New Advancements in Pure and Applied Mathematics via Fractals and Fractional Calculus

New Advancements in Pure and Applied Mathematics via Fractals and Fractional Calculus

Editors

Asifa Tassaddiq
Muhammad Yaseen

MDPI • Basel • Beijing • Wuhan • Barcelona • Belgrade • Manchester • Tokyo • Cluj • Tianjin

Editors
Asifa Tassaddiq
College of Computer and Information Sciences
Majmaah University
Saudi Arabia

Muhammad Yaseen
University of Sargodha
Pakistan

Editorial Office
MDPI
St. Alban-Anlage 66
4052 Basel, Switzerland

This is a reprint of articles from the Special Issue published online in the open access journal *Fractal and Fractional* (ISSN 2504-3110) (available at: https://www.mdpi.com/journal/fractalfract/special_issues/pure_and_applied_math).

For citation purposes, cite each article independently as indicated on the article page online and as indicated below:

LastName, A.A.; LastName, B.B.; LastName, C.C. Article Title. *Journal Name* **Year**, *Volume Number*, Page Range.

ISBN 978-3-0365-4905-7 (Hbk)
ISBN 978-3-0365-4906-4 (PDF)

Contents

Editorial

Editorial for Special Issue "New Advancements in Pure and Applied Mathematics via Fractals and Fractional Calculus"

Asifa Tassaddiq [1,*] and Muhammad Yaseen [2]

1. Department of Basic Sciences and Humanities, College of Computer and Information Sciences, Majmaah University, Al Majmaah 11952, Saudi Arabia
2. Department of Mathematics, University of Sargodha, Sargodha 40100, Pakistan; yaseen.yaqoob@uos.edu.pk
* Correspondence: a.tassaddiq@mu.edu.sa

Citation: Tassaddiq, A.; Yaseen, M. Editorial for Special Issue "New Advancements in Pure and Applied Mathematics via Fractals and Fractional Calculus". *Fractal Fract.* **2022**, *6*, 284. https://doi.org/10.3390/fractalfract6060284

Received: 19 May 2022
Accepted: 24 May 2022
Published: 25 May 2022

Publisher's Note: MDPI stays neutral with regard to jurisdictional claims in published maps and institutional affiliations.

Copyright: © 2022 by the authors. Licensee MDPI, Basel, Switzerland. This article is an open access article distributed under the terms and conditions of the Creative Commons Attribution (CC BY) license (https://creativecommons.org/licenses/by/4.0/).

Fractional calculus has reshaped science and technology since its first appearance in a letter received to Gottfried Wilhelm Leibniz from Guil-laume de l'Hôpital in the year 1695. The existence of fractional behavior in nature cannot be denied. Any phenomenon with a pulse, rhythm, or pattern has the potential to be a fractal. The goal of this Special Issue is to explore new developments in both pure and applied mathematics as a result of fractional behavior. This assertion is supported by the papers in this Special Issue. The variety of topics covered here demonstrates the importance of fractional calculus in various fields and provides adequate coverage to appeal to the interests of each reader. This Special Issue of *Fractal and Fractional* was posted in early 2021 with the goal of exploring the various connections between fractional calculus and its applications in pure and applied mathematics. Initially, a deadline was set and has been extended to 5 April 2022, in consideration of the author's interest. In total, we received 74 submissions. Following a thorough peer-review process, seventeen of them were eventually published and, keeping with the original concept of this Special Issue, have now been compiled into this book. The following are details of the papers published in our Special Issue:

Ali et al. [1] developed a new version of generalized fractional Hadamard and Fejér–Hadamard-type integral inequalities that can be used to investigate the stability and control of corresponding fractional dynamic equations.

Fisher's equation is a precise mathematical result derived from population dynamics and genetics, specifically chemistry. Rashid et al. [2] used a hybrid technique in conjunction with a new iterative transform method to solve the nonlinear fractional Fisher model. Furthermore, while the proposed procedure is highly robust, explicit, and viable for nonlinear fractional PDEs, it has the potential to be consistently applied to other multifaceted physical processes.

It is worth noting that the proposed fuzziness approach is to validate the superiority and dependability of configuring numerical solutions to nonlinear fuzzy fractional partial differential equations arising in physical and complex structures. As a result, in [3], the authors evaluate a semi-analytical method in conjunction with a new hybrid fuzzy integral transform and the Adomian decomposition method using the fuzziness concept known as the Elzaki Adomian decomposition method (EADM).

In [4], the authors analyzed the solutions of a nonlinear div-curl system with fractional derivatives of the Riemann–Liouville or Caputo types. To that end, the fractional-order vector operators of divergence, curl, and gradient were identified as components of the quaternionic fractional Dirac operator. General solutions to some non-homogeneous div-curl systems were derived that consider the presence of fractional-order derivatives of the Riemann–Liouville or Caputo types as one of the most important results of this manuscript.

An integro-differential kinetic equation was derived in [5] by using novel fractional operators and its solution using weighted generalized Laplace transforms. The weighted (k,s)-Riemann–Liouville fractional integral and differential operators are defined by the authors. The paper includes some specific properties of the operators as well as the weighted generalized Laplace transform of the new operators.

The models that include vaccination as a control measure are very important. In light of this, the authors developed and mathematically investigated integer and fractional models of typhoid fever transmission dynamics in [6]. Several numerical simulations were run, allowing us to conclude that such diseases may be combated through vaccination combined with environmental sanitation.

Chemical, electrical, biochemical, geometrical, and meteorological models are examples of nonlinear models used in science and engineering. The authors of [7] investigated the global fractal behavior of a new nonlinear three-step method with tenth-order convergence. Basins of attraction consider various types of complex functions. When compared to other well-known methods, the proposed method achieves the specified tolerance in the smallest number of iterations while assuming different initial guesses.

The authors investigate the existence results for the hybrid Caputo–Hadamard fractional boundary value problem in [8]. The proposed BVP's inclusion version with three-point hybrid Caputo–Hadamard terminal conditions is also considered, and the related existence results are provided. To accomplish these objectives, Dhage's well-known fixed-point theorems for both BVPs are applied. Furthermore, two numerical examples are presented to validate the analytical findings.

The authors of [9] developed a feedback-control strategy to control the chaos caused by bifurcation. The proposed model's fractal dimensions were computed. To further confirm the complexity and chaotic behavior, the maximum Lyapunov exponents and phase portraits were depicted. Finally, numerical simulations were presented to validate the theoretical and analytical results.

Numerical analysis is always necessary to demonstrate the efficacy of proposed schemes. Keeping this in mind, the authors in [10] concentrated on numerically addressing the time fractional Cattaneo equation involving the Caputo–Fabrizio derivative using spline-based numerical techniques. The main advantage of the schemes is that the approximation solution is generated as a smooth piecewise continuous function, which allows to approximate a solution at any point in the domain of interest.

Certain convex and s-convex functions have applications in optimization theory. As a result, in [11], the authors investigated a variety of mean-type integral inequalities for a well-known Hilfer fractional derivative. Some identities were also established in order to infer more interesting mean inequalities. The Caputo fractional derivative consequences were presented as special cases to their general conclusions.

The authors of [12] proposed a numerical method for solving Caputo fractional-order differential equations based on the operational matrices of shifted Vieta–Lucas polynomials (VLPs) (FDEs). A new operational matrix of fractional-order derivatives in the Caputo sense was derived, which was then used in conjunction with the spectral tau and spectral collocation methods to reduce the FDEs to a system of algebraic equations. Numerical examples were provided to demonstrate the accuracy of this method, which demonstrated that the obtained results agree well with the analytical solutions for both linear and nonlinear FDEs.

A semi-analytical analysis of the fractional-order non-linear coupled system of Whitham–Broer–Kaup equations was presented in [13]. The fractional derivative was considered in the Caputo–Fabrizio sense. When the analytical and actual solutions are compared, it is clear that the proposed approaches effectively solve complex nonlinear problems. Furthermore, the proposed methodologies control and manipulate the obtained numerical solutions in an extreme manner in a large acceptable region.

The authors of [14] derived some suitable results for extremal solutions to a class of generalized Caputo-type nonlinear fractional differential equations (FDEs) with nonlinear boundary conditions (NBCs). The aforementioned outcomes were obtained by employing the monotone iterative method, which employs the procedure of upper and lower solutions. There are two sequences of extremal solutions generated, one of which converges to the upper solution and the other to the corresponding lower solution. The method does not require any prior discretization or collocation to generate the aforementioned upper and lower solution sequences.

Q-calculus is a non-trivial and useful generalization of calculus. The authors of [15] presented two new identities involving q-Riemann–Liouville fractional integrals. New q-fractional estimates of trapezoidal-like inequalities were derived using these identities as auxiliary results, in essence of the class of generalized exponential convex functions.

The definition and applicability of new families of polynomials generating function and operational representations are always of great interest. The authors of [16] used operational techniques to investigate a new type of polynomial, specifically the Gould–Hopper–Laguerre–Sheffer matrix polynomials. Furthermore, these particular matrix polynomials were interpreted in terms of quasi-monomiality. The integral transform was used to investigate the properties of the extended versions of the Gould–Hopper–Laguerre–Sheffer matrix polynomials. There were also examples of how these results apply to specific members of the matrix polynomial family.

Laplace transform of the Riemann zeta function using its distributional representation was computed, which played a critical role in applying the operators of generalized fractional calculus to this well-studied function [17]. As a result, as special cases, similar new images can be obtained using various other popular fractional transforms. The Riemann zeta function was used to formulate and solve a new fractional kinetic equation. Following that, a new relationship involving the Laplace transform of the Riemann zeta function and the Fox–Wright function was investigated, which significantly simplified the results.

To summarize, this special selection covers the scope of ongoing activities in the context of fractional calculus by presenting alternative perspectives, viable methods, new derivatives, and strategies to solve practical issues. As editors, we presume that this will be followed by a set of Special Issues and texts to further investigate this theme.

As the guest editors of this Special Issue, we would like to take this opportunity to thank all of the reviewers, editorial board members, and editors who assisted us in perfecting the content of this volume. We would also like to thank Ms. Cecile Zheng from the journal office for her prompt assistance across the Special Issue management process.

All author contributions to this Special Issue are greatly acknowledged with thanks.

Funding: This research received no external funding.

Conflicts of Interest: The authors declare no conflict of interest.

References

1. Ali, R.; Mukheimer, A.; Abdeljawad, T.; Mubeen, S.; Ali, S.; Rahman, G.; Nisar, K. Some New Harmonically Convex Function Type Generalized Fractional Integral Integral Inequalities. *Fractal Fract.* **2021**, *5*, 54. [CrossRef]
2. Rashid, S.; Hammouch, Z.; Aydi, H.; Ahmad, A.; Alsharif, A. Novel Computations of the Time-Fractional Fisher's Model via Generalized Fractional Integral Operators by Means of the Elzaki Transform. *Fractal Fract.* **2021**, *5*, 94. [CrossRef]
3. Rashid, S.; Ashraf, R.; Akdemir, A.; Alqudah, M.; Abdeljawad, T.; Mohamed, M. Analytic Fuzzy Formulation of a Time-Fractional Fornberg–Whitham Model with Power and Mittag–Leffler Kernels. *Fractal Fract.* **2021**, *5*, 113. [CrossRef]
4. Delgado, B.; Macias-Diaz, J. On the General Solutions of Some Non-Homogeneous Div-Curl Systems with Riemann–Liouville and Caputo Fractional Derivatives. *Fractal Fract.* **2021**, *5*, 117. [CrossRef]
5. Samraiz, M.; Umer, M.; Kashuri, A.; Abdeljawad, T.; Iqbal, S.; Mlaiki, N. On Weighted (k, s)-Riemann-Liouville Fractional Operators and Solution of Fractional Kinetic Equation. *Fractal Fract.* **2021**, *5*, 118. [CrossRef]
6. Abboubakar, H.; Kom Regonne, R.; Sooppy Nisar, K. Fractional Dynamics of Typhoid Fever Transmission Models with Mass Vaccination Perspectives. *Fractal Fract.* **2021**, *5*, 149. [CrossRef]
7. Tassaddiq, A.; Qureshi, S.; Soomro, A.; Hincal, E.; Baleanu, D.; Shaikh, A. A New Three-Step Root-Finding Numerical Method and Its Fractal Global Behavior. *Fractal Fract.* **2021**, *5*, 204. [CrossRef]
8. Yaseen, M.; Mumtaz, S.; George, R.; Hussain, A. Existence Results for the Solution of the Hybrid Caputo-Hadamard Fractional Differential Problems Using Dhage's Approach. *Fractal Fract.* **2022**, *6*, 17. [CrossRef]
9. Tassaddiq, A.; Shabbir, M.; Din, Q.; Naaz, H. Discretization, Bifurcation, and Control for a Class of Predator-Prey Interactions. *Fractal Fract.* **2022**, *6*, 31. [CrossRef]
10. Yaseen, M.; Arif, Q.U.N.; George, R.; Khan, S. Comparative Numerical Study of Spline-Based Numerical Techniques for Time Fractional Cattaneo Equation in the Sense of Caputo-Fabrizio. *Fractal Fract.* **2022**, *6*, 50. [CrossRef]

11. Samraiz, M.; Perveen, Z.; Rahman, G.; Adil Khan, M.; Nisar, K. Hermite-Hadamard Fractional Inequalities for Differentiable Functions. *Fractal Fract.* **2022**, *6*, 60. [CrossRef]
12. Noor, Z.; Talib, I.; Abdeljawad, T.; Alqudah, M. Numerical Study of Caputo Fractional-Order Differential Equations by Developing New Operational Matrices of Vieta-Lucas Polynomials. *Fractal Fract.* **2022**, *6*, 79. [CrossRef]
13. Yasmin, H. Numerical Analysis of Time-Fractional Whitham-Broer-Kaup Equations with Exponential-Decay Kernel. *Fractal Fract.* **2022**, *6*, 142. [CrossRef]
14. Derbazi, C.; Baitiche, Z.; Abdo, M.; Shah, K.; Abdalla, B.; Abdeljawad, T. Extremal Solutions of Generalized Caputo-Type Fractional-Order Boundary Value Problems Using Monotone Iterative Method. *Fractal Fract.* **2022**, *6*, 146. [CrossRef]
15. Nonlaopon, K.; Awan, M.; Javed, M.; Budak, H.; Noor, M. Some q-Fractional Estimates of Trapezoid like Inequalities Involving Raina's Function. *Fractal Fract.* **2022**, *6*, 185. [CrossRef]
16. Nahid, T.; Choi, J. Certain Hybrid Matrix Polynomials Related to the Laguerre-Sheffer Family. *Fractal Fract.* **2022**, *6*, 211. [CrossRef]
17. Tassaddiq, A.; Srivastava, R. New Results Involving Riemann Zeta Function Using Its Distributional Representation. *Fractal Fract.* **2022**, *6*, 254. [CrossRef]

fractal and fractional

MDPI

Article

New Results Involving Riemann Zeta Function Using Its Distributional Representation

Asifa Tassaddiq [1,*] and Rekha Srivastava [2,*]

1 Department of Basic Sciences and Humanities, College of Computer and Information Sciences, Majmaah University, Al Majmaah 11952, Saudi Arabia
2 Department of Mathematics and Statistics, University of Victoria, Victoria, BC V8P 5C2, Canada
* Correspondence: a.tassaddiq@mu.edu.sa (A.T.); rekhas@uvic.ca (R.S.)

Abstract: The relation of special functions with fractional integral transforms has a great influence on modern science and research. For example, an old special function, namely, the Mittag–Leffler function, became the queen of fractional calculus because its image under the Laplace transform is known to a large audience only in this century. By taking motivation from these facts, we use distributional representation of the Riemann zeta function to compute its Laplace transform, which has played a fundamental role in applying the operators of generalized fractional calculus to this well-studied function. Hence, similar new images under various other popular fractional transforms can be obtained as special cases. A new fractional kinetic equation involving the Riemann zeta function is formulated and solved. Thereafter, a new relation involving the Laplace transform of the Riemann zeta function and the Fox–Wright function is explored, which proved to significantly simplify the results. Various new distributional properties are also derived.

Keywords: delta function; Riemann zeta-function; fractional transforms; Fox–Wright-function; generalized fractional kinetic equation

Citation: Tassaddiq, A.; Srivastava, R. New Results Involving Riemann Zeta Function Using Its Distributional Representation. *Fractal Fract.* **2022**, *6*, 254. https://doi.org/10.3390/fractalfract6050254

Academic Editor: Ricardo Almeida

Received: 10 March 2022
Accepted: 21 April 2022
Published: 6 May 2022

1. Introduction

In general, the Riemann zeta function and its generalizations have always been of fundamental importance [1–10] due to their widespread applications. For instance, the role of the Riemann zeta function is vital in fractal geometry for studying the complex dimensions of fractal strings [1]. More recently, new representations of special functions are discussed [10–21] in terms of the complex delta function [22,23]. In this article, we use a distributional representation [10], Equation (33), of the Riemann zeta function to obtain further new results. On the one hand, several fractional calculus images involving the Riemann zeta function are obtained under multiple E–K fractional operators, and on the other hand, a non-integer-order kinetic equation including the Riemann zeta function is formulated and solved. The Laplace transform of the Riemann zeta function is computed using its distributional representation, which played a fundamental role in accomplishing the goals of this research. Several new properties and results for this function are also discussed.

Calculation of the images of special functions using the fractional calculus operators has emerged as a popular subject in the data of various newly published papers [24–26]. This number is rising regularly, and such research has commented [24] further in mentioning Kiryakova's unified approach. Taking a cue from these facts, the author has followed the recommendation of [24] and obtained fractional calculus images involving the Riemann zeta function and its simpler cases using the unified approach [24–28]. The Marichev–Saigo–Maeda (M-S-M) operators and the Saigo, Erdélyi–Kober, and Riemann–Liouville (R–L) fractional operators for $m = 3$, $m = 2$, $m = 1$, respectively, are discussed as special cases of generalized fractional calculus operators (namely, multiple E–K operators of the multiplicity m). It is recommended in the conclusion section of [24] to examine whether the special function can be formulated as a general function, namely, the Fox–Wright function ${}_p\Psi_q$, then to use a general result such as [24] and Theorem 3 and 4 therein. It can be observed that it is not possible to apply these theorems for the Riemann zeta function using its classical representations, as already mentioned (see [24], p. 2). It is important to note that the results obtained in this research are completely verifiable with these general results. The corresponding fractional derivatives in the Riemann–Liouville and Caputo sense, as discussed in [24] (p. 9, Definition 6; and p. 17, Theorem 4), can now be used for the Laplace transform of the Riemann zeta function and also straightforwardly using its new representation.

The remaining paper is organized as follows: Necessary preliminaries related to the family of the Fox-H function and the generalized fractional integrals (multiple E–K operators) form part of Section 2. Section 3.1 contains fractional calculus images involving the Riemann zeta-function. The next Section 3.2, is devoted to the formulation and solution of a non-integer-order kinetic equation containing the Riemann zeta-function. Further new properties and results involving the Reimann zeta function are discussed in Section 3.3. The conclusion is given in Section 4. Related special cases of generalized fractional integrals (multiple E–K operators) are listed in Appendix A.

Hence, in order to achieve our purpose, let us first go through the basic definitions and preliminaries in the subsequent section.

2. Materials and Methods

Throughout this article, $\mathbb{C}$ and $\mathbb{R}$ represent the set of complex and real numbers. The real part of any complex number is denoted by $\Re$, $\mathbb{Z}_0^-$ denotes a set of negative integers containing *0*, and $\mathbb{R}^+$ symbolizes the set containing positive reals.

The Riemann zeta function is a classical function investigated by Riemann [2], defined as

$$\zeta(s) := \sum_{n=1}^{\infty} \frac{1}{n^s}; (s = \sigma + i\tau\, ; \Re(s) > 1). \tag{1}$$

With the exception of a simple pole at $s = 1$, the meromorphic continuation of this function extends it to the entire complex s-plane. As shown in [2] (p. 13, Equation (2.1.1)), this function satisfies the following result (also known as Riemann's Functional Equation):

$$\zeta(s) := 2^s \pi^{s-1} \Gamma(1-s)\zeta(1-s), \tag{2}$$

where $\Gamma(s)$ represents the gamma function [3,4] (a generalization of the factorial). The Riemann zeta function has simple zeros at negative even integers that are its trivial zeros. The remaining zeros of the zeta function are known as its nontrivial zeros, which are symmetrically placed on the line $\Re(s) = 1/2$. This unproved fact is also famously known as the "Riemann Hypothesis". Several authors have investigated and analyzed different generalizations of the zeta function. It has different integral representations, for example [2],

$$\begin{aligned} \zeta(s) &:= \tfrac{1}{\Gamma(s)} \int_0^\infty \tfrac{t^{s-1}}{e^t-1} dt; (\Re(s) > 1); \\ \zeta(s) &:= \tfrac{1}{\Gamma(s)} \int_0^\infty \left[\tfrac{1}{e^t-1} - \tfrac{1}{t}\right] t^{s-1} dt; (0 < \Re(s) < 1); \\ \zeta(s) &:= \tfrac{1}{\Gamma(s)\left(1-2^{1-s}\right)} \int_0^\infty \tfrac{t^{s-1}}{e^t+1} dt; (\Re(s) > 0). \end{aligned} \tag{3}$$

For more details about the zeta function, the interested reader is referred to the references [4–9] and the cited bibliographies therein. More recently, the distributional representation of different special functions has been discussed in [10–20]. In this article, for $\Re(s) > 1$, the following representation [10], Equation (33),

$$\Gamma(s)\zeta(s) = 2\pi \sum_{n,l=0}^{\infty} \frac{(-(n+1))^l}{l!} \delta(s+l) \tag{4}$$

is the main focus to achieve the purpose of the current research. For similar studies, the interested reader is referred to [10–20]. For any suitable function f and the number ω, the delta function is a famous generalized function (distribution) defined by [22,23]:

$$\langle \delta(s-\omega), \wp(s)\rangle = \wp(\omega);\ \delta(-s) = \delta(s);\ \delta(\omega s) = \frac{\delta(s)}{|\omega|},\ \text{where } \omega \neq 0. \tag{5}$$

It has several interesting properties, such as the following (see [22,23]):

$$\delta(s+l) = \sum_{p=0}^{\infty} \frac{(l)^p}{p!}\,\delta^{(p)}(s); \tag{6}$$

$$\begin{gathered} \delta(z-c) * \vartheta(z) = \vartheta(z-c); \\ \delta^{(i)}(z-c) * \vartheta(z) = \vartheta^{(i)}(z-c); \\ \left(\sum_{i=0}^{\infty} \delta^{(i)}(z-v)\right) * \left(\sum_{i=0}^{\infty} \delta(z-v)\right) = \sum_{i=0}^{\infty}\sum_{j=0}^{i} \delta^{(j)}(z-v); \\ \left(\sum_{i=0}^{\infty} \delta^{(i)}(z-v)\right) * \left(\sum_{i=0}^{\infty} \delta^{(i)}(z-v)\right) = \left(\sum_{j=0}^{\infty} (v+1)\delta^{(j)}(z-v)\right). \end{gathered} \tag{7}$$

Furthermore, the Laplace transform of an arbitrary function $\varepsilon(t)$ is defined by [23] (Chapter 8):

$$\varepsilon(s) = L[\varepsilon(t) : s] = \int_0^{\infty} e^{-st}(t)dt, \Re(s) > 0 \tag{8}$$

and we will also use [23] (p. 227):

$$L\left\{\delta^{(r)}(z); \xi\right\} = \xi^r. \tag{9}$$

The generalized fractional integrals, namely (multiple) E–K operators of multiplicity m, are defined by [24] (p. 8, Equation (18)):

$$I_{(\beta_k),m}^{(\gamma_k),(\delta_k)} f(z) = \begin{cases} \int_0^1 f(z\sigma)\, H_{m,m}^{m,0}\left[\sigma \middle| \begin{matrix} \left(\gamma_k+\delta_k+1-\frac{1}{\beta_k}, \frac{1}{\beta_k}\right)_1^m \\ \left(\gamma_k+1-\frac{1}{\beta_k}, \frac{1}{\beta_k}\right)_1^m \end{matrix}\right] d\sigma; \sum_k \delta_k > 0 \\ = z^{-1}\int_0^z f(\xi)\, H_{m,m}^{m,0}\left[\frac{\xi}{z} \middle| \begin{matrix} \left(\gamma_k+\delta_k+1-\frac{1}{\beta_k}, \frac{1}{\beta_k}\right)_1^m \\ \left(\gamma_k+1-\frac{1}{\beta_k}, \frac{1}{\beta_k}\right)_1^m \end{matrix}\right] d\xi; \sum_k \delta_k > 0 \\ f(z); \delta_k = 1 \end{cases} \tag{10}$$

where $\delta_k's$ are concerned with the order of integration, $\gamma_k's$ are weights, and $\beta_k's$ are additional parameters. $H_{p,q}^{m,n}$ is the H-function defined in the subsequent paragraph. The limits of integration $(0,\ 1)$ and $(0,\ z)$ in the above equation can be changed to $(0,\infty)$ using the fact that $H_{m,m}^{m,0}$ vanishes for $|\sigma| > 1$ (To avoid prolonging this section, the special cases of (10) in relation to the results of this article are given in Appendix A). However, the corresponding multiple (m-tuple) Erdélyi–Kober fractional derivative of the R–L type of multi-order $\delta = (\delta_1 \geq 0, \ldots, \delta_m \geq 0)$ is defined by [24] (p. 9):

$$D_{(\beta k),m}^{(\gamma k)_1^m,(\delta k)}(f(z)) := D_\eta\, I_{(\beta k),m}^{(\gamma k+\delta k),(\eta k-\delta k)}\, f(z) D_\eta \int_0^1 f(z\sigma) H_{m,m}^{m,0}\left[\sigma \middle| \begin{matrix} \left(\gamma_k+\eta_k+1-\frac{1}{\beta_k}, \frac{1}{\beta_k}\right)_1^m \\ \left(\gamma_k+1-\frac{1}{\beta_k}, \frac{1}{\beta_k}\right)_1^m \end{matrix}\right] d\sigma \tag{11}$$

where D_η, is a polynomial of variable $z\left(\frac{d}{dz}\right)$ of degree $\eta_1 + \ldots + \eta_m$, given by

$$D_\eta = \prod_{r=1}^{m}\prod_{j=1}^{\eta_r}\left(\frac{1}{\beta_r} z\frac{d}{dz} + \gamma_r + j\right);\ \eta_k = \begin{cases} [\delta_k]+1; \delta_k \notin \mathbb{Z} \\ \delta_k; \delta_k \in \mathbb{Z} \end{cases} \tag{12}$$

and the corresponding multiple (m-tuple) Erdélyi–Kober fractional derivative of the Caputo type is given as (see [24] (p. 9) and references therein):

$$ {}^{*}D_{(\beta k),m}^{(\gamma k)_1^m,(\delta k)} f(z) = I_{(\beta k\,),m}^{(\gamma k+\delta k\,),(\eta k-\delta k\,)} D_\eta f(z). \tag{13}$$

The action of the E–K operators on the power function yields [24] (p. 9; Equation (27)):

$$ I_{(\beta k),m}^{(\gamma k),(\delta k)}\{z^p\} = \prod_{i=1}^{m} \frac{\Gamma\left(\gamma_i+1+\frac{p}{\beta_i}\right)}{\Gamma\left(\gamma_i+\delta_i+1+\frac{p}{\beta_i}\right)} z^p;\ \ [-\beta_k(1+\gamma_k)] < p; \delta_k \geq 0; k = 1,\ldots,m. \tag{14}$$

The integrand of (10) involves the Fox H-function defined by [24] (p. 3; see also [25,29]), which is given here in its integral and series form as follows:

$$ \begin{aligned} H_{p,q}^{m,n}(z) &= H_{p,q}^{m,n}\left[z\left|\begin{matrix}(a_i, A_i)\\(b_j,\ B_j)\end{matrix}\right.\right] = H_{p,q}^{m,n}\left[z\left|\begin{matrix}(a_1,A_1),\ldots, & (a_i,A_i)\\(b_1,\ B_1),\ldots, & (b_j,\ B_j)\end{matrix}\right.\right] \\ &= \frac{1}{2\pi i}\int_{\mathcal{L}} \frac{\prod_{j=1}^{m}\Gamma\left(b_j+B_j s\right)\prod_{i=1}^{n}\Gamma\left(1-a_j-A_j s\right)}{\prod_{j=m+1}^{q}\Gamma\left(1-b_j-B_j s\right)\prod_{i=n+1}^{p}\Gamma\left(a_j+A_j s\right)} z^{-s}\, ds, \end{aligned} \tag{15}$$

where $m,\ n, p,$ and q are related as $1 \leqq m \leqq\ q;\ 0 \leqq n \leqq\ p,\ A_i > 0\ (\ i = 1,\ \cdots,\ p);$ $B_j > 0\ (\ j = 1,\ \cdots, q),\ a_i \in \mathbb{C}\ (\ i = 1,\ \cdots,\ p); b_j \in \mathbb{C}\ (\ j = 1,\ \cdots, q)$; and $\mathcal{L}$ is an appropriate Mellin–Barnes type of contour that separates the singularities of $\left\{\Gamma(\ b_j\ + B_j s)\right\}_{j=1}^{m}$ from the singularities of $\left\{\Gamma(\ 1-a_j\ \ - A_j s)\right\}_{j=1}^{n}$. Here, $\Gamma(z)$ denotes the familiar gamma function [4], and when all $A_p = B_q = 1$, then the H-function becomes the Meijer G-function [24] (p.4; see also [25,29]):

$$ \begin{aligned} H_{p,q}^{m,n}\left[z\left|\begin{matrix}(a_1,A_1),\ldots, & (a_i,A_i)\\(b_1,\ B_1),\ldots, & (b_j,\ B_j)\end{matrix}\right.\right] &= \sum_{m=0}^{\infty} \frac{\prod_{j=1}^{m}\Gamma(\ b_j\ +B_j m)\prod_{i=1}^{n}\Gamma(\ 1-a_i\ -A_i m)}{\prod_{j=m+1}^{q}\Gamma(\ 1-b_j\ -B_j m)\prod_{i=n+1}^{p}\Gamma(\ a_i\ +im)} \frac{z^m}{m!} \\ H_{p,q}^{m,n}\left[z\left|\begin{matrix}(a_1,1),\ldots, & (a_i,1)\\(b_1,1),\ldots, & (b_j,1)\end{matrix}\right.\right] &= G_{p,q}^{m,n}\left[z\left|\begin{matrix}a_1\ldots, & a_i\\ b_1,\ldots, & b_j\end{matrix}\right.\right]. \end{aligned} \tag{16}$$

The basic Fox–Wright function denoted by ${}_p\Psi_q$ is defined and related to the H-function:

$$ \begin{aligned} {}_p\Psi_q\left[\begin{matrix}(a_i,\ A_i)\\(b_j, B_j)\end{matrix};z\right] &= \sum_{m=0}^{\infty} \frac{\prod_{i=1}^{p}\Gamma(a_i+A_i m)}{\prod_{l=1}^{q}\Gamma((b_j+B_j m)}\frac{z^m}{m!} = H_{p,q+1}^{1,p}\left[-z\left|\begin{matrix}(1-a_1,A_1),\ldots, & (1-a_i,\ A_i)\\(0,1),(1-b_1,\ B_1),\ldots, & ((1-b_j,B_j)\end{matrix}\right.\right] \\ &\left(a_i \epsilon \mathbb{R}^+ (i = 1,\ldots,p); B_j \epsilon \mathbb{R}^+ (j = 1,\ldots,q); 1+\sum_{i=1}^{q} B_i - \sum_{j=1}^{p} A_j > 0\right) \end{aligned} \tag{17}$$

and contains the hypergeometric and other important functions as [24] (p. 4; see also [25,29]):

$$ \begin{aligned} {}_p\Psi_q\left[\begin{matrix}(a_i,\ 1)\\(b_j,1)\end{matrix};z\right] &= G_{p,q+1}^{1,p}\left[-z\left|\begin{matrix}(1-a_1,1),\ldots, & (1-a_i,1)\\ 0,(1-b_1,1),\ldots, & (1-b_j,1)\end{matrix}\right.\right] \\ &= {}_pF_q\left[\begin{matrix}a_i\\ b_j\end{matrix};z\right]\cdot\frac{\Gamma(a_1)\ldots\Gamma(a_i)}{\Gamma(b_1)\ldots\Gamma(b_j)}. \qquad (a_j > 0; b_j \notin \mathbb{Z}_0^-). \end{aligned} \tag{18}$$

Furthermore, many other special functions studied in the literature are connected with this class of special functions. For example, the Mittag–Leffler function [30] of parameters 1, 2, and 3 is related with the abovementioned special functions as follows:

$$ \begin{aligned} E_{\alpha,\beta}^{\gamma}(z) &= \sum_{r=0}^{\infty}\frac{(\gamma)_r z^r}{\Gamma(\alpha r+\beta)} = \frac{1}{\Gamma(\gamma)}{}_1\Psi_1\left[\begin{matrix}(\gamma,\ 1)\\(\beta,\alpha)\end{matrix};z\right] = H_{1,2}^{1,1}\left[-z\left|\begin{matrix}(1-\gamma,1)\\(0,1),(1-\beta,\alpha)\end{matrix}\right.\right]; \\ E_{\alpha,\beta}^{1}(z) = E_{\alpha,\beta}(z) &= \sum_{r=0}^{\infty}\frac{z^r}{\Gamma(\alpha r+\beta)} = {}_1\Psi_1\left[\begin{matrix}(1,\ 1)\\(\beta,\alpha)\end{matrix};z\right] = H_{1,2}^{1,1}\left[-z\left|\begin{matrix}(0,1)\\(0,1),(1-\beta,\alpha)\end{matrix}\right.\right]; \\ E_{\alpha,1}^{1}(z) = E_{\alpha}(z) &= \sum_{r=0}^{\infty}\frac{z^r}{\Gamma(\alpha r+1)} = {}_1\Psi_1\left[\begin{matrix}(1,\ 1)\\(1,\alpha)\end{matrix};z\right] = H_{1,2}^{1,1}\left[-z\left|\begin{matrix}(0,1)\\(0,1),(0,\alpha)\end{matrix}\right.\right]. \end{aligned} \tag{19}$$

Furthermore, $(s)_k$ is the Pochhammer symbols defined in terms of the gamma function as follows:

$$(s)_\rho = \frac{\Gamma(s+\rho)}{\Gamma(s)} = \begin{cases} 1\ (\rho = 0, s \in C\backslash\{0\}) \\ s(s+1)\ldots(s+k-1)\ (\rho = k \in \mathbb{N}; s \in \mathbb{C}). \end{cases} \tag{20}$$

Furthermore, it is important to mention that if any function can be expressed in the form of the Fox–Wright function, then the generalized (multiple E–K) fractional integrals and derivatives involving this function can be obtained directly using the general results of [24], Theorem 3:

$$I^{(\gamma k)_1^m,(\delta k)}_{(\beta k),m}\left\{z^c{}_p\Psi_q\left[\begin{matrix}(a_k,\alpha_k)_1^p\\(b_k,\beta_k)_1^q\end{matrix};\lambda z^\mu\right]\right\} = z^c\left\{{}_{p+m}\Psi_{q+m}\left[\begin{matrix}(a_k,\alpha_k)_1^p,\left(\gamma_k+1+\frac{c}{\beta_k},\frac{\mu}{\beta_k}\right)_1^m\\(b_k,\beta_k)_1^q,\left(\gamma_k+\delta_k+1+\frac{c}{\beta_k},\frac{\mu}{\beta_k}\right)_1^m\end{matrix};\lambda z^\mu\right]\right\} \tag{21}$$

$$(\delta k \geq 0,\ \gamma_k > -1,\ \beta_k > 0,\ k = 1,\ \ldots,\ m \wedge \mu > 0,\ \lambda \neq 0)$$

and [24], Theorem 4:

$$D^{(\gamma k)_1^m,(\delta k)}_{(\beta k),m}\left\{z^c{}_p\Psi_q\left[\begin{matrix}(a_i,\alpha_i)_1^p\\(b_j,\beta_j)_1^q\end{matrix};\lambda z^\mu\right]\right\} = z^c\left\{{}_{p+m}\Psi_{q+m}\left[\begin{matrix}(a_i,\alpha_i)_1^p,\left(\gamma_k+\delta_k+1+\frac{c}{\beta_k},\frac{\mu}{\beta_k}\right)_1^m\\(b_j,\beta_j)_1^q,\left(\gamma_k+1+\frac{c}{\beta_k},\frac{\mu}{\beta_k}\right)_1^m\end{matrix};\lambda z^\mu\right]\right\}. \tag{22}$$

Unless otherwise stated, the conditions of parameters will remain similar to this Section 2 and references therein.

3. Results

3.1. Fractional Integrals and Derivatives Formulae Involving the Riemann Zeta-Function

The following lemma has significant importance for the application of Equations (21) and (22).

Lemma 1. *Prove that the following result involving the Fox–Wright function holds true:*

$$2\pi\sum_{n=0}^{\infty}{}_0\Psi_0\left[\begin{matrix}-\\-\end{matrix}\middle| -(n+1)e^\omega\right] = \sum_{n,l=0}^{\infty}\frac{(-(n+1))^l}{l!}{}_0\Psi_0\left[\begin{matrix}-\\-\end{matrix}\middle| l\omega\right]. \tag{23}$$

Proof. First of all, let us use (6) in (4) to get the following form:

$$\Gamma(s)\zeta(s) = 2\pi\sum_{n,l,p=0}^{\infty}\frac{(-(n+1))^l(l)^p}{l!p!}\delta^{(p)}(s). \tag{24}$$

Then, by applying the Laplace transform to (24), and by making use of (9), we are led to the following:

$$L(\Gamma(s)\zeta(s);\omega) = 2\pi\sum_{n,l,p=0}^{\infty}\frac{(-(n+1))^l(l)^p}{l!p!}\omega^p = 2\pi\sum_{n,l=0}^{\infty}\frac{(-(n+1))^l}{l!}{}_0\Psi_0\left[\begin{matrix}-\\-\end{matrix}\middle| l\omega\right]. \tag{25}$$

From (25), it can be further noticed that

$$L(\Gamma(s)\zeta(s);\omega) = \frac{2\pi}{\exp(e^\omega)-1} = 2\pi\sum_{n=0}^{\infty}\exp(-(r+1)e^\omega) = 2\pi\sum_{n=0}^{\infty}{}_0\Psi_0\left[\begin{matrix}-\\-\end{matrix}\middle| -(n+1)e^\omega\right]. \tag{26}$$

From (25) and (26), the required result follows. □

Theorem 1. *The multiple E–K fractional transform of the Riemann zeta function is given by:*

$$I^{(\gamma k),(\delta k)}_{(\beta k),m}\left(\omega^{\chi-1}L\{\Gamma(s)\zeta(s);\omega\}\right) = 2\pi x^{\chi-1}\sum_{r=0}^{\infty} {}_m\Psi_m\left[\begin{array}{c}\left(\gamma_i+1+\frac{\chi-1}{\beta_i},\frac{1}{\beta_i}\right)_1^m \\ \left(\gamma_i+\delta_i+1+\frac{\chi-1}{\beta_i},\frac{1}{\beta_i}\right)_1^m\end{array}\middle| -(r+1)e^{\omega}\right] \tag{27}$$
$$[-\beta_k(1+\gamma_k)] < p; \delta_k \geq 0; k = 1,\ldots,m.$$

Proof. Let us first consider multiple E–K's fractional transform using (25):

$$I^{(\gamma k),(\delta k)}_{(\beta k),m}\left(\omega^{\chi-1}L\{\Gamma(s)\zeta(s);\omega\}\right) = I^{(\gamma k),(\delta k)}_{(\beta k),m}\left(\omega^{\chi-1}2\pi\sum_{n,l,p=0}^{\infty}\frac{(-(n+1))^l(m)^p}{l!p!}\omega^p\right), \tag{28}$$

exchanging the summation and integration,

$$I^{(\gamma k),(\delta k)}_{(\beta k),m}\left(\omega^{\chi-1}L\{\Gamma(s)\zeta(s);\omega\}\right) = 2\pi\sum_{n,l,p=0}^{\infty}\frac{(-(n+1))^l(m)^p}{l!p!}I^{(\gamma k),(\delta k)}_{(\beta k),m}\left(\omega^{\chi-1}\omega^p\right), \tag{29}$$

and then using (14) yields

$$\begin{aligned} I^{(\gamma k),(\delta k)}_{(\beta k),m}\left(\omega^{\chi-1}L\{\Gamma(s)\zeta(s);\omega\}\right) &= 2\pi\sum_{n,l,p=0}^{\infty}\frac{(-(n+1))^l(l)^p}{l!p!}\prod_{i=1}^{m}\frac{\Gamma\left(\gamma_i+1+\frac{\chi+p-1}{\beta_i}\right)}{\Gamma\left(\gamma i+\delta i+1+\frac{\chi+p-1}{\beta i}\right)}\omega^{p+\chi-1} \\ &= 2\pi\omega^{\chi-1}\sum_{n,l=0}^{\infty}\frac{(-(n+1))^l}{l!}{}_m\Psi_m\left[\begin{array}{c}\left(\gamma_i+1+\frac{\chi-1}{\beta_i},\frac{1}{\beta_i}\right)_1^m \\ \left(\gamma_i+\delta_i+1+\frac{\chi-1}{\beta_i},\frac{1}{\beta_i}\right)_1^m\end{array}\middle| l\omega\right], \end{aligned} \tag{30}$$

which, after using Lemma 1, leads to the required result. □

Remark 1. *Hence the result (30) is completely verifiable with ([24], Theorem 3) in view of (25). Similarly, the generalized fractional derivatives involving the Riemann zeta function can be obtained using the methodology of theorem 1 or by using directly (22) and (25) as follows:*

$$\begin{aligned} D^{(\gamma k)_1^m,(\delta k)}_{(\beta k),m}\{z^cL(\Gamma(s)\zeta(s);z)\} &= 2\pi z^c\sum_{n,l=0}^{\infty}\frac{(-(n+1))^l}{l!}{}_m\Psi_m\left[\begin{array}{c}\left(\gamma_k+\delta_k+1+\frac{c}{\beta_k},\frac{1}{\beta_k}\right)_1^m \\ \left(\gamma_k+1+\frac{c}{\beta_k},\frac{1}{\beta_k}\right)_1^m\end{array}\middle| l\omega\right] \\ &= 2\pi z^c\sum_{n=0}^{\infty}{}_m\Psi_m\left[\begin{array}{c}\left(\gamma_k+\delta_k+1+\frac{c}{\beta_k},\frac{1}{\beta_k}\right)_1^m \\ \left(\gamma_k+1+\frac{c}{\beta_k},\frac{1}{\beta_k}\right)_1^m\end{array}\middle| -(n+1)e^{\omega}\right]. \end{aligned} \tag{31}$$

Continuing in this way, we obtain the following Table 1 of fractional integrals and derivatives formulae involving the Riemann zeta function by following the methodology of Theorem 1 and using Equations (27), (30), and (31), respectively. (As already mentioned in Section 2, the definitions of the Marichev–Saigo–Maeda, Saigo, Erdélyi–Kober, and Riemann–Liouville (R–L) fractional operators and their relation to (10) for $m = 3$, m = 2, $m = 1$, respectively, are given in Appendix A; see also [31–34]).

Table 1. Fractional integrals and derivatives formulae involving Riemann zeta-function.

$m = 3$	**Marichev–Saigo–Maeda fractional integrals and derivatives [31–34]**
	$I_{0+}^{\gamma_1,\gamma_1',\gamma_2,\gamma_2',\delta}\left(\omega^{\chi-1}L\{\Gamma(s)\zeta(s);\omega\}\right) =$
$2\pi\omega^{\delta+\chi-\gamma_1-\gamma_1'-1}\sum_{n=0}^{\infty}{}_3\Psi_3\Big[$	$\begin{array}{ccc}(\chi,1) & (\chi+\delta-\gamma_1-\gamma_1'-\gamma_2,1) & (\chi+\gamma_2'-\gamma_1',1) \\ (\chi+\gamma_2',1) & (\chi+\delta-\gamma_1-\gamma_1',1) & \chi+\delta-\gamma_1'-\gamma_2\end{array}\Big\vert -(n+1)e^{\omega}\Big]$

Table 1. *Cont.*

$m = 3$	**Marichev–Saigo–Maeda fractional integrals and derivatives [31–34]**
	$I_{0-}^{\gamma_1,\gamma_1',\gamma_2,\gamma_2',\delta}(\omega^{\chi-1}L\{\Gamma(s)\zeta(s);\omega\}) = 2\pi\omega^{\delta+\chi-\gamma_1-\gamma_1'-1}\sum_{n=0}^{\infty}{}_3\Psi_3\left[\begin{matrix}(1-\chi-\delta+\gamma_1+\gamma_1',-1) & (1-\chi+\gamma_1+\gamma_2'-\delta,-1) & (1-\chi-\gamma_1,-1)\\ (1-\chi,-1) & (1-\chi+\gamma_1+\gamma_1'+\gamma_2+\gamma_2'-\delta,-1) & (1-\chi+\gamma_1-\gamma_2,-1)\end{matrix}\middle\vert -(n+1)e^{\omega}\right]$
	$D_{0+}^{\gamma_1,\gamma_1',\gamma_2,\gamma_2',\delta}(\omega^{\chi-1}L(\Gamma(s)\zeta(s);\omega)) = 2\pi\omega^{\chi-1}\sum_{n=0}^{\infty}{}_3\Psi_3\left[\begin{matrix}(\chi,1) & (\chi-\gamma_2+\gamma_1,1) & (\chi+\gamma_1+\gamma_1'+\gamma_2'-\delta,1)\\ (\chi-\gamma_2,1) & (\chi-\delta+\gamma_1++\gamma_2',1) & (\chi-\delta+\gamma_1'+\gamma_1,1)\end{matrix}\middle\vert -(n+1)e^{\omega}\right]$
	$D_{-}^{\gamma_1,\gamma_1',\gamma_2,\gamma_2',\delta}(\omega^{\chi-1}L(\Gamma(s)\zeta(s);\omega)) = 2\pi\omega^{\chi-1}\sum_{n=0}^{\infty}{}_3\Psi_3\left[\begin{matrix}(1-\chi+\gamma_2',1) & (1+\gamma_2'-\chi-\gamma_2+\gamma_1,1) & (1-\chi-\gamma_1-\gamma_1'+\delta,-1)\\ (1-\chi,1) & (1-\chi-\gamma_1'+\gamma_2',1) & (1-\chi+\delta-\gamma_1'-\gamma_1-\gamma_2,-1)\end{matrix}\middle\vert -(n+1)e^{\omega}\right]$
$m = 2$	**Saigo fractional integrals and derivatives [31–34]**
	$I_{0+}^{\gamma_1,\gamma_2,\delta}(\omega^{\chi-1}L\{\Gamma(s)\zeta(s);\omega\}) = 2\pi\omega^{\chi-\gamma_1-1}\sum_{n=0}^{\infty}{}_2\Psi_2\left[\begin{matrix}(\chi,1) & (\chi+\gamma_2-\gamma_1,1)\\ (\chi-\gamma_2,1) & (\chi+\delta+\gamma_2)\end{matrix}\middle\vert -(n+1)e^{\omega}\right]$
	$I_{-}^{\gamma_1,\gamma_2,\delta}\omega^{\chi-1}(L\{\Gamma(s)\zeta(s);\omega\}) = 2\pi\omega^{\chi-\gamma_1-1}\sum_{n=0}^{\infty}{}_2\Psi_2\left[\begin{matrix}(\gamma_1-\chi+1,1) & (\gamma_2-\chi+1,-1)\\ (1-\chi,1) & ((\gamma_1+\gamma_2+\delta-\chi+1,-1)\end{matrix}\middle\vert -(n+1)e^{\omega}\right]$
	$D_{0+}^{\gamma_1,\gamma_2,\delta}(t^{\chi-1}L(\Gamma(s)\zeta(s);\omega)) = 2\pi\sum_{n=0}^{\infty}{}_2\Psi_2\left[\begin{matrix}(\chi,1) & (\chi+\delta+\gamma_2+\gamma_1,1)\\ (\chi+\gamma_2,1) & (\chi+\delta,1)\end{matrix}\middle\vert -(n+1)e^{\omega}\right]$
	$D_{-}^{\gamma_1,\gamma_2,\delta}(t^{\chi-1}L(\Gamma(s)\zeta(s);\omega)) = 2\pi\sum_{n=0}^{\infty}{}_2\Psi_2\left[\begin{matrix}(1-\chi-\gamma_2,1) & (1-\chi+\delta+\gamma_1,-1)\\ (1-\chi+\delta-\gamma_2,1) & (1-\chi,-1)\end{matrix}\middle\vert -(n+1)e^{\omega}\right]$
$m = 1$	**Erdélyi–Kober, Riemann–Liouville (R–L) fractional integrals and derivatives [31–34]**
	$I_{0+}^{\gamma,\delta}(\omega^{\chi-1}L\{\Gamma(s)\zeta(s);\omega\}) = 2\pi\omega^{\chi-1}\sum_{n=0}^{\infty}{}_1\Psi_1\left[\begin{matrix}(\chi+\gamma,1)\\ (\chi+\gamma+\delta,1)\end{matrix}\middle\vert -(n+1)e^{\omega}\right]$
	$I_{0-}^{\gamma,\delta}(\omega^{\chi-1}L\{\Gamma(s)\zeta(s);\omega\}) = 2\pi\omega^{\chi+\delta-1}\sum_{n=0}^{\infty}{}_1\Psi_1\left[\begin{matrix}(\gamma-\chi+1,-1)\\ (\gamma+\delta-\chi+1,-1)\end{matrix}\middle\vert -(n+1)e^{\omega}\right]$
	$D_{0+}^{\gamma,\delta}\{\omega^{\chi-1}L(\Gamma(s)\zeta(s);\omega)\} = 2\pi\omega^{\chi-1}\sum_{n=0}^{\infty}{}_1\Psi_1\left[\begin{matrix}(\gamma+\delta+\chi,1)\\ (\gamma+\chi,1)\end{matrix}\middle\vert -(n+1)e^{\omega}\right]$
	$D_{-}^{\gamma,\delta}\{\omega^{\chi-1}L(\Gamma(s)\zeta(s);\omega)\} = 2\pi\omega^{\chi-1}\sum_{n=0}^{\infty}{}_1\Psi_1\left[\begin{matrix}(1-\chi+\gamma+\delta,-1)\\ (1-\chi+\gamma,-1)\end{matrix}\middle\vert -(n+1)e^{\omega}\right]$
	$I_{+}^{\delta}(\omega^{\chi-1}L\{\Gamma(s)\zeta(s);\omega\}) = 2\pi\omega^{\chi+\delta-1}\sum_{n=0}^{\infty}{}_1\Psi_1\left[\begin{matrix}(\chi,1)\\ (\delta+\chi,1)\end{matrix}\middle\vert -(n+1)e^{\omega}\right]$
	$I_{-}^{\delta}(\omega^{\chi-1}L\{\Gamma(s)\zeta(s);\omega\}) = 2\pi\omega^{\chi+\delta-1}\sum_{n=0}^{\infty}{}_1\Psi_1\left[\begin{matrix}(1-\delta-\chi,-1)\\ (1-\chi,-1)\end{matrix}\middle\vert -(n+1)e^{\omega}\right]$
	$D_{0+}^{\delta}\{\omega^{\chi-1}L(\Gamma(s)\zeta(s);\omega)\} = 2\pi\omega^{\chi-1-\delta}\sum_{n=0}^{\infty}{}_1\Psi_1\left[\begin{matrix}(\chi,1)\\ (\chi-\delta,1)\end{matrix}\middle\vert -(n+1)e^{\omega}\right]$
	$D_{-}^{\delta}\{\omega^{\chi-1}L(\Gamma(s)\zeta(s);\omega)\} = 2\pi\omega^{\chi-1-\delta}\sum_{n=0}^{\infty}{}_1\Psi_1\left[\begin{matrix}(\delta-\chi+1,-1)\\ (1-\chi,-1)\end{matrix}\middle\vert -(n+1)e^{\omega}\right]$

Remark 2. *It is mentionable that the succeeding result involving the products of a large class of special functions is because of (26) and (27):*

$$\int_0^1 \frac{\omega^{\rho-1}}{\exp(e^{\omega})-1} H_{m,m}^{m,0}\left[\omega\middle\vert\begin{matrix}\left(\gamma_k+\delta_k+1-\frac{1}{\beta_k},\frac{1}{\beta_k}\right)_1^m\\ \left(\gamma_k+1-\frac{1}{\beta_k},\frac{1}{\beta_k}\right)_1^m\end{matrix}\right]d\omega = \omega^{\rho-1}\sum_{n=0}^{\infty}{}_m\Psi_m\left[\begin{matrix}\left(\gamma_k+1-\frac{1}{\beta_k},\frac{1}{\beta_k}\right)_1^m\\ \left(\gamma_k+\delta_k+1-\frac{1}{\beta_k},\frac{1}{\beta_k}\right)_1^m\end{matrix}\middle\vert -(n+1)e^{\omega}\right]. \tag{32}$$

Remark 3. *Using the principle of mathematical induction for (23), it can be proved that*

$$\sum_{n=0}^{\infty}{}_p\Psi_q\left[\begin{matrix}(a_i,A_i)\\ (b_j,B_j)\end{matrix};-(n+1)e^{\omega}\right] = \sum_{n,l=0}^{\infty}\frac{(-(n+1))^l}{l!}{}_p\Psi_q\left[\begin{matrix}(a_i,A_i)\\ (b_j,B_j)\end{matrix};l\omega\right]. \tag{33}$$

3.2. Formulation of Fractional Kinetic Equation Involving Riemann Zeta-Function

The use of non-integer operators has emerged recently in the different disciplines of engineering and the physical sciences [35–40]. For instance, the fractional kinetic equation is important to investigate the theory of gases, aerodynamics, and astrophysics [41–48]. By reviewing the literature, it is found that the fractional kinetic equation comprising the Riemann zeta function is not formulated. The main purpose of this section is to formulate and solve this problem.

The change in the rates of production using subsequent kinetic equations to analyze the reaction and destruction is described in [41]:

$$\frac{d\varepsilon}{dt} = -d(\varepsilon_t) + p(\varepsilon_t), \tag{34}$$

where ε_t is given by $\varepsilon_t(t^*) = \varepsilon(t - t^*), t^* > 0$. Further to this $\varepsilon = \varepsilon(t)$ = change in reaction, $d = d(\varepsilon)$ = change in destruction, and $p = p(\varepsilon)$ = change in production. The following is obtainable by ignoring the spatial fluctuation and inhomogeneity of $\varepsilon(t)$ with the concentration of species, $\varepsilon_j(t = 0) = \varepsilon_0$:

$$\frac{d\varepsilon_j}{dt} = -c_j\varepsilon_j(t). \tag{35}$$

Next, ignoring subscript j and integrating (35) yields

$$\varepsilon(t) - \varepsilon_0 = -c\, I_{0+}^{-1}\varepsilon(t).$$

The non-integer-order kinetic equation is due to [41]:

$$\varepsilon(t) - \varepsilon_0 = -c^\delta I_{0+}^\delta \varepsilon(t), \tag{36}$$

where $I_{0+}^\delta, \delta > 0$ is the Riemann–Liouville fractional integral, c is a constant, and its Laplace transform is given by

$$L\Big\{\, I_{0+}^\delta \varepsilon(t); \omega \Big\} = \omega^{-\delta}\varepsilon(\omega). \tag{37}$$

Following to Haubold and Mathai [41], we next formulate and solve the fractional kinetic equation so that for any integrable function *f(t)*, we have

$$\varepsilon(t) - f(t)\varepsilon_0 = -d^\delta\, I_{0+}^\delta \varepsilon(t). \tag{38}$$

In light of this discussion, the fractional kinetic equation involving the Riemann zeta function is formulated and solved in Theorem 2. This becomes possible only due to the Reimann zeta-function's new representation involving the delta function; otherwise, the Laplace transform is not found before the w.r.t variable s (see [49]).

Theorem 2. *For $\delta > 0$, the solution of a given fractional kinetic equation containing the Riemann zeta function is*

$$\varepsilon(t) - \varepsilon_0\Gamma(t)\zeta(t) = -d^\delta\, I_{0+}^\delta \varepsilon(t) \tag{39}$$

$$\varepsilon(t) = \frac{2\pi\varepsilon_0}{t} \sum_{n,l,p=0}^{\infty} \frac{(-(n+1))^l \left(\frac{l}{t}\right)^p}{l!p!} E_{\delta,-p}\Big(-d^\delta t^\delta\Big). \tag{40}$$

Proof. Applying Laplace's transformation to (39) and making use of (25) as well as (37) gives

$$\varepsilon(\omega) = 2\pi\varepsilon_0 \sum_{n,l,p=0}^{\infty} \frac{(-(n+1))^l (l)^p}{l!p!}\, \omega^p - \left(\frac{\omega}{d}\right)^{-\delta} \varepsilon(\omega). \tag{41}$$

Therefore, we have

$$\varepsilon(\omega)\left[1+\left(\frac{\omega}{d}\right)^{-\delta}\right]=2\pi\varepsilon_0\sum_{n,l,p=0}^{\infty}\frac{(-(n+1))^l(l)^p}{l!p!}\omega^p, \quad (42)$$

and

$$\varepsilon(\omega)=2\pi\varepsilon_0\sum_{n,l,p=0}^{\infty}\frac{(-(n+1))^l(l)^p}{l!p!}\omega^p\sum_{m=0}^{\infty}\left[-\left(\frac{\omega}{d}\right)^{-\delta}\right]^m. \quad (43)$$

By considering $\delta m-p>0;\delta>0$ and using $L^{-1}\{\omega^{-\delta};t\}=\frac{t^{\delta-1}}{\Gamma(\delta)}$, the inverse Laplace transform of (43) is given by

$$\varepsilon(t)=2\pi\varepsilon_0\sum_{n,l,p=0}^{\infty}\frac{(-(n+1))^l(l)^p}{l!p!}t^{-p-1}\times\sum_{m=0}^{\infty}\frac{\left(-d^\delta t^\delta\right)^m}{\Gamma(\delta m-p)}. \quad (44)$$

Lastly, making use of (19) in the above equation (44) provides the solution as stated in (39) and (40). □

Remark 4. *It can be noted that the solution methodology of Theorem 2 is in line with the existing methods [41–48], and, as expected, the reaction rate $\varepsilon(t)$ contains the Mittag–Leffler function governed by the non-integer parameter δ. Furthermore, the sum over the coefficients in (40) is well-defined and can be computed as follows:*

$$C(t)=\sum_{n,l,p=0}^{\infty}\frac{(-(n+1))^l\left(\frac{l}{t}\right)^p}{l!p!}=\frac{1}{exp\left(e^{\frac{1}{t}}\right)-1}. \quad (45)$$

Likewise, $\lim_{t\to\infty}C(t)=\frac{1}{\exp(1)-1}$ *and* $\lim_{t\to 0}C(t)=0$.

3.3. Further New Properties of the Riemann Zeta function as a Distribution

The Dirac delta function is a linear functional, which transforms each function to its value at zero. Hence, using (4), we have

$$\int_{s\epsilon\mathbb{C}}\wp(s)\Gamma(s)\zeta(s)ds=2\pi\sum_{n,l=0}^{\infty}\frac{(-(n+1))^l}{l!}\delta(s+l),\wp(s)=2\pi\sum_{n,l=0}^{\infty}\frac{(-(n+1))^l}{l!}\wp(-l), \quad (46)$$

or, for a real t, using (24), we have

$$\int_{t\epsilon\mathbb{R}}\wp(t)\Gamma(t)\zeta(t)dt=2\pi\sum_{n,l,p=0}^{\infty}\frac{(-(n+1))^l(l)^p}{l!p!}\delta^{(p)}(t),\wp(t)=2\pi\sum_{n,l,p=0}^{\infty}\frac{(-(n+1))^l(l)^p}{l!p!}(-1)^p\wp^{(p)}(0) \quad (47)$$

and from the above equations it follows that the most properties that hold for the delta function will also hold for the Riemann zeta-function. It can be noted that the sum over the co-efficient in (46) and (47) is finite and well-defined, as well as rapidly decreasing. This sum also defines a new transform named the zeta transform, and the following formulae given in Table 2 and many others can be obtained using it.

Table 2. Zeta Transform.

Function	Zeta Transform
e^{at}	$\frac{2\pi}{exp(e^{-a})-1}$
$sinat$	$IMG\left(\frac{2\pi}{exp(e^{-ia})-1}\right)$

Table 2. *Cont.*

Function	Zeta Transform
$cosat$	$Re\left(\frac{2\pi}{exp(e^{-ia})-1}\right)$
$E_\alpha(s)$	$2\pi \sum\limits_{n,l=0}^{\infty} \frac{(-(n+1))^l}{l!} E_\alpha(-l)$
$K_v\ (s)$ [McDonald function [4]]	$2\pi \sum\limits_{n,l=0}^{\infty} \frac{(-(n+1))^l}{l!} K_v(-l)$
$H_{r,r}^{r,0}\left[\zeta \middle\vert \begin{matrix} \left(\gamma_k+\delta_k+1-\frac{1}{\beta_k},\frac{1}{\beta_k}\right)_1^r \\ \left(\gamma_k+1-\frac{1}{\beta_k},\frac{1}{\beta_k}\right)_1^r \end{matrix}\right]$	$2\pi \sum\limits_{n,l=0}^{\infty} \frac{(-(n+1))^l}{l!} H_{r,r}^{r,0}\left[-l \middle\vert \begin{matrix} \left(\gamma_k+\delta_k+1-\frac{1}{\beta_k},\frac{1}{\beta_k}\right)_1^r \\ \left(\gamma_k+1-\frac{1}{\beta_k},\frac{1}{\beta_k}\right)_1^r \end{matrix}\right]$

The purpose of the remaining section is to enlist the new properties of the Riemann zeta function as a distribution by following the concepts and methodology of [23] (Chapter 7, pp. 199–207), which is achieved due to the zeta function's new representations (4) and (24) in terms of the delta function. First note that the frequently used test functions [22,23] are of either compact support, or they are rapidly decreasing as well as infinitely differentiable. These domains of test functions are commonly denoted by $\mathcal{D}$ and $\mathcal{S}$, respectively, and their codomains are the spaces $\mathcal{D}'$ and $\mathcal{S}'$ (also known as their dual spaces). Actually, $\mathcal{D}$ and $\mathcal{D}\prime$ are not closed w.r.t Fourier transforms, but $\mathcal{S}$ and $\mathcal{S}'$ are closed. Another space of test functions is denoted by $\mathcal{Z}$, which is the space of the analytic and its entire functions. Hence, Fourier transforms of the elements of $\mathcal{D}'$ to belong to $\mathcal{Z}'$, which is dual to $\mathcal{Z}$, and Fourier transforms of the elements of $-\mathcal{Z}$ into $\mathcal{D}$ [22,23]. Therefore, the Fourier transform as well as its inverse are continuous linear functionals from $\mathcal{D}\prime$ to $\mathcal{Z}'$ ([23], p. 203). Since the complex delta function is an element of $\mathcal{Z}'$, from (4), it is therefore obvious that $\Gamma(s)\zeta(s)$ is also an element of $\mathcal{Z}'$. In light of this discussion, the following theorem follows.

Theorem 3. *Suppose f is a distribution of bounded support; then,*

$$\mathcal{F}\left[f(y) * \frac{\sqrt{2\pi}e^{\sigma y}}{exp(e^{y}) - 1}; s\right] = \mathcal{F}[f(s)]\Gamma(s)\zeta(s). \tag{48}$$

Proof. Because $\Gamma(s)\zeta(s) \in \mathcal{Z}'$ and $\mathcal{F}; \mathcal{F}^{-1}$ are continuous linear functionals from $\mathcal{D}'$ to $\mathcal{Z}'$. Further, we have [14], Equation (42):

$$\mathcal{F}\left[\frac{\sqrt{2\pi}e^{\sigma y}}{exp(e^{y}) - 1}; \tau\right] = \Gamma(s)\zeta(s); (s = \sigma + i\tau). \tag{49}$$

Therefore, $\frac{\sqrt{2\pi}e^{\sigma y}}{exp(e^{y})-1}$ is an element of $\mathcal{D}'$ being a Fourier transform of an element of space $\mathcal{Z}'$. Hence, the proof of the result (48) is complete using ([23], p. 206, Theorem 7.9.1). □

Example 1. *Consider a functionf with bounded support defined by*

$$f(y) = \begin{cases} 1 & |y| < 1 \\ 0 & |y| \geq 1 \end{cases} \tag{50}$$

Then, according to Theorem 3,

$$\mathcal{F}\left[f(y) * \frac{\sqrt{2\pi}e^{\sigma y}}{exp(e^{y}) - 1}\right] = \mathcal{F}[f(y); s]\mathcal{F}\left[\frac{\sqrt{2\pi}e^{\sigma y}}{exp(e^{y}) - 1}; s\right] = \frac{sins}{s}\Gamma(s)\zeta(s), \tag{51}$$

yields a valuable consequence of distributional representation.

Continuing in this way, we can apply the elements of distributions to the Riemann zeta function using its distributional representation (4) and (24). Some of these are listed below in Table 3, and it is mentionable that the proof of all these properties simply follows from the properties of the delta function and are therefore omitted. Here, we restrict these over the space of complex analytic functions $\wp(s)\epsilon\mathcal{Z}$, as defined in [22,23], but these properties may hold for a large space of test functions, and it is supposed that c_1, γ, and c_2 are constants.

Table 3. Properties of Riemann Zeta function as a distribution.

addition with an arbitrary distribution f	$\langle\Gamma(s)\zeta(s)+f,\ \wp(s)\rangle = \langle\Gamma(s)\zeta(s),\wp(s)+f,\ \wp(s)\rangle$
multiplication with an arbitrary constant c_1	$\langle c_1\Gamma(s)\zeta(s),\wp(s)\rangle = \langle\Gamma(s)\zeta(s),c_1\wp(s)\rangle$
shifting by an arbitrary complex constant γ	$\langle\Gamma(s-\gamma)\zeta(s-\gamma),\wp(s)\rangle = \langle\Gamma(s)\zeta(s),\wp(s+\gamma)\rangle$
transposition	$\langle\Gamma(-s)\zeta(-s),\wp(s)\rangle = \langle\Gamma(s)\zeta(s),\wp(-s)\rangle$
multiplication of the independent variable with a positive constant c_1	$\langle\Gamma(c_1s)\zeta(c_1s),\wp(s)\rangle = \langle\Gamma(s)\zeta(s),\frac{1}{c_1}\wp\left(\frac{s}{c_1}\right)\rangle$
distributional differentiation	$\langle\frac{d^k}{ds^k}(\Gamma(s)\zeta(s)),\wp(s)\rangle = \sum\limits_{n,l=0}^{\infty}\frac{(-(n+1))^l}{l!}(-1)^k\wp^k(-l)$
distributional Fourier transform	$\langle\mathcal{F}[\Gamma(s)\zeta(s)],\wp(s)\rangle = \langle\Gamma(s)\zeta(s),\mathcal{F}[\wp](s)\rangle$
duality property of Fourier transform	$\langle\mathcal{F}[\Gamma(s)\zeta(s)],\mathcal{F}[\wp(s)]\rangle = \langle 2\pi\Gamma(s)\zeta(s),\wp(-s)\rangle$
Parseval's identity of Fourier transform	$\langle\mathcal{F}[\Gamma(s)\zeta(s)],\overline{\mathcal{F}[\wp(s)]}\rangle = \langle\overline{\mathcal{F}[\Gamma(s)\zeta(s)]},\mathcal{F}[\wp(s)]\rangle =$ $2\pi\langle[\Gamma(\sigma)\zeta(\sigma)],\overline{[\wp(\sigma)]}\rangle;\sigma=\Re(s)$
differentiation property of Fourier transform	$\langle\mathcal{F}\left[\frac{d^k}{ds^k}(\Gamma(s)\zeta(s))\right],\wp(s)\rangle = \langle(-it)^m\Gamma(s)\zeta(s),\mathcal{F}[\wp](s)\rangle$
Taylor series	$\langle\Gamma(s+c_1)\zeta(s+c_1),\wp(s)\rangle = \langle\sum\limits_{n=0}^{\infty}\frac{(c_1)^n}{n!}\frac{d^n}{ds^n}(\Gamma(s)\zeta(s)),\wp(s)\rangle$
Convolution property	$\Gamma(t)\zeta(t)*f(t)= 2\pi\sum\limits_{n,l=0}^{\infty}\frac{(-(n+1))^l(l)^p}{l!p!}\frac{d^p}{dt^p}(f(t))$
$\Gamma(s)\zeta(s)*exp(as)$	$\frac{2\pi e^{as}}{exp(e^a)-1}$

4. Conclusions

The calculation of the images of special functions using the fractional calculus operators has emerged as a popular subject. In this research, we have obtained fractional calculus images involving Riemann zeta-functions and their simpler cases. Specifications of these results were discussed for $m=3$, $m=2$, and $m=1$. It is reasonable to verify, in view of (25), that Theorems 3 and 4 of [24] are applicable, and the main result (27) and its several special cases are completely verifiable with these theorems. A new fractional kinetic equation involving the Riemann zeta function was formulated and solved. A newly obtained representation of the Riemann zeta function and its Laplace transform has played a crucial role in accomplishing the goals of this research. Certain distributional properties of the Riemann zeta function and examples were also discussed. We hope that this confluence of distribution theory and the function of analytic number theory will have far-reaching applications in the future.

Author Contributions: Conceptualization, A.T.; methodology, A.T.; validation, A.T. and R.S.; formal analysis, A.T. and R.S.; investigation, A.T.; resources, A.T. and R.S.; data curation, R.S.; writing—original draft preparation, A.T.; writing—review and editing, R.S.; visualization, R.S.; supervision, R.S.; project administration, R.S.; funding acquisition, A.T. All authors have read and agreed to the published version of the manuscript.

Funding: This research received no external funding.

Data Availability Statement: Not applicable.

Acknowledgments: Asifa Tassaddiq would like to thank the Deanship of Scientific Research at Majmaah University for supporting this work under Project No. R-2022-133. The authors are also thankful to the worthy reviewers and editors for their useful and valuable suggestions for the improvement of this paper which led to a better presentation.

Conflicts of Interest: The authors declare no conflict of interest.

Appendix A

Related Special Cases to (10)

Case 1: Marichev–Saigo–Maeda fractional integral operator

First of all, let us consider the case $m = 3$ and further take $\beta_1 = \beta_2 = \beta_3 = \beta = 1$ in (10). Then, the kernel of (10) will reduce to a special case of the H-function $H^{3,0}_{3,3}$ that has the following relation with the Meijer G-function $G^{3,0}_{3,3}$–function and the Appel function (Horn function) F_3 ([2], Vol. 1):

$$H^{3,0}_{3,3}\left(\frac{t}{x}\right) = G^{3,0}_{3,3}\left[\frac{t}{x}\left|\begin{matrix}\gamma_1' + \gamma_2', & \delta-\gamma_1, \delta-\gamma_2 \\ \gamma_1', \gamma_2', & \delta-\gamma_1-\gamma_2\end{matrix}\right.\right] = \frac{x^{-\gamma_1}}{\Gamma(\delta)}(x-t)^{\delta-1}t^{-\gamma_1'}F_3\left(\gamma_1,\gamma_1',\gamma_2,\gamma_2',\delta;1-\frac{t}{x};1-\frac{x}{t}\right) \quad (A1)$$

where

$$F_3(\gamma_1,\gamma_1',\gamma_2,\gamma_2',\delta;u;v) = \sum_{k,l=0}^{\infty}\frac{(\gamma_1)_k(\gamma_1')_l(\gamma_2)_k(\gamma_2')_l}{(\delta)_{l+m}}\frac{u^k}{k!}\frac{v^l}{l!}, max(|u|,|v|) < 1. \quad (A2)$$

Hence, due to (A1), for the complex parameters $\gamma_1, \gamma_1', \gamma_2, \gamma_2'$, $\Re(\delta) > 0$, the Marichev–Saigo–Maeda fractional integral operator of integration (see ([2], Vol. 1) is also [31–34]) defined as

$$\left(I^{\gamma_1,\gamma_1',\gamma_2,\gamma_2',\delta}_{0+}f\right)(x) = \frac{x^{-\gamma_1}}{\Gamma(\delta)}\int_0^x (x-t)^{\delta-1}t^{-\gamma_1'}F_3\left(\gamma_1,\gamma_1',\gamma_2,\gamma_2',\delta;1-\frac{t}{x};1-\frac{x}{t}\right)f(t)\mathrm{d}t \quad (A3)$$

and

$$\left(I^{\gamma_1,\gamma_1',\gamma_2,\gamma_2',\delta}_{0-}f\right)(x) = \frac{t^{-\gamma_1'}}{\Gamma(\delta)}\int_x^\infty (x-t)^{\delta-1}t^{-\gamma_1}F_3\left(\gamma_1,\gamma_1',\gamma_2,\gamma_2',\delta;1-\frac{x}{t};1-\frac{t}{x}\right)f(t)\mathrm{d}t. \quad (A4)$$

Both of the above forms have significant importance. Furthermore, it is now obvious from (20)–(23) that the Marichev–Saigo–Maeda fractional integral operator is related to the multiple E–K fractional integral operators as given in (10) for $m = 3$. The Marichev–Saigo–Maeda fractional integral operator can also be expressed as a composition of three commutable classical E–K integrals (see Kiryakova [24,25]) as follows:

$$I^{\gamma_1,\gamma_1',\gamma_2,\gamma_2',\delta}_{0+}f(x) = I^{(0,\delta-\gamma_1-\gamma_1'-\gamma_2,\gamma_2'-\gamma_1'),(\gamma_2',\gamma_2,\delta-\gamma_2-\gamma_2')}_{(1,1,1),3}f(x) = I^{(0,\gamma_2')}_1 I^{(\delta-\gamma_1-\gamma_1'-\gamma_2,\gamma_2)}_1 I^{(\gamma_2'-\gamma_1',\delta-\gamma_2-\gamma_2')}_1 f(x) \quad (A5)$$

Many such representations are found (see Kiryakova [24,25] and cited references) because of the symmetry of variables γ_1, γ_1' and γ_2, γ_2' in F_3, as well as the symmetry in the upper and lower rows of the G-function in (A-1). Hence, the following result holds true in view of (A1)–(A4) and (10) (see also [31–34]).

Let $\gamma_1, \gamma_1', \gamma_2, \gamma_2' \in \mathbb{C}$, $\omega > 0 \wedge \Re(\chi) > \max\{0, \Re(\gamma_1 + \gamma_1' + \gamma_2 - \delta), \Re(\gamma_1' - \gamma_2')\}$, $\Re(\delta) > 0$, then

$$I^{\gamma_1,\gamma_1',\gamma_2,\gamma_2',\delta}_{0+}\left(\omega^{\chi-1}\right) = \frac{\Gamma(\chi)\Gamma(\chi+\delta-\gamma_1-\gamma_1'-\gamma_2)\Gamma(\chi+\gamma_2'-\gamma_1')}{\Gamma(\chi+\gamma_2')\Gamma(\chi+\delta-\gamma_1-\gamma_1')\Gamma(\chi+\delta-\gamma_1'-\gamma_2)}\omega^{\delta+\chi-\gamma_1-\gamma_1'-1} \quad (A6)$$

Similarly, let $\gamma_1, \gamma_1', \gamma_2, \gamma_2' \in \mathbb{C}$, $\omega > 0$, and if $\Re(\delta) > 0$, $\Re(\chi) < 1 +$ min $\{\Re(-\gamma_2), \Re(\gamma_1 + \gamma_1' - \delta), \Re(\gamma_1 + \gamma_2' - \delta)\}$; then, the following image formula holds true in view of (A1)–(A4) and (10) (see also [31–34]):

$$I_{0-}^{\gamma_1,\gamma_1',\gamma_2,\gamma_2',\delta}\left(\omega^{\chi-1}\right) = \frac{\Gamma(1-\chi-\delta+\gamma_1+\gamma_1')\Gamma(1-\chi+\gamma_1+\gamma_2'-\delta)\Gamma(1-\chi-\gamma_1)}{\Gamma(1-\chi)\Gamma(1-\chi+\gamma_1+\gamma_1'+\gamma_2+\gamma_2'-\delta)\Gamma(1-\chi+\gamma_1-\gamma_2)}\omega^{\delta+\chi-\gamma_1-\gamma_1'-1} \quad \text{(A7)}$$

Case 2: Saigo fractional operator

Next, let us consider the case m = 2 with $\beta_1 = \beta_2 = \beta > 0.$; then, the kernel-functions of (10) reduce to the Gauss function [24]:

$$H_{2,2}^{2,0}\left[\sigma\left|\begin{matrix}\left(\gamma_1+\delta_1+1-\frac{1}{\beta},\frac{1}{\beta}\right),\left(\gamma_2+\delta_2+1-\frac{1}{\beta},\frac{1}{\beta}\right)\\ \left(\gamma_1+1-\frac{1}{\beta},\frac{1}{\beta}\right),\left(\gamma_2+1-\frac{1}{\beta},\frac{1}{\beta}\right)\end{matrix}\right.\right] = G_{2,2}^{2,0}\left[\sigma^{\beta}\left|\begin{matrix}\gamma_1+\delta_2,\gamma_2+\delta_2\\ \gamma_1\gamma_2\end{matrix}\right.\right]$$
$$= \beta\,\frac{\sigma^{\beta\gamma_2}\left(1-\sigma^{\beta}\right)^{\delta_1+\delta_2-1}}{\Gamma(\delta_1+\delta_2)}\,{}_2F_1\left(\gamma_2+\delta_2-\gamma_1,\delta_1;\delta_1+\delta_2;1-\sigma^{\beta}\right) \quad \text{(A8)}$$

For the purpose of this investigation, let us focus on two fractional integral operators that are defined for $\gamma_1, \gamma_2, \delta \in \mathbb{C}$ with $x; \Re(\delta) > 0$ by Saigo [33], which can also be obtained by taking $\beta = 1$; $\sigma = \frac{t}{x}$ and then $\sigma = \frac{x}{t}$, also appropriately specifying the other parameter values $\delta_1 + \delta_2 = \delta; \delta_1 = -\gamma_1$ in (A1) and (A8) (see also [31,32]).

$$I_{0+}^{\gamma_1,\gamma_2,\delta} = \frac{x^{-\delta-\gamma_1}}{\Gamma(\delta)}\int_0^x (x-t)^{\delta-1}{}_2F_1\left(\delta+\gamma_2, -\gamma_1; \delta; 1-\frac{t}{x}\right)f(t)\mathrm{d}t \quad \text{(A9)}$$

and

$$I_{-}^{\gamma_1,\gamma_2,\delta}\left(f(x)\right) = \frac{1}{\Gamma(\delta)}\int_x^{\infty} (t-x)^{\delta-1}t^{-\delta-\gamma_1}{}_2F_1\left(\delta+\gamma_2, -\gamma_1; \delta; 1-\frac{x}{t}\right)f(t)\mathrm{d}t \quad \text{(A10)}$$

where ${}_2F_1$ represents the Gauss hypergeometric function given by (see [34]):

$${}_2F_1(\gamma_1, \gamma_2, \gamma_3; u) = \sum_{k=0}^{\infty}\frac{(\gamma_1)_k(\gamma_2)_k}{(\gamma_3)_k}\frac{u^k}{k!}, |u| < 1; |u| = 1(u \neq 1), \Re(\gamma_3 - \gamma_1 - \gamma_2) > 0. \quad \text{(A11)}$$

The Appell function F_3 diminishes to ${}_2F_1$ (Gauss hypergeomatric function) and also contends the following relationships (see [34], p. 301, Equation 9.4):

$$F_3(\gamma_1, \delta-\gamma_1, \gamma_2, \delta-\gamma_2; \delta; u; v) = {}_2F_1(\gamma_1, \gamma_2; \delta; u+v-\mathrm{uv})$$
$$\text{and} \quad \text{(A12)}$$
$$F_3(0, \gamma_1', \gamma_2, \gamma_2', \delta) = {}_2F_1(\gamma_1, \gamma_2; \delta; x); F_3(\gamma_1, 0, \gamma_2, \gamma_2', \delta) = {}_2F_1(\gamma_1', \gamma_2', \delta; y)$$

Hence, the relation of the Marichev–Saigo–Maeda (A3) and (A4) and the Saigo fractional integral operators (A9) and (A10) is obvious using (31) for $\gamma_1 = 0 \vee \gamma_1' = 0$, where both equations also interrelated with (10) in view of (A1) and (A8). Hence using these facts for (10) and (A9), we have (see also [31–34]):

$$I_{0+}^{\gamma_1,\gamma_2,\delta}\left(\omega^{\chi-1}\right) = \frac{\Gamma(\chi)\Gamma(\chi+\gamma_2-\gamma_1)}{\Gamma(\chi-\gamma_2)\Gamma(\chi+\delta+\gamma_2)}\omega^{\chi-\gamma_1-1}, (\gamma_1, \gamma_2, \delta \in \mathbb{C}; \Re(\delta,) > 0, \Re(\chi) > \max[0, \Re(\gamma_1 - \gamma_2)]). \quad \text{(A13)}$$

Similar to (A13), we have the following right-handed formula (see also [31–34]):

$$I_{-}^{\gamma_1,\gamma_2,\delta}\left(\omega^{\chi-1}\right) = \frac{\Gamma(\gamma_1-\chi+1)\Gamma(\gamma_2-\chi+1)}{\Gamma(1-\chi)\Gamma(\gamma_1+\gamma_2+\delta-\chi+1)}\omega^{\chi-\gamma_1-1};$$
$$\gamma_1, \gamma_2, \delta \in \mathbb{C} \wedge \Re(\delta) > 0 \wedge \Re(\chi) < 1 + \min[\Re(\gamma_1), \Re(\gamma_2)]. \quad \text{(A14)}$$

Case 3: Erdélyi–Kober (E–K) and the Riemann–Liouville (R–L) fractional operator

Let us consider $m = 1$ in (10); then, the kernel function of (10) becomes

$$H^{1,1}_{1,0}\left[\sigma\left|\begin{matrix}\left(\gamma+\delta,\frac{1}{\beta}\right)\\ \left(\gamma,\frac{1}{\beta}\right)\end{matrix}\right.\right] = \beta\sigma^{\beta-1}G^{1,1}_{1,0}\left[\sigma^{\beta}\left|\begin{matrix}\gamma+\delta\\ \gamma\end{matrix}\right.\right] = \beta\,\frac{\sigma^{\beta\gamma+\beta-1}\left(1-\sigma^{\beta}\right)^{\delta-1}}{\Gamma(\delta)} \quad \text{(A15)}$$

and one can obtain the classical fractional operators, namely, the Erdélyi–Kober (E–K) operators:

$$I^{\gamma,\delta}_{\beta} f(z) = \frac{1}{\Gamma(\delta)}\int_0^1 \sigma^{\gamma}\,(1-\sigma)^{\delta-1} f\left(z\sigma^{\frac{1}{\beta}}\right)d\sigma, \delta \geq 0,\ \beta > 0,\ \gamma \in \mathbb{R}. \quad \text{(A16)}$$

Further, for $\gamma_1 = 0, \gamma_2 = \gamma$, the Saigo operators (A9) and (A10) reduce the other fractional operators, namely, the Erdélyi–Kober integrals defined for complex $\gamma, \delta \in \mathbb{C}$, $\Re(\delta) > 0$, (see also [31–34]):

$$I^{0,\gamma,\delta}_{0+}(\,f(x)) = \left(I^{\gamma,\delta}_{0+}\,f\right)(x) = \frac{x^{-\delta-\gamma}}{\Gamma(\chi)}\int_0^x (x-t)^{\delta-1}t^{\gamma}f(t)\mathrm{d}t\ (x>0) \quad \text{(A17)}$$

$$I^{0,\gamma,\delta}_{0-}(\,f(x)) = \left(I^{\gamma,\delta}_{0-}\,f\right)(x) = \frac{x^{\gamma}}{\Gamma(\chi)}\int_x^{\infty} (t-x)^{\delta-1}t^{-\delta-\gamma}f(t)\mathrm{d}t\ (x>0) \quad \text{(A18)}$$

It is obvious that (A17) and (A18) are also obtainable from (A16) for specific values of $\beta = 1$; $\sigma = \frac{t}{x}$ and then $\sigma = \frac{x}{t}$. Similarly, the Saigo operators are also related with the E–K and the Riemann–Liouville (R–L) operators:

$$I^{0,\gamma,\delta}_{0+}(\,f(x)) = I^{\gamma_1,0,\delta}_{0+}(\,f(x)); I^{\gamma,\delta}_{0+}(\,f(x)) = I^{\gamma,\delta}_{0-}\,(\,f(x)); \Re(\delta) > 0 \quad \text{(A19)}$$

Continuing in this way, if $\gamma_1 = -\delta$, the Saigo operators (A9) and (A10) reduce to the Riemann–Liouville (R–L) operators (see also [31–34]). The classical left-hand-sided Riemann–Liouville fractional integrals I^{δ}_{0+} and right-hand-sided Riemann–Liouville fractional integrals I^{δ}_{-} of order $\delta \in \mathbb{C}$, $\Re(\delta) > 0$ are defined by [7–9]:

$$I^{\delta}_{0+}(f(x)) = \frac{1}{\Gamma(\delta)}\int_0^x (x-t)^{\delta-1}f(t)dt \qquad (x>0) \quad \text{(A20)}$$

and

$$I^{\delta}_{-}(f(x)) = \frac{1}{\Gamma(\delta)}\int_x^{\infty} (x-t)^{\delta-1}f(t)dt \qquad (x>0) \quad \text{(A21)}$$

respectively. These are also related to the Weyl transform [7–9]. It can be noted that (A20) and (A21) are also obtainable from (A16) for specific values of the involved parameters.

References

1. Lapidus, M.L.; van Frankenhuijsen, M. *Fractal Geometry, Complex Dimensions and Zeta Functions: Geometry and Spectra of Fractal Strings*; Springer: Berlin/Heidelberg, Germany, 2012.
2. Titchmarsh, E.C. *The Theory of the Riemann Zeta Function*; Oxford University Press: Oxford, UK, 1951.
3. Erdélyi, A.; Magnus, W.; Oberhettinger, F.; Tricomi, F. *Higher Transcendental Functions*; McGraw-Hill Book Corp.: New York, NY, USA, 1953; Volume 1–2.
4. Chaudhry, M.A.; Zubair, S.M. *On a Class of Incomplete Gamma Functions with Applications*; Chapman and Hall (CRC Press Company): Boca Raton, FL, USA; London, UK; New York, NY, USA; Washington, DC, USA, 2001.
5. Srivastava, H.M.; Choi, J. *Series Associated with the Zeta and Related Functions*; Kluwer Academic Publishers: Dordrecht, The Netherlands; Boston, MA, USA; London, UK, 2001.
6. Srivastava, H.M. Some parametric and argument variations of the operators of fractional calculus and related special functions and integral transformations. *J. Nonlinear Convex Anal.* **2021**, *22*, 1501–1520.
7. Srivastava, H.M. An introductory overview of fractional-calculus operators based upon the Fox-Wright and related higher transcendental functions. *J. Adv. Engrg. Comput.* **2021**, *5*, 135–166. [CrossRef]
8. Srivastava, H.M. Some general families of the Hurwitz-Lerch Zeta functions and their applications: Recent developments and directions for further researches. *Proc. Inst. Math. Mech. Nat. Acad. Sci. Azerbaijan* **2019**, *45*, 234–269. [CrossRef]

9. Srivastava, H.M. The Zeta and related functions: Recent developments. *J. Adv. Engrg. Comput.* **2019**, *3*, 329–354. [CrossRef]
10. Tassaddiq, A. A New Representation for Srivastava's λ-Generalized Hurwitz-Lerch Zeta Functions. *Symmetry* **2018**, *10*, 733. [CrossRef]
11. Tassaddiq, A. A New Representation of the generalized Krätzel function. *Mathematics* **2020**, *8*, 2009. [CrossRef]
12. Chaudhry, M.A.; Qadir, A. Fourier transform and distributional representation of Gamma function leading to some new identities. *Int. J. Math. Math. Sci.* **2004**, *37*, 2091–2096. [CrossRef]
13. Tassaddiq, A.; Qadir, A. Fourier transform and distributional representation of the generalized gamma function with some applications. *Appl. Math. Comput.* **2011**, *218*, 1084–1088. [CrossRef]
14. Tassaddiq, A.; Qadir, A. Fourier transform representation of the extended Fermi-Dirac and Bose-Einstein functions with applications to the family of the zeta and related functions. *Integral Transform. Spec. Funct.* **2018**, *22*, 453–466. [CrossRef]
15. Tassaddiq, A. Some Representations of the Extended Fermi-Dirac and Bose-Einstein Functions with Applications. Ph.D. Dissertation, National University of Sciences and Technology Islamabad, Islamabad, Pakistan, 2011.
16. Al-Lail, M.H.; Qadir, A. Fourier transform representation of the generalized hypergeometric functions with applications to the confluent and gauss hypergeometric functions. *Appl. Math. Comput.* **2015**, *263*, 392–397. [CrossRef]
17. Tassaddiq, A. A New Representation of the Extended Fermi-Dirac and Bose-Einstein Functions. *Int. J. Math. Appl.* **2017**, *5*, 435–446.
18. Tassaddiq, A.; Safdar, R.; Kanwal, T. A distributional representation of gamma function with generalized complex domain. *Adv. Pure Math.* **2017**, *7*, 441–449. [CrossRef]
19. Tassaddiq, A. A new representation of the k-gamma functions. *Mathematics* **2019**, *10*, 133. [CrossRef]
20. Tassaddiq, A. An application of theory of distributions to the family of λ-generalized gamma function. *AIMS Math.* **2020**, *5*, 5839–5858. [CrossRef]
21. Tassaddiq, A. A new representation of the extended k-gamma function with applications. *Math. Meth. Appl. Sci.* **2021**, *44*, 11174–11195. [CrossRef]
22. Gel'fand, I.M.; Shilov, G.E. *Generalized Functions: Properties and Operations*; Academic Press: New York, NY, USA, 1969; Volume 1.
23. Zamanian, A.H. *Distribution Theory and Transform Analysis*; Dover Publications: New York, NY, USA, 1987.
24. Kiryakova, V. Unified Approach to Fractional Calculus Images of Special Functions—A Survey. *Mathematics* **2020**, *8*, 2260. [CrossRef]
25. Kiryakova, V. A Guide to Special Functions in Fractional Calculus. *Mathematics* **2021**, *9*, 106. [CrossRef]
26. Agarwal, R.; Jain, S.; Agarwal, R.P.; Baleanu, D. A remark on the fractional integral operators and the image formulas of generalized Lommel-Wright function. *Front. Phys.* **2018**, *6*, 79. [CrossRef]
27. Kiryakova, V. Commentary: A remark on the fractional integral operators and the image formulas of generalized Lommel-Wright function. *Front. Phys.* **2019**, *7*, 145. [CrossRef]
28. Agarwal, R.; Jain, S.; Agarwal, R.P.; Baleanu, D. Response: Commentary: A remark on the fractional integral operators and the image formulas of generalized Lommel-Wright function. *Front. Phys.* **2020**, *8*, 72. [CrossRef]
29. Kilbas, A.A. *H-Transforms: Theory and Applications*, 1st ed.; CRC Press: Boca Raton, FL, USA, 2004. [CrossRef]
30. Mittag-Leffler, M.G. Sur la nouvelle fonction E(x). *Comptes Rendus Acad. Sci.* **1903**, *137*, 554–558.
31. Marichev, O.I. Volterra equation of Mellin convolutional type with a Horn function in the kernel. *Izv. AN BSSR Ser. Fiz. Mat. Nauk* **1974**, *1*, 128–129. (In Russian)
32. Saigo, M.; Maeda, N. More generalization of fractional calculus. In *Transformation Methods & Special Functions, Varna '96: Second International Workshop: Proceedings*; Rusev, P., Dimovski, I., Kiryakova, V., Eds.; Science Culture Technology Publishing: Singapore, 1998; pp. 386–400.
33. Saigo, M. A remark on integral operators involving the Gauss hypergeometric functions. *Math. Rep. Coll. Gen. Ed. Kyushu Univ.* **1978**, *11*, 135–143.
34. Srivastava, H.M.; Karlsson, P.W. *Multiple Gaussian Hypergeometric Series (Ellis Horwood Series in Mathematics and Its Applications)*; Halsted Press: Chichester, UK, 1985.
35. Abboubakar, H.; Kom Regonne, R.; Sooppy Nisar, K. Fractional Dynamics of Typhoid Fever Transmission Models with Mass Vaccination Perspectives. *Fractal Fract.* **2021**, *5*, 149. [CrossRef]
36. Yasmin, H. Numerical Analysis of Time-Fractional Whitham-Broer-Kaup Equations with Exponential-Decay Kernel. *Fractal Fract.* **2022**, *6*, 142. [CrossRef]
37. Delgado, B.B.; Macías-Díaz, J.E. On the General Solutions of Some Non-Homogeneous Div-Curl Systems with Riemann–Liouville and Caputo Fractional Derivatives. *Fractal Fract.* **2021**, *5*, 117. [CrossRef]
38. Feng, Z.; Ye, L.; Zhang, Y. On the Fractional Derivative of Dirac Delta Function and Its Application. *Adv. Math. Phys.* **2020**, *2020*, 1842945. [CrossRef]
39. Derbazi, C.; Baitiche, Z.; Abdo, M.S.; Shah, K.; Abdalla, B.; Abdeljawad, T. Extremal Solutions of Generalized Caputo-Type Fractional-Order Boundary Value Problems Using Monotone Iterative Method. *Fractal Fract.* **2022**, *6*, 146. [CrossRef]
40. Srivastava, H.M. Some families of Mittag-Leffler type functions and associated operators of fractional calculus. *TWMS J. Pure Appl. Math.* **2016**, *7*, 123–145.
41. Haubold, H.J.; Mathai, A.M. The fractional kinetic equation and thermonuclear functions. *Astrophys. Space Sci.* **2000**, *273*, 53–63. [CrossRef]

42. Saxena, R.K.; Mathai, A.M.; Haubold, H.J. On fractional kinetic equations. *Astrophys. Space Sci.* **2002**, *282*, 281–287. [CrossRef]
43. Saxena, R.K.; Mathai, A.M.; Haubold, H.J. On generalized fractional kinetic equations. *Phys. A* **2004**, *344*, 653–664. [CrossRef]
44. Saxena, R.K.; Mathai, A.M.; Haubold, H.J. Unified fractional kinetic equations and a fractional diffusion equation. *Astrophys. Space Sci.* **2004**, *290*, 299–310. [CrossRef]
45. Saxena, R.K.; Kalla, S.L. On the solutions of certain fractional kinetic equations. *Appl. Math. Comput.* **2008**, *199*, 504–511. [CrossRef]
46. Chaurasia, V.B.L.; Pandey, S.C. On the new computable solution of the generalized fractional kinetic equations involving the generalized function for the fractional calculus and related functions. *Astrophys. Space Sci.* **2008**, *317*, 213–219. [CrossRef]
47. Chaurasia, V.B.L.; Kumar, D. On the solutions of generalized fractional kinetic equations. *Adv. Stud. Theor. Phys.* **2010**, *4*, 773–780.
48. Chaurasia, V.B.L.; Singh, Y. A novel computable extension of fractional kinetic equations arising in astrophysics. *Int. J. Adv. Appl. Math. Mech.* **2015**, *3*, 1–9.
49. Srivastava, H.M. A survey of some recent developments on higher transcendental functions of analytic number theory and applied mathematics. *Symmetry* **2021**, *13*, 2294. [CrossRef]

fractal and fractional

MDPI

Article

Novel Computations of the Time-Fractional Fisher's Model via Generalized Fractional Integral Operators by Means of the Elzaki Transform

Saima Rashid [1], Zakia Hammouch [2,3,4], Hassen Aydi [5,6,7,*], Abdulaziz Garba Ahmad [8] and Abdullah M. Alsharif [9]

1 Department of Mathematics, Government College University, Faisalabad 38000, Pakistan; saimarashid@gcuf.edu.pk
2 Division of Applied Mathematics, Thu Dau Mo University, Thu Dau Mot 75000, Vietnam; hammouch_zakia@tdmu.edu.vn
3 Ecole Normale Supéerieure de Meknés, Université Moulay Ismail, Meknes 50000, Morocco
4 Department of Mathematics and Science Education, Harran University, Sanliurfa 63510, Turkey
5 Institut Supérieur d'Informatique et des Techniques de Communication, Université de Sousse, Hammam Sousse 4000, Tunisia
6 China Medical University Hospital, China Medical University, Taichung 40402, Taiwan
7 Department of Mathematics and Applied Mathematics, Sefako Makgatho Health Sciences University, P.O. Box 60, Ga-Rankuwa 0208, South Africa
8 Department of Mathematics, National Mathematical Centre Abuja, Abuja 900211, Nigeria; agarbaahmad@yahoo.com
9 Department of Mathematics, Faculty of Science, Taif University, P.O. Box 11099, Taif 21944, Saudi Arabia; a.alshrif@tu.edu.sa
* Correspondence: hassen.aydi@isima.rnu.tn

Citation: Rashid, S.; Hammouch, Z.; Aydi, H.; Ahmad, A.G.; Alsharif, A.M. Novel Computations of the Time-Fractional Fisher's Model via Generalized Fractional Integral Operators by Means of the Elzaki Transform. *Fractal Fract.* **2021**, *5*, 94. https://doi.org/10.3390/fractalfract5030094

Academic Editors: Asifa Tassaddiq and Muhammad Yaseen

Received: 4 July 2021
Accepted: 6 August 2021
Published: 12 August 2021

Abstract: The present investigation dealing with a hybrid technique coupled with a new iterative transform method, namely the iterative Elzaki transform method (IETM), is employed to solve the nonlinear fractional Fisher's model. Fisher's equation is a precise mathematical result that arose in population dynamics and genetics, specifically in chemistry. The Caputo and Antagana-Baleanu fractional derivatives in the Caputo sense are used to test the intricacies of this mechanism numerically. In order to examine the approximate findings of fractional-order Fisher's type equations, the IETM solutions are obtained in series representation. Moreover, the stability of the approach was demonstrated using fixed point theory. Several illustrative cases are described that strongly agree with the precise solutions. Moreover, tables and graphs are included in order to conceptualize the influence of the fractional order and on the previous findings. The projected technique illustrates that only a few terms are sufficient for finding an approximate outcome, which is computationally appealing and accurate to analyze. Additionally, the offered procedure is highly robust, explicit, and viable for nonlinear fractional PDEs, but it could be generalized to other complex physical phenomena.

Keywords: Elzaki transform; Caputo fractional derivative; AB-fractional operator; new iterative transform method; Fisher's equation

1. Introduction

Researchers from various domains have been interested in fractional differential equations (FDEs) due to their wide applicability, and they are considered to be a handy tool for simulating the behaviour of several complex processes that have ramifications in specified disciplines of the physical sciences. Interestingly, it has boosted tremendous applications in autocatalytic reactions, anomalous diffusion process, viscoelastic damping, Maxwell fluid, virology, advection-diffusion process, thermal sciences, kinetics, optics, hydrodynamics, and epidemic diseases; different fractional calculus formulations are implemented in FDEs in order to adequately interpret and analyze memory. Numerous sorts of definitions and notions of fractional operators have been expounded by individuals such as Coimbra, Davison, and Essex; Riesz; Riemann and Liouville; Hadamard; Weyl; Jumarie; Caputo and Fabrizio [1]; Atangana and Baleanu [2]; Grünwald and Letnikov [3]; and Liouville and Caputo [4]. However, the Liouville-Caputo and AB operators are the best fractional filters.

Several studies have been contemplated on the applications of these operators. For example, Morales-Delgado [5] proposed a fractional analysis with and without kernel singularity. The authors of [6] employed AB fractional derivatives for finding the generalized Casson fluid model. Atangana and Alkahtani [7] used the Caputo–Fabrizio derivative for the analysis of groundwater flowing within a confined aquifer. Kumar et al. [8] considered the approximate-analytical solution of the regularized long-wave model by using the AB-fractional operator. Singh et al. [9] use the Mittag-Leffler type function to characterize the kinetics of an AB-fractional operator. The researchers of [10] proposed novel fractional optimal control problems with non-singular Mittage-Leffler functions as a kernel. More specifically, the Mittage-Leffler function is far more effective than the power and exponential functions in expressing physical difficulties. Consequently, the fractional derivative of the AB operator is well suited to unraveling heterogeneities in substances, structures, or media of various sizes.

Fractional PDEs have recently become extremely valuable in a variety of fields, including stochastic models, ground water flow, bacterial growth rates, astrophysics, and many more. Generally, PDEs are classified into conservation laws of energy, momentum, or electric charge (e.g., Fitzhugh-Nagumo equation, Korteweg-de Vties equations, Navier–Stokes equations, and Kawahara equations). The development of accurate and explicit solutions to nonlinear PDEs is a challenging task in applied sciences, and it is one of the most promising and productive research areas. Due to these facts, numerous mathematical methods for configuring approximate solutions have been proposed, such as the Adomian decomposition method (ADM) [11–13], homotopy perturbation method (HPM) [14,15], Laplace iterative transform method (LITM) [16], q-homotopy analysis method (q-HAM) [17], Haar wavelet method (HWM) [18], Lie symmetry analysis (LSA) [19], Chebyshev spectral collocation method (CSCM) [20], and many more.

Consider the generalized time-fractional Burgers–Fisher equation [18] presented as follows:

$$\frac{\partial^{\alpha}\mathbf{f}}{\partial\bar{t}^{\alpha}} + \zeta\mathbf{f}^{\beta}\frac{\partial\mathbf{f}}{\partial\mathbf{x}_1} = \sigma\frac{\partial^2\mathbf{f}}{\partial\mathbf{x}_1^2} + \theta\mathbf{f}(1-\mathbf{f}^{\beta}) \tag{1}$$

where ζ, σ, θ are parameters and $0 < \alpha \leq 1$. (1) plays a vital role in fluid dynamics models, heat conduction, elasticity, and capillary-gravity waves. When $\zeta = 0$ and $\beta = 1$ (1) are transformed into a Fisher's type equation, the derivative in (1) is a Caputo/AB-fractional derivative of order α.

Specifically, if $\zeta = 0$ and $\sigma = \theta = 1$, the generalized time-fractional Fisher's biological population diffusion equation [18] is presented as follows:

$$\frac{\partial^{\alpha}\mathbf{f}}{\partial\bar{t}^{\alpha}} = \frac{\partial^2\mathbf{f}}{\partial\mathbf{x}_1^2} + \mathcal{F}(\mathbf{f}) \tag{2}$$

where $\mathbf{f}(\mathbf{x}_1, \bar{t})$ refers the population density and $\bar{t} > 0$, $\mathbf{x}_1 \in \mathbb{R}$, and $\mathcal{F}(\mathbf{f})$ is a continuous nonlinear function fulfilling the following hypothesis: $\mathcal{F}(0) = \mathcal{F}(1) = 0$, and $\mathcal{F}'(1) < 0 < \mathcal{F}'(0)$.

Equation (1) transformed into the logistic equation if $\alpha = 1$, $\zeta = 0$ and the confluence of the diffusion equation has the diffusion factor σ and the birth rate θ. The coordinates $(\mathbf{x}_1, \bar{t})$ specified by $\mathbf{f}(\mathbf{x}_1, \bar{t})$ provide the state evolution across the spatial-temporal domain. Fisher's equation is used in many fields, including chemical kinetics [21], Neolithic transitions [22], branching Brownian motion [23], epidemics and bacteria [24], and many others. Wazwaz and Gorguis [11] used the ADM to solve Fisher's equation and demonstrated their convergence. Dag et al. [25] contemplated the B-spline Galerkin method for Fisher's equation. Bastani and Salkuyeh [26] adopted the compact finite difference approach in association with the third-order Runge–Kutta method to obtain Fisher's equation. For further investigations into linear and nonlinear Fisher's equations, see [27,28].

In 2001, Elzaki [29] expounded a new transform in order to facilitate the process of solving ODEs and PDEs in the time domain. This novel transform is the generalization of existing transforms (Laplace and Sumudu) that can contribute to in an analogous way to the Laplace and Sumudu transformations in order to determine the analytical solutions to the PDEs.

In [30], Daftardar-Gejji and Jafari suggested a new iterative approach (NITM) for solving functional equations, with the results reported in series form. Decomposing the nonlinear terms constitutes the foundation for the formulation of an iterative technique. Jafari et al. [16], first coupled the Laplace transform in the NITM and then they generated a novel recursive approach, namely ILTM, for obtaining the numerical consequences of FPDEs. Later, this approach has been correlated with different transformations (e.g., Sumudu transform, Aboodh transform, Elzaki transform, and Mohand transform) (see [31–34]). This methodology is incredibly pragmatic, and it does not entail the inclusion of an unconditioned matrix, convoluted integrals, or infinite series expressions. This approach avoids the demand for any explicit problematic configurations. NITM has been employed to solve PDEs in multiple investigations, including the KdV equation [35], Fornberg-Whitham equation [36], and Klein-Gordon equations [37].

Considering the substantial literature on fractional PDE frameworks, determining the analytical results of the underlying PDE is not an inexpensive procedure. In this perspective, we intend to design an appropriate technique for evaluating the numerical solution to Fisher's, the generic Fisher equation, and nonlinear diffusion equations of the Fisher type that depict the complexities of the mechanism under consideration by utilizing NITM. The Elzaki transform (ET) is merged with the NITM, and the proactive concept is said to be the iterative Elzaki transform method (IETM). This novel method is applied to examining fractional-order Fisher's models. In order to illustrate the capability of the recommended methodology, the findings of certain experimental examples were analysed. New strategies are applied to establish the results of the fractional-order and closed form results. An evaluation of IETM's convergence and uniqueness is also supplied. We test the superiority and practicality of the described algorithmic strategies for generating the analytical results in a numerical simulation leveraging fabricated trajectories inferred from Fisher's model. Additionally, other fractional-orders of linear and non-linear PDEs can be handled by the expounded approach.

2. Preliminaries

In this section, we will discus some basic preliminaries, definitions, and fractional frameworks of derivatives with power-law and Mittag-Leffler functions in their kernels, as well as the ET and fractional integrals.

Definition 1 ([35]). *The Caputo fractional derivative (CFD) is defined as follows.*

$$ {}_{0}^{c}\mathcal{D}_{\bar{t}}^{\alpha} = \begin{cases} \frac{1}{\Gamma(\jmath-\alpha)} \int\limits_{0}^{\bar{t}} \frac{\mathbf{f}^{(\jmath)}(\mathbf{x}_1)}{(\bar{t}-\mathbf{x}_1)^{\alpha+1-\jmath}} d\mathbf{x}_1, & \jmath-1 < \alpha < \jmath, \\ \frac{d^{\jmath}}{d\bar{t}^{\jmath}} \mathbf{f}(\bar{t}), & \alpha = \jmath. \end{cases} \tag{3} $$

Definition 2 ([2]). *The AB fractional derivative in the Caputo sense (ABC) is presented as follows:*

$$ {}_{a_1}^{ABC}\mathcal{D}_{\bar{t}}^{\alpha}(\mathbf{f}(\bar{t})) = \frac{\mathbb{N}(\alpha)}{1-\alpha} \int\limits_{a_1}^{\bar{t}} \mathbf{f}'(\bar{t}) E_{\alpha}\Big[-\frac{\alpha(\bar{t}-\mathbf{x}_1)^{\alpha}}{1-\alpha}\Big] d\mathbf{x}_1, \tag{4} $$

where $\mathbf{f} \in \mathcal{H}_1(\check{\alpha}, \check{\beta})$, $\check{\alpha} < \check{\beta}, \alpha \in [0,1]$ *and* $\mathbb{N}(\alpha)$ *indicates a normalization function as* $\mathbb{N}(\alpha) = \mathbb{N}(0) = \mathbb{N}(1) = 1$.

Definition 3 ([2]). *The fractional integral of the ABC-operator is stated as follows.*

$$ {}^{ABC}_{a_1}\mathcal{I}^{\alpha}_{\bar{t}}(\mathbf{f}(\bar{t})) = \frac{1-\alpha}{\mathbb{N}(\alpha)}\mathbf{f}(\bar{t}) + \frac{\alpha}{\Gamma(\alpha)\mathbb{N}(\alpha)} \int_{a_1}^{\bar{t}} \mathbf{f}(\mathbf{x}_1)(\bar{t}-\mathbf{x}_1)^{\alpha-1} d\mathbf{x}_1. \tag{5} $$

Definition 4 ([29]). *A set* $\mathcal{M}$ *containing exponential mapping is presented as follows:*

$$ \mathcal{M} = \left\{ \mathbf{f}(\bar{t}) : \exists z, p_1, p_2 > 0, |\mathbf{f}(\bar{t})| < z e^{\frac{|\bar{t}|}{p_i}},\ if\ \bar{t} \in (-1)^i \times [0,\infty)| \right\}. \tag{6} $$

where z is a finite number, but p_1, p_2 *may be finite or infinite.*

Definition 5 ([29,35]). *The ET of a given mapping* $\mathbf{f}(\bar{t})$ *is stated as follows.*

$$ \mathbb{E}\{\mathbf{f}(\bar{t})\}(\omega) = \tilde{\mathcal{U}}(\omega) = \omega \int_0^{\infty} e^{-\frac{\bar{t}}{\omega}} \mathbf{f}(\bar{t}) d\bar{t}, \quad \bar{t} \geq 0, \quad \omega \in [p_1, p_2]. \tag{7} $$

Definition 6 ([36]). *The Elzaki transform of the CFD is presented as follows.*

$$ \mathbb{E}\left\{ {}^{C}_{0}\mathcal{D}^{\alpha}_{\bar{t}}(\mathbf{f}(\bar{t})) \right\}(\omega) = \omega^{-\alpha}\tilde{\mathcal{U}}(\omega) - \sum_{\kappa=0}^{\jmath-1} \omega^{2-\alpha+\kappa}\mathbf{f}^{(\kappa)}(0), \quad \jmath - 1 < \alpha < \jmath. \tag{8} $$

Definition 7 ([37]). *The ET of the ABC fractional derivative operator is presented as follows:*

$$ \mathbb{E}\left\{ {}^{ABC}_{0}\mathcal{D}^{\alpha}_{\bar{t}}(\mathbf{f}(\bar{t})) \right\}(\omega) = \frac{\mathbb{N}(\alpha)}{\alpha\omega^{\alpha} + 1 - \alpha}\left(\frac{\tilde{\mathcal{U}}(\omega)}{\omega} - \omega \mathbf{f}(0) \right), \tag{9} $$

where $\mathbb{E}\{\mathbf{f}(\bar{t})\}(\omega) = \tilde{\mathcal{U}}(\omega)$.

3. Application of Caputo-Liouville and ABC Fractional Derivatives to the Non-Linear Fisher's Model

In this note, we analyze time fractional Caputo-Liouville and the ABC fractional derivative operator in order to analyze the non-linear Fisher's equation [18]. The model under consideration is presented as follows:

$$ {}^{\otimes}_{0}\mathcal{D}^{\alpha}_{\bar{t}} = \eta \frac{\partial^2 \mathbf{f}(\mathbf{x}_1, \bar{t})}{\partial \mathbf{x}_1^2} - \theta(\mathbf{f}(\mathbf{x}_1, \bar{t}) - \varphi)(1 - \mathbf{f}^{\beta}(\mathbf{x}_1, \bar{t})), \ \beta > 1, \ 0 < \alpha \leq 1, \tag{10} $$

which is subject to the following condition.

$$ \mathbf{f}(\mathbf{x}_1, 0) = 0, \ a \leq \mathbf{x}_1 \leq b. \tag{11} $$

3.1. Description of IETM

Assume the following nonlinear fractional PDE:

$$ {}^{\otimes}\mathcal{D}^{\alpha}_{\bar{t}}\mathbf{f}(\mathbf{x}_1, \bar{t}) + \tilde{\mathcal{L}}\mathbf{f}(\mathbf{x}_1, \bar{t}) + \tilde{\mathcal{N}}\mathbf{f}(\mathbf{x}_1, \bar{t}) = \mathcal{F}(\mathbf{x}_1, \bar{t}), \ \bar{t} > 0, \ 0 < \alpha \leq 1, \ \jmath - 1 < \alpha \leq \jmath, \ \jmath \in \mathbb{N}, \tag{12} $$

subject to the initial condition

$$ \frac{\partial \mathbf{f}^{\kappa}}{\partial \bar{t}^{\kappa}}(\mathbf{x}_1, 0) = \mathcal{G}_{\kappa}(\mathbf{x}_1), \quad \kappa = 0, 1, \ldots, m_1 - 1. \tag{13} $$

where ${}^{\otimes}\mathcal{D}^{\alpha}_{\bar{t}} = \frac{\partial^{\alpha}\mathbf{f}(\mathbf{x}_1, \bar{t})}{\partial \bar{t}^{\alpha}}$ denotes the Caputo or ABC fractional derivative operator with $0 < \alpha \leq 1$, while $\tilde{\mathcal{L}}$ and $\tilde{\mathcal{N}}$ are linear and nonlinear terms, and $\mathcal{F}(\mathbf{x}_1, \bar{t})$ indicates the source term.

By employing the Elzaki transform to (12), we acquire the following.

$$\mathbb{E}\big[{}^{\otimes}\mathcal{D}_{\bar{t}}^{\alpha}\mathbf{f}(\mathbf{x}_1,\bar{t}) + \tilde{\mathcal{L}}\mathbf{f}(\mathbf{x}_1,\bar{t}) + \tilde{\mathcal{N}}\mathbf{f}(\mathbf{x}_1,\bar{t})\big] = \mathbb{E}\big[\mathcal{F}(\mathbf{x}_1,\bar{t})\big].$$

By the virtue of the Elzaki differentiation property for the Caputo fractional derivative operator defined in (6), we have the following.

$$\frac{1}{\omega^{\alpha}}\mathbb{E}\big[\mathbf{f}(\mathbf{x}_1,\bar{t})\big] - \sum_{\kappa=0}^{m_1-1}\mathbf{f}_{(\kappa)}(\mathbf{x}_1,0)\omega^{2-\alpha+\kappa} = -\mathbb{E}\big[\tilde{\mathcal{L}}\mathbf{f}(\mathbf{x}_1,\bar{t}) + \tilde{\mathcal{N}}\mathbf{f}(\mathbf{x}_1,\bar{t})\big] + \mathbb{E}\big[\mathcal{F}(\mathbf{x}_1,\bar{t})\big],$$

$$\mathbb{E}\big[\mathbf{f}(\mathbf{x}_1,\bar{t})\big] = \omega^2\mathbf{f}(\mathbf{x}_1,0) + \omega^{\alpha}\mathbb{E}\big[\mathcal{F}(\mathbf{x}_1,\bar{t})\big] - \omega^{\alpha}\mathbb{E}\big[\tilde{\mathcal{L}}\mathbf{f}(\mathbf{x}_1,\bar{t}) + \tilde{\mathcal{N}}\mathbf{f}(\mathbf{x}_1,\bar{t})\big]. \tag{14}$$

Again, in view of the Elzaki differentiation property for the ABC fractional derivative operator defined in (7), we have the following.

$$\mathbb{E}\big[\mathbf{f}(\mathbf{x}_1,\bar{t})\big] = \omega^2\mathbf{f}(\mathbf{x}_1,0) + \frac{\alpha\omega^{\alpha}+1-\alpha}{\mathbb{N}(\alpha)}\mathbb{E}\big[\mathcal{F}(\mathbf{x}_1,\bar{t})\big] - \frac{\alpha\omega^{\alpha}+1-\alpha}{\mathbb{N}(\alpha)}\mathbb{E}\big[\tilde{\mathcal{L}}\mathbf{f}(\mathbf{x}_1,\bar{t}) + \tilde{\mathcal{N}}\mathbf{f}(\mathbf{x}_1,\bar{t})\big]. \tag{15}$$

Now, by applying the inverse Elzaki transform to (14) and (15), respectively, we have the following:

$$\mathbf{f}(\mathbf{x}_1,\bar{t}) = \mathbb{E}^{-1}\Big\{\omega^2\mathbf{f}(\mathbf{x}_1,0) + \omega^{\alpha}\mathbb{E}\big[\mathcal{F}(\mathbf{x}_1,\bar{t})\big]\Big\} - \mathbb{E}^{-1}\Big\{\omega^{\alpha}\mathbb{E}\big[\tilde{\mathcal{L}}\mathbf{f}(\mathbf{x}_1,\bar{t}) + \tilde{\mathcal{N}}\mathbf{f}(\mathbf{x}_1,\bar{t})\big]\Big\}, \tag{16}$$

and

$$\mathbf{f}(\mathbf{x}_1,\bar{t}) = \mathbb{E}^{-1}\Big\{\omega^2\mathbf{f}(\mathbf{x}_1,0) + \frac{\alpha\omega^{\alpha}+1-\alpha}{\mathbb{N}(\alpha)}\mathbb{E}\big[\mathcal{F}(\mathbf{x}_1,\bar{t})\big]\Big\} - \mathbb{E}^{-1}\Big\{\frac{\alpha\omega^{\alpha}+1-\alpha}{\mathbb{N}(\alpha)}\mathbb{E}\big[\tilde{\mathcal{L}}\mathbf{f}(\mathbf{x}_1,\bar{t}) + \tilde{\mathcal{N}}\mathbf{f}(\mathbf{x}_1,\bar{t})\big]\Big\}. \tag{17}$$

The iterative process in terms of power series is prescribed as follows.

$$\mathbf{f}(\mathbf{x}_1,\bar{t}) = \sum_{m_1=0}^{\infty}\mathbf{f}_{m_1}(\mathbf{x}_1,\bar{t}). \tag{18}$$

Moreover, the linear factor can be stated as the following.

$$\tilde{\mathcal{L}}\Big(\sum_{m_1=0}^{\infty}\mathbf{f}_{m_1}(\mathbf{x}_1,\bar{t})\Big) = \sum_{m_1=0}^{\infty}\tilde{\mathcal{L}}\big[\mathbf{f}_{m_1}(\mathbf{x}_1,\bar{t})\big]. \tag{19}$$

Furthermore, the nonlinear operator $\tilde{\mathcal{N}}$ can be decomposed [30] as follows:

$$\begin{aligned}\tilde{N}\Big(\sum_{m_1=0}^{\infty}\mathbf{f}_{m_1}(\mathbf{x}_1,\bar{t})\Big) &= \tilde{N}(\mathbf{f}_0(\mathbf{x}_1,\bar{t})) + \sum_{m_1=0}^{\infty}\Big[\tilde{N}\Big(\sum_{\kappa=0}^{m_1}\mathbf{f}_{\kappa}(\mathbf{x}_1,\bar{t})\Big) - \tilde{N}\Big(\sum_{\kappa=0}^{m_1-1}\mathbf{f}_{\kappa}(\mathbf{x}_1,\bar{t})\Big)\Big] \\ &= \tilde{N}(\mathbf{f}_0) + \sum_{\kappa=1}^{\infty}D_{m_1},\end{aligned} \tag{20}$$

where $D_m = \tilde{N}\Big(\sum_{\kappa=0}^{m_1}\mathbf{f}_{\kappa}(\mathbf{x}_1,\bar{t})\Big) - \tilde{N}\Big(\sum_{\kappa=0}^{m_1-1}\mathbf{f}_{\kappa}(\mathbf{x}_1,\bar{t})\Big)$.

Substituting (18)–(20) into (16) and (17), respectively, we will obtain the following equations:

$$\begin{aligned}\sum_{m_1=0}^{\infty}\mathbf{f}_{m_1}(\mathbf{x}_1,\bar{t}) &= \mathcal{G}(\mathbf{x}_1) + \mathbb{E}^{-1}\Big\{\omega^{\alpha}\mathbb{E}\big[\mathcal{F}(\mathbf{x}_1,\bar{t})\big]\Big\} \\ &\quad -\mathbb{E}^{-1}\Big\{\omega^{\alpha}\mathbb{E}\Big[\tilde{\mathcal{L}}\sum_{\kappa=0}^{m_1}\mathbf{f}_{\kappa}(\mathbf{x}_1,\bar{t}) + \tilde{N}(\mathbf{f}_0) + \sum_{\kappa=1}^{m_1}D_{m_1}\Big]\Big\},\end{aligned} \tag{21}$$

and

$$\sum_{m_1=0}^{\infty} \mathbf{f}_{m_1}(\mathbf{x}_1,\bar{t}) = \mathcal{G}(\mathbf{x}_1) + \mathbb{E}^{-1}\left\{\frac{\alpha\omega^{\alpha}+1-\alpha}{\mathbb{N}(\alpha)}\mathbb{E}[\mathcal{F}(\mathbf{x}_1,\bar{t})]\right\}$$
$$-\mathbb{E}^{-1}\left\{\frac{\alpha\omega^{\alpha}+1-\alpha}{\mathbb{N}(\alpha)}\mathbb{E}\left[\tilde{\mathcal{L}}\sum_{\kappa=0}^{m_1}\mathbf{f}_{\kappa}(\mathbf{x}_1,\bar{t}) + \tilde{N}(\mathbf{f}_0) + \sum_{\kappa=1}^{m_1} D_{m_1}\right]\right\}. \quad (22)$$

We mention the following iterative scheme for the Caputo fractional derivative operator as follows.

$$\begin{aligned}
\mathbf{f}_0(\mathbf{x}_1,\bar{t}) &= \mathcal{G}(\mathbf{x}_1) + \mathbb{E}^{-1}\left\{\omega^{\alpha}\mathbb{E}[\mathcal{F}(\mathbf{x}_1,\bar{t})]\right\},\\
\mathbf{f}_1(\mathbf{x}_1,\bar{t}) &= \mathbb{E}^{-1}\left\{\omega^{\alpha}\mathbb{E}\left[\tilde{\mathcal{L}}\mathbf{f}_0(\mathbf{x}_1,\bar{t}) + \tilde{\mathcal{N}}(\mathbf{f}_0(\mathbf{x}_1,\bar{t}))\right]\right\},\\
&\vdots\\
\mathbf{f}_{m_1+1}(\mathbf{x}_1,\bar{t}) &= \mathbb{E}^{-1}\left\{\omega^{\alpha}\mathbb{E}\left[\tilde{\mathcal{L}}\mathbf{f}_{m_1}(\mathbf{x}_1,\bar{t}) + D_{m_1}\right]\right\},\ m_1>0,\ m_1\in\mathbb{N}.
\end{aligned} \quad (23)$$

Analogously, the iterative scheme for the ABC fractional derivative operator is presented as follows.

$$\begin{aligned}
\mathbf{f}_0(\mathbf{x}_1,\bar{t}) &= \mathcal{G}(\mathbf{x}_1) + \mathbb{E}^{-1}\left\{\tfrac{\alpha\omega^{\alpha}+1-\alpha}{\mathbb{N}(\alpha)}\mathbb{E}[\mathcal{F}(\mathbf{x}_1,\bar{t})]\right\},\\
\mathbf{f}_1(\mathbf{x}_1,\bar{t}) &= \mathbb{E}^{-1}\left\{\tfrac{\alpha\omega^{\alpha}+1-\alpha}{\mathbb{N}(\alpha)}\mathbb{E}\left[\tilde{\mathcal{L}}\mathbf{f}_0(\mathbf{x}_1,\bar{t}) + \tilde{\mathcal{N}}(\mathbf{f}_0(\mathbf{x}_1,\bar{t}))\right]\right\},\\
&\vdots\\
\mathbf{f}_{m_1+1}(\mathbf{x}_1,\bar{t}) &= \mathbb{E}^{-1}\left\{\tfrac{\alpha\omega^{\alpha}+1-\alpha}{\mathbb{N}(\alpha)}\mathbb{E}\left[\tilde{\mathcal{L}}\mathbf{f}_{m_1}(\mathbf{x}_1,\bar{t}) + D_{m_1}\right]\right\},\ m_1>0,\ m_1\in\mathbb{N}.
\end{aligned} \quad (24)$$

Finally, (12) and (13) yield the m_1-terms solution in series forms as follows.

$$\mathbf{f}(\mathbf{x}_1,\mathbf{f}) = \mathbf{f}_0(\mathbf{x}_1,\bar{t}) + \mathbf{f}_1(\mathbf{x}_1,\bar{t}) + \mathbf{f}_3(\mathbf{x}_1,\bar{t}) + \ldots + \mathbf{f}_{m_1}(\mathbf{x}_1,\bar{t}).,\ m_1 = 1,2,\ldots \quad (25)$$

3.2. Stability Analysis

Let there be a Banach space $(Y, \|.\|)$ and $\mathcal{V}$ be a self-map of Y. Let $\mathbf{x}_{m+1} = h(\mathcal{V},\mathbf{x}_m)$ be a specific iterative scheme. Moreover, let $F(\mathcal{V})$, the fixed-point of $\mathcal{V}$, possess at least one element and that $\mathbf{x}_n$ tends to a point $q \in F(\mathcal{V})$. Consider a sequence $\{\mathbf{y}_m\}$ such that $\{\mathbf{y}_m\} \subseteq Y$ and $\epsilon_m = \|\mathbf{y}_{m+1} - h(\mathcal{V},\mathbf{y}_m)\|$. If $\lim_{m\mapsto\infty}\epsilon^m = 0$ implies that $\lim_{m\mapsto\infty} y^m = q$, then we say that the iterative process $\mathbf{x}_{m+1} = h(\mathcal{H},\mathbf{x}_n)$ is $\mathcal{V}$-stable. Without any loss of generality, we surmise that $\{\mathbf{y}_m\}$ is upper bounded; otherwise convergence cannot be expected. If all hypotheses fulfilled for $\mathbf{x}_{m+1} = \mathcal{V}\mathbf{x}_m$, which is known as Picard's iteration, are satisfied, consequently, the iteration will be $\mathcal{V}$-stable. The following theorem will be presented next.

Theorem 1 ([38]). *Consider a Banach space $(Y, \|.\|)$ and $\mathcal{V}$ a self-map of Y holding the following:*

$$\|\mathcal{V}_{\mathbf{y}} - \mathcal{V}_{\mathbf{x}_1}\| \le \mathcal{K}\|\mathbf{y} - \mathcal{V}_{\mathbf{y}}\| + \hat{k}\|\mathbf{y} - \mathbf{x}_1\|,\ \forall \mathbf{x}_1,\mathbf{y} \in Y, \quad (26)$$

where $\mathcal{K} > 0$, $\hat{k} \in [0,1)$, then $\mathcal{V}$ is Picard $\mathcal{V}$-stable.

Consider the following sequence, which represents the nonlinear fractional Fisher's model as follows:

$$\mathbf{f}_{m+1}(\mathbf{x}_1,\bar{t}) = \mathbf{f}_m(\mathbf{x}_1,\bar{t}) + \mathbb{E}^{-1}\left[\frac{\alpha\omega^{\alpha}+1-\alpha}{\mathbb{N}(\alpha)}\mathbb{E}\left[\sigma\frac{\partial^2\mathbf{f}_m(\mathbf{x}_1,\bar{t})}{\partial\mathbf{x}_1^2} + \theta\mathbf{f}_m(\mathbf{x}_1,\bar{t})(1-\mathbf{f}_m^{\beta}(\mathbf{x}_1,\bar{t}))\right]\right], \quad (27)$$

where $\frac{\alpha\omega^{\alpha}+1-\alpha}{\mathbb{N}(\alpha)}$ is the fractional Langrange multiplier and $\hat{\mathbf{f}}_m^{\beta}$ is a limited variant that denotes $\delta\hat{\mathbf{f}}_m^{\beta} = 0$.

Theorem 2. *Consider $\bar{T}$ as a self-map stated as follows:*

$$\begin{aligned}\bar{T}(\mathbf{f}_m(\mathbf{x}_1,\bar{t})) &= \mathbf{f}_{m+1}(\mathbf{x}_1,\bar{t}) \\ &= \mathbf{f}_n(\mathbf{x}_1,\bar{t}) + \mathbb{E}^{-1}\left[\frac{\alpha\omega^{\alpha}+1-\alpha}{\mathbb{N}(\alpha)}\mathbb{E}\left[\sigma\frac{\partial^2\mathbf{f}_m(\mathbf{x}_1,\bar{t})}{\partial\mathbf{x}_1^2} + \theta\mathbf{f}_m(\mathbf{x}_1,\bar{t})(1-\mathbf{f}_m^{\beta}(\mathbf{x}_1,\bar{t}))\right]\right]\end{aligned} \tag{28}$$

is $\bar{T}$-stable in $L^2(a,b)$ if $\left[1+\left(\frac{\sigma\Omega_1\Omega_2+\theta(1-(\mathcal{K}+\mathcal{H})^{\beta})}{\mathbb{N}(\alpha)}\right)\left(\frac{\alpha\bar{t}^{\alpha}}{\Gamma(\alpha+1)}+(1-\alpha)\right)\right] < \Theta$.

Proof. First, we illustrate that $\bar{T}$ has a fixed point. In order to accomplish this, we examined the following for all $(m,\kappa) \in \mathbb{N}\times\mathbb{N}$.

$$\|\bar{T}(\mathbf{f}_m(\mathbf{x}_1,\bar{t})) - \bar{T}(\mathbf{f}_\kappa(\mathbf{x}_1,\bar{t}))| = \left\|\begin{array}{l}\mathbf{f}_m(\mathbf{x}_1,\bar{t}) - \mathbf{f}_\kappa(\mathbf{x}_1,\bar{t}) \\ +\mathbb{E}^{-1}\left[\frac{\alpha\omega^{\alpha}+1-\alpha}{\mathbb{N}(\alpha)}\mathbb{E}\left[\sigma\frac{\partial^2\mathbf{f}_m(\mathbf{x}_1,\bar{t})}{\partial\mathbf{x}_1^2} + \theta\mathbf{f}_m(\mathbf{x}_1,\bar{t})(1-\mathbf{f}_m^{\beta}(\mathbf{x}_1,\bar{t}))\right]\right] \\ +\mathbb{E}^{-1}\left[\frac{\alpha\omega^{\alpha}+1-\alpha}{\mathbb{N}(\alpha)}\mathbb{E}\left[\sigma\frac{\partial^2\mathbf{f}_\kappa(\mathbf{x}_1,\bar{t})}{\partial\mathbf{x}_1^2} + \theta\mathbf{f}_\kappa(\mathbf{x}_1,\bar{t})(1-\mathbf{f}_\kappa^{\beta}(\mathbf{x}_1,\bar{t}))\right]\right]\end{array}\right\|.$$

Employing the linearity property of the inverse Elzaki transform yields the following.

$$\|\bar{T}(\mathbf{f}_m(\mathbf{x}_1,\bar{t})) - \bar{T}(\mathbf{f}_\kappa(\mathbf{x}_1,\bar{t}))| = \left\|\begin{array}{l}\mathbf{f}_m(\mathbf{x}_1,\bar{t}) - \mathbf{f}_\kappa(\mathbf{x}_1,\bar{t}) \\ +\mathbb{E}^{-1}\left[\frac{\alpha\omega^{\alpha}+1-\alpha}{\mathbb{N}(\alpha)}\mathbb{E}\left\{\begin{array}{l}\left[\sigma\frac{\partial^2\left[\mathbf{f}_m(\mathbf{x}_1,\bar{t})-\mathbf{f}_\kappa(\mathbf{x}_1,\bar{t})\right]}{\partial\mathbf{x}_1^2}\right. \\ +\theta\left[\mathbf{f}_m(\mathbf{x}_1,\bar{t}) - \mathbf{f}_\kappa(\mathbf{x}_1,\bar{t})\right] \\ \left.-\theta\left[\mathbf{f}_m^{\beta+1}(\mathbf{x}_1,\bar{t}) - \mathbf{f}_\kappa^{\beta+1}(\mathbf{x}_1,\bar{t})\right]\right]\end{array}\right.\right.\end{array}\right\|.$$

Utilizing triangular inequality for the norms, we have the following.

$$\begin{aligned}&\|\bar{T}(\mathbf{f}_m(\mathbf{x}_1,\bar{t})) - \bar{T}(\mathbf{f}_\kappa(\mathbf{x}_1,\bar{t}))| \\ &\leq \|\mathbf{f}_m(\mathbf{x}_1,\bar{t}) - \mathbf{f}_\kappa(\mathbf{x}_1,\bar{t})| + \\ &\quad +\mathbb{E}^{-1}\left[\frac{\alpha\omega^{\alpha}+1-\alpha}{\mathbb{N}(\alpha)}\left[\left\|\sigma\frac{\partial^2\left[\mathbf{f}_m(\mathbf{x}_1,\bar{t})-\mathbf{f}_\kappa(\mathbf{x}_1,\bar{t})\right]}{\partial\mathbf{x}_1^2}\right\|\right]\right] \\ &\quad +\mathbb{E}^{-1}\left[\frac{\alpha\omega^{\alpha}+1-\alpha}{\mathbb{N}(\alpha)}\mathbb{E}\left[\left\|\theta\left[\mathbf{f}_m(\mathbf{x}_1,\bar{t}) - \mathbf{f}_\kappa(\mathbf{x}_1,\bar{t})\right]\right\|\right]\right] \\ &\quad +\mathbb{E}^{-1}\left[\frac{\alpha\omega^{\alpha}+1-\alpha}{\mathbb{N}(\alpha)}\mathbb{E}\left[\left\|-\theta\left[\mathbf{f}_m^{\beta+1}(\mathbf{x}_1,\bar{t}) - \mathbf{f}_\kappa^{\beta+1}(\mathbf{x}_1,\bar{t})\right]\right\|\right]\right].\end{aligned} \tag{29}$$

Equation (29) can be examined on a case-by-case basis, beginning with the following.

$$\left\|\sigma\frac{\partial^2\left[\mathbf{f}_m(\mathbf{x}_1,\bar{t}) - \mathbf{f}_\kappa(\mathbf{x}_1,\bar{t})\right]}{\partial\mathbf{x}_1^2}\right\| \leq \sigma\Omega_1\Omega_2. \tag{30}$$

The following results.

$$\left\|\theta\left[\mathbf{f}_m^{\beta+1}(\mathbf{x}_1,\bar{t}) - \mathbf{f}_\kappa^{\beta+1}(\mathbf{x}_1,\bar{t})\right]\right\|$$
$$\leq \left\|\sum_{i=0}^{\beta}\mathcal{K}_\beta^i(\mathbf{f}_m(\mathbf{x}_1,\bar{t}))^i(\mathbf{f}_\kappa(\mathbf{x}_1,\bar{t}))^{\beta-i-1}\right\|.\left\|\mathbf{f}_m(\mathbf{x}_1,\bar{t}) - \mathbf{f}_\kappa(\mathbf{x}_1,\bar{t})\right\|. \quad (31)$$

Since $\mathbf{f}_m(\mathbf{x}_1,\bar{t})$, $\mathbf{f}_\kappa(\mathbf{x}_1,\bar{t})$ are bounded, there are two different positive constants that we can obtain $\mathcal{K},\mathcal{H}$ such that for all $(\mathbf{x}_1,\bar{t})$, we have the following.

$$\left\|\mathbf{f}_m(\mathbf{x}_1,\bar{t})\right\| \leq \mathcal{K}, \ \left\|\mathbf{f}_\kappa(\mathbf{x}_1,\bar{t})\right\| \leq \mathcal{H}, \ (m,\kappa) \in \mathbb{N}\times\mathbb{N}. \quad (32)$$

As a result of combining the triangular inequality with the above inequalities, (31) is obtained as follows.

$$\left\|\theta\left[\mathbf{f}_m^{\beta+1}(\mathbf{x}_1,\bar{t}) - \mathbf{f}_\kappa^{\beta+1}(\mathbf{x}_1,\bar{t})\right]\right\|$$
$$\leq (\mathcal{K}+\mathcal{H})^\beta\left\|\mathbf{f}_m(\mathbf{x}_1,\bar{t}) - \mathbf{f}_\kappa(\mathbf{x}_1,\bar{t})\right\|. \quad (33)$$

Now, by combining (31) and (33) into (34), we obtain the following result:

$$\left\|\bar{T}(\mathbf{f}_m(\mathbf{x}_1,\bar{t})) - \bar{T}(\mathbf{f}_\kappa(\mathbf{x}_1,\bar{t}))\right|$$
$$\leq \left[1+\left(\frac{\sigma\Omega_1\Omega_2+\theta(1-(\mathcal{K}+\mathcal{H})^\beta)}{\mathbb{N}(\alpha)}\right)\left(\frac{\alpha\bar{t}^\alpha}{\Gamma(\alpha+1)}+(1-\alpha)\right)\right]\left\|\mathbf{f}_m(\mathbf{x}_1,\bar{t}) - \mathbf{f}_\kappa(\mathbf{x}_1,\bar{t})\right\| \quad (34)$$

with the following.

$$\left[1+\left(\frac{\sigma\Omega_1\Omega_2+\theta(1-(\mathcal{K}+\mathcal{H})^\beta)}{\mathbb{N}(\alpha)}\right)\left(\frac{\alpha\bar{t}^\alpha}{\Gamma(\alpha+1)}+(1-\alpha)\right)\right] < \Theta.$$

This establishes the existence of a fixed point for the nonlinear $\bar{T}$-self map. As a consequence, the proof is complete. We also proved that $\bar{T}$ fulfills the requirements of Theorem 1. Allow (7) to hold by inserting the following.

$$\mathbf{f} = \left[1+\left(\frac{\sigma\Omega_1\Omega_2+\theta(1-(\mathcal{K}+\mathcal{H})^\beta)}{\mathbb{N}(\alpha)}\right)\left(\frac{\alpha\bar{t}^\alpha}{\Gamma(\alpha+1)}+(1-\alpha)\right)\right], \quad (35)$$

This proves that hypothesis of Theorem 1 fulfills the nonlinear mapping $\bar{T}$. Thus, all assumptions in Theorem 1 satisfies the described nonlinear mapping $\bar{T}$, and $\bar{T}$ is Picard's $\bar{T}$-stable. As a result, the proof of Theorem 2 is complete. □

4. Evaluation of the Fractional Fisher Model via IETM

This section demonstrate the reliability and preciseness of the projected methodology.

Problem 1. *If $\theta = 1, \beta = 1$, and $\varphi = 0$ in (10) with $\mathbf{f}_0(\mathbf{x}_1,0) = \eta$, then the one dimensional time fractional Fisher equation is presented as follows.*

$$\frac{\partial^\alpha\mathbf{f}}{\partial\bar{t}^\alpha} = \frac{\partial^2\mathbf{f}}{\partial\mathbf{x}_1^2} + \mathbf{f}(1-\mathbf{f}). \quad (36)$$

The integer-order solution for the Fisher's Equation (36) is obtained by using the Taylor's series expansion for $\alpha = 1$ as follows.

$$\mathbf{f}(\mathbf{x}_1,\bar{t}) = \frac{\eta\exp(\bar{t})}{1-\eta+\eta\exp(\bar{t})}.$$

Case I. First, we formulate Problem 1 by utilizing the Elzaki transform coupled with the Caputo derivative operator.

By employing the Elzaki transform to (36) with the initial condition, we have the following.

$$\mathbb{E}\Big[\frac{\partial^\alpha \mathbf{f}}{\partial \bar{t}^\alpha}\Big] = \mathbb{E}\Big[\frac{\partial^2 \mathbf{f}}{\partial \mathbf{x}_1^2} + \mathbf{f}(1-\mathbf{f})\Big]. \tag{37}$$

The following is the case:

$$\frac{1}{\omega^\alpha}\mathbb{E}\big[\mathbf{f}(\mathbf{x}_1,\bar{t})\big] - \sum_{\kappa=0}^{m-1} \mathbf{f}_{(\kappa)}(\mathbf{x}_1,0)\omega^{2-\alpha+\kappa} = \mathbb{E}\Big[\frac{\partial^2 \mathbf{f}}{\partial \mathbf{x}_1^2} + \mathbf{f}(1-\mathbf{f})\Big].$$

equivalently, we have

$$\frac{1}{\omega^\alpha}\mathbb{E}\big[\mathbf{f}(\mathbf{x}_1,\bar{t})\big] = \mathbf{f}_{(0)}(\mathbf{x}_1,0)\omega^{2-\alpha} + \mathbb{E}\Big[\frac{\partial^2 \mathbf{f}}{\partial \mathbf{x}_1^2} + \mathbf{f}(1-\mathbf{f})\Big].$$

Using the inverse Elzaki transform, we have the following.

$$\mathbf{f}(\mathbf{x}_1,\bar{t}) = \eta + \mathbb{E}^{-1}\Big[\omega^\alpha \mathbb{E}\Big[\frac{\partial^2 \mathbf{f}}{\partial \mathbf{x}_1^2} + \mathbf{f}(1-\mathbf{f})\Big]\Big].$$

Applying the iterative technique described in Section 3.1, we obtain the following.

$$\begin{aligned}
\mathbf{f}_0(\mathbf{x}_1,\bar{t}) &= \eta, \\
\mathbf{f}_1(\mathbf{x}_1,\bar{t}) &= \mathbb{E}^{-1}\Big[\omega^\alpha \mathbb{E}\Big\{\big(\mathbf{f}_0(\mathbf{x}_1,\bar{t})\big)_{\mathbf{x}_1\mathbf{x}_1} + \mathbf{f}_0(1-\mathbf{f}_0)\Big\}\Big] \\
&= \eta(1-\eta)\frac{\bar{t}^\alpha}{\Gamma(\alpha+1)}, \\
\mathbf{f}_2(\mathbf{x}_1,\bar{t}) &= \mathbb{E}^{-1}\Big[\omega^\alpha \mathbb{E}\Big\{\big(\mathbf{f}_1(\mathbf{x}_1,\bar{t})\big)_{\mathbf{x}_1\mathbf{x}_1} + \mathbf{f}_1(1-\mathbf{f}_1)\Big\}\Big] \\
&= \eta(1-\eta)(1-2\eta)\frac{\bar{t}^{2\alpha}}{\Gamma(2\alpha+1)}, \\
\mathbf{f}_3(\mathbf{x}_1,\bar{t}) &= \mathbb{E}^{-1}\Big[\omega^\alpha \mathbb{E}\Big\{\big(\mathbf{f}_2(\mathbf{x}_1,\bar{t})\big)_{\mathbf{x}_1\mathbf{x}_1} + \mathbf{f}_2(1-\mathbf{f}_2)\Big\}\Big] \\
&= \eta(1-\eta)(1-6\eta+6\eta^2)\frac{\bar{t}^{3\alpha}}{\Gamma(3\alpha+1)}, \\
\mathbf{f}_4(\mathbf{x}_1,\bar{t}) &= \mathbb{E}^{-1}\Big[\omega^\alpha \mathbb{E}\Big\{\big(\mathbf{f}_3(\mathbf{x}_1,\bar{t})\big)_{\mathbf{x}_1\mathbf{x}_1} + \mathbf{f}_3(1-\mathbf{f}_3)\Big\}\Big] \\
&= \eta(1-\eta)(1-2\eta)(1-12\eta+12\eta^2)\frac{\bar{t}^{4\alpha}}{\Gamma(4\alpha+1)}, \\
&\vdots
\end{aligned}$$

Provided that the series form solution is as follows:

$$\mathbf{f}(\mathbf{x}_1,\bar{t}) = \mathbf{f}_0(\mathbf{x}_1,\bar{t}) + \mathbf{f}_1(\mathbf{x}_1,\bar{t}) + \mathbf{f}_2(\mathbf{x}_1,\bar{t}) + \mathbf{f}_3(\mathbf{x}_1,\bar{t}) + \ldots + \mathbf{f}_{m_1}(\mathbf{x}_1,\bar{t}).$$

we consequently have

$$
\begin{aligned}
\mathbf{f}(\mathbf{x}_1,\bar{t}) &= \eta + \eta(1-\eta)\frac{\bar{t}^{\alpha}}{\Gamma(\alpha+1)} + \eta(1-\eta)(1-2\eta)\frac{\bar{t}^{2\alpha}}{\Gamma(2\alpha+1)} + \eta(1-\eta)(1-6\eta+6\eta^2)\frac{\bar{t}^{3\alpha}}{\Gamma(3\alpha+1)} \\
&+\eta(1-\eta)(1-2\eta)(1-12\eta+12\eta^2)\frac{\bar{t}^{4\alpha}}{\Gamma(4\alpha+1)} + \ldots
\end{aligned}
$$

Case II. Now we formulate Problem 1 by utilizing Elzaki transform coupled with the ABC derivative operator.

By employing the Elzaki transform to (36) with the initial condition, we have the following.

$$
\mathbb{E}\Big[\frac{\partial^{\alpha}\mathbf{f}}{\partial \bar{t}^{\alpha}}\Big] = \mathbb{E}\Big[\frac{\partial^2 \mathbf{f}}{\partial \mathbf{x}_1^2} + \mathbf{f}(1-\mathbf{f})\Big]. \tag{38}
$$

The following is then the case.

$$
\mathbb{E}\big[\mathbf{f}(\mathbf{x}_1,\bar{t})\big] = \mathbf{f}_{(0)}(\mathbf{x}_1,0)\omega^2 + \frac{\alpha\omega^{\alpha}+1-\alpha}{\mathbb{N}(\alpha)}\mathbb{E}\Big[\frac{\partial^2 \mathbf{f}}{\partial \mathbf{x}_1^2} + \mathbf{f}(1-\mathbf{f})\Big].
$$

Using the inverse Elzaki transform, we have the following.

$$
\mathbf{f}(\mathbf{x}_1,\bar{t}) = \eta + \mathbb{E}^{-1}\Big[\frac{\alpha\omega^{\alpha}+1-\alpha}{\mathbb{N}(\alpha)}\mathbb{E}\Big[\frac{\partial^2 \mathbf{f}}{\partial \mathbf{x}_1^2} + \mathbf{f}(1-\mathbf{f})\Big]\Big].
$$

Applying the iterative technique described in Section 3.1, we obtain the following results.

$$
\begin{aligned}
\mathbf{f}_0(\mathbf{x}_1,\bar{t}) &= \eta, \\
\mathbf{f}_1(\mathbf{x}_1,\bar{t}) &= \mathbb{E}^{-1}\Big[\frac{\alpha\omega^{\alpha}+1-\alpha}{\mathbb{N}(\alpha)}\mathbb{E}\Big\{\big(\mathbf{f}_0(\mathbf{x}_1,\bar{t})\big)_{\mathbf{x}_1\mathbf{x}_1} + \mathbf{f}_0(1-\mathbf{f}_0)\Big\}\Big] \\
&= \frac{\eta(1-\eta)}{\mathbb{N}(\alpha)}\Big[\frac{\alpha\bar{t}^{\alpha}}{\Gamma(\alpha+1)} + (1-\alpha)\Big], \\
\mathbf{f}_2(\mathbf{x}_1,\bar{t}) &= \mathbb{E}^{-1}\Big[\frac{\alpha\omega^{\alpha}+1-\alpha}{\mathbb{N}(\alpha)}\mathbb{E}\Big\{\big(\mathbf{f}_1(\mathbf{x}_1,\bar{t})\big)_{\mathbf{x}_1\mathbf{x}_1} + \mathbf{f}_1(1-\mathbf{f}_1)\Big\}\Big] \\
&= \frac{\eta(1-\eta)(1-2\eta)}{\mathbb{N}^2(\alpha)}\Big[\frac{\alpha^2\bar{t}^{2\alpha}}{\Gamma(2\alpha+1)} + 2\alpha(1-\alpha)\frac{\bar{t}^{\alpha}}{\Gamma(\alpha+1)} + (1-\alpha)^2\Big], \\
\mathbf{f}_3(\mathbf{x}_1,\bar{t}) &= \mathbb{E}^{-1}\Big[\frac{\alpha\omega^{\alpha}+1-\alpha}{\mathbb{N}(\alpha)}\mathbb{E}\Big\{\big(\mathbf{f}_2(\mathbf{x}_1,\bar{t})\big)_{\mathbf{x}_1\mathbf{x}_1} + \mathbf{f}_2(1-\mathbf{f}_2)\Big\}\Big] \\
&= \frac{\eta(1-\eta)(1-6\eta+6\eta^2)}{\mathbb{N}^3(\alpha)}\Big[\frac{\alpha^3\bar{t}^{3\alpha}}{\Gamma(3\alpha+1)} + 3\alpha^2(1-\alpha)\frac{\bar{t}^{2\alpha}}{\Gamma(2\alpha+1)} + 3\alpha(1-\alpha)^2\frac{\bar{t}^{\alpha}}{\Gamma(\alpha+1)} + (1-\alpha)^3\Big], \\
&\vdots
\end{aligned}
$$

Provided the series form solution is as follows:

$$
\mathbf{f}(\mathbf{x}_1,\bar{t}) = \mathbf{f}_0(\mathbf{x}_1,\bar{t}) + \mathbf{f}_1(\mathbf{x}_1,\bar{t}) + \mathbf{f}_2(\mathbf{x}_1,\bar{t}) + \mathbf{f}_3(\mathbf{x}_1,\bar{t}) + \ldots + \mathbf{f}_{m_1}(\mathbf{x}_1,\bar{t}).
$$

we consequently have

$$
\begin{aligned}
\mathbf{f}(\mathbf{x}_1,\bar{t}) &= \eta + \frac{\eta(1-\eta)}{\mathbb{N}(\alpha)}\Big[\frac{\alpha\bar{t}^{\alpha}}{\Gamma(\alpha+1)} + (1-\alpha)\Big] + \frac{\eta(1-\eta)(1-2\eta)}{\mathbb{N}^2(\alpha)}\Big[\frac{\alpha^2\bar{t}^{2\alpha}}{\Gamma(2\alpha+1)} + 2\alpha(1-\alpha)\frac{\bar{t}^{\alpha}}{\Gamma(\alpha+1)} + (1-\alpha)^2\Big] \\
&+\frac{\eta(1-\eta)(1-6\eta+6\eta^2)}{\mathbb{N}^3(\alpha)}\Big[\frac{\alpha^3\bar{t}^{3\alpha}}{\Gamma(3\alpha+1)} + 3\alpha^2(1-\alpha)\frac{\bar{t}^{2\alpha}}{\Gamma(2\alpha+1)} + 3\alpha(1-\alpha)^2\frac{\bar{t}^{\alpha}}{\Gamma(\alpha+1)} + (1-\alpha)^3\Big] + \ldots
\end{aligned}
$$

For showing the accuracy and compactness of our proposed algorithm (IETM via CFD and ABC fractional derivatives), we compare our results with [39]. It can be observed from Table 1 that the present algorithm is very effective and yields accurate results. The absolute errors of the numerical solution of Fisher's equation obtained by IETM (CFD and ABC fractional derivatives) and the exact solutions for Case 1 are depicted in Table 1 and presents a strong correlation among the proposed technique and rapidly converges to the exact solution very efficiently in a short admissible domain.

Figure 1 compares the exact and approximate solutions to Problem 1 using the CDF operator. The absolute error norm in Figure 2 for (36) with the assumptions of $\eta = 0.05, \theta = 1$, $\beta = 1$, and $\varphi = 0$ ensures the approximation of the numerical results derived by IETM to the exact solution via the CFD and ABC fractional derivative operators, respectively. The results of the graphical representation reveal that the model is highly dependent on fractional order α. The absolute inaccuracy is really small. Two dimensional representations of graphs via Figure 3 show the strong connection between the exact and approximate solutions for various fractional orders. Furthermore, Figure3a,b illustrate that the ABC fractional derivative operator has better harmony than the CFD operator.

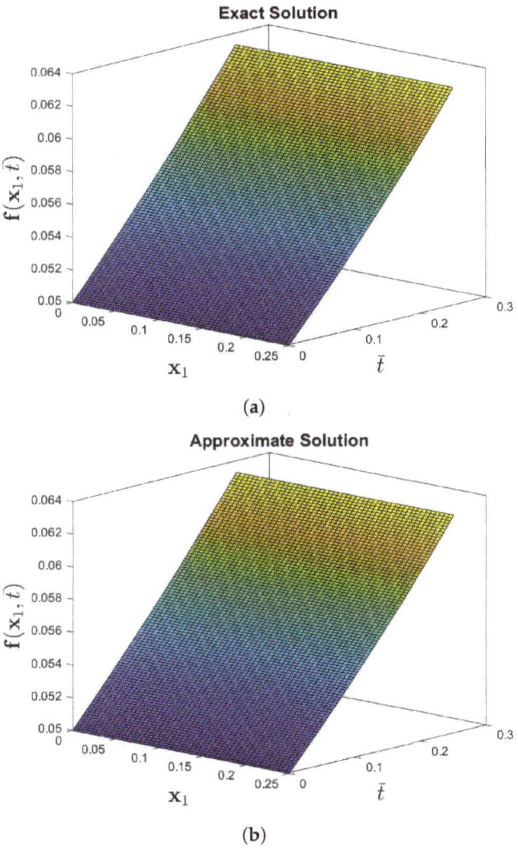

Figure 1. Numerical behavior of exact and approximate solution to the $\mathbf{f}(\mathbf{x}_1, \bar{t})$ for Problem 1 when the parameters are $\eta = 0.05, \theta = 1, \beta = 1$, and $\varphi = 0$.

Table 1. Comparison results with exact ($\mathbf{f}_E$) IETM-numerical solutions ($\mathbf{f}_{Num}$) for CFD and ABC fractional derivative operator of $\mathbf{f}(\mathbf{x}_1, \bar{t})$ of Case 1 with absolute errors and the HPM [39] when $\alpha = 1$, $\bar{t} = 0.01$, $\eta = 0.05$, and $\varphi = 0$ for various values of $\mathbf{x}_1$.

η	$\mathbf{f}_E$	$\mathbf{f}_{Num/CFD}$ sol.	$\mathbf{f}_{Num/ABC}$ sol.	$\lvert\mathbf{f}_E - \mathbf{f}_{Num/CFD}\rvert$	$\lvert\mathbf{f}_E - \mathbf{f}_{Num/ABC}\rvert$	HPM sol. [39]
0.1	1.009×10^{-1}	1.009×10^{-1}	2.329×10^{-1}	0	1.319×10^{-1}	8.999×10^{-1}
0.2	2.016×10^{-1}	2.016×10^{-1}	4.091×10^{-1}	0	2.017×10^{-1}	7.987×10^{-1}
0.3	3.021×10^{-1}	3.021×10^{-1}	5.429×10^{-1}	0	2.408×10^{-1}	9.210×10^{-1}
0.4	4.024×10^{-1}	4.024×10^{-1}	6.464×10^{-1}	1.00×10^{-11}	2.440×10^{-1}	7.540×10^{-1}
0.5	5.025×10^{-1}	5.025×10^{-1}	7.292×10^{-1}	3.00×10^{-10}	2.267×10^{-1}	8.908×10^{-1}
0.6	6.024×10^{-1}	6.024×10^{-1}	7.984×10^{-1}	1.00×10^{-10}	1.960×10^{-1}	9.765×10^{-1}
0.7	7.021×10^{-1}	7.021×10^{-1}	8.589×10^{-1}	0	1.568×10^{-1}	9.344×10^{-1}
0.8	8.016×10^{-1}	8.016×10^{-1}	9.131×10^{-1}	3.00×10^{-10}	1.115×10^{-1}	9.123×10^{-1}
0.9	9.009×10^{-1}	9.009×10^{-1}	9.609×10^{-1}	3.00×10^{-10}	6.000×10^{-2}	9.777×10^{-1}
1.0	1.000×10^{0}	1.000×10^{0}	1.000×10^{0}	0	0	1.000×10^{0}

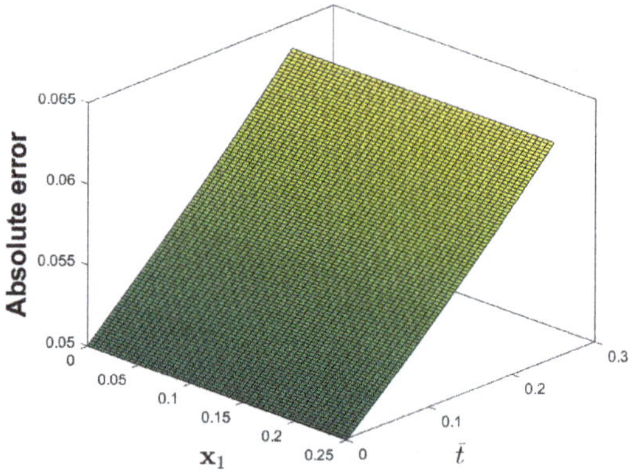

(a)

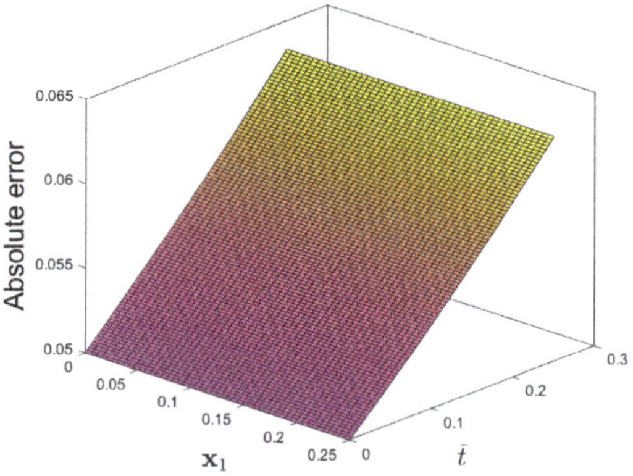

(b)

Figure 2. (**a**) Absolute error plots of $\mathbf{f}(\mathbf{x}_1, \bar{t})$ for Problem 1 for (**a**) CFD and (**b**) ABC when the parameters are $\eta = 0.05, \theta = 1, \beta = 1$, and $\varphi = 0$.

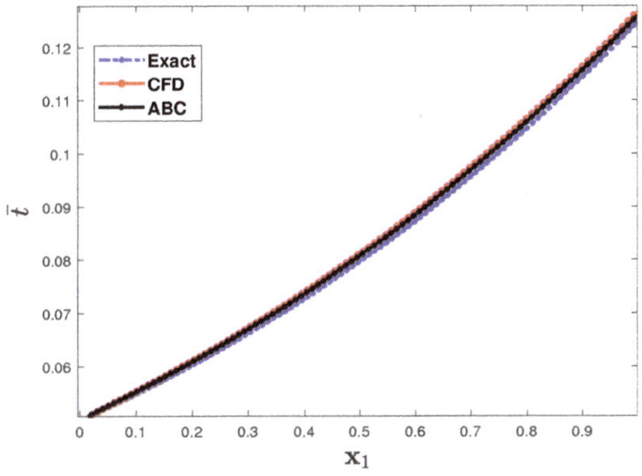

(a)

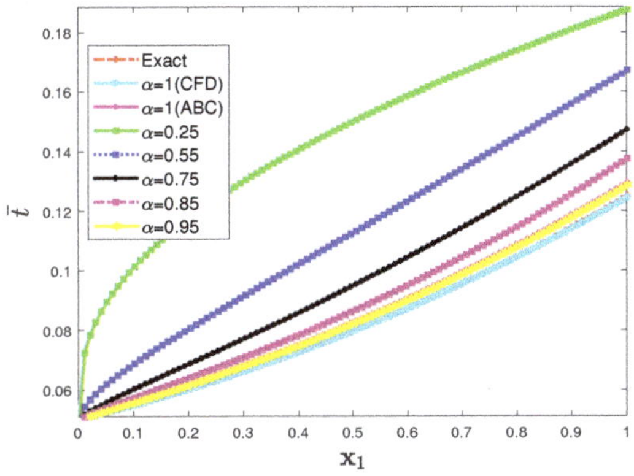

(b)

Figure 3. Two dimensional representation of $\mathbf{f}(\mathbf{x}_1, \bar{t})$ for Problem 1 at different fractional orders when the parameters are $\eta = 0.05, \theta = 1, \beta = 1$, and $\varphi = 0$. (**a**) illustrates the comparison view of CFD and ABC operators with their exact solutions, while (**b**) shows the two dimensional view of the exact-approximate solution with different fractional-order in the CFD and ABC fractional derivative sense.

Problem 2. *If* $\theta = 6, \beta = 1$ *and* $\varphi = 0$ *in (10) with* $\mathbf{f}_0(\mathbf{x}_1, 0) = \frac{1}{(1+\exp(\mathbf{x}_1))^2}$, *then the one dimensional time fractional Fisher equation is presented as follows.*

$$\frac{\partial^\alpha \mathbf{f}}{\partial \bar{t}^\alpha} = \frac{\partial^2 \mathbf{f}}{\partial \mathbf{x}_1^2} + 6\mathbf{f}(1 - \mathbf{f}). \tag{39}$$

The integer-order solution for the Fisher's Equation (39) is obtained by using the Taylor's series expansion for $\alpha = 1$ *as follows.*

$$\mathbf{f}(\mathbf{x}_1, \bar{t}) = \frac{1}{(1 + \exp(\mathbf{x}_1 - 5\bar{t}))^2}.$$

Case I. First, we formulate Problem 2 by utilizing the Elzaki transform coupled with the Caputo derivative operator.

Employing the Elzaki transform to (39) with the initial condition, we have the following.

$$\mathbb{E}\left[\frac{\partial^\alpha \mathbf{f}}{\partial \bar{t}^\alpha}\right] = \mathbb{E}\left[\frac{\partial^2 \mathbf{f}}{\partial \mathbf{x}_1^2} + 6\mathbf{f}(1 - \mathbf{f})\right]. \tag{40}$$

The following results:

$$\frac{1}{\omega^{\alpha}}\mathbb{E}\big[\mathbf{f}(\mathbf{x}_1,\bar{t})\big] - \sum_{\kappa=0}^{m-1}\mathbf{f}_{(\kappa)}(\mathbf{x}_1,0)\omega^{2-\alpha+\kappa} = \mathbb{E}\Big[\frac{\partial^2\mathbf{f}}{\partial\mathbf{x}_1^2} + 6\mathbf{f}(1-\mathbf{f})\Big].$$

equivalently, we have

$$\frac{1}{\omega^{\alpha}}\mathbb{E}\big[\mathbf{f}(\mathbf{x}_1,\bar{t})\big] = \mathbf{f}_{(0)}(\mathbf{x}_1,0)\omega^{2-\alpha} + \mathbb{E}\Big[\frac{\partial^2\mathbf{f}}{\partial\mathbf{x}_1^2} + 6\mathbf{f}(1-\mathbf{f})\Big].$$

Using the inverse Elzaki transform, we have the following.

$$\mathbf{f}(\mathbf{x}_1,\bar{t}) = \eta + \mathbb{E}^{-1}\Big[\omega^{\alpha}\mathbb{E}\Big[\frac{\partial^2\mathbf{f}}{\partial\mathbf{x}_1^2} + 6\mathbf{f}(1-\mathbf{f})\Big]\Big].$$

Applying the iterative technique described in Section 3.1, we obtain the following.

$$\begin{aligned}
\mathbf{f}_0(\mathbf{x}_1,\bar{t}) &= \frac{1}{(1+\exp(\mathbf{x}_1))^2},\\
\mathbf{f}_1(\mathbf{x}_1,\bar{t}) &= \mathbb{E}^{-1}\Big[\omega^{\alpha}\mathbb{E}\Big\{\big(\mathbf{f}_0(\mathbf{x}_1,\bar{t})\big)_{\mathbf{x}_1\mathbf{x}_1} + 6\mathbf{f}_0(1-\mathbf{f}_0)\Big\}\Big]\\
&= \frac{10\exp(\mathbf{x}_1)}{(1+\exp(\mathbf{x}_1))^3}\frac{\bar{t}^{\alpha}}{\Gamma(\alpha+1)},\\
\mathbf{f}_2(\mathbf{x}_1,\bar{t}) &= \mathbb{E}^{-1}\Big[\omega^{\alpha}\mathbb{E}\Big\{\big(\mathbf{f}_1(\mathbf{x}_1,\bar{t})\big)_{\mathbf{x}_1\mathbf{x}_1} + 6\mathbf{f}_1(1-\mathbf{f}_1)\Big\}\Big]\\
&= \frac{50\exp(\mathbf{x}_1)(\exp(2\mathbf{x}_1)-1)}{(1+\exp(\mathbf{x}_1))^4}\frac{\bar{t}^{2\alpha}}{\Gamma(2\alpha+1)},\\
\mathbf{f}_3(\mathbf{x}_1,\bar{t}) &= \mathbb{E}^{-1}\Big[\omega^{\alpha}\mathbb{E}\Big\{\big(\mathbf{f}_2(\mathbf{x}_1,\bar{t})\big)_{\mathbf{x}_1\mathbf{x}_1} + 6\mathbf{f}_2(1-\mathbf{f}_2)\Big\}\Big]\\
&= -\frac{750(7\exp(\mathbf{x}_1)-4\exp(2\mathbf{x}_1)-1)}{3(1+\exp(\mathbf{x}_1))^5}\frac{\bar{t}^{3\alpha}}{\Gamma(3\alpha+1)},\\
&\vdots
\end{aligned}$$

Provided the series form solution is the following:

$$\mathbf{f}(\mathbf{x}_1,\bar{t}) = \mathbf{f}_0(\mathbf{x}_1,\bar{t}) + \mathbf{f}_1(\mathbf{x}_1,\bar{t}) + \mathbf{f}_2(\mathbf{x}_1,\bar{t}) + \mathbf{f}_3(\mathbf{x}_1,\bar{t}) + \ldots + \mathbf{f}_{m_1}(\mathbf{x}_1,\bar{t}).$$

we consequently have the following.

$$\begin{aligned}
\mathbf{f}(\mathbf{x}_1,\bar{t}) &= \frac{1}{(1+\exp(\mathbf{x}_1))^2} + \frac{10\exp(\mathbf{x}_1)}{(1+\exp(\mathbf{x}_1))^3}\frac{\bar{t}^{\alpha}}{\Gamma(\alpha+1)} + \frac{50\exp(\mathbf{x}_1)(\exp(2\mathbf{x}_1)-1)}{(1+\exp(\mathbf{x}_1))^4}\frac{\bar{t}^{2\alpha}}{\Gamma(2\alpha+1)}\\
&\quad -\frac{570(7\exp(\mathbf{x}_1)-4\exp(2\mathbf{x}_1)-1)}{3(1+\exp(\mathbf{x}_1))^5}\frac{\bar{t}^{3\alpha}}{\Gamma(3\alpha+1)} + \ldots
\end{aligned}$$

Case II. Now we formulate Problem 2 by utilizing the Elzaki transform coupled with an ABC derivative operator.

Employing the Elzaki transform to (39) with the initial condition, we have the following.

$$\mathbb{E}\Big[\frac{\partial^{\alpha}\mathbf{f}}{\partial\bar{t}^{\alpha}}\Big] = \mathbb{E}\Big[\frac{\partial^2\mathbf{f}}{\partial\mathbf{x}_1^2} + 6\mathbf{f}(1-\mathbf{f})\Big]. \tag{41}$$

$$\mathbb{E}\big[\mathbf{f}(\mathbf{x}_1,\bar{t})\big] = \mathbf{f}_{(0)}(\mathbf{x}_1,0)\omega^2 + \frac{\alpha\omega^{\alpha} + (1-\alpha)}{\mathbb{N}(\alpha)}\mathbb{E}\Big[\frac{\partial^2\mathbf{f}}{\partial\mathbf{x}_1^2} + 6\mathbf{f}(1-\mathbf{f})\Big].$$

By using the inverse Elzaki transform, we have the following.

$$\mathbf{f}(\mathbf{x}_1,\bar{t}) = \eta + \mathbb{E}^{-1}\left[\frac{\alpha\omega^{\alpha} + (1-\alpha)}{\mathbb{N}(\alpha)}\mathbb{E}\left[\frac{\partial^2 \mathbf{f}}{\partial \mathbf{x}_1^2} + 6\mathbf{f}(1-\mathbf{f})\right]\right].$$

By applying the iterative technique described in Section 3.1, we obtain the following.

$$\begin{aligned}
\mathbf{f}_0(\mathbf{x}_1,\bar{t}) &= \frac{1}{(1+\exp(\mathbf{x}_1))^2},\\
\mathbf{f}_1(\mathbf{x}_1,\bar{t}) &= \mathbb{E}^{-1}\left[\frac{\alpha\omega^{\alpha} + (1-\alpha)}{\mathbb{N}(\alpha)}\mathbb{E}\Big\{\big(\mathbf{f}_0(\mathbf{x}_1,\bar{t})\big)_{\mathbf{x}_1\mathbf{x}_1} + 6\mathbf{f}_0(1-\mathbf{f}_0)\Big\}\right]\\
&= \frac{10\exp(\mathbf{x}_1)}{\mathbb{N}(\alpha)(1+\exp(\mathbf{x}_1))^3}\left[\frac{\alpha\bar{t}^{\alpha}}{\Gamma(\alpha+1)} + (1-\alpha)\right],\\
\mathbf{f}_2(\mathbf{x}_1,\bar{t}) &= \mathbb{E}^{-1}\left[\frac{\alpha\omega^{\alpha} + (1-\alpha)}{\mathbb{N}(\alpha)}\mathbb{E}\Big\{\big(\mathbf{f}_1(\mathbf{x}_1,\bar{t})\big)_{\mathbf{x}_1\mathbf{x}_1} + 6\mathbf{f}_1(1-\mathbf{f}_1)\Big\}\right]\\
&= \frac{50\exp(\mathbf{x}_1)(\exp(2\mathbf{x}_1)-1)}{\mathbb{N}^2(\alpha)(1+\exp(\mathbf{x}_1))^4}\left[\frac{\alpha^2\bar{t}^{2\alpha}}{\Gamma(2\alpha+1)} + 2\alpha(1-\alpha)\frac{\bar{t}^{\alpha}}{\Gamma(1+\alpha)} + (1-\alpha)^2\right],\\
\mathbf{f}_3(\mathbf{x}_1,\bar{t}) &= \mathbb{E}^{-1}\left[\frac{\alpha\omega^{\alpha} + (1-\alpha)}{\mathbb{N}(\alpha)}\mathbb{E}\Big\{\big(\mathbf{f}_2(\mathbf{x}_1,\bar{t})\big)_{\mathbf{x}_1\mathbf{x}_1} + 6\mathbf{f}_2(1-\mathbf{f}_2)\Big\}\right]\\
&= -\frac{750(7\exp(\mathbf{x}_1) - 4\exp(2\mathbf{x}_1) - 1)}{3\mathbb{N}^3(\alpha)(1+\exp(\mathbf{x}_1))^5}\\
&\quad\times\left[\frac{\alpha^3\bar{t}^{3\alpha}}{\Gamma(3\alpha+1)} + 3\alpha^2(1-\alpha)\frac{\bar{t}^{2\alpha}}{\Gamma(1+2\alpha)} + 3\alpha(1-\alpha)^2\frac{\bar{t}^{\alpha}}{\Gamma(1+\alpha)} + (1-\alpha)^3\right],\\
&\vdots
\end{aligned}$$

Provided that the series form solution is as follows:

$$\mathbf{f}(\mathbf{x}_1,\bar{t}) = \mathbf{f}_0(\mathbf{x}_1,\bar{t}) + \mathbf{f}_1(\mathbf{x}_1,\bar{t}) + \mathbf{f}_2(\mathbf{x}_1,\bar{t}) + \mathbf{f}_3(\mathbf{x}_1,\bar{t}) + \ldots + \mathbf{f}_{m_1}(\mathbf{x}_1,\bar{t}).$$

the following consequently results.

$$\begin{aligned}
\mathbf{f}(\mathbf{x}_1,\bar{t}) &= \frac{1}{(1+\exp(\mathbf{x}_1))^2} + \frac{10\exp(\mathbf{x}_1)}{\mathbb{N}(\alpha)(1+\exp(\mathbf{x}_1))^3}\left[\frac{\alpha\bar{t}^{\alpha}}{\Gamma(\alpha+1)} + (1-\alpha)\right]\\
&\quad+\frac{50\exp(\mathbf{x}_1)(\exp(2\mathbf{x}_1)-1)}{\mathbb{N}^2(\alpha)(1+\exp(\mathbf{x}_1))^4}\left[\frac{\alpha^2\bar{t}^{2\alpha}}{\Gamma(2\alpha+1)} + 2\alpha(1-\alpha)\frac{\bar{t}^{\alpha}}{\Gamma(1+\alpha)} + (1-\alpha)^2\right]\\
&\quad-\frac{750(7\exp(\mathbf{x}_1) - 4\exp(2\mathbf{x}_1) - 1)}{3\mathbb{N}^3(\alpha)(1+\exp(\mathbf{x}_1))^5}\\
&\quad\times\left[\frac{\alpha^3\bar{t}^{3\alpha}}{\Gamma(3\alpha+1)} + 3\alpha^2(1-\alpha)\frac{\bar{t}^{2\alpha}}{\Gamma(1+2\alpha)} + 3\alpha(1-\alpha)^2\frac{\bar{t}^{\alpha}}{\Gamma(1+\alpha)} + (1-\alpha)^3\right] + \ldots
\end{aligned}$$

For showing the accuracy and compactness of our proposed algorithm (IETM via CFD and ABC fractional derivatives), we compare our results with [39]. It can be observed from Table 2 that the present algorithm is very effective and yields accurate results. The absolute errors of the numerical solution of Fisher's equation obtained by IETM (CFD and ABC fractional derivatives) and the exact solutions for Case 2 are depicted in Table 2 and presents a strong correlation among the proposed technique and rapidly converges to the exact solution very efficiently in a short admissible domain.

Figure 4 compares the exact and approximate solutions to Problem 2 by using the CDF operator. The absolute error norm in Figure 5 for (39) with the assumptions of $\theta = 6, \beta = 1$, and $\varphi = 0$ ensures the approximation of the numerical results derived by the IETM to the exact solution via the CFD and ABC fractional derivative operators, respectively. The results of the graphical representation reveal that the model is highly dependent on fractional order α. The absolute inaccuracy is really small. Surface and two dimensional representations of graphs via Figure 6 show the strong connection between the exact and approximate solutions for various fractional orders. Furthermore, Figure 6a,b illustrates that the ABC fractional derivative operator has better harmony than the CFD operator.

Table 2. Comparison results with exact ($\mathbf{f}_E$) IETM-numerical solutions ($\mathbf{f}_{Num}$) for CFD and ABC fractional derivative operator of $\mathbf{f}(\mathbf{x}_1, \bar{t})$ of Case 2 with absolute errors and the HPM [39] when $\alpha = 1$, $\bar{t} = 0.01$, and $\varphi = 0$ for various values of $\mathbf{x}_1$.

$\mathbf{x}_1$	$\mathbf{f}_E$	$\mathbf{f}_{Num/CFD}$ sol.	$\mathbf{f}_{Num/ABC}$ sol.	$\lvert\mathbf{f}_E - \mathbf{f}_{Num/CFD}\rvert$	$\lvert\mathbf{f}_E - \mathbf{f}_{Num/ABC}\rvert$	HPM sol. [39]
0.1	2.377×10^{-1}	2.375×10^{-1}	2.432×10^{-1}	1.394×10^{-4}	-5.7000×10^{-4}	8.387×10^{-1}
0.2	2.140×10^{-1}	2.138×10^{-1}	2.159×10^{-1}	1.119×10^{-4}	-0.019×10^{-4}	9.567×10^{-1}
0.3	1.917×10^{-1}	1.915×10^{-1}	2.000×10^{-1}	9.68×10^{-5}	-0.083×10^{-5}	4.534×10^{-1}
0.4	1.709×10^{-1}	1.708×10^{-1}	1.888×10^{-1}	7.32×10^{-5}	-0.179×10^{-5}	8.887×10^{-1}
0.5	1.516×10^{-1}	-0.18×10^{-5}	1.575×10^{-1}	4.85×10^{-5}	-0.059×10^{-5}	7.337×10^{-1}
0.6	1.339×10^{-1}	1.338×10^{-1}	1.958×10^{-1}	2.35×10^{-5}	-0.619×10^{-5}	9.337×10^{-1}
0.7	1.176×10^{-1}	1.175×10^{-1}	1.234×10^{-1}	4.85×10^{-5}	-0.58×10^{-5}	7.337×10^{-1}
0.8	1.029×10^{-1}	1.029×10^{-1}	1.416×10^{-1}	2.51×10^{-5}	-0.055×10^{-5}	9.998×10^{-1}
0.9	8.966×10^{-2}	8.971×10^{-2}	9.516×10^{-1}	4.76×10^{-5}	3.426×10^{-5}	9.001×10^{-1}
1.0	7.778×10^{-1}	7.784×10^{-1}	9.001×10^{0}	6.75×10^{-5}	-1.223×10^{-5}	7.337×10^{-1}

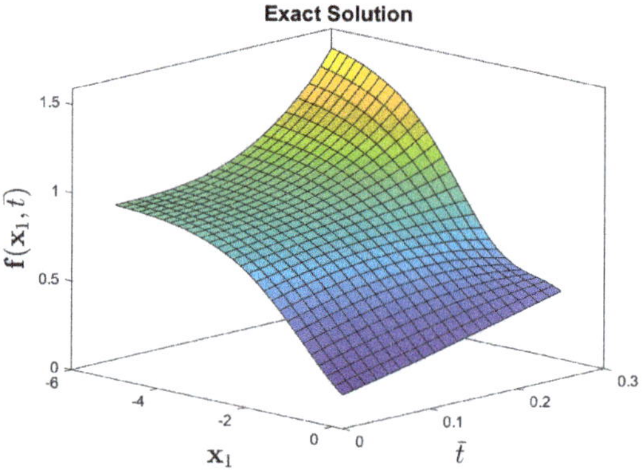

(a)

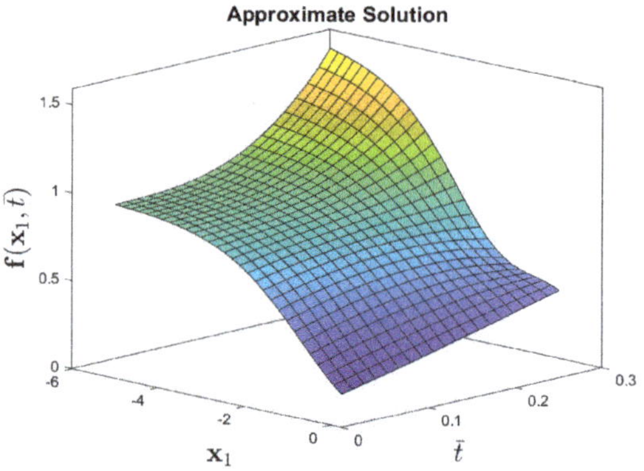

(b)

Figure 4. Numerical-behavior of exact and approximate solution to the $\mathbf{f}(\mathbf{x}_1, \bar{t})$ for Problem 2 when the parameters are $\theta = 6, \beta = 1$, and $\varphi = 0$.

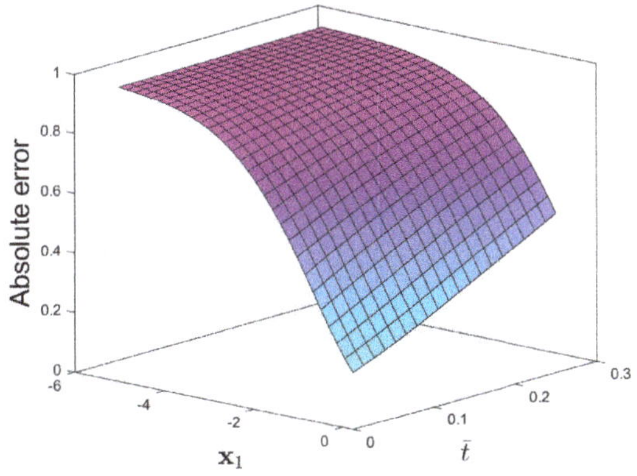

(a)

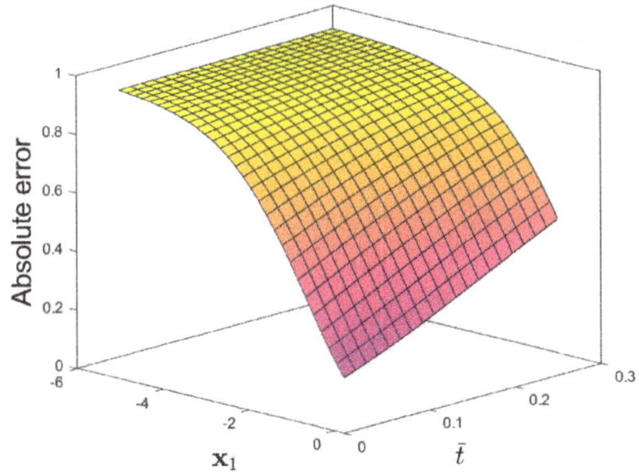

(b)

Figure 5. (a) Absolute-error plots of $\mathbf{f}(\mathbf{x}_1, \bar{t})$ for Problem 2 for (a) CFD and (b) ABC when the parameters are $\theta = 6, \beta = 1$, and $\varphi = 0$.

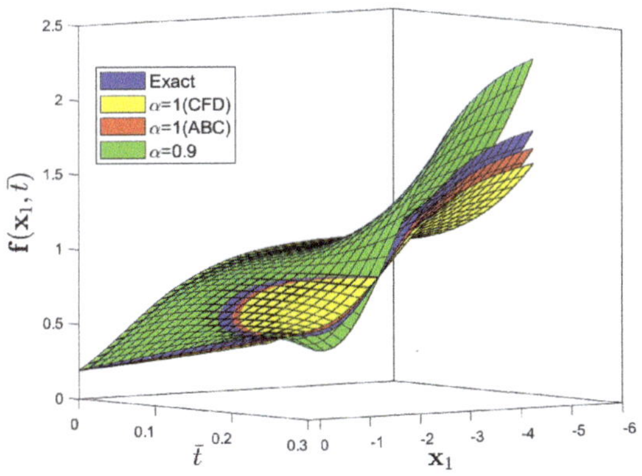

(a)

Figure 6. *Cont.*

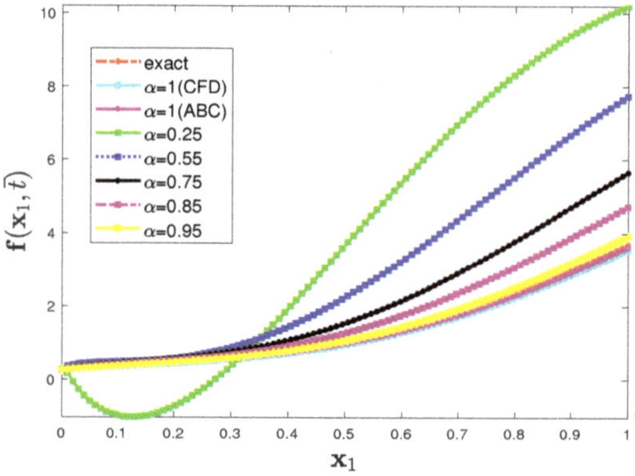

(**b**)

Figure 6. (**a**) Three-dimensional comparison plot among exact, CFD, and ABC fractional derivative operators via the IETM (**b**) Two dimensional representation of $\mathbf{f}(\mathbf{x}_1,\bar{t})$ for Problem 2 when the parameters are $\theta = 6, \beta = 1$, and $\varphi = 0$.

Problem 3. *If* $\theta = 1, \beta = 6$, *and* $\varphi = 0$ *in (10) with* $\mathbf{f}_0(\mathbf{x}_1, 0) = \frac{1}{(1+\exp(\frac{3\mathbf{x}_1}{2}))^{1/3}}$, *then the one dimensional time fractional generalized Fisher's equation is presented as follows.*

$$\frac{\partial^{\alpha}\mathbf{f}}{\partial\bar{t}^{\alpha}} = \frac{\partial^2\mathbf{f}}{\partial\mathbf{x}_1^2} + \mathbf{f}(1-\mathbf{f}^6). \tag{42}$$

The integer-order solution for the Fisher's Equation (42) is obtained by using the Taylor's series expansion for $\alpha = 1$ *as follows.*

$$\mathbf{f}(\mathbf{x}_1,\bar{t}) = \left(\frac{1}{2}\right)^{1/3}\left(\tanh\left(\frac{15}{8}\bar{t} - \frac{3}{4}\mathbf{x}_1\right) + 1\right)^{1/3}.$$

Case I. First, we formulate Problem 3 by utilizing the Elzaki transform coupled with the Caputo derivative operator.

Employing the Elzaki transform to (42) with the initial condition, we have the following.

$$\mathbb{E}\Big[\frac{\partial^{\alpha}\mathbf{f}}{\partial\bar{t}^{\alpha}}\Big] = \mathbb{E}\Big[\frac{\partial^2\mathbf{f}}{\partial\mathbf{x}_1^2} + \mathbf{f}(1-\mathbf{f}^6)\Big]. \tag{43}$$

The following is the case.

$$\frac{1}{\omega^{\alpha}}\mathbb{E}\big[\mathbf{f}(\mathbf{x}_1,\bar{t})\big] - \sum_{\kappa=0}^{m-1}\mathbf{f}_{(\kappa)}(\mathbf{x}_1,0)\omega^{2-\alpha+\kappa} = \mathbb{E}\Big[\frac{\partial^2\mathbf{f}}{\partial\mathbf{x}_1^2} + \mathbf{f}(1-\mathbf{f}^6)\Big].$$

Equivalently, we have the following.

$$\frac{1}{\omega^{\alpha}}\mathbb{E}\big[\mathbf{f}(\mathbf{x}_1,\bar{t})\big] = \mathbf{f}_{(0)}(\mathbf{x}_1,0)\omega^{2-\alpha} + \mathbb{E}\Big[\frac{\partial^2\mathbf{f}}{\partial\mathbf{x}_1^2} + \mathbf{f}(1-\mathbf{f}^6)\Big].$$

By using the inverse Elzaki transform, we have the following.

$$\mathbf{f}(\mathbf{x}_1,\bar{t}) = \frac{1}{(1+\exp(\frac{3\mathbf{x}_1}{2}))^{1/3}} + \mathbb{E}^{-1}\Big[\omega^{\alpha}\mathbb{E}\Big[\frac{\partial^2\mathbf{f}}{\partial\mathbf{x}_1^2} + \mathbf{f}(1-\mathbf{f}^6)\Big]\Big].$$

By applying the iterative technique described in Section 3.1, we obtain the following.

$$\begin{aligned}
\mathbf{f}_0(\mathbf{x}_1,\bar{t}) &= \frac{1}{(1+\exp(\frac{3\mathbf{x}_1}{2}))^{1/3}},\\
\mathbf{f}_1(\mathbf{x}_1,\bar{t}) &= \mathbb{E}^{-1}\Big[\omega^{\alpha}\mathbb{E}\Big\{(\mathbf{f}_0(\mathbf{x}_1,\bar{t}))_{\mathbf{x}_1\mathbf{x}_1}+\mathbf{f}_0(1-\mathbf{f}_0^6)\Big\}\Big]\\
&= \frac{5\exp(\frac{3\mathbf{x}_1}{2})}{4(1+\exp(\frac{3\mathbf{x}_1}{2}))^{4/3}}\frac{\bar{t}^{\alpha}}{\Gamma(\alpha+1)},\\
\mathbf{f}_2(\mathbf{x}_1,\bar{t}) &= \mathbb{E}^{-1}\Big[\omega^{\alpha}\mathbb{E}\Big\{(\mathbf{f}_1(\mathbf{x}_1,\bar{t}))_{\mathbf{x}_1\mathbf{x}_1}+\mathbf{f}_1(1-\mathbf{f}_1^6)\Big\}\Big]\\
&= \frac{50\exp(\frac{3\mathbf{x}_1}{2})(\exp(\frac{3x}{2})-3)}{16(1+\exp(\frac{3\mathbf{x}_1}{2}))^{7/3}}\frac{\bar{t}^{2\alpha}}{\Gamma(2\alpha+1)},\\
\mathbf{f}_3(\mathbf{x}_1,\bar{t}) &= \mathbb{E}^{-1}\Big[\omega^{\alpha}\mathbb{E}\Big\{(\mathbf{f}_2(\mathbf{x}_1,\bar{t}))_{\mathbf{x}_1\mathbf{x}_1}+\mathbf{f}_2(1-\mathbf{f}_2^6)\Big\}\Big]\\
&= \frac{125\exp(\frac{3\mathbf{x}_1}{2})(\exp(3\mathbf{x}_1)-18\exp(\frac{3\mathbf{x}_1}{2})+9)}{16(1+\exp(\frac{3\mathbf{x}_1}{2}))^{10/3}}\frac{\bar{t}^{3\alpha}}{\Gamma(3\alpha+1)},\\
&\vdots
\end{aligned}$$

Provided the series form solution is the following:

$$\mathbf{f}(\mathbf{x}_1,\bar{t}) = \mathbf{f}_0(\mathbf{x}_1,\bar{t})+\mathbf{f}_1(\mathbf{x}_1,\bar{t})+\mathbf{f}_2(\mathbf{x}_1,\bar{t})+\mathbf{f}_3(\mathbf{x}_1,\bar{t})+\ldots+\mathbf{f}_{m_1}(\mathbf{x}_1,\bar{t}).$$

the following consequently results.

$$\begin{aligned}
\mathbf{f}(\mathbf{x}_1,\bar{t}) &= \frac{1}{(1+\exp(\frac{3\mathbf{x}_1}{2}))^{1/3}}+\frac{5\exp(\frac{3\mathbf{x}_1}{2})}{4(1+\exp(\frac{3\mathbf{x}_1}{2}))^{4/3}}\frac{\bar{t}^{\alpha}}{\Gamma(\alpha+1)}+\frac{50\exp(\frac{3\mathbf{x}_1}{2})(\exp(\frac{3x}{2})-3)}{16(1+\exp(\frac{3\mathbf{x}_1}{2}))^{7/3}}\frac{\bar{t}^{2\alpha}}{\Gamma(2\alpha+1)}\\
&+\frac{750\exp(\frac{3\mathbf{x}_1}{2})(\exp(3\mathbf{x}_1)-18\exp(\frac{3\mathbf{x}_1}{2})+9)}{16(1+\exp(\frac{3\mathbf{x}_1}{2}))^{10/3}}\frac{\bar{t}^{3\alpha}}{\Gamma(3\alpha+1)}+\ldots
\end{aligned}$$

Case II. Now we formulate Problem 3 by utilizing the Elzaki transform coupled with an ABC derivative operator.

Employing the Elzaki transform to (42) with the initial condition, we have the following.

$$\mathbb{E}\Big[\frac{\partial^{\alpha}\mathbf{f}}{\partial\bar{t}^{\alpha}}\Big] = \mathbb{E}\Big[\frac{\partial^2\mathbf{f}}{\partial\mathbf{x}_1^2}+\mathbf{f}(1-\mathbf{f}^6)\Big]. \tag{44}$$

$$\mathbb{E}\big[\mathbf{f}(\mathbf{x}_1,\bar{t})\big] = \mathbf{f}_{(0)}(\mathbf{x}_1,0)\omega^2+\frac{\alpha\omega^{\alpha}+(1-\alpha)}{\mathbb{N}(\alpha)}\mathbb{E}\Big[\frac{\partial^2\mathbf{f}}{\partial\mathbf{x}_1^2}+\mathbf{f}(1-\mathbf{f}^6)\Big].$$

By using the inverse Elzaki transform, we have the following.

$$\mathbf{f}(\mathbf{x}_1,\bar{t}) = \frac{1}{(1+\exp(\frac{3\mathbf{x}_1}{2}))^{1/3}}+\mathbb{E}^{-1}\Big[\frac{\alpha\omega^{\alpha}+(1-\alpha)}{\mathbb{N}(\alpha)}\mathbb{E}\Big[\frac{\partial^2\mathbf{f}}{\partial\mathbf{x}_1^2}+\mathbf{f}(1-\mathbf{f}^6)\Big]\Big].$$

By applying the iterative technique described in Section 3.1, we obtain the following.

$$
\begin{aligned}
\mathbf{f}_0(\mathbf{x}_1,\bar{t}) &= \frac{1}{(1+\exp(\frac{3\mathbf{x}_1}{2}))^{1/3}},\\
\mathbf{f}_1(\mathbf{x}_1,\bar{t}) &= \mathbb{E}^{-1}\left[\frac{\alpha\omega^{\alpha}+(1-\alpha)}{\mathbb{N}(\alpha)}\mathbb{E}\left\{\left(\mathbf{f}_0(\mathbf{x}_1,\bar{t})\right)_{\mathbf{x}_1\mathbf{x}_1}+\mathbf{f}_0(1-\mathbf{f}_0^6)\right\}\right]\\
&= \frac{5\exp(\frac{3\mathbf{x}_1}{2})}{4\mathbb{N}(\alpha)(1+\exp(\frac{3\mathbf{x}_1}{2}))^{4/3}}\left[\frac{\alpha\bar{t}^{\alpha}}{\Gamma(\alpha+1)}+(1-\alpha)\right],\\
\mathbf{f}_2(\mathbf{x}_1,\bar{t}) &= \mathbb{E}^{-1}\left[\frac{\alpha\omega^{\alpha}+(1-\alpha)}{\mathbb{N}(\alpha)}\mathbb{E}\left\{\left(\mathbf{f}_1(\mathbf{x}_1,\bar{t})\right)_{\mathbf{x}_1\mathbf{x}_1}+\mathbf{f}_1(1-\mathbf{f}_1^6)\right\}\right]\\
&= \frac{50\exp(\frac{3\mathbf{x}_1}{2})(\exp(\frac{3x}{2})-3)}{16\mathbb{N}^2(\alpha)(1+\exp(\frac{3\mathbf{x}_1}{2}))^{7/3}}\left[\frac{\alpha^2\bar{t}^{2\alpha}}{\Gamma(2\alpha+1)}+2\alpha(1-\alpha)\frac{\bar{t}^{\alpha}}{\Gamma(\alpha+1)}+(1-\alpha)^2\right],\\
\mathbf{f}_3(\mathbf{x}_1,\bar{t}) &= \mathbb{E}^{-1}\left[\frac{\alpha\omega^{\alpha}+(1-\alpha)}{\mathbb{N}(\alpha)}\mathbb{E}\left\{\left(\mathbf{f}_2(\mathbf{x}_1,\bar{t})\right)_{\mathbf{x}_1\mathbf{x}_1}+\mathbf{f}_2(1-\mathbf{f}_2^6)\right\}\right]\\
&= \frac{750\exp(\frac{3\mathbf{x}_1}{2})(\exp(3\mathbf{x}_1)-18\exp(\frac{3\mathbf{x}_1}{2})+9)}{16\mathbb{N}^3(\alpha)(1+\exp(\frac{3\mathbf{x}_1}{2}))^{10/3}}\\
&\quad\times\left[\frac{\alpha^3\bar{t}^{3\alpha}}{\Gamma(3\alpha+1)}+3\alpha^2(1-\alpha)\frac{\bar{t}^{2\alpha}}{\Gamma(2\alpha+1)}+3\alpha(1-\alpha)^2\frac{\bar{t}^{\alpha}}{\Gamma(\alpha+1)}+(1-\alpha)^3\right],\\
&\vdots
\end{aligned}
$$

Provided the series form solution is as follows:

$$
\mathbf{f}(\mathbf{x}_1,\bar{t}) = \mathbf{f}_0(\mathbf{x}_1,\bar{t})+\mathbf{f}_1(\mathbf{x}_1,\bar{t})+\mathbf{f}_2(\mathbf{x}_1,\bar{t})+\mathbf{f}_3(\mathbf{x}_1,\bar{t})+\ldots+\mathbf{f}_{m_1}(\mathbf{x}_1,\bar{t}).
$$

the following consequently results.

$$
\begin{aligned}
\mathbf{f}(\mathbf{x}_1,\bar{t}) &= \frac{1}{(1+\exp(\frac{3\mathbf{x}_1}{2}))^{1/3}}+\frac{5\exp(\frac{3\mathbf{x}_1}{2})}{4\mathbb{N}(\alpha)(1+\exp(\frac{3\mathbf{x}_1}{2}))^{4/3}}\left[\frac{\alpha\bar{t}^{\alpha}}{\Gamma(\alpha+1)}+(1-\alpha)\right]\\
&+\frac{50\exp(\frac{3\mathbf{x}_1}{2})(\exp(\frac{3x}{2})-3)}{16\mathbb{N}^2(\alpha)(1+\exp(\frac{3\mathbf{x}_1}{2}))^{7/3}}\left[\frac{\alpha^2\bar{t}^{2\alpha}}{\Gamma(2\alpha+1)}+2\alpha(1-\alpha)\frac{\bar{t}^{\alpha}}{\Gamma(\alpha+1)}+(1-\alpha)^2\right]\\
&+\frac{750\exp(\frac{3\mathbf{x}_1}{2})(\exp(3\mathbf{x}_1)-18\exp(\frac{3\mathbf{x}_1}{2})+9)}{16\mathbb{N}^3(\alpha)(1+\exp(\frac{3\mathbf{x}_1}{2}))^{10/3}}\\
&\times\left[\frac{\alpha^3\bar{t}^{3\alpha}}{\Gamma(3\alpha+1)}+3\alpha^2(1-\alpha)\frac{\bar{t}^{2\alpha}}{\Gamma(2\alpha+1)}+3\alpha(1-\alpha)^2\frac{\bar{t}^{\alpha}}{\Gamma(\alpha+1)}+(1-\alpha)^3\right]+\ldots
\end{aligned}
$$

Figure 7 compares the exact and approximate solutions to Problem 3 using the CDF operator. The absolute error norm in Figure 8 for (42) with the assumptions of $\theta = 1, \beta = 6$, and $\varphi = 0$ ensures the approximation of the numerical results derived by the IETM to the exact solution via the CFD and ABC fractional derivative operators, respectively. The results of the graphical representation reveal that the model is highly dependent on fractional order α. The absolute inaccuracy is really small. Surface and two dimensional representations of graphs via Figure 9 show the strong connection between the exact and approximate solutions for various fractional orders. Furthermore, Figure 9a,b illustrate that the ABC fractional derivative operator has better harmony than the CFD operator.

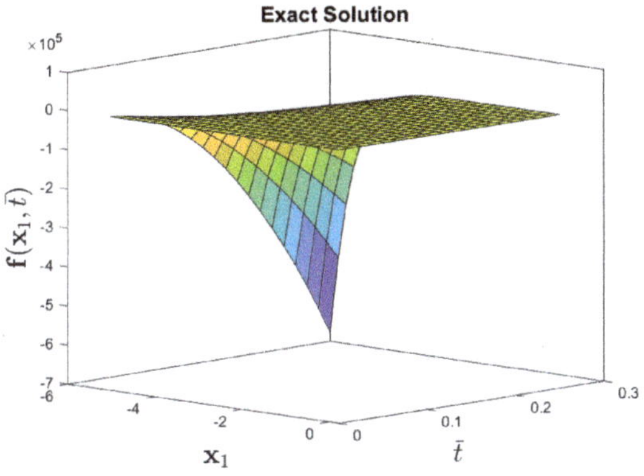

(**a**)

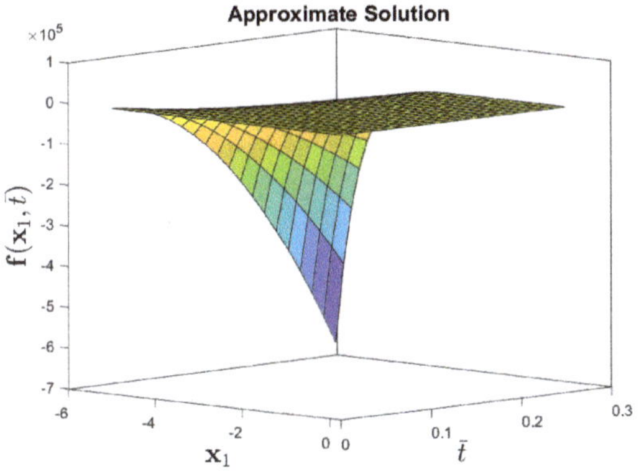

(**b**)

Figure 7. Numerical-behavior of exact and approximate solution to the $\mathbf{f}(\mathbf{x}_1, \bar{t})$ for Problem 3 when the parameters are $\theta = 1, \beta = 6$, and $\varphi = 0$.

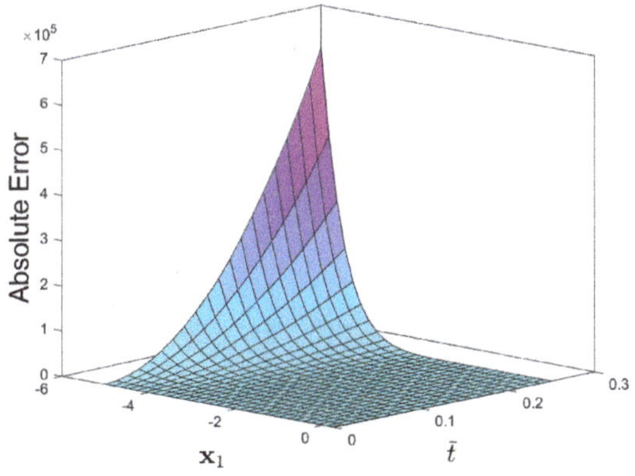

(**a**)

Figure 8. *Cont.*

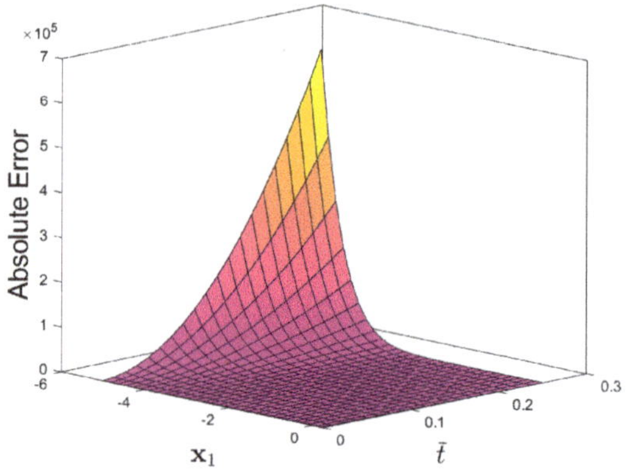

(b)

Figure 8. (**a**) Absolute-error plots of $\mathbf{f}(\mathbf{x}_1, \bar{t})$ for Problem 3 for (**a**) CFD and (**b**) ABC when the parameters are $\theta = 1, \beta = 6$ and $\varphi = 0$.

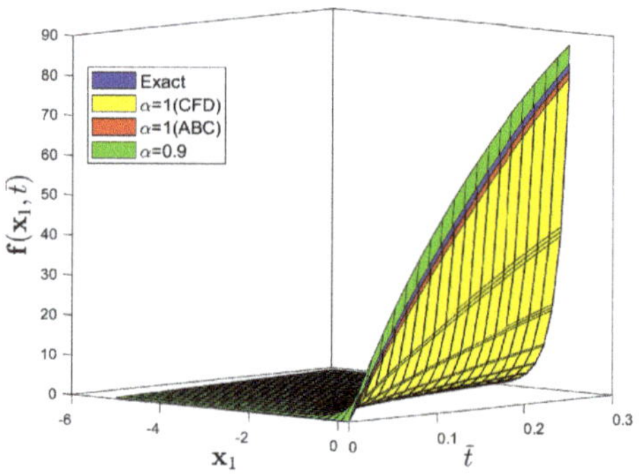

(a)

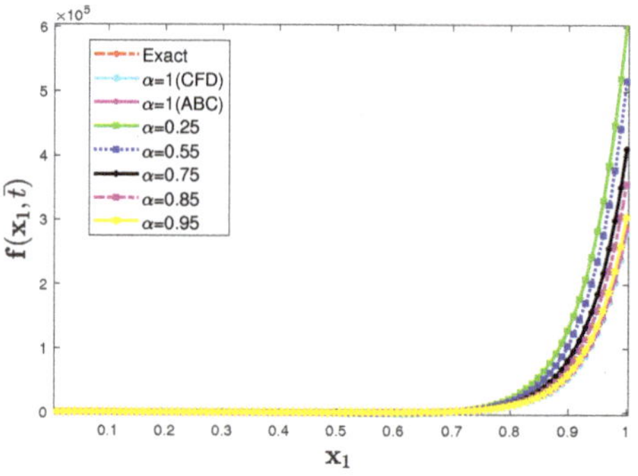

(b)

Figure 9. (**a**) Three-dimensional comparison plot among exact, CFD, and ABC fractional derivative operators via the IETM. (**b**) Two dimensional representation of $\mathbf{f}(\mathbf{x}_1, \bar{t})$ for Problem 3 at different fractional orders when $\theta = 1, \beta = 6$, and $\varphi = 0$.

Problem 4. *If* $\theta = 1, \beta = 1$, *and* $0 < \varphi < 1$ *in (10) with* $\mathbf{f}_0(\mathbf{x}_1, 0) = \frac{1}{(1+\exp(-\frac{1}{\sqrt{2}})\mathbf{x}_1)}$, *then the nonlinear diffusion equation of the Fisher's type is as follows.*

$$\frac{\partial^{\alpha}\mathbf{f}}{\partial\bar{t}^{\alpha}} = \frac{\partial^{2}\mathbf{f}}{\partial\mathbf{x}_1^{2}} + \mathbf{f}(1-\mathbf{f})(\mathbf{f}-\varphi). \tag{45}$$

The integer-order solution for the the nonlinear diffusion equation of the Fisher's type (45) is obtained by using the Taylor's series expansion for $\alpha = 1$ as follows.

$$\mathbf{f}(\mathbf{x}_1,\bar{t}) = \frac{1}{1+\exp(-\zeta/\sqrt{2})}.$$

Case I. First, we formulate Problem 4 by utilizing the Elzaki transform coupled with the Caputo derivative operator.

By employing the Elzaki transform to (45) with the initial condition, we have the following.

$$\mathbb{E}\Big[\frac{\partial^{\alpha}\mathbf{f}}{\partial\bar{t}^{\alpha}}\Big] = \mathbb{E}\Big[\frac{\partial^{2}\mathbf{f}}{\partial\mathbf{x}_1^{2}} + \mathbf{f}(1-\mathbf{f})(\mathbf{f}-\varphi)\Big]. \tag{46}$$

The following results.

$$\frac{1}{\omega^{\alpha}}\mathbb{E}\big[\mathbf{f}(\mathbf{x}_1,\bar{t})\big] - \sum_{\kappa=0}^{m-1}\mathbf{f}_{(\kappa)}(\mathbf{x}_1,0)\omega^{2-\alpha+\kappa} = \mathbb{E}\Big[\frac{\partial^{2}\mathbf{f}}{\partial\mathbf{x}_1^{2}} + \mathbf{f}(1-\mathbf{f})(\mathbf{f}-\varphi)\Big].$$

Equivalently, we also have the following.

$$\frac{1}{\omega^{\alpha}}\mathbb{E}\big[\mathbf{f}(\mathbf{x}_1,\bar{t})\big] = \mathbf{f}_{(0)}(\mathbf{x}_1,0)\omega^{2-\alpha} + \mathbb{E}\Big[\frac{\partial^{2}\mathbf{f}}{\partial\mathbf{x}_1^{2}} + \mathbf{f}(1-\mathbf{f})(\mathbf{f}-\varphi)\Big].$$

By using the inverse Elzaki transform, the following results.

$$\mathbf{f}(\mathbf{x}_1,\bar{t}) = \frac{1}{(1+\exp(-\frac{1}{\sqrt{2}})\mathbf{x}_1)} + \mathbb{E}^{-1}\Big[\omega^{\alpha}\mathbb{E}\Big[\frac{\partial^{2}\mathbf{f}}{\partial\mathbf{x}_1^{2}} - \varphi\mathbf{f} + (\varphi+1)\mathbf{f}^{2} - \mathbf{f}^{3}\Big]\Big].$$

By applying the iterative technique described in Section 3.1, we obtain the following.

$$\begin{aligned}
\mathbf{f}_0(\mathbf{x}_1,\bar{t}) &= \frac{1}{(1+\exp(-\frac{1}{\sqrt{2}})\mathbf{x}_1)},\\
\mathbf{f}_1(\mathbf{x}_1,\bar{t}) &= \mathbb{E}^{-1}\Big[\omega^{\alpha}\mathbb{E}\Big\{\big(\mathbf{f}_0(\mathbf{x}_1,\bar{t})\big)_{\mathbf{x}_1\mathbf{x}_1} + (\varphi+1)\mathbf{f}_0^{2} - \mathbf{f}_0^{3} - \varphi\mathbf{f}_0\Big\}\Big]\\
&= \frac{(1-2\varphi)\exp(\frac{-\mathbf{x}_1}{\sqrt{2}})}{2(1+\exp(\frac{-\mathbf{x}_1}{\sqrt{2}}))^{2}}\frac{\bar{t}^{\alpha}}{\Gamma(\alpha+1)},\\
\mathbf{f}_2(\mathbf{x}_1,\bar{t}) &= \mathbb{E}^{-1}\Big[\omega^{\alpha}\mathbb{E}\Big\{\big(\mathbf{f}_1(\mathbf{x}_1,\bar{t})\big)_{\mathbf{x}_1\mathbf{x}_1} + (\varphi+1)\mathbf{f}_1^{2} - \mathbf{f}_1^{3} - \varphi\mathbf{f}_1\Big\}\Big]\\
&= -\frac{(1-2\varphi)^{2}\exp(\frac{-\mathbf{x}_1}{\sqrt{2}})(\exp(\frac{-\mathbf{x}_1}{\sqrt{2}})-1)}{4(1+\exp(-\frac{\mathbf{x}_1}{\sqrt{2}}))^{3}}\frac{\bar{t}^{2\alpha}}{\Gamma(2\alpha+1)},\\
\mathbf{f}_3(\mathbf{x}_1,\bar{t}) &= \mathbb{E}^{-1}\Big[\omega^{\alpha}\mathbb{E}\Big\{\big(\mathbf{f}_2(\mathbf{x}_1,\bar{t})\big)_{\mathbf{x}_1\mathbf{x}_1} + (\varphi+1)\mathbf{f}_2^{2} - \mathbf{f}_2^{3} - \varphi\mathbf{f}_2\Big\}\Big]\\
&= -\frac{(1-2\varphi)^{3}\exp(\frac{-\mathbf{x}_1}{\sqrt{2}})(\exp(\sqrt{2}x) - 4\exp(\frac{-\mathbf{x}_1}{\sqrt{2}})+1)}{8(1+\exp(\frac{-\mathbf{x}_1}{\sqrt{2}}))^{4}}\frac{\bar{t}^{3\alpha}}{\Gamma(3\alpha+1)},\\
&\vdots
\end{aligned}$$

Provided that the series form solution is the following:

$$\mathbf{f}(\mathbf{x}_1,\bar{t}) = \mathbf{f}_0(\mathbf{x}_1,\bar{t}) + \mathbf{f}_1(\mathbf{x}_1,\bar{t}) + \mathbf{f}_2(\mathbf{x}_1,\bar{t}) + \mathbf{f}_3(\mathbf{x}_1,\bar{t}) + \ldots + \mathbf{f}_{m_1}(\mathbf{x}_1,\bar{t}).$$

the following consequently results.

$$\begin{aligned}\mathbf{f}(\mathbf{x}_1,\bar{t}) &= \frac{1}{(1+\exp(-\frac{1}{\sqrt{2}})\mathbf{x}_1)} + \frac{(1-2\varphi)\exp(-\frac{\mathbf{x}_1}{\sqrt{2}})}{2(1+\exp(\frac{-\mathbf{x}_1}{\sqrt{2}}))^2}\frac{\bar{t}^{\alpha}}{\Gamma(\alpha+1)} - \frac{(1-2\varphi)^2\exp(\frac{-\mathbf{x}_1}{\sqrt{2}})(\exp(\frac{-\mathbf{x}_1}{\sqrt{2}})-1)}{4(1+\exp(\frac{-\mathbf{x}_1}{\sqrt{2}}))^3}\frac{\bar{t}^{2\alpha}}{\Gamma(2\alpha+1)} \\ &- \frac{(1-2\varphi)^3\exp(\frac{-\mathbf{x}_1}{\sqrt{2}})(\exp(\sqrt{2}\mathbf{x}_1)-4\exp(\frac{-\mathbf{x}_1}{\sqrt{2}})+1)}{8(1+\exp(\frac{-\mathbf{x}_1}{\sqrt{2}}))^4}\frac{\bar{t}^{3\alpha}}{\Gamma(3\alpha+1)} + \ldots\end{aligned}$$

Case II. We now formulate Problem 4 by utilizing the Elzaki transform coupled with the ABC derivative operator.

By employing the Elzaki transform to (45) with the initial condition, we have the following.

$$\mathbb{E}\Big[\frac{\partial^{\alpha}\mathbf{f}}{\partial\bar{t}^{\alpha}}\Big] = \mathbb{E}\Big[\frac{\partial^2\mathbf{f}}{\partial\mathbf{x}_1^2} + \mathbf{f}(1-\mathbf{f})(\mathbf{f}-\varphi)\Big]. \quad (47)$$

$$\mathbb{E}\big[\mathbf{f}(\mathbf{x}_1,\bar{t})\big] = \mathbf{f}_{(0)}(\mathbf{x}_1,0)\omega^2 + \frac{\alpha\omega^{\alpha}+1-\alpha}{\mathbb{N}(\alpha)}\mathbb{E}\Big[\frac{\partial^2\mathbf{f}}{\partial\mathbf{x}_1^2} + \mathbf{f}(1-\mathbf{f})(\mathbf{f}-\varphi)\Big].$$

By using the inverse Elzaki transform, we have the following.

$$\mathbf{f}(\mathbf{x}_1,\bar{t}) = \frac{1}{(1+\exp(-\frac{1}{\sqrt{2}})\mathbf{x}_1)} + \mathbb{E}^{-1}\Big[\frac{\alpha\omega^{\alpha}+1-\alpha}{\mathbb{N}(\alpha)}\mathbb{E}\Big[\frac{\partial^2\mathbf{f}}{\partial\mathbf{x}_1^2} - \varphi\mathbf{f} + (\varphi+1)\mathbf{f}^2 - \mathbf{f}^3\Big]\Big].$$

Applying the iterative technique described in Section 3.1, we obtain the following.

$$\begin{aligned}\mathbf{f}_0(\mathbf{x}_1,\bar{t}) &= \frac{1}{(1+\exp(-\frac{1}{\sqrt{2}})\mathbf{x}_1)}, \\ \mathbf{f}_1(\mathbf{x}_1,\bar{t}) &= \mathbb{E}^{-1}\Big[\frac{\alpha\omega^{\alpha}+1-\alpha}{\mathbb{N}(\alpha)}\mathbb{E}\Big\{(\mathbf{f}_0(\mathbf{x}_1,\bar{t}))_{\mathbf{x}_1\mathbf{x}_1} + (\varphi+1)\mathbf{f}_0^2 - \mathbf{f}_0^3 - \varphi\mathbf{f}_0\Big\}\Big] \\ &= \frac{(1-2\varphi)\exp(\frac{\mathbf{x}_1}{\sqrt{2}})}{2\mathbb{N}(\alpha)(1+\exp(\frac{\mathbf{x}_1}{\sqrt{2}}))^2}\Big[\frac{\alpha\bar{t}^{\alpha}}{\Gamma(\alpha+1)} + (1-\alpha)\Big], \\ \mathbf{f}_2(\mathbf{x}_1,\bar{t}) &= \mathbb{E}^{-1}\Big[\frac{\alpha\omega^{\alpha}+1-\alpha}{\mathbb{N}(\alpha)}\mathbb{E}\Big\{(\mathbf{f}_1(\mathbf{x}_1,\bar{t}))_{\mathbf{x}_1\mathbf{x}_1} + (\varphi+1)\mathbf{f}_1^2 - \mathbf{f}_1^3 - \varphi\mathbf{f}_1\Big\}\Big] \\ &= -\frac{(1-2\varphi)^2\exp(\frac{\mathbf{x}_1}{\sqrt{2}})(\exp(\frac{\mathbf{x}_1}{\sqrt{2}})-1)}{4\mathbb{N}^2(\alpha)(1+\exp(\frac{\mathbf{x}_1}{\sqrt{2}}))^3}\Big[\frac{\alpha^2\bar{t}^{2\alpha}}{\Gamma(2\alpha+1)} + 2\alpha(1-\alpha)\frac{\bar{t}^{\alpha}}{\Gamma(\alpha+1)} + (1-\alpha)^2\Big], \\ \mathbf{f}_3(\mathbf{x}_1,\bar{t}) &= \mathbb{E}^{-1}\Big[\frac{\alpha\omega^{\alpha}+1-\alpha}{\mathbb{N}(\alpha)}\mathbb{E}\Big\{(\mathbf{f}_2(\mathbf{x}_1,\bar{t}))_{\mathbf{x}_1\mathbf{x}_1} + (\varphi+1)\mathbf{f}_2^2 - \mathbf{f}_2^3 - \varphi\mathbf{f}_2\Big\}\Big] \\ &= -\frac{(1-2\varphi)^3\exp(\frac{\mathbf{x}_1}{\sqrt{2}})(\exp(\sqrt{2}x)-4\exp(\frac{\mathbf{x}_1}{\sqrt{2}})+1)}{8\mathbb{N}^3(\alpha)(1+\exp(\frac{\mathbf{x}_1}{\sqrt{2}}))^4} \\ &\times\Big[\frac{\alpha^3\bar{t}^{3\alpha}}{\Gamma(3\alpha+1)} + 3\alpha^2(1-\alpha)\frac{\bar{t}^{2\alpha}}{\Gamma(2\alpha+1)} + 3\alpha(1-\alpha)^2\frac{\bar{t}^{\alpha}}{\Gamma(\alpha+1)} + (1-\alpha)^3\Big], \\ &\vdots\end{aligned}$$

Provided that the series form solution is the following:

$$\mathbf{f}(\mathbf{x}_1,\bar{t}) = \mathbf{f}_0(\mathbf{x}_1,\bar{t}) + \mathbf{f}_1(\mathbf{x}_1,\bar{t}) + \mathbf{f}_2(\mathbf{x}_1,\bar{t}) + \mathbf{f}_3(\mathbf{x}_1,\bar{t}) + \ldots + \mathbf{f}_{m_1}(\mathbf{x}_1,\bar{t}).$$

the following consequently results.

$$\begin{aligned}
\mathbf{f}(\mathbf{x}_1,\bar{t}) \quad &= \frac{1}{(1+\exp(-\frac{1}{\sqrt{2}})\mathbf{x}_1)} + \frac{(1-2\varphi)\exp(\frac{\mathbf{x}_1}{\sqrt{2}})}{2\mathbb{N}(\alpha)(1+\exp(\frac{\mathbf{x}_1}{\sqrt{2}}))^2}\left[\frac{\alpha\bar{t}^{\alpha}}{\Gamma(\alpha+1)} + (1-\alpha)\right] \\
&- \frac{(1-2\varphi)^2\exp(\frac{\mathbf{x}_1}{\sqrt{2}})(\exp(\frac{\mathbf{x}_1}{\sqrt{2}})-1)}{4\mathbb{N}^2(\alpha)(1+\exp(\frac{\mathbf{x}_1}{\sqrt{2}}))^3}\left[\frac{\alpha^2\bar{t}^{2\alpha}}{\Gamma(2\alpha+1)} + 2\alpha(1-\alpha)\frac{\bar{t}^{\alpha}}{\Gamma(\alpha+1)} + (1-\alpha)^2\right] \\
&- \frac{(1-2\varphi)^3\exp(\frac{\mathbf{x}_1}{\sqrt{2}})(\exp(\sqrt{2}\mathbf{x}_1) - 4\exp(\frac{\mathbf{x}_1}{\sqrt{2}})+1)}{8\mathbb{N}^3(\alpha)(1+\exp(\frac{\mathbf{x}_1}{\sqrt{2}}))^4} \\
&\times\left[\frac{\alpha^3\bar{t}^{3\alpha}}{\Gamma(3\alpha+1)} + 3\alpha^2(1-\alpha)\frac{\bar{t}^{2\alpha}}{\Gamma(2\alpha+1)} + 3\alpha(1-\alpha)^2\frac{\bar{t}^{\alpha}}{\Gamma(\alpha+1)} + (1-\alpha)^3\right] + \ldots
\end{aligned}$$

Figure 10 compares the exact and approximate solutions to Problem 4 by using the CDF operator. The absolute error norm in Figure 11 for (45) with the assumptions of $\theta = 1, \beta = 1$, and $0 < \varphi < 1$ ensures the approximation of the numerical results derived by the IETM to the exact solution via the CFD and ABC fractional derivative operators, respectively. The results of the graphical representation reveal that the model is highly dependent on fractional order α. The absolute inaccuracy is really small. Surface and two dimensional representations of graphs via Figure 12 show the strong connection between the exact and approximate solutions for various fractional orders. Furthermore, Figure 12a,b illustrate that the ABC fractional derivative operator has better harmony than the CFD operator.

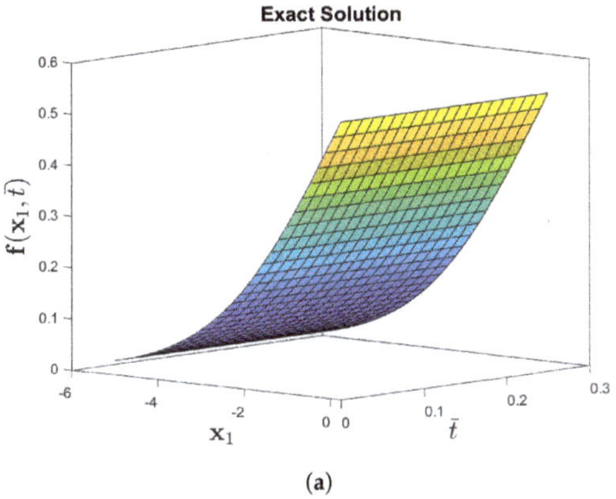

(a)

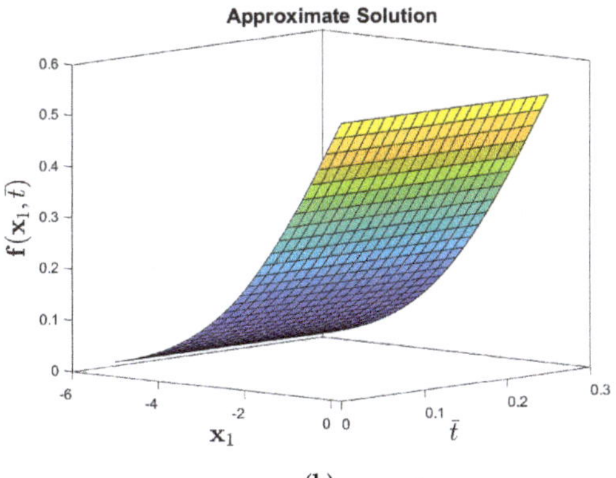

(b)

Figure 10. Numerical-behavior of exact and approximate solution to the $\mathbf{f}(\mathbf{x}_1,\bar{t})$ for Problem 4 when the parameters are $\theta = 1, \beta = 1$, and $\varphi = 1/10$.

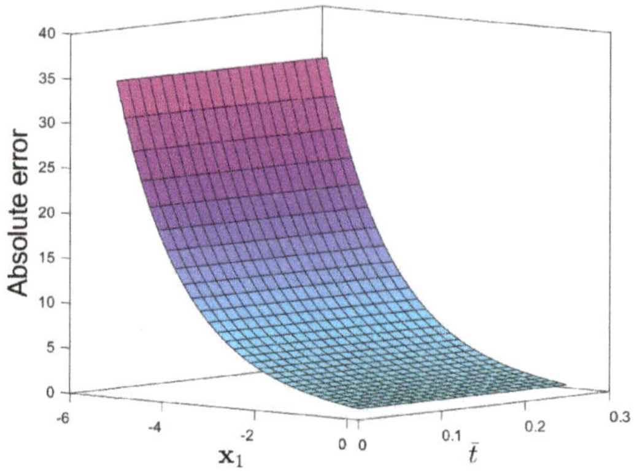

(**a**)

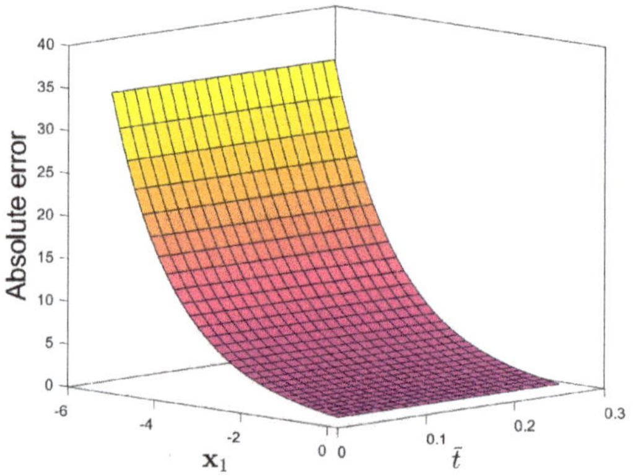

(**b**)

Figure 11. (**a**) Absolute-error plots of $\mathbf{f}(\mathbf{x}_1, \bar{t})$ for Problem 3 for (textbfa) CFD and (**b**) ABC when the parameters are $\theta = 1, \beta = 1$, and $\varphi = 1/10$.

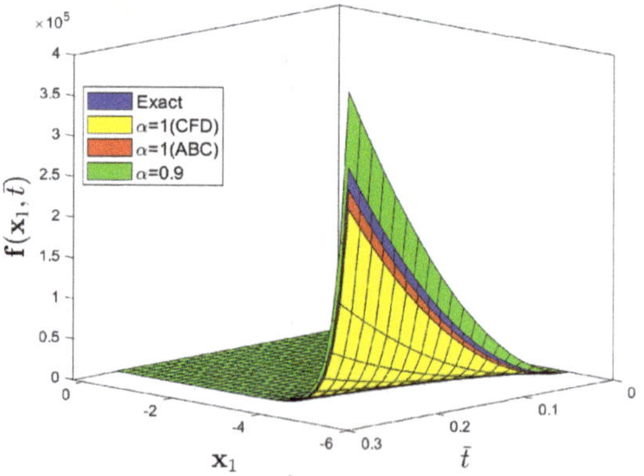

(a)

Figure 12. *Cont.*

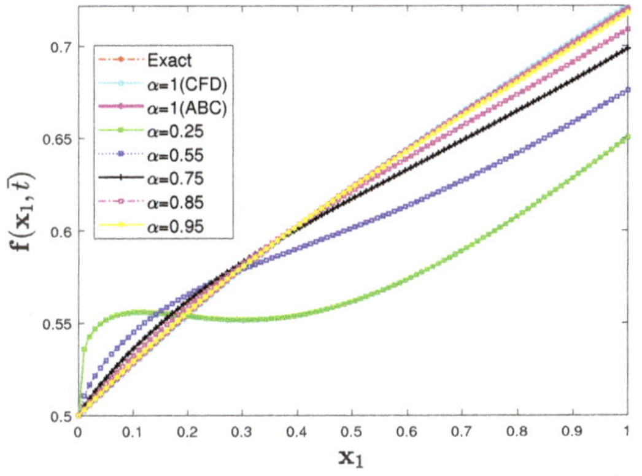

(b)

Figure 12. (**a**) Three-dimensional comparison plot among exact, CFD, and ABC fractional derivative operators via the IETM. (**b**) Two dimensional representation of $\mathbf{f}(\mathbf{x}_1, \bar{t})$ for Problem 4 at different fractional orders when the parameters are $\theta = 1, \beta = 1$, and $\varphi = 1/10$.

Remark 1. *The integer-order ($\alpha = 1$) solution of Problem 4 is the following:*

$$\mathbf{f}(\mathbf{x}_1, \bar{t}) = \frac{1}{1 + \exp(-\zeta/\sqrt{2})},$$

which agrees completely with the findings [40], where $\zeta = \mathbf{x}_1 + c_1\bar{t}$ and $c_1 = \frac{1}{2\sqrt{2}} - \sqrt{2}\varphi$. It is obvious that $\mathbf{f}(-\infty) = 0$. $\mathbf{f}(\infty) = 1$, and hence $\mathbf{f}(\zeta)$ in this scenario is a wave front traveling from right to left with speed $c_1 = \frac{1}{2\sqrt{2}} - \sqrt{2}\varphi$. It is worth noting that (45) enables steady travelling wave solutions.

$$\begin{aligned}
\mathbf{f}(\mathbf{x}_1, \bar{t}) &= \frac{1}{2}\left(1 + \tanh\left(\pm \frac{\mathbf{x}_1}{2\sqrt{2}} + \frac{(1-2\varphi)}{4}\bar{t}\right)\right), \\
\mathbf{f}(\mathbf{x}_1, \bar{t}) &= \frac{\varphi}{2}\left(1 + \tanh\left(\pm \frac{\varphi\mathbf{x}_1}{2\sqrt{2}} + \frac{(\varphi^2-2\varphi)}{4}\bar{t}\right)\right), \\
\mathbf{f}(\mathbf{x}_1, \bar{t}) &= \frac{(1+\varphi)}{2} + \frac{(1-\varphi)}{2}\left(1 + \tanh\left(\pm \frac{(1-\varphi)\mathbf{x}_1}{2\sqrt{2}} + \frac{(1-\varphi^2)}{4}\bar{t}\right)\right),
\end{aligned} \tag{48}$$

The above is proposed by Khawara and Tanaka [41].

5. Conclusions

In this paper, the new iterative Elzaki transform method is used to efficiently solve the nonlinear Fisher's equation by using the Caputo and the AB fractional derivative operators, which possess a fractional Lagrange multiplier. In addition, the concept of $\mathcal{V}$-stable mapping and the fixed point theorem demonstrates the stability of the proposed technique in the sense of the AB-fractional operator. Several illustrative cases were carried out to verify the efficacy and reliability of the proposed technique. The findings indicate that the ABC fractional derivative is completely accurate and has a wide spectrum of uses as compared to CFD. In comparison to existing numerical algorithms, the suggested method has a lower processing complexity. Furthermore, the approach interprets and regulates the series of solutions, which converge swiftly to the exact solution in a short admissible domain. In this process, we do not require rectification functionals, stationary constraints, or hefty integrals since the findings are noise-free, which addresses the drawbacks of earlier techniques. Furthermore, we believe that this technique will be adopted to contend with other non-linear fractional order systems of equations that are extremely complex. In the future, we will investigate a similar problem by utilizing the double Laplace transform and generalized Kudryashov method, which will be a useful mechanism for solving nonlinear PDEs and other FDEs.

Author Contributions: Conceptualization and data curation—S.R., Z.H.; formal analysis—S.R., H.A., A.M.A.; funding acquisition—S.R., H.A., Z.H.; investigation—S.R., A.G.A., A.M.A.; methodology—S.R., Z.H., A.G.A., A.M.A.; project administration—S.R., Z.H., H.A.; supervision—S.R.; resources, software—H.A., A.M.A.; validation—Z.H., H.A., A.G.A., A.M.A.; visualization—S.R., Z.H., H.A., A.G.A., A.M.A.; writing—original draft—S.R., H.A.; writing—review & editing—S.R. All authors have read and agreed to the published version of the manuscript.

Funding: This research received no external funding.

Institutional Review Board Statement: Not Applicable.

Informed Consent Statement: Not Applicable.

Data Availability Statement: Not Applicable.

Acknowledgments: The authors would like to express their sincere thanks for the support of Taif University Researchers Supporting Project Number (TURSP-2020/96), Taif University, Taif, Saudi Arabia.

Conflicts of Interest: The authors declare no conflict of interest.

References

1. Caputo, M.; Fabrizio, M. A new definition of fractional derivative without singular kernel. *Prog. Fract. Differ. Appl.* **2015**, *73*, 1–13.
2. Atangana, A.; Baleanu, D. New fractional derivatives with non-local and non-singular kernel. Theory and Application to Heat Transfer Model. *arXiv* **2016**, arXiv:1602.03408.
3. Scherer, R.; Kalla, S.L.; Tang, Y.; Huanget, J. The Grünwald-Letnikov method for fractional differential equations. *Comput. Math. Appl.* **2011**, *62*, 902–917. [CrossRef]
4. Li, C.; Qian, D.; Chen, Y.Q. On Riemann-Liouville and Caputo Derivatives. *Discret. Dyn. Nat. Soc.* **2011**, *2011*, 562494. [CrossRef]
5. Morales-Delgado, V.F.; Gómez-Aguilar, J.F.; Yepez-Martínez, H.; Baleanu, D.; Escobar-Jimenez, R.F.; Olivares-Peregrino, V.H. Laplace homotopy analysis method for solving linear partial differential equations using a fractional derivative with and without kernel singular. *Adv. Differ. Equ.* **2016**, *2016*, 164. [CrossRef]
6. Sheikh, N.A.; Ali, F.; Saqib, M.; Khan, I.; Jan, S.A.A.; Alshomrani, A.S.; Alghamdi, M.S. Comparison and analysis of the Atangana-Baleanu and Caputo-Fabrizio fractional derivatives for generalized Casson fluid model with heat generation and chemical reaction. *Results Phys.* **2017**, *7*, 789. [CrossRef]
7. Atangana, A.; Alkahtani, B.S.T. New model of groundwater flowing within a confine aquifer: Application of Caputo-Fabrizio derivative. *Arab. J. Geo-Sci.* **2016**, *9*, 8. [CrossRef]
8. Kumar, D.; Singh, J.; Baleanu, D.; Sushila. Analysis of regularized long-wave equation associated with a new fractional operator with Mittag-Leffler type kernel. *Physica A* **2018**, *492*, 155–167. [CrossRef]
9. Singh, J.; Kumar, D.; Baleanu, D. On the analysis of chemical kinetics system pertaining to a fractional derivative with Mittag-Leffler type kernel. *Chaos* **2017**, *27*, 103113. [CrossRef]
10. Baleanu, D.; Jajarmi, A.; Hajipour, M. A new formulation of the fractional optimal control problems involving Mittag-Leffler nonsingular kernel. *J. Optim. Theory Appl.* **2017**, *175*, 718–737. [CrossRef]
11. Wazwaz, A.M.; Gorguis, A. An analytic study of Fisher's equation by using Adomian decomposition method. *Appl. Math. Comput.* **2004**, *154*, 609–620. [CrossRef]
12. Rashid, S.; Khalid, A.; Sultana, S.; Hammouch, Z.; Shah, R.; Alsharif, A.M. A novel analytical view of time-fractional Korteweg-De Vries equations via a new integral transform. *Symmetry* **2021**, *13*, 1254. [CrossRef]
13. Rashid, S.; Kubra, K.T.; Rauf, A.; Chu, Y.-M.; Hamed, Y.S. New numerical approach for time-fractional partial differential equations arising in physical system involving natural decomposition method. *Phys. Scr.* **2021**, *96*, 105204. [CrossRef]
14. Abedle-Rady, A.S.; Rida, S.Z.; Arafa, A.A.M.; Adedl-Rahim, H.R. Approximate analytical solutions of the fractional nonlinear dispersive equations using homotopy perturbation Sumudu transform method. *Int. J. Innov. Sci. Eng. Technol.* **2014**, *1*, 257–267.
15. Rashid, S.; Kubra, K.T.; Lehre, S.U. Fractional spatial diffusion of a biological population model via a new integral transform in the settings of power and Mittag-Leffler nonsingular kernel. *Phys. Scr. A* **2021**, *96*, 114003. [CrossRef]
16. Jafari, H.; Nazari, M.; Baleanu, D.; Khalique, C.M. A new approach for solving a system of fractional partial differential equations. *Comput. Math. Appl.* **2013**, *66*, 838–843. [CrossRef]
17. Veeresha, P.; Prakasha, D.G.; Baskonus, H.M. Novel simulations to the time-fractional Fisher's equation. *Math. Sci.* **2019**, *13*, 33–42. [CrossRef]
18. Gupta, A.K.; Ray, S.S. On the solutions of fractional Burgers-Fisher and generalized Fisher's equations using two reliable methods. *Int. J. Math. Math. Sci.* **2014**, *2014*, 682910. [CrossRef]
19. Li, C.; Zhang, J. Lie symmetry analysis and exact solutions of generalized fractional Zakharov-Kuznetsov equations. *Symmetry* **2019**, *11*, 601. [CrossRef]
20. Khader, M.M.; Saad, K.M. A numerical approach for solving the fractional Fisher equation using Chebyshev spectral collocation method. *Chaos Solitons Fract.* **2018**, *110*, 169–177. [CrossRef]

21. Rossa, J.; Villaverdeb, A.F.; Bangab, J.R.; Vazquezc, S.; Moranc, F. A generalized Fisher equation and its utility in chemical kinetics. *Proc. Natl. Acad. Sci. USA* **2010**, *107*, 12777–12781. [CrossRef]
22. Ammerman, A.J.; Cavalli-Sforza, L.L. *The Neolithic Transition and the Genetics of Population in Europe*; Princeton University Press: Princeton, NJ, USA, 1984.
23. Merdan, M. Solutions of time-fractional reaction-diffusion equation with modified Riemann-Liouville derivative. *Int. J. Phys. Sci.* **2012**, *7*, 2317–2326. [CrossRef]
24. Kerke, V.M. Results from variants of the Fisher equation in the study of epidemics and bacteria. *Physica A* **2004**, *342*, 242–248.
25. Dag, I.; Sahin, A.; Korkmaz, A. Numerical investigation of the solution of Fisher's equation via the B-spline galerkin method. *Numer. Meth. Partial Differ. Equ.* **2010**, *26*, 1483–1503. [CrossRef]
26. Bastani, M.; Salkuyeh, D.K. A highly accurate method to solve Fisher's equation. *Pramana* **2012**, *78*, 335–346. [CrossRef]
27. Gazdag, J.; Canosa, J. Numerical solution of Fisher's equation. *J. Appl. Probab.* **1974**, *11*, 445–457. [CrossRef]
28. Zhao, T.; Li, C.; Zang, Z.; Wu, Y. Chebyshev-Legendre pseudo-spectral method for the generalised Burgers-Fisher equation. *Appl. Math. Model.* **2012**, *36*, 1046–1056. [CrossRef]
29. Elzaki, T.M. The new integral transform Elzaki transform. *Glob. J. Pure Appl. Math.* **2011**, *7*, 57–64.
30. Daftardar-Gejji, V.; Jafari, H. An iterative method for solving nonlinear functional equations. *J. Math. Anal. Appl.* **2006**, *316*, 753–763. [CrossRef]
31. Belgacem, F.B.M.; Karaballi, A.A.; Kalla, S.L. Analytical investigations of the sumudu transform and applications to integral production equations. *Math. Probab. Eng.* **2003**, *2003*, 103–118. [CrossRef]
32. Aboodh, K.S. The new integral transform "Aboodh Transform". *Glob. J. Pure Appl. Math.* **2013**, *9*, 35–43.
33. Elzaki, T.M. Application of new transform Elzaki transform to partial differential equations. *Glob. J. Pure Appl. Math.* **2011**, *7*, 65–70.
34. Mahgoub, M.M.A. The new integral transform "Mohand Transform". *Adv. Theor. Appl. Math.* **2017**, *12*, 113–120.
35. Alderremy, A.A.; Elzaki, T.M.; Chamekh, M. New transform iterative method for solving some Klein-Gordon equations. *Results Phys.* **2018**, *10*, 655–659. [CrossRef]
36. Sedeeg, A.H. A coupling Elzaki transform and homotopy perturbation method for solving nonlinear fractional heat-like equations. *Am. J. Math. Comput. Model* **2016**, *1*, 15–20.
37. Yavuz, M.; Abdeljawad, T. Nonlinear regularized long-wave models with a new integral transformation applied to the fractional derivative with power and Mittag-Leffler kernel. *Adv. Differ. Equ.* **2020**, *2020*, 367. [CrossRef]
38. Odibat, Z.M.; Momani, S. Application of variational iteration method to nonlinear differential equation of fractional order. *Int. J. Nonlinear Sci. Numer. Simul.* **2006**, *7*, 27–34. [CrossRef]
39. Ağrseven, D.; Öziş, T. An analytical study for Fisher type equations by using homotopy perturbation method. *Comput. Math. Appl.* **2010**, *60*, 602–609. [CrossRef]
40. Jone, D.S.; Sleeman, B.D. *Differential Equations and Mathematical Biology*; Chapman & Hall/CRC: New York, NY, USA, 2003.
41. Kawahara, T.; Tanaka, M. Interactions of traveling fronts: An exact solution of a nonlinear diffusion equation. *Phys. Lett. A* **1983**, *97*, 311–314. [CrossRef]

fractal and fractional

MDPI

Article

Analytic Fuzzy Formulation of a Time-Fractional Fornberg–Whitham Model with Power and Mittag–Leffler Kernels

Saima Rashid [1], Rehana Ashraf [2], Ahmet Ocak Akdemir [3], Manar A. Alqudah [4,*], Thabet Abdeljawad [5,6,*] and Mohamed S. Mohamed [7]

1 Department of Mathematics, Government College University, Faisalabad 38000, Pakistan; saimarashid@gcuf.edu.pk
2 Department of Mathematics, Lahore College Women University, Lahore 54000, Pakistan; rehana.ashraf@jhang.lcwu.edu.pk
3 Department of Mathematics, Faculty of Science and Letters, Agri Ibrahim Cecen University, Agri 04100, Turkey; aocakakdemir@gmail.com
4 Department Mathematical Sciences, Faculty of Sciences, Princess Nourah Bint Abdulrahman University, P.O. Box 84428, Riyadh 11671, Saudi Arabia
5 Department of Mathematics and General Sciences, Prince Sultan University, Riyadh 11586, Saudi Arabia
6 China Medical University Hospital, China Medical University, Taichung 40402, Taiwan
7 Department of Mathematics, Faculty of Science, Taif University, P.O. Box 11099, Taif 21944, Saudi Arabia; m.saaad@tu.edu.sa
* Correspondence: maalqudah@pnu.edu.sa (M.A.A.); tabdeljawad@psu.edu.sa (T.A.)

Citation: Rashid, S.; Ashraf, R.; Akdemir, A.O.; Alqudah, M.A.; Abdeljawad, T.; Mohamed, M.S. Analytic Fuzzy Formulation of a Time-Fractional Fornberg–Whitham Model with Power and Mittag–Leffler Kernels. *Fractal Fract.* **2021**, *5*, 113. https://doi.org/10.3390/fractalfract5030113

Academic Editors: Muhammad Yaseen and Asifa Tassaddiq

Received: 12 August 2021
Accepted: 2 September 2021
Published: 8 September 2021

Abstract: This manuscript assesses a semi-analytical method in connection with a new hybrid fuzzy integral transform and the Adomian decomposition method via the notion of fuzziness known as the Elzaki Adomian decomposition method (briefly, EADM). Moreover, we use the aforesaid strategy to address the time-fractional Fornberg–Whitham equation (FWE) under $g\mathcal{H}$-differentiability by employing different initial conditions (IC). Several algebraic aspects of the fuzzy Caputo fractional derivative (CFD) and fuzzy Atangana–Baleanu (AB) fractional derivative operator in the Caputo sense, with respect to the Elzaki transform, are presented to validate their utilities. Apart from that, a general algorithm for fuzzy Caputo and AB fractional derivatives in the Caputo sense is proposed. Some illustrative cases are demonstrated to understand the algorithmic approach of FWE. Taking into consideration the uncertainty parameter $\zeta \in [0,1]$ and various fractional orders, the convergence and error analysis are reported by graphical representations of FWE that have close harmony with the closed form solutions. It is worth mentioning that the projected approach to fuzziness is to verify the supremacy and reliability of configuring numerical solutions to nonlinear fuzzy fractional partial differential equations arising in physical and complex structures.

Keywords: Elzaki transform; Hukuhara difference; Caputo fractional derivative; Atangana–Baleanu fractional derivative operator; Mittag–Leffler kernel; Fornberg–Whitham equation

1. Introduction

Recently, fractional calculus (FC) theory has shown incredible capabilities for describing the dynamical behavior and memory-related properties of scientific structures and procedures. Fractional differential equations (FDEs) have been developed by researchers to investigate and interpret natural phenomena in a variety of domains. FC theory comprises numerous generalizations in terms of non-local properties of fractional operators, expanded degree of independence, and maximum information application, and these features only arise in fractional order processes, not in integer-order mechanisms. Some scholars have recently investigated a series of innovative mathematical models using distinct local and non-local fractional derivative operators (see, [1–12]).

Recently, many innovative fractional derivative operators beyond the singular kernel have been explored, such as the Mittag–Leffler function [13] and exponential function [14]. In particular, researchers who would like to develop and address a real-life problem have recommended fractional operators, see [15]. Problems involving these operators can be solved quickly and reliably because they include a non-singular kernel. Numerical algorithms can also be conducted conveniently regarding the integral transforms of these fractional formulations. Many authors have investigated fractional operators, as evidenced by the references [16,17] and those cited therein.

Modeling uncertain problems with fuzzy set theory is a useful method. As a consequence, fuzzy notions have been employed to model a wide range of natural processes. Specifically, fuzzy partial differential equations (PDEs) have been exploited in a wide range of scientific domains, including patteren formation theory, engineering, population dynamics, control systems, knowledge-based systems, image processing, power engineering, industrial automation, robotics, consumer electronics, artificial intelligence/expert systems, management, and operations research. However, the notion of fuzzy set theory has a strong connection with fractional calculus, due to its crucial aspects in various scientific disciplines [18]. In 1978, Kandel and Byatt [19] contemplated the idea of fuzzy DEs, then Agarwal et al. [20] were the first to address fuzziness and the Riemann–Liouville differentiability notion under the Hukuhara differentiability. Fuzzy set theory and FC incorporate several numerical approaches that enable a more in-depth understanding of complicated systems while also reducing the amount of computational cost involved in the solution process. In the case of FPDEs, finding accurate analytical solutions is a difficult task. To cope with this challenge, several numerical methods have been expounded to obtain the analytical solutions of PDES and ODEs, such as the Adomian decomposition method (ADM) [21], q-homotopy analysis method (q-HAM) [22], pseudo spectral method (PSM) [23], Laguerre wavelets collocation method (LWCM) [24], new Legendre-Wavelets decomposition method (NLWDM) [25], etc. Fuzzy FPDEs have expanded in prominence over the last decade as a result of their vast applicability and significance in analyzing the behavior of complex geometries. Recently, Hoa et al. [26,27] investigated the gH-differentiability with a Katugampola fractional derivative in the Caputo sense and employed fuzzy FDEs. Salahshour et al. [28] expounded the H-differentiability with the Laplace transform to solve the FDEs. Ahmad et al. [29] studied the third order fuzzy dispersive PDEs in the Caputo, Caputo–Fabrizio, and Atangana–Baleanu fractional operator frameworks. Shah et al. [30] presented the evolution of one dimensional fuzzy fractional PDEs. For more details, see [31–34] and the references cited therein.

Accessing the influence of PDEs for external potential has been extensively applied as a model for the evaluation of multiple challenges. The Fornberg–Whitham (FW) model is an important complex formulation in mathematical physics. The FWE [35,36] is presented as

$$\frac{\partial \mathbf{f}}{\partial \vartheta} - \frac{\partial^3 \mathbf{f}}{\partial \vartheta \partial \ell_1^2} + \frac{\partial \mathbf{f}}{\partial \ell_1} = \mathbf{f}\frac{\partial^3 \mathbf{f}}{\partial \ell_1^3} - \mathbf{f}\frac{\partial \mathbf{f}}{\partial \ell_1} + 3\frac{\partial \mathbf{f}}{\partial \ell_1}\frac{\partial^2 \mathbf{f}}{\partial \ell_1^2}, \tag{1}$$

where the fluid velocity is expressed by $\mathbf{f}(\ell_1, \vartheta)$ along with ℓ_1 as the spatial co-ordinate and ϑ denoting time. In 1978, Fornberg and Whitham [35,36] contemplated a solution $\mathbf{f}(\ell_1, \vartheta) = \delta \exp\left(\frac{\ell_1}{2} - \frac{4\vartheta}{3}\right)$ with an arbitrarily defined constant of α. The FWE has been discovered to need peakon outcomes as a model for controlling wave heights and wave break frequency. Recently, various sorts of FWE models in physics have been investigated by Abidi and Omrani [37], Gupta and Singh [38], Lu [39], Sakar et al. [40], Chen et al. [41], Yin et al. [42], Zhou and Tian [43], He et al. [44], Fan et al. [45], Jiang and Bi [46].

This research creates a modified fuzzy EADM framework to assess the fuzzy time fractional FWE. The approximate analytical solutions for various fractional Brownian movements, as well as standard motion, are derived using the uncertainty parameter in ICs. Graphically, the diversity of approximate results is illustrated, and the error estimate demonstrates the validity of the approximate analytical solutions. In the time fractional operator form, this equation can be expressed as

$$\frac{\partial^\alpha \mathbf{f}}{\partial \vartheta^\alpha} - \frac{\partial^3 \mathbf{f}}{\partial \vartheta \partial \ell_1^2} + \frac{\partial \mathbf{f}}{\partial \ell_1} = \mathbf{f}\frac{\partial^3 \mathbf{f}}{\partial \ell_1^3} - \mathbf{f}\frac{\partial \mathbf{f}}{\partial \ell_1} + 3\frac{\partial \mathbf{f}}{\partial \ell_1}\frac{\partial^2 \mathbf{f}}{\partial \ell_1^2}, \tag{2}$$

subject to ICs $\mathbf{f}(\ell_1, 0) = \exp\left(\frac{\ell_1}{2}\right)$ and $\cosh^2\left(\frac{\ell_1}{4}\right)$, and $\alpha \in (0,1]$ is the order of the time fractional derivative. It is remarkable that the exact traveling wave solution of FWE subject to IC $\mathbf{f}(\ell_1, \vartheta) = 0.75\exp\left(\frac{3\ell_1 - 2\vartheta}{6}\right)$ has been investigated in [38].

In order to simplify the approach to solving ODEs and PDEs in the time domain, Tarig Elzaki [47] introduced a new transform known as ET in 2001. This innovative transform is a refinement of existing transforms (Laplace and Sumudu) that can help determine the analytical solutions of PDEs in a similar fashion to the Laplace and Sumudu transformations.

The ADM is a semi-analytical approach to solving linear-nonlinear FDEs by advantageously creating a functional series solution, initially presented by Adomian [48]. Later, this approach was used with numerous transformations (such as the Sumudu, Aboodh, Laplace, and Mohand transforms), as shown in [49–58].

Owing to the above propensity, configuring the exact solution of nonlinear fuzzy fractional PDEs is an ever challenging task. In this paper, our intention is to establish an efficacious algorithm for generating estimated solutions of fuzzy fractional FWE, the general FWE arising in wave breaking subject to uncertainty in IC by EADM that models the dynamics of the system being analyzed. The EADM is merged with the Elzaki transform (ET), and the ADM is known as the Elzaki Adomian decomposition method (EADM). This novel method is applied to examining fractional-order FW models. The findings of a particular test case are evaluated in terms of showing that the proposed technique is viable. The findings of the fractional-order with an uncertainty factor are determined by advanced tools and methods. The convergence analysis for EADM was also briefly discussed. The FW model was leveraged to generate synthesized trajectories. In a simulation study, we illustrate the applicability and effectiveness of the offered algorithmic strategies for determining numerical solutions. Several fuzzy fractional orders of linear and non-linear PDEs can be addressed using the proposed method.

2. Preliminaries

This section clearly exhibits some major features connected to the stream of fuzzy (**F**) set theory and FC, as well as certain key findings about ET. For more details, we refer to [59].

Definition 1 ([60,61]). *We say that $\Omega : \mathbb{R} \mapsto [0,1]$ is a* **F** *number, if it holds the subsequent assumptions:*

1. *Ω is normal (for some $\ell_{10} \in \mathbb{R}; \Omega(\ell_{10}) = 1$);*
2. *Ω is upper semi continuous;*
3. *$\Omega(\ell_1\vartheta + (1-\vartheta)\ell_2) \geq (\Omega(\ell_1) \wedge \Omega(\ell_2))$ $\forall \vartheta \in [0,1]$, $\ell_1, \ell_2 \in \mathbb{R}$,, i.e., Ω is convex;*
4. *$cl\{\ell_1 \in \mathbb{R}, \Omega(\ell_1) > 0\}$ is compact.*

Definition 2 ([60]). *We say that a* **F** *number Ω is a ζ-level set described as*

$$[\Omega]^\zeta = \{\mathbf{f} \in \mathbb{R} : \Omega(\mathbf{f}) \geq \zeta\}, \tag{3}$$

where $\zeta \in [0,1]$ and $\mathbf{f} \in \mathbb{R}$.

Definition 3 ([60]). *The parameterized version of a* **F** *number is denoted by $[\underline{\Omega}(\zeta), \bar{\Omega}(\zeta)]$ such that $\zeta \in [0,1]$ satisfies the subsequent assumptions:*

1. *$\underline{\Omega}(\zeta)$ is non-decreasing, left continuous, bounded over $(0,1]$ and left continuous at 0.*

2. $\bar{\Omega}(\zeta)$ *is non-increasing, right continuous, bounded over* $(0,1]$ *and right continuous at* 0.
3. $\underline{\Omega}(\zeta) \leq \bar{\Omega}(\zeta)$.

   Moreover, ζ *is known to be a crisp number if* $\underline{\Omega}(\zeta) = \bar{\Omega}(\zeta) = \zeta$.

Definition 4 ([59]). *For* $\zeta \in [0,1]$ *and* χ *to be a scalar, assume that there are two* **F** *numbers* $\tilde{\mathbf{f}} = (\underline{\mathbf{f}}, \bar{\mathbf{f}})$, $\tilde{\phi} = (\underline{\phi}, \bar{\phi})$, *then the addition, subtraction and scalar multiplication, respectively, are stated as*

1. $\tilde{\mathbf{f}} \oplus \tilde{\phi} = (\underline{\mathbf{f}}(\zeta) \oplus \underline{\phi}(\zeta), \bar{\mathbf{f}}(\zeta) \oplus \bar{\phi}(\zeta))$;
2. $\tilde{\mathbf{f}} \ominus \tilde{\phi} = (\underline{\mathbf{f}}(\zeta) \ominus \underline{\phi}(\zeta), \bar{\mathbf{f}}(\zeta) \ominus \bar{\phi}(\zeta))$;
3. $\chi \odot \tilde{\mathbf{f}} = \begin{cases} (\chi \odot \underline{\mathbf{f}}, \chi \odot \bar{\mathbf{f}}) & \chi \geq 0, \\ (\chi \odot \bar{\mathbf{f}}, \chi \odot \underline{\mathbf{f}}) & \chi < 0. \end{cases}$

Suppose the set $\tilde{D}$ is a domain of **F**-valued mapping Θ. Let us introduce the mappings $\underline{\Theta}(.,.,\zeta), \bar{\Theta}(.,.,\zeta)$: $\tilde{D} \mapsto \mathbb{R}$, $\forall$ *zeta* $\in [0,1]$. These mappings are said to be the left and right $\wp$-level mappings of the map Θ.

Definition 5 ([28]). *Suppose a* **F** *mapping* $\Theta : \tilde{E} \times \tilde{E} \mapsto \mathbb{R}$ *with two* **F** *numbers* $\tilde{\mathbf{f}} = (\underline{\mathbf{f}}, \bar{\mathbf{f}})$, $\tilde{\phi} = (\underline{\phi}, \bar{\phi})$, *then the* Θ*-distance between* $\tilde{\mathbf{f}}$ *and* $\tilde{\phi}$ *is represented as*

$$\Theta(\tilde{\mathbf{f}}, \tilde{\phi}) = \sup_{\zeta \in [0,1]} \left[\max\left\{ |\underline{\mathbf{f}}(\zeta) - \underline{\phi}(\zeta)|, |\bar{\mathbf{f}}(\zeta) - \bar{\phi}(\zeta)| \right\} \right]. \tag{4}$$

Definition 6 ([28]). *Consider a* **F** *mapping* $\Theta : \tilde{D} \mapsto \tilde{E}$, *is said to be continuous at* $(a_0, b_0) \in \tilde{D}$ *if for every* $\epsilon > 0$ *and there is* $\delta > 0$ *such that*

$$d(\Theta(a,b), \Theta(a_0, b_0)) < \epsilon; \ \ whenever |a - a_0| + |b - b_0| < \delta. \tag{5}$$

If Θ *is continuous for each* $(a,b) \in \tilde{D}$, *we say that* Θ *is continuous on* $\tilde{D}$.

Definition 7 ([62]). *Let* $\beta_1, \beta_2 \in \tilde{E}$, *if* $\beta_3 \in \tilde{E}$ *and* $\beta_1 = \beta_2 + \beta_3$. *The* $\mathcal{H}$*-difference* β_3 *of* β_1 *and* β_2 *is denoted as* $\beta_1 \ominus^{\mathcal{H}} \beta_2$. *Observe that* $\beta_1 \ominus^{\mathcal{H}} \beta_2 \neq \beta_1 + (-1)\beta_2$.

Now suppose $\beta_1, \beta_2 \in \tilde{E}$, *then* $\beta_1 \ominus \, _{g\mathcal{H}}\beta_2 = \beta_3 \Leftrightarrow$

(i) $\beta_3 = (\underline{\beta_1}(\zeta) - \underline{\beta}_2(\zeta), \bar{\beta}_1(\zeta) - \bar{\beta}_2(\zeta))$.

or

(ii) $\beta_3 = (\bar{\beta}_1(\zeta) - \bar{\beta}_2(\zeta), \underline{\beta_1}(\zeta) - \underline{\beta}_2(\zeta))$.

The following Lemma demonstrates the link between the $g\mathcal{H}$-difference and the Housdroff distance.

Lemma 1 ([63]). *For all* $\beta_1, \beta_2 \in \tilde{E}$, *then*

$$d(\beta_1, \beta_2) = \sup_{\zeta \in [0,1]} \|[\beta_1]^{\zeta} \ominus \, _{g\mathcal{H}}[\beta_2]^{\zeta}\|, \tag{6}$$

where, for an interval $[a,b]$, *the norm is* $\|[a,b]\| = \max\{|a|, |b|\}$.

Definition 8 ([64]). *Suppose* $\Theta : \tilde{D} \mapsto \tilde{E}$ *and* $(\ell_0, \vartheta) \in \tilde{D}$. *Then* Θ *is said to be strongly generalized Hukuhara differentiable on* (ζ_0, ϑ) *(*$g\mathcal{H}$*-differentiable) if* $\exists$ *an element* $\frac{\partial\theta(\ell_0,\vartheta)}{\partial\ell} \in \tilde{E}$ *such that the following holds:*

(i) *For all* $\hbar > 0$ *of sufficiently small size, the subsequent* $g\mathcal{H}$*-differences exist:*

$$\Theta(\ell_0 + \hbar, \vartheta) \ominus \, _{g\mathcal{H}}\Theta(\ell_0, \vartheta), \ \ \Theta(\ell_0, \vartheta) \ominus \, _{g\mathcal{H}}\Theta(\ell_0 - \hbar, \vartheta)$$

and the following limits hold:

$$\lim_{\hbar \mapsto 0} \frac{\Theta(\ell_0 + \hbar, \vartheta) \ominus_{gH} \Theta(\ell_0, \vartheta)}{\hbar} = \lim_{\hbar \mapsto 0} \frac{\Theta(\ell_0, \vartheta) \ominus_{gH} \Theta(\ell_0 - \hbar, \vartheta)}{\hbar} = \frac{\partial \theta(\ell_0, \vartheta)}{\partial \ell}.$$

(ii) *For all $\hbar > 0$ of sufficiently small size, the subsequent gH-differences exist:*

$$\Theta(\ell_0, \vartheta) \ominus_{gH} \Theta(\ell_0 + \hbar, \vartheta), \;\; \Theta(\ell_0 - \hbar, \vartheta) \ominus_{gH} \Theta(\ell_0, \vartheta)$$

and the following limits hold:

$$\lim_{\hbar \mapsto 0} \frac{\Theta(\ell_0, \vartheta) \ominus_{gH} \Theta(\ell_0 + \hbar, \vartheta)}{-\hbar} = \lim_{\hbar \mapsto 0} \frac{\Theta(\ell_0 - \hbar, \vartheta) \ominus_{gH} \Theta(\ell_0, \vartheta)}{-\hbar} = \frac{\partial \theta(\ell_0, \vartheta)}{\partial \ell}.$$

Lemma 2 ([65]). *Consider that $\Theta:\tilde{D} \mapsto \tilde{E}$ is a continuous* **F** *-valued mapping and $\Theta(\ell, \vartheta) = (\underline{\Theta}(\ell, \vartheta, \zeta), \bar{\Theta}(\ell, \vartheta, \zeta))$, $\forall \zeta \in [0,1]$. Then*

(i) *If $\Theta(\ell, \vartheta)$ is* **(i)***-partial differentiable for ℓ (i.e., Θ is partial differentiable for ℓ under the meaning of Definition 8* **(i)***), then*

$$\frac{\partial \Theta(\ell, \vartheta)}{\partial \ell} = \left(\frac{\partial \underline{\Theta}(\ell, \vartheta)}{\partial \ell}, \frac{\partial \bar{\Theta}(\ell, \vartheta)}{\partial \ell} \right), \tag{7}$$

(ii) *If $\Theta(\ell, \vartheta)$ is* **(i)***-partial differentiable for ℓ (i.e., Θ is partial differentiable for ℓ under the meaning of Definition 8* **(ii)***), then*

$$\frac{\partial \Theta(\ell, \vartheta)}{\partial \ell} = \left(\frac{\partial \bar{\Theta}(\ell, \vartheta)}{\partial \ell}, \frac{\partial \underline{\Theta}(\ell, \vartheta)}{\partial \ell} \right). \tag{8}$$

Definition 9 ([28]). *Assume that a* **F** *mapping $\mathbf{f} \in \mathbb{C}^F[0, d_1] \bigcap \mathbb{L}^F[0, d_1]$ is represented in parameterized version as $\mathbf{f} = [\underline{\mathbf{f}}_\zeta(\vartheta), \bar{\mathbf{f}}_\zeta(\vartheta)]$, $\zeta \in [0,1]$ and $\vartheta_0 \in (0, d_1)$, then CFD in the* **F** *sense is stated as*

$$\left[\mathcal{D}^\alpha \mathbf{f}(\vartheta_0) \right]_\zeta = \left[\mathcal{D}^\alpha \underline{\mathbf{f}}(\vartheta_0), \mathcal{D}^\alpha \bar{\mathbf{f}}(\vartheta_0) \right], \; \zeta \in (0, \zeta], \tag{9}$$

where $q = [\zeta]$.

$$\begin{aligned} \left[\mathcal{D}^\alpha \underline{\mathbf{f}}(\vartheta_0) \right] &= \frac{1}{\Gamma(q-\alpha)} \left[\int_0^\vartheta (\vartheta - \ell_1)^{q-\alpha-1} \frac{d^q}{d\ell_1^q} \underline{\mathbf{f}}(\ell_1) d\ell_1 \right]_{\vartheta = \vartheta_0}, \\ \left[\mathcal{D}^\alpha \bar{\mathbf{f}}(\vartheta_0) \right] &= \frac{1}{\Gamma(q-\alpha)} \left[\int_0^\vartheta (\vartheta - \ell_1)^{q-\alpha-1} \frac{d^q}{d\ell_1^q} \bar{\mathbf{f}}(\ell_1) d\ell_1 \right]_{\vartheta = \vartheta_0}. \end{aligned} \tag{10}$$

2.1. A Fuzzy Elzaki Transform for Fuzzy Caputo Fractional Derivative and a Fuzzy Atangana–Baleanu Fractional Derivative Operator

Definition 10. *Consider $\tilde{\mathbf{f}}$ to be continuous* **F***-valued mapping and assume that $\exp\left(\frac{-\vartheta}{\omega}\right) \odot \tilde{\mathbf{f}}(\vartheta)$ is an improper fuzzy Riemann-integrable on $[0, \infty)$ and then $\int_0^\infty \exp\left(\frac{-\vartheta}{\omega}\right) \odot \tilde{\mathbf{f}}(\vartheta) d\vartheta$ is said to be the fuzzy Elzaki transform and is described over the set of mappings:*

$$\mathcal{M} = \left\{ \mathbf{f}(\vartheta) : \exists z, p_1, p_2 > 0, \left| \mathbf{f}(\vartheta) \right| < z e^{\frac{|\vartheta|}{p_i}}, \; if \; \vartheta \in (-1)^i \times [0, \infty) | \right\}. \tag{11}$$

where z is a finite number, but p_1, p_2 may be finite or infinite, then the fuzzy Elzaki transform is described as

$$\mathbb{E}\{\tilde{\mathbf{f}}(\vartheta)\} = \mathbf{Q}(\omega) = \omega \int_0^\infty e^{-\frac{\vartheta}{\omega}} \odot \tilde{\mathbf{f}}(\vartheta)d\vartheta, \quad \vartheta \geq 0, \quad \omega \in [p_1, p_2]. \tag{12}$$

Remark 1. *In (12), $\tilde{\mathbf{f}}$ hold the cases of the decreasing diameter $\underline{\mathbf{f}}$ and the increasing diameter $\bar{\mathbf{f}}$ of a fuzzy mapping $\mathbf{f}$. Moreover, when $\omega = 1$, then the fuzzy Elzaki transform reduces to a Laplace transform.*

The parameterized version of $\tilde{\mathbf{f}}(\vartheta)$ is defined as

$$\omega \int_0^\infty e^{-\frac{\vartheta}{\omega}} \tilde{\mathbf{f}}(\vartheta)d\vartheta = \left[\omega \int_0^\infty e^{-\frac{\vartheta}{\omega}} \underline{\mathbf{f}}(\vartheta)d\vartheta, \omega \int_0^\infty e^{-\frac{\vartheta}{\omega}} \bar{\mathbf{f}}(\vartheta)d\vartheta\right]. \tag{13}$$

Thus,

$$\mathbb{E}[\mathbf{f}(\vartheta,\zeta)] = \left[\mathbb{E}[\underline{\mathbf{f}}(\vartheta,\zeta)], \mathbb{E}[\bar{\mathbf{f}}(\vartheta,\zeta)]\right]. \tag{14}$$

2.2. Some Algebraic Properties of Fuzzy Elzaki Transform

Here, our first result is the following theorem.

Theorem 1. *Assume that an integrable fuzzy valued mapping $\tilde{\mathbf{f}}^{(q)}(\vartheta)$ and $\tilde{\mathbf{f}}(\vartheta)$ is the primitive of $\tilde{\mathbf{f}}^{(q)}(\vartheta)$ on $[0,\infty)$, then*

$$\mathbb{E}\left[\tilde{\mathbf{f}}^{(q)}(\vartheta)\right] = (\frac{1}{\omega})^q \odot \mathbb{E}[\tilde{\mathbf{f}}(\vartheta)] \ominus \sum_{\kappa=0}^{q-1} \omega^{2-q+\kappa} \odot \tilde{\mathbf{f}}^{(\kappa)}(0). \tag{15}$$

The first few terms of (15) are represented as follows:

$$\begin{aligned}
\mathbb{E}\left[\tilde{\mathbf{f}}'(\vartheta)\right] &= (\frac{1}{\omega}) \odot \mathbf{Q}(\omega) \ominus \omega \odot \tilde{\mathbf{f}}(0), \\
\mathbb{E}\left[\tilde{\mathbf{f}}''(\vartheta)\right] &= (\frac{1}{\omega})^2 \odot \mathbf{Q}(\omega) - \omega \odot \tilde{\mathbf{f}}'(0) \ominus \tilde{\mathbf{f}}(0).
\end{aligned} \tag{16}$$

Proof. Assume that $\zeta \in [0,1]$ is arbitrary, and then we deduce

$$\begin{aligned}
&(\frac{1}{\omega})^q \odot \mathbb{E}[\tilde{\mathbf{f}}(\vartheta)] \ominus \sum_{\kappa=0}^{q-1} \omega^{2-q+\kappa} \odot \tilde{\mathbf{f}}^{(\kappa)}(0) \\
&= \left((\frac{1}{\omega})^q \odot \mathbb{E}[\bar{\mathbf{f}}(\vartheta;\zeta)] \ominus \sum_{\kappa=0}^{q-1} \omega^{2-q+\kappa} \odot \bar{\mathbf{f}}^{(\kappa)}(0;\zeta), (\frac{1}{\omega})^q \odot \mathbb{E}[\underline{\mathbf{f}}(\vartheta;\zeta)] \ominus \sum_{\kappa=0}^{q-1} \omega^{2-q+\kappa} \odot \underline{\mathbf{f}}^{(\kappa)}(0;\zeta)\right).
\end{aligned} \tag{17}$$

In view of (14), we have

$$\begin{aligned}
&(\frac{1}{\omega})^n \odot \mathbb{E}[\tilde{\mathbf{f}}(\vartheta)] \ominus \sum_{\kappa=0}^{q-1} \omega^{2-q+\kappa} \odot \tilde{\mathbf{f}}^{(\kappa)}(0) \\
&= \left(\mathbb{E}\left[\bar{\mathbf{f}}^{(q)}(\vartheta;\zeta)\right], \mathbb{E}\left[\underline{\mathbf{f}}^{(q)}(\vartheta;\zeta)\right]\right).
\end{aligned} \tag{18}$$

By mathematical induction, (15) holds for $q = \kappa$ and, utilizing (18), we have

$$\begin{aligned}\mathbb{E}\Big[(\tilde{\mathbf{f}}^{\kappa}(\vartheta))'\Big] &= \frac{1}{\omega} \odot \mathbb{E}\Big[\tilde{\mathbf{f}}^{\kappa}(\vartheta)\Big] \ominus \tilde{\mathbf{f}}^{\kappa}(0) \\ &= \frac{1}{\omega} \odot \Big[(\frac{1}{\omega})^{\kappa} \odot \mathbb{E}[\tilde{\mathbf{f}}(\vartheta)] \ominus \sum_{j=0}^{\kappa-1} \omega^{2-\kappa+j} \odot \tilde{\mathbf{f}}^{(j)}(0)\Big] \ominus \mathbf{f}^{\tilde{(\kappa)}}(0) \\ &= (\frac{1}{\omega})^{\kappa+1} \odot \mathbb{E}[\tilde{\mathbf{f}}(\vartheta)] \ominus \sum_{j=0}^{\kappa} \omega^{1-j+\kappa} \odot \tilde{\mathbf{f}}^{(j)}(0). \end{aligned} \tag{19}$$

Consequently, (15) is truewhen $q = \kappa + 1$. This completes the proof. □

Our next result is the convolution property of the fuzzy Elzaki transform.

Theorem 2. *Assume that two integrable fuzzy-valued mappings $\tilde{\mathbf{f}}_1(\vartheta)$ and $\tilde{\mathbf{f}}_2(\vartheta)$, with their respective fuzzy Elzaki transforms $\mathbf{Q}_1(\omega)$ and $\mathbf{Q}_2(\omega)$, respectively, then*

$$\mathbb{E}\Big[(\tilde{\mathbf{f}}_1 * \tilde{\mathbf{f}}_2)(\vartheta)\Big] = \omega^{-1} \odot \mathbf{Q}_1(\omega) \odot \mathbf{Q}_2(\omega), \tag{20}$$

where the convolution of $\tilde{\mathbf{f}}_1 * \tilde{\mathbf{f}}_2$ is defined as

$$\int_0^{\vartheta} \tilde{\mathbf{f}}_1(\tau) \odot \tilde{\mathbf{f}}_2(\vartheta - \tau) d\tau = \int_0^{\vartheta} \tilde{\mathbf{f}}_1(\vartheta - \tau) \odot \tilde{\mathbf{f}}_2(\tau) d\tau. \tag{21}$$

Proof. Utilizing (13), (20) and (21), we have

$$\mathbb{E}\Big[\int_0^{\vartheta} \tilde{\mathbf{f}}_1(\tau) \odot \tilde{\mathbf{f}}_2(\vartheta - \tau) d\tau\Big] = \omega \int_0^{\infty} \exp\big(-\frac{\vartheta}{\omega}\big) \odot \Big(\int_0^{\vartheta} \tilde{\mathbf{f}}_1(\tau) \odot \tilde{\mathbf{f}}_2(\vartheta - \tau) d\tau\Big) d\tau. \tag{22}$$

Exchanging the order and limit of the integrals, we have

$$\mathbb{E}\Big[\int_0^{\vartheta} \tilde{\mathbf{f}}_1(\tau) \odot \tilde{\mathbf{f}}_2(\vartheta - \tau) d\tau\Big] = \omega \int_0^{\infty} \Big(\tilde{\mathbf{f}}(\tau) \odot \int_{\tau}^{\infty} \exp\big(-\frac{\vartheta}{\omega}\big) \odot \tilde{\mathbf{f}}_2(\vartheta - \tau) d\vartheta\Big) d\tau. \tag{23}$$

Substituting $\theta = \vartheta - \tau$, we have

$$\begin{aligned}\int_{\tau}^{\infty} \exp\big(\frac{-\vartheta}{\omega}\big) \odot \tilde{\mathbf{f}}_2(\vartheta - \tau) d\vartheta &= \int_0^{\infty} \exp\big(\frac{-(\theta + \tau)}{\omega}\big) \odot \tilde{\mathbf{f}}_2(\theta) d\theta \\ &= \exp\Big(-\frac{\tau}{\omega}\Big) \odot \int_0^{\infty} \exp\big(-\frac{\theta}{\omega}\big) \odot \tilde{\mathbf{f}}_2(\theta) d\theta \\ &= \exp\Big(-\frac{\tau}{\omega}\Big) \odot \frac{1}{\omega} \odot \mathbf{Q}_2(\omega). \end{aligned} \tag{24}$$

Thus, we conclude

$$\begin{aligned}\mathbb{E}\Big[\int_0^{\vartheta} \tilde{\mathbf{f}}_1(\tau) \odot \tilde{\mathbf{f}}_2(\vartheta - \tau) d\tau\Big] &= \int_0^{\infty} \tilde{\mathbf{f}}_1(\tau) \odot \exp\Big(-\frac{\tau}{\omega}\Big) \odot \mathbf{Q}_2(\omega) d\tau \\ &= \omega^{-1} \odot \mathbf{Q}_1(\omega) \odot \Big[\int_0^{\infty} \tilde{\mathbf{f}}_1(\tau) \odot \exp\Big(-\frac{\tau}{\omega}\Big) d\tau\Big] \\ &= \omega^{-1} \odot \mathbf{Q}_2(\omega) \odot \mathbf{Q}_1(\omega). \end{aligned} \tag{25}$$

□

Theorem 3. *(Inverse fuzzy Elzaki transform.) Consider the mapping* $\mathbf{f}(\vartheta) \in \mathcal{M}$ *and* $\mathbf{Q}(\omega)$ *to be the fuzzy Elzaki transform of the mapping* $\mathbf{f}(\vartheta)$*, then the inverse transforms* $\mathbb{E}^{-1}$ *are presented as follows:*

$$\begin{aligned}\mathbb{E}^{-1}[\mathbf{Q}(\omega)] &= \lim_{b_1 \mapsto \infty} \frac{1}{2\pi\iota} \odot \int_{a_1-\iota b_1}^{a_1+\iota b_1} \mathbf{Q}(\frac{1}{\omega}) \odot \exp(\vartheta\omega) \odot \omega d\omega \\ &= \sum residues\ of \left[\mathbf{Q}(\frac{1}{\omega}) \odot \exp(\vartheta\omega) \odot \omega\right]. \end{aligned} \tag{26}$$

Adopting the idea of Allahviranloo et al. [66], here, we illustrate the fuzzy Elzaki transform of Caputo and generalize the Hukuhara derivative ${}_{g\mathcal{H}}\mathcal{D}^{\alpha}_{\vartheta}\mathbf{f}(\vartheta)$.

Theorem 4. *Consider an integrable fuzzy-valued mapping* ${}_{g\mathcal{H}}\mathcal{D}^{\alpha}_{\vartheta}\tilde{\mathbf{f}}(\vartheta)$ *and* $\mathbf{f}(\vartheta)$ *is the primitive of* ${}_{g\mathcal{H}}\mathcal{D}^{\alpha}_{\vartheta}\tilde{\mathbf{f}}(\vartheta)$ *defined on* $[0, \infty)$*, then the CFD operator of order* α *satisfies*

$$\mathbb{E}\left[{}_{g\mathcal{H}}\mathcal{D}^{\alpha}_{\vartheta}\tilde{\mathbf{f}}(\vartheta)\right] = \left(\frac{1}{\omega}\right)^{\alpha} \odot \mathbb{E}[\tilde{\mathbf{f}}(\vartheta)] \ominus \sum_{\kappa=0}^{q-1} \omega^{2-\alpha+\kappa} \odot \tilde{\mathbf{f}}^{(\kappa)}(0), \quad q-1 < \alpha \leq 1. \tag{27}$$

Proof. By means of Definition 10 and Theorem 2, we have

$$\begin{aligned}{}_{g\mathcal{H}}\mathcal{D}^{\alpha}_{\vartheta}\tilde{\mathbf{f}}(\vartheta) &= \frac{1}{\Gamma(q-\alpha)} \odot \int_0^{\vartheta} (\vartheta-\tau)^{q-\alpha-1} \odot \frac{\partial^q \tilde{f}(\tau)}{\partial \tau^q} d\tau \\ &= \frac{1}{\Gamma(q-\alpha)} \odot \tilde{\mathbf{f}}^{(q)} \odot \vartheta^{q-\alpha-1}. \end{aligned} \tag{28}$$

Again, in view of Definition 10 and Theorem 1, we obtain

$$\begin{aligned}{}_{g\mathcal{H}}\mathcal{D}^{\alpha}_{\vartheta}\tilde{\mathbf{f}}(\vartheta) &= \frac{1}{\Gamma(q-\alpha)} \odot \mathbb{E}\left[\vartheta^{q-\alpha-1} \odot \tilde{\mathbf{f}^{(q)}}(\vartheta)\right] \\ &= \left(\frac{1}{\omega}\right)^{\alpha} \odot \mathbb{E}[\tilde{\mathbf{f}}(\vartheta)] \ominus \sum_{\kappa=0}^{q-1} \omega^{2-\alpha+\kappa} \odot \tilde{\mathbf{f}}^{(\kappa)}(0). \end{aligned} \tag{29}$$

Using the fact that $\zeta \in [0,1]$, and the result provided by Salahhshour et al. [67], we have

$$\begin{aligned}&\left(\frac{1}{\omega}\right)^{\alpha} \odot \mathbb{E}[\tilde{\mathbf{f}}(\vartheta)] \ominus \sum_{\kappa=0}^{q-1} \omega^{2-\alpha+\kappa} \odot \tilde{\mathbf{f}}^{(\kappa)}(0) \\ &= \left(\left(\frac{1}{\omega}\right)^{\alpha} \mathbb{E}[\underline{\mathbf{f}}(\vartheta;\zeta)] - \sum_{\kappa=0}^{q-1} \omega^{2-\alpha+\kappa} \underline{\mathbf{f}}^{(\kappa)}(0;\zeta), \left(\frac{1}{\omega}\right)^{\alpha} \mathbb{E}[\bar{\mathbf{f}}(\vartheta\ wp)] - \sum_{\kappa=0}^{q-1} \omega^{2-\alpha+\kappa} \bar{\mathbf{f}}^{(\kappa)}(0;\zeta)\right). \end{aligned} \tag{30}$$

This completes the proof. □

Next we illustrate the linearity property of yjr fuzzy Elzaki transform.

Theorem 5. *Assume that there are two continuous fuzzy valued-mappings* $\tilde{\mathbf{f}}_1(\vartheta)$ *and* $\tilde{\mathbf{f}}_2(\vartheta)$ *with real constants* c_1 *and* c_2*,, then*

$$\mathbb{E}\left[c_1 \odot \tilde{\mathbf{f}}_1(\vartheta) \oplus c_2 \odot \tilde{\mathbf{f}}_2(\vartheta)\right] = c_1 \odot \mathbb{E}\left[\tilde{\mathbf{f}}_1(\vartheta)\right] + c_2 \odot \mathbb{E}\left[\tilde{\mathbf{f}}_2(\vartheta)\right]. \tag{31}$$

Proof. Consider $\zeta \in [0,1]$ to be arbitrarily fixed. Then, by means of (13), we have

$$\begin{aligned}
\mathbb{E}\Big[c_1 \odot \tilde{\mathbf{f}}_1(\vartheta) \oplus c_2 \odot \tilde{\mathbf{f}}_1(\vartheta)\Big] &= \omega \int_0^\infty (a_1 \odot \tilde{\mathbf{f}}_1(\vartheta) \oplus c_2 \odot \tilde{\mathbf{f}}_2(\vartheta)) \odot \exp\big(\frac{-\vartheta}{\omega}\big) d\vartheta \\
&= \omega \int_0^\infty c_1 \odot \tilde{\mathbf{f}}_1(\vartheta) \odot \exp\big(\frac{-\vartheta}{\omega}\big) d\vartheta \oplus \omega \int_0^\infty c_2 \odot \tilde{\mathbf{f}}_2(\vartheta) \odot \exp\big(\frac{-\vartheta}{\omega}\big) d\vartheta \\
&= \Big(c_1 \odot \omega \int_0^\infty \tilde{\mathbf{f}}_1(\vartheta) \odot \exp\big(\frac{-\vartheta}{\omega}\big) d\vartheta\Big) \oplus \Big(c_2 \odot \omega \int_0^\infty \tilde{\mathbf{f}}_1(\vartheta) \odot \exp\big(\frac{-\vartheta}{\omega}\big) d\vartheta\Big) \\
&= c_1 \odot \Big(\omega \int_0^\infty \underline{\mathbf{f}}_1(\vartheta;\zeta) \odot \exp\big(\frac{-\vartheta}{\omega}\big) d\vartheta, \omega \int_0^\infty \bar{\mathbf{f}}_1(\vartheta;\zeta) \odot \exp\big(\frac{-\vartheta}{\omega}\big) d\vartheta\Big) \\
&\oplus \ c_2 \odot \Big(\omega \int_0^\infty \underline{\mathbf{f}}_2(\vartheta;\zeta) \odot \exp\big(\frac{-\vartheta}{\omega}\big) d\vartheta, \omega \int_0^\infty \bar{\mathbf{f}}_2(\vartheta;\zeta) \odot \exp\big(\frac{-\vartheta}{\omega}\big) d\vartheta\Big) \\
&= c_1 \odot \mathbb{E}\Big[\tilde{\mathbf{f}}_1(\vartheta)\Big] + c_2 \odot \mathbb{E}\Big[\tilde{\mathbf{f}}_2(\vartheta)\Big],
\end{aligned} \tag{32}$$

This completes the proof. □

Definition 11. *Consider* $\mathbf{f} \in \mathbb{C}^F[0,\bar{d}_1] \bigcap \mathbb{L}^F[0,\bar{d}_1]$ *such that* $\mathbf{f}(\vartheta) = \big[\underline{\mathbf{f}}(\vartheta,\zeta), \bar{\mathbf{f}}(\vartheta,\zeta)\big]$, $\zeta \in [0,1]$, *then the Elzaki transform of fuzzy CFD of order* $\alpha \in (0,1]$ *is described as follows:*

$$\mathbb{E}\big[(\mathcal{D}^\alpha \mathbf{f}(\vartheta))_\zeta\big] = \big[\mathbb{E}[\mathcal{D}^\alpha \underline{\mathbf{f}}(\vartheta,\zeta)], \mathbb{E}[\mathcal{D}^\alpha \bar{\mathbf{f}}(\vartheta,\zeta)]\big], \tag{33}$$

where

$$\begin{aligned}
\mathbb{E}[\mathcal{D}^\alpha \underline{\mathbf{f}}(\vartheta,\zeta)] &= \frac{1}{\omega^\alpha}\mathbb{E}[\underline{\mathbf{f}}(\vartheta,\zeta)] - \sum_{\kappa=0}^{q-1} \underline{\mathbf{f}}_{(\kappa)}(\ell_1;\zeta)\omega^{2-\alpha+\kappa}, \quad \alpha \in (q-1,q], \\
\mathbb{E}[\mathcal{D}^\alpha \bar{\mathbf{f}}(\vartheta,\zeta)] &= \frac{1}{\omega^\alpha}\mathbb{E}[\bar{\mathbf{f}}(\vartheta,\zeta)] - \sum_{\kappa=0}^{q-1} \bar{\mathbf{f}}_{(\kappa)}(\ell_1;\zeta)\omega^{2-\alpha+\kappa}, \quad \alpha \in (q-1,q].
\end{aligned} \tag{34}$$

Definition 12. *Consider* $\tilde{\mathbf{f}}(\vartheta) \in \tilde{\mathbb{H}}^1(0,T)$ *and* $\alpha \in [0,1]$, *then the* α*th-order variable Atangana–Baleanu derivative under (i)—*$g\mathcal{H}$ *differentiability of* $\tilde{\mathbf{f}}$ *in the Caputo sense is stated as*

$$\Big[{}^{ABC}\mathcal{D}^\alpha_{(i)-g\mathcal{H}}\tilde{\mathbf{f}}(\vartheta_0;\zeta)\Big] = \Big[{}^{ABC}\mathcal{D}^\alpha_{(i)-g\mathcal{H}}\underline{\mathbf{f}}(\vartheta_0;\zeta),\ {}^{ABC}\mathcal{D}^\alpha_{(i)-g\mathcal{H}}\bar{\mathbf{f}}(\vartheta_0;\zeta)\Big], \ \zeta \in [0,1], \tag{35}$$

where

$$\begin{aligned}
{}^{ABC}\mathcal{D}^\alpha_{(i)-g\mathcal{H}}\underline{\mathbf{f}}(\vartheta_0;\zeta) &= \frac{\mathcal{N}(\alpha)}{1-\alpha}\Big[\int_0^\vartheta \underline{\mathbf{f}}'_{(i)-g\mathcal{H}}(\ell_1)E_\alpha\Big[\frac{-\alpha(\vartheta-\ell_1)^\alpha}{1-\alpha}\Big]d\ell_1\Big]_{\vartheta=\vartheta_0}, \\
{}^{ABC}\mathcal{D}^\alpha_{(i)-g\mathcal{H}}\bar{\mathbf{f}}(\vartheta_0;\zeta) &= \frac{\mathcal{N}(\alpha)}{1-\alpha}\Big[\int_0^\vartheta \bar{\mathbf{f}}'_{(i)-g\mathcal{H}}(\ell_1)E_\alpha\Big[\frac{-\alpha(\vartheta-\ell_1)^\alpha}{1-\alpha}\Big]d\ell_1\Big]_{\vartheta=\vartheta_0},
\end{aligned} \tag{36}$$

where $\mathcal{N}(\alpha)$ *denotes the normalize function that equals 1 when* α *is assumed to be 0 and 1. Furthermore, we suppose that type (i)—*$g\mathcal{H}$ *exists. So here is no need to consider (ii)—*$g\mathcal{H}$ *differentiability.*

Yauvaz and Abdeljawad [68] defined the ABC fractional derivative operator in the Elzaki sense. Furthermore, we extend the idea of a fuzzy ABC fractional derivative in the Elzaki transform sense as follows:

Definition 13. *Consider* $\mathbf{f} \in \mathbb{C}^F[0, \bar{d}_1] \bigcap \mathbb{L}^F[0, \bar{d}_1]$ *such that* $\mathbf{f}(\vartheta) = [\underline{\mathbf{f}}(\vartheta, \zeta), \bar{\mathbf{f}}(\vartheta, \zeta)]$, $\zeta \in [0,1]$, *then the Elzaki transform of fuzzy ABC of order* $\alpha \in [0,1]$ *is described as follows:*

$$\mathbb{E}[({}^{ABC}\mathcal{D}^{\alpha}\mathbf{f}(\vartheta))_{\zeta}] = [\mathbb{E}[{}^{ABC}\mathcal{D}^{\alpha}\underline{\mathbf{f}}(\vartheta,\zeta)], \mathbb{E}[{}^{ABC}\mathcal{D}^{\alpha}\bar{\mathbf{f}}(\vartheta,\zeta)]], \tag{37}$$

where

$$\begin{aligned} \mathbb{E}[{}^{ABC}\mathcal{D}^{\alpha}\underline{\mathbf{f}}(\vartheta,\zeta)] &= \frac{\omega\mathcal{N}(\alpha)}{\alpha\omega^{\alpha}+1-\alpha}\left[\frac{\mathbb{E}[\underline{\mathbf{f}}(\vartheta,\zeta)]}{\omega} - \omega\underline{\mathbf{f}}(0)\right], \\ \mathbb{E}[{}^{ABC}\mathcal{D}^{\alpha}\bar{\mathbf{f}}(\vartheta,\zeta)] &= \frac{\omega\mathcal{N}(\alpha)}{\alpha\omega^{\alpha}+1-\alpha}\left[\frac{\mathbb{E}[\bar{\mathbf{f}}(\vartheta,\zeta)]}{\omega} - \omega\bar{\mathbf{f}}(0)\right]. \end{aligned} \tag{38}$$

3. Proposed Algorithm

Here, the general methodology of obtaining the numerical results of one-dimensional fractional FWE involving the CFD and ABC fractional derivative operator in the fuzzy ET is investigated.

The parameterized version of (2) is presented as

$$\begin{cases} \frac{\partial^{\alpha}}{\partial\vartheta^{\alpha}}\underline{\mathbf{f}}(\ell_1,\vartheta;\zeta) = \frac{\partial^3}{\partial\ell_1^2\partial\vartheta}\underline{\mathbf{f}}(\ell_1,\vartheta;\zeta) - \frac{\partial}{\partial\ell_1}\underline{\mathbf{f}}(\ell_1,\vartheta;\zeta) + \underline{\mathbf{f}}(\ell_1,\vartheta;\zeta)\frac{\partial^3}{\partial\ell_1^3}\underline{\mathbf{f}}(\ell_1,\vartheta;\zeta) \\ \quad -\underline{\mathbf{f}}(\ell_1,\vartheta;\zeta)\frac{\partial}{\partial\ell_1}\underline{\mathbf{f}}(\ell_1,\vartheta;\zeta) + 3\frac{\partial}{\partial\ell_1}\underline{\mathbf{f}}(\ell_1,\vartheta;\zeta)\frac{\partial^2}{\partial\ell_1^2}\underline{\mathbf{f}}(\ell_1,\vartheta;\zeta), \\ \underline{\mathbf{f}}(\ell_1,0) = \underline{g}(\ell_1;\zeta), \\ \frac{\partial^{\alpha}}{\partial\vartheta^{\alpha}}\bar{\mathbf{f}}(\ell_1,\vartheta;\zeta) = \frac{\partial^3}{\partial\ell_1^2\partial\vartheta}\bar{\mathbf{f}}(\ell_1,\vartheta;\zeta) - \frac{\partial}{\partial\ell_1}\bar{\mathbf{f}}(\ell_1,\vartheta;\zeta) + \bar{\mathbf{f}}(\ell_1,\vartheta;\zeta)\frac{\partial^3}{\partial\ell_1^3}\bar{\mathbf{f}}(\ell_1,\vartheta;\zeta) \\ \quad -\bar{\mathbf{f}}(\ell_1,\vartheta;\zeta)\frac{\partial}{\partial\ell_1}\bar{\mathbf{f}}(\ell_1,\vartheta;\zeta) + 3\frac{\partial}{\partial\ell_1}\bar{\mathbf{f}}(\ell_1,\vartheta;\zeta)\frac{\partial^2}{\partial\ell_1^2}\bar{\mathbf{f}}(\ell_1,\vartheta;\zeta), \\ \bar{\mathbf{f}}(\ell_1,0) = \bar{g}(\ell_1;\zeta). \end{cases} \tag{39}$$

Employing ET on both sides of the first preceding case of (39) by utilizing the fuzzy CFD, we have

$$\begin{aligned} \mathbb{E}[\underline{\mathbf{f}}(\ell_1,\vartheta;\zeta)] = \mathbb{E}\Big[&\frac{\partial^3}{\partial\ell_1^2\partial\vartheta}\underline{\mathbf{f}}(\ell_1,\vartheta;\zeta) - \frac{\partial}{\partial\ell_1}\underline{\mathbf{f}}(\ell_1,\vartheta;\zeta) + \underline{\mathbf{f}}(\ell_1,\vartheta;\zeta)\frac{\partial^3}{\partial\ell_1^3}\underline{\mathbf{f}}(\ell_1,\vartheta;\zeta) \\ &-\underline{\mathbf{f}}(\ell_1,\vartheta;\zeta)\frac{\partial}{\partial\ell_1}\underline{\mathbf{f}}(\ell_1,\vartheta;\zeta) + 3\frac{\partial}{\partial\ell_1}\underline{\mathbf{f}}(\ell_1,\vartheta;\zeta)\frac{\partial^2}{\partial\ell_1^2}\underline{\mathbf{f}}(\ell_1,\vartheta;\zeta)\Big] \end{aligned} \tag{40}$$

subject to the IC $\underline{\mathbf{f}}(\ell_1,0) = \underline{g}(\ell_1)$, we have

$$\begin{aligned} \frac{1}{\omega^{\alpha}}\mathbb{E}[\underline{\mathbf{f}}(\ell_1,\vartheta;\zeta)] - \sum_{\kappa=0}^{q-1}\underline{\mathbf{f}}_{(\kappa)}(\ell_1;\zeta)\omega^{2-\alpha+\kappa} = \mathbb{E}\Big[&\frac{\partial^3}{\partial\ell_1^2\partial\vartheta}\underline{\mathbf{f}}(\ell_1,\vartheta;\zeta) - \frac{\partial}{\partial\ell_1}\underline{\mathbf{f}}(\ell_1,\vartheta;\zeta) + \underline{\mathbf{f}}(\ell_1,\vartheta;\zeta)\frac{\partial^3}{\partial\ell_1^3}\underline{\mathbf{f}}(\ell_1,\vartheta;\zeta) \\ &-\underline{\mathbf{f}}(\ell_1,\vartheta;\zeta)\frac{\partial}{\partial\ell_1}\underline{\mathbf{f}}(\ell_1,\vartheta;\zeta) + 3\frac{\partial}{\partial\ell_1}\underline{\mathbf{f}}(\ell_1,\vartheta;\zeta)\frac{\partial^2}{\partial\ell_1^2}\underline{\mathbf{f}}(\ell_1,\vartheta;\zeta)\Big], \end{aligned}$$

or, accordingly, we have

$$\mathbb{E}\big[\underline{\mathbf{f}}(\ell_1,\vartheta;\zeta)\big] = \omega^2\underline{g}(\ell_1;\zeta) + \omega^\alpha\mathbb{E}\bigg[\frac{\partial^3}{\partial\ell_1^2\partial\vartheta}\underline{\mathbf{f}}(\ell_1,\vartheta;\zeta) - \frac{\partial}{\partial\ell_1}\underline{\mathbf{f}}(\ell_1,\vartheta;\zeta) + \underline{\mathbf{f}}(\ell_1,\vartheta;\zeta)\frac{\partial^3}{\partial\ell_1^3}\underline{\mathbf{f}}(\ell_1,\vartheta;\zeta) \\ -\underline{\mathbf{f}}(\ell_1,\vartheta;\zeta)\frac{\partial}{\partial\ell_1}\underline{\mathbf{f}}(\ell_1,\vartheta;\zeta) + 3\frac{\partial}{\partial\ell_1}\underline{\mathbf{f}}(\ell_1,\vartheta;\zeta)\frac{\partial^2}{\partial\ell_1^2}\underline{\mathbf{f}}(\ell_1,\vartheta;\zeta)\bigg]. \quad (41)$$

Again, applying ET on both sides of the first preceding case of (39) by utilizing the fuzzy ABC fractional derivative, we have

$$\mathbb{E}\big[\underline{\mathbf{f}}(\ell_1,\vartheta;\zeta)\big] = \omega^2\underline{g}(\ell_1;\zeta) + \left(\frac{\alpha\omega^\alpha + 1 - \alpha}{\mathcal{N}(\alpha)}\right)\mathbb{E}\bigg[\frac{\partial^3}{\partial\ell_1^2\partial\vartheta}\underline{\mathbf{f}}(\ell_1,\vartheta;\zeta) - \frac{\partial}{\partial\ell_1}\underline{\mathbf{f}}(\ell_1,\vartheta;\zeta) \\ +\underline{\mathbf{f}}(\ell_1,\vartheta;\zeta)\frac{\partial^3}{\partial\ell_1^3}\underline{\mathbf{f}}(\ell_1,\vartheta;\zeta) - \underline{\mathbf{f}}(\ell_1,\vartheta;\zeta)\frac{\partial}{\partial\ell_1}\underline{\mathbf{f}}(\ell_1,\vartheta;\zeta) + 3\frac{\partial}{\partial\ell_1}\underline{\mathbf{f}}(\ell_1,\vartheta;\zeta)\frac{\partial^2}{\partial\ell_1^2}\underline{\mathbf{f}}(\ell_1,\vartheta;\zeta)\bigg]. \quad (42)$$

The unknown series solution is expressed as

$$\underline{\mathbf{f}}(\ell_1,\vartheta;\zeta) = \sum_{q=0}^{\infty}\underline{\mathbf{f}}(\ell_1,\vartheta;\zeta), \quad (43)$$

and the nonlinear terms are decomposed as

$$\begin{aligned} \underline{\mathcal{N}}_1(\ell_1,\vartheta;\zeta) &= \sum_{q=0}^{\infty}\underline{\mathcal{A}}_q(\ell_1,\vartheta;\zeta) = \underline{\mathbf{f}}(\ell_1,\vartheta;\zeta)\frac{\partial^3}{\partial\ell_1^3}\underline{\mathbf{f}}(\ell_1,\vartheta;\zeta), \\ \underline{\mathcal{N}}_2(\ell_1,\vartheta;\zeta) &= \sum_{q=0}^{\infty}\underline{\mathcal{B}}_q(\ell_1,\vartheta;\zeta) = \underline{\mathbf{f}}(\ell_1,\vartheta;\zeta)\frac{\partial}{\partial\ell_1}\underline{\mathbf{f}}(\ell_1,\vartheta;\zeta), \\ \underline{\mathcal{N}}_3(\ell_1,\vartheta;\zeta) &= \sum_{q=0}^{\infty}\underline{\mathcal{C}}_q(\ell_1,\vartheta;\zeta) = \frac{\partial}{\partial\ell_1}\underline{\mathbf{f}}(\ell_1,\vartheta;\zeta)\frac{\partial^2}{\partial\ell_1^2}\underline{\mathbf{f}}(\ell_1,\vartheta;\zeta), \end{aligned} \quad (44)$$

where $\underline{\mathcal{A}}_q, \underline{\mathcal{B}}_q$ and $\underline{\mathcal{C}}_q$ are known to be the Adomian polynomial are presented as

$$\begin{aligned} \underline{\mathcal{A}}_q &= \frac{1}{q!}\frac{d^q}{d\lambda^q}\left[\underline{\mathcal{N}}_1\left(\sum_{q=0}^{\infty}\lambda^q\underline{\mathbf{f}}_q(\ell_1,\vartheta;\zeta)\right)\right]_{\lambda=0}, \\ \underline{\mathcal{B}}_q &= \frac{1}{q!}\frac{d^q}{d\lambda^q}\left[\underline{\mathcal{N}}_2\left(\sum_{q=0}^{\infty}\lambda^q\underline{\mathbf{f}}_q(\ell_1,\vartheta;\zeta)\right)\right]_{\lambda=0}, \\ \underline{\mathcal{C}}_q &= \frac{1}{q!}\frac{d^q}{d\lambda^q}\left[\underline{\mathcal{N}}_3\left(\sum_{q=0}^{\infty}\lambda^q\underline{\mathbf{f}}_q(\ell_1,\vartheta;\zeta)\right)\right]_{\lambda=0}. \end{aligned} \quad (45)$$

Now, (41) and (42), respectively, can be expressed as

$$\mathbb{E}\left[\sum_{q=0}^{\infty}\underline{\mathbf{f}}(\ell_1,\vartheta;\zeta)\right] = \omega^2\underline{g}(\ell_1;\zeta) + \omega^\alpha\mathbb{E}\left[\sum_{q=0}^{\infty}\left(\frac{\partial^3}{\partial\ell_1^2\partial\vartheta}\underline{\mathbf{f}}(\ell_1,\vartheta;\zeta)\right)_q - \sum_{q=0}^{\infty}\left(\frac{\partial}{\partial\ell_1}\underline{\mathbf{f}}(\ell_1,\vartheta;\zeta)\right)_q \\ + \sum_{q=0}^{\infty}\underline{\mathcal{A}}_q(\ell_1,\vartheta;\zeta) - \sum_{q=0}^{\infty}\underline{\mathcal{B}}_q(\ell_1,\vartheta;\zeta) + 3\sum_{q=0}^{\infty}\underline{\mathcal{C}}_q(\ell_1,\vartheta;\zeta)\right] \quad (46)$$

and

$$\mathbb{E}\Big[\sum_{q=0}^{\infty}\underline{\mathbf{f}}(\ell_1,\vartheta;\zeta)\Big] = \omega^2\underline{g}(\ell_1;\zeta)+\Big(\frac{\alpha\omega^\alpha+1-\alpha}{\mathcal{N}(\alpha)}\Big)\mathbb{E}\Big[\sum_{q=0}^{\infty}\Big(\frac{\partial^3}{\partial\ell_1^2\partial\vartheta}\underline{\mathbf{f}}(\ell_1,\vartheta;\zeta)\Big)_q-\sum_{q=0}^{\infty}\Big(\frac{\partial}{\partial\ell_1}\underline{\mathbf{f}}(\ell_1,\vartheta;\zeta)\Big)_q + \sum_{q=0}^{\infty}\underline{\mathcal{A}}_q(\ell_1,\vartheta;\zeta)-\sum_{q=0}^{\infty}\underline{\mathcal{B}}_q(\ell_1,\vartheta;\zeta)+3\sum_{q=0}^{\infty}\underline{\mathcal{C}}_q(\ell_1,\vartheta;\zeta)\Big]. \tag{47}$$

Applying the inverse ET on (46) and comparing terms by terms on both sides, we have

$$\begin{aligned}
\underline{\mathbf{f}}_0(\ell_1,\vartheta;\zeta) &= \mathbb{E}^{-1}\Big[\omega^2\underline{g}(\ell_1;\zeta)\Big],\\
\underline{\mathbf{f}}_1(\ell_1,\vartheta;\zeta) &= \mathbb{E}^{-1}\Big[\omega^\alpha\mathbb{E}\Big[\Big(\frac{\partial^3}{\partial\ell_1^2\partial\vartheta}\underline{\mathbf{f}}(\ell_1,\vartheta;\zeta)\Big)_1-\Big(\frac{\partial}{\partial\ell_1}\underline{\mathbf{f}}(\ell_1,\vartheta;\zeta)\Big)_1+\underline{\mathcal{A}}_1(\ell_1,\vartheta;\zeta)-\underline{\mathcal{B}}_1(\ell_1,\vartheta;\zeta)\\
&\quad +3\underline{\mathcal{C}}_1(\ell_1,\vartheta;\zeta)\Big]\Big],\\
\underline{\mathbf{f}}_2(\ell_1,\vartheta;\zeta) &= \mathbb{E}^{-1}\Big[\omega^\alpha\mathbb{E}\Big[\Big(\frac{\partial^3}{\partial\ell_1^2\partial\vartheta}\underline{\mathbf{f}}(\ell_1,\vartheta;\zeta)\Big)_2-\Big(\frac{\partial}{\partial\ell_1}\underline{\mathbf{f}}(\ell_1,\vartheta;\zeta)\Big)_2+\underline{\mathcal{A}}_2(\ell_1,\vartheta;\zeta)-\underline{\mathcal{B}}_2(\ell_1,\vartheta;\zeta)\\
&\quad +3\underline{\mathcal{C}}_2(\ell_1,\vartheta;\zeta)\Big]\Big],\\
&\ \vdots\\
\underline{\mathbf{f}}_{q+1}(\ell_1,\vartheta;\zeta) &= \mathbb{E}^{-1}\Big[\omega^\alpha\mathbb{E}\Big[\Big(\frac{\partial^3}{\partial\ell_1^2\partial\vartheta}\underline{\mathbf{f}}(\ell_1,\vartheta;\zeta)\Big)_q-\Big(\frac{\partial}{\partial\ell_1}\underline{\mathbf{f}}(\ell_1,\vartheta;\zeta)\Big)_q+\underline{\mathcal{A}}_q(\ell_1,\vartheta;\zeta)-\underline{\mathcal{B}}_q(\ell_1,\vartheta;\zeta)\\
&\quad +3\underline{\mathcal{C}}_q(\ell_1,\vartheta;\zeta)\Big]\Big].
\end{aligned}$$

Again, applying the inverse ET on (47) and comparing terms by terms on both sides, we have

$$\begin{aligned}
\underline{\mathbf{f}}_0(\ell_1,\vartheta;\zeta) &= \mathbb{E}^{-1}\Big[\omega^2\underline{g}(\ell_1;\zeta)\Big],\\
\underline{\mathbf{f}}_1(\ell_1,\vartheta;\zeta) &= \mathbb{E}^{-1}\Big[\Big(\frac{\alpha\omega^\alpha+1-\alpha}{\mathcal{N}(\alpha)}\Big)\mathbb{E}\Big[\Big(\frac{\partial^3}{\partial\ell_1^2\partial\vartheta}\underline{\mathbf{f}}(\ell_1,\vartheta;\zeta)\Big)_0-\Big(\frac{\partial}{\partial\ell_1}\underline{\mathbf{f}}(\ell_1,\vartheta;\zeta)\Big)_0+\underline{\mathcal{A}}_0(\ell_1,\vartheta;\zeta)\\
&\quad -\underline{\mathcal{B}}_0(\ell_1,\vartheta;\zeta)+3\underline{\mathcal{C}}_0(\ell_1,\vartheta;\zeta)\Big]\Big],\\
\underline{\mathbf{f}}_2(\ell_1,\vartheta;\zeta) &= \mathbb{E}^{-1}\Big[\Big(\frac{\alpha\omega^\alpha+1-\alpha}{\mathcal{N}(\alpha)}\Big)\mathbb{E}\Big[\Big(\frac{\partial^3}{\partial\ell_1^2\partial\vartheta}\underline{\mathbf{f}}(\ell_1,\vartheta;\zeta)\Big)_1-\Big(\frac{\partial}{\partial\ell_1}\underline{\mathbf{f}}(\ell_1,\vartheta;\zeta)\Big)_1+\underline{\mathcal{A}}_1(\ell_1,\vartheta;\zeta)\\
&\quad -\underline{\mathcal{B}}_1(\ell_1,\vartheta;\zeta)+3\underline{\mathcal{C}}_1(\ell_1,\vartheta;\zeta)\Big]\Big],\\
&\ \vdots\\
\underline{\mathbf{f}}_{q+1}(\ell_1,\vartheta;\zeta) &= \mathbb{E}^{-1}\Big[\Big(\frac{\alpha\omega^\alpha+1-\alpha}{\mathcal{N}(\alpha)}\Big)\mathbb{E}\Big[\Big(\frac{\partial^3}{\partial\ell_1^2\partial\vartheta}\underline{\mathbf{f}}(\ell_1,\vartheta;\zeta)\Big)_q-\Big(\frac{\partial}{\partial\ell_1}\underline{\mathbf{f}}(\ell_1,\vartheta;\zeta)\Big)_q+\underline{\mathcal{A}}_q(\ell_1,\vartheta;\zeta)\\
&\quad -\underline{\mathcal{B}}_q(\ell_1,\vartheta;\zeta)+3\underline{\mathcal{C}}_q(\ell_1,\vartheta;\zeta)\Big]\Big].
\end{aligned}$$

Hence, the required series solution is expressed as

$$\underline{\mathbf{f}}(\ell_1,\vartheta;\zeta)=\underline{\mathbf{f}}_0(\ell_1,\vartheta;\zeta)+\underline{\mathbf{f}}_1(\ell_1,\vartheta;\zeta)+\dots. \tag{48}$$

Repeating the same procedure for the upper case of (39). Therefore, we mention the solution in the parameterized version as follows:

$$\begin{cases} \underline{\mathbf{f}}(\ell_1,\vartheta;\zeta) = \underline{\mathbf{f}}_0(\ell_1,\vartheta;\zeta) + \underline{\mathbf{f}}_1(\ell_1,\vartheta;\zeta) + \dots, \\ \bar{\mathbf{f}}(\ell_1,\vartheta;\zeta) = \bar{\mathbf{f}}_0(\ell_1,\vartheta;\zeta) + \bar{\mathbf{f}}_1(\ell_1,\vartheta;\zeta) + \dots. \end{cases} \tag{49}$$

4. Test Examples and Their Physical Interpretation

In this note, we demonstrate the series solutions with the aid of EADM concerning different initial conditions by employing fuzzy Caputo and ABC fractional derivative operators, respectively.
Firstly, we surmise the FW model (2) by considering EADM.

Problem 1. *Assume the one-dimension fuzzy fractional FW model with fuzzy ICs is represented as follows:*

$$\begin{aligned} \frac{\partial^\alpha}{\partial\vartheta^\alpha}\tilde{\mathbf{f}}(\ell_1,\vartheta;\zeta) &= \frac{\partial^3}{\partial\ell_1^2\partial\vartheta}\tilde{\mathbf{f}}(\ell_1,\vartheta;\zeta) \ominus \frac{\partial}{\partial\ell_1}\tilde{\mathbf{f}}(\ell_1,\vartheta;\zeta) \oplus \tilde{\mathbf{f}}(\ell_1,\vartheta;\zeta) \odot \frac{\partial^3}{\partial\ell_1^3}\tilde{\mathbf{f}}(\ell_1,\vartheta;\zeta) \\ &\ominus\tilde{\mathbf{f}}(\ell_1,\vartheta;\zeta) \odot \frac{\partial}{\partial\ell_1}\tilde{\mathbf{f}}(\ell_1,\vartheta;\zeta) \oplus \frac{\partial}{\partial\ell_1}\tilde{\mathbf{f}}(\ell_1,\vartheta;\zeta) \odot \frac{\partial^2}{\partial\ell_1^2}\tilde{\mathbf{f}}(\ell_1,\vartheta;\zeta), \\ &\tilde{\mathbf{f}}(\ell_1,0) = \tilde{\chi}(\zeta) \odot \exp\left(\frac{\ell_1}{2}\right), \end{aligned} \tag{50}$$

where $\tilde{\chi}(\zeta) = [\underline{\chi}(\zeta), \bar{\chi}(\zeta)] = [\zeta - 1, 1 - \zeta]$ *for* $\zeta \in [0,1]$ *is a fuzzy number.*

Proof. The parameterized version of the problem (50) is expressed as follows

$$\begin{cases} \frac{\partial^\alpha}{\partial\vartheta^\alpha}\underline{\mathbf{f}}(\ell_1,\vartheta;\zeta) = \frac{\partial^3}{\partial\ell_1^2\partial\vartheta}\underline{\mathbf{f}}(\ell_1,\vartheta;\zeta) - \frac{\partial}{\partial\ell_1}\underline{\mathbf{f}}(\ell_1,\vartheta;\zeta) + \underline{\mathbf{f}}(\ell_1,\vartheta;\zeta)\frac{\partial^3}{\partial\ell_1^3}\underline{\mathbf{f}}(\ell_1,\vartheta;\zeta) \\ \quad -\underline{\mathbf{f}}(\ell_1,\vartheta;\zeta)\frac{\partial}{\partial\ell_1}\underline{\mathbf{f}}(\ell_1,\vartheta;\zeta) + \frac{\partial}{\partial\ell_1}\underline{\mathbf{f}}(\ell_1,\vartheta;\zeta)\frac{\partial^2}{\partial\ell_1^2}\underline{\mathbf{f}}(\ell_1,\vartheta;\zeta), \\ \underline{\mathbf{f}}(\ell_1,0) = \underline{\chi}(\zeta)\exp\left(\frac{\ell_1}{2}\right), \\ \frac{\partial^\alpha}{\partial\vartheta^\alpha}\bar{\mathbf{f}}(\ell_1,\vartheta;\zeta) = \frac{\partial^3}{\partial\ell_1^2\partial\vartheta}\bar{\mathbf{f}}(\ell_1,\vartheta;\zeta) - \frac{\partial}{\partial\ell_1}\bar{\mathbf{f}}(\ell_1,\vartheta;\zeta) + \bar{\mathbf{f}}(\ell_1,\vartheta;\zeta)\frac{\partial^3}{\partial\ell_1^3}\bar{\mathbf{f}}(\ell_1,\vartheta;\zeta) \\ \quad -\bar{\mathbf{f}}(\ell_1,\vartheta;\zeta)\frac{\partial}{\partial\ell_1}\bar{\mathbf{f}}(\ell_1,\vartheta;\zeta) + \frac{\partial}{\partial\ell_1}\bar{\mathbf{f}}(\ell_1,\vartheta;\zeta)\frac{\partial^2}{\partial\ell_1^2}\bar{\mathbf{f}}(\ell_1,\vartheta;\zeta), \\ \bar{\mathbf{f}}(\ell_1,0) = \bar{\chi}(\zeta)\exp\left(\frac{\ell_1}{2}\right). \end{cases} \tag{51}$$

Case I. (For the fuzzy Caputo fractional derivative)

Here, we obtain the EADM solution for the first case of (51) by using the fuzzy Caputo fractional derivative operator.

Taking into consideration the procedure described in Section 3, we have

$$\begin{aligned} &\frac{1}{\omega^\alpha}\mathbb{E}[\underline{\mathbf{f}}(\ell_1,\vartheta;\zeta)] - \sum_{\kappa=0}^{q-1}\underline{\mathbf{f}}_{(\kappa)}(\ell_1;\zeta)\omega^{2-\alpha+\kappa} \\ &= \mathbb{E}\Bigg[\frac{\partial^3}{\partial\ell_1^2\partial\vartheta}\underline{\mathbf{f}}(\ell_1,\vartheta;\zeta) - \frac{\partial}{\partial\ell_1}\underline{\mathbf{f}}(\ell_1,\vartheta;\zeta) + \underline{\mathbf{f}}(\ell_1,\vartheta;\zeta)\frac{\partial^3}{\partial\ell_1^3}\underline{\mathbf{f}}(\ell_1,\vartheta;\zeta) \\ &\quad -\underline{\mathbf{f}}(\ell_1,\vartheta;\zeta)\frac{\partial}{\partial\ell_1}\underline{\mathbf{f}}(\ell_1,\vartheta;\zeta) + \frac{\partial}{\partial\ell_1}\underline{\mathbf{f}}(\ell_1,\vartheta;\zeta)\frac{\partial^2}{\partial\ell_1^2}\underline{\mathbf{f}}(\ell_1,\vartheta;\zeta)\Bigg]. \end{aligned}$$

Simple computations result in

$$\begin{aligned} \underline{\mathbf{f}}(\ell_1,\vartheta;\zeta) = (\zeta-1)\exp\left(\frac{\ell_1}{2}\right) + \mathbb{E}^{-1}\Bigg[\omega^\alpha\mathbb{E}\Bigg[&\frac{\partial^3}{\partial\ell_1^2\partial\vartheta}\underline{\mathbf{f}}(\ell_1,\vartheta;\zeta) - \frac{\partial}{\partial\ell_1}\underline{\mathbf{f}}(\ell_1,\vartheta;\zeta) + \underline{\mathbf{f}}(\ell_1,\vartheta;\zeta)\frac{\partial^3}{\partial\ell_1^3}\underline{\mathbf{f}}(\ell_1,\vartheta;\zeta) \\ &-\underline{\mathbf{f}}(\ell_1,\vartheta;\zeta)\frac{\partial}{\partial\ell_1}\underline{\mathbf{f}}(\ell_1,\vartheta;\zeta) + \frac{\partial}{\partial\ell_1}\underline{\mathbf{f}}(\ell_1,\vartheta;\zeta)\frac{\partial^2}{\partial\ell_1^2}\underline{\mathbf{f}}(\ell_1,\vartheta;\zeta)\Bigg]\Bigg]. \end{aligned} \tag{52}$$

Let us surmise the infinite sum $\underline{\mathbf{f}}(\ell_1,\vartheta;\zeta) = \sum_{q=0}^{\infty} \underline{\mathbf{f}}_q(\ell_1,\vartheta;\zeta)$, $(q = 0,1,2,\ldots)$ accompanying it with (45) and affirming the non-linearity. Therefore, (52) takes the form

$$\begin{aligned}\sum_{q=0}^{\infty} \underline{\mathbf{f}}_q(\ell_1,\vartheta;\zeta) &= (\zeta-1)\exp\left(\frac{\ell_1}{2}\right) + \mathbb{E}^{-1}\left[\omega^{\alpha}\mathbb{E}\left[\sum_{q=0}^{\infty}\left(\frac{\partial^3}{\partial\ell_1^2\partial\vartheta}\underline{\mathbf{f}}(\ell_1,\vartheta;\zeta)\right)_q - \sum_{q=0}^{\infty}\left(\frac{\partial}{\partial\ell_1}\underline{\mathbf{f}}(\ell_1,\vartheta;\zeta)\right)_q \right.\right.\\ &\left.\left. + \sum_{q=0}^{\infty}\underline{\mathcal{A}}_q - \sum_{q=0}^{\infty}\underline{\mathcal{B}}_q + 3\sum_{q=0}^{\infty}\underline{\mathcal{C}}_q\right]\right].\end{aligned} \tag{53}$$

The first few Adomian polynomials are

$$\begin{aligned}\underline{\mathcal{A}}_q\left(\underline{\mathbf{f}}\frac{\partial^3}{\partial\ell_1^3}\underline{\mathbf{f}}\right) &= \begin{cases}\underline{\mathbf{f}}_0\frac{\partial^3}{\partial\ell_1^3}\underline{\mathbf{f}}_0, & q=0,\\ \underline{\mathbf{f}}_0\frac{\partial^3}{\partial\ell_1^3}\underline{\mathbf{f}}_1 + \underline{\mathbf{f}}_1\frac{\partial^3}{\partial\ell_1^3}\underline{\mathbf{f}}_0, & q=1,\\ \underline{\mathbf{f}}_1\frac{\partial^3}{\partial\ell_1^3}\underline{\mathbf{f}}_2 + \underline{\mathbf{f}}_1\frac{\partial^3}{\partial\ell_1^3}\underline{\mathbf{f}}_1 + \underline{\mathbf{f}}_2\frac{\partial^3}{\partial\ell_1^3}\underline{\mathbf{f}}_0, & q=2,\end{cases}\\ \underline{\mathcal{B}}_q\left(\underline{\mathbf{f}}\frac{\partial}{\partial\ell_1}\underline{\mathbf{f}}\right) &= \begin{cases}\underline{\mathbf{f}}_0\frac{\partial}{\partial\ell_1}\underline{\mathbf{f}}_0, & q=0,\\ \underline{\mathbf{f}}_0\frac{\partial}{\partial\ell_1}\underline{\mathbf{f}}_1 + \underline{\mathbf{f}}_1\frac{\partial}{\partial\ell_1}\underline{\mathbf{f}}_0, & q=1,\\ \underline{\mathbf{f}}_1\frac{\partial}{\partial\ell_1}\underline{\mathbf{f}}_2 + \underline{\mathbf{f}}_1\frac{\partial}{\partial\ell_1}\underline{\mathbf{f}}_1 + \underline{\mathbf{f}}_2\frac{\partial}{\partial\ell_1}\underline{\mathbf{f}}_0, & q=2,\end{cases}\\ \underline{\mathcal{C}}_q\left(\frac{\partial}{\partial\ell_1}\underline{\mathbf{f}}\frac{\partial^2}{\partial\ell_1^2}\underline{\mathbf{f}}\right) &= \begin{cases}\frac{\partial}{\partial\ell_1}\underline{\mathbf{f}}_0\frac{\partial^2}{\partial\ell_1^2}\underline{\mathbf{f}}_0, & q=0,\\ \frac{\partial}{\partial\ell_1}\underline{\mathbf{f}}_0\frac{\partial^2}{\partial\ell_1^2}\underline{\mathbf{f}}_1 + \frac{\partial}{\partial\ell_1}\underline{\mathbf{f}}_1\frac{\partial^2}{\partial\ell_1^2}\underline{\mathbf{f}}_0, & q=1,\\ \frac{\partial}{\partial\ell_1}\underline{\mathbf{f}}_1\frac{\partial^2}{\partial\ell_1^2}\underline{\mathbf{f}}_2 + \frac{\partial}{\partial\ell_1}\underline{\mathbf{f}}_1\frac{\partial^2}{\partial\ell_1^2}\underline{\mathbf{f}}_1 + \frac{\partial}{\partial\ell_1}\underline{\mathbf{f}}_2\frac{\partial^2}{\partial\ell_1^2}\underline{\mathbf{f}}_0, & q=2.\end{cases}\end{aligned} \tag{54}$$

then (53) simplifies to

$$\begin{aligned}\underline{\mathbf{f}}_0(\ell_1,\vartheta;\zeta) &= (\zeta-1)\exp\left(\frac{\ell_1}{2}\right),\\ \underline{\mathbf{f}}_1(\ell_1,\vartheta;\zeta) &= \mathbb{E}^{-1}\left[\omega^{\alpha}\mathbb{E}\left[\left(\frac{\partial^3}{\partial\ell_1^2\partial\vartheta}\underline{\mathbf{f}}(\ell_1,\vartheta;\zeta)\right)_0 - \left(\frac{\partial}{\partial\ell_1}\underline{\mathbf{f}}(\ell_1,\vartheta;\zeta)\right)_0 + \underline{\mathcal{A}}_0 - \underline{\mathcal{B}}_0 + 3\underline{\mathcal{C}}_0\right]\right]\\ &= -\frac{\zeta-1}{2}\exp\left(\frac{\ell_1}{2}\right)\frac{\vartheta^{\alpha}}{\Gamma(\alpha+1)},\\ \underline{\mathbf{f}}_2(\ell_1,\vartheta;\zeta) &= \mathbb{E}^{-1}\left[\omega^{\alpha}\mathbb{E}\left[\left(\frac{\partial^3}{\partial\ell_1^2\partial\vartheta}\underline{\mathbf{f}}(\ell_1,\vartheta;\zeta)\right)_1 - \left(\frac{\partial}{\partial\ell_1}\underline{\mathbf{f}}(\ell_1,\vartheta;\zeta)\right)_1 + \underline{\mathcal{A}}_1 - \underline{\mathcal{B}}_1 + 3\underline{\mathcal{C}}_1\right]\right]\\ &= \frac{-(\zeta-1)}{8}\exp\left(\frac{\ell_1}{2}\right)\frac{\vartheta^{2\alpha-1}}{\Gamma(2\alpha)} + \frac{(\zeta-1)}{4}\exp\left(\frac{\ell_1}{2}\right)\frac{\vartheta^{2\alpha}}{\Gamma(2\alpha+1)},\\ \underline{\mathbf{f}}_3(\ell_1,\vartheta;\zeta) &= \mathbb{E}^{-1}\left[\omega^{\alpha}\mathbb{E}\left[\left(\frac{\partial^3}{\partial\ell_1^2\partial\vartheta}\underline{\mathbf{f}}(\ell_1,\vartheta;\zeta)\right)_2 - \left(\frac{\partial}{\partial\ell_1}\underline{\mathbf{f}}(\ell_1,\vartheta;\zeta)\right)_2 + \underline{\mathcal{A}}_2 - \underline{\mathcal{B}}_2 + 3\underline{\mathcal{C}}_2\right]\right]\\ &= \frac{-(\zeta-1)}{32}\exp\left(\frac{\ell_1}{2}\right)\frac{\vartheta^{3\alpha-2}}{\Gamma(3\alpha-1)} + \frac{(\zeta-1)}{8}\exp\left(\frac{\ell_1}{2}\right)\frac{\vartheta^{3\alpha-1}}{\Gamma(3\alpha)} - \frac{(\zeta-1)}{8}\exp\left(\frac{\ell_1}{2}\right)\frac{\vartheta^{3\alpha}}{\Gamma(3\alpha+1)},\\ &\vdots\end{aligned}$$

By implementing a similar technique, the remaining terms of $\underline{\mathbf{f}}_q$ $(q \geq 4)$ of the EADM solution can be simply determined. Furthermore, when the iterative process expands, the accuracy of the obtained solution improves dramatically, and the deduced solution moves closer to the precise result. Finally, we have come up with the following answers in a series form

$$\underline{\mathbf{f}}(\ell_1,\vartheta,\zeta) = \underline{\mathbf{f}}_0(\ell_1,\vartheta,\zeta) + \underline{\mathbf{f}}_1(\ell_1,\vartheta,\zeta) + \underline{\mathbf{f}}_1(\ell_1,\vartheta,\zeta) + \ldots,$$

such that

$$\begin{aligned}\underline{\mathbf{f}}(\ell_1,\vartheta,\zeta) &= \underline{\mathbf{f}}_0(\ell_1,\vartheta,\zeta) + \underline{\mathbf{f}}_1(\ell_1,\vartheta,\zeta) + \underline{\mathbf{f}}_1(\ell_1,\vartheta,\zeta) + \ldots,\\ \bar{\mathbf{f}}(\ell_1,\vartheta,\zeta) &= \bar{\mathbf{f}}_0(\ell_1,\vartheta,\zeta) + \bar{\mathbf{f}}_1(\ell_1,\vartheta,\zeta) + \bar{\mathbf{f}}_1(\ell_1,\vartheta,\zeta) + \ldots.\end{aligned}$$

Consequently, we have

$$\begin{aligned}
\underline{\mathbf{f}}(\ell_1,\vartheta,\zeta) &= (\zeta-1)\exp\left(\frac{\ell_1}{2}\right) - \frac{\zeta-1}{2}\exp\left(\frac{\ell_1}{2}\right)\frac{\vartheta^{\alpha}}{\Gamma(\alpha+1)} \\
&- \frac{(\zeta-1)}{8}\exp\left(\frac{\ell_1}{2}\right)\frac{\vartheta^{2\alpha-1}}{\Gamma(2\alpha)} + \frac{(\zeta-1)}{4}\exp\left(\frac{\ell_1}{2}\right)\frac{\vartheta^{2\alpha}}{\Gamma(2\alpha+1)} \\
&- \frac{(\zeta-1)}{32}\exp\left(\frac{\ell_1}{2}\right)\frac{\vartheta^{3\alpha-2}}{\Gamma(3\alpha-1)} + \frac{(\zeta-1)}{8}\exp\left(\frac{\ell_1}{2}\right)\frac{\vartheta^{3\alpha-1}}{\Gamma(3\alpha)} - \frac{(\zeta-1)}{8}\exp\left(\frac{\ell_1}{2}\right)\frac{\vartheta^{3\alpha}}{\Gamma(3\alpha+1)}\cdots, \\
\overline{\mathbf{f}}(\ell_1,\vartheta,\zeta) &= (1-\zeta)\exp\left(\frac{\ell_1}{2}\right) - \frac{1-\zeta}{2}\exp\left(\frac{\ell_1}{2}\right)\frac{\vartheta^{\alpha}}{\Gamma(\alpha+1)} \\
&- \frac{(1-\zeta)}{8}\exp\left(\frac{\ell_1}{2}\right)\frac{\vartheta^{2\alpha-1}}{\Gamma(2\alpha)} + \frac{(1-\zeta)}{4}\exp\left(\frac{\ell_1}{2}\right)\frac{\vartheta^{2\alpha}}{\Gamma(2\alpha+1)} \\
&- \frac{(1-\zeta)}{32}\exp\left(\frac{\ell_1}{2}\right)\frac{\vartheta^{3\alpha-2}}{\Gamma(3\alpha-1)} + \frac{(1-\zeta)}{8}\exp\left(\frac{\ell_1}{2}\right)\frac{\vartheta^{3\alpha-1}}{\Gamma(3\alpha)} - \frac{(1-\zeta)}{8}\exp\left(\frac{\ell_1}{2}\right)\frac{\vartheta^{3\alpha}}{\Gamma(3\alpha+1)}\cdots. \qquad (55)
\end{aligned}$$

Case II. (For the fuzzy Atangana–Baleanu Caputo fractional derivative)

Here, we obtain the EADM solution for the first case of (51) by the using fuzzy ABC fractional derivative operator.

Taking into consideration the procedure described in Section 3, we have

$$\mathbb{E}\left[\underline{\mathbf{f}}(\ell_1,\vartheta;\zeta)\right] = \omega^2\underline{\mathbf{f}}_{(\kappa)}(\ell_1;\zeta) + \left(\frac{\alpha\omega^{\alpha}+1-\alpha}{\mathcal{N}(\alpha)}\right)\mathbb{E}\left[\frac{\partial^3}{\partial\ell_1^2\partial\vartheta}\underline{\mathbf{f}}(\ell_1,\vartheta;\zeta) - \frac{\partial}{\partial\ell_1}\underline{\mathbf{f}}(\ell_1,\vartheta;\zeta)\right.$$
$$\left. + \underline{\mathbf{f}}(\ell_1,\vartheta;\zeta)\frac{\partial^3}{\partial\ell_1^3}\underline{\mathbf{f}}(\ell_1,\vartheta;\zeta) - \underline{\mathbf{f}}(\ell_1,\vartheta;\zeta)\frac{\partial}{\partial\ell_1}\underline{\mathbf{f}}(\ell_1,\vartheta;\zeta) + \frac{\partial}{\partial\ell_1}\underline{\mathbf{f}}(\ell_1,\vartheta;\zeta)\frac{\partial^2}{\partial\ell_1^2}\underline{\mathbf{f}}(\ell_1,\vartheta;\zeta)\right].$$

Simple computations result in

$$\begin{aligned}
\underline{\mathbf{f}}(\ell_1,\vartheta;\zeta) &= (\zeta-1)\exp\left(\frac{\ell_1}{2}\right) + \mathbb{E}^{-1}\left[\left(\frac{\alpha\omega^{\alpha}+1-\alpha}{\mathcal{N}(\alpha)}\right)\mathbb{E}\left[\frac{\partial^3}{\partial\ell_1^2\partial\vartheta}\underline{\mathbf{f}}(\ell_1,\vartheta;\zeta) - \frac{\partial}{\partial\ell_1}\underline{\mathbf{f}}(\ell_1,\vartheta;\zeta)\right.\right. \\
&\left.\left. + \underline{\mathbf{f}}(\ell_1,\vartheta;\zeta)\frac{\partial^3}{\partial\ell_1^3}\underline{\mathbf{f}}(\ell_1,\vartheta;\zeta) - \underline{\mathbf{f}}(\ell_1,\vartheta;\zeta)\frac{\partial}{\partial\ell_1}\underline{\mathbf{f}}(\ell_1,\vartheta;\zeta) + \frac{\partial}{\partial\ell_1}\underline{\mathbf{f}}(\ell_1,\vartheta;\zeta)\frac{\partial^2}{\partial\ell_1^2}\underline{\mathbf{f}}(\ell_1,\vartheta;\zeta)\right]\right]. \qquad (56)
\end{aligned}$$

Let us surmise the infinite sum $\underline{\mathbf{f}}(\ell_1,\vartheta;\zeta) = \sum_{q=0}^{\infty}\underline{\mathbf{f}}_q(\ell_1,\vartheta;\zeta)$, $(q=0,1,2,...)$ accompanying it with (45) and affirming the non-linearity. Therefore, (52) takes the form

$$\begin{aligned}
\sum_{q=0}^{\infty}\underline{\mathbf{f}}_q(\ell_1,\vartheta;\zeta) &= (\zeta-1)\exp\left(\frac{\ell_1}{2}\right) \\
&+ \mathbb{E}^{-1}\left[\left(\frac{\alpha\omega^{\alpha}+1-\alpha}{\mathcal{N}(\alpha)}\right)\mathbb{E}\left[\sum_{q=0}^{\infty}\left(\frac{\partial^3}{\partial\ell_1^2\partial\vartheta}\underline{\mathbf{f}}(\ell_1,\vartheta;\zeta)\right)_q - \sum_{q=0}^{\infty}\left(\frac{\partial}{\partial\ell_1}\underline{\mathbf{f}}(\ell_1,\vartheta;\zeta)\right)_q\right.\right. \\
&\left.\left. + \sum_{q=0}^{\infty}\mathcal{A}_q - \sum_{q=0}^{\infty}\mathcal{B}_q + 3\sum_{q=0}^{\infty}\mathcal{C}_q\right]\right]. \qquad (57)
\end{aligned}$$

Utilizing the Adomian polynomials described in (54), then (57) simplifies to

$$\begin{aligned}
\underline{\mathbf{f}}_0(\ell_1,\vartheta;\zeta) &= (\zeta-1)\exp\left(\frac{\ell_1}{2}\right),\\
\underline{\mathbf{f}}_1(\ell_1,\vartheta;\zeta) &= \mathbb{E}^{-1}\left[\left(\frac{\alpha\omega^\alpha+1-\alpha}{\mathcal{N}(\alpha)}\right)\mathbb{E}\left[\left(\frac{\partial^3}{\partial\ell_1^2\partial\vartheta}\underline{\mathbf{f}}(\ell_1,\vartheta;\zeta)\right)_0-\left(\frac{\partial}{\partial\ell_1}\underline{\mathbf{f}}(\ell_1,\vartheta;\zeta)\right)_0+\underline{\mathcal{A}}_0-\underline{\mathcal{B}}_0+3\underline{\mathcal{C}}_0\right]\right]\\
&= -\frac{\zeta-1}{2\mathcal{N}(\alpha)}\exp\left(\frac{\ell_1}{2}\right)\left[\frac{\alpha\vartheta^\alpha}{\Gamma(\alpha+1)}+(1-\alpha)\right],\\
\underline{\mathbf{f}}_2(\ell_1,\vartheta;\zeta) &= \mathbb{E}^{-1}\left[\left(\frac{\alpha\omega^\alpha+1-\alpha}{\mathcal{N}(\alpha)}\right)\mathbb{E}\left[\left(\frac{\partial^3}{\partial\ell_1^2\partial\vartheta}\underline{\mathbf{f}}(\ell_1,\vartheta;\zeta)\right)_1-\left(\frac{\partial}{\partial\ell_1}\underline{\mathbf{f}}(\ell_1,\vartheta;\zeta)\right)_1+\underline{\mathcal{A}}_1-\underline{\mathcal{B}}_1+3\underline{\mathcal{C}}_1\right]\right]\\
&= \frac{-(\zeta-1)}{8\mathcal{N}^2(\alpha)}\exp\left(\frac{\ell_1}{2}\right)\left[\frac{\alpha^2\vartheta^{2\alpha-1}}{\Gamma(2\alpha)}+\alpha(\alpha-1)\frac{\vartheta^{\alpha-1}}{\Gamma(\alpha)}+\alpha\frac{\vartheta^\alpha}{\Gamma(\alpha+1)}+(1-\alpha)\right]\\
&\quad+\frac{(\zeta-1)}{4\mathcal{N}^2(\alpha)}\exp\left(\frac{\ell_1}{2}\right)\left[\frac{\alpha^2\vartheta^{2\alpha}}{\Gamma(2\alpha+1)}+2\alpha(1-\alpha)\frac{\vartheta^\alpha}{\Gamma(\alpha+1)}+(1-\alpha)^2\right],\\
\underline{\mathbf{f}}_3(\ell_1,\vartheta;\zeta) &= \mathbb{E}^{-1}\left[\left(\frac{\alpha\omega^\alpha+1-\alpha}{\mathcal{N}(\alpha)}\right)\mathbb{E}\left[\left(\frac{\partial^3}{\partial\ell_1^2\partial\vartheta}\underline{\mathbf{f}}(\ell_1,\vartheta;\zeta)\right)_2-\left(\frac{\partial}{\partial\ell_1}\underline{\mathbf{f}}(\ell_1,\vartheta;\zeta)\right)_2+\underline{\mathcal{A}}_2-\underline{\mathcal{B}}_2+3\underline{\mathcal{C}}_2\right]\right]\\
&= \frac{-(\zeta-1)}{32\mathcal{N}^3(\alpha)}\exp\left(\frac{\ell_1}{2}\right)\begin{cases}\frac{4\alpha^3\vartheta^{3\alpha}}{\Gamma(3\alpha+1)}-\frac{2\alpha^3\vartheta^{3\alpha-1}}{\Gamma(3\alpha)}-\frac{\alpha^3\vartheta^{3\alpha-2}}{\Gamma(3\alpha-1)}+2\alpha^2(5-2\alpha)\frac{\vartheta^{2\alpha}}{\Gamma(2\alpha+1)}\\ -2\alpha^2(1-\alpha)\frac{\vartheta^{2\alpha-1}}{\Gamma(2\alpha)}-\alpha^2(1-\alpha)\frac{\vartheta^{2\alpha-2}}{\Gamma(2\alpha-1)}+\alpha(1-\alpha)(7-6\alpha)\frac{\vartheta^{\alpha-1}}{\Gamma(\alpha)}\\ +2(1-\alpha)(1-2\alpha),\end{cases}\\
&\ \ \vdots
\end{aligned}$$

By implementing a similar technique, the remaining terms of $\underline{\mathbf{f}}_q$ $(q\geq 3)$ of the EADM solution can be simply determined. Furthermore, when the iterative process expands, the accuracy of the obtained solution improves dramatically, and the deduced solution moves closer to the precise result. Finally, we have come up with the following answers in a series form

$$\underline{\mathbf{f}}(\ell_1,\vartheta,\zeta)=\underline{\mathbf{f}}_0(\ell_1,\vartheta,\zeta)+\underline{\mathbf{f}}_1(\ell_1,\vartheta,\zeta)+\underline{\mathbf{f}}_1(\ell_1,\vartheta,\zeta)+\ldots,$$

such that

$$\begin{aligned}
\underline{\mathbf{f}}(\ell_1,\vartheta,\zeta) &= \underline{\mathbf{f}}_0(\ell_1,\vartheta,\zeta)+\underline{\mathbf{f}}_1(\ell_1,\vartheta,\zeta)+\underline{\mathbf{f}}_1(\ell_1,\vartheta,\zeta)+\ldots,\\
\bar{\mathbf{f}}(\ell_1,\vartheta,\zeta) &= \bar{\mathbf{f}}_0(\ell_1,\vartheta,\zeta)+\bar{\mathbf{f}}_1(\ell_1,\vartheta,\zeta)+\bar{\mathbf{f}}_1(\ell_1,\vartheta,\zeta)+\ldots.
\end{aligned}$$

Consequently, we have

$$\begin{aligned}
\underline{\mathbf{f}}(\ell_1,\vartheta,\zeta) &= (\zeta-1)\exp\left(\frac{\ell_1}{2}\right)-\frac{\zeta-1}{2\mathcal{N}(\alpha)}\exp\left(\frac{\ell_1}{2}\right)\left[\frac{\alpha\vartheta^\alpha}{\Gamma(\alpha+1)}+(1-\alpha)\right]\\
&\quad-\frac{(\zeta-1)}{8\mathcal{N}^2(\alpha)}\exp\left(\frac{\ell_1}{2}\right)\left[\frac{\alpha^2\vartheta^{2\alpha-1}}{\Gamma(2\alpha)}+\alpha(\alpha-1)\frac{\vartheta^{\alpha-1}}{\Gamma(\alpha)}+\alpha\frac{\vartheta^\alpha}{\Gamma(\alpha+1)}+(1-\alpha)\right]\\
&\quad+\frac{(\zeta-1)}{4\mathcal{N}^2(\alpha)}\exp\left(\frac{\ell_1}{2}\right)\left[\frac{\alpha^2\vartheta^{2\alpha}}{\Gamma(2\alpha+1)}+2\alpha(1-\alpha)\frac{\vartheta^\alpha}{\Gamma(\alpha+1)}+(1-\alpha)^2\right]\\
&\quad-\frac{(\zeta-1)}{32\mathcal{N}^3(\alpha)}\exp\left(\frac{\ell_1}{2}\right)\times\begin{cases}\frac{4\alpha^3\vartheta^{3\alpha}}{\Gamma(3\alpha+1)}-\frac{2\alpha^3\vartheta^{3\alpha-1}}{\Gamma(3\alpha)}-\frac{\alpha^3\vartheta^{3\alpha-2}}{\Gamma(3\alpha-1)}+2\alpha^2(5-2\alpha)\frac{\vartheta^{2\alpha}}{\Gamma(2\alpha+1)}\\ -2\alpha^2(1-\alpha)\frac{\vartheta^{2\alpha-1}}{\Gamma(2\alpha)}-\alpha^2(1-\alpha)\frac{\vartheta^{2\alpha-2}}{\Gamma(2\alpha-1)}+\alpha(1-\alpha)(7-6\alpha)\frac{\vartheta^{\alpha-1}}{\Gamma(\alpha)}\\ +2(1-\alpha)(1-2\alpha)\end{cases}+\ldots,\\
\bar{\mathbf{f}}(\ell_1,\vartheta,\zeta) &= (1-\zeta)\exp\left(\frac{\ell_1}{2}\right)-\frac{1-\zeta}{2\mathcal{N}(\alpha)}\exp\left(\frac{\ell_1}{2}\right)\left[\frac{\alpha\vartheta^\alpha}{\Gamma(\alpha+1)}+(1-\alpha)\right]\\
&\quad-\frac{(1-\zeta)}{8\mathcal{N}^2(\alpha)}\exp\left(\frac{\ell_1}{2}\right)\left[\frac{\alpha^2\vartheta^{2\alpha-1}}{\Gamma(2\alpha)}+\alpha(\alpha-1)\frac{\vartheta^{\alpha-1}}{\Gamma(\alpha)}+\alpha\frac{\vartheta^\alpha}{\Gamma(\alpha+1)}+(1-\alpha)\right]\\
&\quad+\frac{(1-\zeta)}{4\mathcal{N}^2(\alpha)}\exp\left(\frac{\ell_1}{2}\right)\left[\frac{\alpha^2\vartheta^{2\alpha}}{\Gamma(2\alpha+1)}+2\alpha(1-\alpha)\frac{\vartheta^\alpha}{\Gamma(\alpha+1)}+(1-\alpha)^2\right]\\
&\quad-\frac{(1-\zeta)}{32\mathcal{N}^3(\alpha)}\exp\left(\frac{\ell_1}{2}\right)\times\begin{cases}\frac{4\alpha^3\vartheta^{3\alpha}}{\Gamma(3\alpha+1)}-\frac{2\alpha^3\vartheta^{3\alpha-1}}{\Gamma(3\alpha)}-\frac{\alpha^3\vartheta^{3\alpha-2}}{\Gamma(3\alpha-1)}+2\alpha^2(5-2\alpha)\frac{\vartheta^{2\alpha}}{\Gamma(2\alpha+1)}\\ -2\alpha^2(1-\alpha)\frac{\vartheta^{2\alpha-1}}{\Gamma(2\alpha)}-\alpha^2(1-\alpha)\frac{\vartheta^{2\alpha-2}}{\Gamma(2\alpha-1)}+\alpha(1-\alpha)(7-6\alpha)\frac{\vartheta^{\alpha-1}}{\Gamma(\alpha)}\\ +2(1-\alpha)(1-2\alpha)\end{cases}+\ldots. \qquad (58)
\end{aligned}$$

In this analysis, Figure 1 demonstrates the insight into the influence of multiple layer surface plots for Problem 1 correlated with the CFD and Elzaki transform in the fuzzy sense. It is worth mentioning that the profile identifies the variation in the mapping $\mathbf{f}(\ell_1, \vartheta; \zeta)$ on space co-ordinate ℓ_1 with respect to ϑ and uncertainty parameter $\zeta \in [0, 1]$.

The graph illustrates that as, time progresses, the mapping $\mathbf{f}(\ell_1, \vartheta; \zeta)$ will also increase.

- The effect of the proposed methodology on the mapping $\mathbf{f}(\ell_1, \vartheta; \zeta)$ is displayed in Figure 2a for the varying fractional orders $\alpha = 1, 0.85, 0.75, 0.55$ by considering CFD operator. It exhibits a relatively small increase in the mapping $\underline{\mathbf{f}}(\ell_1, \vartheta; \zeta)$ with the decrease in $\bar{\mathbf{f}}(\ell_1, \vartheta; \zeta)$.
- The profile graph of Figure 2b demonstrates the lower and upper solution of varying uncertainty when the fractional order is assumed to be $\alpha = 0.2$ by proposing CFD operator. It emphasizes a relatively small variation in the mapping $\underline{\mathbf{f}}(\ell_1, \vartheta; \zeta)$ with the increase in $\bar{\mathbf{f}}(\ell_1, \vartheta; \zeta)$.
- The effect of the proposed methodology on the mapping $\mathbf{f}(\ell_1, \vartheta; \zeta)$ is displayed in Figure 3a for the varying fractional orders $\alpha = 1, 0.85, 0.75, 0.55$ by considering the ABC fractional derivative operator. It exhibits a relatively small increase in the mapping $\underline{\mathbf{f}}(\ell_1, \vartheta; \zeta)$ with the decrease in $\bar{\mathbf{f}}(\ell_1, \vartheta; \zeta)$.
- The Profile graph of Figure 3b demonstrates the lower and upper solutions of varying uncertainty when the fractional order is assumed to be $\alpha = 0.2$ by proposing the ABC fractional derivative operator. It emphasizes a relatively small variation in the mapping $\underline{\mathbf{f}}(\ell_1, \vartheta; \zeta)$ with the increase in $\bar{\mathbf{f}}(\ell_1, \vartheta; \zeta)$.
- Figure 4 demonstrates the comparison analysis between the CFD operator and the ABC fractional derivative operator for varying fractional order with uncertainty $\kappa \in [0, 1]$, exhibits that lower the solution profile for the ABC fractional operator has strong ties with the upper solution as compared to the CFD operator.
- Figure 5 shows the comparison analysis between ($\underline{\mathbf{f}}(\ell_1, \vartheta; \zeta)$ and the exact solution), ($\bar{\mathbf{f}}(\ell_1, \vartheta; \zeta)$ and exact solution), respectively, for three dimensional error plots by considering the CFD operator.

Furthermore, the offered approach does not provide a unique solution but will aid scientists in selecting the best approximate solution. It is remarkable that the fuzzy ABC fractional derivative operator has better performance than the CFD operators, because the curves have a strong harmony with the integer-order graph in the ABC operator case.

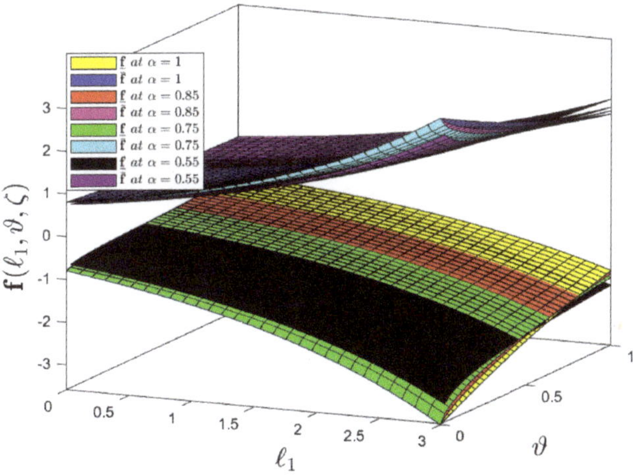

Figure 1. A three-dimensional surface plot indicates the lower and upper solution $\mathbf{f}(\ell_1, \vartheta, \zeta)$ taking into consideration the fuzzy Caputo fractional derivative operator for Problem 1 when $\alpha = 1, 0.85, 0.75, 0.55$ with uncertainty $\zeta \in [0, 1]$.

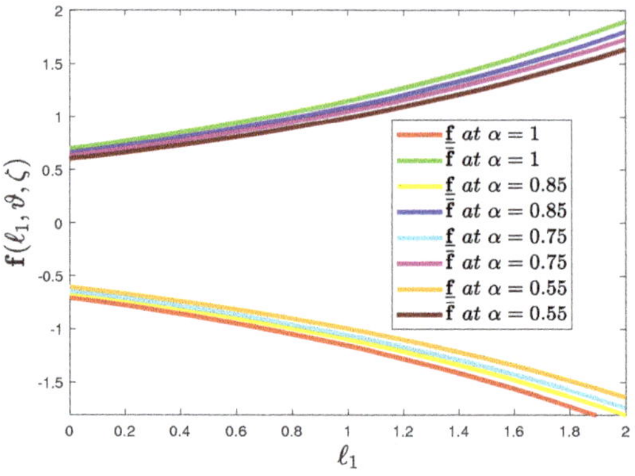

(**a**)

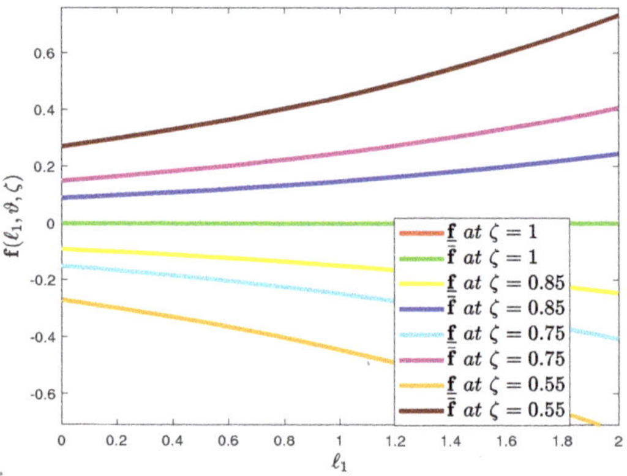

(**b**)

Figure 2. (**a**) A two-dimensional plot indicates the lower and upper solution $\mathbf{f}(\ell_1, \vartheta, \zeta)$ taking into consideration the fuzzy Caputo fractional derivative operator for Problem 1 when $\alpha = 1, 0.85, 0.75, 0.55$ with uncertainty $\zeta \in [0, 1]$. (**b**) A two-dimensional plot indicates the lower and upper solution $\mathbf{f}(\ell_1, \vartheta, \zeta)$ taking into consideration the fuzzy Caputo fractional derivative operator for Problem 1 when $\zeta = 1, 0.85, 0.75, 0.55$ with the fractional order $\alpha = 0.2$.

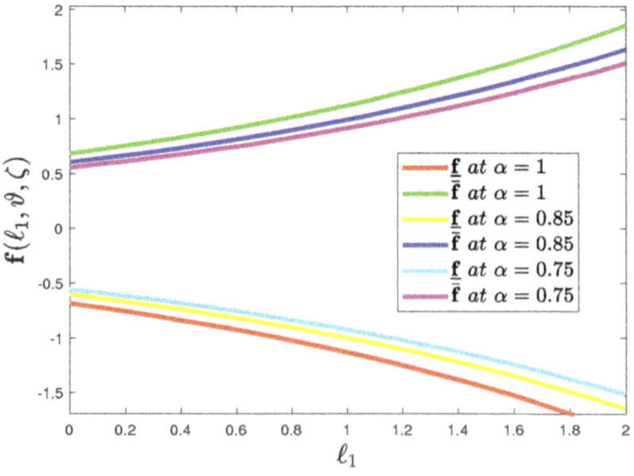

(**a**)

Figure 3. *Cont.*

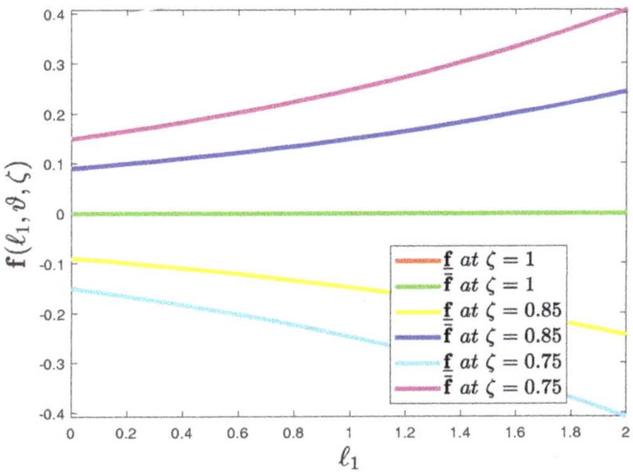

(**b**)

Figure 3. (**a**) A two-dimensional plot indicates the lower and upper solution $\mathbf{f}(\ell_1, \vartheta, \zeta)$ taking into consideration the ABC fractional derivative operator for Problem 1 when $\alpha = 1, 0.85, 0.75, 0.55$ with uncertainty $\zeta \in [0,1]$. (**b**) A two-dimensional plot indicates the lower and upper solution $\mathbf{f}(\ell_1, \vartheta, \zeta)$ taking into consideration the ABC fractional derivative operator for Problem 1 when $\zeta = 1, 0.85, 0.75, 0.55$ with fractional order $\alpha = 0.2$.

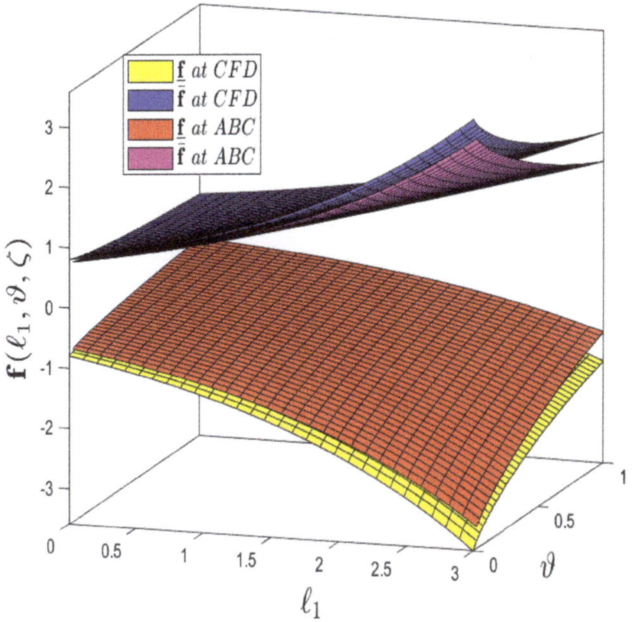

Figure 4. A comparison three-dimensional plot indicates the lower and upper solution $\mathbf{f}(\ell_1, \vartheta, \zeta)$ taking into consideration the fuzzy Caputo and fuzzy ABC fractional derivative operators for Problem 1 when $\zeta = 0.2$ with fractional order $\alpha = 0.85$.

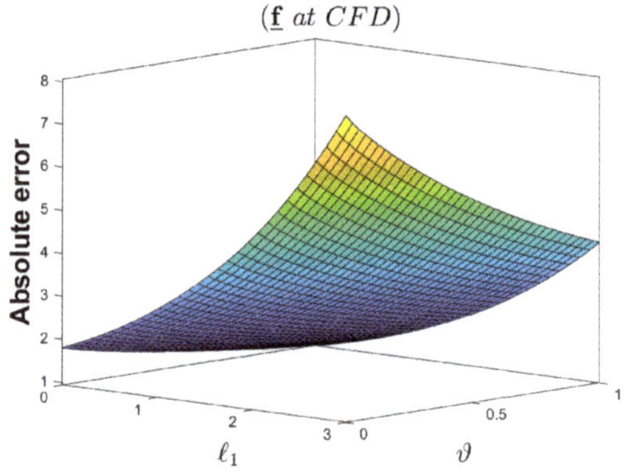

(**a**)

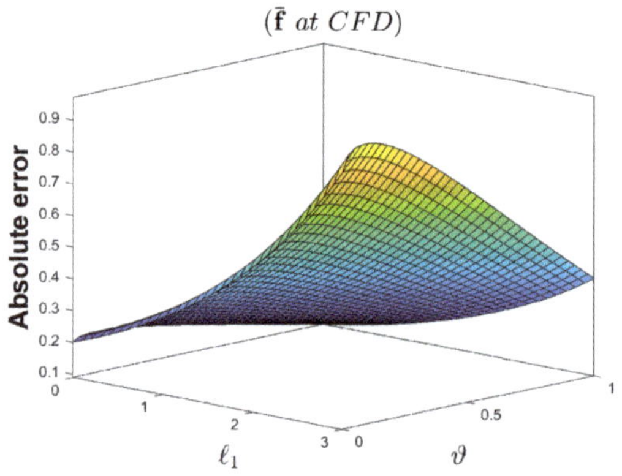

(**b**)

Figure 5. (**a**) A three-dimensional absolute error plot indicates the lower and exact solution (Remark 2) of $\mathbf{f}(\ell_1, \vartheta, \zeta)$ taking into consideration the fuzzy Caputo fractional derivative operator for Problem 1 when $\alpha = 0.85$ with uncertainty $\zeta \in [0, 1]$. (**b**)A three-dimensional absolute error plot indicates the upper and exact solution (Remark 2) of $\mathbf{f}(\ell_1, \vartheta, \zeta)$ taking into consideration the fuzzy Caputo fractional derivative operator for Problem 1 when $\alpha = 0.85$ with uncertainty $\zeta \in [0, 1]$.

□

Remark 2. *When $\underline{\chi}(\zeta) = \bar{\chi}(\zeta) = 1$ along with $\alpha = 1$, then both solutions of Problem 1 converge to the integer-order solution $\mathbf{f}(\ell_1, \vartheta) = \exp\left(\frac{\ell_1}{2} - \frac{2\vartheta}{3}\right)$.*

Problem 2. *Assume the one-dimension fuzzy fractional FW model with fuzzy ICs is represented as follows:*

$$\begin{aligned} \frac{\partial^\alpha}{\partial \vartheta^\alpha}\tilde{\mathbf{f}}(\ell_1, \vartheta; \zeta) &= \frac{\partial^3}{\partial \ell_1^2 \partial \vartheta}\tilde{\mathbf{f}}(\ell_1, \vartheta; \zeta) \ominus \frac{\partial}{\partial \ell_1}\tilde{\mathbf{f}}(\ell_1, \vartheta; \zeta) \oplus \tilde{\mathbf{f}}(\ell_1, \vartheta; \zeta) \odot \frac{\partial^3}{\partial \ell_1^3}\tilde{\mathbf{f}}(\ell_1, \vartheta; \zeta) \\ &\ominus \tilde{\mathbf{f}}(\ell_1, \vartheta; \zeta) \odot \frac{\partial}{\partial \ell_1}\tilde{\mathbf{f}}(\ell_1, \vartheta; \zeta) \oplus \frac{\partial}{\partial \ell_1}\tilde{\mathbf{f}}(\ell_1, \vartheta; \zeta) \odot \frac{\partial^2}{\partial \ell_1^2}\tilde{\mathbf{f}}(\ell_1, \vartheta; \zeta), \\ \tilde{\mathbf{f}}(\ell_1, 0) &= \tilde{\chi}(\zeta) \odot \cosh^2\left(\frac{\ell_1}{4}\right), \end{aligned} \tag{59}$$

where $\tilde{\chi}(\zeta) = [\underline{\chi}(\zeta), \bar{\chi}(\zeta)] = [\zeta - 1, 1 - \zeta]$ for $\zeta \in [0, 1]$ is a fuzzy number.

Proof. The parameterized version of the problem (50) is expressed as follows

$$\begin{cases} \frac{\partial^\alpha}{\partial\vartheta^\alpha}\underline{\mathbf{f}}(\ell_1,\vartheta;\zeta) = \frac{\partial^3}{\partial\ell_1^2\partial\vartheta}\underline{\mathbf{f}}(\ell_1,\vartheta;\zeta) - \frac{\partial}{\partial\ell_1}\underline{\mathbf{f}}(\ell_1,\vartheta;\zeta) + \underline{\mathbf{f}}(\ell_1,\vartheta;\zeta)\frac{\partial^3}{\partial\ell_1^3}\underline{\mathbf{f}}(\ell_1,\vartheta;\zeta) \\ \quad -\underline{\mathbf{f}}(\ell_1,\vartheta;\zeta)\frac{\partial}{\partial\ell_1}\underline{\mathbf{f}}(\ell_1,\vartheta;\zeta) + \frac{\partial}{\partial\ell_1}\underline{\mathbf{f}}(\ell_1,\vartheta;\zeta)\frac{\partial^2}{\partial\ell_1^2}\underline{\mathbf{f}}(\ell_1,\vartheta;\zeta), \\ \underline{\mathbf{f}}(\ell_1,0) = \underline{\chi}(\zeta)\cosh^2\left(\frac{\ell_1}{4}\right), \\ \frac{\partial^\alpha}{\partial\vartheta^\alpha}\overline{\mathbf{f}}(\ell_1,\vartheta;\zeta) = \frac{\partial^3}{\partial\ell_1^2\partial\vartheta}\overline{\mathbf{f}}(\ell_1,\vartheta;\zeta) - \frac{\partial}{\partial\ell_1}\overline{\mathbf{f}}(\ell_1,\vartheta;\zeta) + \overline{\mathbf{f}}(\ell_1,\vartheta;\zeta)\frac{\partial^3}{\partial\ell_1^3}\overline{\mathbf{f}}(\ell_1,\vartheta;\zeta) \\ \quad -\overline{\mathbf{f}}(\ell_1,\vartheta;\zeta)\frac{\partial}{\partial\ell_1}\overline{\mathbf{f}}(\ell_1,\vartheta;\zeta) + \frac{\partial}{\partial\ell_1}\overline{\mathbf{f}}(\ell_1,\vartheta;\zeta)\frac{\partial^2}{\partial\ell_1^2}\overline{\mathbf{f}}(\ell_1,\vartheta;\zeta), \\ \overline{\mathbf{f}}(\ell_1,0) = \overline{\chi}(\zeta)\cosh^2\left(\frac{\ell_1}{4}\right). \end{cases} \tag{60}$$

Case I. (For the fuzzy Caputo fractional derivative)

Here, we obtain the EADM solution for the first case of (51) by using the fuzzy CFD operator. Taking into consideration the procedure described in Section 3, we have

$$\begin{aligned} &\frac{1}{\omega^\alpha}\mathbb{E}\left[\underline{\mathbf{f}}(\ell_1,\vartheta;\zeta)\right] - \sum_{\kappa=0}^{q-1}\underline{\mathbf{f}}_{(\kappa)}(\ell_1;\zeta)\omega^{2-\alpha+\kappa} \\ &= \mathbb{E}\left[\frac{\partial^3}{\partial\ell_1^2\partial\vartheta}\underline{\mathbf{f}}(\ell_1,\vartheta;\zeta) - \frac{\partial}{\partial\ell_1}\underline{\mathbf{f}}(\ell_1,\vartheta;\zeta) + \underline{\mathbf{f}}(\ell_1,\vartheta;\zeta)\frac{\partial^3}{\partial\ell_1^3}\underline{\mathbf{f}}(\ell_1,\vartheta;\zeta) \right. \\ &\quad \left. -\underline{\mathbf{f}}(\ell_1,\vartheta;\zeta)\frac{\partial}{\partial\ell_1}\underline{\mathbf{f}}(\ell_1,\vartheta;\zeta) + \frac{\partial}{\partial\ell_1}\underline{\mathbf{f}}(\ell_1,\vartheta;\zeta)\frac{\partial^2}{\partial\ell_1^2}\underline{\mathbf{f}}(\ell_1,\vartheta;\zeta)\right]. \end{aligned}$$

Simple computations result in

$$\begin{aligned} \underline{\mathbf{f}}(\ell_1,\vartheta;\zeta) = (\zeta-1)\cosh^2\left(\frac{\ell_1}{4}\right) + \mathbb{E}^{-1}\Bigg[\omega^\alpha\mathbb{E}\Bigg[&\frac{\partial^3}{\partial\ell_1^2\partial\vartheta}\underline{\mathbf{f}}(\ell_1,\vartheta;\zeta) - \frac{\partial}{\partial\ell_1}\underline{\mathbf{f}}(\ell_1,\vartheta;\zeta) + \underline{\mathbf{f}}(\ell_1,\vartheta;\zeta)\frac{\partial^3}{\partial\ell_1^3}\underline{\mathbf{f}}(\ell_1,\vartheta;\zeta) \\ &-\underline{\mathbf{f}}(\ell_1,\vartheta;\zeta)\frac{\partial}{\partial\ell_1}\underline{\mathbf{f}}(\ell_1,\vartheta;\zeta) + \frac{\partial}{\partial\ell_1}\underline{\mathbf{f}}(\ell_1,\vartheta;\zeta)\frac{\partial^2}{\partial\ell_1^2}\underline{\mathbf{f}}(\ell_1,\vartheta;\zeta)\Bigg]\Bigg]. \end{aligned} \tag{61}$$

Let us surmise the infinite sum $\underline{\mathbf{f}}(\ell_1,\vartheta;\zeta) = \sum_{q=0}^{\infty}\underline{\mathbf{f}}_q(\ell_1,\vartheta;\zeta)$ accompanying it with (45) and affirming the non-linearity. Therefore, (63) takes the form

$$\begin{aligned} \sum_{q=0}^{\infty}\underline{\mathbf{f}}_q(\ell_1,\vartheta;\zeta) &= (\zeta-1)\cosh^2\left(\frac{\ell_1}{4}\right) + \mathbb{E}^{-1}\Bigg[\omega^\alpha\mathbb{E}\Bigg[\sum_{q=0}^{\infty}\left(\frac{\partial^3}{\partial\ell_1^2\partial\vartheta}\underline{\mathbf{f}}(\ell_1,\vartheta;\zeta)\right)_q - \sum_{q=0}^{\infty}\left(\frac{\partial}{\partial\ell_1}\underline{\mathbf{f}}(\ell_1,\vartheta;\zeta)\right)_q \\ &\quad + \sum_{q=0}^{\infty}\mathcal{A}_q - \sum_{q=0}^{\infty}\mathcal{B}_q + 3\sum_{q=0}^{\infty}\mathcal{C}_q\Bigg]\Bigg]. \end{aligned} \tag{62}$$

Utilizing the Adomian polynomials described in (54), then (64) simplifies to

$$
\begin{aligned}
\underline{\mathbf{f}_0}(\ell_1,\vartheta;\zeta) &= (\zeta-1)\Big(\frac{1}{2}+\frac{1}{2}\cosh\Big(\frac{\ell_1}{2}\Big)\Big),\\
\underline{\mathbf{f}_1}(\ell_1,\vartheta;\zeta) &= \mathbb{E}^{-1}\Big[\omega^{\alpha}\mathbb{E}\Big[\Big(\frac{\partial^3}{\partial\ell_1^2\partial\vartheta}\underline{\mathbf{f}}(\ell_1,\vartheta;\zeta)\Big)_0-\Big(\frac{\partial}{\partial\ell_1}\underline{\mathbf{f}}(\ell_1,\vartheta;\zeta)\Big)_0+\underline{\mathcal{A}}_0-\underline{\mathcal{B}}_0+3\underline{\mathcal{C}}_0\Big]\Big]\\
&= -\frac{11}{32}(\zeta-1)\sinh\Big(\frac{\ell_1}{2}\Big)\frac{\vartheta^{\alpha}}{\Gamma(\alpha+1)},\\
\underline{\mathbf{f}_2}(\ell_1,\vartheta;\zeta) &= \mathbb{E}^{-1}\Big[\omega^{\alpha}\mathbb{E}\Big[\Big(\frac{\partial^3}{\partial\ell_1^2\partial\vartheta}\underline{\mathbf{f}}(\ell_1,\vartheta;\zeta)\Big)_1-\Big(\frac{\partial}{\partial\ell_1}\underline{\mathbf{f}}(\ell_1,\vartheta;\zeta)\Big)_1+\underline{\mathcal{A}}_1-\underline{\mathcal{B}}_1+3\underline{\mathcal{C}}_1\Big]\Big]\\
&= \frac{-11}{128}(\zeta-1)\sinh\Big(\frac{\ell_1}{2}\Big)\frac{\vartheta^{\alpha}}{\Gamma(\alpha+1)}+\frac{242(\zeta-1)}{1024}\cosh\Big(\frac{\ell_1}{2}\Big)\frac{\vartheta^{2\alpha}}{\Gamma(2\alpha+1)},\\
\underline{\mathbf{f}_3}(\ell_1,\vartheta;\zeta) &= \mathbb{E}^{-1}\Big[\omega^{\alpha}\mathbb{E}\Big[\Big(\frac{\partial^3}{\partial\ell_1^2\partial\vartheta}\underline{\mathbf{f}}(\ell_1,\vartheta;\zeta)\Big)_2-\Big(\frac{\partial}{\partial\ell_1}\underline{\mathbf{f}}(\ell_1,\vartheta;\zeta)\Big)_2+\underline{\mathcal{A}}_2-\underline{\mathcal{B}}_2+3\underline{\mathcal{C}}_2\Big]\Big]\\
&= \frac{-11(\zeta-1)}{512}\sinh\Big(\frac{\ell_1}{2}\Big)\frac{\vartheta^{\alpha}}{\Gamma(\alpha+1)}+\frac{242(\zeta-1)}{2048}\cosh\Big(\frac{\ell_1}{2}\Big)\frac{\vartheta^{2\alpha}}{\Gamma(2\alpha+1)}\\
&\quad -\frac{7986(\zeta-1)}{49152}\sinh\Big(\frac{\ell_1}{2}\Big)\frac{\vartheta^{3\alpha}}{\Gamma(3\alpha+1)},\\
&\ \vdots
\end{aligned}
$$

By implementing a similar technique, the remaining terms of $\underline{\mathbf{f}}_q$ $(q\geq 4)$ of EADM solution can be simply determined. Furthermore, when the iterative process expands, the accuracy of the obtained solution improves dramatically, and the deduced solution moves closer to the precise result. Finally, we have come up with the following answers in a series form

$$
\underline{\mathbf{f}}(\ell_1,\vartheta,\zeta)=\underline{\mathbf{f}}_0(\ell_1,\vartheta,\zeta)+\underline{\mathbf{f}}_1(\ell_1,\vartheta,\zeta)+\underline{\mathbf{f}}_1(\ell_1,\vartheta,\zeta)+\dots,
$$

such that

$$
\begin{aligned}
\underline{\mathbf{f}}(\ell_1,\vartheta,\zeta) &= \underline{\mathbf{f}}_0(\ell_1,\vartheta,\zeta)+\underline{\mathbf{f}}_1(\ell_1,\vartheta,\zeta)+\underline{\mathbf{f}}_1(\ell_1,\vartheta,\zeta)+\dots,\\
\bar{\mathbf{f}}(\ell_1,\vartheta,\zeta) &= \bar{\mathbf{f}}_0(\ell_1,\vartheta,\zeta)+\bar{\mathbf{f}}_1(\ell_1,\vartheta,\zeta)+\bar{\mathbf{f}}_1(\ell_1,\vartheta,\zeta)+\dots.
\end{aligned}
$$

Consequently, we have

$$
\begin{aligned}
\underline{\mathbf{f}}(\ell_1,\vartheta,\zeta) &= (\zeta-1)\Big(\frac{1}{2}+\frac{1}{2}\cosh\Big(\frac{\ell_1}{2}\Big)\Big)-\frac{231}{32}(\zeta-1)\sinh\Big(\frac{\ell_1}{2}\Big)\frac{\vartheta^{\alpha}}{\Gamma(\alpha+1)}\\
&\quad+\frac{363(\zeta-1)}{1024}\cosh\Big(\frac{\ell_1}{2}\Big)\frac{\vartheta^{2\alpha}}{\Gamma(2\alpha+1)}-\frac{7986(\zeta-1)}{49152}\sinh\Big(\frac{\ell_1}{2}\Big)\frac{\vartheta^{3\alpha}}{\Gamma(3\alpha+1)}+\dots,\\
\bar{\mathbf{f}}(\ell_1,\vartheta,\zeta) &= (1-\zeta)\Big(\frac{1}{2}+\frac{1}{2}\cosh\Big(\frac{\ell_1}{2}\Big)\Big)-\frac{231}{32}(1-\zeta)\sinh\Big(\frac{\ell_1}{2}\Big)\frac{\vartheta^{\alpha}}{\Gamma(\alpha+1)}\\
&\quad+\frac{363(1-\zeta)}{1024}\cosh\Big(\frac{\ell_1}{2}\Big)\frac{\vartheta^{2\alpha}}{\Gamma(2\alpha+1)}-\frac{7986(1-\zeta)}{49152}\sinh\Big(\frac{\ell_1}{2}\Big)\frac{\vartheta^{3\alpha}}{\Gamma(3\alpha+1)}+\dots.
\end{aligned}
$$

Case II. (For the fuzzy Atangana–Baleanu Caputo fractional derivative)

Here, we obtain the EADM solution for the first case of (51) by using the fuzzy ABC fractional derivative operator.

Taking into consideration the procedure described in Section 3, we have

$$
\begin{aligned}
\mathbb{E}[\underline{\mathbf{f}}(\ell_1,\vartheta;\zeta)] &= \omega^2\underline{\mathbf{f}}_{(\kappa)}(\ell_1;\zeta)+\Big(\frac{\alpha\omega^{\alpha}+1-\alpha}{\mathcal{N}(\alpha)}\Big)\mathbb{E}\Big[\frac{\partial^3}{\partial\ell_1^2\partial\vartheta}\underline{\mathbf{f}}(\ell_1,\vartheta;\zeta)-\frac{\partial}{\partial\ell_1}\underline{\mathbf{f}}(\ell_1,\vartheta;\zeta)\\
&\quad+\underline{\mathbf{f}}(\ell_1,\vartheta;\zeta)\frac{\partial^3}{\partial\ell_1^3}\underline{\mathbf{f}}(\ell_1,\vartheta;\zeta)-\underline{\mathbf{f}}(\ell_1,\vartheta;\zeta)\frac{\partial}{\partial\ell_1}\underline{\mathbf{f}}(\ell_1,\vartheta;\zeta)+\frac{\partial}{\partial\ell_1}\underline{\mathbf{f}}(\ell_1,\vartheta;\zeta)\frac{\partial^2}{\partial\ell_1^2}\underline{\mathbf{f}}(\ell_1,\vartheta;\zeta)\Big].
\end{aligned}
$$

Simple computations result in

$$\underline{\mathbf{f}}(\ell_1,\vartheta;\zeta) = (\zeta-1)\cosh^2\left(\frac{\ell_1}{4}\right) + \mathbb{E}^{-1}\left[\left(\frac{\alpha\omega^\alpha+1-\alpha}{\mathcal{N}(\alpha)}\right)\mathbb{E}\left[\frac{\partial^3}{\partial\ell_1^2\partial\vartheta}\underline{\mathbf{f}}(\ell_1,\vartheta;\zeta) - \frac{\partial}{\partial\ell_1}\underline{\mathbf{f}}(\ell_1,\vartheta;\zeta) + \underline{\mathbf{f}}(\ell_1,\vartheta;\zeta)\frac{\partial^3}{\partial\ell_1^3}\underline{\mathbf{f}}(\ell_1,\vartheta;\zeta) - \underline{\mathbf{f}}(\ell_1,\vartheta;\zeta)\frac{\partial}{\partial\ell_1}\underline{\mathbf{f}}(\ell_1,\vartheta;\zeta) + \frac{\partial}{\partial\ell_1}\underline{\mathbf{f}}(\ell_1,\vartheta;\zeta)\frac{\partial^2}{\partial\ell_1^2}\underline{\mathbf{f}}(\ell_1,\vartheta;\zeta)\right]\right]. \quad (63)$$

Let us surmise the infinite sum $\underline{\mathbf{f}}(\ell_1,\vartheta;\zeta) = \sum_{q=0}^{\infty}\underline{\mathbf{f}}_q(\ell_1,\vartheta;\zeta)$ accompanying it with (45) and affirming the non-linearity. Therefore, (63) takes the form

$$\sum_{q=0}^{\infty}\underline{\mathbf{f}}_q(\ell_1,\vartheta;\zeta) = (\zeta-1)\cosh^2\left(\frac{\ell_1}{4}\right) + \mathbb{E}^{-1}\left[\left(\frac{\alpha\omega^\alpha+1-\alpha}{\mathcal{N}(\alpha)}\right)\mathbb{E}\left[\sum_{q=0}^{\infty}\left(\frac{\partial^3}{\partial\ell_1^2\partial\vartheta}\underline{\mathbf{f}}(\ell_1,\vartheta;\zeta)\right)_q - \sum_{q=0}^{\infty}\left(\frac{\partial}{\partial\ell_1}\underline{\mathbf{f}}(\ell_1,\vartheta;\zeta)\right)_q + \sum_{q=0}^{\infty}\mathcal{A}_q - \sum_{q=0}^{\infty}\mathcal{B}_q + 3\sum_{q=0}^{\infty}\mathcal{C}_q\right]\right]. \quad (64)$$

Utilizing the Adomian polynomials described in (54), then (64) simplifies to

$$\begin{aligned}
\underline{\mathbf{f}}_0(\ell_1,\vartheta;\zeta) &= (\zeta-1)\left(\frac{1}{2}+\frac{1}{2}\cosh\left(\frac{\ell_1}{2}\right)\right),\\
\underline{\mathbf{f}}_1(\ell_1,\vartheta;\zeta) &= \mathbb{E}^{-1}\left[\left(\frac{\alpha\omega^\alpha+1-\alpha}{\mathcal{N}(\alpha)}\right)\mathbb{E}\left[\left(\frac{\partial^3}{\partial\ell_1^2\partial\vartheta}\underline{\mathbf{f}}(\ell_1,\vartheta;\zeta)\right)_0 - \left(\frac{\partial}{\partial\ell_1}\underline{\mathbf{f}}(\ell_1,\vartheta;\zeta)\right)_0 + \mathcal{A}_0 - \mathcal{B}_0 + 3\mathcal{C}_0\right]\right]\\
&= -\frac{11}{32}(\zeta-1)\sinh\left(\frac{\ell_1}{2}\right)\left[\frac{\alpha\vartheta^\alpha}{\Gamma(\alpha+1)} + (1-\alpha)\right],\\
\underline{\mathbf{f}}_2(\ell_1,\vartheta;\zeta) &= \mathbb{E}^{-1}\left[\left(\frac{\alpha\omega^\alpha+1-\alpha}{\mathcal{N}(\alpha)}\right)\mathbb{E}\left[\left(\frac{\partial^3}{\partial\ell_1^2\partial\vartheta}\underline{\mathbf{f}}(\ell_1,\vartheta;\zeta)\right)_1 - \left(\frac{\partial}{\partial\ell_1}\underline{\mathbf{f}}(\ell_1,\vartheta;\zeta)\right)_1 + \mathcal{A}_1 - \mathcal{B}_1 + 3\mathcal{C}_1\right]\right]\\
&= \frac{-11}{128}(\zeta-1)\sinh\left(\frac{\ell_1}{2}\right)\left[\frac{\alpha\vartheta^\alpha}{\Gamma(\alpha+1)} + (1-\alpha)\right]\\
&\quad + \frac{242(\zeta-1)}{1024}\cosh\left(\frac{\ell_1}{2}\right)\left[\frac{\alpha^2\vartheta^{2\alpha}}{\Gamma(2\alpha+1)} + 2\alpha(1-\alpha)\frac{\vartheta^\alpha}{\Gamma(\alpha+1)} + (1-\alpha)^2\right],\\
\underline{\mathbf{f}}_3(\ell_1,\vartheta;\zeta) &= \mathbb{E}^{-1}\left[\left(\frac{\alpha\omega^\alpha+1-\alpha}{\mathcal{N}(\alpha)}\right)\mathbb{E}\left[\left(\frac{\partial^3}{\partial\ell_1^2\partial\vartheta}\underline{\mathbf{f}}(\ell_1,\vartheta;\zeta)\right)_2 - \left(\frac{\partial}{\partial\ell_1}\underline{\mathbf{f}}(\ell_1,\vartheta;\zeta)\right)_2 + \mathcal{A}_2 - \mathcal{B}_2 + 3\mathcal{C}_2\right]\right]\\
&= \frac{-11(\zeta-1)}{512}\sinh\left(\frac{\ell_1}{2}\right)\left[\frac{\alpha\vartheta^\alpha}{\Gamma(\alpha+1)} + (1-\alpha)\right]\\
&\quad + \frac{242(\zeta-1)}{2048}\cosh\left(\frac{\ell_1}{2}\right)\left[\frac{\alpha^2\vartheta^{2\alpha}}{\Gamma(2\alpha+1)} + 2\alpha(1-\alpha)\frac{\vartheta^\alpha}{\Gamma(\alpha+1)} + (1-\alpha)^2\right]\\
&\quad - \frac{7986(\zeta-1)}{49152}\sinh\left(\frac{\ell_1}{2}\right)\left[\frac{\alpha^3\vartheta^{3\alpha}}{\Gamma(3\alpha+1)} + 3\alpha^2(1-\alpha)\frac{\vartheta^{2\alpha}}{\Gamma(2\alpha+1)} + 3\alpha(1-\alpha)^2\frac{\vartheta^\alpha}{\Gamma(\alpha+1)} + (1-\alpha)^3\right],\\
&\;\;\vdots
\end{aligned}$$

By implementing a similar technique, the remaining terms of $\underline{\mathbf{f}}_q$ $(q \geq 4)$ of the EADM solution can be simply determined. Furthermore, when the iterative process expands, the accuracy of the obtained solution improves dramatically, and the deduced solution moves closer to the precise result. Finally, we have come up with the following answers in a series form

$$\underline{\mathbf{f}}(\ell_1,\vartheta,\zeta) = \underline{\mathbf{f}}_0(\ell_1,\vartheta,\zeta) + \underline{\mathbf{f}}_1(\ell_1,\vartheta,\zeta) + \underline{\mathbf{f}}_1(\ell_1,\vartheta,\zeta) + \dots,$$

such that

$$\begin{aligned}
\underline{\mathbf{f}}(\ell_1,\vartheta,\zeta) &= \underline{\mathbf{f}}_0(\ell_1,\vartheta,\zeta) + \underline{\mathbf{f}}_1(\ell_1,\vartheta,\zeta) + \underline{\mathbf{f}}_1(\ell_1,\vartheta,\zeta) + \dots,\\
\bar{\mathbf{f}}(\ell_1,\vartheta,\zeta) &= \bar{\mathbf{f}}_0(\ell_1,\vartheta,\zeta) + \bar{\mathbf{f}}_1(\ell_1,\vartheta,\zeta) + \bar{\mathbf{f}}_1(\ell_1,\vartheta,\zeta) + \dots.
\end{aligned}$$

Consequently, we have

$$\begin{aligned}
\underline{\mathbf{f}}(\ell_1,\vartheta,\zeta) &= (\zeta-1)\Big(\frac{1}{2}+\frac{1}{2}\cosh\Big(\frac{\ell_1}{2}\Big)\Big)-\frac{231}{32}(\zeta-1)\sinh\Big(\frac{\ell_1}{2}\Big)\Big[\frac{\alpha\vartheta^{\alpha}}{\Gamma(\alpha+1)}+(1-\alpha)\Big] \\
&+\frac{363(\zeta-1)}{1024}\cosh\Big(\frac{\ell_1}{2}\Big)\Big[\frac{\alpha^2\vartheta^{2\alpha}}{\Gamma(2\alpha+1)}+2\alpha(1-\alpha)\frac{\vartheta^{\alpha}}{\Gamma(\alpha+1)}+(1-\alpha)^2\Big] \\
&-\frac{7986(\zeta-1)}{49152}\sinh\Big(\frac{\ell_1}{2}\Big)\Big[\frac{\alpha^3\vartheta^{3\alpha}}{\Gamma(3\alpha+1)}+3\alpha^2(1-\alpha)\frac{\vartheta^{2\alpha}}{\Gamma(2\alpha+1)}+3\alpha(1-\alpha)^2\frac{\vartheta^{\alpha}}{\Gamma(\alpha+1)} \\
&+(1-\alpha)^3\Big]+\dots, \\
\bar{\mathbf{f}}(\ell_1,\vartheta,\zeta) &= (1-\zeta)\Big(\frac{1}{2}+\frac{1}{2}\cosh\Big(\frac{\ell_1}{2}\Big)\Big)-\frac{231}{32}(1-\zeta)\sinh\Big(\frac{\ell_1}{2}\Big)\Big[\frac{\alpha\vartheta^{\alpha}}{\Gamma(\alpha+1)}+(1-\alpha)\Big] \\
&+\frac{363(1-\zeta)}{1024}\cosh\Big(\frac{\ell_1}{2}\Big)\Big[\frac{\alpha^2\vartheta^{2\alpha}}{\Gamma(2\alpha+1)}+2\alpha(1-\alpha)\frac{\vartheta^{\alpha}}{\Gamma(\alpha+1)}+(1-\alpha)^2\Big] \\
&-\frac{7986(1-\zeta)}{49152}\sinh\Big(\frac{\ell_1}{2}\Big)\Big[\frac{\alpha^3\vartheta^{3\alpha}}{\Gamma(3\alpha+1)}+3\alpha^2(1-\alpha)\frac{\vartheta^{2\alpha}}{\Gamma(2\alpha+1)}+3\alpha(1-\alpha)^2\frac{\vartheta^{\alpha}}{\Gamma(\alpha+1)} \\
&+(1-\alpha)^3\Big]+\dots.
\end{aligned}$$

In this analysis, Figure 6 demonstrates the insight into the influence of multiple-layer surface plots for Problem 2 correlated with the CFD and Elzaki transform in the fuzzy sense. It is worth mentioning that the profile identifies the variation in the mapping $\mathbf{f}(\ell_1,\vartheta;\zeta)$ on space co-ordinate ℓ_1 with respect to ϑ and the uncertainty parameter $\zeta\in[0,1]$.

The graph illustrates that, as time progresses, the mapping $\mathbf{f}(\ell_1,\vartheta;\zeta)$ will also increase.

- The effect of the proposed methodology on the mapping $\mathbf{f}(\ell_1,\vartheta;\zeta)$ is displayed in Figure 7a for the varying fractional-orders $\alpha=1,0.85,0.75,0.55$ by the considering CFD operator. It exhibits a relatively small increase in the mapping $\underline{\mathbf{f}}(\ell_1,\vartheta;\zeta)$ with the decrease in $\bar{\mathbf{f}}(\ell_1,\vartheta;\zeta)$.
- The profile graph of Figure 7b demonstrates the lower and upper solution of varying uncertainty when the fractional order is assumed to be $\alpha=0.2$ by proposing the CFD operator. It emphasizes a relatively small variation in the mapping $\underline{\mathbf{f}}(\ell_1,\vartheta;\zeta)$ with the increase in $\bar{\mathbf{f}}(\ell_1,\vartheta;\zeta)$.
- The effect of the proposed methodology on the mapping $\mathbf{f}(\ell_1,\vartheta;\zeta)$ is displayed in Figure 8a for the varying fractional orders $\alpha=1,0.85,0.75,0.55$ by considering ABC fractional derivative operator. It exhibits a relatively small increase in the mapping $\underline{\mathbf{f}}(\ell_1,\vartheta;\zeta)$ with the decrease in $\bar{\mathbf{f}}(\ell_1,\vartheta;\zeta)$.
- The profile graph of Figure 8b demonstrates the lower and upper solutions of varying uncertainty when the fractional order is assumed to be $\alpha=0.2$ by proposing the ABC fractional derivative operator. It emphasizes a relatively small variation in the mapping $\underline{\mathbf{f}}(\ell_1,\vartheta;\zeta)$ with the increase in $\bar{\mathbf{f}}(\ell_1,\vartheta;\zeta)$.
- Figure 9 demonstrates the comparison analysis between CFD operator and ABC fractional derivative operator for varying fractional order with uncertainty $\kappa\in[0,1]$, exhibits that the lower solution profile for ABC fractional operator has strong ties with the upper solution as compared to the CFD operator.
- Figure 10 shows the comparison analysis between ($\underline{\mathbf{f}}(\ell_1,\vartheta;\zeta)$ and the exact solution), ($\bar{\mathbf{f}}(\ell_1,\vartheta;\zeta)$ and the exact solution), respectively, for three dimensional error plots by considering the CFD operator.

Furthermore, the offered approach does not provide a unique solution but will aid scientists in selecting the best approximate solution. It is remarkable that the fuzzy ABC fractional derivative operator has better performance than the CFD operators, because the curves have a strong harmony with the integer-order graph in the ABC operator case.

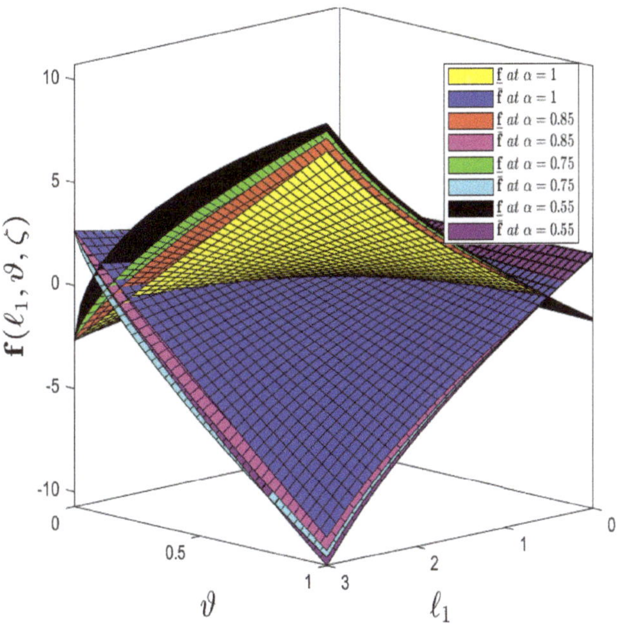

Figure 6. A three-dimensional surface plot indicates the lower and upper solution $\mathbf{f}(\ell_1, \vartheta, \zeta)$ taking into consideration the fuzzy Caputo fractional derivative operator for Problem 2 when $\alpha = 1, 0.85, 0.75, 0.55$ with uncertainty $\zeta \in [0, 1]$.

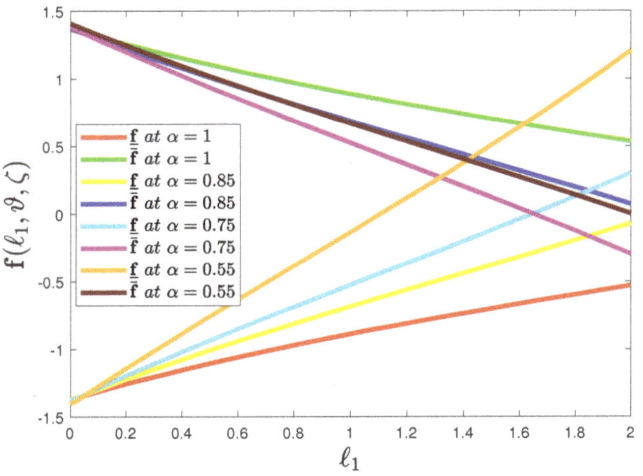

(a)

Figure 7. *Cont.*

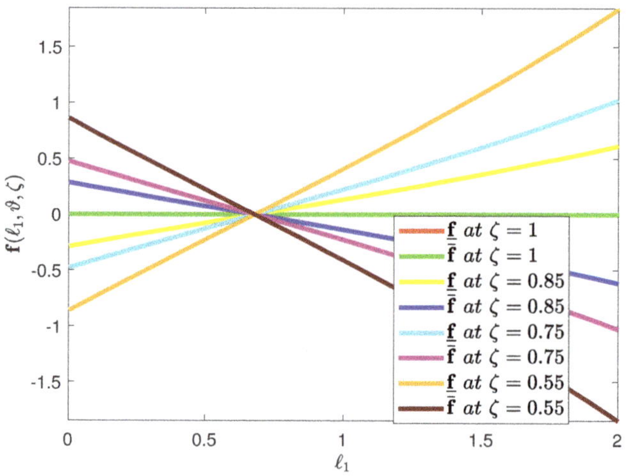

(**b**)

Figure 7. (**a**) A two-dimensional plot indicates the lower and upper solution $\mathbf{f}(\ell_1, \vartheta, \zeta)$ taking into consideration the fuzzy Caputo fractional derivative operator for Problem 2 when $\alpha = 1, 0.85, 0.75, 0.55$ with uncertainty $\zeta \in [0, 1]$. (**b**) A two-dimensional plot indicates the lower and upper solution $\mathbf{f}(\ell_1, \vartheta, \zeta)$ taking into consideration the fuzzy Caputo fractional derivative operator for Problem 2 when $\zeta = 1, 0.85, 0.75, 0.55$ with the fractional order $\alpha = 0.2$.

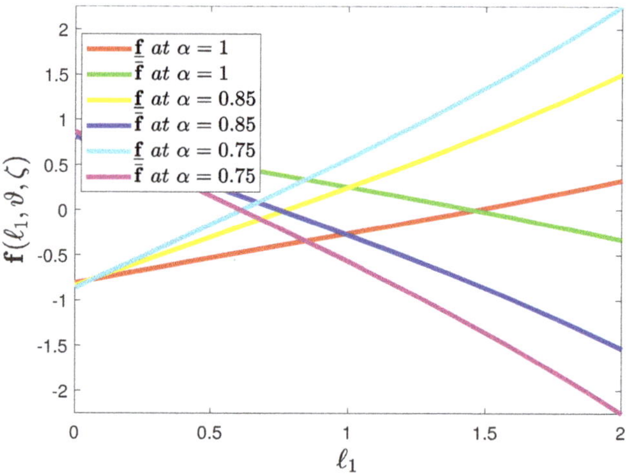

(**a**)

Figure 8. *Cont.*

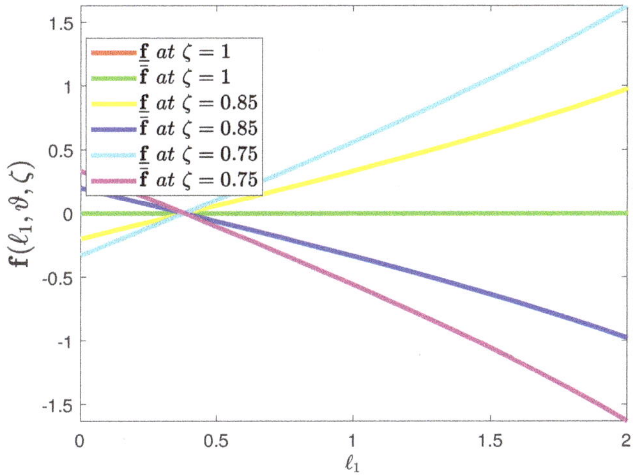

(**b**)

Figure 8. (**a**) A two-dimensional plot indicates the lower and upper solution $\mathbf{f}(\ell_1, \vartheta, \zeta)$ taking into consideration the fuzzy ABC fractional derivative operator for Problem 2 when $\alpha = 1, 0.85, 0.75, 0.55$ with uncertainty $\zeta \in [0, 1]$. (**b**) A two-dimensional plot indicates the lower and upper solution $\mathbf{f}(\ell_1, \vartheta, \zeta)$ taking into consideration the fuzzy ABC fractional derivative operator for Problem 2 when $\zeta = 1, 0.85, 0.75, 0.55$ with the fractional order $\alpha = 0.2$.

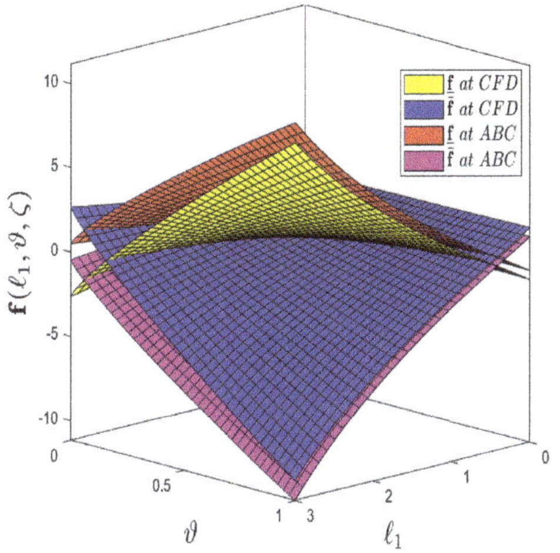

Figure 9. A comparison three-dimensional plot indicates the lower and upper solution $\mathbf{f}(\ell_1, \vartheta, \zeta)$, taking into consideration the fuzzy Caputo and fuzzy ABC fractional derivative operators for Problem 2 when $\zeta = 0.2$ with the fractional order $\alpha = 0.85$.

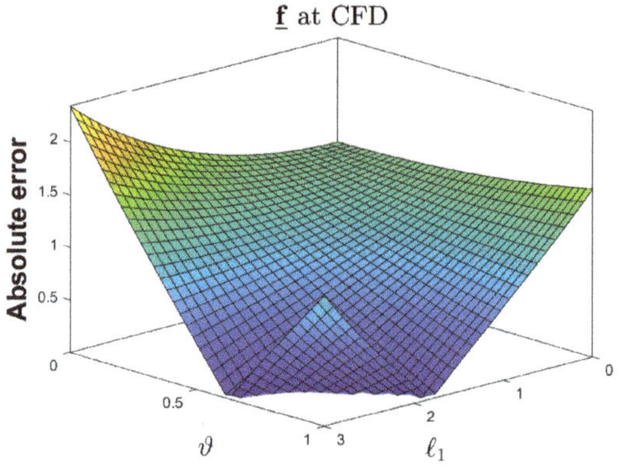

(a)

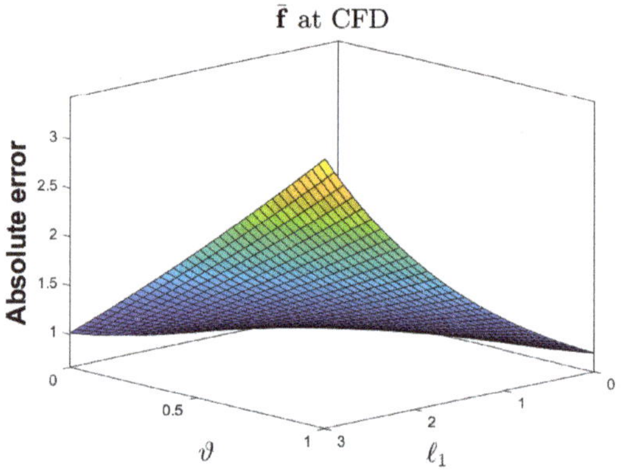

(b)

Figure 10. (**a**) A three-dimensional absolute error plot indicates the lower and exact solution (Remark 3) of $\mathbf{f}(\ell_1, \vartheta, \zeta)$ taking into consideration the fuzzy Caputo fractional derivative operator for Problem 2 when $\alpha = 0.85$ with uncertainty $\zeta \in [0, 1]$. (**b**)A three-dimensional absolute error plot indicates the upper and exact solution (Remark 3) of $\mathbf{f}(\ell_1, \vartheta, \zeta)$, taking into consideration the fuzzy Caputo fractional derivative operator for Problem 2 when $\alpha = 0.85$ with uncertainty $\zeta \in [0, 1]$. □

Remark 3. *When* $\underline{\chi}(\zeta) = \bar{\chi}(\zeta) = 1$ *along with* $\alpha = 1$, *then both solutions of Problem 2 converge to the integer-order solution* $\mathbf{f}(\ell_1, \vartheta) = \cosh^2\left(\frac{\ell_1}{4} - \frac{11\vartheta}{24}\right)$.

5. Conclusions

The paper has demonstrated families of approximate solutions to the FWE under $g\mathcal{H} - (i)$ differentiability taking into consideration the Elzaki and ADM. Fractional operators (Caputo and ABC) describing fuzzy characteristics have been separately discussed. The fuzzy solutions of FWE proposed for such flows are characterized by EADM. Nevertheless, the crisp operators are unable to simulate any physical mechanism in an unpredictable setting. Therefore, fuzzy operators are a preferable means to describe the physical phenomenon in such a scenario. Specifically, we illustrated two test examples of the evolutionary method to gain deeper insight into the exact–approximate solutions to validate the projected technique to attain a parametric solution for each case of the fuzzy (Caputo and ABC) fractional derivative operator. It has been demonstrated that the solution graphs predict the fuzzy solution since they satisfy the fuzzy number conditions. As for applications of this framework, the convergence and error analysis can be predicated by the simulation study that specified that fractional-order plots have a strong correlation with the evolutionary trajectories of FWE. It has also been shown that fuzzy EADM represents two solutions, which often leads to an advantage in selecting the best one possible for a governing model. As a consequence, the fuzzy theory connected with FC allows a model to improve performance in an uncertain domain. In the future, we will investigate a similar problem by defining the Henstock integrals (fuzzy integrals in the sense of Lebesgue) at infinite intervals [69,70].

Author Contributions: Conceptualization and data curation—S.R., R.A., A.O.A., M.S.M.; formal analysis—S.R., R.A., A.O.A., M.A.A., T.A., M.S.M.; funding acquisition—S.R., R.A., A.O.A., M.A.A., T.A., M.S.M.; investigation—S.R., R.A., A.O.A., M.A.A., T.A.; methodology— S.R., R.A., A.O.A., M.A.A., T.A.; project administration—S.R., R.A., A.O.A., M.A.A., T. A, M.S.M.: supervision—S.R.; resources, software—R.A., A.O.A., M.A.A., T.A., M.S.M.; validation—S.R., R.A., A.O.A,: visualization—S.R., R.A., A.O.A., M.A.A., T.A, M.S.M.; writing—original draft—S.R., R.A., A.O.A,; writing—review & editing—S.R., M.A.A., T.A., M.S.M.; All authors have read and agreed to the published version of the manuscript.

Funding: This research received no external funding.

Institutional Review Board Statement: Not Applicable.

Informed Consent Statement: Not Applicable.

Data Availability Statement: Not Applicable.

Acknowledgments: This research was funded by the Deanship of Scientific Research at Princess Nourah bint Abdulrahman University, Saudi Arabia through the Fast-track Research Funding Program.

Conflicts of Interest: The authors declare no conflict of interest.

References

1. Podlubny, I. *Fractional Differential Equations*; Academic Press: San Diego, CA, USA, 1999.
2. Hilfer, R. *Applications of Fractional Calculus in Physics*; Word Scientific: Singapore, 2000.
3. Kilbas, A.; Srivastava, H.M.; Trujillo, J.J. *Theory and Application of Fractional Differential Equations*; Elsevier: Amsterdam, The Netherlands, 2006; Volume 204, pp. 1–523.
4. Magin, R.L. *Fractional Calculus in Bioengineering*; Begell House Publishers: Redding, CT, USA, 2006.
5. Samko, S.G.; Kilbas, A.A.; Marichev, O.I. *Fractional Integrals and Derivatives: Theory and Applications*; Gordon and Breach Science Publishers, Philadelphia, PA, USA, 1993.
6. Jajarmi, A.; Baleanu, D. On the fractional optimal control problems with a general derivative operator. *Asian J. Cont.* **2021**, *23*, 1062–1071. [CrossRef]
7. Jajarmi, A.; Baleanu, D. Suboptimal control of fractional-order dynamic systems with delay argument. *J. Vib. Control* **2018**, *24*, 2430–2446. [CrossRef]
8. Baleanu, D.; Zibaei, S.; Namjoo, M.; Jajarmi, A. A nonstandard finite difference scheme for the modeling and nonidentical synchronization of a novel fractional chaotic system. *Adv. Diff. Eqs.* **2021**, *2021*, 308. [CrossRef]
9. Tuan, N.H.; Ganji, R.M.; Jafari, H. A numerical study of fractional rheological models and fractional Newell-Whitehead-Segel equation with non-local and non-singular kernel. *Chin. J. Phys.* **2020**, *68*, 308–320. [CrossRef]
10. Ganji, R.M.; Jafari, H.; Baleanu, D. A new approach for solving multi variable orders differential equations with Mittag–Leffler kernel. *Chaos Solitons Fract.* **2020**, *130*, 109405. [CrossRef]
11. Baleanu, D.; Jajarmi, A.; Mohammadi, H.; Rezapour, S. A new study on the mathematical modelling of human liver with Caputo–Fabrizio fractional derivative. *Chaos Solitons Fract.* **2020**, *134*, 109705. [CrossRef]
12. Rashid, S.; Khalid, A.; Bazighifan, O.; Oros, G.I. New modifications of integral inequalities via $\wp$-convexity pertaining to fractional calculus and their applications. *Mathematics* **2021**, *9*, 1753. [CrossRef]
13. Atangana, A.; Baleanu D. New fractional derivatives with nonlocal and non-singular kernel: Theory and application to heat transfer model. *Therm. Sci.* **2016**, *20*, 763–769. [CrossRef]
14. Caputo, M.; Fabrizio, M. A new definition of fractional derivative without singular kernel. *Prog. Fract. Differ. Appl.* **2015**, *1*, 1–13.
15. Baleanu, D.; Sajjadi, S.S.; Jajarmi, A.; Defterli, O. On a nonlinear dynamical system with both chaotic and non-chaotic behaviours: A new fractional analysis and control. *Adv. Differ. Equ.* **2021**, *2021*, 234. [CrossRef]
16. Abdeljawad, T.; Baleanu, D. Integration by parts and its applications of a new nonlocal fractional derivative with Mittag-Leffler nonsingular kernel. *Nonlinear Anal. Theory Methods Appl.* **2017**, *10*, 1098–1107. [CrossRef]

17. Abdeljawad, T. Fractional operators with generalized Mittag-Leffler kernels and their iterated differintegrals. *Chaos* **2019**, *29*, 023102. [CrossRef]
18. Li, Z.; Wang, C.; Agarwal, R.P.; Sakthivel, R. Hyers-Ulam-Rassias stability of quaternion multidimensional fuzzy nonlinear difference equations with impulses. *Iran. J. Fuzzy Syst.* **2021**, *18*, 143–160.
19. Kandel, A.; Byatt, W.J. Fuzzy differential equations. In Proceedings of the International Conference Cybernetics and Society, Tokyo, Japan, 3–7 November 1978; pp. 1213–1216.
20. Agarwal, R.P.; Lakshmikantham, V.; Nieto, J.J. On the concept of solution for fractional differential equations with uncertainty. *Nonlin. Anal. Theory Meth Appl.* **2010**, *72*, 2859–2862. [CrossRef]
21. El-Sayed, S.; Kaya, D. An application of the ADM to seven-order Sawada-Kotara equations. *Appl. Math. Comput.* **2004**, *157*, 93–101. [CrossRef]
22. El-Tawil, M.A.; Huseen, S. On convergence of the q-homotopy analysis method. *Int. J. Contemp. Math. Sci.* **2013**, *8*, 481–497. [CrossRef]
23. Darvishia, M.T.; Kheybaria, S.; Khanib, F. A numerical solution of the Lax's 7th-order KdV equation by Pseudo spectral method and Darvishi's Preconditioning. *Int. J. Contemp. Math. Sci.* **2007**, *2*, 1097–1106. [CrossRef]
24. Shiralashetti, S.C.; Kumbinarasaiah, S. Laguerre wavelets collocation method for the numerical solution of the Benjamina–Bona–Mohany equations. *J. Taibah Univ. Sci.* **2019**, *13*, 9–15. [CrossRef]
25. Lahmar, N.A.; Belhamitib, O.; Bahric, S.M. A new Legendre-Wavelets decomposition method for solving PDEs. *Malaya J. Mat.* **2014**, *1*, 72–81.
26. Hoa, N.V.; Vu, H.; Duc, T.M. Fuzzy fractional differential equations under Caputo Katugampola fractional derivative approach. *Fuzzy Sets Syst.* **2019**, *375*, 70–99. [CrossRef]
27. Hoa, N.V. Fuzzy fractional functional differential equations under Caputo gH-differentiability. *Commun. Nonlinear Sci. Numer. Simul.* **2015**, *22*, 1134–1157. [CrossRef]
28. Salahshour, S.; Ahmadian, A.; Senu, N.; Baleanu, D.; Agarwal, P. On analytical aolutions of the fractional differential equation with uncertainty: Application to the Basset problem. *Entropy* **2015**, *17*, 885–902. [CrossRef]
29. Ahmad, S.; Ullah, A.; Akgül, A.; Abdeljawad, T. Semi-analytical solutions of the 3rd order fuzzy dispersive partial differential equations under fractional operators. *Alex. Eng. J.* **2021**, *60*, 5861–5878. [CrossRef]
30. Shah, K.; Seadawy, A.R.; Arfan, M. Evaluation of one dimensional fuzzy fractional partial differential equations. *Alex. Eng. J.* **2020**, *59*, 3347–3353. [CrossRef]
31. Allahviranloo, T. An analytic approximation to the solution of fuzzy heat equation by Adomian decomposition method. *Int. J. Contemp. Math. Sci.* **2009**, *4*, 105–114.
32. Allahviranloo, T. The Adomian decomposition method for fuzzy system of linear equations. *Appl. Math. Comput.* **2005**, *163*, 553–563. [CrossRef]
33. Biswas, S.; Roy, T.K. Adomian decomposition method for fuzzy differential equations with linear differential operator. *J. Comput. Inf. Sci. Eng.* **2016**, *11*, 243–250.
34. Hamoud, A.; Ghadle, K. Modified Adomian decomposition method for solving fuzzy Volterra–Fredholm integral equations. *J. Indian Math. Soc.* **2018**, *85*, 52–69.
35. Whitham, G.B. Variational methods and applications to water wave. *Proc. R. Soc. Lond. Ser. A* **1967**, *299*, 6–25.
36. Fornberg, B.; Whitham, G.B. A numerical and theoretical study of certain nonlinear wave phenomena. *Philos. Trans. R. Soc. Lond. Ser. A* **1978**, *289*, 373–404.
37. Abidi, F.; Omrani, K. The homotopy analysis method for solving the Fornberg—Whitham equation and comparison with Adomian's decomposition method. *Comput. Math. Appl.* **2010**, *59*, 2743–2750. [CrossRef]
38. Gupta, P.K.; Singh, M. Homotopy perturbation method fractional Fornberg–Whitham equation. *Comput. Math. Appl.* **2011**, *61*, 250–254. [CrossRef]
39. Lu, J. An analytical approach to the Fornberg–Whitham type equations by using the variational iteration method. *Comput. Math. Appl.* **2011**, *61*, 2010–2013. [CrossRef]
40. Sakar, M.G.; Erdogan, F.; Yildirim, A. Variational iteration method for the time-fractional Fornberg–Whitham equation. *Comput. Math. Appl.* **2012**, *63*, 1382–1388. [CrossRef]
41. Chen, A.; Li, J. Deng, X.; Huang, W. Travelling wave solutions of the Fornberg–Whitham equation. *Appl. Math. Comput.* **2009**, *215*, 3068–3075.
42. Yin, J.; Tian, L.; Fan, X. Classification of travelling waves in the Fornberg–Whitham equation. *J. Math. Anal. Appl.* **2010**, *368*, 133–143. [CrossRef]
43. Zhou, J.; Tian, L. A type of bounded traveling wave solutions for the Fornberg–Whitham equation. *J. Math. Anal. Appl.* **2008**, *346*, 255–261. [CrossRef]
44. He, B.; Meng, Q.; Li, S. Explicit peakon and solitary wave solutions for the modified Fornberg–Whitham equation. *Appl. Math. Comput.* **2010**, *5*, 1976–1982. [CrossRef]
45. Fan, X.; Yang, S.; Yin, J.; Tian, L. Bifurcations of traveling wave solutions for a two-component Fornberg–Whitham equation. *Commun. Nonlinear Sci. Numer. Simul.* **2011**, *16*, 3956–3963. [CrossRef]
46. Jiang, B.; Bi, B. Smooth and non-smooth traveling wave solutions of the Fornberg–Whitham equation with linear dispersion term. *Appl. Math. Comput.* **2010**, *216*, 2155–2162. [CrossRef]

47. Adomian, G. A review of the decomposition method in applied mathematics. *J. Math. Anal. Appl.* **1988**, *135*, 501–544. [CrossRef]
48. Belgacem, F.B.M.; Karaballi, A.A.; Kalla, S.L. Analytical investigations of the sumudu transform and applications to integral production equations. *Math. Prob. Eng.* **2003**, *2003*, 103–118. [CrossRef]
49. Aboodh, K.S. The new integral transform "Aboodh Transform". *Glob. J. Pure Appl. Math.* **2013**, *9*, 35–43.
50. Elzaki, T.M. Application of new transform Elzaki transform to partial differential equations. *Glob. J. Pure Appl. Math.* **2011**, *7*, 65–70.
51. Mahgoub, M.M.A. The new integral transform "Mohand Transform". *Adv. Theor. Appl. Math.* **2017**, *12*, 113–120.
52. Rashid, S.; Hammouch, Z.; Aydi, H.; Ahmad, A.G.; Alsharif, A.M. Novel computations of the time-fractional Fisher's model via generalized fractional integral operators by means of the Elzaki transform. *Fractal Fract.* **2021**, *5*, 94. [CrossRef]
53. Rashid, S.; Kubra, K.T.; Guirao, J.L.G. Construction of an approximate analytical solution for multi-dimensional fractional Zakharov-Kuznetsov equation via Aboodh adomian decomposition method. *Symmetry* **2021**, *13*, 1542. [CrossRef]
54. Rashid, S.; Kubra, K.T.; Abualnaja, K.M. Fractional view of heat-like equations via the Elzaki transform in the settings of the Mittag-Leffler function. *Math. Methods Appl. Sci.* **2021**. [CrossRef]
55. Rashid, S.; Khalid, A.; Sultana, S.; Hammouch, Z.; Shah, R.; Alsharif, A.M. A novel analytical view of time-fractional Korteweg-De Vries equations via a new integral transform. *Symmetry* **2021**, *13*, 1254. [CrossRef]
56. Rashid, S.; Kubra, K.T.; Ullah, S. Fractional spatial diffusion of a biological population model via a new integral transform in the settings of power and Mittag-Leffler nonsingular kernel. *Phys. Scr.* **2021**, *96*, 114003. [CrossRef]
57. Rashid, S.; Jarad, F.; Abualnaja, K.M. On fuzzy Volterra-Fredholm integrodifferential equation associated with Hilfer-generalized proportional fractional derivative. *AIMS Math.* **2021**, *6*, 10920–10946. [CrossRef]
58. Rashid, S.; Kubra, K.T.; Rauf, A.; Chu, Y.-M.; Hamed, Y.S. New numerical approach for time-fractional partial differential equations arising in physical system involving natural decomposition method. *Phys. Scr.* **2021**, *96*, 105204. [CrossRef]
59. Allahviranloo, T. *Fuzzy Fractional Differential Operators and Equation Studies in Fuzziness and Soft Computing*; Springer: Berlin, Germany, 2021.
60. Zimmermann, H.J. *Fuzzy Set Theory and Its Applications*; Kluwer Academic Publishers: Dordrecht, The Netherlands, 1991.
61. Zadeh, L.A. Fuzzy sets. *Inform. Contr.* **1965**, *8*, 338–353. [CrossRef]
62. Allahviranloo, T.; Ahmadi, M.B. Fuzzy Lapalce Transform. *Soft Comput.* **2010**, *14*, 235–243. [CrossRef]
63. Georgieva, A. Double Fuzzy Sumudu transform to solve partial Volterra fuzzy integro-differential equations. *Mathematics* **2020**, *8*, 692. [CrossRef]
64. Bede, B.; Stefanini, L. Generalized differentiability of fuzzy-valued functions. *Fuzzy Sets Syst.* **2013**, *230*, 119–141. [CrossRef]
65. Bede, B.; Gal, S.G. Generalizations of the differentiability of fuzzy-number-valued functions with applications to fuzzy differential equations. *Fuzzy Sets Syst.* **2005**, *151*, 581–599. [CrossRef]
66. Allahviranloo, T.; Armand, A.; Gouyandeh, Z. Fuzzy fractional differential equations under generalized fuzzy Caputo derivative. *J. Intell. Fuzzy Syst.* **2014**, *26*, 1481–1490. [CrossRef]
67. Salahshour, S.; Allahviranloo, T.; Abbasbandy, S. Solving fuzzy fractional differential equations by fuzzy Laplace transforms. *Commun. Nonlinear Sci. Numer. Simul.* **2012**, *17*, 1372–1381. [CrossRef]
68. Yavuz, M.; Abdeljawad, T. Nonlinear regularized long-wave models with a new integral transformation applied to the fractional derivative with power and Mittag-Leffler kernel. *Adv. Differ. Equ.* **2020**, *2020*, 367. [CrossRef]
69. Henstock, R. *Theory of Integration*; Butterworth: London, UK, 1963.
70. Gong, Z.T.; Wang, L.L. The Henstock–Stieltjes integral for fuzzy-number-valued functions. *Inf. Sci.* **2012**, *188*, 276–297. [CrossRef]

fractal and fractional

MDPI

Article

On the General Solutions of Some Non-Homogeneous Div-Curl Systems with Riemann–Liouville and Caputo Fractional Derivatives

Briceyda B. Delgado [1] and Jorge E. Macías-Díaz [1,2,*]

[1] Departamento de Matemáticas y Física, Universidad Autónoma de Aguascalientes, Aguascalientes 20100, Mexico; profesor.invitado368@edu.uaa.mx
[2] Department of Mathematics and Didactics of Mathematics, Tallinn University, 10120 Tallinn, Estonia
* Correspondence: jemacias@correo.uaa.mx; Tel.: +52-449-9108400

Abstract: In this work, we investigate analytically the solutions of a nonlinear div-curl system with fractional derivatives of the Riemann–Liouville or Caputo types. To this end, the fractional-order vector operators of divergence, curl and gradient are identified as components of the fractional Dirac operator in quaternionic form. As one of the most important results of this manuscript, we derive general solutions of some non-homogeneous div-curl systems that consider the presence of fractional-order derivatives of the Riemann–Liouville or Caputo types. A fractional analogous to the Teodorescu transform is presented in this work, and we employ some properties of its component operators, developed in this work to establish a generalization of the Helmholtz decomposition theorem in fractional space. Additionally, right inverses of the fractional-order curl, divergence and gradient vector operators are obtained using Riemann–Liouville and Caputo fractional operators. Finally, some consequences of these results are provided as applications at the end of this work.

Keywords: fractional div-curl systems; Helmholtz decomposition theorem; Riemann–Liouville derivative; Caputo derivative; fractional vector operators

MSC: 34Axx; 91Bxx

Citation: Delgado, B.B.; Macías-Díaz, J.E. On the General Solutions of Some Non-Homogeneous Div-Curl Systems with Riemann–Liouville and Caputo Fractional Derivatives. *Fractal Fract.* **2021**, *5*, 117. https://doi.org/10.3390/fractalfract5030117

Academic Editors: Asifa Tassaddiq and Muhammad Yaseen

Received: 9 August 2021
Accepted: 4 September 2021
Published: 10 September 2021

1. Introduction

From an analytical point of view, the literature offers a wide range of reports that focus on the extension of integer-order methods and results for the fractional case. From a more particular point of view, the fractional generalization of the classical vector calculus operators (that is, the gradient, divergence, curl and Laplacian operators) has been also an active topic of research, which has been developed from different approaches. Some of the first attempts to extend these operators to the fractional scenario are described in [1,2] using the Nishimoto fractional derivative. These operators were used later on in [3] to provide a physical interpretation for the fractional advection-dispersion equation for flow in heterogeneous porous media. In 2008, Vasily E. Tarasov described different approaches to formulate a fractional form of vector calculus with physical applications in [4] (see also references therein). More recently, a new generalization of the Helmholtz decomposition theorem for both fractional time and space was proposed in [5,6] using the discrete Grünwald–Letnikov fractional derivative. Another related work is [7], where the authors investigate the dynamic creation of fractionalized half-quantum vortices in Bose–Einstein condensates of sodium atoms.

In this work, we consider fractional derivatives of the Riemann–Liouville and the Caputo types and provide extensions of the definitions of the main differential operators from vector calculus using these fractional operators. In such a way, we present fractional forms of the divergence, the rotational and the gradient operators. Moreover, we also consider generalized forms of the Dirac and the Laplace operators using fractional derivatives.

Our goal in this work is to extend quaternionic analysis to consider fractional forms of the classical differential operators. Analogues of the properties satisfied by the classical operators will be mathematically established in this work. For instance, we will show that

$$^{RL}\mathcal{D}^{\alpha}_{a^+}[w] = -^{RL}\operatorname{div}^{\alpha}_{a^+}\vec{w} + ^{RL}\operatorname{grad}^{\alpha}_{a^+} w_0 + ^{RL}\operatorname{curl}^{\alpha}_{a^+}\vec{w}, \tag{1}$$

$$^{C}\mathcal{D}^{\alpha}_{a^+}[w] = -^{C}\operatorname{div}^{\alpha}_{a^+}\vec{w} + ^{C}\operatorname{grad}^{\alpha}_{a^+} w_0 + ^{C}\operatorname{curl}^{\alpha}_{a^+}\vec{w}. \tag{2}$$

In other words, when we apply the fractional Dirac operator to a quaternion valued function $w = w_0 + \vec{w}$, this expression can be decomposed in terms of a fractional divergence, a fractional gradient and a fractional rotational operator. Based on this, we will also provide explicit expressions of general weak solutions for some fractional forms of the div-curl system, considering various analytical hypotheses. More precisely, we will prove that if we restrict ourselves to the class of functions with fractional divergence zero and whose Riemann–Liouville fractional integral has zero normal trace, then the fractional Teodorescu transform represents a solution of the above-mentioned div-curl systems (see Theorems 2 and 3).

This manuscript is organized as follows. In Section 2, we recall some important definitions from the literature, including those of the left and right Riemann–Liouville fractional derivatives, the left and right Caputo fractional derivatives and the two-parameter Mittag–Leffler function. A useful characterization of the functions with summable fractional derivatives is also recalled. In Section 2.2, the Riemann–Liouville and the Caputo Dirac and Laplace operators are introduced. Moreover, some fundamental solutions for the fractional Dirac operators are recalled in Section 2.3. In Section 3, we introduce fractional extensions of the divergence, rotational and gradient differential operators. Some properties among these operators are established, and a useful factorization theorem for the fractional Laplace operators is proven. It is worth mentioning that this factorization is not new; however, we were able to derive it only using the identities preserving the fractional gradient, divergence and rotational operators. Among the most important results, we establish that the fractional Teodorescu transform is a right inverse of the fractional Dirac operator under suitable analytical conditions, and we prove a fractional form of the Divergence Theorem.

Section 4 is devoted to establishing the existence of weak solutions for Riemann–Liouville and Caputo fractional div-curl systems. The explicit form of the operators involved in the solution, as well as some of their properties, allow the solution to be re-expressed as the sum of the fractional gradient of a scalar potential plus a fractional curl of a vector potential; we can say that our solutions preserve a Helmholtz-type decomposition (see Propositions 6 and 7). As a consequence, right inverses of the fractional rotational and divergence operators are provided in a subclass of the fractional divergence-free vector fields. In turn, Section 5 provides some consequences of the factorization results proven in Section 2 to the construction of fractional hyper-conjugate pairs. A theorem providing necessary and sufficient conditions for the existence of Caputo fractional hyper-conjugate pairs is proven in this stage, along with a result of the existence of a right inverse for the fractional gradient. Finally, this manuscript closes with a section of concluding remarks.

2. Background

2.1. Fractional Calculus

The present section is devoted to recalling some useful definitions from fractional calculus. Throughout, we assume that $a, b, \alpha \in \mathbb{R}$ satisfy $\alpha > 0$ and $a < b$. Meanwhile, we suppose that $f : \mathbb{R} \to \mathbb{R}$ is a sufficiently smooth function, with the property that f is identically equal to zero outside of the interval $[a, b]$.

Definition 1. *The* left *and* right Riemann–Liouville fractional integrals *of f of order α with respect to the interval $[a, b]$ (whenever they exist) are the functions $I^{\alpha}_{a^+}[f]$ and $I^{\alpha}_{b^-}[f]$, defined, respectively, by (see [8])*

$$I^{\alpha}_{a^+}[f](x) = \frac{1}{\Gamma(\alpha)} \int_a^x \frac{f(t)}{(x-t)^{1-\alpha}}\, dt, \quad x > a, \tag{3}$$

$$I^{\alpha}_{b^-}[f](x) = \frac{1}{\Gamma(\alpha)} \int_x^b \frac{f(t)}{(t-x)^{1-\alpha}}\, dt, \quad x < b. \tag{4}$$

Let $m = [\alpha] + 1 \in \mathbb{Z}$. *The* left Riemann–Liouville *and the* left Caputo fractional derivatives *of order* α *with respect to the interval* $[a, b]$ *are, respectively, defined as follows:*

$$^{RL}D^{\alpha}_{a^+}[f](x) = D^m I^{m-\alpha}_{a^+}[f](x) = \frac{1}{\Gamma(m-\alpha)} \frac{d^m}{dx^m} \int_a^x \frac{f(t)}{(x-t)^{\alpha-m+1}}\, dt, \quad x > a, \tag{5}$$

$$^{C}D^{\alpha}_{a^+}[f](x) = I^{m-\alpha}_{a^+} D^m[f](x) = \frac{1}{\Gamma(m-\alpha)} \int_a^x \frac{f^{(m)}(t)}{(x-t)^{\alpha-m+1}}\, dt, \quad x > a. \tag{6}$$

Finally, we define the right Riemann–Liouville *and the* right Caputo fractional derivatives *of order* α *with respect to the point* a, *respectively, as the functions*

$$^{RL}D^{\alpha}_{b^-}[f](x) = D^m I^{m-\alpha}_{b^-}[f](x) = \frac{(-1)^m}{\Gamma(m-\alpha)} \frac{d^m}{dx^m} \int_x^b \frac{f(t)}{(t-x)^{\alpha-m+1}}\, dt, \quad x < b, \tag{7}$$

$$^{C}D^{\alpha}_{b^-}[f](x) = I^{m-\alpha}_{b^-} D^m[f](x) = \frac{(-1)^m}{\Gamma(m-\alpha)} \int_x^b \frac{f^{(m)}(t)}{(t-x)^{\alpha-m+1}}\, dt, \quad x < b. \tag{8}$$

For the sake of convenience, we will employ the notation $D^{\alpha}_{a^\pm}$ when we present properties satisfied by both fractional derivatives $^{RL}D^{\alpha}_{a^\pm}$ and $^{C}D^{\alpha}_{a^\pm}$. Throughout, we let $I^{\alpha}_{a^+}(L_1)$ denote the class of all functions f that are represented by the fractional integral (3) of some integrable function, i.e., $f = I^{\alpha}_{a^+}[g]$, for some $g \in L_1(a, b)$. Using this notation, the following result provides a characterization of these functions.

Theorem 1 (Samko et al. [9]). *Let* $\alpha > 0$ *and* $m = [\alpha] + 1$. *Then, the function* f *belongs to* $I^{\alpha}_{a^+}(L_1)$ *if and only if* $I^{m-\alpha}_{a^+}[f] \in AC^m([a, b])$, *and* $(I^{m-\alpha}_{a^+}[f])^k(a) = 0$, *for each* $k \in \{0, 1, \ldots, m-1\}$.

Definition 2. *If* $(I^{m-\alpha}_{a^+}[f])^{(k)}(a) = 0$, *for each* $k \in \{0, 1 \ldots, m-1\}$, *then it follows that* $f^{(k)}(a) = 0$ *holds, for each* $k \in \{0, \ldots, m-1\}$ *(see [8,9]). In light of this fact, we will say that* f *has a* summable fractional derivative $D^{\alpha}_{a^+}[f]$ *of order* α *on* $[a, b]$ *if* $I^{m-\alpha}_{a^+}[f] \in AC^m([a, b])$.

In the following discussion, suppose that $\alpha > 0$, f admits a summable fractional derivative of order $\alpha > 0$ on $[a, b]$, and let $m = [\alpha] + 1$. Then, the following composition rules are satisfied:

$$I^{\alpha}_{a^+}\, {}^{RL}D^{\alpha}_{a^+}[f](x) = f(x) - \sum_{k=0}^{m-1} \frac{(x-a)^{\alpha-k-1}}{\Gamma(\alpha-k)} (I^{m-\alpha}_{a^+}[f])^{m-k-1}(a), \tag{9}$$

$$I^{\alpha}_{a^+}\, {}^{C}D^{\alpha}_{a^+}[f](x) = f(x) - \sum_{k=0}^{m-1} \frac{f^{(k)}(a)}{k!} (x-a)^k. \tag{10}$$

On the other hand, we know that both fractional derivatives $^{RL}D^{\alpha}_{a^+}$ and $^{C}D^{\alpha}_{a^+}$ satisfy the one-sided invertibility property $D^{\alpha}_{a^+} I^{\alpha}_{a^+}[f] = f$. It is worth noting that this identity is a particular case of the property $D^{\alpha}_{a^+} I^{\beta}_{a^+}[f] = D^{\alpha-\beta}_{a^+}[f]$, which holds for each $\alpha, \beta \in \mathbb{R}$ satisfying $\alpha \geq \beta > 0$.

It is important to recall also that the following semi-group property for the composition of fractional derivatives is not generally satisfied:

$$^{C}D^{\alpha}_{a^+}\, {}^{C}D^{\beta}_{a^+}[f] = {}^{C}D^{\beta}_{a^+}\, {}^{C}D^{\alpha}_{a^+}[f] = {}^{C}D^{\alpha+\beta}_{a^+}[f]. \tag{11}$$

However, if $f^{(k)}(a) = 0$ for $k = 0, 1, \ldots, \max\{[\alpha] + 1, [\beta] + 1\} - 1$, then (11) does hold; see Section 2.2.6 [8]. An analogous condition for the semi-group property in the context of the Riemann–Liouville derivative is found in Section 2.3.6 [8]. Finally, the following relation between the Riemann–Liouville and Caputo fractional derivatives is valid:

$$ {}^{C}D^{\alpha}_{a^+}[f](x) = {}^{RL}D^{\alpha}_{a^+}[f](x) - \sum_{j=0}^{m-1} \frac{(x-a)^{j-\alpha}}{\Gamma(j-\alpha+1)} D^{(j)}_x[f](a). \tag{12} $$

Definition 3 (Gorenflo et al. [10]). *Let $\mu, \nu \in \mathbb{R}$ be such that $\mu > 0$. We define the two-parameter Mittag–Leffler function $E_{\mu,\nu} : \mathbb{C} \to \mathbb{C}$ with parameters μ and ν in terms of the following power series:*

$$ E_{\mu,\nu}(z) = \sum_{n=0}^{\infty} \frac{z^n}{\Gamma(\mu n + \nu)}, \quad \forall z \in \mathbb{C}. \tag{13} $$

2.2. Fractional Quaternionic Analysis

In this section, we will mention some recent results in fractional Clifford analysis. The Dirac operator in Clifford analysis, also known as the Moisil–Teodorescu differential operator, represents the cornerstone of the analysis in higher dimensions. A remarkable number of systems of differential equations have been analyzed using this operator or a perturbation of it, and the monographs [11–13] of the authors Gürlebeck and Sprößig contain many examples of the applications that have been made over the years. See also [14], where the authors introduced the fractional Dirac operator with Caputo derivatives as well as the basic tool of a fractional function theory in more dimensions.

More precisely, this section is devoted to the collection of some recent results of the authors Ferreira et al. [15–17], by whom fundamental solutions of the fractional Laplacian were found, where the derivatives are of Riemann–Liouville and Caputo types, as well as of the fractional Dirac operators.

For the remainder, let $a_i, b_i \in \mathbb{R}$ satisfy $a_i < b_i$, for each $i = 1,2,3$. We will suppose that $\Omega = \Pi_{i=1}^{3}(a_i, b_i)$ is a bounded open rectangular domain in $\mathbb{R}^3$, and let $\alpha = (\alpha_1, \alpha_2, \alpha_3)$, with $\alpha_i \in (0,1]$, for all $i = 1,2,3$.

Definition 4 (Ferreira et al. [15–17]). *The* fractional Riemann–Liouville *and* fractional Caputo Dirac operators *are represented by ${}^{RL}\mathcal{D}^{\alpha}_{a^+}$ and ${}^{C}\mathcal{D}^{\alpha}_{a^+}$, respectively, and they are defined as*

$$ {}^{RL}\mathcal{D}^{\alpha}_{a^+} = \sum_{i=1}^{3} e_i \, {}^{RL}\partial^{\frac{1+\alpha_i}{2}}_{x_i,a_i}, \tag{14} $$

$$ {}^{C}\mathcal{D}^{\alpha}_{a^+} = \sum_{i=1}^{3} e_i \, {}^{C}\partial^{\frac{1+\alpha_i}{2}}_{x_i,a_i}. \tag{15} $$

Here, ${}^{RL}\partial^{\frac{1+\alpha_i}{2}}_{x_i,a_i}$ and ${}^{C}\partial^{\frac{1+\alpha_i}{2}}_{x_i,a_i}$ are, respectively, the Riemann–Liouville and the Caputo fractional derivative operators of order $(1+\alpha_i)/2$ with respect to the variable $x_i \in (a_i, b_i)$, for each $i = 1,2,3$. We define the fractional Laplace operators *${}^{RL}\Delta^{\alpha}_{a^+}$ and ${}^{RL}\Delta^{\alpha}_{a^+}$, respectively, by*

$$ {}^{RL}\Delta^{\alpha}_{a^+} = \sum_{i=1}^{3} {}^{RL}\partial^{1+\alpha_i}_{x_i,a_i}, \tag{16} $$

$$ {}^{C}\Delta^{\alpha}_{a^+} = \sum_{i=1}^{3} {}^{C}\partial^{1+\alpha_i}_{x_i,a_i}, \tag{17} $$

where ${}^{RL}\partial^{1+\alpha_i}_{x_i,a_i}$ and ${}^{C}\partial^{1+\alpha_i}_{x_i,a_i}$ are, respectively, the Riemann–Liouville and Caputo fractional derivatives of order $1+\alpha_i$ with respect to the variable $x_i \in (a_i, b_i)$, for each $i = 1,2,3$.

2.3. Fundamental Solutions

The purpose of this subsection is to determine fundamental solutions for the fractional Dirac operator and use their properties in the investigation of the solutions of fractional div-curl systems. Beforehand, it is worth recalling that a family of fundamental solutions for the fractional Laplace operators ${}^{RL}\Delta^{\alpha}_{a^+}$ and ${}^{C}\Delta^{\alpha}_{a^+}$, and a family of fundamental solutions for the fractional Dirac operators ${}^{RL}\mathcal{D}^{\alpha}_{a^+}$ and ${}^{C}\mathcal{D}^{\alpha}_{a^+}$, were obtained in [15,16] for the Riemann–Liouville and Caputo case, respectively. In the case of Riemann–Liouville fractional operators, the authors employed some properties of the Mittag–Leffler function and the Laplace transform in two dimensions.

We will begin this section by recalling some relevant results derived in [15]. To this end, let u be an eigenfunction of the fractional Laplace operator, i.e., suppose that ${}^{RL}\Delta^{\alpha}_{a^+} u = \lambda u$ for some $\lambda \in \mathbb{C}$, and assume that $u(\vec{x})$ admits a summable fractional derivative ${}^{RL}\partial^{\frac{1+\alpha_1}{2}}_{x_1,a_1^+}$ in the variable x_1, and that it belongs to $I^{1+\alpha_i}_{a_i^+}(L_1)$ in the variables x_2 and x_3. In what follows, we will consider the following integral and differential conditions of Cauchy type:

$$\begin{cases} f_0(x_2,x_3) = I^{1-\alpha_1}_{a_1^+}[u](a_1,x_2,x_3), \\ f_1(x_2,x_3) = \partial^{\alpha_1}_{x_1,a_1^+}[u](a_1,x_2,x_3). \end{cases} \tag{18}$$

Lemma 1 (Ferreira et al. [15,17]). *A family of eigenfunctions of the fractional Laplace operator ${}^{RL}\Delta^{\alpha}_{a^+}$ is given by the function*

$$\begin{aligned} u_\lambda(\vec{x}) = {}& (x_1-a_1)^{\alpha_1-1} E_{1+\alpha_1,\alpha_1}\left(-(x_1-a_1)^{1+\alpha_1}\left({}^{RL}\partial^{1+\alpha_2}_{x_2,a_2^+} + {}^{RL}\partial^{1+\alpha_3}_{x_3,a_3^+} - \lambda\right)\right) f_0(x_2,x_3) \\ &+ (x_1-a_1)^{\alpha_1} E_{1+\alpha_1,1+\alpha_1}\left(-(x_1-a_1)^{1+\alpha_1}\left({}^{RL}\partial^{1+\alpha_2}_{x_2,a_2^+} + {}^{RL}\partial^{1+\alpha_3}_{x_3,a_3^+} - \lambda\right)\right) f_1(x_2,x_3). \end{aligned} \tag{19}$$

Meanwhile, a family of fundamental solutions of the fractional Dirac operator ${}^{RL}\mathcal{D}^{\alpha}_{a^+}$ is obtained by considering $\lambda \equiv 0$ in (19). More precisely, this family of solutions is given by

$${}^{RL}\mathcal{E}^{\alpha}_{a^+}(\vec{x}) = -{}^{RL}\mathcal{D}^{\alpha}_{a^+}[u_0](\vec{x}), \tag{20}$$

where u_0 is a fundamental solution of ${}^{RL}\Delta^{\alpha}_{a^+}$, i.e.,

$$\begin{aligned} u_0(\vec{x}) = {}& (x_1-a_1)^{\alpha_1-1} E_{1+\alpha_1,\alpha_1}\left(-(x_1-a_1)^{1+\alpha_1}\left({}^{RL}\partial^{1+\alpha_2}_{x_2,a_2^+} + {}^{RL}\partial^{1+\alpha_3}_{x_3,a_3^+}\right)\right) f_0(x_2,x_3) \\ &+ (x_1-a_1)^{\alpha_1} E_{1+\alpha_1,1+\alpha_1}\left(-(x_1-a_1)^{1+\alpha_1}\left({}^{RL}\partial^{1+\alpha_2}_{x_2,a_2^+} + {}^{RL}\partial^{1+\alpha_3}_{x_3,a_3^+}\right)\right) f_1(x_2,x_3). \end{aligned} \tag{21}$$

Here, f_0 and f_1 satisfy the conditions (18).

Let v be an eigenfunction of the fractional Laplace operator, i.e., suppose that ${}^{C}\Delta^{\alpha}_{a^+} v = \lambda v$, for some $\lambda \in \mathbb{C}$. Assume that $v(\vec{x})$ admits a summable fractional derivative ${}^{RL}\partial^{\frac{1+\alpha_1}{2}}_{x_1,a_1^+}$ in the variable x_1, and that it belongs to $I^{1+\alpha_i}_{a_i^+}(L_1)$ in the variables x_2 and x_3. By Theorem 1, $v(x_1,a_2,x_3) = v(x_1,x_2,a_3) = 0$. In what follows, we will consider the following Cauchy conditions:

$$\begin{cases} g_0(x_2,x_3) = v(a_1,x_2,x_3), \\ g_1(x_2,x_3) = v'_{x_1}(a_1,x_2,x_3). \end{cases} \tag{22}$$

As a consequence, $g_0(a_2,x_3) = g_0(x_2,a_3) = g_1(a_2,x_3) = g_1(x_2,a_3) = 0$.

Lemma 2 (Ferreira et al. [16,17]). *A family of eigenfunctions for the fractional Laplace operator* ${}^{C}\Delta^{\alpha}_{a^+}$ *is given by the function*

$$\begin{aligned} v_\lambda(\vec{x}) = {} & E_{1+\alpha_1,1}\left(-(x_1-a_1)^{1+\alpha_1}\left({}^{C}\partial^{1+\alpha_2}_{x_2,a_2^+} + {}^{C}\partial^{1+\alpha_3}_{x_3,a_3^+} - \lambda\right)\right)g_0(x_2,x_3) \\ & + (x_1-a_1)E_{1+\alpha_1,2}\left(-(x_1-a_1)^{1+\alpha_1}\left({}^{C}\partial^{1+\alpha_2}_{x_2,a_2^+} + {}^{C}\partial^{1+\alpha_3}_{x_3,a_3^+} - \lambda\right)\right)g_1(x_2,x_3). \end{aligned} \tag{23}$$

Meanwhile, a family of fundamental solutions of the fractional Dirac operator ${}^{C}\mathcal{D}^{\alpha}_{a^+}$ *is obtained by considering* $\lambda \equiv 0$ *in* (23)*. More precisely, this family of solutions is given by*

$${}^{C}\mathcal{E}^{\alpha}_{a^+}(\vec{x}) = -{}^{C}\mathcal{D}^{\alpha}_{a^+}[v_0](\vec{x}), \tag{24}$$

where v_0 *is a fundamental solution of* ${}^{C}\Delta^{\alpha}_{a^+}$*, i.e.,*

$$\begin{aligned} v_0(\vec{x}) = {} & E_{1+\alpha_1,1}\left(-(x_1-a_1)^{1+\alpha_1}\left({}^{C}\partial^{1+\alpha_2}_{x_2,a_2^+} + {}^{C}\partial^{1+\alpha_3}_{x_3,a_3^+}\right)\right)g_0(x_2,x_3) \\ & + (x_1-a_1)E_{1+\alpha_1,2}\left(-(x_1-a_1)^{1+\alpha_1}\left({}^{C}\partial^{1+\alpha_2}_{x_2,a_2^+} + {}^{C}\partial^{1+\alpha_3}_{x_3,a_3^+}\right)\right)g_1(x_2,x_3), \end{aligned} \tag{25}$$

where $g_0(x_2,x_3)$ *and* $g_1(x_2,x_3)$ *satisfy* (22)*.*

3. Fractional Vector Calculus

For the remainder of this section, we will study the fractional divergence, gradient and rotational operators as parts of a decomposition of the fractional Dirac operator in three dimensions. More precisely, recall that if $w = w_0 + \vec{w}$ is a quaternionic-valued function, then the following decomposition in quaternionic form is satisfied:

$$Dw = -\operatorname{div}\vec{w} + \operatorname{grad} w_0 + \operatorname{curl}\vec{w}. \tag{26}$$

Here, $D = \sum_{i=1}^{3} e_i\partial_i$ is the classical Dirac operator, which is also called the Moisil–Teodorescu differential operator. For more details about quaternionic analysis, see [11,18,19]. Our goal in this section is to provide an extension of this decomposition (27) using fractional operators of the Riemann–Liouville and Caputo types.

Let $w = w_0 + \sum_{i=1}^{3} e_i w_i$ be a quaternionic-valued function in $\mathrm{AC}(\Omega)$, whose scalar part is denoted by $\mathrm{Sc}[w] = w_0$ and its vector part by $\mathrm{Vec}[w] = \vec{w} = \sum_{i=1}^{3} e_i w_i$. Then, the action of the operator ${}^{RL}\mathcal{D}^{\alpha}_{a^+}$ on w reduces to

$$\begin{aligned} {}^{RL}\mathcal{D}^{\alpha}_{a^+}[w] &= \sum_{i=1}^{3} e_i\, {}^{RL}\partial^{\frac{1+\alpha_i}{2}}_{x_i,a_i}[w] \\ &= -\left(\sum_{i=1}^{3} {}^{RL}\partial^{\frac{1+\alpha_i}{2}}_{x_i,a_i}[w_i]\right) + \left(\sum_{i=1}^{3} e_i\, {}^{RL}\partial^{\frac{1+\alpha_i}{2}}_{x_i,a_i}[w_0]\right) + e_1\left({}^{RL}\partial^{\frac{1+\alpha_2}{2}}_{x_2,a_2}[w_3] - {}^{RL}\partial^{\frac{1+\alpha_3}{2}}_{x_3,a_3}[w_2]\right) \\ &\quad + e_2\left({}^{RL}\partial^{\frac{1+\alpha_3}{2}}_{x_3,a_3}[w_1] - {}^{RL}\partial^{\frac{1+\alpha_1}{2}}_{x_1,a_1}[w_3]\right) + e_3\left({}^{RL}\partial^{\frac{1+\alpha_2}{2}}_{x_1,a_1}[w_2] - {}^{RL}\partial^{\frac{1+\alpha_2}{2}}_{x_2,a_2}[w_1]\right). \end{aligned} \tag{27}$$

The above decomposition of Equation (27) originates a fractional version of the classical divergence, rotational and gradient differential operators from vector calculus. These operators are, respectively, the scalar component of (27), the vector term acting over $\vec{w}$ and the vector term of the equation acting over w_0. These facts are the motivation to analyze the following fractional differential operators.

Definition 5. *Define the fractional divergence, curl and gradient operators in the Riemann–Liouville sense by*

$$ {}^{RL}\operatorname{div}^{\alpha}_{a^+}\,\vec{w} = \sum_{i=1}^{3} {}^{RL}\partial^{\frac{1+\alpha_i}{2}}_{x_i,a_i}[w_i], \tag{28}$$

$$ {}^{RL}\operatorname{curl}^{\alpha}_{a^+}\,\vec{w} = e_1\left({}^{RL}\partial^{\frac{1+\alpha_2}{2}}_{x_2,a_2}[w_3] - {}^{RL}\partial^{\frac{1+\alpha_3}{2}}_{x_3,a_3}[w_2]\right) + e_2\left({}^{RL}\partial^{\frac{1+\alpha_3}{2}}_{x_3,a_3}[w_1] - {}^{RL}\partial^{\frac{1+\alpha_1}{2}}_{x_1,a_1}[w_3]\right) \tag{29}$$
$$ + e_3\left({}^{RL}\partial^{\frac{1+\alpha_2}{2}}_{x_1,a_1}[w_2] - {}^{RL}\partial^{\frac{1+\alpha_2}{2}}_{x_2,a_2}[w_1]\right),$$

$$ {}^{RL}\operatorname{grad}^{\alpha}_{a^+}[w_0] = \sum_{i=1}^{3} e_i\,{}^{RL}\partial^{\frac{1+\alpha_i}{2}}_{x_i,a_i}[w_0]. \tag{30}$$

It is important to point out that the fractional operators (28)–(30) reduce, respectively, to the classical div, curl, and grad operators from vector calculus when $\alpha_i = 1$, for each $i = 1,2,3$. Moreover, if $\alpha^* > 0$ and $\alpha_i = \alpha^*$, for each $i = 1,2,3$, then the above fractional operators coincide with the divergence, curl and gradient operators defined in [1,2] up to a constant factor. See also [4] and references therein for a historic account of the efforts to formulate a fractional form of vector calculus. Unlike the classical vector calculus operators, these fractional operators are non-local. Consequently, the fractional divergence, curl and gradient depend on the domain Ω.

Notice now that (27) can be rewritten as the following decomposition

$$ {}^{RL}\mathcal{D}^{\alpha}_{a^+}[w] = -{}^{RL}\operatorname{div}^{\alpha}_{a^+}\,\vec{w} + {}^{RL}\operatorname{grad}^{\alpha}_{a^+}\,w_0 + {}^{RL}\operatorname{curl}^{\alpha}_{a^+}\,\vec{w}. \tag{31}$$

Since the specific form of the Riemann–Liouville fractional derivative does not affect the above decomposition, we analogously obtain the following decomposition in terms of Caputo fractional derivatives:

$$ {}^{C}\mathcal{D}^{\alpha}_{a^+}[w] = -{}^{C}\operatorname{div}^{\alpha}_{a^+}\,\vec{w} + {}^{C}\operatorname{grad}^{\alpha}_{a^+}\,w_0 + {}^{C}\operatorname{curl}^{\alpha}_{a^+}\,\vec{w}. \tag{32}$$

Here, the operators ${}^{C}\operatorname{div}^{\alpha}_{a^+}$, ${}^{C}\operatorname{curl}^{\alpha}_{a^+}$ and ${}^{C}\operatorname{grad}^{\alpha}_{a^+}$ are defined as in (28)–(30), respectively, using Caputo fractional derivatives instead of Riemann–Liouville operators.

We define now a class of functions in $AC^1(\Omega)$ where we can apply the semi-group property (11).

Definition 6. *We set* $\mathcal{Z}^{\alpha}_{a^+}(\Omega) = \{f \in AC^1(\Omega) \colon f(a_1,x_2,x_3) = f(x_1,a_2,x_3) = f(x_1,x_2,a_3) = 0\}$.

Proposition 1. *If* $f = f_0 + \vec{f} \in \mathcal{Z}^{\alpha}_{a^+}(\Omega)$, *then*

(i) ${}^{RL}\operatorname{div}^{\alpha}_{a^+}\,{}^{RL}\operatorname{curl}^{\alpha}_{a^+}[\vec{f}] = 0$,
(ii) ${}^{RL}\operatorname{curl}^{\alpha}_{a^+}\,{}^{RL}\operatorname{grad}^{\alpha}_{a^+}[f_0] = 0$,
(iii) ${}^{RL}\operatorname{div}^{\alpha}_{a^+}\,{}^{RL}\operatorname{grad}^{\alpha}_{a^+}[f_0] = {}^{RL}\Delta^{\alpha}_{a^+}[f_0]$,
(iv) ${}^{RL}\operatorname{grad}^{\alpha}_{a^+}\,{}^{RL}\operatorname{div}^{\alpha}_{a^+}[\vec{f}] - {}^{RL}\operatorname{curl}^{\alpha}_{a^+}\,{}^{RL}\operatorname{curl}^{\alpha}_{a^+}[\vec{f}] = {}^{RL}\Delta^{\alpha}_{a^+}[\vec{f}]$.

Moreover, the identities (i)–(iv) *also hold for the Caputo fractional operators.*

Proof. The results readily follow from the identities

$$ {}^{RL}\partial^{\frac{1+\alpha_i}{2}}_{x_i,a_i}\,{}^{RL}\partial^{\frac{1+\alpha_i}{2}}_{x_i,a_i} = {}^{RL}\partial^{1+\alpha_i}_{x_i,a_i}, \quad \forall i = 1,2,3. \tag{33}$$

$$ {}^{RL}\partial^{\frac{1+\alpha_i}{2}}_{x_i,a_i}\,{}^{RL}\partial^{\frac{1+\alpha_j}{2}}_{x_j,a_j} = {}^{RL}\partial^{\frac{1+\alpha_j}{2}}_{x_j,a_j}\,{}^{RL}\partial^{\frac{1+\alpha_i}{2}}_{x_i,a_i}, \quad \forall i,j = 1,2,3, \tag{34}$$

which are trivially satisfied in $\mathcal{Z}^{\alpha}_{a^+}(\Omega)$. The identities with Caputo fractional operators are established in a similar fashion. □

Similar identities using Caputo fractional derivatives were proven in [4] when all the orders of the fractional derivatives are equal, i.e., when there exists $\alpha^* > 0$ such that $\alpha_i = \alpha^*$, for each $i = 1, 2, 3$. On the other hand, a direct consequence of Proposition 1 is that the fractional Dirac operator ${}^{RL}\mathcal{D}^{\alpha}_{a^+}$ factorizes the fractional Laplace operator ${}^{RL}\Delta^{\alpha}_{a^+}$. A more general factorization for functions taking values in $Cl_{0,3}$ was proven in Section 4 [16]. In light of these remarks, the following result is a direct consequence of Proposition 1.

Corollary 1. *If $f = f_0 + \vec{f} \in \mathcal{Z}^{\alpha}_{a^+}(\Omega)$, then the following factorizations of the fractional Laplace operators are satisfied:*

$$ {}^{RL}\Delta^{\alpha}_{a^+}[f] = -{}^{RL}\mathcal{D}^{\alpha}_{a^+}\,{}^{RL}\mathcal{D}^{\alpha}_{a^+}[f], \tag{35} $$

$$ {}^{C}\Delta^{\alpha}_{a^+}[f] = -{}^{C}\mathcal{D}^{\alpha}_{a^+}\,{}^{C}\mathcal{D}^{\alpha}_{a^+}[f]. \tag{36} $$

We turn our attention now to the right Caputo fractional Dirac operator, which is given by the expression

$$ {}^{C}\mathcal{D}^{\alpha}_{b^-} = \sum_{i=1}^{3} e_i\,{}^{C}\partial^{\frac{1+\alpha_i}{2}}_{x_i, b_i^-}. \tag{37} $$

The following *fractional Stokes formula* was proven in Theorem 10 [17]:

$$ \int_{\Omega} \left(-([h]^{C}\mathcal{D}^{\alpha}_{b^-})(\vec{y})f(\vec{y}) + h(\vec{y})({}^{RL}\mathcal{D}^{\alpha}_{a^+}[f])(\vec{y})\right) d\vec{y} = \int_{\partial\Omega} h(\vec{y})\eta(\vec{y})ds_{\vec{y}} I^{\alpha}_{a^+}[f](\vec{y}). \tag{38} $$

Here, we require that $f, h \in AC^1(\Omega) \cap AC(\overline{\Omega})$ and $I^{\alpha}_{a^+}[f] = \sum_{i=1}^{3} I^{\frac{1-\alpha_i}{2}}_{a_i^+}[f]$. It is worth noting here that the operator ${}^{C}\mathcal{D}^{\alpha}_{b^-}$ acts on the right, while ${}^{RL}\mathcal{D}^{\alpha}_{a^+}$ acts on the left. Intuitively, the last formula shows that the left Riemann–Liouville and right Caputo fractional Dirac operators act by 'intertwining' to obtain the fractional analogue of the Stokes formula.

The following result is a fractional form of the well-known Divergence Theorem.

Proposition 2 (Fractional Divergence Theorem). *If $\vec{f} \in AC^1(\Omega) \cap AC(\overline{\Omega})$, then*

$$ \int_{\Omega} {}^{RL}\mathrm{div}^{\alpha}_{a^+}[\vec{f}](\vec{y})\, d\vec{y} = \int_{\partial\Omega} \eta(\vec{y}) \cdot I^{\alpha}_{a^+}[\vec{f}](\vec{y})\, ds_{\vec{y}}, \tag{39} $$

$$ \int_{\Omega} {}^{RL}\mathrm{curl}^{\alpha}_{a^+}[\vec{f}](\vec{y})\, d\vec{y} = \int_{\partial\Omega} \eta(\vec{y}) \times I^{\alpha}_{a^+}[\vec{f}](\vec{y})\, ds_{\vec{y}}. \tag{40} $$

Proof. Taking $h \equiv 1$ in (38) and using that $[1]^{C}\mathcal{D}^{\alpha}_{b^-} = 0$ yields that

$$ \int_{\Omega} {}^{RL}\mathcal{D}^{\alpha}_{a^+}[\vec{f}](\vec{y})\, d\vec{y} = \int_{\partial\Omega} \eta(\vec{x}) I^{\alpha}_{a^+}[\vec{f}](\vec{y})\, ds_{\vec{y}}. \tag{41} $$

Due to the decomposition (31) and because $I^{\alpha}_{a^+}[\vec{f}]$ is purely vectorial, we can readily calculate their scalar and vector parts, respectively. As a consequence, we readily achieve formulas (39) and (40), respectively. □

Proposition 3. *Let $\vec{f} \in \mathcal{Z}^{\alpha}_{a^+}(\Omega)$, $c_0 \in AC(\Omega)$ and $0 < \alpha_i < 1$, for all $i = 1, 2, 3$. Then, the following identities hold:*

$$ {}^{RL}\mathrm{div}^{\alpha}_{a^+}[c_0\vec{f}] = \vec{f} \cdot {}^{RL}\mathrm{grad}^{\alpha}_{a^+}[c_0], \tag{42} $$

$$ {}^{RL}\mathrm{curl}^{\alpha}_{a^+}[c_0\vec{f}] = \vec{f} \times {}^{RL}\mathrm{grad}^{\alpha}_{a^+}[c_0]. \tag{43} $$

Proof. We will only calculate ${}^{RL}\partial^{\frac{1+\alpha_1}{2}}_{x_1, a_1^+}[f_i c_0]$ using integration by parts and Leibniz' rule, the determination of ${}^{RL}\partial^{\frac{1+\alpha_2}{2}}_{x_2, a_2^+}[f_i c_0]$ and ${}^{RL}\partial^{\frac{1+\alpha_3}{2}}_{x_3, a_3^+}[f_i c_0]$ being similar. Beforehand, note that

$m_i = 1$, for all $i = 1,2,3$. Recall now that $D^{\alpha}_{a^+}$ is a left inverse of $I^{\alpha}_{a^+}$; and by hypothesis $f_i(a_1, x_2, x_3) = 0$, for all $i = 1,2,3$. It follows then that

$$\begin{aligned} {}^{\mathrm{RL}}\partial^{\frac{1+\alpha_1}{2}}_{x_1,a_1^+}[f_i c_0](\vec{x}) &= \frac{\partial}{\partial x_1}\frac{1}{\Gamma(\frac{1-\alpha_1}{2})}\int_{a_1}^{x_1}\frac{f_i(t,x_2,x_3)c_0(t,x_2,x_3)}{(x_1-t)^{\frac{1+\alpha_1}{2}}}dt \\ &= \frac{\partial}{\partial x_1}\left(f_i(t,x_2,x_3)I^{\frac{1-\alpha_1}{2}}_{a_1^+}[c_0](t,x_2,x_3)\Big|_{a_1}^{x_1} - I^1_{a_1^+}\left[I^{\frac{1-\alpha_1}{2}}_{a_1^+}[c_0]\frac{\partial}{\partial x_1}f_i\right]\right) \\ &= \frac{\partial}{\partial x_1}\left(f_i(x_1,x_2,x_3)I^{\frac{1-\alpha_1}{2}}_{a_1^+}[c_0](x_1,x_2,x_3)\right) - I^{\frac{1-\alpha_1}{2}}_{a_1^+}[c_0]\frac{\partial}{\partial x_1}f_i \\ &= f_i(\vec{x})\,{}^{\mathrm{RL}}\partial^{\frac{1+\alpha_1}{2}}_{x_1,a_1^+}[c_0](\vec{x}). \end{aligned} \tag{44}$$

Analogously, ${}^{\mathrm{RL}}\partial^{\frac{1+\alpha_j}{2}}_{x_j,a_j^+}[f_i c_0] = f_i\,{}^{\mathrm{RL}}\partial^{\frac{1+\alpha_j}{2}}_{x_j,a_j^+}[c_0]$, for each $j = 2,3$ and $i = 1,2,3$. As a consequence,

$$ {}^{\mathrm{RL}}\operatorname{div}^{\alpha}_{a^+}[c_0\vec{f}] = \sum_{i=1}^{3}{}^{\mathrm{RL}}\partial^{\frac{1+\alpha_i}{2}}_{x_i,a_i^+}[c_0 f_i] = \sum_{i=1}^{3} f_i\,{}^{\mathrm{RL}}\partial^{\frac{1+\alpha_i}{2}}_{x_i,a_i^+}[c_0] = \vec{f}\cdot{}^{\mathrm{RL}}\operatorname{grad}^{\alpha}_{a^+}[c_0]. \tag{45}$$

This establishes the first identity of the conclusion. The proof of the second equation is analogous. □

Before closing this section, it is natural to compare qualitatively the results obtained in traditional vector calculus against those in the fractional case. In classical vector calculus, the following product rules are satisfied:

$$\operatorname{div}(c_0\vec{f}) = \vec{f}\cdot\operatorname{grad}c_0 + c_0\operatorname{div}\vec{f}, \tag{46}$$

$$\operatorname{curl}(v_0\vec{f}) = \vec{f}\times\operatorname{grad}c_0 + c_0\operatorname{curl}\vec{f}. \tag{47}$$

On the other hand, in the fully fractional case considered in Proposition 3, when we restrict $\vec{f}$ to the class of functions $\mathcal{Z}^{\alpha}_{a^+}(\Omega)$, the first part of these identities is also satisfied, except that the second terms on the right-hand sides of (46) and (47) are not present anymore. Notice that it is not difficult to construct a family of functions belonging to $\mathcal{Z}^{\alpha}_{a^+}(\Omega)$, for instance $\vec{f}(x) = (x_1-a_1)^{\gamma_1}(x_2-a_2)^{\gamma_2}(x_3-a_3)^{\gamma_3}\vec{g}(x)$ for all $\vec{g}\in AC^1(\Omega)$ and $\gamma_i \geq 0$ for all $i = 1,2,3$.

4. Fractional Div-Curl Systems

4.1. Properties of the Fractional Teodorescu Transform

As a derivation of the fractional Borel–Pompeiu formula [17], the authors defined the Caputo-type Teodorescu transform in a very similar way to the following definition, the difference being that the kernel is now a fundamental solution of the fractional Dirac operator defined in (20).

Definition 7. *Let $\vec{x} = (x_1, x_2, x_3)$ with $x_i > a_i$, for all $i = 1,2,3$. Define the* Riemann–Liouville and Caputo fractional Teodorescu transform *by*

$${}^{RL}T^{\alpha}_{a^+}[w](\vec{x}) = \int_{\Omega}{}^{RL}\mathcal{E}^{\alpha}_{a^+}(\vec{x}+a-\vec{y})w(\vec{y})\,d\vec{y}, \tag{48}$$

$${}^{C}T^{\alpha}_{a^+}[w](\vec{x}) = \int_{\Omega}{}^{C}\mathcal{E}^{\alpha}_{a^+}(\vec{x}+a-\vec{y})w(\vec{y})\,d\vec{y}. \tag{49}$$

Here, we follow the nomenclature and conventions of Lemma 1. Moreover, the derivatives ${}^{RL}\mathcal{E}^{\alpha}_{a^+} = -{}^{RL}\mathcal{D}^{\alpha}_{a^+}[u_0]$ and ${}^{C}\mathcal{E}^{\alpha}_{a^+} = -{}^{C}\mathcal{D}^{\alpha}_{a^+}[v_0]$ that appear in the kernel of (48) and (49), respecrively, are with respect to the variable $\vec{x}$.

In the following, it will be convenient to employ the translation operator, which is defined by $\mathcal{T}_{\vec{z}} f(\vec{y}) = f(\vec{y}+\vec{z})$. Similarly, we will use the reflection operator given by $\mathcal{R}_{\vec{y}} f(\vec{y}) = f(-\vec{y})$. An important relation between $\mathcal{D}^{\alpha}_{a^+}$ and $\mathcal{T}_\theta$ is

$$ {}^{RL}\mathcal{D}^{\alpha}_{(-\theta+a)^+}[\mathcal{T}_\theta f](\vec{x}) = {}^{RL}\mathcal{D}^{\alpha}_{a^+}[f](\vec{x}+\theta), \tag{50} $$

where the derivative is taken with respect to the variable $\vec{x}$. The proof is analogous to Theorem 11 [17], which is for the Caputo case, but we will give it here for completeness.

Proposition 4. *Let $\alpha_i \in (0,1)$, for each $i = 1,2,3$, and let $\alpha^* = \min_{1\le i\le 3}\{\alpha_i\}$. The fractional Teodorescu transform ${}^{RL}T^{\alpha}_{a^+}$ is a right inverse of the fractional Dirac operator ${}^{RL}\mathcal{D}^{\alpha}_{a^+}$ in $L^p(\Omega)$, for all $p \in \mathbb{R}$ satisfying $1 < p < \dfrac{2}{1-\alpha^*}$.*

Proof. Observe firstly that the fundamental solution ${}^{RL}\mathcal{E}^{\alpha}_{a^+}(\vec{x})$ of the fractional Dirac operator ${}^{RL}\mathcal{D}^{\alpha}_{a^+}$ is defined only for $x_i > a_i$, for each $i = 1,2,3$. Moreover, it satisfies the identity ${}^{RL}\mathcal{D}^{\alpha}_{a^+}\,{}^{RL}\mathcal{E}^{\alpha}_{a^+}(\vec{x}) = \delta(\vec{x}-a)$ or, equivalently, ${}^{RL}\Delta^{\alpha}_{a^+}[u_0](\vec{x}) = -{}^{RL}\mathcal{D}^{\alpha}_{a^+}\,{}^{RL}\mathcal{D}^{\alpha}_{a^+}[u_0](\vec{x}) = \delta(\vec{x}-a)$. In the following, the derivatives ${}^{RL}\mathcal{D}^{\alpha}_{\vec{y}^+}$ and ${}^{RL}\mathcal{D}^{\alpha}_{a^+}$ are with respect to the variable $\vec{x}$. Using (50) with $\theta = a - \vec{y}$ yields

$$ \begin{aligned} {}^{RL}\mathcal{D}^{\alpha}_{a^+}\,{}^{RL}T^{\alpha}_{a^+}[w](\vec{x}) &= \int_\Omega {}^{RL}\mathcal{D}^{\alpha}_{\vec{y}^+}\,{}^{RL}\mathcal{E}^{\alpha}_{a^+}(\vec{x}+a-\vec{y})w(\vec{y})\,d\vec{y} \\ &= -\int_\Omega {}^{RL}\mathcal{D}^{\alpha}_{\vec{y}^+}\Big(\mathcal{T}_{a-\vec{y}}({}^{RL}\mathcal{D}^{\alpha}_{a^+}[u_0](\vec{x}))\Big)w(\vec{y})\,d\vec{y} \\ &= -\int_\Omega \mathcal{T}_{a-\vec{y}}({}^{RL}\mathcal{D}^{\alpha}_{a^+}\,{}^{RL}\mathcal{D}^{\alpha}_{a^+}[u_0](\vec{x}))w(\vec{y})\,dy \\ &= \int_\Omega \mathcal{T}_{a-\vec{y}}\delta(\vec{x}-a)w(\vec{y})\,dy \\ &= \int_\Omega \delta(\vec{x}-\vec{y})\,w(\vec{y})\,d\vec{y} = w(\vec{x}), \end{aligned} \tag{51} $$

which we wished to prove. □

In the following, we will employ a key decomposition of the classical Teodorescu operator used in [20–23] for different kinds of bounded or unbounded domains in $\mathbb{R}^3$. For the fractional version, we denote the component operators of the fractional Teodorescu transform as follows:

$$ {}^{RL}T^{\alpha}_{a^+}[w_0+\vec{w}] := {}^{RL}T^{\alpha}_{0,a^+}[\vec{w}] + {}^{RL}\overrightarrow{T}^{\alpha}_{1,a^+}[w_0] + {}^{RL}\overrightarrow{T}^{\alpha}_{2,a^+}[\vec{w}]. \tag{52} $$

The first term on the right-hand side of Equation (52) is the scalar part, while the last two summands represent the vector part, and it has been split into two components for the sake of convenience. These three terms are given by

$$ {}^{RL}T^{\alpha}_{0,a^+}[\vec{w}](\vec{x}) = -\int_\Omega {}^{RL}\mathcal{E}^{\alpha}_{a^+}(\vec{x}+a-\vec{y})\cdot\vec{w}(\vec{y})\,d\vec{y}, \tag{53} $$

$$ {}^{RL}\overrightarrow{T}^{\alpha}_{1,a^+}[w_0](\vec{x}) = \int_\Omega {}^{RL}\mathcal{E}^{\alpha}_{a^+}(\vec{x}+a-\vec{y})w_0(\vec{y})\,d\vec{y}, \tag{54} $$

$$ {}^{RL}\overrightarrow{T}^{\alpha}_{2,a^+}[\vec{w}](\vec{x}) = \int_\Omega {}^{RL}\mathcal{E}^{\alpha}_{a^+}(\vec{x}+a-\vec{y})\times\vec{w}(\vec{y})\,d\vec{y}. \tag{55} $$

We will see later in Corollary 2 how some of these component operators themselves represent right inverse operators of the fractional divergence and rotational operators, under certain conditions. Moreover, ${}^{RL}T^{\alpha}_{a^+}$, as a good generalization of the classical Teodorescu operator in quaternionic analysis, preserves many of its properties. To this end, we use the above decomposition (52), ${}^{RL}T^{\alpha}_{a^+}[\vec{g}] = {}^{RL}T^{\alpha}_{0,a^+}[\vec{g}] + {}^{RL}\overrightarrow{T}^{\alpha}_{2,a^+}[\vec{g}]$, to see necessary and sufficient conditions to guarantee that both its scalar part and its vector part belong to the kernel of the fractional Laplacian ${}^{RL}\Delta^{\alpha}_{a^+}$. In order to apply the factorization of Corollary 1 to the fractional Teodorescu transform, we will need to guarantee that the condition ${}^{RL}T^{\alpha}_{a^+}[\vec{g}] \in \mathcal{Z}^{\alpha}_{a^+}(\Omega)$ can be satisfied.

Proposition 5. *If ${}^{RL}T^{\alpha}_{a^+}[\vec{g}] \in \mathcal{Z}^{\alpha}_{a^+}(\Omega)$, then the following hold for the fractional Teodorescu transform ${}^{RL}T^{\alpha}_{a^+}[\vec{g}]$:*

(i) *The scalar part of ${}^{RL}T^{\alpha}_{a^+}[\vec{g}]$, ${}^{RL}T^{\alpha}_{0,a^+}[\vec{g}]$, belongs to the kernel of ${}^{RL}\Delta^{\alpha}_{a^+}$ if and only if ${}^{RL}\mathrm{div}^{\alpha}_{a^+}\, \vec{g} = 0$.*

(ii) *The vector part of ${}^{RL}T^{\alpha}_{a^+}[\vec{g}]$, ${}^{RL}\overrightarrow{T}^{\alpha}_{2,a^+}[\vec{g}]$, belongs to the kernel of ${}^{RL}\Delta^{\alpha}_{a^+}$ if and only if ${}^{RL}\mathrm{curl}^{\alpha}_{a^+}\, \vec{g} = 0$.*

Moreover, the statements (i) *and* (ii) *also hold for the Caputo fractional Teodorescu transform.*

Proof. Using the factorization in Corollary 1, Proposition 4 and the decomposition (31), it readily follows that

$$ {}^{RL}\Delta^{\alpha}_{a^+}\, {}^{RL}T^{\alpha}_{a^+}[\vec{g}] = -{}^{RL}\mathcal{D}^{\alpha}_{a^+}\, {}^{RL}\mathcal{D}^{\alpha}_{a^+}\, {}^{RL}T^{\alpha}_{a^+}[\vec{g}] = {}^{RL}\mathrm{div}^{\alpha}_{a^+}\, \vec{g} - {}^{RL}\mathrm{curl}^{\alpha}_{a^+}\, \vec{g}. \tag{56} $$

Taking its scalar part or vector part, respectively, we obtain the desired result. □

4.2. Riemann–Liouville System

In the following, we will study a fractional form of the classical div-curl system and construct its solution. More precisely, we fix the domain Ω, and consider the fractional system

$$ \begin{cases} {}^{RL}\mathrm{div}^{\alpha}_{a^+}\, \vec{w} = g_0, \\ {}^{RL}\mathrm{curl}^{\alpha}_{a^+}\, \vec{w} = \vec{g}, \end{cases} \tag{57} $$

where $g_0 \in L^p(\Omega, \mathbb{R})$ and $\vec{g} \in L^p(\Omega, \mathbb{R}^3)$, for some $1 < p < 2/(1-\alpha^*)$. Notice that if the solution is such that $\vec{w} \in \mathcal{Z}^{\alpha}_{a^+}(\Omega)$, then $\vec{g}$ satisfies ${}^{RL}\mathrm{div}^{\alpha}_{a^+}\, \vec{g} = 0$, i.e., $\vec{g}$ is a 'fractional divergence-free' vector field. Let $f_0 \in AC(\Omega)$. The following relations will be fundamental in the sequel:

$$ \begin{aligned} {}^{RL}\mathrm{grad}^{\alpha}_{a^+}[f_0](\theta - \vec{y}) &= -{}^{RL}\mathrm{grad}^{\alpha}_{(\theta-a)^-}[\mathcal{T}_\theta \mathcal{R}_{\vec{y}}[f_0]](\vec{y}), \\ {}^{C}\mathrm{grad}^{\alpha}_{a^+}[f_0](\theta - \vec{y}) &= -{}^{C}\mathrm{grad}^{\alpha}_{(\theta-a)^-}[\mathcal{T}_\theta \mathcal{R}_{\vec{y}}[f_0]](\vec{y}). \end{aligned} \tag{58} $$

Here, the derivatives are taken with respect to the variable $\vec{y}$. In the sequel and for the sake of convenience, we will employ Ker to denote the kernel of operators. As in the previous section, we will let $\Omega = \Pi_{i=1}^{3}(a_i, b_i)$ be a bounded open rectangular domain in $\mathbb{R}^3$, and assume that $\alpha = (\alpha_1, \alpha_2, \alpha_3)$, with $\alpha_i \in (0,1)$, for each $i = 1,2,3$. Using this notation, we have the following result.

Theorem 2. *Let $g = g_0 + \vec{g} \in L^p(\Omega)$ with $1 < p < 2/(1-\alpha^*)$. If ${}^{RL}\mathrm{div}^{\alpha}_{a^+}[\vec{g}] = 0$ and the normal trace of $I^{\alpha}_{a^+}[\vec{g}]$ vanishes, then a general weak solution of the fractional Riemann–Liouville div-curl system (57) is given by*

$$ \vec{w} = -{}^{RL}\overrightarrow{T}^{\alpha}_{1,a^+}[g_0] + {}^{RL}\overrightarrow{T}^{\alpha}_{2,a^+}[\vec{g}] + {}^{RL}\mathrm{grad}^{\alpha}_{a^+}[h], \tag{59} $$

where $h \in \mathrm{Ker}({}^{RL}\Delta^{\alpha}_{a^+}) \cap \mathcal{Z}^{\alpha}_{a^+}(\Omega)$ is an arbitrary scalar function.

Proof. By Proposition 4 and the decomposition (31), the fractional Teodorescu transform ${}^{RL}T^{\alpha}_{a^+}[-g_0 + \vec{g}]$ is a quaternionic solution of the system (57). To obtain a pure-vector solution, note that the decomposition (52) yields

$$ {}^{RL}T^{\alpha}_{a^+}[-g_0 + \vec{g}] = {}^{RL}T^{\alpha}_{0,a^+}[\vec{g}] - {}^{RL}\overrightarrow{T}^{\alpha}_{1,a^+}[g_0] + {}^{RL}\overrightarrow{T}^{\alpha}_{2,a^+}[\vec{g}]. \tag{60} $$

Taking $h(\vec{y}) = \mathcal{T}_{\vec{x}+a}\mathcal{R}_{\vec{y}}u_0(\vec{y})$ and $f(\vec{y}) = \vec{g}(\vec{y})$ in the Stokes formula (38) (all derivatives with respect to the variable $\vec{y}$) and letting u_0 be a fundamental solution of ${}^{\mathrm{RL}}\Delta^{\alpha}_{a^+}$ given in Lemma 1, we obtain

$$\begin{aligned}&\int_{\partial\Omega} \mathcal{T}_{\vec{x}+a}\mathcal{R}_{\vec{y}}[u_0](\vec{y})\eta(\vec{y})I^{\alpha}_{a^+}[\vec{g}]\,ds_{\vec{y}} \\ &= \int_{\Omega}\Big(\Big(-[\mathcal{T}_{\vec{x}+a}\mathcal{R}_{\vec{y}}[u_0]]^{\mathrm{C}}\mathcal{D}^{\alpha}_{b^-}\Big)(\vec{y})\vec{g}(\vec{y}) + \mathcal{T}_{\vec{x}+a}\mathcal{R}_{\vec{y}}[u_0](\vec{y})^{\mathrm{RL}}\mathcal{D}^{\alpha}_{a^+}[\vec{g}](\vec{y})\Big)\,d\vec{y}.\end{aligned} \tag{61}$$

Calculate the scalar part of (61) and use the hypotheses $-\mathrm{Sc}\,\mathcal{D}^{\alpha}_{a^+}[\vec{g}] = {}^{\mathrm{RL}}\mathrm{div}^{\alpha}_{a^+}[\vec{g}] = 0$ in Ω and $I^{\alpha}_{a^+}[\vec{g}]| \cdot \eta = 0$ on $\partial\Omega$ to reach

$$\int_{\Omega} {}^{\mathrm{C}}\mathrm{grad}^{\alpha}_{b^-}\Big[\mathcal{T}_{\vec{x}+a}\mathcal{R}_{\vec{y}}[u_0]\Big](\vec{y}) \cdot \vec{g}(\vec{y})\,d\vec{y} = 0. \tag{62}$$

By differentiating now under the integral sign and using the traditional Leibniz' rule, we readily obtain the following fundamental relation:

$${}^{\mathrm{RL}}\mathrm{grad}^{\alpha}_{a^+,\vec{x}}[u_0](\vec{x}+a-\vec{y}) = -{}^{\mathrm{RL}}\mathrm{grad}^{\alpha}_{a^+,\vec{y}}[u_0](\vec{x}+a-\vec{y}). \tag{63}$$

Here, the second sub-index indicates whether we are taking derivatives with respect to the variable $\vec{x}$ or $\vec{y}$. Recall that ${}^{\mathrm{RL}}\mathcal{E}^{\alpha}_{a^+} = -{}^{\mathrm{RL}}\mathrm{grad}^{\alpha}_{a^+,\vec{x}}[u_0]$ with respect to the variable $\vec{x}$. On the other hand, due to $u_0 \in \mathcal{Z}^{\alpha}_{a^+}(\Omega)$ and relations (12), (63) and (58), we obtain

$${}^{\mathrm{RL}}\mathcal{E}^{\alpha}_{a^+}(\vec{x}+a-\vec{y}) = {}^{\mathrm{C}}\mathrm{grad}^{\alpha}_{a^+,\vec{y}}[u_0](\vec{x}+a-\vec{y}) = -{}^{\mathrm{C}}\mathrm{grad}^{\alpha}_{x^-,\vec{y}}[\mathcal{T}_{\vec{x}+a}\mathcal{R}_{\vec{y}}[u_0]](\vec{y}). \tag{64}$$

However, we know that ${}^{\mathrm{RL}}\mathcal{E}^{\alpha}_{a^+}(\vec{x}+a-\vec{y})$ is only defined when $x_i > y_i$, for all $i = 1,2,3$. As a consequence, we can readily replace the operator ${}^{\mathrm{C}}\mathrm{grad}^{\alpha}_{b^-}$ by ${}^{\mathrm{C}}\mathrm{grad}^{\alpha}_{x^-}$ in (62). Finally, by employing (64) and (62), we readily reach

$${}^{\mathrm{RL}}T^{\alpha}_{0,a^+}[\vec{g}](\vec{x}) = -\int_{\Omega} {}^{\mathrm{RL}}\mathcal{E}^{\alpha}_{a^+}(\vec{x}+a-\vec{y}) \cdot \vec{g}(\vec{y})\,d\vec{y} = 0, \quad \text{in } \Omega. \tag{65}$$

This means that ${}^{\mathrm{RL}}T^{\alpha}_{a^+}[\vec{g}] = -{}^{\mathrm{RL}}\overrightarrow{T}^{\alpha}_{1,a^+}[g_0] + {}^{\mathrm{RL}}\overrightarrow{T}^{\alpha}_{2,a^+}[\vec{g}]$ is purely vectorial and, moreover,

$$(-{}^{\mathrm{RL}}\mathrm{div}^{\alpha}_{a^+} + {}^{\mathrm{RL}}\mathrm{curl}^{\alpha}_{a^+})\,{}^{\mathrm{RL}}T^{\alpha}_{a^+}[-g_0+\vec{g}] = {}^{\mathrm{RL}}\mathcal{D}^{\alpha}_{a^+}\,{}^{\mathrm{RL}}T^{\alpha}_{a^+}[-g_0+\vec{g}] = -g_0+\vec{g}. \tag{66}$$

Setting the scalar and vector parts equal to each other, we obtain that ${}^{\mathrm{RL}}T^{\alpha}_{a^+}[\vec{g}] = -{}^{\mathrm{RL}}\overrightarrow{T}^{\alpha}_{1,a^+}[g_0] + {}^{\mathrm{RL}}\overrightarrow{T}^{\alpha}_{2,a^+}[\vec{g}]$ is a solution of the fractional div-curl system (57). Finally, the fact that the solution is not unique is a consequence of the identities (ii) and (iii) of Proposition 1. □

In the limit $\alpha_i \to 1^-$, the fractional div-curl system (57) reduces to the well-known integer-order system from vector calculus, and the hypothesis ${}^{\mathrm{RL}}\mathrm{div}^{\alpha}_{a^+}[\vec{g}] = 0$ reduces to the evident requirement that $\vec{g}$ be a divergence-free vector field. Moreover, $I^{\frac{1-\alpha_i}{2}}_{a_i^+}[u_0\vec{g}] \to u_0\vec{g}$. This means that it is sufficient to require that $\vec{g}$ has zero normal trace. Taking $g_0 \equiv 0$ or $\vec{g} \equiv 0$ in (57), we readily obtain

Corollary 2. *Under the same assumptions of Theorem 2, ${}^{RL}\overrightarrow{T}^{\alpha}_{2,a^+}$ is a right inverse operator of ${}^{RL}\mathrm{curl}^{\alpha}_{a^+}$ in the class of functions considered in Theorem 2. Meanwhile, $-{}^{RL}\overrightarrow{T}^{\alpha}_{1,a^+}$ is always a right inverse operator of ${}^{RL}\mathrm{div}^{\alpha}_{a^+}$ in $L^p(\Omega)$.*

Another important analogy with the classical vector calculus is that the solution of the non-linear fractional system (57) also admits a Helmholtz-type fractional decomposition as follows.

Proposition 6. *Under the same assumptions of Theorem 2, the general weak solution of the fractional Riemann–Liouville div-curl system (57) admits a fractional Helmholtz decomposition as follows*

$$\vec{w} = {}^{RL}\mathrm{grad}^{\alpha}_{\vec{x}^-}\,\varphi_0 - {}^{RL}\mathrm{curl}^{\alpha}_{\vec{x}^-}\,\vec{\varphi}, \quad \text{in } \Omega, \tag{67}$$

where the scalar potential φ_0 and the vector potential $\vec{\varphi}$ are given by

$$\varphi_0(\vec{x}) = \int_{\Omega} \mathcal{T}_{\vec{x}+a}\mathcal{R}_{\vec{y}}[u_0](\vec{y})g_0(\vec{y})\,d\vec{y}, \quad \vec{\varphi}(\vec{x}) = \int_{\Omega} \mathcal{T}_{\vec{x}+a}\mathcal{R}_{\vec{y}}[u_0](\vec{y}) \times \vec{g}(\vec{y})\,d\vec{y}.$$

Proof. Due to ${}^{RL}\mathcal{E}^{\alpha}_{a^+} = -{}^{RL}\mathrm{grad}^{\alpha}_{a^+}[u_0]$ and by relation (58), we can write

$$\begin{aligned} {}^{RL}\vec{T}^{\alpha}_{1,a^+}[g_0](\vec{x}) &= \int_{\Omega} {}^{RL}\mathrm{grad}^{\alpha}_{\vec{x}^-}\,\mathcal{T}_{\vec{x}+a}\mathcal{R}_{\vec{y}}[u_0](\vec{y})g_0(\vec{y})\,d\vec{y}, \\ {}^{RL}\vec{T}^{\alpha}_{2,a^+}[\vec{g}](\vec{x}) &= \int_{\Omega} {}^{RL}\mathrm{grad}^{\alpha}_{\vec{x}^-}\,\mathcal{T}_{\vec{x}+a}\mathcal{R}_{\vec{y}}[u_0](\vec{y}) \times \vec{g}(\vec{y})\,d\vec{y}. \end{aligned}$$

Finally, since the fractional gradient involved in the above re-expressions of the component operators ${}^{RL}\vec{T}^{\alpha}_{1,a^+}$ and ${}^{RL}\vec{T}^{\alpha}_{2,a^+}$ is taken with respect to the variable $\vec{x}$, we readily obtain (67). □

4.3. Caputo System

For the corresponding fractional div-curl system in the sense of Caputo derivatives, we will follow the same approach as that used with the Riemann–Liouville div-curl system (57). Let us consider the system

$$\begin{cases} {}^{C}\mathrm{div}^{\alpha}_{a^+}\,\vec{w} = g_0, \\ {}^{C}\mathrm{curl}^{\alpha}_{a^+}\,\vec{w} = \vec{g}, \end{cases} \tag{68}$$

where $g_0 \in L^p(\Omega, \mathbb{R})$ and $\vec{g} \in L^p(\Omega, \mathbb{R}^3)$, for some $1 < p < 2/(1-p^*)$. The cornerstone in our analysis will be again the fractional Teodorescu transform associated with Caputo derivatives, which means that its kernel is a fundamental solution of the fractional Dirac operator ${}^{C}\mathcal{D}^{\alpha}_{a^+}$. As in the case of the Riemann–Liouville fractional Teodorescu operator, we define the decomposition

$${}^{C}T^{\alpha}_{a^+}[w_0 + \vec{w}] := {}^{C}T^{\alpha}_{0,a^+}[\vec{w}] + {}^{C}\vec{T}^{\alpha}_{1,a^+}[w_0] + {}^{C}\vec{T}^{\alpha}_{2,a^+}[\vec{w}], \tag{69}$$

where ${}^{C}T^{\alpha}_{0,a^+}$, ${}^{C}\vec{T}^{\alpha}_{1,a^+}$ and ${}^{C}\vec{T}^{\alpha}_{2,a^+}$ are given by (53), (54) and (55), respectively, but using now the Caputo kernel ${}^{C}\mathcal{E}^{\alpha}_{a^+}$ in the integrand, instead of the Riemann–Liouville kernel ${}^{RL}\mathcal{E}^{\alpha}_{a^+}$.

Analogously to the proof of Theorem 2, it follows that ${}^{C}T^{\alpha}_{a^+}[-g_0 + \vec{g}]$ is a quaternionic solution of the fractional div-curl system (68). This is a direct consequence of the fact that ${}^{C}T^{\alpha}_{a^+}$ is a right inverse of ${}^{C}\mathcal{D}^{\alpha}_{a^+}$ in L^p(see [17], Theorem 11). Apply the decomposition (69) to the quaternion-valued function $g = g_0 + \vec{g}$, where g_0 and $\vec{g}$ are the known data provided by the fractional div-curl system (68). In this way, we notice that

$${}^{C}T^{\alpha}_{a^+}[-g_0 + \vec{g}] = {}^{C}T^{\alpha}_{0,a^+}[\vec{g}] - {}^{C}\vec{T}^{\alpha}_{1,a^+}[g_0] + {}^{C}\vec{T}^{\alpha}_{2,a^+}[\vec{g}].$$

To obtain a purely vectorial solution, we will impose suitable conditions over $\vec{g}$ in order to guarantee that the scalar part of ${}^{C}T^{\alpha}_{a^+}[-g_0 + \vec{g}]$ vanishes in Ω, i.e., that ${}^{C}T^{\alpha}_{0,a^+}[\vec{g}] \equiv 0$ is satisfied in Ω. We will see that the fractional divergence-free functions of the Riemann–Liouville type, whose Riemann–Liouville fractional integral has zero normal trace, belong to the kernel of the operator ${}^{C}T^{\alpha}_{0,a^+}$, in the same way as seen in Theorem 2 for the operator ${}^{RL}T^{\alpha}_{0,a^+}$. Nevertheless, the upcoming proof is relatively more straightforward.

Theorem 3. *Let $g = g_0 + \vec{g} \in L^p(\Omega)$ with $1 < p < 2/(1-\alpha^*)$. If ${}^{RL}\text{div}^{\alpha}_{a^+}[\vec{g}] = 0$ and the normal trace of $I^{\alpha}_{a^+}[\vec{g}]$ vanishes, then a general weak solution of the fractional Caputo div-curl system (68) is given by*

$$\vec{w} = -{}^{C}\overrightarrow{T}^{\alpha}_{1,a^+}[g_0] + {}^{C}\overrightarrow{T}^{\alpha}_{2,a^+}[\vec{g}] + {}^{C}\text{grad}^{\alpha}_{a^+}[h], \tag{70}$$

where $h \in \text{Ker}({}^{C}\Delta^{\alpha}_{a^+}) \cap \mathcal{Z}^{\alpha}_{a^+}(\Omega)$ is an arbitrary scalar function.

Proof. It suffices to prove that ${}^{C}T^{\alpha}_{0,a^+}[\vec{g}] = 0$ in Ω, in light of the previous discussion. Take $h(\vec{y}) = \mathcal{T}_{\vec{x}+a}\mathcal{R}_{\vec{y}}v_0(\vec{y})$ and $f(\vec{y}) = \vec{g}(\vec{y})$ in the Stokes formula (38), and let v_0 be the fundamental solution of ${}^{C}\Delta^{\alpha}_{a^+}$ (25). It follows that

$$\begin{aligned}&\int_{\partial\Omega} \mathcal{T}_{\vec{x}+a}\mathcal{R}_{\vec{y}}[v_0](\vec{y})\eta(\vec{y})I^{\alpha}_{a^+}[\vec{g}]\, ds_{\vec{y}} \\ &= \int_{\Omega}\Big(\Big(-[\mathcal{T}_{\vec{x}+a}\mathcal{R}_{\vec{y}}[v_0]]^{C}\mathcal{D}^{\alpha}_{b^-}\Big)(\vec{y})\vec{g}(\vec{y}) + \mathcal{T}_{\vec{x}+a}\mathcal{R}_{\vec{y}}[v_0](\vec{y})^{RL}\mathcal{D}^{\alpha}_{a^+}[\vec{g}](\vec{y})\Big)\, d\vec{y}.\end{aligned} \tag{71}$$

Taking now the scalar part of (71) and using the hypotheses, we readily obtain

$$\int_{\Omega} {}^{C}\text{grad}^{\alpha}_{b^-}\Big[\mathcal{T}_{\vec{x}+a}\mathcal{R}_{\vec{y}}[v_0]\Big](\vec{y}) \cdot \vec{g}(\vec{y})\, d\vec{y} = 0. \tag{72}$$

As a consequence of (58), we reach ${}^{C}\mathcal{E}^{\alpha}_{a^+}(\vec{x}+a-\vec{y}) = -{}^{C}\text{grad}^{\alpha}_{x^-}[\mathcal{T}_{\vec{x}+a}\mathcal{R}_{\vec{y}}[v_0]](\vec{y})$. However, we know that ${}^{C}\mathcal{E}^{\alpha}_{a^+}(\vec{x}+a-\vec{y})$ is only defined for $x_i > y_i$, for all $i = 1,2,3$. This implies that we can replace the operator ${}^{C}\text{grad}^{\alpha}_{b^-}$ with ${}^{C}\text{grad}^{\alpha}_{x^-}$ in (72). It is easy to see then that

$${}^{C}T^{\alpha}_{0,a^+}[\vec{g}](\vec{x}) = -\int_{\Omega} {}^{C}\mathcal{E}^{\alpha}_{a^+}(\vec{x}+a-\vec{y}) \cdot \vec{g}(\vec{y})\, d\vec{y} = 0, \quad \text{in } \Omega, \tag{73}$$

whence the conclusion readily follows. □

Corollary 3. *Under the same assumptions of Theorem 3, ${}^{C}\overrightarrow{T}^{\alpha}_{2,a^+}$ is a right inverse operator of ${}^{C}\text{curl}^{\alpha}_{a^+}$ in the class of functions considered in Theorem 3. Moreover, $-{}^{C}\overrightarrow{T}^{\alpha}_{1,a^+}$ is a right inverse operator of ${}^{C}\text{div}^{\alpha}_{a^+}$ in $L^p(\Omega)$.*

Analogously to the Riemann–Liouville case, the fractional Caputo div-curl system (68) also admits a fractional Helmholtz decomposition, but now the potential is in terms of v_0 defined in (25), which is a fundamental solution of the fractional Laplace operator of Caputo type.

Proposition 7. *Under the same assumptions of Theorem 3, the general weak solution of the fractional Caputo div-curl system (68) admits a fractional Helmholtz decomposition as follows*

$$\vec{w} = {}^{C}\text{grad}^{\alpha}_{\vec{x}^-}\psi_0 - {}^{C}\text{curl}^{\alpha}_{\vec{x}^-}\vec{\psi}, \quad \text{in } \Omega, \tag{74}$$

where the scalar potential ψ_0 and the vector potential $\vec{\psi}$ are given by

$$\psi_0(\vec{x}) = \int_{\Omega} \mathcal{T}_{\vec{x}+a}\mathcal{R}_{\vec{y}}[v_0](\vec{y})g_0(\vec{y})\, d\vec{y}, \quad \vec{\psi}(\vec{x}) = \int_{\Omega} \mathcal{T}_{\vec{x}+a}\mathcal{R}_{\vec{y}}[v_0](\vec{y}) \times \vec{g}(\vec{y})\, d\vec{y}.$$

Proof. The proof is analogous to that of Proposition 6 for the Riemann–Liouville case. □

5. Application

Let $\Omega = \Pi^3_{i=1}(a_i, b_i)$ be as in the previous sections, and assume $0 < \alpha_i < 1$, for all $i = 1,2,3$. The present section provides some consequences of the factorization provided by Corollary 1 to the construction of fractional hyper-conjugate pairs. In addition, we will give an explicit expression of a right inverse of the fractional gradient of Caputo type considering different derivative orders.

Definition 8. *Let* $w = w_0 + \vec{w} \in AC(\Omega)$*. We say that* $(w_0, \vec{w})$ *is a Riemann–Liouville fractional hyper-conjugate pair if* $w \in \mathrm{Ker}({}^{RL}\mathcal{D}^{\alpha}_{a^+})$*. Analogously,* $(w_0, \vec{w})$ *is a Caputo fractional hyper-conjugate pair if* $w \in \mathrm{Ker}({}^{C}\mathcal{D}^{\alpha}_{a^+})$*.*

The following result is a straightforward consequence of the factorization of the fractional Laplace operator in the class $\mathcal{Z}^{\alpha}_{a^+}(\Omega)$ provided by Corollary 1. For this reason, we omit the proof.

Corollary 4. *Let* $w = \sum_{i=0}^{3} w_i$*, and suppose that* $w \in \mathrm{Ker}({}^{RL}\mathcal{D}^{\alpha}_{a^+}) \cap \mathcal{Z}^{\alpha}_{a^+}(\Omega)$ *(respectively,* $w \in \mathrm{Ker}({}^{C}\mathcal{D}^{\alpha}_{a^+}) \cap \mathcal{Z}^{\alpha}_{a^+}(\Omega)$*). Then,* $w_i \in \mathrm{Ker}({}^{RL}\Delta^{\alpha}_{a^+})$ *(respectively,* $w_i \in \mathrm{Ker}({}^{C}\Delta^{\alpha}_{a^+})$*), for all* $i = 0, 1, 2, 3$.

By Definition 8, it is obvious that $(w_0, \vec{w})$ forms a Riemann–Liouville fractional hyper-conjugate pair if and only if the following fractional div-curl system is satisfied:

$$\begin{aligned} {}^{\mathrm{RL}}\,\mathrm{div}^{\alpha}_{a^+}\,\vec{w} &= 0, \\ {}^{\mathrm{RL}}\,\mathrm{curl}^{\alpha}_{a^+}\,\vec{w} &= -{}^{\mathrm{RL}}\,\mathrm{grad}^{\alpha}_{a^+}\,w_0. \end{aligned} \tag{75}$$

Similarly, $(w_0, \vec{w})$ is a Caputo fractional hyper-conjugate pair if and only if

$$\begin{aligned} {}^{\mathrm{C}}\,\mathrm{div}^{\alpha}_{a^+}\,\vec{w} &= 0, \\ {}^{\mathrm{C}}\,\mathrm{curl}^{\alpha}_{a^+}\,\vec{w} &= -{}^{\mathrm{C}}\,\mathrm{grad}^{\alpha}_{a^+}\,w_0. \end{aligned} \tag{76}$$

The above systems (75) and (76) can be considered fractional generalizations of the Moisil–Teodorescu system studied for the first time in [24].

Let us define the following integral operator in terms of the Riemann–Liouville fractional integrals $I^{\frac{1+\alpha_i}{2}}_{a_i^+}$ as

$$\mathcal{A}^{\alpha}_{a^+}[\vec{f}](\vec{x}) = I^{\frac{1+\alpha_1}{2}}_{a_1^+}[f_1](x_1, a_2, a_3) + I^{\frac{1+\alpha_2}{2}}_{a_2^+}[f_2](x_1, x_2, a_3) + I^{\frac{1+\alpha_3}{2}}_{a_3^+}[f_3](x_1, x_2, x_3). \tag{77}$$

As the following result shows, it turns out that $\mathcal{A}^{\alpha}_{a^+}$ behaves as a right-inverse operator of ${}^{\mathrm{C}}\,\mathrm{grad}^{\alpha}_{a^+}$ in the class of functions satisfying ${}^{\mathrm{C}}\,\mathrm{curl}^{\alpha}_{a^+}\,\vec{f} = 0$. For this reason, $\mathcal{A}^{\alpha}_{a^+}$ is called the *fractional anti-gradient operator*.

Proposition 8. *If* ${}^{\mathrm{C}}\,\mathrm{curl}^{\alpha}_{a^+}\,\vec{f} = 0$*, then* ${}^{\mathrm{C}}\,\mathrm{grad}^{\alpha}_{a^+}\,\mathcal{A}^{\alpha}_{a^+}[\vec{f}] = \vec{f}$*.*

Proof. Using the characterization of Caputo fractional hyper-conjugate pairs given under Corollary 4 and differentiating under the integral sign, we readily obtain

$$\begin{aligned} {}^{\mathrm{C}}\partial^{\frac{1+\alpha_1}{2}}_{x_1,a_1^+}\mathcal{A}^{\alpha}_{a^+}[\vec{f}](\vec{x}) &= {}^{\mathrm{C}}\,\partial^{\frac{1+\alpha_1}{2}}_{x_1,a_1^+}\left(I^{\frac{1+\alpha_1}{2}}_{a_1^+}[f_1](x_1, a_2, a_3) + I^{\frac{1+\alpha_2}{2}}_{a_2^+}[f_2](x_1, x_2, a_3) + I^{\frac{1+\alpha_3}{2}}_{a_3^+}[f_3](x_1, x_2, x_3)\right) \\ &= f_1(x_1, a_2, a_3) + I^{\frac{1+\alpha_2}{2}}_{a_2^+}\,{}^{\mathrm{C}}\partial^{\frac{1+\alpha_1}{2}}_{x_1,a_1^+}[f_2](x_1, x_2, a_3) + I^{\frac{1+\alpha_3}{2}}_{a_3^+}\,{}^{\mathrm{C}}\partial^{\frac{1+\alpha_1}{2}}_{x_1,a_1^+}[f_3](x_1, x_2, x_3). \end{aligned} \tag{78}$$

Now, by hypothesis ${}^{\mathrm{C}}\,\mathrm{curl}^{\alpha}_{a^+}\,\vec{f} = 0$ or, equivalently, the following identities are satisfied:

$${}^{\mathrm{C}}\partial^{\frac{1+\alpha_2}{2}}_{x_2,a_2}[f_3] - {}^{\mathrm{C}}\,\partial^{\frac{1+\alpha_3}{2}}_{x_3,a_3}[f_2] = {}^{\mathrm{C}}\partial^{\frac{1+\alpha_3}{2}}_{x_3,a_3}[f_1] - {}^{\mathrm{C}}\,\partial^{\frac{1+\alpha_1}{2}}_{x_1,a_1}[f_3] = {}^{\mathrm{C}}\partial^{\frac{1+\alpha_1}{2}}_{x_1,a_1}[f_2] - {}^{\mathrm{C}}\,\partial^{\frac{1+\alpha_2}{2}}_{x_2,a_2}[f_1] = 0. \tag{79}$$

Substituting (79) into (78) and using the composition rule (10), it follows that

$$ {}^{C}\partial^{\frac{1+\alpha_1}{2}}_{x_1,a_1^+} \mathcal{A}^{\alpha}_{a^+}[\vec{f}](\vec{x}) = f_1(x_1,a_2,a_3) + f_1(x_1,x_2,a_3) - f_1(x_1,a_2,a_3) + f_1(x_1,x_2,x_3) - f_1(x_1,x_2,a_3) = f_1(\vec{x}). \quad (80) $$

Analogously, one can establish that ${}^{C}\partial^{\frac{1+\alpha_i}{2}}_{x_i,a_i^+} \mathcal{A}^{\alpha}_{a^+}[\vec{f}](\vec{x}) = f_i(\vec{x})$, for $i = 2,3$. The conclusion readily follows now. □

The next proposition shows that the fractional anti-gradient operator $\mathcal{A}^{\alpha}_{a^+}$ allows us to construct a Caputo fractional hyper-conjugate pair w_0 when $\vec{w} \in \mathrm{Ker}({}^{C}\Delta^{\alpha}_{a^+})$ is known beforehand. The proposition is clearly a generalization of [20], Proposition 2.1.

Proposition 9. *Let $\vec{w} \in \mathrm{Ker}({}^{C}\Delta^{\alpha}_{a^+}) \cap \mathcal{Z}^{\alpha}_{a^+}(\Omega)$. A necessary and sufficient condition for the existence of a Caputo fractional hyper-conjugate pair of $\vec{w}$ is that ${}^{C}\operatorname{div}^{\alpha}_{a^+} \vec{w} = 0$. In that case, there is w_0 such that $w = w_0 + \vec{w} \in \mathrm{Ker}({}^{C}\mathcal{D}^{\alpha}_{a^+})$.*

Proof. The necessity is clear due to the characterization of Caputo fractional hyper-conjugate pairs provided by (76). Suppose now that ${}^{C}\operatorname{div}^{\alpha}_{a^+} \vec{w} = 0$. By Proposition 1(iv), it follows that ${}^{C}\operatorname{curl}^{\alpha}_{a^+}\, {}^{C}\operatorname{curl}^{\alpha}_{a^+} \vec{w} = 0$. On the other hand, Proposition 8 ensures that

$$ {}^{C}\operatorname{grad}^{\alpha}_{a^+} \mathcal{A}^{\alpha}_{a^+}[{}^{C}\operatorname{curl}^{\alpha}_{a^+} \vec{w}] = {}^{C}\operatorname{curl}^{\alpha}_{a^+} \vec{w}. \quad (81) $$

The conclusion of this result follows from (76) if we let $w_0 = -\mathcal{A}^{\alpha}_{a^+}[{}^{C}\operatorname{curl}^{\alpha}_{a^+} \vec{w}]$. □

6. Conclusions

In this work, we extend some results from vector calculus to the fractional case—for instance, the space fractional Helmholtz Decomposition Theorem provided by Propositions 6 and 7. The key tools used are the decompositions of the fractional Teodorescu transform in the Riemann–Liouville case (52) and in the Caputo case (69) as well as various properties associated with these fractional operators, which are thoroughly established in this manuscript. To this end, we consider fractional derivatives in the senses of Riemann–Liouville and Caputo, and we analyze fractional forms of various integer-order differential operators, including the divergence, the rotational, the gradient, the Dirac and the Laplace operators. As the most important result, we prove an existence theorem for the solutions of a div-curl system, considering fractional differential operators of the Riemann–Liouville and Caputo types (see Theorems 2 and 3). Other important generalizations of well-known theorems from vector calculus are proven in this way. More precisely, we present fractional versions of the classical Divergence and Stokes Theorems for vector fields (see Proposition 2). Furthermore, we focus on the construction of fractional hyper-conjugate pairs, which represent a fractional generalization of the well-known Moisil–Teodorescu system in quaternionic analysis. Finally, we note that we are also able to provide an explicit expression for an inverse of the fractional gradient operator when we restrict ourselves to vector fields whose fractional rotational is zero, when we consider fractional derivatives of the Caputo type (see Proposition 8).

Author Contributions: Conceptualization, B.B.D. and J.E.M.-D.; methodology, B.B.D. and J.E.M.-D.; software, B.B.D. and J.E.M.-D.; validation, B.B.D. and J.E.M.-D.; formal analysis, B.B.D. and J.E.M.-D.; investigation, B.B.D. and J.E.M.-D.; resources, B.B.D. and J.E.M.-D.; data curation, B.B.D. and J.E.M.-D.; writing—original draft preparation, B.B.D. and J.E.M.-D.; writing—review and editing, B.B.D. and J.E.M.-D.; visualization, B.B.D. and J.E.M.-D.; supervision, J.E.M.-D.; project administration, J.E.M.-D.; funding acquisition, B.B.D. and J.E.M.-D. All authors have read and agreed to the published version of the manuscript.

Funding: The corresponding author (J.E.M.-D.) wishes to acknowledge the financial support of the National Council for Science and Technology of Mexico (CONACYT) through grant A1-S-45928.

Institutional Review Board Statement: Not applicable.

Informed Consent Statement: Not applicable.

Data Availability Statement: No new data were created or analyzed in this study. Data sharing is not applicable to this article.

Acknowledgments: The authors wish to thank the anonymous reviewers for their comments and criticisms. All of their comments were taken into account in the revised version of the paper, resulting in a substantial improvement with respect to the original submission.

Conflicts of Interest: The authors declare no conflict of interest.

References

1. Adda, F.B. Geometric interpretation of the differentiability and gradient of real order. *C. R. L'Acad. Des Sci. Ser. Math.* **1998**, *8*, 931–934.
2. Adda, F.B. The differentiability in the fractional calculus. *Nonlinear Anal.* **2001**, *47*, 5423–5428. [CrossRef]
3. Meerschaert, M.M.; Mortensen, J.; Wheatcraft, S.W. Fractional vector calculus for fractional advection–dispersion. *Phys. A Stat. Mech. Its Appl.* **2006**, *367*, 181–190. [CrossRef]
4. Tarasov, V.E. Fractional vector calculus and fractional Maxwell's equations. *Ann. Phys.* **2008**, *323*, 2756–2778. [CrossRef]
5. Ortigueira, M.D.; Rivero, M.; Trujillo, J.J. From a generalised Helmholtz decomposition theorem to fractional Maxwell equations. *Commun. Nonlinear Sci. Numer. Simul.* **2015**, *22*, 1036–1049. [CrossRef]
6. Ortigueira, M.; Machado, J. On fractional vectorial calculus. *Bull. Pol. Acad. Sci. Tech. Sci.* **2018**, *66*, 389–402.
7. Ji, An-Chun, L.W.M.S.J.L.; Zhou, F. Dynamical creation of fractionalized vortices and vortex lattices. *Phys. Rev. Lett.* **2008**, *101*, 4. [CrossRef] [PubMed]
8. Podlubny, I. *Fractional Differential Equations: An Introduction to Fractional Derivatives, Fractional Differential Equations, to Methods of Their Solution and Some of Their Applications*; Elsevier: Amsterdam, The Netherlands, 1998; Volume 198.
9. Samko, S.G.; Kilbas, A.A.; Marichev, O.I. *Fractional Integrals and Derivatives*; Gordon and Breach Science Publishers: Yverdon Yverdon-les-Bains, Switzerland, 1993; Volume 1.
10. Gorenflo, R.; Mainardi, F.; Rogosin, S. Mittag–Leffler function: Properties and applications. *Handb. Fract. Calc. Appl.* **2019**, *1*, 269–296.
11. Gürlebeck, K.; Sprössig, W. *Quaternionic Analysis and Elliptic Boundary Value Problems*; Birkhäuser: Basel, Switzerland, 1990.
12. Gürlebeck, K.; Sprößig, W. *Quaternionic and Clifford Calculus for Physicists and Engineers*; John Wiley & Sons: Chicheste, UK, 1997.
13. Gürlebeck, K.; Habetha, K.; Sprößig, W. *Holomorphic Functions in the Plane and n-Dimensional Space*; Birkhäuser: Basel, Switzerland, 2008.
14. Kähler, U.; Vieira, N. Fractional Clifford analysis. In *Hypercomplex Analysis: New Perspectives and Applications*; Bernstein, S., Kähler, U., Sabadini, I., Sommen, F., Eds.; Trends in Mathematics; Springer: Birkhäuser, Germany, 2014; pp. 191–201.
15. Ferreira, M.; Vieira, N. Eigenfunctions and fundamental solutions of the fractional Laplace and Dirac operators: The Riemann–Liouville case. *Complex Anal. Oper. Theory* **2016**, *10*, 1081–1100. [CrossRef]
16. Ferreira, M.; Vieira, N. Eigenfunctions and fundamental solutions of the fractional Laplace and Dirac operators using Caputo derivatives. *Complex Var. Elliptic Equ.* **2017**, *62*, 1237–1253. [CrossRef]
17. Ferreira, M.; Kraußhar, R.S.; Rodrigues, M.M.; Vieira, N. A higher dimensional fractional Borel–Pompeiu formula and a related hypercomplex fractional operator calculus. *Math. Methods Appl. Sci.* **2019**, *42*, 3633–3653. [CrossRef]
18. Sudbery, A. Quaternionic analysis. *Math. Proc. Camb. Phil. Soc.* **1979**, *85*, 99–225. [CrossRef]
19. Kravchenko, V.V. *Applied Quaternionic Analysis*; Heldermann Verlag: Lemgo, Germany, 2003.
20. Delgado, B.B.; Porter, R.M. General solution of the inhomogeneous div-curl system and consequences. *Adv. Appl. Clifford Algebr.* **2017**, *27*, 3015–3037. [CrossRef]
21. Delgado, B.B.; Porter, R.M. Hilbert transform for the three-dimensional Vekua equation. *Complex Var. Elliptic Equ.* **2019**, *64*, 1797–1824. [CrossRef]
22. Delgado, B.B.; Kravchneko, V.V. A Right Inverse Operator for curl+λ and Applications. *Adv. Appl. Clifford Algebr.* **2019**, *29*, 40. [CrossRef]
23. Delgado, B.B.; Macías-Díaz, J. An Exterior Neumann Boundary-Value Problem for the Div-Curl System and Applications. *Mathematics* **2021**, *9*, 1609. [CrossRef]
24. Moisil, G.; Theodorescu, N. Functions holomorphes dans l'espace. *Mathematica* **1931**, *5*, 142–159.

fractal and fractional

MDPI

Article

On Weighted (k, s)-Riemann-Liouville Fractional Operators and Solution of Fractional Kinetic Equation

Muhammad Samraiz [1], Muhammad Umer [1], Artion Kashuri [2], Thabet Abdeljawad [3,4,5,*], Sajid Iqbal [6] and Nabil Mlaiki [3]

[1] Department of Mathematics, University of Sargodha, Sargodha 40100, Pakistan; muhammad.samraiz@uos.edu.pk (M.S.); msamraizuos@gmail.com or mianumerlink4u99@gmail.com (M.U.)

[2] Department of Mathematics, Faculty of Technical Science, University Ismail Qemali, 9400 Vlora, Albania; artionkashuri@gmail.com or artion.kashuri@univlora.edu.al

[3] Department of Mathematics and General Sciences, Prince Sultan University, Riyadh 12345, Saudi Arabia; nmlaiki@psu.edu.sa

[4] Department of Medical Research, China Medical University, Taichung 40402, Taiwan

[5] Department of Computer Science and Information Engineering, Asia University, Taichung 40402, Taiwan

[6] Department of Mathematics, Riphah International University, Faisalabad Campus, Satyana Road, Faisalabad 38000, Pakistan; sajid_uos2000@yahoo.com

* Correspondence: tabdeljawad@psu.edu.sa

Abstract: In this article, we establish the weighted (k, s)-Riemann-Liouville fractional integral and differential operators. Some certain properties of the operators and the weighted generalized Laplace transform of the new operators are part of the paper. The article consists of Chebyshev-type inequalities involving a weighted fractional integral. We propose an integro-differential kinetic equation using the novel fractional operators and find its solution by applying weighted generalized Laplace transforms.

Keywords: weighted (k, s) fractional integral operator; weighted (k, s) fractional derivative; weighted generalized Laplace transform; fractional kinetic equation

Citation: Samraiz, M.; Umer, M.; Kashuri, A.; Abdeljawad, T.; Iqbal, S.; Mlaiki, N. On Weighted (k, s)-Riemann-Liouville Fractional Operators and Solution of Fractional Kinetic Equation. *Fractal Fract.* **2021**, *5*, 118. https://doi.org/10.3390/fractalfract5030118

Academic Editor: Renato Spigler

Received: 26 August 2021
Accepted: 8 September 2021
Published: 13 September 2021

1. Introduction

Fractional calculus history dates back to the 17th century, when the derivative of order $\alpha = 1/2$ was defined by Leibnitz in 1695. Fractional calculus has gained broad significance in the last few decades due to its applications in various fields of science and engineering. The Tautocrone problem can be solved using fractional calculus, as shown by Abel [1]. It also has applications in group theory, field theory, polymers, continuum mechanics, wave theory, quantum mechanics, biophysics, spectroscopy, Lie theory, and in several other fields [2–6]. Despite the fact that this calculus is ancient, it has gained attention over the last few decades because of the interesting results derived when this calculus is applied to the models of some real-world problems [7–14]. The fact that there are various fractional operators is what makes fractional calculus special. Thus, any scientist working on modeling real global phenomena can choose the operator that best suits the model.

The Riemann-Liouville, Grünwald-Letnikov, and Caputo and Hadamard definitions [7,15,16] are some of the most well-known definitions of fractional operators, such that their formulations include single-kernel integrals, and they are used to explore and analyze memory effect problems, for example [17]. The fractional derivatives are represented by the fractional integrals [7,10,15,18] in fractional calculus. There are several varieties of fractional integrals, of which two have been studied extensively for their applications. The first one is the Riemann-Liouville fractional integral defined for parameter $\beta \in \mathbb{R}^+$ by

$$(\mathfrak{I}^{\beta}_{a^+} f)(\xi) = \frac{1}{\Gamma(\beta)} \int_a^{\xi} (\xi - t)^{\beta - 1} \varphi(t) dt, \;\; \beta > 0, \xi > a,$$

inspired by Cauchy's integral formula

$$\int_a^{\xi} dt_1 \int_a^{t_1} dt_2 \cdots \int_a^{t_{n-1}} dt_n = \frac{1}{\Gamma(n)} \int_a^{\xi} (\xi - t)^{n-1} \varphi(t) dt,$$

well-defined for $n \in N$. The second is Hadamard's fractional integral, which is defined by Hadamard [19]

$$(\mathfrak{J}_a^{\beta} \varphi)(\xi) = \frac{1}{\Gamma(\beta)} \int_a^{\xi} (log\frac{\xi}{t})^{\beta-1} \frac{\varphi(t)}{t} dt, \quad \beta > 0, \xi > a,$$

and is derived by the following integral:

$$\int_a^{\xi} \frac{dt_1}{t_1} \int_a^{t_1} \frac{dt_2}{t_2} \cdots \int_a^{t_{n-1}} \frac{dt_n}{t_n} = \frac{1}{\Gamma(n)} \int_a^{\xi} (log\frac{\xi}{t})^{n-1} \frac{\varphi(t)}{t} dt.$$

We start by recalling some related results and notions.

Definition 1 ([20])**.** *The integral form of the k-gamma function is defined by*

$$\Gamma_k(\alpha) = \int_0^{\infty} \xi^{\alpha-1} e^{\frac{-\xi^k}{k}} d\xi, \ \Re(\alpha) > 0.$$

Clearly, $\Gamma(\alpha) = \lim_{k \to 1} \Gamma_k(\alpha)$ *and* $\Gamma_k(\alpha) = k^{\frac{\alpha}{k}-1} \Gamma(\frac{\alpha}{k})$.

Definition 2. *Let* $\Re(\alpha)$, $\Re(\beta) > 0$ *and* $k > 0$, *where we have the following k-beta function*

$$B_k(\alpha, \beta) = \frac{1}{k} \int_0^1 \tau^{\frac{\alpha}{k}-1} (1-\tau)^{\frac{\beta}{k}-1} d\tau.$$

Note that the relation between Γ_k *and* B_k *functions is given by* $B_k(\alpha, \beta) = \frac{\Gamma_k(\alpha)\Gamma_k(\beta)}{\Gamma_k(\alpha+\beta)}$.

The (k, s)-Riemann-Liouville fractional integral (RLFI) [21] is given in the following definition.

Definition 3. *Suppose* $\varphi \in C[a, b]$, *then* (k, s)*-RLFI of order* α *is defined by*

$$({}^s_k\mathfrak{J}^{\alpha}_{a^+} \varphi)(\xi) = \frac{(s+1)^{1-\frac{\alpha}{k}}}{k\Gamma_k(\alpha)} \int_a^{\xi} (\xi^{s+1} - t^{s+1})^{\frac{\alpha}{k}-1} t^s \varphi(t) dt, \ \ \xi \in [a, b], \tag{1}$$

where $\alpha, k > 0$ *and* $s \in \mathbb{R} \backslash \{-1\}$.

Definition 4 ([22])**.** *Suppose* φ *is a continuous function on* $[0, \infty)$ *and* $s, \alpha \in \mathbb{R}^+$. *Then for all* $0 < t < \xi < \infty$

$$({}^s_k\mathfrak{D}^{\alpha}_{a^+} \varphi)(\xi) = \frac{(s)^{\frac{\alpha-nk+k}{k}}}{k\Gamma_k(nk-\alpha)} (\xi^{1-s} \frac{d}{d\xi})^n \int_a^{\xi} (\xi^s - t^s)^{\frac{nk-\alpha}{k}-1} t^{s-1} \varphi(t) dt, \tag{2}$$

where $n = [\alpha] + 1$ *and* $k > 0$, *is called a weighted* (k, s)*-Riemann Liouville fractional derivative, provided it exists.*

Definition 5 ([23]). *Let φ, $\psi \in [a,\infty)$ be a real valued function such that $\psi(\xi)$ is continuous and $\psi'(\xi) > 0$ on $[a,\infty)$. The generalized weighted Laplace transform of φ with weight function ω defined on $[a,\infty)$ is given by*

$$L_{\psi}^{\omega}\{\varphi(x)\}(u) = \int_a^{\infty} e^{-u(\psi(x)-\psi(a))}\omega(x)\varphi(x)\psi'(x)dx, \tag{3}$$

holds for all values of u.

Theorem 1 ([23]). *The generalized weighted Laplace transform of $\mathfrak{D}_{\omega}^{n}\varphi$ exists and is given by*

$$\mathfrak{L}_{\Psi}^{\omega}\{\mathfrak{D}_{\omega}^{n}\Phi\}(u) = u^n \mathfrak{L}_{\Psi}^{\omega}\{\Phi(\xi)\}(u) - \sum_{k=0}^{n-1} u^{n-k-1}\Phi_k(a).$$

Definition 6 ([23]). *The generalized weighted convolution of φ and ψ is defined by*

$$\begin{aligned}(\Phi *_{\Psi}^{\omega} h)(\xi) =& \omega^{-1}(\xi)\int_a^{\xi} \omega(\Psi^{-1}(\Psi(\xi)+\Psi(a)-\Psi(t))) \\ &\times \Phi(\Psi^{-1}(\Psi(\xi)+\Psi(a)-\Psi(t)))\omega(t)h(t)\Psi'(t)dt.\end{aligned}$$

2. Weighted (k,s)-Riemann Liouville Fractional Operators

In the present section, we define the weighted (k,s)-Riemann Liouville fractional operators and discuss some of their properties.

Definition 7. *Let φ be a continuous function on $[a,b]$. Then, the weighted (k,s)-RLFI of order α is defined by*

$$({}_{k}^{s}\mathfrak{J}_{a^+,\omega}^{\alpha}\varphi)(\xi) = \frac{(s+1)^{1-\frac{\alpha}{k}}\omega^{-1}(\xi)}{k\Gamma_k(\alpha)}\int_a^{\xi}(\xi^{s+1}-t^{s+1})^{\frac{\alpha}{k}-1}t^s\omega(t)\varphi(t)dt,\ \xi \in [a,b], \tag{4}$$

where $\alpha, k > 0$, $\omega(\xi) \neq 0$ and $s \in \mathbb{R}\backslash\{-1\}$.

Remark 1. *It should be noted that this integral operator covers many fractional integral operators.*

(i) *If we choose $\omega(\xi) = 1$, we obtain (k,s)-RLFI [21].*
(ii) *If we choose $s = 0$ and $\omega(\xi) = 1$, k-RLFI is obtained [24].*
(iii) *For $k = 1$, $s = 0$ and $\omega(\xi) = 1$, it gives RLFI [7].*
(iv) *For $s \to -1^+$ and $\omega(\xi) = 1$, it is converted to the k-Hadamard fractional integral [25].*

The following modification of Definition 4 is required to prove the claimed results.

Definition 8. *The (k,s)-Riemann Liouville fractional derivative is defined as follows:*
Let φ be a continuous function on $[0,\infty)$ and $s \in \mathbb{R}\backslash\{-1\}$. Then for all $0 < t < \xi < \infty$

$$({}_{k}^{s}\mathfrak{D}_{a^+}^{\alpha}\varphi)(\xi) = \frac{k^{n-1}(s+1)^{\frac{\alpha-nk+k}{k}}}{\Gamma_k(nk-\alpha)}(\xi^{-s}\frac{d}{d\xi})^n\int_a^{\xi}(\xi^{s+1}-t^{s+1})^{\frac{nk-\alpha}{k}-1}t^s\varphi(t)dt,$$

where $n = [\alpha]+1$ and $\alpha, k > 0$, is called the (k,s)-Riemann Liouville fractional derivative, provided it exists.

Definition 9. *Let φ be a continuous function on $[0,\infty)$, $s \in \mathbb{R}\backslash\{-1\}$, $n = [\alpha]+1$, $\alpha, k > 0$ and $\omega(\xi) \neq 0$. Then for all $0 < t < \xi < \infty$*

$$({}^{s}_{k}\mathfrak{D}^{\alpha}_{a^+,\omega}\varphi)(\xi) = \omega^{-1}(\xi)(k\xi^{-s}\frac{d}{d\xi})^{n}\omega(\xi)({}^{s}_{k}\mathfrak{J}^{nk-\alpha}_{a^+,\omega}\varphi)(t), \tag{5}$$

where ${}^{s}_{k}\mathfrak{J}^{nk-\alpha}_{a^+,\omega}$ is a weighted (k,s)-RLFI.

It can also be written as

$$({}^{s}_{k}\mathfrak{D}^{\alpha}_{a^+,\omega}\varphi)(\xi) = \frac{k^{n-1}(s+1)^{\frac{\alpha-nk+k}{k}}\omega^{-1}(\xi)}{\Gamma_k(nk-\alpha)}(\xi^{-s}\frac{d}{d\xi})^{n} \times \int_a^{\xi}(\xi^{s+1}-t^{s+1})^{\frac{nk-\alpha}{k}-1}t^{s}\omega(t)\varphi(t)dt. \tag{6}$$

Remark 2. *It is worth mentioning that many other derivative operators can be represented as special cases of (6).*

(i) If $\omega(\xi) = 1$ is chosen, we obtain the (k,s)-Riemann-Liouville fractional derivative [22].
(ii) Let $s = 0$ and $\omega(\xi) = 1$, where it gives the k-Riemann-Liouville fractional derivative [26].
(iii) For $k = 1$, $s = 0$ and $\omega(\xi) = 1$, it reduces to the Riemann-Liouville fractional derivative [27].
(iv) It reduces to the k-Hadamard fractional derivative for $s \to -1^+$, $\omega(\xi) = 1$ [25].

Next, we present the space where the weighted (k,s)-Riemann-Liouville fractional integrals are bounded.

Definition 10. *Let φ be a function defined on $[a,b]$. The space $X^p_\omega(a,b)$, $1 \leq p \leq \infty$ is the space of all Lebesgue measurable functions for which $\| \varphi \|_{X^p_\omega} < \infty$, where*

$$\| \varphi \|_{X^p_\omega} = \Big[(s+1)\int_a^b | \omega(\xi)\varphi(\xi) |^p \xi^s d\xi\Big]^{\frac{1}{p}}, \; 1 \leq p < \infty,$$

$\omega(\xi) \neq 0$, $s \in \mathbb{R}$ and

$$\| \varphi \|_{X^\infty_\omega} = ess\,sup_{a\leq\xi\leq b} | \omega(\xi)\varphi(\xi) | < \infty.$$

Noted that $\varphi \in X^p_\omega(a,b) \Leftrightarrow \omega(\xi)\varphi(\xi)(\xi^s)^{\frac{1}{p}} \in L_p(a,b)$ for $1 \leq p < \infty$ and $\varphi \in X^\infty_\omega(a,b)$ $\Leftrightarrow \omega(\xi)\varphi(\xi) \in L_\infty(a,b)$.

Theorem 2. *Let $\alpha > 0$, $k > 0$, $1 \leq p \leq \infty$ and $\varphi \in X^p_\omega(a,b)$. Then ${}^{s}_{k}\mathfrak{J}^{\alpha}_{a^+,\omega}\varphi$ is bounded in $X^p_\omega(a,b)$ and*

$$\|{}^{s}_{k}\,\mathfrak{J}^{\alpha}_{a^+,\omega}\varphi\,\|_{X^p_\omega} \leq \frac{(s+1)^{-\frac{\alpha}{k}}(b^{s+1}-a^{s+1})^{\frac{\alpha}{k}}}{\Gamma_k(\alpha+1)} \| \varphi \|_{X^p_\omega}.$$

Proof. For $1 \leq p < \infty$, we have

$$\begin{aligned}\|{}^{s}_{k}\,\mathfrak{J}^{\alpha}_{a^+,\omega}\varphi\,\|_{X^p_\omega} &= \Big[(s+1)\int_a^b \Big|\omega(\xi)\frac{(s+1)^{1-\frac{\alpha}{k}}\omega^{-1}(\xi)}{k\Gamma_k(\alpha)} \\ &\times \int_a^{\xi}(\xi^{s+1}-t^{s+1})^{\frac{\alpha}{k}-1}t^{s}\omega(t)\varphi(t)dt\Big|^p \xi^s d\xi\Big]^{\frac{1}{p}} \\ &= \frac{(s+1)^{2-\frac{\alpha}{k}}}{k\Gamma_k(\alpha)}\Big[\int_a^b\Big|\int_a^{\xi}(\xi^{s+1}-t^{s+1})^{\frac{\alpha}{k}-1}t^{s}\omega(t)\varphi(t)dt\Big|^p \xi^s d\xi\Big]^{\frac{1}{p}}.\end{aligned} \tag{7}$$

Substituting $\xi^{s+1} = v$ and $t^{s+1} = u$ on the right side of (7), we obtain

$$\|{}_k^s\mathfrak{J}^{\alpha}_{a^+,\omega}\varphi\|_{X^p_\omega} = \frac{(s+1)^{2-\frac{\alpha}{k}}}{k\Gamma_k(\alpha)}\Big[\int_{a^{s+1}}^{b^{s+1}}\Big|\int_{a^{s+1}}^{v}(v-u)^{\frac{\alpha}{k}-1}\omega(u^{\frac{1}{s+1}})\varphi(u^{\frac{1}{s+1}})du\Big|^p dv\Big]^{\frac{1}{p}}.$$

By using Minkowski's inequality, we have

$$\begin{aligned}\|{}_k^s\mathfrak{J}^{\alpha}_{a^+,\omega}\varphi\|_{X^p_\omega} &\leq \frac{(s+1)^{-\frac{\alpha}{k}}}{k\Gamma_k(\alpha)}\int_{a^{s+1}}^{b^{s+1}}\Big[\int_{u}^{a^{s+1}}\Big|(v-u)^{\frac{\alpha}{k}-1}\omega(u^{\frac{1}{s+1}})\varphi(u^{\frac{1}{s+1}})dv\Big|^p du\Big]^{\frac{1}{p}}\\ &\leq \frac{(s+1)^{-\frac{\alpha}{k}}}{k\Gamma_k(\alpha)}\int_{a^{s+1}}^{b^{s+1}}\Big|\omega(u^{\frac{1}{s+1}})\varphi(u^{\frac{1}{s+1}})\Big|\Big[\frac{(b^{s+1}-u)^{(\frac{\alpha}{k}-1)p+1}}{(\frac{\alpha}{k}-1)p+1}\Big]^{\frac{1}{p}}du.\end{aligned}$$

Applying Hölder's inequality, we obtain

$$\begin{aligned}\|{}_k^s\mathfrak{J}^{\alpha}_{a^+,\omega}\varphi\|_{X^p_\omega} \leq{}& \frac{(s+1)^{-\frac{\alpha}{k}}}{k\Gamma_k(\alpha)}\Big[\int_{a^{s+1}}^{b^{s+1}}\Big|\omega(u^{\frac{1}{s+1}})\varphi(u^{\frac{1}{s+1}})\Big|^p du\Big]^{\frac{1}{p}}\\ &\times\Big[\int_{a^{s+1}}^{b^{s+1}}\Big(\frac{(b^{s+1}-u)^{(\frac{\alpha}{k}-1)p+1}}{(\frac{\alpha}{k}-1)p+1}\Big)^{\frac{q}{p}}du\Big]^{\frac{1}{q}},\end{aligned}$$

where $\frac{1}{p}+\frac{1}{q}=1$. Further,

$$\begin{aligned}\|{}_k^s\mathfrak{J}^{\alpha}_{a^+,\omega}\varphi\|_{X^p_\omega} \leq{}& \frac{(s+1)^{-\frac{\alpha}{k}}}{k\Gamma_k(\alpha)}\Big[\int_a^b\Big|\omega(t)\varphi(t)\Big|^p(s+1)dt\Big]^{\frac{1}{p}}\\ &\times\Big[\int_{a^{s+1}}^{b^{s+1}}\Big(\frac{(b^{s+1}-u)^{(\frac{\alpha}{k}-1)p+1}}{(\frac{\alpha}{k}-1)p+1}\Big)^{\frac{q}{p}}du\Big]^{\frac{1}{q}}\\ \leq{}& \frac{(s+1)^{-\frac{\alpha}{k}}(b^{s+1}-a^{s+1})^{\frac{\alpha}{k}}}{k\Gamma_k(\alpha)\frac{\alpha}{k}}\|\varphi\|_{X^p_\omega}\\ ={}& \frac{(s+1)^{-\frac{\alpha}{k}}(b^{s+1}-a^{s+1})^{\frac{\alpha}{k}}}{\Gamma_k(\alpha+1)}\|\varphi\|_{X^p_\omega}.\end{aligned}$$

For $p=\infty$, we obtain

$$\Big|\omega(\xi){}_k^s\mathfrak{J}^{\alpha}_{a^+,\omega}\varphi(\xi)\Big| = \frac{(s+1)^{-\frac{\alpha}{k}}(b^{s+1}-a^{s+1})^{\frac{\alpha}{k}}}{\Gamma_k(\alpha+1)}\|\varphi\|_{X^\infty_\omega}.$$

Hence, we obtain the desired result. □

Theorem 3. *Let φ be a continuous function on $[0,\infty)$ and $s \in \mathbb{R}\backslash\{-1\}$ and $\omega(\xi) \neq 0$, $n = [\alpha]+1$. Then for all $0 < a < \xi$, we obtain*

$${}_k^s\mathfrak{D}^{\alpha}_{a,\omega}({}_k^s\mathfrak{J}^{\alpha}_{a^+,\omega}\varphi)(\xi) = \varphi(\xi),$$

where $\alpha, k > 0$.

Proof. Consider

$$
\begin{aligned}
&{}^{s}_{k}\mathfrak{D}^{\alpha}_{a^+,\omega}({}^{s}_{k}\mathfrak{J}^{\alpha}_{a^+,\omega}\varphi)(\xi)\\
&= \frac{(s+1)^{\frac{\alpha-nk+k}{k}}\omega^{-1}(\xi)}{k\Gamma_k(nk-\alpha)}(\xi^{-s}\frac{d}{d\xi})^n k^n\\
&\times \int_a^{\xi}(\xi^{s+1}-y^{s+1})^{\frac{nk-\alpha}{k}-1}y^s\omega(y)({}^{s}_{k}\mathfrak{J}^{\alpha}_{a^+,\omega}\varphi)(y)dy\\
&= \frac{(s+1)^{\frac{\alpha-nk+k}{k}}\omega^{-1}(\xi)}{k\Gamma_k(nk-\alpha)}(\xi^{-s}\frac{d}{d\xi})^n k^n \int_a^{\xi}(\xi^{s+1}-y^{s+1})^{\frac{nk-\alpha}{k}-1}y^s\omega(y)\\
&\times \frac{(s+1)^{\frac{1-\alpha}{k}}\omega^{-1}(\xi)}{k\Gamma_k(\alpha)}\int_a^{y}(y^{s+1}-t^{s+1})^{\frac{\alpha}{k}-1}t^s\omega(t)(t)dt\\
&= \frac{(s+1)^{2-n}\omega^{-1}(\xi)}{k^2\Gamma_k(\alpha)\Gamma_k(nk-\alpha)}(\xi^{-s}\frac{d}{d\xi})^n k^n\\
&\times \int_a^{\xi} t^s\omega(t)\varphi(t)\Big[\int_t^{\xi}(y^{s+1}-t^{s+1})^{\frac{\alpha}{k}-1}(\xi^{s+1}-y^{s+1})^{\frac{nk-\alpha}{k}-1}y^s dy\Big]dt.
\end{aligned}
\tag{8}
$$

By substituting $z=\frac{y^{s+1}-t^{s+1}}{\xi^{s+1}-t^{s+1}}$ on the right side of (8), we obtain

$$
\begin{aligned}
&{}^{s}_{k}\mathfrak{D}^{\alpha}_{a^+,\omega}({}^{s}_{k}\mathfrak{J}^{\alpha}_{a^+,\omega}\varphi)(\xi)\\
&= \frac{(s+1)^{1-n}\omega^{-1}(\xi)}{k^2\Gamma_k(\alpha)\Gamma_k(nk-\alpha)}(\xi^{-s}\frac{d}{d\xi})^n k^n\\
&\times \int_a^{\xi} t^s\omega(t)\varphi(t)(\xi^{s+1}-t^{s+1})^{n-1}\Big[\int_t^{\xi}(1-z)^{\frac{\alpha}{k}-1}(z)^{\frac{nk-\alpha}{k}-1}dz\Big]dt\\
&= \frac{(s+1)^{1-n}\omega^{-1}(\xi)}{k^2\Gamma_k(\alpha)\Gamma_k(nk-\alpha)}(\xi^{-s}\frac{d}{d\xi})^n k^n\\
&\times \int_a^{\xi} t^s\omega(t)\varphi(t)(\xi^{s+1}-t^{s+1})^{n-1}[kB_k(\alpha,nk-\alpha)]dt\\
&= \frac{(s+1)^{1-n}\omega^{-1}(\xi)}{k\Gamma_k(nk)}(\xi^{-s}\frac{d}{d\xi})^n k^n\int_a^{\xi} t^s\omega(t)\varphi(t)(\xi^{s+1}-t^{s+1})^{n-1}dt\\
&= \frac{(s+1)^{1-n}\omega^{-1}(\xi)}{k^n\Gamma(n)}(\xi^{-s}\frac{d}{d\xi})^n k^n\int_a^{\xi} t^s\omega(t)\varphi(t)(\xi^{s+1}-t^{s+1})^{n-1}dt,
\end{aligned}
$$

which gives

$$
{}^{s}_{k}\mathfrak{D}^{\alpha}_{a^+,\omega}({}^{s}_{k}\mathfrak{J}^{\alpha}_{a^+,\omega}\varphi)(\xi)=\varphi(\xi).
$$

The inverse property is proved. □

Corollary 1. *Let φ be a continuous function on $[0,\infty)$ and $s\in\mathbb{R}\backslash\{-1\}$ and $\omega(\xi)\neq 0$, $m=[\beta]+1$, $n=[\alpha]+1$. Then for all $0<a<\xi$*

$$
{}^{s}_{k}\mathfrak{D}^{\alpha}_{a^+,\omega}({}^{s}_{k}\mathfrak{J}^{\beta}_{a^+,\omega}\varphi)(\xi)=({}^{s}_{k}\mathfrak{D}^{\alpha-\beta}_{a^+,\omega}\varphi)(\xi),
$$

where $\alpha,\beta,k>0$.

Corollary 2. *(Semi-group property) Let φ be a continuous function on $[0,\infty)$ and $s\in\mathbb{R}\backslash\{-1\}$, $\omega(\xi)\neq 0$, $n=[\alpha]+1$, $m=[\beta]+1$ and $\alpha+\beta<nk$. Then for all $0<a<\xi$*

$$
{}^{s}_{k}\mathfrak{D}^{\alpha}_{a^+,\omega}({}^{s}_{k}\mathfrak{D}^{\beta}_{a^+,\omega}\varphi)(\xi)=({}^{s}_{k}\mathfrak{D}^{\alpha+\beta}_{a^+,\omega}\varphi)(\xi),
$$

where $\alpha,\beta,k>0$.

Proof. By using Definition 9, we have

$$
\begin{aligned}
{}^{s}_{k}\mathfrak{D}^{\alpha}_{a^+,\omega}({}^{s}_{k}\mathfrak{D}^{\beta}_{a^+,\omega}\varphi)(\xi) &= \omega^{-1}(\xi)(k\xi^{-s}\frac{d}{d\xi})^{n}\omega(\xi)({}^{s}_{k}\mathfrak{J}^{nk-\alpha}_{a^+,\omega})({}^{s}_{k}\mathfrak{D}^{\beta}_{a^+,\omega}\varphi)(\xi) \\
&= \omega^{-1}(\xi)(k\xi^{-s}\frac{d}{d\xi})^{n}\omega(\xi)({}^{s}_{k}\mathfrak{J}^{nk-\alpha}_{a^+,\omega})({}^{s}_{k}\mathfrak{D}^{\beta}_{a^+,\omega}\varphi)(\xi)({}^{s}_{k}\mathfrak{J}^{\beta}_{a^+,\omega})({}^{s}_{k}\mathfrak{J}^{-\beta}_{a^+,\omega}).
\end{aligned}
$$

By using Theorem 3, we have

$$
\begin{aligned}
{}^{s}_{k}\mathfrak{D}^{\alpha}_{a^+,\omega}({}^{s}_{k}\mathfrak{D}^{\beta}_{a^+,\omega}\varphi)(\xi) &= \omega^{-1}(\xi)(k\xi^{-s}\frac{d}{d\xi})^{n}\omega(\xi)({}^{s}_{k}\mathfrak{J}^{nk-\alpha}_{a^+,\omega})({}^{s}_{k}\mathfrak{J}^{-\beta}_{a^+,\omega}) \\
&= \omega^{-1}(\xi)(k\xi^{-s}\frac{d}{d\xi})^{n}\omega(\xi)({}^{s}_{k}\mathfrak{J}^{nk-(\alpha+\beta)}_{a^+,\omega}),
\end{aligned}
$$

which implies

$$
{}^{s}_{k}\mathfrak{D}^{\alpha}_{a^+,\omega}({}^{s}_{k}\mathfrak{D}^{\beta}_{a^+,\omega}\varphi)(\xi) = ({}^{s}_{k}\mathfrak{D}^{\alpha+\beta}_{a^+,\omega}\varphi)(\xi),
$$

which is the required result. □

Corollary 3 (Commutative property). *Let φ be a continuous function on $[0,\infty)$ and $\alpha,\beta \in \mathbb{R}^+$, $\omega(\xi) \neq 0$ and $s \in \mathbb{R}\backslash\{-1\}$. Then for all $0 < a < \xi$*

$$
{}^{s}_{k}\mathfrak{D}^{\alpha}_{a^+,\omega}({}^{s}_{k}\mathfrak{D}^{\beta}_{a^+,\omega}\varphi)(\xi) = {}^{s}_{k}\mathfrak{D}^{\beta}_{a^+,\omega}({}^{s}_{k}\mathfrak{D}^{\alpha}_{a^+,\omega}\varphi)(\xi).
$$

Corollary 4 (Linearity property). *Let φ be a continuous function on $[0,\infty)$, $k,\alpha \in \mathbb{R}^+$, $\omega(\xi) \neq 0$ and $s \in \mathbb{R}\backslash\{-1\}$. Then for all $0 < a < \xi$*

$$
{}^{s}_{k}\mathfrak{D}^{\alpha}_{a^+,\omega}[\psi(\xi) + \mu h(\xi)] = {}^{s}_{k}\mathfrak{D}^{\alpha}_{a^+,\omega}\psi(\xi) + \mu\, {}^{s}_{k}\mathfrak{D}^{\alpha}_{a^+,\omega}h(\xi),
$$

where $n \in N$ and $n = [\alpha] + 1$.

Theorem 4. *Let φ be a continuous function on $[a,b]$, $k > 0$, $\omega(\xi) \neq 0$ and $s \in \mathbb{R}\backslash\{-1\}$*

$$
{}^{s}_{k}\mathfrak{J}^{\beta}_{a^+,\omega}[{}^{s}_{k}\mathfrak{J}^{\alpha}_{a^+,\omega}\varphi(\xi)] = {}^{s}_{k}\mathfrak{J}^{\alpha}_{a^+,\omega}[{}^{s}_{k}\mathfrak{J}^{\beta}_{a^+,\omega}\varphi(\xi)] = {}^{s}_{k}\mathfrak{J}^{\alpha+\beta}_{a^+,\omega}\varphi(\xi),
$$

for all α, $\beta > 0$ and $\xi \in [a,b]$.

Proof. By using Definition 7 and Dirichlet's formula, we obtain

$$
\begin{aligned}
&{}^{s}_{k}\mathfrak{J}^{\alpha}_{a^+,\omega}[{}^{s}_{k}\mathfrak{J}^{\beta}_{a^+,\omega}\varphi(\xi)] \\
&= \frac{(s+1)^{1-\frac{\alpha}{k}}\omega^{-1}(\xi)}{k\Gamma_k(\alpha)}\int_a^{\xi}(\xi^{s+1}-t^{s+1})^{\frac{\alpha}{k}-1}t^{s}\omega(t)\,{}^{s}_{k}\mathfrak{J}^{\beta}_{a^+,\omega}\varphi(t)dt \\
&= \frac{(s+1)^{1-\frac{\alpha}{k}}\omega^{-1}(\xi)}{k\Gamma_k(\alpha)}\int_a^{\xi}(\xi^{s+1}-t^{s+1})^{\frac{\alpha}{k}-1}t^{s}\omega(t)\varphi(\tau) \\
&\times\Big[\frac{(s+1)^{1-\frac{\beta}{k}}\omega^{-1}(t)}{k\Gamma_k(\beta)}\int_a^{t}(t^{s+1}-\tau^{s+1})^{\frac{\beta}{k}-1}\tau^{s}\omega(\tau)d\tau\Big]dt \\
&= \frac{(s+1)^{2-\frac{\alpha+\beta}{k}}\omega^{-1}(\xi)}{k^2\Gamma_k(\alpha)\Gamma_k(\beta)}\int_a^{\xi}\tau^{s}\omega(\tau)\varphi(\tau) \\
&\times\int_{\tau}^{\xi}(\xi^{s+1}-t^{s+1})^{\frac{\alpha}{k}-1}(t^{s+1}-\tau^{s+1})^{\frac{\beta}{k}-1}t^{s}dtd\tau. \qquad (9)
\end{aligned}
$$

By substituting $y = \frac{t^{s+1}-\tau^{s+1}}{\xi^{s+1}-\tau^{s+1}}$ on the right side of (9), we obtain

$$
\begin{aligned}
&{}^{s}_{k}\mathfrak{J}^{\alpha}_{a^+,\omega}[{}^{s}_{k}\mathfrak{J}^{\beta}_{a^+,\omega}\varphi(\xi)] \\
&= \frac{(s+1)^{2-\frac{\alpha+\beta}{k}}\omega^{-1}(\xi)}{k^2\Gamma_k(\alpha)\Gamma_k(\beta)} \\
&\times \int_a^\xi \frac{(\xi^{s+1}-\tau^{s+1})^{\frac{\alpha+\beta}{k}-1}}{(s+1)} kB_k(\alpha,\beta)\tau^s\omega(\tau)\varphi(\tau)d\tau \\
&= \frac{(s+1)^{1-\frac{\alpha+\beta}{k}}\omega^{-1}(\xi)}{k\Gamma_k(\alpha+\beta)} \int_a^\xi (\xi^{s+1}-\tau^{s+1})^{\frac{\alpha+\beta}{k}-1}\tau^s\omega(\tau)\varphi(\tau)d\tau \\
&= {}^{s}_{k}\mathfrak{J}^{\alpha+\beta}_{a^+,\omega}\varphi(\xi).
\end{aligned}
$$

The proof is completed. □

Theorem 5. *Let $\alpha, \beta, k > 0$, $\omega(\xi) \neq 0$ and $s \in \mathbb{R}\backslash\{-1\}$. Then we have*

$$
{}^{s}_{k}\mathfrak{J}^{\beta}_{a^+,\omega}[\omega^{-1}(\xi)(\xi^{s+1}-a^{s+1})^{\frac{\beta}{k}-1}] = \frac{\Gamma_k(\beta)(\xi^{s+1}-a^{s+1})^{\frac{\alpha+\beta}{k}-1}\omega^{-1}(\xi)}{(s+1)^{\frac{\alpha}{k}}\Gamma_k(\alpha+\beta)},
$$

where Γ_k denotes the k-Gamma function.

Proof. By using Definition 7, we obtain

$$
\begin{aligned}
&{}^{s}_{k}\mathfrak{J}^{\beta}_{a^+,\omega}[\omega^{-1}(\xi)(\xi^{s+1}-a^{s+1})^{\frac{\beta}{k}-1}] \\
&= \frac{(s+1)^{1-\frac{\alpha}{k}}\omega^{-1}(\xi)}{k\Gamma_k(\alpha)} \int_a^\xi (\xi^{s+1}-t^{s+1})^{\frac{\alpha}{k}-1}t^s \\
&\times (\xi^{s+1}-a^{s+1})^{\frac{\beta}{k}-1}\omega^{-1}(t)\omega(t)\varphi(t)dt.
\end{aligned} \tag{10}
$$

By substituting $y = \frac{\xi^{s+1}-t^{s+1}}{\xi^{s+1}-a^{s+1}}$ on the right side of (10), we obtain

$$
\begin{aligned}
&{}^{s}_{k}\mathfrak{J}^{\beta}_{a^+,\omega}[\omega^{-1}(\xi)(\xi^{s+1}-a^{s+1})^{\frac{\beta}{k}-1}] \\
&= \frac{(s+1)^{\frac{-\alpha}{k}}\omega^{-1}(\xi)(\xi^{s+1}-a^{s+1})^{\frac{\alpha+\beta}{k}-1}}{k\Gamma_k(\alpha)} \\
&\times \int_0^1 (1-y)^{\frac{\alpha}{k}-1}(y)^{\frac{\beta}{k}-1}dy \\
&= \frac{(s+1)^{\frac{-\alpha}{k}}(\xi^{s+1}-a^{s+1})^{\frac{\alpha+\beta}{k}-1}\omega^{-1}(\xi)}{k\Gamma_k(\alpha)} kB_k(\alpha,\beta) \\
&= \frac{\Gamma_k(\beta)(\xi^{s+1}-a^{s+1})^{\frac{\alpha+\beta}{k}-1}\omega^{-1}(\xi)}{(s+1)^{\frac{\alpha}{k}}\Gamma_k(\alpha+\beta)}.
\end{aligned}
$$

This completes the proof. □

Corollary 5. *Let $k > 0$, $\omega(\xi) \neq 0$ and $s \in \mathbb{R}\backslash\{-1\}$. Then, we have*

$$
{}^{s}_{k}\mathfrak{J}^{\alpha}_{a^+,\omega}[\omega^{-1}(\xi)(1)] = \frac{(\xi^{s+1}-a^{s+1})^{\frac{\alpha}{k}-2}\omega^{-1}(\xi)}{(s+1)^{\frac{\alpha}{k}}\Gamma_k(\alpha+\beta)}. \tag{11}
$$

Remark 3. *Taking $\omega(\xi) = 1$ in Theorem 5 and Corollary 5, we obtain results of [21].*

Remark 4. *If we choose $s = 0$, $k = 1$ and $\omega(\xi) = 1$ in Theorem 5 and Corollary 5, we obtain results for Riemann Liouville.*

3. Some New Chebyshev Inequalities Involving Weighted (k, s)-RLFI

Weighted (k, s)-RLFI formulations of Chebyshev-type inequalities are as follows:

Theorem 6. *Let φ and ψ be two synchronous functions on $[0, \infty)$. Then for all $t > a \geq 0$ and the weighted function $\omega(\xi) \neq 0$, the following inequalities for weighted (k, s)-RLFI hold:*

$$ {}^{s}_{k}\mathfrak{J}^{\alpha}_{a^+,\omega}\varphi\psi(t) \geq \frac{1}{{}^{s}_{k}\mathfrak{J}^{\alpha}_{a^+,\omega}(1)}\, {}^{s}_{k}\mathfrak{J}^{\alpha}_{a^+,\omega}\varphi(t)\, {}^{s}_{k}\mathfrak{J}^{\alpha}_{a^+,\omega}\psi(t) \tag{12} $$

and

$$ \begin{aligned} & {}^{s}_{k}\mathfrak{J}^{\alpha}_{a^+,\omega}\varphi\psi(t)\, {}^{s}_{k}\mathfrak{J}^{\beta}_{a^+,\omega}(1) + {}^{s}_{k}\mathfrak{J}^{\beta}_{a^+,\omega}\varphi\psi(t)\, {}^{s}_{k}\mathfrak{J}^{\alpha}_{a^+,\omega}(1) \\ \geq\; & {}^{s}_{k}\mathfrak{J}^{\alpha}_{a^+,\omega}\varphi(t)\, {}^{s}_{k}\mathfrak{J}^{\beta}_{a^+,\omega}\psi(t) + {}^{s}_{k}\mathfrak{J}^{\alpha}_{a^+,\omega}\psi(t)\, {}^{s}_{k}\mathfrak{J}^{\beta}_{a^+,\omega}\varphi(t), \end{aligned} \tag{13} $$

where $\alpha, \beta > 0$.

Proof. Since φ and ψ are synchronous on $[0, \infty)$, for all $\xi, y \geq 0$, we have

$$ \begin{aligned} (\varphi(\xi) - \varphi(y))(\psi(\xi) - \psi(y)) &\geq 0 \\ \varphi(\xi)\psi(\xi) + \varphi(y)\psi(y) &\geq \varphi(\xi)\psi(y) + \varphi(y)\psi(\xi). \end{aligned} \tag{14} $$

Both sides of (14) are multiplied by $\frac{(s+1)^{1-\frac{\alpha}{k}}\omega^{-1}(t)}{k\Gamma_k(\alpha)}(t^{s+1} - \xi^{s+1})^{\frac{\alpha}{k}-1}\omega(\xi)\xi^s$ and integrating w.r.t ξ over (a,t), we obtain

$$ \begin{aligned} & \frac{(s+1)^{1-\frac{\alpha}{k}}\omega^{-1}(t)}{k\Gamma_k(\alpha)} \int_a^t (t^{s+1} - \xi^{s+1})^{\frac{\alpha}{k}-1}\xi^s\omega(\xi)\varphi(\xi)\psi(\xi)d\xi \\ & + \varphi(y)\psi(y)\frac{(s+1)^{1-\frac{\alpha}{k}}\omega^{-1}(t)}{k\Gamma_k(\alpha)} \int_a^t (t^{s+1} - \xi^{s+1})^{\frac{\alpha}{k}-1}\xi^s\omega(\xi)d\xi \\ & \geq \psi(y)\frac{(s+1)^{1-\frac{\alpha}{k}}\omega^{-1}(t)}{k\Gamma_k(\alpha)} \int_a^t (t^{s+1} - \xi^{s+1})^{\frac{\alpha}{k}-1}\xi^s\omega(\xi)\varphi(\xi)d\xi \\ & + \varphi(y)\frac{(s+1)^{1-\frac{\alpha}{k}}\omega^{-1}(t)}{k\Gamma_k(\alpha)} \int_a^t (t^{s+1} - \xi^{s+1})^{\frac{\alpha}{k}-1}\xi^s\omega(\xi)\psi(\xi)d\xi, \end{aligned} \tag{15} $$

which gives

$$ {}^{s}_{k}\mathfrak{J}^{\alpha}_{a^+,\omega}\varphi\psi(t) + \varphi(y)\psi(y)\, {}^{s}_{k}\mathfrak{J}^{\alpha}_{a^+,\omega}(1) \geq \psi(y)\, {}^{s}_{k}\mathfrak{J}^{\alpha}_{a^+,\omega}\varphi(t) + \varphi(y)\, {}^{s}_{k}\mathfrak{J}^{\alpha}_{a^+,\omega}\psi(t). $$

Both sides of (15) are multiplied by $\frac{(s+1)^{1-\frac{\alpha}{k}}\omega^{-1}(t)}{k\Gamma_k(\alpha)}(t^{s+1}-y^{s+1})^{\frac{\alpha}{k}-1}\omega(y)y^s$ and integrating w.r.t y over (a,t), we obtain

$$\begin{aligned}&{}_k^s\mathfrak{J}^{\alpha}_{a^+,\omega}\varphi\psi(t)\frac{(s+1)^{1-\frac{\alpha}{k}}\omega^{-1}(t)}{k\Gamma_k(\alpha)}\int_a^t(t^{s+1}-y^{s+1})^{\frac{\alpha}{k}-1}y^s\omega(y)dy\\&+{}_k^s\mathfrak{J}^{\alpha}_{a^+,\omega}(1)\frac{(s+1)^{1-\frac{\alpha}{k}}\omega^{-1}(t)}{k\Gamma_k(\alpha)}\int_a^t(t^{s+1}-y^{s+1})^{\frac{\alpha}{k}-1}y^s\omega(y)\varphi(y)\psi(y)dy\\&\geq{}_k^s\mathfrak{J}^{\alpha}_{a^+,\omega}\varphi(t)\frac{(s+1)^{1-\frac{\alpha}{k}}\omega^{-1}(t)}{k\Gamma_k(\alpha)}\int_a^t(t^{s+1}-y^{s+1})^{\frac{\alpha}{k}-1}y^s\omega(y)\psi(y)dy\\&+{}_k^s\mathfrak{J}^{\alpha}_{a^+,\omega}\psi(t)\frac{(s+1)^{1-\frac{\alpha}{k}}\omega^{-1}(t)}{k\Gamma_k(\alpha)}\int_a^t(t^{s+1}-y^{s+1})^{\frac{\alpha}{k}-1}y^s\omega(y)\varphi(y)dy.\end{aligned}$$

This can be written as

$$\begin{aligned}&{}_k^s\mathfrak{J}^{\alpha}_{a^+,\omega}\varphi\psi(t){}_k^s\mathfrak{J}^{\alpha}_{a^+,\omega}(1)+{}_k^s\mathfrak{J}^{\alpha}_{a^+,\omega}\varphi\psi(t){}_k^s\mathfrak{J}^{\alpha}_{a^+,\omega}(1)\\&\geq{}_k^s\mathfrak{J}^{\alpha}_{a^+,\omega}\varphi(t){}_k^s\mathfrak{J}^{\alpha}_{a^+,\omega}\psi(t)+{}_k^s\mathfrak{J}^{\alpha}_{a^+,\omega}\varphi(t){}_k^s\mathfrak{J}^{\alpha}_{a^+,\omega}\psi(t).\end{aligned}\tag{16}$$

On simplification, we obtain

$$2{}_k^s\mathfrak{J}^{\alpha}_{a^+,\omega}\varphi\psi(t){}_k^s\mathfrak{J}^{\alpha}_{a^+,\omega}(1)\geq 2{}_k^s\mathfrak{J}^{\alpha}_{a^+,\omega}\varphi(t){}_k^s\mathfrak{J}^{\alpha}_{a^+,\omega}\psi(t),$$

which can be written as

$${}_k^s\mathfrak{J}^{\alpha}_{a^+,\omega}\varphi\psi(t)\geq\frac{1}{{}_k^s\mathfrak{J}^{\alpha}_{a^+,\omega}(1)}{}_k^s\mathfrak{J}^{\alpha}_{a^+,\omega}\varphi(t){}_k^s\mathfrak{J}^{\alpha}_{a^+,\omega}\psi(t).$$

This completes the proof of (12).

Both sides of (16) are multiplied by $\frac{(s+1)^{1-\frac{\beta}{k}}\omega^{-1}(t)}{k\Gamma_k(\alpha)}(t^{s+1}-y^{s+1})^{\frac{\beta}{k}-1}\omega(y)y^s$ and integrating w.r.t y over (a,t), we obtain

$$\begin{aligned}&{}_k^s\mathfrak{J}^{\alpha}_{a^+,\omega}\varphi\psi(t)\frac{(s+1)^{1-\frac{\beta}{k}}\omega^{-1}(t)}{k\Gamma_k(\beta)}\int_a^t(t^{s+1}-y^{s+1})^{\frac{\beta}{k}-1}y^s\omega(y)dy\\&+{}_k^s\mathfrak{J}^{\alpha}_{a^+,\omega}(1)\frac{(s+1)^{1-\frac{\beta}{k}}\omega^{-1}(t)}{k\Gamma_k(\beta)}\int_a^t(t^{s+1}-y^{s+1})^{\frac{\beta}{k}-1}y^s\varphi(y))\psi(y)\omega(y)dy\\\geq\;&{}_k^s\mathfrak{J}^{\alpha}_{a^+,\omega}\varphi(t)\frac{(s+1)^{1-\frac{\beta}{k}}\omega^{-1}(t)}{k\Gamma_k(\beta)}\int_a^t(t^{s+1}-y^{s+1})^{\frac{\beta}{k}-1}y^s\omega(y)\psi(y)dy\\&+{}_k^s\mathfrak{J}^{\alpha}_{a^+,\omega}\psi(t)\frac{(s+1)^{1-\frac{\beta}{k}}\omega^{-1}(t)}{k\Gamma_k(\beta)}\int_a^t(t^{s+1}-y^{s+1})^{\frac{\beta}{k}-1}y^s\omega(y)\varphi(y)dy,\end{aligned}$$

which gives

$$\begin{aligned}&{}_k^s\mathfrak{J}^{\alpha}_{a^+,\omega}\varphi\psi(t){}_k^s\mathfrak{J}^{\beta}_{a^+,\omega}(1)+{}_k^s\mathfrak{J}^{\beta}_{a^+,\omega}\varphi\psi(t){}_k^s\mathfrak{J}^{\alpha}_{a^+,\omega}(1)\\&\geq{}_k^s\mathfrak{J}^{\alpha}_{a^+,\omega}\varphi(t){}_k^s\mathfrak{J}^{\beta}_{a^+,\omega}\psi(t)+{}_k^s\mathfrak{J}^{\alpha}_{a^+,\omega}\psi(t){}_k^s\mathfrak{J}^{\beta}_{a^+,\omega}\varphi(t).\end{aligned}$$

The proof of (13) is done. □

Theorem 7. *Let φ and ψ be two synchronous functions on $[0,\infty)$ and $h(t) \geq 0$. Then for all $t > a \geq 0$, the following inequality holds:*

$$\begin{aligned}
&\frac{\omega^{-1}(\xi)}{(s+1)^{1-\frac{\beta}{k}}\Gamma_k(\beta+k)}(t^{s+1}-a^{s+1})^{\frac{\beta}{k}-2s}{}_k^s\mathfrak{J}^{\alpha}_{a^+,\omega}\varphi\psi h(t) \\
&+\frac{\omega^{-1}(\xi)}{(s+1)^{1-\frac{\alpha}{k}}\Gamma_k(\alpha+k)}(t^{s+1}-a^{s+1})^{\frac{\alpha}{k}-2s}{}_k^s\mathfrak{J}^{\beta}_{a^+,\omega}\varphi\psi h(t) \\
&\geq{}_k^s\mathfrak{J}^{\alpha}_{a^+,\omega}\varphi h(t){}_k^s\mathfrak{J}^{\beta}_{a^+,\omega}\psi(t)+{}_k^s\mathfrak{J}^{\alpha}_{a^+,\omega}\psi h(t){}_k^s\mathfrak{J}^{\beta}_{a^+,\omega}\varphi(t) \\
&-{}_k^s\mathfrak{J}^{\alpha}_{a^+,\omega}h(t){}_k^s\mathfrak{J}^{\beta}_{a^+,\omega}\varphi\psi(t)-{}_k^s\mathfrak{J}^{\alpha}_{a^+,\omega}\varphi\psi(t){}_k^s\mathfrak{J}^{\beta}_{a^+,\omega}h(t) \\
&+{}_k^s\mathfrak{J}^{\alpha}_{a^+,\omega}\varphi(t){}_k^s\mathfrak{J}^{\beta}_{a^+,\omega}\psi h(t)+{}_k^s\mathfrak{J}^{\alpha}_{a^+,\omega}\psi(t){}_k^s\mathfrak{J}^{\beta}_{a^+,\omega}\varphi h(t), \qquad (17)
\end{aligned}$$

where $\alpha,\beta > 0$ and $\omega(\xi) \neq 0$.

Proof. Since the function φ and ψ are synchronous on $[0,\infty)$, $h \geq 0$, for all α, $\beta > 0$, we have

$$(\varphi(\xi)-\varphi(y))(\psi(\xi)-\psi(y))(h(\xi)+h(y)) \geq 0.$$

This gives

$$\begin{aligned}
&\varphi(\xi)\psi(\xi)h(\xi)+\varphi(y)\psi(y)h(y) \\
\geq\ &\varphi(\xi)\psi(y)h(\xi)+\varphi(y)\psi(\xi)h(\xi)-\varphi(y)\psi(y)h(\xi) \\
&-\varphi(\xi)\psi(\xi)h(y)+\varphi(\xi)\psi(y)h(y)+\varphi(y)\psi(\xi)h(y). \qquad (18)
\end{aligned}$$

Both sides of (18) are multiplied by $\frac{(s+1)^{1-\frac{\alpha}{k}}\omega^{-1}(\xi)}{k\Gamma_k(\alpha)}(t^{s+1}-\xi^{s+1})^{\frac{\beta}{k}-1}\omega(\xi)\xi^s$ and integrating w.r.t ξ over (a,t), we obtain

$$\begin{aligned}
&\frac{(s+1)^{1-\frac{\alpha}{k}}\omega^{-1}(t)}{k\Gamma_k(\alpha)}\int_a^t(t^{s+1}-\xi^{s+1})^{\frac{\alpha}{k}-1}\xi^s\omega(\xi)\varphi(\xi)\psi(\xi)h(\xi)d\xi \\
&+\varphi(y)\psi(y)h(y)\frac{(s+1)^{1-\frac{\alpha}{k}}\omega^{-1}(t)}{k\Gamma_k(\alpha)}\int_a^t(t^{s+1}-\xi^{s+1})^{\frac{\alpha}{k}-1}\xi^s\omega(\xi)d\xi \\
\geq\ &\psi(y)\frac{(s+1)^{1-\frac{\alpha}{k}}\omega^{-1}(t)}{k\Gamma_k(\alpha)}\int_a^t(t^{s+1}-\xi^{s+1})^{\frac{\alpha}{k}-1}\xi^s\omega(x)\varphi(x)h(x)dx+ \\
&\varphi(y)\frac{(s+1)^{1-\frac{\alpha}{k}}\omega^{-1}(t)}{k\Gamma_k(\alpha)}\int_a^t(t^{s+1}-\xi^{s+1})^{\frac{\alpha}{k}-1}\xi^s\omega(\xi)\psi(\xi)h(\xi)d\xi \\
&-\varphi(y)\psi(y)\frac{(s+1)^{1-\frac{\alpha}{k}}\omega^{-1}(t)}{k\Gamma_k(\alpha)}\int_a^t(t^{s+1}-\xi^{s+1})^{\frac{\alpha}{k}-1}\xi^s\omega(\xi)h(\xi)d\xi \\
&-h(y)\frac{(s+1)^{1-\frac{\alpha}{k}}\omega^{-1}(t)}{k\Gamma_k(\alpha)}\int_a^t(t^{s+1}-\xi^{s+1})^{\frac{\alpha}{k}-1}\xi^s\omega(\xi)\psi(\xi)f(\xi)d\xi \\
&+\psi(y)h(y)\frac{(s+1)^{1-\frac{\alpha}{k}}\omega^{-1}(t)}{k\Gamma_k(\alpha)}\int_a^t(t^{s+1}-\xi^{s+1})^{\frac{\alpha}{k}-1}\xi^s\omega(\xi)\varphi(\xi)d\xi \\
&+\varphi(y)h(y)\frac{(s+1)^{1-\frac{\alpha}{k}}\omega^{-1}(t)}{k\Gamma_k(\alpha)}\int_a^t(t^{s+1}-\xi^{s+1})^{\frac{\alpha}{k}-1}\xi^s\omega(\xi)\psi(\xi)d\xi \qquad (19)
\end{aligned}$$

After multiplying both sides of (19) by $\frac{(s+1)^{1-\frac{\beta}{k}}\omega^{-1}(t)}{k\Gamma_k(\beta)}(t^{s+1}-y^{s+1})^{\frac{\beta}{k}-1}\omega(y)y^s$ and integrating w.r.t y over (a,t), we obtain

$$\begin{aligned}
&{}_k^s\mathfrak{J}_{a^+,\omega}^{\alpha}\varphi\psi h(t)\,{}_k^s\mathfrak{J}_{a^+,\omega}^{\beta}[\omega^{-1}(\xi)(1)] + {}_k^s\mathfrak{J}_{a^+,\omega}^{\alpha}[\omega^{-1}(\xi)(1)]\,{}_k^s\mathfrak{J}_{a^+,\omega}^{\beta}\varphi\psi h(t)\\
&\geq {}_k^s\mathfrak{J}_{a^+,\omega}^{\alpha}\varphi h(t)\,{}_k^s\mathfrak{J}_{a^+,\omega}^{\beta}\psi(t) + {}_k^s\mathfrak{J}_{a^+,\omega}^{\alpha}g\psi h(t)\,{}_k^s\mathfrak{J}_{a^+,\omega}^{\beta}\varphi(t) - {}_k^s\mathfrak{J}_{a^+,\omega}^{\alpha}h(t)\,{}_k^s\mathfrak{J}_{a^+,\omega}^{\beta}\varphi\psi(t)\\
&-{}_k^s\mathfrak{J}_{a^+,\omega}^{\alpha}\varphi\psi(t)\,{}_k^s\mathfrak{J}_{a^+,\omega}^{\beta}h(t) + {}_k^s\mathfrak{J}_{a^+,\omega}^{\alpha}\varphi(t)\,{}_k^s\mathfrak{J}_{a^+,\omega}^{\beta}\psi h(t) + {}_k^s\mathfrak{J}_{a^+,\omega}^{\alpha}\psi(t)\,{}_k^s\mathfrak{J}_{a^+,\omega}^{\beta}\varphi h(t),
\end{aligned}$$

which implies

$$\begin{aligned}
&\frac{\omega^{-1}(\xi)}{(s+1)^{1-\frac{\beta}{k}}\Gamma_k(\beta+k)}(t^{s+1}-a^{s+1})^{\frac{\beta}{k}-2s}\,{}_k^s\mathfrak{J}_{a^+,\omega}^{\alpha}\varphi\psi h(t)\\
&+\frac{\omega^{-1}(\xi)}{(s+1)^{1-\frac{\alpha}{k}}\Gamma_k(\alpha+k)}(t^{s+1}-a^{s+1})^{\frac{\alpha}{k}-2s}\,{}_k^s\mathfrak{J}_{a^+,\omega}^{\beta}\varphi\psi h(t)\\
&\geq {}_k^s\mathfrak{J}_{a^+,\omega}^{\alpha}\varphi h(t)\,{}_k^s\mathfrak{J}_{a^+,\omega}^{\beta}\psi(t) + {}_k^s\mathfrak{J}_{a^+,\omega}^{\alpha}\psi h(t)\,{}_k^s\mathfrak{J}_{a^+,\omega}^{\beta}\varphi(t) - {}_k^s\mathfrak{J}_{a^+,\omega}^{\alpha}h(t)\,{}_k^s\mathfrak{J}_{a^+,\omega}^{\beta}\varphi\psi(t)\\
&-{}_k^s\mathfrak{J}_{a^+,\omega}^{\alpha}\varphi\psi(t)\,{}_k^s\mathfrak{J}_{a^+,\omega}^{\beta}h(t) + {}_k^s\mathfrak{J}_{a^+,\omega}^{\alpha}\varphi(t)\,{}_k^s\mathfrak{J}_{a^+,\omega}^{\beta}\psi h(t) + {}_k^s\mathfrak{J}_{a^+,\omega}^{\alpha}\psi(t)\,{}_k^s\mathfrak{J}_{a^+,\omega}^{\beta}\varphi h(t).
\end{aligned}$$

Hence, the result is proved. □

Corollary 6. *Let φ and ψ be two synchronous functions on $[0,\infty]$ and $h \geq 0$. Then for all $t > a \geq 0$, the following inequality holds:*

$$\begin{aligned}
&\frac{\omega^{-1}(\xi)}{(s+1)^{1-\frac{\alpha}{k}}\Gamma_k(\alpha+k)}(t^{s+1}-a^{s+1})^{\frac{\alpha}{k}-2s}\,{}_k^s\mathfrak{J}_{a^+,\omega}^{\alpha}\varphi\psi h(t)\\
&\geq {}_k^s\mathfrak{J}_{a^+,\omega}^{\alpha}\varphi h(t)\,{}_k^s\mathfrak{J}_{a^+,\omega}^{\alpha}\psi(t) + {}_k^s\mathfrak{J}_{a^+,\omega}^{\alpha}\psi h(t)\,{}_k^s\mathfrak{J}_{a^+,\omega}^{\alpha}\varphi(t) - {}_k^s\mathfrak{J}_{a^+,\omega}^{\alpha}h(t)\,{}_k^s\mathfrak{J}_{a^+,\omega}^{\alpha}\varphi\psi(t), \quad (20)
\end{aligned}$$

where $\alpha,\beta > 0$ and $\omega(\xi) \neq 0$.

Proof. If we replace β to α in Theorem 7, we obtain the result (20). □

Theorem 8. *Let φ ψ and h be three monotonic functions defined on $[0,\infty]$ and satisfying the following*

$$(\varphi(\xi)-\varphi(y))(\psi(\xi)-\psi(y))(h(\xi)-h(y)) \geq 0.$$

Then for all $t > a \geq 0$, the following inequality holds:

$$\begin{aligned}
&\frac{\omega^{-1}(\xi)}{(s+1)^{1-\frac{\beta}{k}}\Gamma_k(\beta+k)}(t^{s+1}-a^{s+1})^{\frac{\beta}{k}-2s}\,{}_k^s\mathfrak{J}_{a^+,\omega}^{\alpha}\varphi\psi h(t)\\
&-\frac{\omega^{-1}(\xi)}{(s+1)^{1-\frac{\alpha}{k}}\Gamma_k(\alpha+k)}(t^{s+1}-a^{s+1})^{\frac{\alpha}{k}-2s}\,{}_k^s\mathfrak{J}_{a^+,\omega}^{\beta}\varphi\psi h(t)\\
&\geq {}_k^s\mathfrak{J}_{a^+,\omega}^{\alpha}\varphi h(t)\,{}_k^s\mathfrak{J}_{a^+,\omega}^{\beta}\psi(t) + {}_k^s\mathfrak{J}_{a^+,\omega}^{\alpha}\psi h(t)\,{}_k^s\mathfrak{J}_{a^+,\omega}^{\beta}\varphi(t) - {}_k^s\mathfrak{J}_{a^+,\omega}^{\alpha}h(t)\,{}_k^s\mathfrak{J}_{a^+,\omega}^{\beta}\varphi\psi(t)\\
&+{}_k^s\mathfrak{J}_{a^+,\omega}^{\alpha}\varphi\psi(t)\,{}_k^s\mathfrak{J}_{a^+,\omega}^{\beta}h(t) - {}_k^s\mathfrak{J}_{a^+,\omega}^{\alpha}\varphi(t)\,{}_k^s\mathfrak{J}_{a^+,\omega}^{\beta}\psi h(t) - {}_k^s\mathfrak{J}_{a^+,\omega}^{\alpha}\psi(t)\,{}_k^s\mathfrak{J}_{a^+,\omega}^{\beta}\varphi h(t).
\end{aligned}$$

where $\alpha,\beta > 0$ and $\omega(\xi) \neq 0$.

Proof. Use the same argument as in the proof of Theorem 7. □

Theorem 9. *Let φ and ψ be defined on $[0,\infty]$. Then for all $t > a \geq 0$ $\omega(\xi) \neq 0$, $\alpha, \beta > 0$, the following inequalities for weighted (k,s)-RLFI hold:*

$$\begin{aligned}
&\frac{\omega^{-1}(\xi)}{(s+1)^{1-\frac{\beta}{k}}\Gamma_k(\beta+k)}(t^{s+1}-a^{s+1})^{\frac{\beta}{k}-2}{}^s_k\mathfrak{J}^{\alpha}_{a^+,\omega}\varphi^2(t)\\
&+\frac{\omega^{-1}(\xi)}{(s+1)^{1-\frac{\alpha}{k}}\Gamma_k(\alpha+k)}(t^{s+1}-a^{s+1})^{\frac{\alpha}{k}-2}{}^s_k\mathfrak{J}^{\beta}_{a^+,\omega}\psi^2(t)\\
&\geq 2{}^s_k\mathfrak{J}^{\alpha}_{a^+,\omega}\varphi(t){}^s_k\mathfrak{J}^{\beta}_{a^+,\omega}\psi(t)
\end{aligned}\tag{21}$$

and

$${}^s_k\mathfrak{J}^{\alpha}_{a^+,\omega}\varphi^2(t){}^s_k\mathfrak{J}^{\beta}_{a^+,\omega}\psi^2(t)+{}^s_k\mathfrak{J}^{\beta}_{a^+,\omega}\varphi^2(t){}^s_k\mathfrak{J}^{\alpha}_{a^+,\omega}\psi^2(t) \geq 2{}^s_k\mathfrak{J}^{\alpha}_{a^+,\omega}\varphi\psi(t){}^s_k\mathfrak{J}^{\beta}_{a^+,\omega}\varphi\psi(t) \tag{22}$$

Proof. Since $(\varphi(\xi)-\psi(y))^2 \geq 0$ and $(\varphi(\xi)\psi(y)-\varphi(y)\psi(\xi))^2 \geq 0$ using the same argument as the proof in Theorem 7, we obtain (22) and (21). □

Corollary 7. *We have*

$$\begin{aligned}
&\frac{\omega^{-1}(\xi)}{(s+1)^{1-\frac{\alpha}{k}}\Gamma_k(\alpha+k)}(t^{s+1}-a^{s+1})^{\frac{\alpha}{k}-2}[{}^s_k\mathfrak{J}^{\alpha}_{a^+,\omega}\varphi^2(t)+{}^s_k\mathfrak{J}^{\beta}_{a^+,\omega}\psi^2(t)]\\
&\geq 2{}^s_k\mathfrak{J}^{\alpha}_{a^+,\omega}\varphi(t){}^s_k\mathfrak{J}^{\beta}_{a^+,\omega}\psi(t)
\end{aligned}\tag{23}$$

and

$${}^s_k\mathfrak{J}^{\alpha}_{a^+,\omega}\varphi^2(t){}^s_k\mathfrak{J}^{\alpha}_{a^+,\omega}\psi^2(t) \geq [{}^s_k\mathfrak{J}^{\alpha}_{a^+,\omega}\varphi\psi(t)]^2. \tag{24}$$

Proof. If we replace β to α in Theorem 9, we obtain the inequalities (23) and (24). □

Remark 5. *If we set $\omega(\xi) = 1$ in Theorems 6–9, then we obtain the inequalities of Theorems 3.1, 3.2, 3.4, and 3.5, respectively, given in [21].*

Theorem 10. *Let $\varphi : \mathbb{R} \to \mathbb{R}$ with $\overline{\varphi}(\xi) := \int_a^t \omega(t)t^s\varphi(t)dt$, for all $\xi > a \geq 0$, $s \in \mathbb{R}\backslash\{-1\}$. Then $\alpha \geq k > 0$ and $\omega(\xi) \neq 0$, we have*

$${}^s_k\mathfrak{J}^{\alpha+k}_{a^+,\omega}\varphi(\xi) = \frac{1}{k}{}^s_k\mathfrak{J}^{\alpha}_{a^+,\omega}[\omega^{-1}(\xi)\overline{\varphi}(\xi)]. \tag{25}$$

Proof. By using Definition 7 and the Dirichlet's formula, we have

$$\begin{aligned}
&{}^s_k\mathfrak{J}^{\alpha}_{a^+,\omega}[\omega^{-1}(\xi)\overline{\varphi}(\xi)]\\
&=\frac{(s+1)^{1-\frac{\alpha}{k}}\omega^{-1}(\xi)}{k\Gamma_k(\alpha)}\int_a^{\xi}(\xi^{s+1}-t^{s+1})^{\frac{\alpha}{k}-1}t^s\omega^{-1}(t)\omega(t)\overline{\varphi}(t)dt\\
&=\frac{(s+1)^{1-\frac{\alpha}{k}}\omega^{-1}(\xi)}{k\Gamma_k(\alpha)}\int_a^{\xi}(\xi^{s+1}-t^{s+1})^{\frac{\alpha}{k}-1}t^s[\int_a^t u^s\varphi(u)\omega(u)du]dt\\
&=\frac{(s+1)^{1-\frac{\alpha}{k}}\omega^{-1}(\xi)}{k\Gamma_k(\alpha)}\int_a^{\xi}u^s\varphi(u)\omega(u)[\int_u^{\xi}(\xi^{s+1}-t^{s+1})^{\frac{\alpha}{k}-1}t^s dt]du\\
&=\frac{(s+1)^{-\frac{\alpha}{k}}\omega^{-1}(\xi)}{\Gamma_k(\alpha+k)}\int_a^{\xi}(\xi^{s+1}-u^{s+1})^{\frac{\alpha}{k}}u^s\omega(u)\varphi(u)du\\
&=k{}^s_k\mathfrak{J}^{\alpha+k}_{a^+,\omega}\varphi(\xi).
\end{aligned}$$

Hence, we obtained the desired result. □

4. The Weighted Laplace Transform of the Weighted Fractional Operators

In this section, we apply the weighted laplace transformation to the new fractional operators. For this purpose we need to substitute $\psi(t) = t^{s+1}$ on the right side of (3), where we have

$$\mathfrak{L}_{\psi}^{\omega}\{\varphi(t)\}(u) = (s+1)\int_{a}^{\infty} e^{-u(t^{s+1}-a^{s+1})}\omega(t)t^{s}\varphi(t)dt, \tag{26}$$

which holds for all values of u.

Proposition 1.

$$\mathfrak{L}_{\psi}^{\omega}\{\omega^{-1}(\xi)(\xi^{s+1}-a^{s+1})^{\frac{\alpha}{k}-1}\}(u) = \frac{\Gamma(\frac{\alpha}{k})}{u^{\frac{\alpha}{k}}}, \quad u > 0.$$

Proof. By using (26), we have

$$\begin{aligned}
&\mathfrak{L}_{\psi}^{\omega}\{\omega^{-1}(\xi)(\xi^{s+1}-a^{s+1})^{\frac{\alpha}{k}-1}\}(u)\\
&= (s+1)\int_{a}^{\infty} e^{-u(\xi^{s+1}-a^{s+1})}(\xi^{s+1}-a^{s+1})^{\frac{\alpha}{k}-1}\xi^{s}d\xi.
\end{aligned} \tag{27}$$

By substituting $t = (\xi^{s+1} - a^{s+1})$ on the right side of (27), we obtain

$$\begin{aligned}
&\mathfrak{L}_{\psi}^{\omega}\{\omega^{-1}(\xi)(\xi^{s+1}-a^{s+1})^{\frac{\alpha}{k}-1}\}(u)\\
&= \int_{0}^{\infty} e^{-ut}t^{\frac{\alpha}{k}-1}dt\\
&= \int_{0}^{\infty} e^{-ut}\frac{(ut)^{\frac{\alpha}{k}-1}}{(u)^{\frac{\alpha}{k}-1}}\frac{u}{u}dt\\
&= \frac{1}{u^{\frac{\alpha}{k}}}\int_{0}^{\infty} e^{-ut}(ut)^{\frac{\alpha}{k}-1}udt,
\end{aligned}$$

which gives the required result. □

Theorem 11. *Let φ be a piecewise continuous function on each interval $[a, \xi]$ and of weighted ψ-exponential order. Then*

$$\mathfrak{L}_{\psi}^{\omega}\{({}_{k}^{s}\mathfrak{J}_{a^{+},\omega}^{\alpha}\varphi)(\xi)\}(u) = ((s+1)uk)^{\frac{-\alpha}{k}}\mathfrak{L}_{\psi}^{\omega}\{\varphi(\xi)\}(u),$$

where $k > 0$, $\omega(\xi) \neq 0$, $s \in \mathbb{R}\backslash\{-1\}$.

Proof. By using Definitions 6 and 7 and Proposition 1, we have

$$\begin{aligned}
&\mathfrak{L}_{\psi}^{\omega}\{({}_{k}^{s}\mathfrak{J}_{a^{+},\omega}^{\alpha}\varphi)(\xi)\}(u)\\
&= \mathfrak{L}_{\psi}^{\omega}\Big\{\frac{(s+1)^{1-\frac{\alpha}{k}}\omega^{-1}(\xi)}{k\Gamma_{k}(\alpha)}\int_{a}^{\xi}(\xi^{s+1}-t^{s+1})^{\frac{\alpha}{k}-1}t^{s}\omega(t)\varphi(t)dt\Big\}(u)\\
&= \frac{(s+1)^{-\frac{\alpha}{k}}}{k\Gamma_{k}(\alpha)}\mathfrak{L}_{\psi}^{\omega}\{\omega^{-1}(\xi)(\xi^{s+1}-t^{s+1})^{\frac{\alpha}{k}-1} *_{\psi}^{\omega} \varphi(\xi)\}(u)\\
&= \frac{(s+1)^{-\frac{\alpha}{k}}}{k\Gamma_{k}(\alpha)}\mathfrak{L}_{\psi}^{\omega}\{\omega^{-1}(\xi)(\xi^{s+1}-t^{s+1})^{\frac{\alpha}{k}-1}\}(u)\mathfrak{L}_{\psi}^{\omega}\{\varphi(\xi)\}(u)\\
&= \frac{(s+1)^{-\frac{\alpha}{k}}}{k\Gamma_{k}(\alpha)}\frac{\Gamma(\frac{\alpha}{k})}{u^{\frac{\alpha}{k}}}\mathfrak{L}_{\psi}^{\omega}\{\varphi(\xi)\}(u)\\
&= ((s+1)uk)^{-\frac{\alpha}{k}}\mathfrak{L}_{\psi}^{\omega}\{\varphi(\xi)\}(u).
\end{aligned} \tag{28}$$

This proves the claimed result. □

Theorem 12. *The Laplace transform of the weighted (k,s)-Riemann Liouville derivative is given by*

$$\begin{aligned}&\mathfrak{L}_{\psi}^{\omega}\{({}_{k}^{s}\mathfrak{D}_{a^+,\omega}^{\alpha}\varphi)(\xi)\}(u)\\&=(s+1)^{-\frac{nk-\alpha}{k}}(ku)^{\frac{\alpha}{k}}\mathfrak{L}_{\psi}^{\omega}\{\varphi(\xi)\}(u)\\&-k^{n}\sum_{m=0}^{n-1}u^{n-m-1}({}_{k}^{s}\mathfrak{J}_{a^+,\omega}^{nk-\alpha}\varphi)_{m}(a^{+}).\end{aligned} \tag{29}$$

Proof. By using Definition 9, Theorems 1 and 11, we obtain

$$\begin{aligned}&\mathfrak{L}_{\psi}^{\omega}\{({}_{k}^{s}\mathfrak{D}_{a^+,\omega}^{\alpha}\varphi)(\xi)\}(u)\\&=\mathfrak{L}_{\psi}^{\omega}\{(\xi^{1-s}\frac{d}{d\xi})^{n}k^{n}({}_{k}^{s}\mathfrak{J}_{a^+,\omega}^{nk-\alpha}\varphi)(t)\}(u)\\&=k^{n}u^{n}\mathfrak{L}_{\psi}^{\omega}\{({}_{k}^{s}\mathfrak{J}_{a^+,\omega}^{nk-\alpha}\varphi)(t)\}(u)\\&-k^{n}\sum_{m=0}^{n-1}u^{n-m-1}({}_{k}^{s}\mathfrak{J}_{a^+,\omega}^{nk-\alpha}\varphi)_{k}(a^{+})\\&=(uk)^{n}((s+1)uk)^{\frac{nk-\alpha}{k}}\mathfrak{L}_{\psi}^{\omega}\{\varphi(\xi)\}(u)\\&-k^{n}\sum_{m=0}^{n-1}u^{n-m-1}({}_{k}^{s}\mathfrak{J}_{a^+,\omega}^{nk-\alpha}\varphi)_{k}(a^{+})\\&=(s+1)^{\frac{nk-\alpha}{k}}(ku)^{\frac{\alpha}{k}}\mathfrak{L}_{\psi}^{\omega}\{\varphi(\xi)\}(u)\\&-k^{n}\sum_{m=0}^{n-1}u^{n-m-1}({}_{k}^{s}\mathfrak{J}_{a^+,\omega}^{nk-\alpha}\varphi)_{k}(a^{+}),\end{aligned}$$

which gives the required series solution. □

5. Fractional Kinetic Differ-Integral Equation

The fractional differential equations are significant in the field of applied science and have gained interest in dynamic systems, physics, and engineering. In the previous decade, the fractional kinetic equation has gained interest due to the discovery of its relationship with the CTRW theory [28]. The kinetic equations are essential in natural sciences and mathematical physics that explain the continuation of motion of the material. The generalized weighted fractional kinetic equation and its solution related to novel operators are discussed in this section. Consider the fractional kinetic equation given by

$$a({}_{k}^{s}\mathfrak{D}_{0^+,\omega}^{\alpha}N)(t)-N_{0}\varphi(t)=b({}_{k}^{s}\mathfrak{J}_{0^+,\omega}^{\beta}N)(t),\ \ \varphi\in L^{1}[0,\infty), \tag{30}$$

with initial condition

$$\omega(0)({}_{k}^{s}\mathfrak{J}_{0^+,\omega}^{nk-\alpha}N)(0)=d,\ \ d\geq 0, \tag{31}$$

where $\alpha\geq 0, a,b\in R(a\neq 0), k>0, n=[\frac{\alpha}{k}]=1$.

Theorem 13. *The solution of (30) with initial condition (31) is*

$$\begin{aligned}N(t)=d\omega^{-1}(t)\sum_{m=0}^{\infty}(\frac{a}{b})^{n}\frac{(s+1)^{\beta+(1+n)k}}{\Gamma_{k}(\alpha+(\alpha+\beta)n)}(\xi^{s+1}-a^{s+1})^{\frac{\alpha+(\alpha+\beta)n}{k}}\\+\frac{N_{0}}{a}\sum_{m=0}^{\infty}(s+1)^{\beta+(1+n)k}({}_{k}^{s}\mathfrak{J}_{0^+,\omega}^{(\alpha+\beta)n+\alpha}\varphi)(t).\end{aligned} \tag{32}$$

Proof. Applying the modified weighted Laplace transform on both side of (30), we obtain

$$a\mathfrak{L}_{\psi}^{\omega}\{({}_{k}^{s}\mathfrak{D}_{0^+,\omega}^{\alpha}N)(t)\}(u) - \mathfrak{L}_{\psi}^{\omega}\{N_0\varphi(t)\}(u) = b\mathfrak{L}_{\psi}^{\omega}\{({}_{k}^{s}\mathfrak{J}_{0^+,\omega}^{\beta}N)(t)\}(u).$$

Using Theorems 11 and 12, we obtain

$$\begin{aligned} a(s+1)^{-\frac{k-\alpha}{k}}(ku)^{\frac{\alpha}{k}}\mathfrak{L}_{\psi}^{\omega}\{N(t)\}(u) - kw(0)({}_{k}^{s}\mathfrak{J}_{a^+,\omega}^{k-\alpha}N)(0) - N_0\mathfrak{L}_{\psi}^{\omega}\{\varphi(t)\}(u) \\ = b(s+1)^{\frac{-\alpha}{k}}(uk)^{\frac{-\alpha}{k}}\mathfrak{L}_{\psi}^{\omega}\{N(t)\}(u) \end{aligned}$$

$$[\frac{a - b(s+1)^{-\frac{\alpha-k+\beta}{k}}(ku)^{-\frac{\alpha+\beta}{k}}}{(s+1)^{-\frac{\alpha-k}{k}}(ku)^{-\frac{\alpha}{k}}}]\mathfrak{L}_{\psi}^{\omega}\{N(t)\} = akd + N_0\mathfrak{L}_{\psi}^{\omega}\{\varphi(t)\}(u)$$

$$\begin{aligned} \mathfrak{L}_{\psi}^{\omega}\{N(t)\} = akd[&\frac{(s+1)^{-\frac{\alpha-k}{k}}(ku)^{-\frac{\alpha}{k}}}{a - b(s+1)^{-\frac{\alpha-k+\beta}{k}}(ku)^{-\frac{\alpha+\beta}{k}}}] \\ &+ [\frac{(s+1)^{-\frac{\alpha-k}{k}}(ku)^{-\frac{\alpha}{k}}}{a - b(s+1)^{-\frac{\alpha-k+\beta}{k}}(ku)^{-\frac{\alpha+\beta}{k}}}] \\ &\times N_0\mathfrak{L}_{\psi}^{\omega}\{\varphi(t)\}(u). \end{aligned}$$

Taking $|\frac{b}{a}(s+1)^{-\frac{\alpha-k+\beta}{k}}(ku)^{-\frac{\alpha+\beta}{k}}| < 1$, we obtain

$$\begin{aligned} \mathfrak{L}_{\psi}^{\omega}\{N(t)\} =& \big[kd\big[(s+1)^{-\frac{\alpha-k}{k}}(ku)^{-\frac{\alpha}{k}}\big] + a^{-1}N_0\big[(s+1)^{-\frac{\alpha-k}{k}}(ku)^{-\frac{\alpha}{k}}\big]\big] \\ &\times \sum_{n=0}^{\infty}(\frac{b}{a})^n(s+1)^{-\frac{(\alpha-k+\beta)n}{k}}(ku)^{-\frac{(\alpha+\beta)n}{k}}\mathfrak{L}_{\psi}^{\omega}\{\varphi(t)\}(u) \\ =& kd\big[(s+1)^{-\frac{\alpha-k}{k}}(ku)^{-\frac{\alpha}{k}}\big]\sum_{n=0}^{\infty}(\frac{b}{a})^n(s+1)^{-\frac{(\alpha-k+\beta)n}{k}}(ku)^{-\frac{(\alpha+\beta)n}{k}} \\ &+ a^{-1}N_0\big[(s+1)^{-\frac{\alpha-k}{k}}(ku)^{-\frac{\alpha}{k}}\big] \\ &\times \sum_{n=0}^{\infty}(\frac{b}{a})^n(s+1)^{-\frac{(\alpha-k+\beta)n}{k}}(ku)^{-\frac{(\alpha+\beta)n}{k}}\mathfrak{L}_{\psi}^{\omega}\{\varphi(t)\}(u) \\ =& kd\sum_{n=0}^{\infty}(\frac{b}{a})^n(s+1)^{-\frac{(\alpha-k)(n+1)+n\beta}{k}}(ku)^{-\frac{(\alpha+\beta)n+\alpha}{k}} \\ &+ \frac{N_0}{a}\sum_{n=0}^{\infty}(\frac{b}{a})^n(s+1)^{-\frac{(\alpha-k)(n+1)+n\beta}{k}}(ku)^{-\frac{(\alpha+\beta)n+\alpha}{k}}\mathfrak{L}_{\psi}^{\omega}\{\varphi(t)\}(u) \\ =& kd\sum_{n=0}^{\infty}(\frac{b}{a})^n(s+1)^{-\frac{(\alpha-k)(n+1)+n\beta}{k}}(ku)^{-\frac{(\alpha+\beta)n+\alpha}{k}} \\ &+ \frac{N_0}{a}\sum_{n=0}^{\infty}(\frac{b}{a})^n(s+1)^{-\frac{(\alpha+\beta)n+\alpha}{k}}(s+1)^{(n+1)}(ku)^{-\frac{(\alpha+\beta)n+\alpha}{k}}\mathfrak{L}_{\psi}^{\omega}\{\varphi(t)\}(u). \end{aligned}$$

Applying inverse Laplace transform, we obtain

$$\begin{aligned} N(t) =& dw^{-1}(t)\sum_{n=0}^{\infty}(\frac{b}{a})^n\frac{(s+1)^{-\frac{(\alpha-k)(n+1)+n\beta}{k}}}{\Gamma_k((\alpha+\beta)n+\alpha)}(\zeta^{s+1} - a^{s+1})^{\frac{(\alpha+\beta)n+\alpha}{k}-1} \\ &+ \frac{N_0}{a}\sum_{n=0}^{\infty}(\frac{b}{a})^n(s+1)^{(n+1)}({}_{k}^{s}\mathfrak{J}_{0^+,\omega}^{(\alpha+\beta)n+\alpha}\varphi)(t). \end{aligned}$$

The proof of the result is completed. □

6. Conclusions and Discussion

Fractional calculus is currently one of the most widely debated topics. In the present article, we introduced the weighted versions of the (k, s)-RLF operators. We then investigated and examined their properties and found the weighted Laplace transform of the new operators. Significantly, these operators reduce to notable fractional operators in the literature. Other fractional operators, such as the Riemann-Liouville fractional operators and Hadamard fractional operators, show up as special cases of these weighted fractional operators with specific choices of weighted functions and operator functions. We have developed the Chebyshev inequalities by involving the introduced fractional integral operator. We developed a fractional kinetic equation and the weighted Laplace transform used to find the solution of the said model. The presented results motivate scientists to stimulate more work in such directions.

Author Contributions: Conceptualization, M.S.; Formal analysis, M.U.; Funding acquisition, T.A.; Investigation, S.I.; Methodology, A.K.; Project administration, T.A.; Supervision, M.S.; Writing—review and editing, N.M., All authors jointly worked on the results and they read and approved the final manuscript.

Funding: There is no funding available for this work.

Institutional Review Board Statement: Not applicable.

Informed Consent Statement: Not applicable.

Data Availability Statement: No data were used to support this study.

Acknowledgments: The authors Nabil Mlaiki and Thabet Abdeljawad would like to thank Prince Sultan University (PSU) for the support through the TAS research lab. The authors would like to thank for paying the article processing charges.

Conflicts of Interest: The authors declare no conflict of interest.

References

1. Abel, N. Solution de quelques problemes l'aide d'integrales definies. *Mag. Naturv.* **1823**, *1*, 11–27.
2. Herrmann, R. *Fractional Calculus, an Introduction for Physicists*; World Scientific Publishing Company: Singapore; Hackensack, NJ, USA; London, UK; Hong Kong, China, 2018.
3. Hilfer, R. *Applications of Fractional Calculus in Physics*; World Scientific Publishing Co. Pte. Ltd.: Singapore, 2000.
4. Hilfer, R. Threefold introduction to fractional derivatives. In *Anomalous Transport Foundations and Applications*; Wiley-VCH Verlag: Weinheim, Germany, 2008.
5. Mohammed, P.O.; Abdeljawad, T.; Kashuri, A. Fractional Hermite-Hadamard-Fejer Inequalities for a Convex Function with Respect to an Increasing Function Involving a Positive Weighted Symmetric Function. *Symmetry* **2020**, *12*, 1503. [CrossRef]
6. Baleanu, D.; Samraiz, M.; Perveen, Z.; Iqbal, S.; Nisar, K.S.; Rahman, G. Hermite-Hadamard-Fejer type inequalities via fractional integral of a function concerning another function. *AIMS Math.* **2020**, *6*, 4280–4295. [CrossRef]
7. Kilbas, A.A.; Srivastava, H.M.; Trujillo, J.J. *Theory and Applications of Fractional Differential Equations, North-Holland Mathematics Studies*; Elsevier: New York, NY, USA; London, UK, 2006; Volume 204.
8. Lorenzoand, C.F.; Hartley, T.T. Variable order and distributed order fractional operators. *Nonlinear Dyn.* **2002**, 29, 57–98. [CrossRef]
9. Magin, R.L. *Fractional Calculus in Bioengineering*; Begell House Publishers: Danbury, CT, USA, 2006.
10. Podlubny, I. *Fractional Differential Equations*; Academic Press: New York, NY, USA, 1999.
11. Samraiz, M.; Perveen, Z.; Abdeljawad, T.; Iqbal, S.; Naheed, S. On certain fractional calculus operators and applications in mathematical physics. *Phy. Scr.* **2002**, *95*, 115210. [CrossRef]
12. Garra, R.; Goreno, R.; Polito, F.; Tomovski, Z. Hilfer-Prabhakar derivative and some applications. *Appl. Math. Comput.* **2014**, *242*, 576–589. [CrossRef]
13. Samraiz, M.; Perveen, Z.; Rahman, G.; Nisar, K.S.; Kumar, D. On the (k, s)-Hilfer-Prabhakar Fractional Derivative with Applications to Mathematical Physics. *Front. Phy.* **2020**, *8*, 309. [CrossRef]
14. Jarad, F.; Abdeljawad, T.; Shah, K. On the weighted fractional operators of a function with respect to another function. *World Sci.* **2020**, *28*, 2040011. [CrossRef]
15. Oldham, K.B.; Spanier, J. *The Fractional Calculus*; Academic Press: New York, NY, USA, 1974.
16. Baleanu, D.; Diethelm, K.; Scalas, E.; Trujillo, J.J. *Fractional Calculus, Models and Numerical Methods*; World Scientific Publishing Company: Singapore, 2012.
17. Mainardi, F. *Fractional Calculus and Waves in Linear Viscoelasticity, an Introduction to Mathematical Models*; World Scientific Publishing Company: Singapore, 2010.

18. Samko, S.G.; Kilbas, A.A.; Marichev, O.I. *Fractional Integrals and Derivatives: Theory and Applications*; Gordon and Breach Science Publishers: Langhorne, PA, USA, 1993.
19. Hadamard, J. Essai sur l'etude des fonctions donnees par leur developpment de Taylor. *J. Pure Appl. Math.* **1892**, *4*, 101–186.
20. Diaz, R.; Pariguan, E. On hypergeometric functions and Pochhammer k-symbol. *Divulg. Mat.* **2007**, *15*, 179–192.
21. Sarikaya, Z.M.; Dahmani, Z.; Kiris, E.M.; Ahmad, F. (k,s)-Riemann-Liouville fractioinal integral and applications. *J. Math. Stat.* **2016**, *45*, 77–89.
22. Azam, M.K.; Farid, G.; Rehman, M.A. Study of generalized type k-fractional derivatives. *Adv. Differ. Equ.* **2017**, *249*, 1328. [CrossRef]
23. Jarad, F.; Abdeljawad, T. Generalized fractional derivatives and Laplace transform. *Discret. Contin. Syst. Ser.* **2020**, *13*, 709–722. [CrossRef]
24. Mubeen, S.; Habibullah, G.M. k-Fractional integrals and application. *Int. J. Contemp. Math. Sci.* **2012**, *7*, 89–94.
25. Farid, G.; Habibullah, G.M. An extension of Hadamard fractional integral. *Int. J. Math. Anal.* **2015**, *9*, 471–482. [CrossRef]
26. Romero, L.G.; Luque, L.L.; Dorrego, G.A.; Cerutti, R.A. On the k-Riemann-Liouville fractional derivative. *Int. J. Contemp. Math. Sci.* **2013**, *8*, 41–51. [CrossRef]
27. Katugampola, U.N. A new approach to generalized fractional derivative. *Bull. Math. Anal. Appl.* **2014**, *6*, 1–15.
28. Hilfer, R.; Anton, L. Fractional master equations and fractal time random walks. *Phys. Rev. E* **1995**, *51*, 848–851. [CrossRef]

fractal and fractional

MDPI

Article

Fractional Dynamics of Typhoid Fever Transmission Models with Mass Vaccination Perspectives

Hamadjam Abboubakar [1,*], Raissa Kom Regonne [2,†] and Kottakkaran Sooppy Nisar [3,*,†]

1 Department of Computer Engineering, University Institute of Technology of Ngaoundéré, University of Ngaoundéré, Ngaoundéré P.O. Box 455, Cameroon
2 National School of Agro-Industrial Sciences, University of Ngaoundéré, Ngaoundéré P.O. Box 455, Cameroon; rkregonne@yahoo.fr
3 Department of Mathematics, College of Arts and Science, Prince Sattam bin Abdulaziz University, Wadi Aldawaser 11991, Saudi Arabia
* Correspondence: h.abboubakar@gmail.com or hamadjam.abboubakar@univ-ndere.cm (H.A.); n.sooppy@psau.edu.sa (K.S.N.); Tel.: +237-694-52-3111 (H.A.)
† These authors contributed equally to this work.

Abstract: In this work, we formulate and mathematically study integer and fractional models of typhoid fever transmission dynamics. The models include vaccination as a control measure. After recalling some preliminary results for the integer model (determination of the epidemiological threshold denoted by $\mathcal{R}_c$, asymptotic stability of the equilibrium point without disease whenever $\mathcal{R}_c < 1$, the existence of an equilibrium point with disease whenever $\mathcal{R}_c > 1$), we replace the integer derivative with the Caputo derivative. We perform a stability analysis of the disease-free equilibrium and prove the existence and uniqueness of the solution of the fractional model using fixed point theory. We construct the numerical scheme and prove its stability. Simulation results show that when the fractional-order η decreases, the peak of infected humans is delayed. To reduce the proliferation of the disease, mass vaccination combined with environmental sanitation is recommended. We then extend the previous model by replacing the mass action incidences with standard incidences. We compute the corresponding epidemiological threshold denoted by $\mathcal{R}_{c\star}$ and ensure the uniform stability of the disease-free equilibrium, for both new models, when $\mathcal{R}_{c\star} < 1$. A new calibration of the new model is conducted with real data of Mbandjock, Cameroon, to estimate $\mathcal{R}_{c\star} = 1.4348$. We finally perform several numerical simulations that permit us to conclude that such diseases can possibly be tackled through vaccination combined with environmental sanitation.

Keywords: typhoid fever disease; vaccination; model calibration; Caputo derivative; asymptotic stability; fixed point theory

MSC: 26A33; 93D20; 47H10; 93E24; 92D30

Citation: Abboubakar, H.; Kom Regonne, R.; Sooppy Nisar, K. Fractional Dynamics of Typhoid Fever Transmission Models with Mass Vaccination Perspectives. *Fractal Fract.* **2021**, *5*, 149. https://doi.org/10.3390/fractalfract5040149

Academic Editor: Vasily E. Tarasov

Received: 7 August 2021
Accepted: 20 September 2021
Published: 30 September 2021

1. Introduction

Typhoid fever, caused by a salmonella bacterium (*Salmonella typhi*), is a tropical disease transmitted by the ingestion of food or/and water contaminated with feces. It is most prevalent in countries located below the equator, in Southeast Asia, and in the Indian subcontinent, where hygiene conditions are poor [1,2]. The principal symptoms of typhoid fever are insomnia, fever, generalized fatigue, headaches, stomach ache, anorexia, constipation or diarrhea, and vomiting. These symptoms can last several weeks. Without effective treatment, typhoid fever can lead to death. According to the World Health Organization (WHO), the number of cases of typhoid fever is estimated to be between 11 and 21 million, with 128,000 to 161,000 deaths annually due to the severity of the disease [1,2]. Vaccination, sanitary measures, and hygiene measures are the best ways to prevent the spread of the disease [2].

Since the work of Sir Ronald Ross on malaria [3], mathematical tools such as differential equations have usually been used to understand and describe the dynamics of infectious diseases [4–12]. In [7], the authors proposed a $SVII_cR$ that takes into account some control mechanisms (treatment, education campaigns, and vaccination). Quarantining the infected individuals and their treatment are the main control measures studied in [8]. The author used optimal control methods to conclude that the outbreak can be eliminated or controlled if the control strategies are combined to their highest levels. The same conclusions are given in [10,11]. Recently, Olumuyiwa James et al. [9] formulated and studied an optimal control model for typhoid fever that takes into account both indirect and direct transmission. They compared various proposed strategies using numerical simulations and concluded that the disease burden can be controlled if all the available control measures are combined.

Very recently, many authors have proposed fractional-order models in mathematical epidemiology (ref. [13]), ecology (ref. [14]), plant epidemiology (refs. [15,16]), and psychology (ref. [17]). Indeed, the necessity of the use of fractional derivatives in epidemiology, for example, comes from the fact that these operators have many properties, such as their different types of kernels and the crossover behavior in the model, which can only be described using these operators. Moreover, any real data that have zigzag dynamics (mostly many) that cannot be projected by an integer-order derivative can be solved by a fractional model more clearly.

The principal disadvantage of models with integer derivatives is that they do not permit the definition of memory effects. Replacing integer derivatives with fractional derivatives makes it possible to remedy this problem. Indeed, they offer different ways to forecast data by varying the fractional-order parameter [12,18,19]. Several fractional operators have been defined so far. The most popular are the fractional operators of Caputo, the fractional Caputo–Fabrizio operator, and the fractional operator of Atangana–Baleanu. Each operator explores the dynamics of the studied phenomenon differently, thus helping us to predict more variations in the evolution of the phenomenon. The advantages and disadvantages of each fractional operator and their application domains can be found in [20–22].

To the best of our knowledge, there are few mathematical works on typhoid fever using fractional operators [6,12]. In [12], the authors used the Caputo–Fabrizio operator to extend the model proposed in [23]. They provided existence, uniqueness, and stability criteria for the proposed fractional-order typhoid model. More recently, Abboubakar et al. [6] formulated a SIR-B-type compartmental model with both integer and Caputo derivatives. The only control measure was vaccination. They computed the control reproduction number, R_c, and performed stability analysis of the disease-free equilibrium point for both models. The present contributions are listed as follows:

1. Using a fractional derivative in place of an integer derivative, as used in our previous model [5], we formulate a new model. To prove the existence and uniqueness of the solutions, we use fixed point theory. The corresponding numerical scheme is obtained through the Adams–Bashforth–Moulton method [24,25]. The stability of this numerical scheme is also proven. Finally, several numerical simulations are carried out from the real values of parameters estimated with real data of Mbandjock, in Cameroon (see [5]).
2. Secondly, we extend the previous models by replacing the mass action incidence law with the standard incidence law. For these new models, we compute the corresponding control reproduction number, $\mathcal{R}_{c\star}$, and ensure the uniform stability of the equilibrium point without disease. As in [6], model parameters are estimated. With these new parameter values, we finally perform several numerical simulations that permit us to compare the quantitative dynamics of the two types of models.

The rest of the work is organized as follows. We devote Section 2 to preliminary definitions of the fractional derivative in the sense of Caputo and useful results. Formulation of the models with mass incidence law and standard incidence, as well as their mathematical analysis (computation of control reproduction numbers, asymptotic stability analysis of the disease-free equilibrium, existence as well as the uniqueness of solutions, construction of the numerical scheme with its stability), is also described in this section. The calibration of the model with standard incidences and several numerical results is given in Section 3. We end the paper with a discussion and conclusions.

2. Materials and Methods

2.1. Useful Definitions and Results

For over ten years, fractional derivatives have captured the attention of researchers, who use these fractional operators to model physical, chemical, and biological processes. One can cite the dynamics of transmissible diseases [26–28]. Before the formulation of the fractional models, it is important to recall their definition, as well as two results that will be used later in the fractional model analysis.

Definition 1. *Let $f \in C^{l}_{-1}$, and we have the following relation:*

$$D^{\nu}_{\tau}f(\tau) = \begin{cases} \frac{d^r f(\tau)}{d\tau^r}, & \nu = r \in \mathbb{N} \\ \frac{1}{\Gamma(r-\nu)} \int_0^{\tau} (\tau - \iota)^{r-\nu-1} f^{(r)}(\iota) d\iota, & -1 + r < \nu < r, r \in \mathbb{N}, \end{cases} \quad (1)$$

which represents the Caputo derivative of f.

Lemma 1. *Assume that $\chi, \mathbf{Q}, \mathbf{h}, \mathbf{Y} > 0$, $\mathbf{kh} \leq \mathbf{Y}$ with $\mathbf{k} \in \mathbb{N}$, and*

$$\mathbf{y}_{q,m} = \begin{cases} (m-q)^{\chi-1} & q = 1, 2, \ldots, m-1, \\ 0 & q \geq m. \end{cases}$$

Let $\sum_{q=\mathbf{k}}^{q=i} \mathbf{y}_{q,m} |\mathbf{e}_q| = 0$ for $\mathbf{k} > m \geq 1$.
If

$$|\mathbf{e}_m| \leq \mathbf{Q}\mathbf{h}^{\chi} \sum_{q=1}^{m-1} \mathbf{y}_{q,m} |\mathbf{e}_q| + |\eta_0|, \quad m = 1, 2, \ldots, \mathbf{k},$$

then

$$|\mathbf{e}_k| \leq \mathcal{M} |\eta_0|, \quad \mathbf{k} \in \{1, 2, \ldots\}$$

where $\mathcal{M} \in \mathbb{R}_+$ does not depend on $\mathbf{h}$ and $\mathbf{k}$.

Lemma 2. *If $0 < \chi < 1$ and $\mathbf{d} \in \mathbb{N}$, then there exist positive constants $W_{\chi,1}$ and $W_{\chi,2}$ only dependent on χ, such that*

$$(1+\mathbf{v})^{\chi} - \mathbf{v}^{\chi} \leq \mathcal{W}_{\chi,1}(1+\mathbf{v})^{\chi-1} \quad \text{and} \quad (2+\mathbf{v})^{\chi+1} - 2(1+\mathbf{v})^{\chi+1} + \mathbf{v}^{\chi+1} \leq \mathcal{W}_{\chi,2}(1+\mathbf{v})^{\chi-1}.$$

2.2. Model Dynamics with Mass Action Incidence Law

2.2.1. Model Formulation in ODE Sense and Its Analysis

In a previous work [5], we formulated and analyzed a new mathematical model for the transmission dynamics of typhoid fever with application to the town of Mbandjock, in the central region of Cameroon. The model is divided into seven compartments: susceptible individuals $S(t)$, vaccinated individuals $V(t)$, infected individuals in latent stage $E(t)$, infected individuals without any sign of the disease $C(t)$, symptomatic infected individuals $I(t)$, recovered individuals $R(t)$, and the density of bacteria in the environment $B(t)$. Each individual in each compartment naturally dies at a rate μ_h. Susceptible humans are recruited at a constant rate Λ_h. The population in compartment S_h decreases either by vaccination at a rate ξ, or by infection at an incidence rate $\nu B(t)S(t)$, where ν is the contact rate. We denote by θ the rate at which vaccinated individuals lose their immunity. The vaccine efficacy is denoted by ϵ. The compartment E of latent individuals, which include infected susceptible individuals and vaccinated individuals, progresses either to the carriers compartment C at a rate $q\gamma_1$ or to the compartment I at the rate $(1-q)\gamma_1$. Asymptomatic individuals become symptomatic at the rate $(1-p)\gamma_2$ or recover at the rate $p\gamma_2$. δ denotes the disease-induced death rate of symptomatic individuals. Recovered individuals become susceptible at a rate α. With this brief description, the mathematical formulation of the model studied in [5] is presented as follows:

$$\dot{S}(t) = \Lambda_h - (k_1 + \nu B(t))S(t) + \theta V(t) + \alpha R(t), \tag{2a}$$
$$\dot{V}(t) = -[k_2 + (1-\epsilon)\nu B(t)]V(t) + \xi S(t), \tag{2b}$$
$$\dot{E}(t) = -k_3 E(t) + \nu[\pi V(t) + S(t)]B(t), \tag{2c}$$
$$\dot{C}(t) = q\gamma_1 E(t) - k_4 C(t), \tag{2d}$$
$$\dot{I}(t) = q_1\gamma_1 E(t) + p_1\gamma_2 C(t) - [k_5 + \sigma]I(t), \tag{2e}$$
$$\dot{R}(t) = p\gamma_2 C(t) + \sigma I(t) - k_6 R(t), \tag{2f}$$
$$\dot{B}(t) = p_c C(t) + p_i I(t) - \mu_b B(t), \tag{2g}$$

where $k_1 = \xi + \mu_h$, $k_2 = \theta + \mu_h$, $k_3 = \gamma_1 + \mu_h$, $k_4 = \mu_h + \gamma_2$, $k_5 = \delta + \mu_h$, $k_6 = \alpha + \mu_h$, $\pi = -1 + \epsilon$, $q_1 = 1 - q$, $p_1 = -p + 1$, $k_7 = k_1 k_2 - \theta\xi = \mu_h(k_2 + \xi) > 0$, $k_8 = k_5 + \sigma$.

Model (2) is defined in the following set:

$$\mathcal{W} = \left\{(S, V, E, C, I, R, B)' \in \mathbb{R}^7_+ : N = V + S + C + E + I + R \le \frac{\Lambda_h}{\mu_h}; B \le \frac{(p_i + p_c)\Lambda_h}{\mu_h \mu_b}\right\},$$

in which a dynamical system is defined, and where N denotes the human population.

Without disease, model (2) has the following equilibrium: $\mathcal{Q}_0 = (S_0, V_0, 0, 0, 0, 0, 0)'$, where $S_0 = \Lambda_h k_2/(\mu_h(k_2 + \xi))$ and $V_0 = \Lambda_h \xi/(\mu_h(k_2 + \xi))$. Using the same approach developed in [29], we obtain the control reproduction number given by

$$\mathcal{R}_c = \sqrt{\frac{\nu\Lambda_h(k_2 + \pi\xi)\gamma_1[p_c q(\sigma + k_5) + p_i(k_4(1-q) + \gamma_2 q(1-p))]}{\mu_b\mu_h k_3 k_4(k_2 + \xi)(\sigma + k_5)}}. \tag{3}$$

Considering the model without vaccination, R_c is equal to the basic reproduction number:

$$\mathcal{R}_0 = \sqrt{\frac{\nu\Lambda_h\gamma_1[p_c q(\sigma + k_5) + p_i(k_4(1-q) + \gamma_2 q(1-p))]}{\mu_b\mu_h k_3 k_4(\sigma + k_5)}}. \tag{4}$$

Thus, it follows that

$$\mathcal{R}_c = \mathcal{R}_0\sqrt{\frac{(k_2 + \pi\xi)}{(k_2 + \xi)}}.$$

Since $\pi\xi = (1-\epsilon)\xi \le \xi$, we have $\dfrac{(k_2 + \pi\xi)}{(k_2 + \xi)} \le 1$, which means that $\mathcal{R}_c \le \mathcal{R}_0$. This proves that mass vaccination is a useful tool that can be used to effectively tackle this kind of tropical disease.

For typhoid model (2) in the ODE sense, the following results were proven in [5].

Proposition 1 ([5])**.** *For model (2), $\mathcal{Q}_0$ is locally and globally asymptotically stable in $\mathcal{W}$ if $\mathcal{R}_c < 1$ and unstable if $\mathcal{R}_c > 1$.*

Proposition 2 ([5]). *Let us define the following coefficients:*

$$\begin{aligned}
\mathbf{a}_2 &= \mathcal{R}_c^4 k_3^2 k_4^2 k_8^2 \mu_h^2 (\xi + k_2)^2 \pi \times \\
&\quad \times (\mu_h \gamma_1 \alpha (k_5 + \sigma q) + \gamma_2 \mu_h k_8 (\alpha + \gamma_1) + \gamma_1 \gamma_2 \alpha (1 - pq) k_5 + [\mu_h^2 + (\alpha + \gamma_2 + \gamma_1)\mu_h]\mu_h k_8), \\
\mathbf{a}_1 &= -\mathcal{R}_c^2 k_3^2 k_4^2 k_8^2 \mu_h^2 (\xi + k_2)^2 \gamma_1 \Lambda_h k_6 \pi (k_4 q_1 + \gamma_2 p_1 q)(\mathcal{R}_c^2 - \mathcal{R}_b^2), \\
\mathbf{a}_0 &= -\gamma_1^2 \mu_h \Lambda_h^2 k_3 k_4 k_6 k_8 (k_4 q_1 + \gamma_2 p_1 q)^2 (\pi \xi + k_2)^2 (\xi + k_2)(\mathcal{R}_c^2 - 1).
\end{aligned}$$

Model (2) with the integer derivative either has (1) only one endemic equilibrium whenever ($\mathbf{a}_0 < 0 \iff \mathcal{R}_c > 1$) or ($\mathbf{a}_1 < 0$ and $\mathbf{a}_0 = 0$ or $\mathbf{a}_1^2 - 4\mathbf{a}_2\mathbf{a}_0 = 0$), (2) two endemic equilibrium points if ($\mathbf{a}_0 > 0$ ($\mathcal{R}_c < 1$), $\mathbf{a}_1 < 0$ ($\mathcal{R}_c > \mathcal{R}_b$) and $a_1^2 - 4\mathbf{a}_2\mathbf{a}_0 > 0$), or (3) no equilibrium otherwise.

Theorem 1 ([5]). *Model (2) exhibits a supercritical bifurcation at $\mathcal{R}_c = 1$, which implies that whenever $\mathcal{R}_c > 1$, the endemic equilibrium is locally asymptotically stable.*

Remark 1. *Proposition 1 combined with Theorem 1 implies that Proposition 2 (iii) will never hold true for model (2). Thus, the condition $\mathcal{R}_c < 1$ is sufficient to eradicate the disease.*

2.2.2. Fractional-Order Typhoid Model

The following model is obtained when we replace the integer derivative operator in (2) with the non-integer operator in the Caputo sense.

$$\begin{aligned}
{}_{t_0}^{C}D_t^{\eta} S(t) &= \Lambda_h - (k_1 + \nu B(t))S(t) + \theta V(t) + \alpha R(t), && \text{(5a)} \\
{}_{t_0}^{C}D_t^{\eta} V(t) &= -[k_2 + (1 - \epsilon)\nu B(t)]V(t) + \xi S(t), && \text{(5b)} \\
{}_{t_0}^{C}D_t^{\eta} E(t) &= -k_3 E(t) + \nu[\pi V(t) + S(t)]B(t), && \text{(5c)} \\
{}_{t_0}^{C}D_t^{\eta} C(t) &= q\gamma_1 E(t) - k_4 C(t), && \text{(5d)} \\
{}_{t_0}^{C}D_t^{\eta} I(t) &= q_1\gamma_1 E(t) + p_1\gamma_2 C(t) - [k_5 + \sigma]I(t), && \text{(5e)} \\
{}_{t_0}^{C}D_t^{\eta} R(t) &= p\gamma_2 C(t) + \sigma I(t) - k_6 R(t), && \text{(5f)} \\
{}_{t_0}^{C}D_t^{\eta} B(t) &= p_c C(t) + p_i I(t) - \mu_b B(t). && \text{(5g)}
\end{aligned}$$

with $S(0) \geq 0$, $V(0) \geq 0$, $B(0) \geq 0$, $E(0) \geq 0$, $I(0) \geq 0$, $C(0) \geq 0$, and $R(0) \geq 0$.

Asymptotic Stability of the Disease-Free Equilibrium

Before investigating the stability of the disease-free equilibrium point, we consider the fractional-order system (5) as follows:

$${}_{t_0}^{C}D_t^{\eta} \mathbf{x}(t) = \mathcal{F}(\mathbf{x}(t)), \tag{6}$$

where $\mathbf{x}(\zeta) \in \mathbb{R}^7$, $\mathcal{F} \in \mathbb{R}^7 \times \mathbb{R}^7$, $0 < \eta < 1$. The characteristic equation of the matrix $\mathcal{F}$ evaluated at any equilibrium (see [30]) is given by

$$\det(s(\mathbf{I} - (1 - \eta)\mathcal{F}) - \eta\mathcal{F}) = 0. \tag{7}$$

For the fractional model (5), the asymptotic stability of $\mathcal{Q}_0$ is claimed in the following result.

Theorem 2. *The disease-free equilibrium $\mathcal{Q}_0$ is uniformly asymptotically stable if $\mathcal{R}_c < 1$, and unstable otherwise.*

Proof. The Jacobian matrix of system (5) evaluated at the disease–free equilibrium $\mathcal{Q}_0$ is given by

$$J(\mathcal{Q}_0)=\begin{pmatrix} -k_1 & \theta & 0 & 0 & 0 & \alpha & -\nu S_0 \\ \xi & -k_2 & 0 & 0 & 0 & 0 & -\nu\pi V_0 \\ 0 & 0 & -k_3 & 0 & 0 & 0 & \nu(S_0+\pi V_0) \\ 0 & 0 & q\gamma_1 & -k_4 & 0 & 0 & 0 \\ 0 & 0 & q_1\gamma_1 & p_1\gamma_2 & -(k_5+\sigma) & 0 & 0 \\ 0 & 0 & 0 & p\gamma_2 & \sigma & -k_6 & 0 \\ 0 & 0 & 0 & p_c & p_i & 0 & -\mu_b \end{pmatrix}=\begin{pmatrix} J_1 & J_2 \\ J_3 & J_4 \end{pmatrix},$$

where $J_1=\begin{pmatrix} -k_1 & \theta \\ \xi & -k_2 \end{pmatrix}$, $J_2=\begin{pmatrix} 0 & 0 & 0 & \alpha & -\nu S_0 \\ 0 & 0 & 0 & 0 & -\nu\pi V_0 \end{pmatrix}$, $J_3=O_{\mathbb{R}^{5\times 2}}$, and

$$J_4=\begin{pmatrix} -k_3 & 0 & 0 & 0 & \nu(S_0+\pi V_0) \\ q\gamma_1 & -k_4 & 0 & 0 & 0 \\ q_1\gamma_1 & p_1\gamma_2 & -(k_5+\sigma) & 0 & 0 \\ 0 & p\gamma_2 & \sigma & -k_6 & 0 \\ 0 & p_c & p_i & 0 & -\mu_b \end{pmatrix}.$$

The characteristic Equation (7) of the typhoid fractional model (5) becomes

$$\begin{cases} \det(s(\mathbf{I}_2-(1-\eta)J_1)-\eta J_1)=0 \\ \det(s(\mathbf{I}_5-(1-\eta)J_4)-\eta J_4)=0, \end{cases} \tag{8}$$

which is equivalent to

$$\begin{cases} a_2s^2+a_1s+a_0=0, \\ \{s[(1-\eta)k_6+1]+k_6\eta\}\left(s^4+\frac{A_1}{A_5}s^3+\frac{A_2}{A_5}s^2+\frac{A_3}{A_5}s+\frac{A_4}{A_5}\right)=0, \end{cases} \tag{9}$$

where $a_2=(1-\eta)^2k_7+(1-\eta)(k_1+k_2)+1$, $a_1=2\eta(1-\eta)k_7+\eta(k_1+k_2)$, $a_0=\eta^2k_7$,

$$\begin{aligned} A_5 &= \eta_1^3k_3k_4k_8\mu_b(\eta_1k_9+q_1p_i+p_cq)(1-\mathcal{R}_c^2) \\ &\quad+\eta_1^3\mu_b[k_3k_8(k_8p_cq+p_1\gamma_2p_iq)+k_3k_4(q_1k_4p_i+p_1\gamma_2p_iq)+k_4k_8k_9] \\ &\quad+\eta_1^2\mu_b(k_8k_9+k_4k_9+k_3k_9)+\eta_1k_9\mu_b \\ &\quad+\left(\left(\left(\eta_1^3k_3+\eta_1^2\right)k_4+\eta_1^2k_3+\eta_1\right)k_8+\left(\eta_1^2k_3+\eta_1\right)k_4+\eta_1k_3+1\right)k_9, \end{aligned}$$

$$\begin{aligned} A_1 &= \left\{4k_3k_4k_8k_9\mu_b\eta_1^3\eta+3\eta_1^2k_3k_8\mu_b\eta q_1k_4p_i+3\eta_1^2\eta k_3k_4\mu_bk_8p_cq\right\}(1-R_c^2) \\ &\quad+3\eta_1^2\eta k_3k_4\mu_b(q_1k_4p_i+p_1\gamma_2p_iq)+3\eta_1^2k_3k_8\mu_b\eta[k_8p_cq+p_1\gamma_2p_iq] \\ &\quad+2\eta_1k_4k_9\mu_b\eta+(2\eta_1k_3+1)k_9\mu_b\eta+3\eta_1^2k_4k_8k_9\mu_b\eta+2\eta_1k_8k_9\mu_b\eta \\ &\quad+\left(\left(\left(3\eta_1^2k_3+2\eta_1\right)k_4+2\eta_1k_3+1\right)k_8+(2\eta_1k_3+1)k_4+k_3\right)k_9\eta, \end{aligned}$$

$$\begin{aligned} A_2 &= \left[6\eta_1^2k_3k_4k_8k_9\mu_b\eta^2+3\eta_1k_3k_4k_8\mu_b\eta^2(p_cq+q_1p_i)\right](1-\mathcal{R}_c^2) \\ &\quad+3\eta_1\eta^2[k_3k_8\mu_b(k_8p_cq+p_1\gamma_2p_iq)+k_4k_8k_9(\mu_b+k_3)+k_3k_4\mu_b(q_1k_4p_i+p_1\gamma_2p_iq)] \\ &\quad+(k_4k_8+k_3k_8+k_3k_4+k_8\mu_b+k_4\mu_b+k_3\mu_b)k_9\eta^2, \end{aligned}$$

$$\begin{aligned} A_3 &= \mu_b\eta^3k_3k_4k_8[4k_9\eta_1+q_1p_i+p_cq](1-\mathcal{R}_c^2) \\ &\quad+(\mu_b+k_3)k_4k_8k_9\eta^3+k_3\mu_b\eta^3\{k_8q(k_8p_c+p_1\gamma_2p_i)+k_4p_i(q_1k_4+p_1\gamma_2q)\}, \end{aligned}$$

$$A_4=\left(1-\mathcal{R}_c^2\right)\eta^4k_3k_4k_8\mu_b\overbrace{[q_1k_4p_i+k_8p_cq+p_1\gamma_2p_iq]}^{k_9},$$

and $\eta_1=1-\eta$.

Since both coefficients of the first equation of (9) are positive, it follows that the real parts of the solution of (9) are negative. From the second equation of (9), we have that

$s_1 = -\frac{k_6\eta}{[(1-\eta)k_6+1]}$ is one solution, and the other solutions are the root of $\mathcal{T}(s) := A_5s^4 + A_1s^3 + A_2s^2 + A_3s + A_4$. Note that $A_4 > 0 \Longleftrightarrow \mathcal{R}_c < 1$, and $\mathcal{R}_c < 1 \Longrightarrow (A_1 > 0, A_2 > 0, A_3 > 0,$ and $A_5 > 0)$. Then, it follows that, if $\mathcal{R}_c < 1$, then the disease-free equilibrium $\mathcal{Q}_0$ is asymptotically stable whenever the following Routh–Hurwitz criteria $\frac{A_1A_2}{A_5^2} - \frac{A_3}{A_5} > 0$ and $\frac{A_1A_2A_3}{A_5^3} - \frac{A_1^2A_4}{A_5^3} - \frac{A_3^2}{A_5^2} > 0$ are satisfied for polynomial $\mathcal{T}(s)$.

For uniform stability, we use the Lyapunov function:

$$\mathcal{M}(E, C, I, B) = \mathbf{b}_1E(t) + \mathbf{b}_2C(t) + \mathbf{b}_3I(t) + \mathbf{b}_4B(t), \tag{10}$$

where $\mathbf{b}_1 = 1$, $\mathbf{b}_2 = k_3(k_8p_c + p_ip_1\gamma_2)/[k_4p_iq_1\gamma_1 + q\gamma_1(k_8p_c + p_ip_1\gamma_2)]$, and $\mathbf{b}_3 = k_3k_4p_i/[k_4p_iq_1\gamma_1 + q\gamma_1(k_8p_c + p_ip_1\gamma_2)]$ and $\mathbf{b}_4 = k_3k_4k_8/[k_4p_iq_1\gamma_1 + q\gamma_1(k_8p_c + p_ip_1\gamma_2)]$.

The fractional derivative of $\mathcal{M}$ is given by

$$\begin{aligned}
{}^C_{t_0}D^\eta_t\mathcal{M}(E, C, I, B) \leq\ & {}^C_{t_0}D^\eta_t E\mathbf{b}_1 + {}^C_{t_0}D^\eta_t C\mathbf{b}_2 + {}^C_{t_0}D^\eta_t I\mathbf{b}_3 + {}^C_{t_0}D^\eta_t B\mathbf{b}_4 \\
&= \mathbf{b}_1(\nu B[S + \pi V] - k_3E) + \mathbf{b}_2(q\gamma_1E - k_4C) + \mathbf{b}_3(q_1\gamma_1E + p_1\gamma_2C - k_8I) \\
&+ \mathbf{b}_4(p_cC + p_iI - \mu_bB) \\
&\leq \mathbf{b}_1\left(\nu B\left[S^0 + \pi V^0\right] - k_3E\right) + \mathbf{b}_2(q\gamma_1E - k_4C) + \mathbf{b}_3(q_1\gamma_1E + p_1\gamma_2C - k_8I) \\
&+ \mathbf{b}_4(p_cC + p_iI - \mu_bB) \\
&= \mathbf{b}_1\nu B\left(S^0 + \pi V^0\right) - \mathbf{b}_1k_3E + \mathbf{b}_2q\gamma_1E - \mathbf{b}_2k_4C + \mathbf{b}_3q_1\gamma_1E + \mathbf{b}_3p_1\gamma_2C \\
&- \mathbf{b}_3k_8I + \mathbf{b}_4p_cC + \mathbf{b}_4p_iI - \mathbf{b}_4\mu_bB \\
&= \mathbf{b}_1\nu B\left(S^0 + \pi V^0\right) - \mathbf{b}_4\mu_bB + (\mathbf{b}_3q_1\gamma_1 + \mathbf{b}_2q\gamma_1 - \mathbf{b}_1k_3)E \\
&+ (\mathbf{b}_4p_c + \mathbf{b}_3p_1\gamma_2 - \mathbf{b}_2k_4)C + (\mathbf{b}_4p_i - \mathbf{b}_3k_8)I \\
&= \frac{\mu_bk_3k_4(k_5 + \sigma)}{p_iq_1\gamma_1k_4 + q\gamma_1(p_1p_i\gamma_2 + p_c(k_5 + \sigma))}\left(\mathcal{R}_c^2 - 1\right)B.
\end{aligned}$$

Thus, ${}^C_{t_0}D^\eta_t\mathcal{M}(E, C, I, B) < 0$ whenever $\mathcal{R}_c < 1$, and ${}^C_{t_0}D^\eta_t\mathcal{M}(E, C, I, B) = 0$ if and only if $\mathcal{R}_c = 1$ or $B(\zeta) = 0$. Setting $B = 0$ in (5), we obtain $S = S_0$, $V = V_0$, and $E = C = I = R = 0$. Thus, $\lim_{t\to\infty}(S(t), V(t), E(t), C(t), I(t), R(t))' \to (S_0, V_0, 0, 0, 0, 0)' := \mathcal{Q}_0$. Consequently, by [31] (Theorem 2.5), it follows that if $\mathcal{R}_c < 1$, then $\mathcal{Q}_0$ is uniformly asymptotically stable in $\mathcal{W}$. □

Existence and Uniqueness Analysis

Before describing the existence and uniqueness of solutions for the fractional model using the fixed point theorem, we define $\mathcal{T}$ as a Banach space of continuous functions defined on an interval $\mathcal{P}$ with the norm

$$\| \mathbf{X} \| = \sum_{i=1}^{i=7} \| \mathbf{X}_i \|,$$

where $\mathbf{X} = (S, V, E, C, I, R, B)$, $\| \mathbf{X}_i \| = \sup\{|\mathbf{X}_i(t)| : t \in \mathcal{P}\}$, and $\mathcal{T} = \mathcal{M}(\mathcal{P}) \times \mathcal{M}(\mathcal{P}) \times \mathcal{M}(\mathcal{P}) \times \mathcal{M}(\mathcal{P}) \times \mathcal{M}(\mathcal{P}) \times \mathcal{M}(\mathcal{P}) \times \mathcal{M}(\mathcal{P})$.

Let us write system (5) as follows:

$$\begin{aligned}
{}^{C}D_t^{\eta}S(t) &= \mathbf{H}_1(t,S), \\
{}^{C}D_t^{\eta}V(t) &= \mathbf{H}_2(t,V), \\
{}^{C}D_t^{\eta}E(t) &= \mathbf{H}_3(t,E), \\
{}^{C}D_t^{\eta}C(t) &= \mathbf{H}_4(t,C) \\
{}^{C}D_t^{\eta}I(t) &= \mathbf{H}_5(t,I) \\
{}^{C}D_t^{\eta}R(t) &= \mathbf{H}_6(t,R) \\
{}^{C}D_t^{\eta}B(t) &= \mathbf{H}_7(t,B)
\end{aligned} \tag{11}$$

Application of the Caputo fractional integral operator permits us to reduce (11) to the following system:

$$\begin{aligned}
-S(0)+S(t) &= \left[\int_0^t (t-\varsigma)^{\eta-1}\mathbf{H}_1(\varsigma,S)d\varsigma\right]\frac{1}{\Gamma(\eta)}, \\
-V(0)+V(t) &= \left[\int_0^t (t-\varsigma)^{\eta-1}\mathbf{H}_2(\varsigma,V)d\varsigma\right]\frac{1}{\Gamma(\eta)}, \\
-E(0)+E(t) &= \left[\int_0^t (t-\varsigma)^{\eta-1}\mathbf{H}_3(\varsigma,E)d\varsigma\right]\frac{1}{\Gamma(\eta)}, \\
-C(0)+C(t) &= \left[\int_0^t (t-\varsigma)^{\eta-1}\mathbf{H}_4(\varsigma,C)d\varsigma\right]\frac{1}{\Gamma(\eta)}, \\
-I(0)+I(t) &= \left[\int_0^t (t-\varsigma)^{\eta-1}\mathbf{H}_5(\varsigma,I)d\varsigma\right]\frac{1}{\Gamma(\eta)}, \\
-R(0)+R(t) &= \left[\int_0^t (t-\varsigma)^{\eta-1}\mathbf{H}_6(\varsigma,R)d\varsigma\right]\frac{1}{\Gamma(\eta)}, \\
-B(0)+B(t) &= \left[\int_0^t (t-\varsigma)^{\eta-1}\mathbf{H}_7(\varsigma,R)d\varsigma\right]\frac{1}{\Gamma(\eta)},
\end{aligned} \tag{12}$$

with $0 < \eta < 1$.

Now, we will provide the Lipschitz conditions fulfilled by $\mathbf{H}_i$, for $i = 1, 2, \ldots, 7$, as well as the contraction conditions. In the following theorem, we only provide the condition for $\mathbf{H}_1$, the rest being similar.

Theorem 3. *The kernel* $\mathbf{H}_1$ *satisfies the Lipschitz and contraction conditions provided that* $0 \leq \nu\kappa_7 + k_1 < 1$.

Proof. For S, we proceed as below:

$$\begin{aligned}
\| \mathbf{H}_1(t,S) - \mathbf{H}_1(t,S_1) \| &= \| -\nu B(S-S_1) - k_1(S-S_1) \| \\
&= \| \nu B + k_1 \| \| (S(t) - S_1(t)) \| \\
&\leq \| \nu\kappa_7 + k_1 \| \| S(t) - S_1(t) \|
\end{aligned}$$

where κ_7 is the upper bound of the function $B(t)$. Now, setting $W_1 = \nu\kappa_7 + k_1$, we finally obtain

$$\| \mathbf{H}_1(t,S) - \mathbf{H}_1(t,S_1) \| \leq W_1 \| S(t) - S(t_1) \|, \tag{13}$$

which provide the Lipschitz condition. If, additionally, we can have $0 < W_1 = \nu\kappa_7 + k_1 < 1$, then the contraction is also obtained.

As in the case of $\mathbf{H}_1$, it is easy to obtain the Lipschitz condition for the other kernels. Thus, we have

$$\begin{aligned}
\| \mathbf{H}_2(t,C) - \mathbf{H}_2(t,C_1) \| &\leq W_2 \| V(t) - V(t_1) \|, \\
\| \mathbf{H}_3(t,E) - \mathbf{H}_3(t,E_1) \| &\leq W_3 \| E(t) - E(t_1) \|, \\
\| \mathbf{H}_4(t,C) - \mathbf{H}_4(t,C_1) \| &\leq W_4 \| C(t) - C(t_1) \|, \\
\| \mathbf{H}_5(t,I) - \mathbf{H}_5(t,I_1) \| &\leq W_5 \| I(t) - I(t_1) \|, \\
\| \mathbf{H}_6(t,R) - \mathbf{H}_6(t,R_1) \| &\leq W_6 \| R(t) - R(t_1) \|, \\
\| \mathbf{H}_7(t,B) - \mathbf{H}_7(t,B_1) \| &\leq W_7 \| B(t) - B(t_1) \| .
\end{aligned} \tag{14}$$

□

Recursively, Equation (12) can be rewritten as follows:

$$\begin{aligned}
S_n(t) - S(0) &= \frac{1}{\Gamma(\eta)} \int_0^t (t-\varsigma)^{\eta-1} \mathbf{H}_1(\varsigma, S_{n-1}) d\varsigma, \\
V_n(t) - V(0) &= \frac{1}{\Gamma(\eta)} \int_0^t (t-\varsigma)^{\eta-1} \mathbf{H}_2(\varsigma, V_{n-1}) d\varsigma, \\
E_n(t) - E(0) &= \frac{1}{\Gamma(\eta)} \int_0^t (t-\varsigma)^{\eta-1} \mathbf{H}_3(\varsigma, E_{n-1}) d\varsigma, \\
C_n(t) - C(0) &= \frac{1}{\Gamma(\eta)} \int_0^t (t-\varsigma)^{\eta-1} \mathbf{H}_4(\varsigma, C_{n-1}) d\varsigma, \\
I_n(t) - I(0) &= \frac{1}{\Gamma(\eta)} \int_0^t (t-\varsigma)^{\eta-1} \mathbf{H}_5(\varsigma, I_{n-1}) d\varsigma, \\
R_n(t) - R(0) &= \frac{1}{\Gamma(\eta)} \int_0^t (t-\varsigma)^{\eta-1} \mathbf{H}_6(\varsigma, R_{n-1}) d\varsigma, \\
B_n(t) - B(0) &= \frac{1}{\Gamma(\eta)} \int_0^t (t-\varsigma)^{\eta-1} \mathbf{H}_7(\varsigma, B_{n-1}) d\varsigma.
\end{aligned} \tag{15}$$

By determing, in a recursive manner, the difference between the successive terms of (11), we obtain

$$\begin{aligned}
\psi_{1n}(t) = S_n(t) - S_{n-1}(t) &= \frac{1}{\Gamma(\eta)} \int_0^t (t-\varsigma)^{\eta-1} (\mathbf{H}_1(\varsigma, S_{n-1}) - \mathbf{H}_1(\varsigma, S_{n-2})) d\varsigma, \\
\psi_{2n}(t) = V_n(t) - V_{n-1}(t) &= \frac{1}{\Gamma(\eta)} \int_0^t (t-\varsigma)^{\eta-1} (\mathbf{H}_2(\varsigma, V_{n-1}) - \mathbf{H}_2(\varsigma, V_{n-2})) d\varsigma, \\
\psi_{3n}(t) = E_n(t) - E_{n-1}(t) &= \frac{1}{\Gamma(\eta)} \int_0^t (t-\varsigma)^{\eta-1} (\mathbf{H}_3(\varsigma, E_{n-1}) - \mathbf{H}_3(\varsigma, E_{n-2})) d\varsigma, \\
\psi_{4n}(t) = C_n(t) - C_{n-1}(t) &= \frac{1}{\Gamma(\eta)} \int_0^t (t-\varsigma)^{\eta-1} (\mathbf{H}_4(\varsigma, C_{n-1}) - \mathbf{H}_4(\varsigma, C_{n-2})) d\varsigma, \\
\psi_{5n}(t) = I_n(t) - I_{n-1}(t) &= \frac{1}{\Gamma(\eta)} \int_0^t (t-\varsigma)^{\eta-1} (\mathbf{H}_5(\varsigma, I_{n-1}) - \mathbf{H}_5(\varsigma, I_{n-2})) d\varsigma, \\
\psi_{6n}(t) = R_n(t) - R_{n-1}(t) &= \frac{1}{\Gamma(\eta)} \int_0^t (t-\varsigma)^{\eta-1} (\mathbf{H}_6(\varsigma, R_{n-1}) - \mathbf{H}_6(\varsigma, R_{n-2})) d\varsigma, \\
\psi_{7n}(t) = B_n(t) - B_{n-1}(t) &= \frac{1}{\Gamma(\eta)} \int_0^t (t-\varsigma)^{\eta-1} (\mathbf{H}_7(\varsigma, B_{n-1}) - \mathbf{H}_7(\varsigma, B_{n-2})) d\varsigma,
\end{aligned} \tag{16}$$

with $S_0(t) = S(0)$, $V_0(t) = V(0)$, $E_0(t) = E(0)$, $C_0(t) = C(0)$, $I_0(t) = I(0)$, $R_0(t) = R(0)$, and $B_0(t) = B(0)$.

The norm of $\phi_{1n}(t)$ gives

$$\begin{aligned} \| \psi_{1n}(t) \| = \| S_n(t) - S_{n-1}(t) \| &= \| \frac{1}{\Gamma(\eta)} \int_0^t (t-\varsigma)^{\eta-1} (\mathbf{H}_1(\varsigma, S_{n-1}) - \mathbf{H}_1(\varsigma, S_{n-2})) d\varsigma \| \\ &\leq \frac{1}{\Gamma(\eta)} \| \int_0^t (t-\varsigma)^{\eta-1} (\mathbf{H}_1(\varsigma, S_{n-1}) - \mathbf{H}_1(\varsigma, S_{n-2})) d\varsigma \| . \end{aligned} \tag{17}$$

With the Lipschitz condition (13), we obtain

$$\| \psi_{1n}(t) \| = \| S_n(t) - S_{n-1}(t) \| \leq \frac{1}{\Gamma(\eta)} W_1 \int_0^t (t-\varsigma)^{\eta-1} \| S_{n-1} - S_{n-2} \| d\varsigma. \tag{18}$$

Thus, we have

$$\| \psi_{1n}(t) \| \leq \frac{1}{\Gamma(\eta)} W_1 \int_0^t (t-\varsigma)^{\eta-1} \| \psi_{1(n-1)}(\varsigma) \| d\varsigma. \tag{19}$$

By proceeding in a similar way, we obtain, for the other $\phi_{in}(t), i = 2, \ldots, 7$,

$$\begin{aligned} \| \psi_{2n}(t) \| &\leq \frac{1}{\Gamma(\eta)} W_2 \int_0^t (t-\varsigma)^{\eta-1} \| \psi_{2(n-1)}(\varsigma) \| d\varsigma, \\ \| \psi_{3n}(t) \| &\leq \frac{1}{\Gamma(\eta)} W_3 \int_0^t (t-\varsigma)^{\eta-1} \| \psi_{3(n-1)}(\varsigma) \| d\varsigma, \\ \| \psi_{4n}(t) \| &\leq \frac{1}{\Gamma(\eta)} W_4 \int_0^t (t-\varsigma)^{\eta-1} \| \psi_{4(n-1)}(\varsigma) \| d\varsigma, \\ \| \psi_{5n}(t) \| &\leq \frac{1}{\Gamma(\eta)} W_5 \int_0^t (t-\varsigma)^{\eta-1} \| \psi_{5(n-1)}(\varsigma) \| d\varsigma, \\ \| \psi_{6n}(t) \| &\leq \frac{1}{\Gamma(\eta)} W_6 \int_0^t (t-\varsigma)^{\eta-1} \| \psi_{6(n-1)}(\varsigma) \| d\varsigma, \\ \| \psi_{7n}(t) \| &\leq \frac{1}{\Gamma(\eta)} W_7 \int_0^t (t-\varsigma)^{\eta-1} \| \psi_{7(n-1)}(\varsigma) \| d\varsigma. \end{aligned} \tag{20}$$

Each nth term of the state variables of (5) is given by

$$\begin{cases} S_n(t) = \sum\limits_{i=1}^{n} \psi_{1i}(t), V_n(t) = \sum\limits_{i=1}^{n} \psi_{2i}(t), E_n(t) = \sum\limits_{i=1}^{n} \psi_{3i}(t), \\ C_n(t) = \sum\limits_{i=1}^{n} \psi_{4i}(t), I_n(t) = \sum\limits_{i=1}^{n} \psi_{5i}(t), R_n(t) = \sum\limits_{i=1}^{n} \psi_{6i}(t), \\ B_n(t) = \sum\limits_{i=1}^{n} \psi_{7i}(t). \end{cases} \tag{21}$$

The following result guarantees that the solution of the fractional model (5) is unique.

Theorem 4. *If the following inequality holds,*

$$\frac{1}{\Gamma(\eta)} \mathbf{x}^{\eta} W_i < 1, \quad \textit{for} \quad i = 1, 2, \ldots, 7, \tag{22}$$

then the solution of the fractional model (5), for $t \in [0, T]$, is unique.

Proof. With Equation (13), inequalities (19) and (20), combined with a recursive technique, we obtain:

$$\begin{aligned}
&\| \psi_{1n}(t) \| \leq \| S_0(t) \| \left[\frac{W_1}{\Gamma(\eta)}\mathbf{x}^\eta\right]^n, \| \psi_{2n}(t) \| \leq \| W_0(t) \| \left[\frac{W_2}{\Gamma(\eta)}\mathbf{x}^\eta\right]^n, \\
&\| \psi_{3n}(t) \| \leq \| E_0(t) \| \left[\frac{W_3}{\Gamma(\eta)}\mathbf{x}^\eta\right]^n, \| \psi_{4n}(t) \| \leq \| C_0(t) \| \left[\frac{W_4}{\Gamma(\eta)}\mathbf{x}^\eta\right]^n, \\
&\| \psi_{5n}(t) \| \leq \| I_0(t) \| \left[\frac{W_5}{\Gamma(\eta)}\mathbf{x}^\eta\right]^n, \| \psi_{6n}(t) \| \leq \| R_0(t) \| \left[\frac{W_6}{\Gamma(\eta)}\mathbf{x}^\eta\right]^n, \\
&\| \psi_{7n}(t) \| \leq \| B_0(t) \| \left[\frac{W_7}{\Gamma(\eta)}\mathbf{x}^\eta\right]^n.
\end{aligned} \tag{23}$$

Therefore, the above sequences satisfy $\lim_{n\to\infty} \| \psi_{in}(t) \| \to 0,\ j = 1,2,\ldots,7$. The triangle inequality applied to (23) permits us to obtain

$$\begin{aligned}
\| S_{k+n}(t) - S_n(t) \| &\leq \sum_{j=n+1}^{k+n} \mathbf{Z}_1^j = \frac{\mathbf{Z}_1^{n+1} - \mathbf{Z}_1^{k+n+1}}{1-\mathbf{Z}_1}, \\
\| V_{k+n}(t) - V_n(t) \| &\leq \sum_{j=n+1}^{k+n} \mathbf{Z}_2^j = \frac{\mathbf{Z}_2^{n+1} - \mathbf{Z}_2^{k+n+1}}{1-\mathbf{Z}_2}, \\
\| E_{k+n}(t) - E_n(t) \| &\leq \sum_{j=n+1}^{k+n} \mathbf{Z}_3^j = \frac{\mathbf{Z}_3^{n+1} - \mathbf{Z}_3^{k+n+1}}{1-\mathbf{Z}_3}, \\
\| C_{k+n}(t) - C_n(t) \| &\leq \sum_{j=n+1}^{k+n} \mathbf{Z}_4^j = \frac{\mathbf{Z}_4^{n+1} - \mathbf{Z}_4^{k+n+1}}{1-\mathbf{Z}_4}, \\
\| I_{k+n}(t) - I_n(t) \| &\leq \sum_{j=n+1}^{k+n} \mathbf{Z}_5^j = \frac{\mathbf{Z}_5^{n+1} - \mathbf{Z}_5^{k+n+1}}{1-\mathbf{Z}_5}, \\
\| R_{k+n}(t) - R_n(t) \| &\leq \sum_{j=n+1}^{k+n} \mathbf{Z}_6^j = \frac{\mathbf{Z}_6^{n+1} - \mathbf{Z}_6^{k+n+1}}{1-\mathbf{Z}_6}, \\
\| B_{k+n}(t) - B_n(t) \| &\leq \sum_{j=n+1}^{k+n} \mathbf{Z}_7^j = \frac{\mathbf{Z}_7^{n+1} - \mathbf{Z}_7^{n+k+1}}{1-\mathbf{Z}_7},
\end{aligned} \tag{24}$$

with $\frac{1}{\Gamma(\eta)} b^\eta W_l < 1$ and $\mathbf{Z}_l = \left(\frac{1}{\Gamma(\eta)} W_l \mathbf{x}^\eta\right)^n, l = 1,2,\ldots,7.$

Therefore, S_n, V_n, E_n, C_n, I_n, R_n, and B_n are uniformly convergent Cauchy sequences (see [32]). With $n \to \infty$, it follows that the limit of these sequences represents the unique solution to model (5). □

Numerical Scheme of the Fractional Model and Its Stability Analysis

Several methods have been developed to construct numerical schemes for fractional models. One can cite, among others, the Implicit Quadrature method [24], the Approximate Mittag–Leffler method [33], the Predictor Corrector method [34], and the Adams–Bashforth–Moulton method [25]. The choice of the method depends on several factors, such as the amount of information treated [35] and the accuracy order [36]. The numerical scheme proposed in this work is constructed using the Adams–Bashforth–Moulton method.

Let us consider the following general form of a fractional differential equation [37]:

$$\begin{cases} \mathbb{D}_t^\eta \varphi(t) &= h(t, \varphi(t)), \quad 0 \leq t \leq T, \\ \varphi^{(l)}(0) &= \varphi_0^l, \qquad\qquad l = 0,1,2,\ldots,n-1, where\ n = [\eta], \end{cases} \tag{25}$$

which is equivalent to

$$\varphi(t) = \sum_{l=0}^{n-1} h_0^l \frac{t^l}{l!} + \frac{1}{\Gamma(\eta)} \int_0^t (t-\zeta)^{\eta-1} h(\zeta, \varphi(\zeta)) d\zeta. \quad (26)$$

For $\eta \in [0,1], 0 \le t \le T$ and setting $\kappa = T/N$ and $t_m = m\kappa$, for $m = 0,1,2,\ldots,N \in \mathbb{Z}^+$, the solution of the fractional model is

$$\begin{aligned}
S_{1+m} &= S_0 + \frac{\kappa^\eta}{\Gamma(\eta+2)} \Big(\Lambda_h + \eta R^p_{1+m} + \theta V^p_{1+m} - (\nu B^p_{1+m} + k_1) S^p_{1+m} \Big) \\
&\quad + \frac{\kappa^\eta}{\Gamma(\eta+2)} \sum_{j=0}^{m} a_{j,1+m} \Big(\Lambda_h + \eta R_j + \theta V_j - (\nu B_j + k_1) S_j \Big), \\
V_{1+m} &= V_0 + \frac{\kappa^\eta}{\Gamma(\eta+2)} \Big(\xi S^p_{1+m} - \Big[(1-\epsilon)\nu B^p_{1+m} + k_2 \Big] V^p_{1+m} \Big) \\
&\quad + \frac{\kappa^\eta}{\Gamma(\eta+2)} \sum_{j=0}^{m} a_{j,1+m} \Big(\xi S_j - [(1-\epsilon)\nu B_j + k_2] V_j \Big), \\
E_{1+m} &= E_0 + \frac{\kappa^\eta}{\Gamma(\eta+2)} \Big(\nu B^p_{1+m} \Big[S^p_{1+m} + \pi V^p_{1+m} \Big] - k_3 E^p_{1+m} \Big) \\
&\quad + \frac{\kappa^\eta}{\Gamma(\eta+2)} \sum_{j=0}^{m} a_{j,1+m} \Big(\nu B_j [S_j + \pi V_j] - k_3 E_j \Big), \\
C_{1+m} &= C_0 + \frac{\kappa^\eta}{\Gamma(\eta+2)} \Big(q\gamma_1 E^p_{1+m} - k_4 C^p_{1+m} \Big) + \frac{\kappa^\eta}{\Gamma(\eta+2)} \sum_{j=0}^{m} a_{j,1+m} \Big(q\gamma_1 E_j - k_4 C_j \Big), \\
I_{1+m} &= I_0 + \frac{\kappa^\eta}{\Gamma(\eta+2)} \Big(q_1\gamma_1 E^p_{1+m} + p_1\gamma_2 C^p_{1+m} - [k_5 + \sigma] I^p_{1+m} \Big) \\
&\quad + \frac{\kappa^\eta}{\Gamma(\eta+2)} \sum_{j=0}^{m} a_{j,1+m} \Big(q_1\gamma_1 E_j + p_1\gamma_2 C_j - [k_5 + \sigma] I_j \Big), \\
R_{1+m} &= R_0 + \frac{\kappa^\eta}{\Gamma(\eta+2)} \Big(p\gamma_2 C^p_{1+m} + \sigma I^p_{1+m} - k_6 R^p_{1+m} \Big) + \frac{\kappa^\eta}{\Gamma(\eta+2)} \sum_{j=0}^{m} a_{j,1+m} \Big(p\gamma_2 C_j + \sigma I_j - k_6 R_j \Big), \\
B_{1+m} &= B_0 + \frac{\kappa^\eta}{\Gamma(\eta+2)} \Big(p_c C^p_{1+m} + p_i I^p_{1+m} - \mu_b B^p_{1+m} \Big) + \frac{\kappa^\eta}{\Gamma(\eta+2)} \sum_{j=0}^{m} a_{j,1+m} \Big(p_c C_j + p_i I_j - \mu_b B_j \Big),
\end{aligned} \quad (27)$$

where

$$
\begin{aligned}
S^p_{1+m} &= S_0 + \frac{1}{\Gamma(\eta)}\sum_{j=0}^{m} b_{j,1+m}\Big(\Lambda_h + \eta R_j + \theta V_j - (\nu B_j + k_1)S_j\Big),\\
V^p_{1+m} &= V_0 + \frac{1}{\Gamma(\eta)}\sum_{j=0}^{m} b_{j,1+m}\Big(\xi S_j - \big[(1-\epsilon)\nu B_j + k_2\big]V_j\Big),\\
E^p_{1+m} &= E_0 + \frac{1}{\Gamma(\eta)}\sum_{j=0}^{m} b_{j,1+m}\Big(\nu B_j[S_j + \pi V_j] - k_3 E_j\Big),\\
C^p_{1+m} &= C_0 + \frac{1}{\Gamma(\eta)}\sum_{j=0}^{m} b_{j,1+m}\Big(q\gamma_1 E_j - k_4 C_j\Big),\\
I^p_{1+m} &= I_0 + \frac{1}{\Gamma(\eta)}\sum_{j=0}^{m} b_{j,1+m}\Big(= q_1\gamma_1 E_j + p_1\gamma_2 C_j - [k_5 + \sigma]I_j\Big),\\
R^p_{1+m} &= R_0 + \frac{1}{\Gamma(\eta)}\sum_{j=0}^{m} b_{j,1+m}\Big(p\gamma_2 C_j + \sigma I_j - k_6 R_j\Big),\\
B^p_{1+m} &= B_0 + \frac{1}{\Gamma(\eta)}\sum_{j=0}^{m} b_{j,1+m}\Big(p_c C_j + p_i I_j - \mu_b B_j\Big),
\end{aligned} \tag{28}
$$

and

$$
a_{j,1+m} = \begin{cases} m^{\eta+1} - (m+1)(-\eta+m), & j = 0,\\ (2-j+m)^{\eta+1} - 2(m-j+1)^{1+\eta} + (-j+m)^{1+\eta} & 1 \le j \le m,\\ 1, \quad j = 1+m, \end{cases}
$$

$$
b_{j,1+m} = \frac{\kappa^\eta}{\eta}\left((m+1-j)^\eta - (m-j)^\eta\right),\ 0 \le j \le m.
$$

We then claim the following result.

Theorem 5. *Under some conditions, the above numerical scheme (see Equations (27) and (28)) is stable.*

Proof. Let $S^\star_0$, $S^\star_j (j = 0, \dots, 1+m)$ and $S^{\star p}_{1+m} (m = 0, \dots, N-1)$ be perturbations of S_0, S_j, and S^p_{1+m}, respectively. By using Equations (19) and (28), the following perturbation equations are obtained:

$$
S^{\star p}_{1+m} = S^\star_0 + \frac{1}{\Gamma(\eta)}\sum_{j=0}^{m} b_{j,1+m}(\mathcal{G}_1(t_j, S_j + S^\star_j) - \mathcal{G}_1(t_j, S_j)), \tag{29}
$$

$$
\begin{aligned}
S^\star_{1+m} = S^\star_0 &+ \frac{\kappa^\eta}{\Gamma(\eta+2)}(\mathcal{G}_1(t_{1+m}, S^p_{1+m} + S^P_{1+m}) - \mathcal{G}_1(t_{1+m}, S^p_{1+m}))\\
&+ \frac{\kappa^\eta}{\Gamma(\eta+2)}\sum_{j=0}^{m} a_{j,1+m}(\mathcal{G}_1(t_j, S_j + S^\star_j) - \mathcal{G}_1(t_j, S_j)).
\end{aligned} \tag{30}
$$

The Lipschitz condition permits us to obtain

$$
|S^\star_{1+m}| \le \phi_0 + \frac{\kappa^\eta M}{\Gamma(\eta+2)}\left(|S^{\star p}_{1+m}| + \sum_{j=1}^{m} a_{j,1+m}|S^\star_j|\right), \tag{31}
$$

where $\phi_0 = \max_{0 \le m \le N} \left\{ |S_0^\star| + \frac{\kappa^\eta M a_{m,0}}{\Gamma(\eta+2)} |S_0^\star| \right\}$. From [32] (Equation (3.18)), it follows that

$$|S_{1+m}^{\star p}| \le \Theta_0 + \frac{M}{\Gamma(\eta)} \sum_{j=1}^{m} b_{j,1+m} |S_j^\star|, \tag{32}$$

where $\Theta_0 = \max_{0 \le m \le N} \left\{ |S_0^\star| + \frac{M b_{m,0}}{\Gamma(\eta)} |S_0^\star| \right\}$. Substituting $|S_{1+m}^{\star p}|$ from Equation (32) into Equation (31) gives

$$\begin{aligned} |S_{1+m}^\star| &\le \varrho_0 + \frac{\kappa^\eta M}{\Gamma(\eta+2)} \left(\frac{M}{\Gamma(\eta)} \sum_{j=1}^{m} b_{j,1+m} |S_j^\star| + \sum_{j=1}^{m} a_{j,1+m} |S_j^\star| \right) \\ &\le \varrho_0 + \frac{\kappa^\eta M}{\Gamma(\eta+2)} \sum_{j=1}^{m} \left(\frac{M}{\Gamma(\eta)} b_{j,1+m} + a_{j,1+m} \right) |S_j^\star| \\ &\le \varrho_0 + \frac{\kappa^\eta M C_{\eta,2}}{\Gamma(\eta+2)} \sum_{j=1}^{m} (m+1-j)^{\eta-1} |S_j^\star|, \end{aligned} \tag{33}$$

where $\varrho_0 = \max\{\phi_0 + \frac{\kappa^\eta M a_{1+m,1+m}}{\Gamma(\eta+2)} \Theta_0\}$.

Thanks to Lemma 2, we have that $C_{\eta,2} > 0$ and depends only on η, and κ is assumed to be small enough. A direct application of Lemma 1 implies $|S_{1+m}^\star| \le C\varrho_0$. The proof for the other variables is obtained in the same way. This ends the proof. □

2.3. Model Dynamics with the Standard Incidence Law

In this section, we extend model (2) by replacing the mass action incidence law with the standard incidence law, and considering direct transmission (human to human). The new typhoid fever transmission dynamics model is thus presented as follows:

$$\dot{S}(t) = \Lambda_h + \theta V(t) + \alpha R(t) - k_1 S(t) - \beta \frac{(I+C)}{N(t)} S(t) - \nu \frac{B(t)}{B(t)+K} S(t), \tag{34a}$$

$$\dot{V}(t) = -\left[k_2 + \pi\beta \frac{(C+I)}{N(t)} + \pi\nu \frac{B(t)}{K+B(t)} \right] V(t) + \xi S(t), \tag{34b}$$

$$\dot{E}(t) = \left[\beta \frac{(C+I)}{N(t)} + \nu \frac{B(t)}{B(t)+K} \right] (\pi V(t) + S(t)) - k_3 E(t), \tag{34c}$$

$$\dot{C}(t) = q\gamma_1 E(t) - k_4 C(t), \tag{34d}$$

$$\dot{I}(t) = q_1\gamma_1 E(t) + p_1\gamma_2 C(t) - k_8 I(t), \tag{34e}$$

$$\dot{R}(t) = p\gamma_2 C(t) + \sigma I(t) - k_6 R(t), \tag{34f}$$

$$\dot{B}(t) = p_c C(t) + p_i I(t) - \mu_b B(t), \tag{34g}$$

where β is the direct transmission rate, and K represents the half-saturation constant.

The corresponding fractional model is given by

$$ {}_{t_0}^{C}D_t^{\eta} S(t) = \Lambda_h + \theta V(t) + \alpha R(t) - k_1 S(t) - \beta \frac{(I+C)}{N(t)} S(t) - \nu \frac{B(t)}{K+B(t)} S(t), \quad (35a)$$

$$ {}_{t_0}^{C}D_t^{\eta} V(t) = -\left[k_2 + \pi\beta \frac{(I+C)}{N(t)} + \pi\nu \frac{B(t)}{K+B(t)} \right] V(t) + \xi S(t), \quad (35b)$$

$$ {}_{t_0}^{C}D_t^{\eta} E(t) = \left[\beta \frac{(I+C)}{N(t)} + \nu \frac{B(t)}{K+B(t)} \right] (\pi V(t) + S(t)) - k_3 E(t), \quad (35c)$$

$$ {}_{t_0}^{C}D_t^{\eta} C(t) = q\gamma_1 E(t) - k_4 C(t), \quad (35d)$$

$$ {}_{t_0}^{C}D_t^{\eta} I(t) = q_1\gamma_1 E(t) + p_1\gamma_2 C(t) - k_8 I(t), \quad (35e)$$

$$ {}_{t_0}^{C}D_t^{\eta} R(t) = p\gamma_2 C(t) + \sigma I(t) - k_6 R(t), \quad (35f)$$

$$ {}_{t_0}^{C}D_t^{\eta} B(t) = p_c C(t) + p_i I(t) - \mu_b B(t). \quad (35g)$$

Without loss of generality, it is evident that the new model (34) (resp. (35)) is also defined in $\mathcal{W}$.

Model (34) has the same disease-free equilibrium as model system (2). Using the same approach developed in [29], we define the next-generation matrix of model system (34) as

$$NGM = \begin{pmatrix} R_1 & \frac{H_0 p_1 \gamma_2 \beta}{N_0 k_4 k_8} + \frac{H_0 \beta}{N_0 k_4} & \frac{H_0 \beta}{N_0 k_8} & R_4 \\ 0 & 0 & 0 & 0 \\ 0 & 0 & 0 & 0 \\ R_5 & \frac{p_1 \gamma_2 p_i}{k_4 k_8} + \frac{p_c}{k_4} & \frac{p_i}{k_8} & 0 \end{pmatrix}$$

where $R_1 = \dfrac{H_0 \beta \gamma_1}{N_0 k_3 k_4} \left[\dfrac{(p_1\gamma_2 q + q_1 k_4)}{k_8} + q \right]$, $R_4 = \dfrac{H_0 \nu}{K \mu_b}$, $R_5 = \dfrac{\gamma_1}{k_3 k_4} \left[\dfrac{p_i(p_1\gamma_2 q + q_1 k_4)}{k_8} + p_c q \right]$, and $H_0 = S_0 + \pi V_0$.

Thus, the control reproduction number of model (34), which is the spectral radius of NGM, is given by

$$\mathcal{R}_{c\star} = \frac{R_1 + \sqrt{R_1^2 + 4R_4R_5}}{2}. \quad (36)$$

The following result is a direct consequence of Theorem 2 in [29] (see Appendix A for the proof).

Proposition 3. *For model (34) (resp. (35)), $\mathcal{Q}_0$ is locally asymptotically stable in $\mathcal{W}$ if $\mathcal{R}_{c\star} < 1$ and unstable if $\mathcal{R}_{c\star} > 1$.*

Theorem 6. *For the new typhoid model (34) (resp. (35)), the disease-free equilibrium $\mathcal{Q}_0$ is globally asymptotically stable if $\mathcal{R}_{c\star} < 1$ and unstable otherwise.*

Proof. See Appendix B. □

3. Results

3.1. Numerical Results of the Fractional Model with Mass Action Incidence Law

We perform several simulations with the parameter values listed in Table 1.

Table 1. Parameter values of (2) taken from [5].

Parameter	Value	Parameter	Value	Parameter	Value
Λ_h	1	μ_b	0.149990	ν	3.2618×10^{-6}
γ_1	0.2145	ϵ	0.9495	δ	0.1499
γ_2	0.1498	ξ	0.3221	σ	0.49992
α	0.0834	θ	0.0833	$\mathcal{R}_c$	2.4750

The step size is $\kappa = 10^{-8}$, the initial time is $T = 0$, and the final time is $T > 0$. We begin by showing the impact of fractional order on the dynamics of the disease. To this aim, the fractional-order parameter varies between $\eta = 1$ and $\eta = 0.5$.

Figure 1 displays the impact of the Caputo fractional operator on the model dynamics. For different values of the fractional-order parameter η, the infected state profiles are drawn. From Figure 1, it follows that when the fractional order η decreases, the solutions of our fractional model (5) have different behaviors. Indeed, when the fractional order decreases, the number of infected humans in latent, carrier, and symptomatic states increases. This is the same for the compartment B. This phenomenon was also observed in a malaria fractional model studied by [38]. It is important to note that for $\eta = 1$, the solutions of the fractional model converge to the solutions of the integer model.

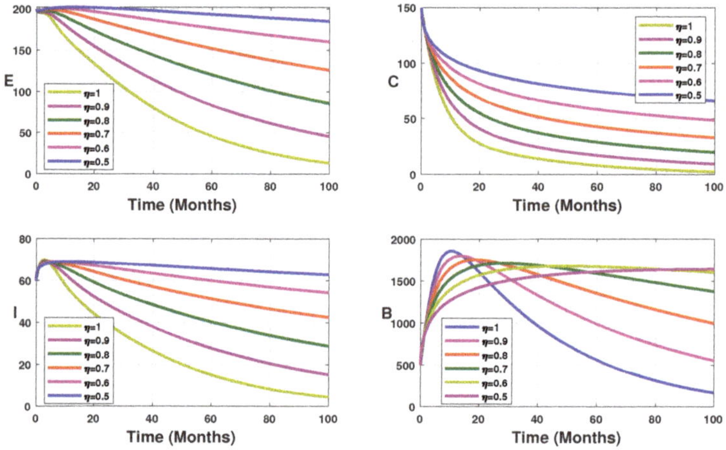

Figure 1. Simulation results showing the fractional dynamics on the infected state variable profiles for different values of the fractional-order parameter η.

To evaluate the impact of vaccination on typhoid fever transmission dynamics, we fix the vaccine efficacy at $\epsilon = 70\%$ while the vaccine coverage parameter varies between $\xi = 0\%$ and $\xi = 90\%$ ($\xi \in \{0.90, 0.50, 0.20, 0\}$), with different values of the fractional-order η ($\eta \in \{1, 0.90, 0.80, 0.70, 0.60, 0.50\}$). The results are displayed in Figures 2–5.

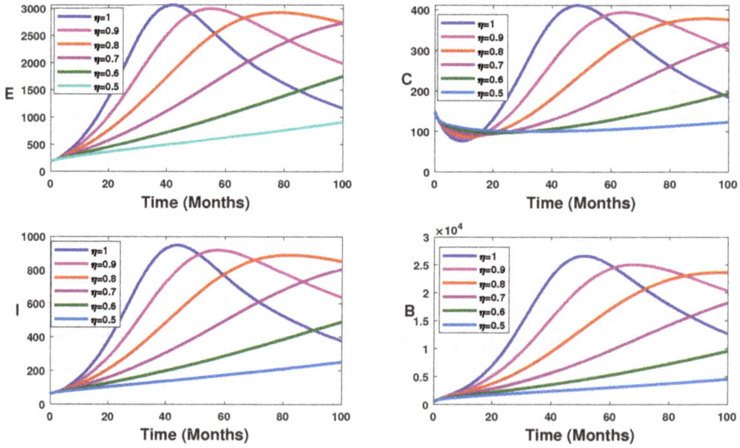

Figure 2. Simulation results showing the infected state variable profiles without vaccination ($\xi = 0$).

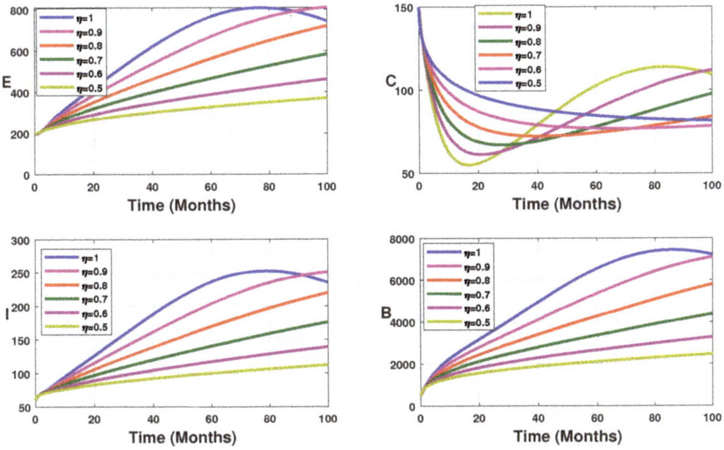

Figure 3. Simulation results showing the infected state variable profiles when the vaccination coverage $\xi = 20\%$ with different values of the fractional order.

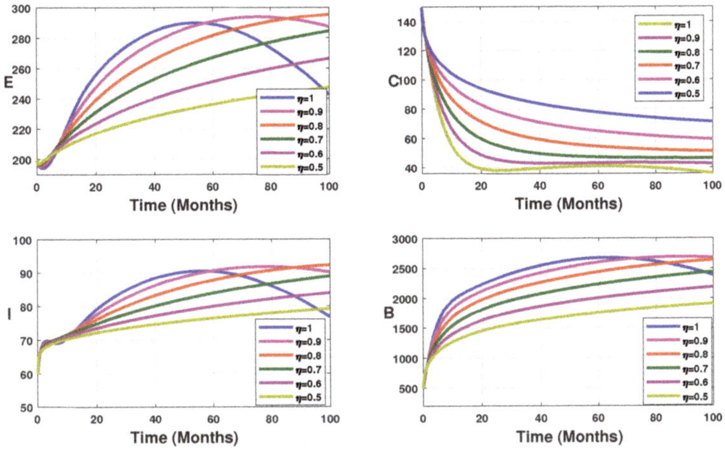

Figure 4. Simulation results showing the infected state variable profiles when the vaccination coverage $\xi = 50\%$ with different values of the fractional order.

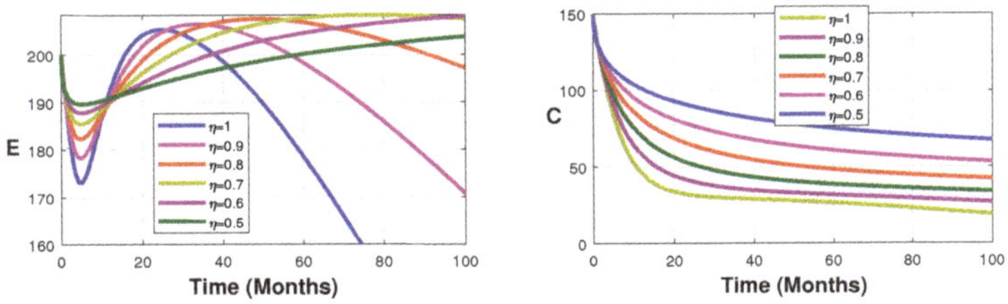

Figure 5. *Cont.*

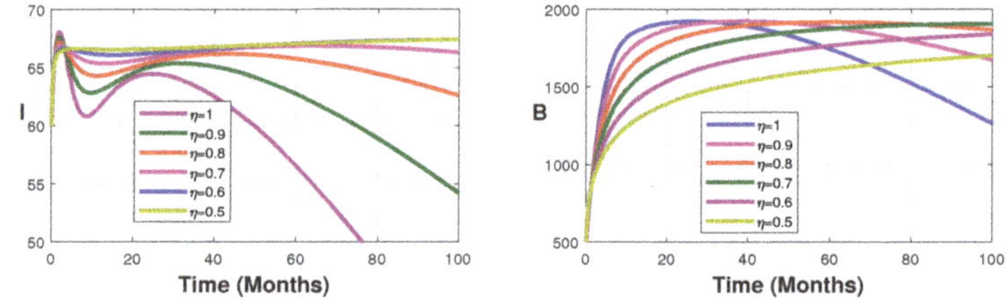

Figure 5. Simulation results showing the infected state variable profiles when the vaccination coverage $\xi = 90\%$ with different values of the fractional order.

Without vaccination coverage $\xi = 0$ (Figure 2), the peak delayed phenomenon is observable. Indeed, for $\eta = 1$, the peak date corresponds to $T = 45$ months, with approximately 3000 infected individuals in the latent stage, 400 asymptomatic individuals, 950 symptomatic individuals, and 26,000 free salmonella in the environment. This peak date is delayed when the fractional-order parameter η decreases. Thus, the peak dates are beyond $T = 45$ months. From Figures 3–5, we note that this peak date is forward delayed beyond $T = 45$ months whenever the fractional-order parameter η decreases. Moreover, the number of infected individuals decreases with the decrease in the fractional-order parameter.

Now, in addition to vaccination, we consider environmental sanitation. To this aim, the bacterial decay rate μ_b is modified to $\mu_b := \mu_b + \omega$, where $\omega \in \{0, 0.1, 0.2.0.3, 0.4\}$ represents the additional decay rate of free salmonella due to environmental sanitation [6]. The vaccination coverage is fixed at $\xi = 32.21\%$ as reported in Table 1. From Figures 6–9, it is evident that mass vaccination combined with environmental sanitation has a positive impact, reducing the disease burden.

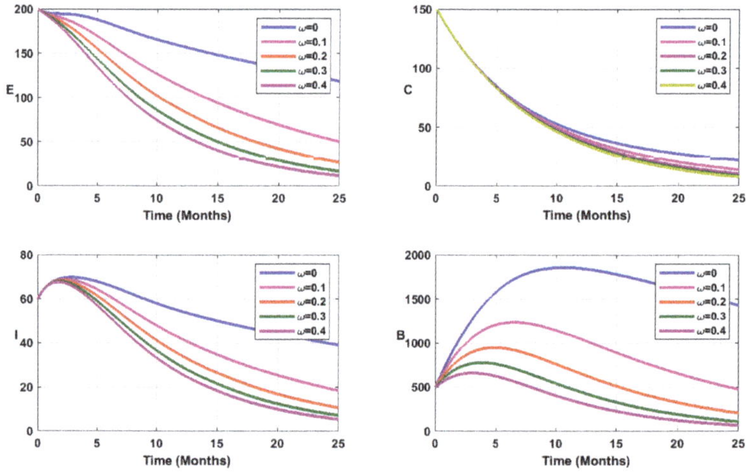

Figure 6. Simulation results showing the infected state variable profiles when vaccination is combined with environmental sanitation, for fractional order $\eta = 1$.

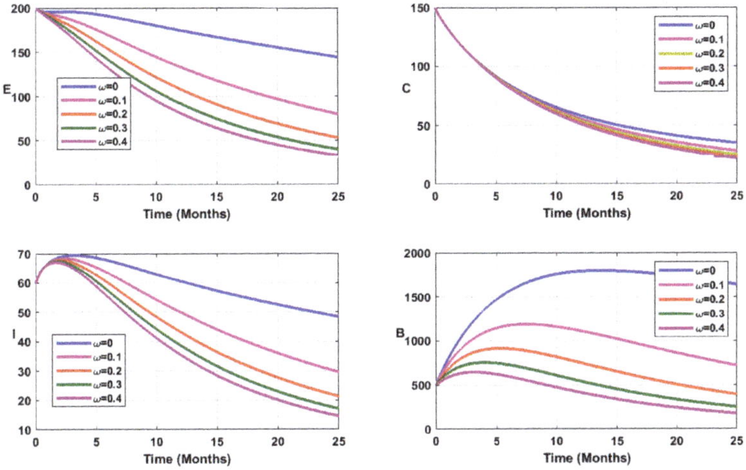

Figure 7. Simulation results showing the infected state variable profiles when vaccination is combined with environmental sanitation, for fractional order $\eta = 0.90$.

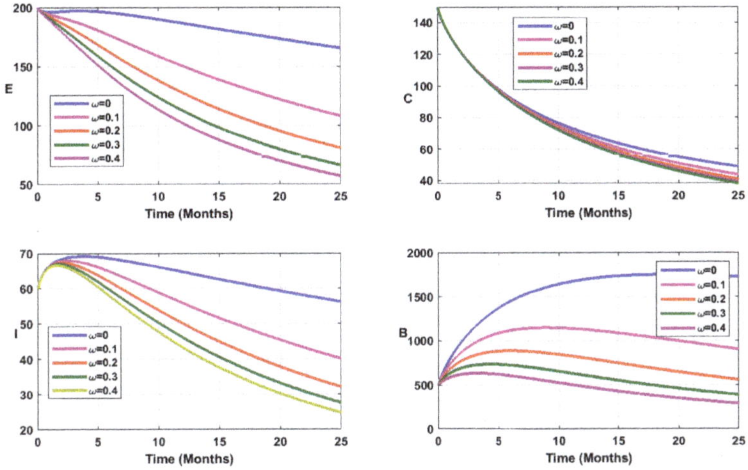

Figure 8. Simulation results showing the infected state variable profiles when vaccination is combined with environmental sanitation, for fractional order $\eta = 0.80$.

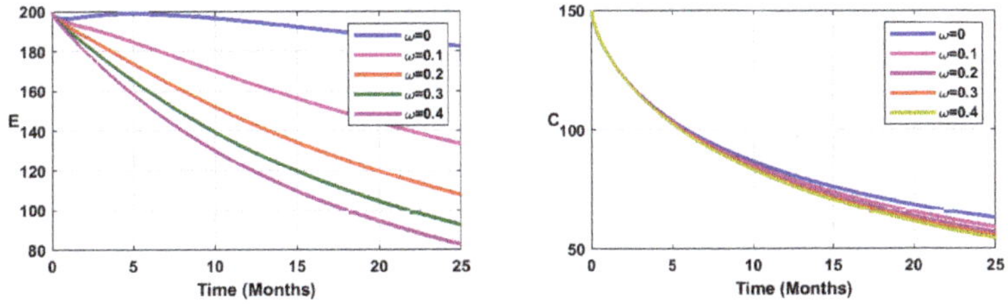

Figure 9. *Cont.*

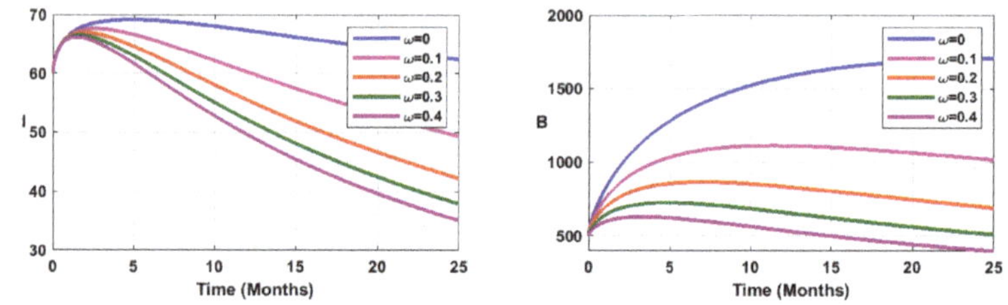

Figure 9. Simulation results showing the infected state variable profiles when vaccination is combined with environmental sanitation, for fractional order $\eta = 0.70$.

3.2. Numerical Results of the Fractional Model with Standard Incidence Law

To simulate the new fractional model (35), we use the parameter values listed in Table 2. Note that the new typhoid model has been calibrated using real data from Mbandjock, Cameroon (see [5,6]). In Figure 10, panel (a) shows the cumulative typhoid cases versus fitted confirmed cases (infectious individuals tested positive), which is equal to $(1-q)\gamma_1 E(t) + (1-p)\gamma_2 C(t)$, while panel (b) presents the cumulative estimated cases for the next year. The following fractions are used as initial conditions $S(0) = 20{,}950/32{,}000$, $V(0) = 20/32{,}000$, $E(0) = 200/32{,}000$, $C(0) = 150/32{,}000$, $I(0) = 60/32{,}000$, $R(0) = 1/32{,}000$, and $B(0) = 500/10^6$. The relative change is $r = 1.83 \times 10^{-7}$ and the function tolerance is equal to 10^{-6}.

Table 2. Estimated parameter values of the new typhoid model (35).

Parameter	Values	Source	Parameter	Values	Source
Λ_h	3	Fitted	μ_b	0.0015	Fitted
γ_1	0.1512	Fitted	ϵ	0.9497	Fitted
γ_2	0.3039	Fitted	ξ	0.1538	Fitted
δ	0.1382	Fitted	β	0.60	Fitted
ν	0.00050	Fitted	K	995.7957	Fitted
σ	0.4992	Fitted	$\mathcal{R}_{c\star}$	1.4348	Estimated

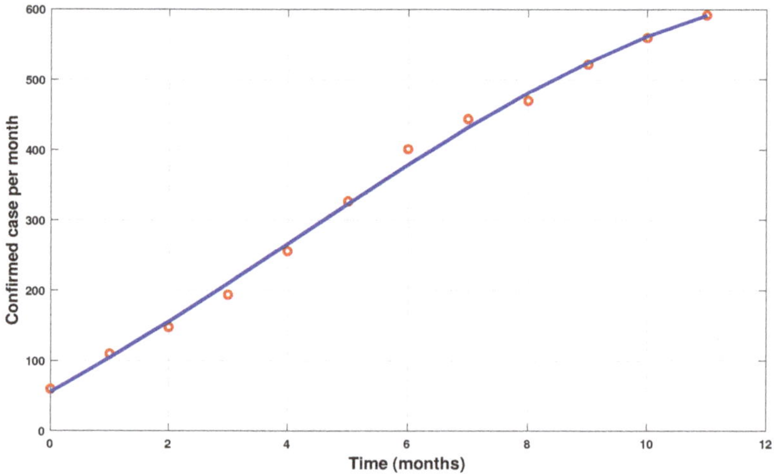

(a)

Figure 10. *Cont.*

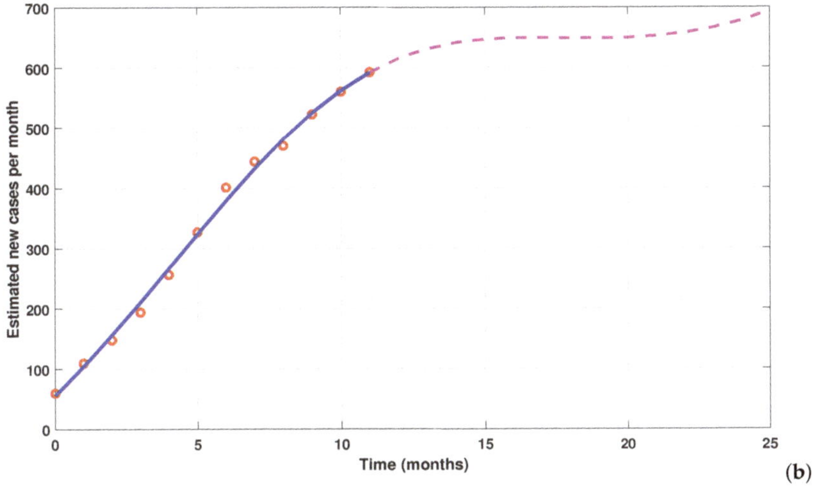

Figure 10. Parameter estimation and forecasting of cumulative new cases of typhoid fever in Mbandjock, Cameroon, from 1 July 2019 to 31 August 2020, for the new model (35). Red bullets denote real data (see [5,6]). The fitted model is represented with the blue line, and new forecasted cases are represented by the dotted line.

First, we observe the general dynamics of the new fractional model. The results are displayed in Figure 11. As for the case of the fractional model with mass incidence law (5), Figure 11 reveals that when the fractional order η decreases, the solutions of our fractional model (35) have different behaviors. The number of typhoid cases decreases and the peak is delayed when the fractional order decreases.

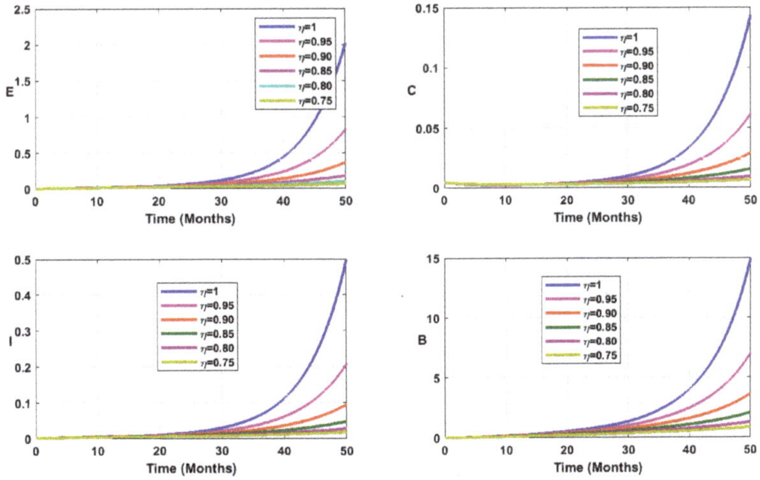

Figure 11. Simulation results showing the fractional dynamics on the infected state variable profiles for different values of the fractional-order parameter η.

Vaccination coverage impact is studied numerically. From Figures 12–15, it follows that the more ξ increases, the fewer individuals are infected. This shows that mass vaccination plays a important role in reducing the spread of the disease.

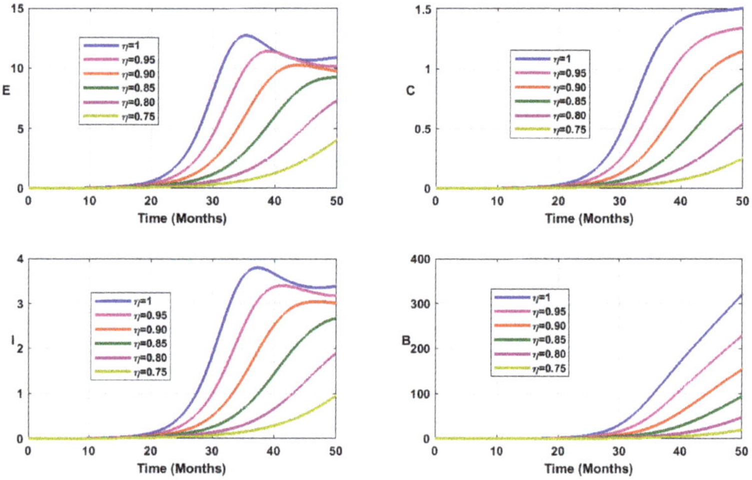

Figure 12. Simulation results showing the infected state variable profiles when the vaccination coverage $\xi = 0\%$ with different values of the fractional order.

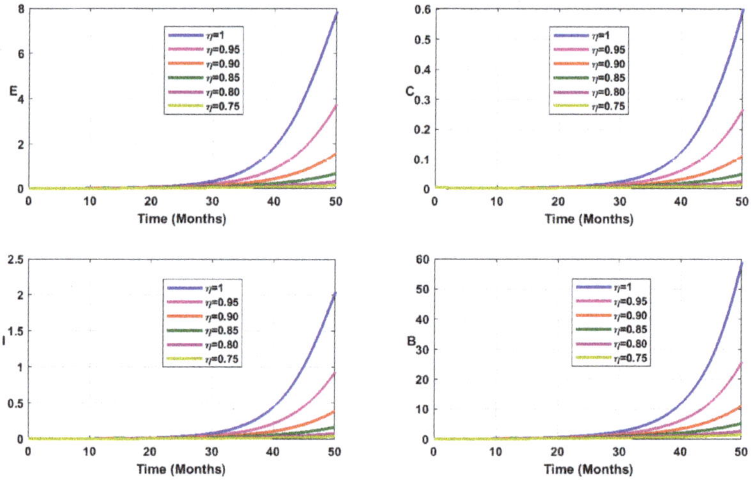

Figure 13. Simulation results showing the infected state variable profiles when the vaccination coverage $\xi = 20\%$ with different values of the fractional order.

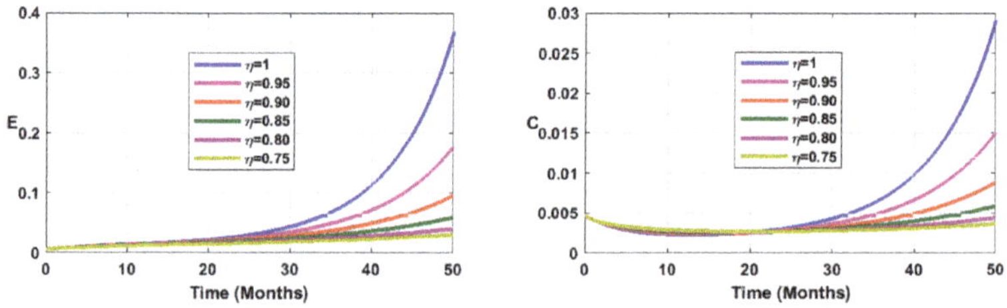

Figure 14. *Cont.*

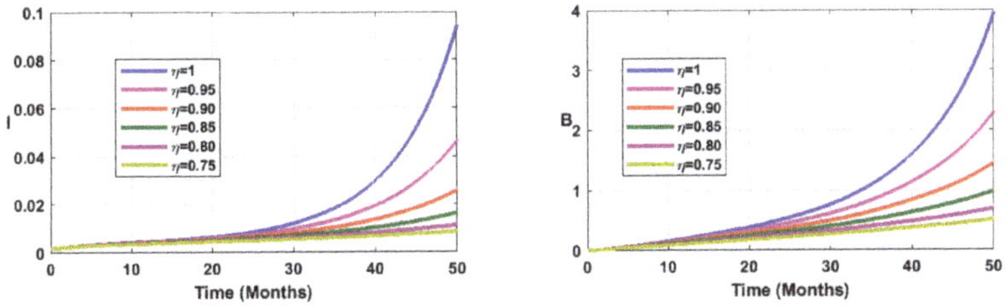

Figure 14. Simulation results showing the infected state variable profiles when the vaccination coverage $\xi = 50\%$ with different values of the fractional order.

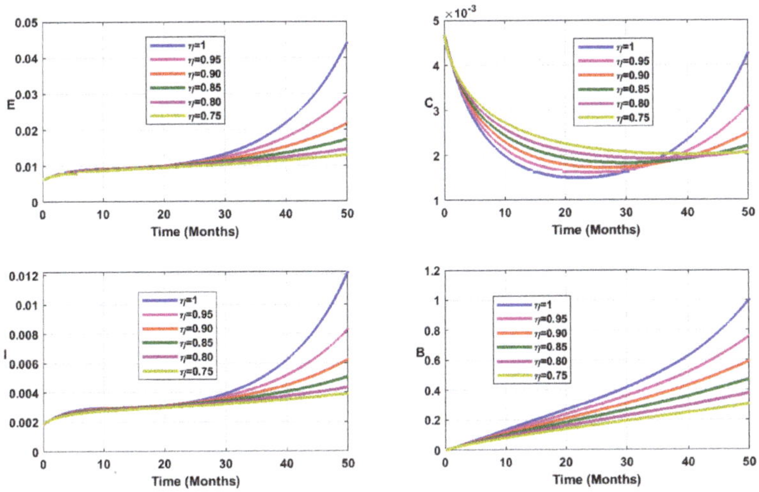

Figure 15. Simulation results showing the infected state variable profiles when the vaccination coverage $\xi = 90\%$ with different values of the fractional order.

4. Discussion and Conclusions

In this work, we extended our previous $SVEIR$-B compartmental model [5] by replacing the integer derivative with fractional derivatives, to evaluate the memory effect on the transmission dynamics of typhoid fever. We began by recalling some previous results on the integer model (the control reproduction number $\mathcal{R}_c$, existence and stability of equilibrium points). In order to describe the non-local character as well as long-term memory effects in the typhoid fever transmission dynamics, we replaced the integer derivative with the fractional derivative in the Caputo sense and studied the asymptotic stability of the disease-free equilibrium. Using fixed point theory, we proved the existence as well as the uniqueness of the solutions of the fractional model. We used the Adams–Bashforth method to construct the numerical scheme of the proposed fractional model. We then established the stability of this proposed numerical scheme. We simulated our fractional model using the Adams–Bashforth–Moulton scheme implemented by [39]. Using parameter values for Mbandjock, a city in the central region of Cameroon, we simulates the model by varying the fractional-order parameter, the vaccination coverage, and the bacterial decay rate. Apart from the fact that the solutions of the fractional model converged to the solutions of the integer model when the fractional-order approached one ($\eta = 1$), the simulation results showed that the expected date of the disease peak was forward delayed when the fractional-order parameter decreased. In addition, combining vaccination with environmental sanitation can permit a considerable reduction in the disease's spread.

We then extended the previous models by replacing the mass action incidence law with the standard incidence. The analysis of the models showed that the disease-free equilibrium is also globally asymptotically stable whenever the corresponding reproduction number $\mathcal{R}_{c\star}$ is less than one. Due to the complexity of the newly proposed models, we could not prove the existence and uniqueness of the endemic equilibrium. However, numerical simulations showed that it is possible that the new typhoid fever models permit a unique endemic equilibrium that is globally stable whenever $\mathcal{R}_{c\star} > 1$, and no equilibrium otherwise. We also found that, from a quantitative point of view, the disease burden was overestimated with the models the with mass incidence law compared to the one with the standard incidence law. Indeed, for the models with mass incidence, the control reproduction number was estimated at 2.4750, while the one with the standard incidence was estimated at 1.4348. This was in accordance with our previous work in which we considered the standard incidence law. In [6], the control reproduction number was estimated at 1.3722. As for the models with mass action incidences, we observed a delay in the disease peaks whenever the fractional-order derivative decreased.

It was observed that mass vaccination can overcome this disease. In fact, if the means are put in place to finance and implement vaccination campaigns in rural areas, it is possible to eradicate typhoid fever. Moreover, these vaccination campaigns must be accompanied by awareness campaigns among the population in order to combat this type of disease, as well as instructing citizens on ways to protect their environment against the proliferation of salmonella.

Our main contribution in this paper consisted in the formulation, using both integer and fractional derivatives, of new transmission dynamics typhoid fever models that incorporate the standard incidence rates and mass vaccination. The values of the control reproductive number differ from the model with mass action incidence and those with the standard incidences. Indeed, for the model with mass action incidences, $\mathcal{R}_c = 2.4750$, while, for those with standard incidences, $\mathcal{R}_{c*} = 1.4348$. This proves that mass action incidence overestimates the disease burden.

Author Contributions: Conceptualization, H.A. and R.K.R.; methodology, H.A. and R.K.R.; software, H.A. and R.K.R.; validation, H.A., R.K.R. and K.S.N.; formal analysis, H.A.; investigation, H.A. and R.K.R.; resources, H.A., R.K.R. and K.S.N.; data curation, H.A. and R.K.R.; writing—original draft preparation, H.A.; writing—review and editing, H.A., R.K.R. and K.S.N.; visualization, H.A.; supervision, H.A.; project administration, H.A.; funding acquisition, K.S.N. All authors have read and agreed to the published version of the manuscript.

Funding: The APC was funded by Kottakkaran Sooppy Nisar. No other funds have been received for this work.

Institutional Review Board Statement: Not applicable.

Informed Consent Statement: Not applicable.

Data Availability Statement: The data used to calibrate our model were taken from Mbandjock district hospital.

Acknowledgments: The authors thank the LESIA laboratory manager of the National School of Agro-industrial Sciences of the University of Ngaoundéré for their hospitality during the performance of the numerical simulations. The authors thank the Handling Editor and the anonymous reviewers for their comments and suggestions, which enabled us to improve the manuscript.

Conflicts of Interest: The authors declare no conflict of interest.

Appendix A. Proof of Proposition 3

Proof. The Jacobian matrix of (34) (resp. (35)) evaluated at the disease-free equilibrium $\mathcal{Q}_0$ is given by

$$\mathcal{J}(\mathcal{Q}_0)=\begin{pmatrix} -k_1 & \theta & 0 & -\frac{S_0\beta}{N_0} & -\frac{S_0\beta}{N_0} & \alpha & -\frac{S_0\nu}{K} \\ \xi & -k_2 & 0 & -\frac{\pi V_0\beta}{N_0} & -\frac{\pi V_0\beta}{N_0} & 0 & -\frac{\pi V_0\nu}{K} \\ 0 & 0 & -k_3 & \frac{H_0\beta}{N_0} & \frac{H_0\beta}{N_0} & 0 & \frac{H_0\nu}{K} \\ 0 & 0 & \gamma_1 q & -k_4 & 0 & 0 & 0 \\ 0 & 0 & q_1\gamma_1 & p_1\gamma_2 & -k_8 & 0 & 0 \\ 0 & 0 & 0 & p\gamma_2 & \sigma & -k_6 & 0 \\ 0 & 0 & 0 & p_c & p_i & 0 & -\mu_b \end{pmatrix}=\begin{pmatrix} \mathcal{J}_1 & \mathcal{J}_2 \\ \mathcal{J}_3 & \mathcal{J}_4 \end{pmatrix},$$

where $\mathcal{J}_1=\begin{pmatrix} -k_1 & \theta \\ \xi & -k_2 \end{pmatrix}$, $\mathcal{J}_4=\begin{pmatrix} -k_3 & \frac{H_0\beta}{N_0} & \frac{H_0\beta}{N_0} & 0 & \frac{H_0\nu}{K} \\ \gamma_1 q & -k_4 & 0 & 0 & 0 \\ q_1\gamma_1 & p_1\gamma_2 & -k_8 & 0 & 0 \\ 0 & p\gamma_2 & \sigma & -k_6 & 0 \\ 0 & p_c & p_i & 0 & -\mu_b \end{pmatrix}$,

$\mathcal{J}_2=\begin{pmatrix} 0 & -\frac{S_0\beta}{N_0} & -\frac{S_0\beta}{N_0} & \alpha & -\frac{S_0\nu}{K} \\ 0 & -\frac{\pi V_0\beta}{N_0} & -\frac{\pi V_0\beta}{N_0} & 0 & -\frac{\pi V_0\nu}{K} \end{pmatrix}$, and $\mathcal{J}_3=\mathbf{0}_{\mathbb{R}^{5\times 2}}$. The eigenvalues of $\mathcal{J}(\mathcal{Q}_0)$ are those of $\mathcal{J}_1$ and $\mathcal{J}_4$. It is evident that the eigenvalues of $\mathcal{J}_1$ have negative real parts. Indeed, the characteristic polynomial of $\mathcal{J}_1$ is $\mathcal{T}(x)=\det(\mathcal{J}_1-xI_2)=x^2+(k_1+k_2)x+k_7$. Since all its coefficients are positive, it follows that all its roots have negative real parts. A trivial eigenvalue of $\mathcal{J}_4$ is $x=-k_6$. The others are the roots of the following polynomial: $\mathcal{I}(x)=x^4+a_1x^3+a_2x^2+a_3x+a_4$, with $a_1=\mu_b+k_8+k_4+k_3$, $a_4=k_3k_4k_8\mu_b(1-\mathcal{R}_c^\star)(\mathcal{R}_c^\star-R_1+1)$,

$$\begin{aligned} a_2=&\frac{1}{k_8q+p_1\gamma_2q+q_1k_4}\Big[k_8^2\mu_bq+k_4k_8\mu_bq+k_3k_8\mu_bq+p_1\gamma_2k_8\mu_bq+p_1\gamma_2k_4\mu_bq+p_1\gamma_2k_3\mu_bq\\ &+k_4k_8^2q+k_3k_8^2q+k_3k_4k_8q(1-R_1)+p_1\gamma_2k_4k_8q+p_1\gamma_2k_3k_8q+p_1\gamma_2k_3k_4q+q_1k_4k_8\mu_b\\ &+q_1k_4^2\mu_b+q_1k_3k_4\mu_b+q_1k_4^2k_8+q_1k_3k_4k_8(1-R_1)+q_1k_3k_4^2\Big], \end{aligned}$$

and

$$\begin{aligned} a_3=&\frac{1}{((p_1\gamma_2k_8+p_1^2\gamma_2^2)p_i+(k_8^2+p_1\gamma_2k_8)p_c)q^2+((q_1k_4k_8+2p_1q_1\gamma_2k_4)p_i+q_1k_4k_8p_c)q+q_1^2k_4^2p_i}\times\\ &\times\Big[((((p_1\gamma_2k_4+p_1\gamma_2k_3)k_8^2+(((1-R_1)p_1\gamma_2k_3+p_1^2\gamma_2^2)k_4+p_1^2\gamma_2^2k_3)k_8+p_1^2\gamma_2^2k_3k_4)\mu_b\\ &+(1-R_1)p_1\gamma_2k_3k_4k_8^2+(1-R_1)p_1^2\gamma_2^2k_3k_4k_8)p_i+\mathbf{K}_1p_c)q^2\\ &+(\mathbf{K}_2p_i+(q_1k_3k_4^2k_8\mu_b\Big(1-\mathcal{R}_{c\star}^2+R_1\mathcal{R}_{c\star}\Big)+q_1k_4^2k_8^2\mu_b+(1-R_1)q_1k_3k_4k_8^2\mu_b+(1-R_1)q_1k_3k_4^2k_8^2)p_c)q\\ &+(q_1^2k_3k_4^2k_8\mu_b(1-\mathcal{R}_{c\star})(1-R_1+\mathcal{R}_{c\star})+q_1^2k_4^3k_8\mu_b+q_1^2k_3k_4^3\mu_b+(1-R_1)q_1^2k_3k_4^3k_8)p_i\Big]. \end{aligned}$$

$$\begin{aligned} \mathbf{K}_1=&\Big\{k_8(1-\mathcal{R}_{c\star})(1-R_1+\mathcal{R}_{c\star})+p_1\gamma_2\Big(1-\mathcal{R}_{c\star}^2+R_1\mathcal{R}_{c\star}\Big)\Big\}k_3k_4k_8\mu_b\\ &+(k_4+k_3)(k_8+p_1\gamma_2)k_8^2\mu_b+(1-R_1)(k_8+p_1\gamma_2)k_3k_4k_8^2\\ \mathbf{K}_2=&(k_8+p_1\gamma_2)q_1k_3k_4k_8\mu_b\Big(1-\mathcal{R}_{c\star}^2+R_1\mathcal{R}_{c\star}\Big)\\ &+(1-R_1)p_1\gamma_2q_1k_3k_4k_8\mu_b+((1-R_1)q_1k_3+2p_1\gamma_2q_1)k_4^2k_8\mu_b^2+2p_1\gamma_2q_1k_3k_4^2\mu_b\\ &+q_1k_4^2k_8^2\mu_b+q_1k_3k_4^2k_8(1-R_1)(k_8+2p_1\gamma_2) \end{aligned}$$

It is clear that a_1 is always positive, and a_i, $i \in \{2, 3, 4\}$ are positive if $\mathcal{R}_{c\star} < 1$. Indeed, it is important to note that

$$\mathcal{R}_{c\star} < 1 \Longrightarrow R_1 < 1, \tag{A1}$$

which implies that $\mathbf{K}_1 > 0$ and $\mathbf{K}_2 > 0$.

Thus, all coefficients of the polynomial $\mathcal{I}(x)$ are always positive whenever $\mathcal{R}_{c\star} < 1$. It follows that, if $\mathcal{R}_{c\star} < 1$, then the disease-free equilibrium is locally asymptotically stable if and only if the following conditions hold (because of the length of the expressions, we omit them here):

$$a_1 a_2 - a_3 > 0 \quad \text{and} \quad a_1 a_2 a_3 - a_1^2 a_4 - a_3^2 > 0. \tag{A2}$$

This ends the proof. □

It remains now to prove the corresponding result for the new fractional model (35). To this aim, let us define the following equation:

$$\det[r(I - (1-\eta)\mathcal{J}(\mathcal{Q}_0)) - \eta \mathcal{J}(\mathcal{Q}_0)] = 0, \tag{A3}$$

which is the characteristic equation of

$$\mathcal{J}(\mathcal{Q}_0) = \begin{pmatrix} -k_1 & \vartheta & 0 & -\frac{S_0\beta}{N_0} & -\frac{S_0\beta}{N_0} & 0 & -\frac{S_0\nu}{K} \\ \xi & -k_2 & 0 & -\frac{V_0\beta\pi}{N_0} & -\frac{V_0\beta\pi}{N_0} & 0 & -\frac{V_0\nu\pi}{N_0} \\ 0 & 0 & -k_3 & \frac{H_0\beta}{N_0} & \frac{H_0\beta}{N_0} & 0 & \frac{H_0\nu}{K} \\ 0 & 0 & \gamma_1 q & -k_4 & 0 & 0 & 0 \\ 0 & 0 & q_1\gamma_1 & p_1\gamma_2 & -k_8 & 0 & 0 \\ 0 & 0 & 0 & \gamma_2 p & \sigma & -k_6 & 0 \\ 0 & 0 & 0 & p_c & p_i & 0 & -\mu_b \end{pmatrix}.$$

From [4,30], it follows that $\mathcal{Q}_0$ is asymptotically stable, for the new fractional model, if all solutions of (A3) have negative real parts.

Setting $\mathcal{D} := [r(I - (1-\eta)\mathcal{J}(\mathcal{Q}_0)) - \eta \mathcal{Q}_0] = \begin{pmatrix} D_1 & \bullet \\ \mathbf{0}_{\mathbb{R}^{5\times 2}} & D_4 \end{pmatrix}$, with

$$D_1 := \begin{pmatrix} (\eta_1 k_1 + 1)r + k_1\eta & -\eta_1 r\vartheta - \eta\vartheta \\ -\eta_1 r\xi - \eta\xi & (\eta_1 k_2 + 1)r + k_2\eta \end{pmatrix} \text{ and}$$

$$D_4 := \begin{pmatrix} (\eta_1 k_3 + 1)r + k_3\eta & -\frac{H_0\eta_1\beta r}{N_0} - \frac{H_0\beta\eta}{N_0} & -\frac{H_0\eta_1\beta r}{N_0} - \frac{H_0\beta\eta}{N_0} & 0 & -\frac{H_0\eta_1\nu r}{K} - \frac{H_0\eta\nu}{K} \\ -\eta_1\gamma_1 q r - \gamma_1\eta q & (\eta_1 k_4 + 1)r + k_4\eta & 0 & 0 & 0 \\ -\eta_1 q_1\gamma_1 r - q_1\gamma_1\eta & -\eta_1 p_1\gamma_2 r - p_1\gamma_2\eta & (\eta_1 k_8 + 1)r + k_8\eta & 0 & 0 \\ 0 & -\eta_1\gamma_2 p r - \gamma_2\eta p & -\eta_1 r\sigma - \eta\sigma & (\eta_1 k_6 + 1)r + k_6\eta & 0 \\ 0 & -\eta_1 p_c r - p_c\eta & -\eta_1 p_i r - \eta p_i & 0 & (\eta_1\mu_b + 1)r + \mu_b\eta \end{pmatrix},$$

it follows that the solutions of (A3) are the solutions of $\det(D_1) = 0$ and $\det(D_4) = 0$. From the Proof of Theorem 2, it follows that the solutions of $\det(D_1) = 0$ have negative real parts. It thus remains to show that the same is true for $\det(D_4) = 0$. Note that $r = -\dfrac{k_8\eta}{\eta_1 k_8 + 1} < 0$ is a solution of $\det(D_4) = 0$. The others are the solutions of $\det(D_4^\star) = 0$, where

$$D_4^\star := \begin{pmatrix} (\eta_1 k_3 + 1)r + k_3\eta & -\frac{H_0\eta_1\beta r}{N_0} - \frac{H_0\beta\eta}{N_0} & -\frac{H_0\eta_1\beta r}{N_0} - \frac{H_0\beta\eta}{N_0} & -\frac{H_0\eta_1\nu r}{K} - \frac{H_0\eta\nu}{K} \\ -\eta_1\gamma_1 q r - \gamma_1\eta q & (\eta_1 k_4 + 1)r + k_4\eta & 0 & 0 \\ -\eta_1 q_1\gamma_1 r - q_1\gamma_1\eta & -\eta_1 p_1\gamma_2 r - p_1\gamma_2\eta & (\eta_1 k_8 + 1)r + k_8\eta & 0 \\ 0 & -\eta_1 p_c r - p_c\eta & -\eta_1 p_i r - \eta p_i & (\eta_1\mu_b + 1)r + \mu_b\eta \end{pmatrix}.$$

After some straightforward algebraic computations, we obtain that

$$\det(D_4^\star) = 0 \Longleftrightarrow r^4 + \frac{A_2}{A_1}r^3 + \frac{A_3}{A_1}r^2 + \frac{A_4}{A_1}r + \frac{A_5}{A_1} = 0, \tag{A4}$$

where

$$
\begin{aligned}
A_1 = {} & ((((-p_1\gamma_2k_3k_4k_8^2 - p_1^2\gamma_2^2k_3k_4k_8)\mu_b\mathcal{R}_{c\star}^2 + (R_1p_1\gamma_2k_3k_4k_8^2 + R_1p_1^2\gamma_2^2k_3k_4k_8)\mu_b\mathcal{R}_{c\star} \\
& + ((1-R_1)p_1\gamma_2k_3k_4k_8^2 + (1-R_1)p_1^2\gamma_2^2k_3k_4k_8)\mu_b)\eta^4 \\
& + (((p_1\gamma_2k_4 + p_1\gamma_2k_3)k_8^2 + (((1-R_1)p_1\gamma_2k_3 + p_1^2\gamma_2^2)k_4 + p_1^2\gamma_2^2k_3)k_8 + p_1^2\gamma_2^2k_3k_4)\mu_b \\
& + (1-R_1)p_1\gamma_2k_3k_4k_8^2 + (1-R_1)p_1^2\gamma_2^2k_3k_4k_8)\eta^3 + ((p_1\gamma_2k_8^2 + (p_1\gamma_2k_4 + p_1\gamma_2k_3 + p_1^2\gamma_2^2)k_8 + p_1^2\gamma_2^2k_4 + p_1^2\gamma_2^2k_3)\mu_b \\
& + (p_1\gamma_2k_4 + p_1\gamma_2k_3)k_8^2 + (((1-R_1)p_1\gamma_2k_3 + p_1^2\gamma_2^2)k_4 + p_1^2\gamma_2^2k_3)k_8 + p_1^2\gamma_2^2k_3k_4)\eta^2 \\
& + ((p_1\gamma_2k_8 + p_1^2\gamma_2^2)\mu_b + p_1\gamma_2k_8^2 + (p_1\gamma_2k_4 + p_1\gamma_2k_3 + p_1^2\gamma_2^2)k_8 + p_1^2\gamma_2^2k_4 + p_1^2\gamma_2^2k_3)\eta + p_1\gamma_2k_8 + p_1^2\gamma_2^2)p_i \\
& + ((-k_3k_4k_8^3 - p_1\gamma_2k_3k_4k_8^2)\mu_bR_c^2 + (R_1k_3k_4k_8^3 + R_1p_1\gamma_2k_3k_4k_8^2)\mu_b\mathcal{R}_{c\star} + ((1-R_1)k_3k_4k_8^3 + (1-R_1)p_1\gamma_2k_3k_4k_8^2)\mu_b)p_c\eta^4 \\
& + ((-k_3k_4k_8^2 - p_1\gamma_2k_3k_4k_8)\mu_b\mathcal{R}_{c\star}^2 + (R_1k_3k_4k_8^2 + R_1p_1\gamma_2k_3k_4k_8)\mu_b\mathcal{R}_{c\star} + ((k_4+k_3)k_8^3 + (((1-R_1)k_3 + p_1\gamma_2)k_4 + p_1\gamma_2k_3)k_8^2 \\
& + p_1\gamma_2k_3k_4k_8)\mu_b + (1-R_1)k_3k_4k_8^3 + (1-R_1)p_1\gamma_2k_3k_4k_8^2)p_c\eta^3 + ((k_8^3 + (k_4+k_3+p_1\gamma_2)k_8^2 + (p_1\gamma_2k_4 + p_1\gamma_2k_3)k_8)\mu_b \\
& + (k_4+k_3)k_8^3 + (((1-R_1)k_3 + p_1\gamma_2)k_4 + p_1\gamma_2k_3)k_8^2 + p_1\gamma_2k_3k_4k_8)p_c\eta^2 + ((k_8^2 + p_1\gamma_2k_8)\mu_b + k_8^3 + (k_4+k_3+p_1\gamma_2)k_8^2 \\
& + (p_1\gamma_2k_4 + p_1\gamma_2k_3)k_8)p_c\eta + (k_8^2 + p_1\gamma_2k_8)p_c)q^2 \\
& + ((((-q_1k_3k_4^2k_8^2 - 2p_1q_1\gamma_2k_3k_4^2k_8)\mu_b\mathcal{R}_{c\star}^2 + (R_1q_1k_3k_4^2k_8^2 + 2R_1p_1q_1\gamma_2k_3k_4^2k_8)\mu_b\mathcal{R}_{c\star} \\
& + ((1-R_1)q_1k_3k_4^2k_8^2 + (2-2R_1)p_1q_1\gamma_2k_3k_4^2k_8)\mu_b)\eta^4 \\
& + ((-q_1k_3k_4k_8^2 - p_1q_1\gamma_2k_3k_4k_8)\mu_b\mathcal{R}_{c\star}^2 + (R_1q_1k_3k_4k_8^2 + R_1p_1q_1\gamma_2k_3k_4k_8)\mu_b\mathcal{R}_{c\star} \\
& + ((q_1k_4^2 + q_1k_3k_4)k_8^2 + (((1-R_1)q_1k_3 + 2p_1q_1\gamma_2)k_4^2 + (2-R_1)p_1q_1\gamma_2k_3k_4)k_8 + 2p_1q_1\gamma_2k_3k_4^2)\mu_b \\
& + (1-R_1)q_1k_3k_4^2k_8^2 + (2-2R_1)p_1q_1\gamma_2k_3k_4^2k_8)\eta^3 + ((q_1k_4k_8^2 + (q_1k_4^2 + (q_1k_3 + 2p_1q_1\gamma_2)k_4)k_8 + 2p_1q_1\gamma_2k_4^2 + 2p_1q_1\gamma_2k_3k_4)\mu_b \\
& + (q_1k_4^2 + q_1k_3k_4)k_8^2 + (((1-R_1)q_1k_3 + 2p_1q_1\gamma_2)k_4^2 + (2-R_1)p_1q_1\gamma_2k_3k_4)k_8 + 2p_1q_1\gamma_2k_3k_4^2)\eta^2 + ((q_1k_4k_8 + 2p_1q_1\gamma_2k_4)\mu_b \\
& + q_1k_4k_8^2 + (q_1k_4^2 + (q_1k_3 + 2p_1q_1\gamma_2)k_4)k_8 + 2p_1q_1\gamma_2k_4^2 + 2p_1q_1\gamma_2k_3k_4)\eta + q_1k_4k_8 + 2p_1q_1\gamma_2k_4)p_i \\
& + (-q_1k_3k_4^2k_8^2\mu_b\mathcal{R}_{c\star}^2 + R_1q_1k_3k_4^2k_8^2\mu_b\mathcal{R}_{c\star} + (1-R_1)q_1k_3k_4^2k_8^2\mu_b)p_c\eta^4 \\
& + (-q_1k_3k_4^2k_8\mu_b\mathcal{R}_{c\star}^2 + R_1q_1k_3k_4^2k_8\mu_b\mathcal{R}_{c\star} + ((q_1k_4^2 + (1-R_1)q_1k_3k_4)k_8^2 + q_1k_3k_4^2k_8)\mu_b + (1-R_1)q_1k_3k_4^2k_8^2)p_c\eta^3 \\
& + ((q_1k_4k_8^2 + (q_1k_4^2 + q_1k_3k_4)k_8)\mu_b + (q_1k_4^2 + (1-R_1)q_1k_3k_4)k_8^2 + q_1k_3k_4^2k_8)p_c\eta^2 \\
& + (q_1k_4k_8\mu_b + q_1k_4k_8^2 + (q_1k_4^2 + q_1k_3k_4)k_8)p_c\eta + q_1k_4k_8p_c)q \\
& + ((-q_1^2k_3k_4^3k_8\mu_b\mathcal{R}_{c\star}^2 + R_1q_1^2k_3k_4^3k_8\mu_b\mathcal{R}_{c\star} + (1-R_1)q_1^2k_3k_4^3k_8\mu_b)\eta^4 \\
& + (-q_1^2k_3k_4^2k_8\mu_b\mathcal{R}_{c\star}^2 + R_1q_1^2k_3k_4^2k_8\mu_b\mathcal{R}_{c\star} + ((q_1^2k_4^3 + (1-R_1)q_1^2k_3k_4^2)k_8 + q_1^2k_3k_4^3)\mu_b + (1-R_1)q_1^2k_3k_4^3k_8)\eta^3 \\
& + ((q_1^2k_4^2k_8 + q_1^2k_4^3 + q_1^2k_3k_4^2)\mu_b + (q_1^2k_4^3 + (1-R_1)q_1^2k_3k_4^2)k_8 + q_1^2k_3k_4^3)\eta^2 + (q_1^2k_4^2\mu_b + q_1^2k_4^2k_8 + q_1^2k_4^3 + q_1^2k_3k_4^2)\eta + q_1^2k_4^2)p_i,
\end{aligned}
$$

$$
\begin{aligned}
A_4 = {} & ((4(k_8 + p_1\gamma_2)p_1\gamma_2k_3k_4k_8\mu_b(1-\mathcal{R}_{c\star})(1+\mathcal{R}_{c\star}-R_1)\eta^4 \\
& + (((p_1\gamma_2k_4 + p_1\gamma_2k_3)k_8^2 + (((1-R_1)p_1\gamma_2k_3 + p_1^2\gamma_2^2)k_4 + p_1^2\gamma_2^2k_3)k_8 + p_1^2\gamma_2^2k_3k_4)\mu_b \\
& + (1-R_1)p_1\gamma_2k_3k_4k_8^2 + (1-R_1)p_1^2\gamma_2^2k_3k_4k_8)\eta^3)p_i + 4(k_8+p_1\gamma_2)(1-\mathcal{R}_{c\star})(1+\mathcal{R}_{c\star}-R_1)k_3k_4k_8^2\mu_bp_c\eta^4 \\
& + ((k_8 + p_1\gamma_2)k_3k_4k_8\mu_b\mathcal{R}_{c\star}(R_1 - \mathcal{R}_{c\star}) + ((k_4+k_3)k_8^3 \\
& + (((1-R_1)k_3 + p_1\gamma_2)k_4 + p_1\gamma_2k_3)k_8^2 + p_1\gamma_2k_3k_4k_8)\mu_b + (1-R_1)k_3k_4k_8^3 + (1-R_1)p_1\gamma_2k_3k_4k_8^2)p_c\eta^3)q^2 \\
& + ((((-4q_1k_3k_4^2k_8^2 - 8p_1q_1\gamma_2k_3k_4^2k_8)\mu_b\mathcal{R}_{c\star}^2 + (4R_1q_1k_3k_4^2k_8^2 + 8R_1p_1q_1\gamma_2k_3k_4^2k_8)\mu_b\mathcal{R}_{c\star} \\
& + ((4-4R_1)q_1k_3k_4^2k_8^2 + (8-8R_1)p_1q_1\gamma_2k_3k_4^2k_8)\mu_b)\eta^4 \\
& + ((-q_1k_3k_4k_8^2 - p_1q_1\gamma_2k_3k_4k_8)\mu_b\mathcal{R}_{c\star}^2 + (R_1q_1k_3k_4k_8^2 + R_1p_1q_1\gamma_2k_3k_4k_8)\mu_b\mathcal{R}_{c\star} + ((q_1k_4^2 + q_1k_3k_4)k_8^2 \\
& + (((1-R_1)q_1k_3 + 2p_1q_1\gamma_2)k_4^2 + (2-R_1)p_1q_1\gamma_2k_3k_4)k_8 + 2p_1q_1\gamma_2k_3k_4^2)\mu_b + (1-R_1)q_1k_3k_4^2k_8^2 \\
& + (2-2R_1)p_1q_1\gamma_2k_3k_4^2k_8)\eta^3)p_i + (-4q_1k_3k_4^2k_8^2\mu_b\mathcal{R}_{c\star}^2 + 4R_1q_1k_3k_4^2k_8^2\mu_b\mathcal{R}_{c\star} + (4-4R_1)q_1k_3k_4^2k_8^2\mu_b)p_c\eta^4 \\
& + (-q_1k_3k_4^2k_8\mu_b\mathcal{R}_{c\star}^2 + R_1q_1k_3k_4^2k_8\mu_b\mathcal{R}_{c\star} + ((q_1k_4^2 + (1-R_1)q_1k_3k_4)k_8^2 + q_1k_3k_4^2k_8)\mu_b \\
& + (1-R_1)q_1k_3k_4^2k_8^2)p_c\eta^3)q + ((-4q_1^2k_3k_4^3k_8\mu_b\mathcal{R}_{c\star}^2 + 4R_1q_1^2k_3k_4^3k_8\mu_b\mathcal{R}_{c\star} + (4-4R_1)q_1^2k_3k_4^3k_8\mu_b)\eta^4 \\
& + (-q_1^2k_3k_4^2k_8\mu_b\mathcal{R}_{c\star}^2 + R_1q_1^2k_3k_4^2k_8\mu_b\mathcal{R}_{c\star} + ((q_1^2k_4^3 + (1-R_1)q_1^2k_3k_4^2)k_8 + q_1^2k_3k_4^3)\mu_b + (1-R_1)q_1^2k_3k_4^3k_8)\eta^3)p_i,
\end{aligned}
$$

$$A_5 = k_3k_4k_8\mu_b(\mathcal{R}_{c\star} - R_1 + 1)\eta^4(k_8q + p_1\gamma_2q + q_1k_4)(p_1\gamma_2p_iq + k_8p_cq + q_1k_4p_i)(1 - \mathcal{R}_{c\star}),$$

$$\begin{aligned}
A_2 = {} & ((((-4p_1\gamma_2k_3k_4k_8^2 - 4p_1^2\gamma_2^2k_3k_4k_8)\mu_b\mathcal{R}_{c\star}^2 + (4R_1p_1\gamma_2k_3k_4k_8^2 + 4R_1p_1^2\gamma_2^2k_3k_4k_8)\mu_b\mathcal{R}_{c\star} \\
& + ((4-4R_1)p_1\gamma_2k_3k_4k_8^2 + (4-4R_1)p_1^2\gamma_2^2k_3k_4k_8)\mu_b)\eta^4 \\
& + (((3p_1\gamma_2k_4 + 3p_1\gamma_2k_3)k_8^2 + (((3-3R_1)p_1\gamma_2k_3 + 3p_1^2\gamma_2^2)k_4 + 3p_1^2\gamma_2^2k_3)k_8 + 3p_1^2\gamma_2^2k_3k_4)\mu_b \\
& + (3-3R_1)p_1\gamma_2k_3k_4k_8^2 + (3-3R_1)p_1^2\gamma_2^2k_3k_4k_8)\eta^3 + ((2p_1\gamma_2k_8^2 + (2p_1\gamma_2k_4 + 2p_1\gamma_2k_3 + 2p_1^2\gamma_2^2)k_8 + 2p_1^2\gamma_2^2k_4 + 2p_1^2\gamma_2^2k_3)\mu_b \\
& + (2p_1\gamma_2k_4 + 2p_1\gamma_2k_3)k_8^2 + (((2-2R_1)p_1\gamma_2k_3 + 2p_1^2\gamma_2^2)k_4 + 2p_1^2\gamma_2^2k_3)k_8 + 2p_1^2\gamma_2^2k_3k_4)\eta^2 \\
& + ((p_1\gamma_2k_8 + p_1^2\gamma_2^2)\mu_b + p_1\gamma_2k_8^2 + (p_1\gamma_2k_4 + p_1\gamma_2k_3 + p_1^2\gamma_2^2)k_8 + p_1^2\gamma_2^2k_4 + p_1^2\gamma_2^2k_3)\eta)p_i \\
& + ((-4k_3k_4k_8^3 - 4p_1\gamma_2k_3k_4k_8^2)\mu_b\mathcal{R}_{c\star}^2 + (4R_1k_3k_4k_8^3 + 4R_1p_1\gamma_2k_3k_4k_8^2)\mu_b\mathcal{R}_{c\star} + 4(1-R_1)(k_8 + p_1\gamma_2)k_3k_4k_8^2\mu_b)p_c\eta^4 \\
& + ((-3k_3k_4k_8^2 - 3p_1\gamma_2k_3k_4k_8)\mu_b\mathcal{R}_{c\star}^2 + (3R_1k_3k_4k_8^2 + 3R_1p_1\gamma_2k_3k_4k_8)\mu_b\mathcal{R}_{c\star} \\
& + (3(k_4 + k_3)k_8^3 + ((3(1-R_1)k_3 + 3p_1\gamma_2)k_4 + 3p_1\gamma_2k_3)k_8^2 + 3p_1\gamma_2k_3k_4k_8)\mu_b + 3(1-R_1)k_3k_4k_8^3 + 3(1-R_1)p_1\gamma_2k_3k_4k_8^2)p_c\eta^3 \\
& + ((2k_8^3 + (2k_4 + 2k_3 + 2p_1\gamma_2)k_8^2 + (2p_1\gamma_2k_4 + 2p_1\gamma_2k_3)k_8)\mu_b + (2k_4 + 2k_3)k_8^3 + (((2-2R_1)k_3 + 2p_1\gamma_2)k_4 + 2p_1\gamma_2k_3)k_8^2 \\
& + 2p_1\gamma_2k_3k_4k_8)p_c\eta^2 + ((k_8^2 + p_1\gamma_2k_8)\mu_b + k_8^3 + (k_4 + k_3 + p_1\gamma_2)k_8^2 + (p_1\gamma_2k_4 + p_1\gamma_2k_3)k_8)p_c\eta)q^2 \\
& + ((((-4q_1k_3k_4^2k_8^2 - 8p_1q_1\gamma_2k_3k_4^2k_8)\mu_b\mathcal{R}_{c\star}^2 + (4R_1q_1k_3k_4^2k_8^2 + 8R_1p_1q_1\gamma_2k_3k_4^2k_8)\mu_b\mathcal{R}_{c\star} \\
& + ((4-4R_1)q_1k_3k_4^2k_8^2 + (8-8R_1)p_1q_1\gamma_2k_3k_4^2k_8)\mu_b)\eta^4 \\
& + ((-3q_1k_3k_4k_8^2 - 3p_1q_1\gamma_2k_3k_4k_8)\mu_b\mathcal{R}_{c\star}^2 + (3R_1q_1k_3k_4k_8^2 + 3R_1p_1q_1\gamma_2k_3k_4k_8)\mu_b\mathcal{R}_{c\star} \\
& + ((3q_1k_4^2 + 3q_1k_3k_4)k_8^2 + (((3-3R_1)q_1k_3 + 6p_1q_1\gamma_2)k_4^2 + (6-3R_1)p_1q_1\gamma_2k_3k_4)k_8 + 6p_1q_1\gamma_2k_3k_4^2)\mu_b + (3-3R_1)q_1k_3k_4^2k_8^2 \\
& + (6-6R_1)p_1q_1\gamma_2k_3k_4^2k_8)\eta^3 + ((2q_1k_4k_8^2 + (2q_1k_4^2 + (2q_1k_3 + 4p_1q_1\gamma_2)k_4)k_8 + 4p_1q_1\gamma_2k_4^2 + 4p_1q_1\gamma_2k_3k_4)\mu_b \\
& + (2q_1k_4^2 + 2q_1k_3k_4)k_8^2 + (((2-2R_1)q_1k_3 + 4p_1q_1\gamma_2)k_4^2 + (4-2R_1)p_1q_1\gamma_2k_3k_4)k_8 + 4p_1q_1\gamma_2k_3k_4^2)\eta^2 \\
& + ((q_1k_4k_8 + 2p_1q_1\gamma_2k_4)\mu_b + q_1k_4k_8^2 + (q_1k_4^2 + (q_1k_3 + 2p_1q_1\gamma_2)k_4)k_8 + 2p_1q_1\gamma_2k_4^2 + 2p_1q_1\gamma_2k_3k_4)\eta)p_i \\
& + 4(1 - R_1 + R_1\mathcal{R}_{c\star} - \mathcal{R}_{c\star}^2)q_1k_3k_4^2k_8^2\mu_bp_c\eta^4 \\
& + (-3q_1k_3k_4^2k_8\mu_b\mathcal{R}_{c\star}^2 + 3R_1q_1k_3k_4^2k_8\mu_b\mathcal{R}_{c\star} + ((3q_1k_4^2 + (3-3R_1)q_1k_3k_4)k_8^2 + 3q_1k_3k_4^2k_8)\mu_b + (3-3R_1)q_1k_3k_4^2k_8^2)p_c\eta^3 \\
& + ((2q_1k_4k_8^2 + (2q_1k_4^2 + 2q_1k_3k_4)k_8)\mu_b + (2q_1k_4^2 + (2-2R_1)q_1k_3k_4)k_8^2 + 2q_1k_3k_4^2k_8)p_c\eta^2 \\
& + (q_1k_4k_8\mu_b + q_1k_4k_8^2 + (q_1k_4^2 + q_1k_3k_4)k_8)p_c\eta)q \\
& + ((-4q_1^2k_3k_4^3k_8\mu_b\mathcal{R}_{c\star}^2 + 4R_1q_1^2k_3k_4^3k_8\mu_b\mathcal{R}_{c\star} + (4-4R_1)q_1^2k_3k_4^3k_8\mu_b)\eta^4 \\
& + (-3q_1^2k_3k_4^2k_8\mu_bR_c^2 + 3R_1q_1^2k_3k_4^2k_8\mu_b\mathcal{R}_{c\star} + ((3q_1^2k_4^3 + (3-3R_1)q_1^2k_3k_4^2)k_8 + 3q_1^2k_3k_4^3)\mu_b + (3-3R_1)q_1^2k_3k_4^3k_8)\eta^3 \\
& + ((2q_1^2k_4^2k_8 + 2q_1^2k_4^3 + 2q_1^2k_3k_4^2)\mu_b + (2q_1^2k_4^3 + (2-2R_1)q_1^2k_3k_4^2)k_8 + 2q_1^2k_3k_4^3)\eta^2 + (q_1^2k_4^2\mu_b + q_1^2k_4^2k_8 + q_1^2k_4^3 + q_1^2k_3k_4^2)\eta)p_i,
\end{aligned}$$

$$
\begin{aligned}
A_3 = {} & ((((-6p_1\gamma_2k_3k_4k_8^2 - 6p_1^2\gamma_2^2k_3k_4k_8)\mu_bR_c^2 + (6R_1p_1\gamma_2k_3k_4k_8^2 + 6R_1p_1^2\gamma_2^2k_3k_4k_8)\mu_bR_c \\
& + ((6-6R_1)p_1\gamma_2k_3k_4k_8^2 + (6-6R_1)p_1^2\gamma_2^2k_3k_4k_8)\mu_b)\eta^4 + (((3p_1\gamma_2k_4 + 3p_1\gamma_2k_3)k_8^2 \\
& + (((3-3R_1)p_1\gamma_2k_3 + 3p_1^2\gamma_2^2)k_4 + 3p_1^2\gamma_2^2k_3)k_8 + 3p_1^2\gamma_2^2k_3k_4)\mu_b + (3-3R_1)p_1\gamma_2k_3k_4k_8^2 + (3-3R_1)p_1^2\gamma_2^2k_3k_4k_8)\eta^3 \\
& + ((p_1\gamma_2k_8^2 + (p_1\gamma_2k_4 + p_1\gamma_2k_3 + p_1^2\gamma_2^2)k_8 + p_1^2\gamma_2^2k_4 + p_1^2\gamma_2^2k_3)\mu_b + (p_1\gamma_2k_4 + p_1\gamma_2k_3)k_8^2 \\
& + (((1-R_1)p_1\gamma_2k_3 + p_1^2\gamma_2^2)k_4 + p_1^2\gamma_2^2k_3)k_8 + p_1^2\gamma_2^2k_3k_4)\eta^2)p_i \\
& + ((-6k_3k_4k_8^3 - 6p_1\gamma_2k_3k_4k_8^2)\mu_b\mathcal{R}_{c\star}^2 + (6R_1k_3k_4k_8^3 + 6R_1p_1\gamma_2k_3k_4k_8^2)\mu_b\mathcal{R}_{c\star} + (6-6R_1)(k_3k_4k_8^3 + p_1\gamma_2k_3k_4k_8^2)\mu_b)p_c\eta^4 \\
& + ((-3k_3k_4k_8^2 - 3p_1\gamma_2k_3k_4k_8)\mu_b\mathcal{R}_{c\star}^2 + (3R_1k_3k_4k_8^2 + 3R_1p_1\gamma_2k_3k_4k_8)\mu_bR_c \\
& + ((3k_4 + 3k_3)k_8^3 + (((3-3R_1)k_3 + 3p_1\gamma_2)k_4 + 3p_1\gamma_2k_3)k_8^2 + 3p_1\gamma_2k_3k_4k_8)\mu_b + 3(1-R_1)k_3k_4k_8^2(k_8 + p_1\gamma_2))p_c\eta^3 \\
& + ((k_8^3 + (k_4 + k_3 + p_1\gamma_2)k_8^2 + p_1\gamma_2(k_4 + k_3)k_8)\mu_b + (k_4 + k_3)k_8^3 + (((1-R_1)k_3 + p_1\gamma_2)k_4 + p_1\gamma_2k_3)k_8^2 + p_1\gamma_2k_3k_4k_8)p_c\eta^2)q^2 \\
& + ((((-6q_1k_3k_4^2k_8^2 - 12p_1q_1\gamma_2k_3k_4^2k_8)\mu_bR_c^2 + (6R_1q_1k_3k_4^2k_8^2 + 12R_1p_1q_1\gamma_2k_3k_4^2k_8)\mu_b\mathcal{R}_{c\star} \\
& + ((6-6R_1)q_1k_3k_4^2k_8^2 + (12-12R_1)p_1q_1\gamma_2k_3k_4^2k_8)\mu_b)\eta^4 \\
& + ((-3q_1k_3k_4k_8^2 - 3p_1q_1\gamma_2k_3k_4k_8)\mu_b\mathcal{R}_{c\star}^2 + (3R_1q_1k_3k_4k_8^2 + 3R_1p_1q_1\gamma_2k_3k_4k_8)\mu_bR_c \\
& + ((3q_1k_4^2 + 3q_1k_3k_4)k_8^2 + (((3-3R_1)q_1k_3 + 6p_1q_1\gamma_2)k_4^2 + (6-3R_1)p_1q_1\gamma_2k_3k_4)k_8 + 6p_1q_1\gamma_2k_3k_4^2)\mu_b \\
& + (3-3R_1)q_1k_3k_4^2k_8^2 + (6-6R_1)p_1q_1\gamma_2k_3k_4^2k_8)\eta^3 \\
& + ((q_1k_4k_8^2 + (q_1k_4^2 + (q_1k_3 + 2p_1q_1\gamma_2)k_4)k_8 + 2p_1q_1\gamma_2k_4^2 + 2p_1q_1\gamma_2k_3k_4)\mu_b + (q_1k_4^2 + q_1k_3k_4)k_8^2 \\
& + (((1-R_1)q_1k_3 + 2p_1q_1\gamma_2)k_4^2 + (2-R_1)p_1q_1\gamma_2k_3k_4)k_8 + 2p_1q_1\gamma_2k_3k_4^2)\eta^2)p_i \\
& + (-6q_1k_3k_4^2k_8^2\mu_b\mathcal{R}_{c\star}^2 + 6R_1q_1k_3k_4^2k_8^2\mu_b\mathcal{R}_{c\star} + (6-6R_1)q_1k_3k_4^2k_8^2\mu_b)p_c\eta^4 \\
& + (-3q_1k_3k_4^2k_8\mu_b\mathcal{R}_{c\star}^2 + 3R_1q_1k_3k_4^2k_8\mu_b\mathcal{R}_{c\star} + ((3q_1k_4^2 + (3-3R_1)q_1k_3k_4)k_8^2 + 3q_1k_3k_4^2k_8)\mu_b + (3-3R_1)q_1k_3k_4^2k_8^2)p_c\eta^3 \\
& + ((q_1k_4k_8^2 + (q_1k_4^2 + q_1k_3k_4)k_8)\mu_b + (q_1k_4^2 + (1-R_1)q_1k_3k_4)k_8^2 + q_1k_3k_4^2k_8)p_c\eta^2)q \\
& + ((-6q_1^2k_3k_4^3k_8\mu_b\mathcal{R}_{c\star}^2 + 6R_1q_1^2k_3k_4^3k_8\mu_b\mathcal{R}_{c\star} + (6-6R_1)q_1^2k_3k_4^3k_8\mu_b)\eta^4 \\
& + (-3q_1^2k_3k_4^2k_8\mu_b\mathcal{R}_{c\star}^2 + 3R_1q_1^2k_3k_4^2k_8\mu_b\mathcal{R}_{c\star} + ((3q_1^2k_4^3 + (3-3R_1)q_1^2k_3k_4^2)k_8 + 3q_1^2k_3k_4^3)\mu_b + (3-3R_1)q_1^2k_3k_4^3k_8)\eta^3 \\
& + ((q_1^2k_4^2k_8 + q_1^2k_4^3 + q_1^2k_3k_4^2)\mu_b + (q_1^2k_4^3 + (1-R_1)q_1^2k_3k_4^2)k_8 + q_1^2k_3k_4^3)\eta^2)p_i.
\end{aligned}
$$

It is possible to show that all the above coefficients are positive whenever $\mathcal{R}_{c\star} < 1$. Then, it follows that, if $\mathcal{R}_c < 1$, then the disease-free equilibrium $\mathcal{Q}_0$ is asymptotically stable whenever the following Routh–Hurwitz criteria, $\frac{A_1A_2}{A_5^2} - \frac{A_3}{A_5} > 0$ and $\frac{A_1A_2A_3}{A_5^3} - \frac{A_1^2A_4}{A_5^3} - \frac{A_3^2}{A_5^2} > 0$, are satisfied for polynomial $\det(D_4^\star)$. (Given the heaviness of these coefficients, we do not present the Routh–Hurwitz conditions here.)

Appendix B. Proof of Theorem 6

Let us rewrite system (34) as

$$
\begin{pmatrix} \frac{dE}{dt} \\ \frac{dC}{dt} \\ \frac{dI}{dt} \\ \frac{dB}{dt} \end{pmatrix} = \begin{pmatrix} -k_3 & \frac{H_0\beta}{N_0} & \frac{H_0\beta}{N_0} & \frac{H_0\nu}{K} \\ \gamma_1 q & -k_4 & 0 & 0 \\ q_1\gamma_1 & p_1\gamma_2 & -k_8 & 0 \\ 0 & p_c & p_i & -\mu_b \end{pmatrix} \begin{pmatrix} E \\ C \\ I \\ B \end{pmatrix} - \mathcal{N}(S,V,E,C,I,R,B), \tag{A5}
$$

where $\mathcal{N}(S,V,E,C,I,R,B) = \begin{pmatrix} \beta(C+I)\left(\frac{H_0}{N_0} - \frac{H}{N}\right) + \nu B\left(\frac{H_0}{K} - \frac{H}{K+B}\right) \\ 0 \\ 0 \\ 0 \end{pmatrix}.$

In $\mathcal{W}$, $H = S + \pi V < H_0 = S_0 + \pi V_0$ for $t > 0$; thus, $\mathcal{N}(S, V, E, C, I, R, B) \geq \mathbf{O}_{\mathbb{R}^4}$. Note that Proposition 3 ensures that the following matrix

$$\mathcal{J}(\mathcal{Q}_0) = \begin{pmatrix} -k_3 & \frac{H_0\beta}{N_0} & \frac{H_0\beta}{N_0} & \frac{H_0\nu}{K} \\ \gamma_1 q & -k_4 & 0 & 0 \\ q_1\gamma_1 & p_1\gamma_2 & -k_8 & 0 \\ 0 & p_c & p_i & -\mu_b \end{pmatrix}$$

has all its eigenvalues with negative real parts. It follows that from the comparison theorem [40], $(E, C, I, B) \longrightarrow (0, 0, 0, 0)$ and $(S, V, R) \longrightarrow (S_0, V_0, 0)$ as $t \longrightarrow +\infty$. Thus, $(S, V, E, C, I, R, B) \longrightarrow \mathcal{Q}_0 = (S_0, V_0, 0, 0, 0, 0, 0)$ as $t \longrightarrow +\infty$. We finally conclude that the disease–free equilibrium is globally asymptotically stable in $\mathcal{W}$ if $\mathcal{R}_{c\star} < 1$.

References

1. World Health Organization. Typhoid vaccines: WHO position paper. *Wkly. Epidemiol. Rec.* **2008**, *83*, 49–59.
2. World Health Organization. Typhoid vaccines: WHO position paper–March 2018–Vaccins antityphoïdiques: Note de synthèse de l'OMS–mars 2018. *Wkly. Epidemiol. Rec.* **2018**, *93*, 153–172.
3. Ross, R. Some quantitative studies in epidemiology. *Nature* **1911**, *87*, 466–467. [CrossRef]
4. Abboubakar, H.; Kumar, P.; Erturk, V.S.; Kumar, A. A mathematical study of a tuberculosis model with fractional derivatives. *Int. J. Model. Simul. Sci. Comput.* **2021**, *12*, 2150037. [CrossRef]
5. Abboubakar, H.; Racke, R. Mathematical modeling, forecasting, and optimal control of typhoid fever transmission dynamics. *Chaos Solitons Fractals* **2021**, *149*, 111074. [CrossRef]
6. Abboubakar, H.; Kombou, L.K.; Koko, A.D.; Fouda, H.P.E.; Kumar, A. Projections and fractional dynamics of the typhoid fever: A case study of Mbandjock in the Centre Region of Cameroon. *Chaos Solitons Fractals* **2021**, *150*, 111129. [CrossRef]
7. Edward, S.; Nyerere, N. Modelling typhoid fever with education, vaccination and treatment. *Eng. Math.* **2016**, *1*, 44–52.
8. Mushayabasa, S. Modeling the impact of optimal screening on typhoid dynamics. *Int. J. Dyn. Control* **2016**, *4*, 330–338. [CrossRef]
9. Peter, O.J.; Ibrahim, M.O.; Edogbanya, H.O.; Oguntolu, F.A.; Oshinubi, K.; Ibrahim, A.A.; Ayoola, T.A.; Lawal, J.O. Direct and Indirect Transmission of Typhoid Fever Model with Optimal Control. *Results Phys.* **2021**, *27*, 104463. [CrossRef]
10. Tilahun, G.T.; Makinde, O.D.; Malonza, D. Modelling and optimal control of typhoid fever disease with cost-effective strategies. *Comput. Math. Methods Med.* **2017**, *2017*, 2324518. [CrossRef]
11. Tilahun, G.T.; Makinde, O.D.; Malonza, D. Co-dynamics of pneumonia and typhoid fever diseases with cost effective optimal control analysis. *Appl. Math. Comput.* **2018**, *316*, 438–459. [CrossRef]
12. Shaikh, A.S.; Nisar, K.S. Transmission dynamics of fractional order Typhoid fever model using Caputo–Fabrizio operator. *Chaos Solitons Fractals* **2019**, *128*, 355–365. [CrossRef]
13. Erturk, V.S.; Kumar, P. Solution of a COVID-19 model via new generalized Caputo-type fractional derivatives. *Chaos Solitons Fractals* **2020**, *139*, 110280. [CrossRef]
14. Kumar, P.; Erturk, V.S. Environmental persistence influences infection dynamics for a butterfly pathogen via new generalised Caputo type fractional derivative. *Chaos Solitons Fractals* **2021**, *144*, 110672. [CrossRef]
15. Kumar, P.; Erturk, V.S.; Almusawa, H. Mathematical structure of mosaic disease using microbial biostimulants via Caputo and Atangana–Baleanu derivatives. *Results Phys.* **2021**, *24*, 104186. [CrossRef]
16. Kumar, P.; Suat Ertürk, V.; Nisar, K.S. Fractional dynamics of huanglongbing transmission within a citrus tree. *Math. Methods Appl. Sci.* **2021**, *44*, 11404–11424. [CrossRef]
17. Kumar, P.; Erturk, V.S.; Murillo-Arcila, M. A complex fractional mathematical modeling for the love story of Layla and Majnun. *Chaos Solitons Fractals* **2021**, *150*, 111091. [CrossRef]
18. Loverro, A. *Fractional Calculus: History, Definitions and Applications for the Engineer*; Rapport Technique; Department of Aerospace and Mechanical Engineering, Univeristy of Notre Dame: Notre Dame, IN, USA, 2004; pp. 1–28.
19. Nabi, K.N.; Abboubakar, H.; Kumar, P. Forecasting of COVID-19 pandemic: From integer derivatives to fractional derivatives. *Chaos Solitons Fractals* **2020**, *141*, 110283. [CrossRef] [PubMed]
20. Angstmann, C.N.; Jacobs, B.A.; Henry, B.I.; Xu, Z. Intrinsic Discontinuities in Solutions of Evolution Equations Involving Fractional Caputo–Fabrizio and Atangana–Baleanu Operators. *Mathematics* **2020**, *8*, 2023. [CrossRef]
21. Tarasov, V.E. On chain rule for fractional derivatives. *Commun. Nonlinear Sci. Numer. Simul.* **2016**, *30*, 1–4. [CrossRef]
22. Tarasov, V.E. No violation of the Leibniz rule. No fractional derivative. *Commun. Nonlinear Sci. Numer. Simul.* **2013**, *18*, 2945–2948. [CrossRef]
23. Peter, O.; Ibrahim, M.; Oguntolu, F.; Akinduko, O.; Akinyemi, S. Direct and indirect transmission dynamics of typhoid fever model by differential transform method. *ATBU J. Sci. Technol. Educ. (JOSTE)* **2018**, *6*, 167–177.
24. Diethelm, K. An algorithm for the numerical solution of differential equations of fractional order. *Electron. Trans. Numer. Anal* **1997**, *5*, 1–6.
25. Diethelm, K.; Ford, N.J. Analysis of fractional differential equations. *J. Math. Anal. Appl.* **2002**, *265*, 229–248. [CrossRef]

26. Boccaletti, S.; Ditto, W.; Mindlin, G.; Atangana, A. Modeling and forecasting of epidemic spreading: The case of Covid-19 and beyond. *Chaos Solitons Fractals* **2020**, *135*, 109794. [CrossRef]
27. Khan, M.A.; Atangana, A.; Alzahrani, E. The dynamics of COVID-19 with quarantined and isolation. *Adv. Differ. Equ.* **2020**, *2020*, 425. [CrossRef]
28. Schmidt, A.; Gaul, L. On the numerical evaluation of fractional derivatives in multi-degree-of-freedom systems. *Signal Process.* **2006**, *86*, 2592–2601. [CrossRef]
29. van den Driessche, P.; Watmough, J. Reproduction numbers and the sub-threshold endemic equilibria for compartmental models of disease transmission. *Math. Biosci.* **2002**, *180*, 29–48. [CrossRef]
30. Li, H.; Cheng, J.; Li, H.B.; Zhong, S.M. Stability analysis of a fractional-order linear system described by the Caputo–Fabrizio derivative. *Mathematics* **2019**, *7*, 200. [CrossRef]
31. Wojtak, W.; Silva, C.J.; Torres, D.F. Uniform asymptotic stability of a fractional tuberculosis model. *Math. Model. Nat. Phenom.* **2018**, *13*, 9. [CrossRef]
32. Li, C.; Zeng, F. The finite difference methods for fractional ordinary differential equations. *Numer. Funct. Anal. Optim.* **2013**, *34*, 149–179. [CrossRef]
33. Diethelm, K.; Luchko, Y. Numerical solution of linear multi-term initial value problems of fractional order. *J. Comput. Anal. Appl.* **2004**, *6*, 243–263.
34. Diethelm, K.; Ford, N.J.; Freed, A.D. A predictor-corrector approach for the numerical solution of fractional differential equations. *Nonlinear Dyn.* **2002**, *29*, 3–22. [CrossRef]
35. Pieskä, J.; Laitinen, E.; Lapin, A. Predictor-corrector methods for solving continuous casting problem. In *Domain Decomposition Methods in Science and Engineering;* Springer: Berlin/Heidelberg, Germany, 2005; pp. 677–684.
36. Butcher, J.C. Numerical methods for ordinary differential equations in the 20th century. *J. Comput. Appl. Math.* **2000**, *125*, 1–29. [CrossRef]
37. Li, C.; Tao, C. On the fractional Adams method. *Comput. Math. Appl.* **2009**, *58*, 1573–1588. [CrossRef]
38. Abboubakar, H.; Kumar, P.; Rangaig, N.A.; Kumar, S. A Malaria Model with Caputo-Fabrizio and Atangana-Baleanu Derivatives. *Int. J. Model. Simul. Sci. Comput.* **2021**, *12*, 2150013. [CrossRef]
39. Garrappa, R. Predictor-Corrector PECE Method for Fractional Differential Equations. MATLAB, Central File Exchange [File ID: 32918]. Available online: https://www.mathworks.com/matlabcentral/fileexchange/32918-predictor-corrector-pece-method-for-fractional-differential-equations (accessed on 23 September 2021).
40. Lakshmikantham, V.; Leela, S.; Martynyuk, A.A. *Stability Analysis of Non Linear Systems;* Springer: Berlin/Heidelberg, Germany, 1989.

fractal and fractional

MDPI

Article

A New Three-Step Root-Finding Numerical Method and Its Fractal Global Behavior

Asifa Tassaddiq [1,†], Sania Qureshi [2,3,*,†], Amanullah Soomro [2,†], Evren Hincal [3,†], Dumitru Baleanu [4,5,†] and Asif Ali Shaikh [2,†]

1 Department of Basic Sciences and Humanities, College of Computer and Information Sciences, Majmaah University, Al-Majmaah 11952, Saudi Arabia; a.tassaddiq@mu.edu.sa
2 Department of Basic Sciences and Related Studies, Mehran University of Engineering and Technology, Jamshoro 76062, Pakistan; soomroamanullah820@gmail.com (A.S.); asif.shaikh@faculty.muet.edu.pk (A.A.S.)
3 Department of Mathematics, Near East University TRCN, Mersin 10, 99138 Nicosia, Turkey; evren.hincal@neu.edu.tr
4 Department of Mathematics, Cankaya University, 06530 Ankara, Turkey; dumitru.baleanu@gmail.com
5 Institute of Space Sciences, Magurele, R76900 Bucharest, Romania
* Correspondence: sania.qureshi@faculty.muet.edu.pk
† These authors contributed equally to this work.

Abstract: There is an increasing demand for numerical methods to obtain accurate approximate solutions for nonlinear models based upon polynomials and transcendental equations under both single and multivariate variables. Keeping in mind the high demand within the scientific literature, we attempt to devise a new nonlinear three-step method with tenth-order convergence while using six functional evaluations (three functions and three first-order derivatives) per iteration. The method has an efficiency index of about 1.4678, which is higher than most optimal methods. Convergence analysis for single and systems of nonlinear equations is also carried out. The same is verified with the approximated computational order of convergence in the absence of an exact solution. To observe the global fractal behavior of the proposed method, different types of complex functions are considered under basins of attraction. When compared with various well-known methods, it is observed that the proposed method achieves prespecified tolerance in the minimum number of iterations while assuming different initial guesses. Nonlinear models include those employed in science and engineering, including chemical, electrical, biochemical, geometrical, and meteorological models.

Keywords: nonlinear models; efficiency index; computational cost; Halley's method; basin of attraction; computational order of convergence

MSC: 65H04; 68W05

Citation: Tassaddiq, A.; Qureshi, S.; Soomro, A.; Hincal, E.; Baleanu, D; Shaikh, A.A. A New Three-Step Root-Finding Numerical Method and Its Fractal Global Behavior. *Fractal Fract.* **2021**, *5*, 204. https://doi.org/10.3390/fractalfract5040204

Academic Editor: Carlo Cattani

Received: 6 September 2021
Accepted: 27 October 2021
Published: 8 November 2021

1. Introduction

The study of iterative methods for solving nonlinear equations and systems appears to be a very important area in theory and practice. Such problems appear not only in applied mathematics but also in many branches of science including engineering (design of an electric circuit), physics (pipe friction), chemistry (greenhouse gases and rainwater), biology (steady-state of Lotka–Volterra system), fluid dynamics (combined conductive–radiative heat transfer), environmental engineering (oxygen level in a river downstream from a sewage discharge), finance (option pricing), and many more. The study of nonlinear models is a vital area of research in numerical analysis. Interesting applications in pure and applied sciences can be studied in the general construction of the nonlinear equations expressed in the form $g(x) = 0$. Due to their significance, several iterative methods have been devised under certain situations since it is near to impossible to obtain exact solutions of models that are nonlinear in nature. These iterative methods have been constructed using different existing methods such as Taylor expansion, the perturbation method, the homotopy perturbation method, Adomian decomposition method, quadrature formula, multi-point iterative methods, the Steffensen-type methods adapted to multidimensional cases, and the variational iteration method. For detailed information, see [1–8]. Among existing iterative methods, the optimal methods are considered those that satisfy the condition for an order of convergence of 2^{k-1}, where k stands for the number of function evaluations per iteration as suggested in [9]. In this way, Newton's classical method $x_{n+1} = x_n - \frac{g(x_n)}{g'(x_n)}$ is the optimal method with quadratic convergence. Various attempts have been made to improve the efficiency of Newton's classical method in past and recent research, as can be seen in [10–15] and most of the references cited therein.

2. Materials and Methods

In a general form, the uni-variate nonlinear equation can be expressed as $g(x) = 0$, where x is the desired quantity. It is extremely difficult to solve the nonlinear equation to find the value of x. Therefore, we attempt to devise a new, highly convergent iterative method to obtain an accurate approximate x that could yield the smallest possible error in the numerical solution. Before we continue with a discussion and derivation of the proposed method, we present some of the methods that are frequently used in the available literature. Later, we use these methods to compare the results obtained under these methods and the results obtained via the method we plan to propose.

2.1. Some Existing Methods

The iterative method ,called the Newton Rahpson method NR_2 [1,16,17] with quadratic convergence, is shown below and uses two function evaluations: one for the function $g(x)$ itself and 1 for the first derivative $g'(x)$:

$$x_{n+1} = x_n - \frac{g(x_n)}{g'(x_n)}, \; n = 0, 1, 2, \ldots, \tag{1}$$

where $g'(x_n) \neq 0$.

In [2], the authors proposed an iterative method with fifth-order convergence as abbreviated by MHM_5. The method requires four function evaluations per iteration: two for the function itself and two first derivatives. The computational steps for the two-step method MHM_5 is described below:

$$\begin{aligned} y_n &= x_n - \frac{g(x_n)}{g'(x_n)}, \\ x_{n+1} &= y_n - \frac{g'(y_n) + 3g'(x_n)}{5g'(y_n) - g'(x_n)} \cdot \frac{g(y_n)}{g'(x_n)}, \end{aligned} \quad n = 0, 1, 2, \ldots, \tag{2}$$

where $g'(x_n) \neq 0$ and $5g'(y_n) \neq g'(x_n)$.

An efficient three-step iterative method with sixth-order convergence is proposed in [3]. This method is the combination of two different methods from [1,18] with second and third-order convergence, respectively. The method requires five function evaluations: three evaluations of the function itself and two evaluations of the first-order derivative per iteration. We represent the method as HM_6. The computational steps of the method can be described as follows:

$$\begin{aligned} y_n &= x_n - \frac{g(x_n)}{g'(x_n)}, \\ z_n &= y_n - \frac{g(y_n)}{g'(y_n)}, \\ x_{x+1} &= y_n - \frac{g(y_n) + g(z_n)}{g'(y_n)}. \end{aligned} \quad n = 0, 1, 2, \ldots, \tag{3}$$

The three-step method with eighth-order convergence as denoted by WO_8 is proposed in [19]. The method requires four function evaluations: three evaluations of the function itself and one evaluation of the first-order derivative per iteration. The computational steps of WO_8 can be described as follows:

$$\begin{aligned} y_n &= x_n - \frac{g(x_n)}{g'(x_n)}, \\ z_n &= x_n - \frac{g(x_n)}{g'(x_n)} \frac{4g(x_n)^2 - 5g(x_n)g(y_n) - g(y_n)^2}{4g(x_n)^2 - 9g(x_n)g(y_n)}, \qquad n = 0,1,2,\ldots, \\ x_{n+1} &= z_n - \frac{g(z_n)}{g'(x_n)}[1 + 4\frac{g(z_n)}{g(x_n)}][\frac{8g(y_n)}{4g(x_n) - 11g(y_n)} + 1 + \frac{g(z_n)}{g(y_n)}]. \end{aligned} \tag{4}$$

The iterative method with ninth-order convergence can be seen in [20]. The method requires five function evaluations: three evaluations of function itself and two evaluations of the first-order derivative per iteration. This method is abbreviated as NM_9. The computational steps of the method are described as follows:

$$\begin{aligned} y_n &= x_n - \frac{g(x_n)}{g'(x_n)}, \\ z_n &= y_n - [1 + (\frac{g(y_n)}{g(x_n)})^2]\frac{g(y_n)}{g'(y_n)}, \qquad n = 0,1,2,\ldots, \\ x_{n+1} &= z_n - [1 + 2(\frac{g(y_n)}{g(x_n)})^2 + 2\frac{g(z_n)}{g(y_n)}]\frac{g(z_n)}{g'(y_n)}. \end{aligned} \tag{5}$$

The predictor–corrector modified Householder's method with tenth order convergence, as denoted by MH_{10}, is proposed in [21]. The method is free from the second derivative and requires only five function evaluations per iteration: three evaluations of the function itself and two evaluations of the first-order derivative. The computational steps of MH_{10} can be described as

$$\begin{aligned} y_n &= x_n - \frac{g(x_n)}{g'(x_n)}, \\ z_n &= y_n - \frac{g(y_n)}{g'(y_n)} - \frac{g^2(y_n)P(y_n)}{2g'^3(y_n)}, \qquad n = 0,1,2,\ldots, \\ x_{n+1} &= z_n - \frac{g(z_n)}{g[z_n, y_n] + (z_n - y_n)g[z_n, y_n, y_n]}, \end{aligned} \tag{6}$$

where

$$\begin{aligned} P(y_n) &= g''(y_n) = \frac{2}{x_n - y_n}[3\frac{g(x_n) - g(y_n)}{x_n - y_n} - 2g'(y_n) - g'(x_n)], \\ g[z_n, y_n] &= \frac{g(z_n) - g(y_n)}{z_n - y_n}, \\ g[z_n, y_n, y_n] &= \frac{g[z_n, y_n] - g'(y_n)}{z_n - y_n}. \end{aligned} \tag{7}$$

2.2. Proposed Iterative Method

There are many recent research studies wherein researchers have presented modified iterative methods to solve nonlinear models of the form $g(x) = 0$. In some of these methods, modification is based on the idea of combining two existing methods to develop a new method with a better order of convergence. After being motivated by such an idea as observed in [3,22–24], we have developed a new method with tenth-order convergence via the blending of two different iterative methods with second (Newton method) and fifth-order (modified Halley method) convergence as given in [1,2], respectively. We recommended the higher-order convergent method of the convergence order $2 \times 5 = 10$. The choice of the methods in the present work is suitable because the resultant iterative method with tenth order convergence uses only 6 function evaluations (3 function evaluations and three evaluations of the first-order derivative) per iteration. It may be noted that the choice of blending of methods is extremely important to avoid extra function evaluations that could bring additional computational cost, as can be seen in [25,26]. Hence, the proposed iterative method not only confirms higher-order convergence but also employs fewer function evaluations, as can be described by the following proposed three-step method abbreviated as PM_{10}:

$$\begin{aligned} y_n &= x_n - \frac{g(x_n)}{g'(x_n)}, \\ z_n &= y_n - \frac{g(y_n)}{g'(y_n)}, \qquad\qquad n = 0, 1, 2, \ldots, \\ x_{x+1} &= z_n - \frac{g'(z_n) + 3g'(y_n)}{5g'(z_n) - g'(y_n)} \left(\frac{g(z_n)}{g'(y_n)} \right). \end{aligned} \tag{8}$$

The proposed iterative three-step method given in (8) is discussed in the flowchart presented in Figure 1.

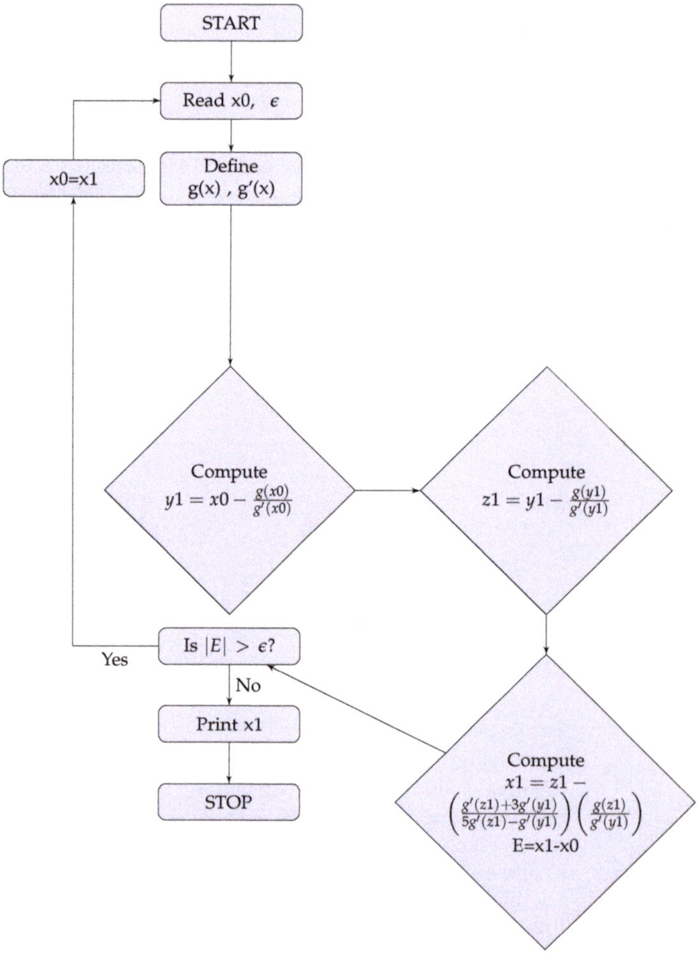

Figure 1. Flowchart of the proposed three-step method given in (8).

Furthermore, the efficiency index e is also computed for the proposed iterative method as $10^{1/6} \approx 1.4678$, whereas it would generally be shown as $10^{1/3(n+n^2)}$ for $n \geq 1$. The following Table 1 and the Figure 2 can be consulted for the computation and comparison of all iterative methods taken for comparison in the present research work. Although the function evaluations (FV) of PM_{10} in the Table 1 seem to be more than some of the methods under consideration, to achieve the desired accuracy regarding the performance of the latter under different initial guesses (IG), the number of iterations (N) and CPU time (seconds) are better than most of the methods. This discussion is presented in Section 5.

Table 1. Comparison of efficiency indices for methods under consideration.

Method	Order	FV	EI	New Function Evaluations per Iteration for $n \geq 1$.
PM_{10}	10	6	1.4678×10^0	$3(n+n^2)$
MH_{10}	10	5	1.5849×10^0	$3n+2n^2$
NM_9	9	5	1.5518×10^0	$3n+2n^2$
WO_8	8	4	1.6818×10^0	$3n+n^2$
HM_6	6	5	1.4310×10^0	$3n+2n^2$
MHM_5	5	4	1.4953×10^0	$2(n+n^2)$
NR_2	2	2	1.4142×10^0	$n+n^2$

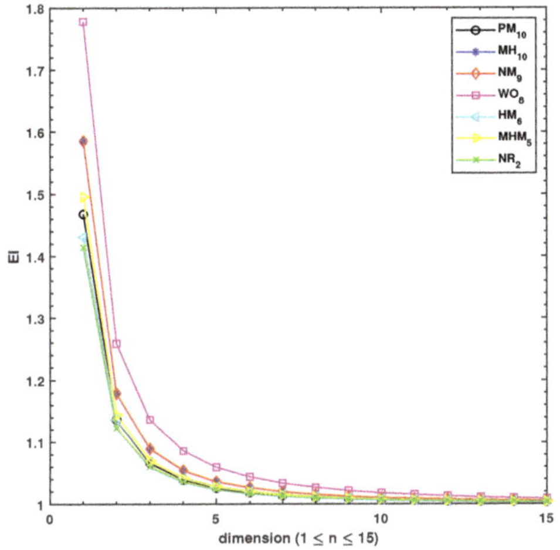

Figure 2. Behavior of efficiency index of various iterative methods for increasing dimensions of the nonlinear problem.

3. Convergence Analysis

This section has been devoted to the proof of local convergence analysis for the proposed tenth-order method under both scalar and vector (systems) cases. Single and multivariate Taylor's series expansion have been used to obtain the required order of local convergence. It is worth noting that the convergence analysis is addressed in a similar manner to many other existing articles, and the main interest in developing higher-order methods is of the academic type. Even if higher-order methods are more complicated, the efficiency can be measured, and this is why we have included the CPU time, as found in Section 5. The theorems stated below are later used for the theoretical analysis of the convergence.

Theorem 1. *(Single real variable Taylor's series expansion): Suppose that $r \geq 1$ is an integer and further suppose that $g : \mathbb{R} \to \mathbb{R}$ is an r-times differentiable function at some finite point $\alpha \in \mathbb{R}$. Then, there exists the following expression:*

$$g(x_n) = g(\alpha) + g'(\alpha)\delta_n + \frac{1}{2!}g''(\alpha)\delta_n^2 + \cdot + \frac{1}{r!}g^{(r)}(\alpha)\delta_n^r + R_r(x_n), \tag{9}$$

where $R_r(x)$ is the remainder term, whose integral form is

$$R_r(x_n) = \int_{\alpha}^{x_n} \frac{1}{(r+1)!}g^{(r+1)}(t)(x_n - t)^r dt. \tag{10}$$

Theorem 2. *(Multivariable Taylor's series expansion): Suppose that $\mathbf{G} : P \subseteq \mathbb{R}^n \to \mathbb{R}^n$ is an r-times Frechet differentiable system of functions in a convex set $P \subseteq \mathbb{R}^n$; then, for any $\boldsymbol{x}$ and $k \in \mathbb{R}^n$ the equation given below is true:*

$$\mathbf{G}(\boldsymbol{x}+k) = \mathbf{G}(\boldsymbol{x}) + k\mathbf{G}'(\boldsymbol{x}) + \frac{k^2}{2!}\mathbf{G}''(\boldsymbol{x}) + \frac{k^3}{3!}\mathbf{G}''(\boldsymbol{x}) + \ldots + \frac{k^{r-1}}{r!}\mathbf{G}^{(r-1)}(\boldsymbol{x}) + R_r, \tag{11}$$

where $||R_r|| \leq \frac{1}{r!}\sup_{0<t<1}||\mathbf{G}^{(r)}(\boldsymbol{x}+kt)||||k||^r$ and $\mathbf{G}^{(q)}(\boldsymbol{x})k^q = (\ldots(\mathbf{G}^{(q)}(\boldsymbol{x})k).\overset{q}{.}.)k \in \mathbb{R}^n$.

3.1. Convergence under Scalar Case

In this subsection, we theoretically prove the local order of convergence for PM_{10}.

Theorem 3. *Suppose that $\alpha \in P$ is the required simple root for a differentiable function $g : P \subseteq \mathbb{R} \to \mathbb{R}$ within an open real interval P. Then, the proposed three-step numerical method (8) possesses tenth-order convergence, and the asymptotic error term is given by*

$$\epsilon_{x_{n+1}} = \frac{g''^7(\alpha)}{4096g'^9(\alpha)}\Big(52g'(\alpha)g'''(\alpha) - 35g''^2(\alpha)\Big)\epsilon_{x_n}^{10} + \mathcal{O}(\epsilon_{x_n}^{11}). \tag{12}$$

Proof. Suppose α is the root of $g(x_n)$, where x_n is the nth approximation for the root by the proposed method (8), and $\epsilon_{x_n} = x_n - \alpha$ is the error term in variable x at the nth iteration step. Employing the single real variable Taylor's series given in the theorem (1) for $g(x_n)$ around α, we obtain

$$g(x_n) = g'(\alpha)\epsilon_{x_n} + \frac{1}{2!}g''(\alpha)\epsilon_{x_n}^2 + \mathcal{O}(\epsilon_{x_n}^3). \tag{13}$$

Similarly, using the Taylor's series for $1/g'(x_n)$ around α, we obtain

$$\frac{1}{g'(x_n)} = \frac{1}{g'(\alpha)} - \frac{g''(\alpha)}{g'^2(\alpha)}\epsilon_{x_n} + \frac{1}{g'(\alpha)}\Big(\frac{g''^2(\alpha)}{g'^2(\alpha)} - \frac{g'''(\alpha)}{2g'(\alpha)}\Big)\epsilon_{x_n}^2 + \mathcal{O}(\epsilon_{x_n}^3). \tag{14}$$

Multiplying (13) and (14), we obtain

$$\frac{g(x_n)}{g'(x_n)} = \epsilon_{x_n} - \frac{g''(\alpha)}{2g'(\alpha)}\epsilon_{x_n}^2 + \Big(\frac{g''^2(\alpha)}{2g'^2(\alpha)} - \frac{g'''(\alpha)}{2g'(\alpha)}\Big)\epsilon_{x_n}^3 + \mathcal{O}(\epsilon_{x_n}^4). \tag{15}$$

Now, substituting (15) in the first step of (8), we obtain

$$\epsilon_{y_n} = \frac{g''(\alpha)}{2g'(\alpha)}\epsilon_{x_n}^2 + \frac{1}{2g'^2(\alpha)}\Big(g'(\alpha)g'''(\alpha) - g''^2(\alpha)\Big)\epsilon_{x_n}^3 + \mathcal{O}(\epsilon_{x_n}^4), \tag{16}$$

where $\epsilon_{y_n} = y_n - \alpha$. Using the Taylor's series (1) for $g(y_n)$ around α, we obtain

$$g(y_n) = g'(\alpha)\epsilon_{y_n} + \frac{1}{2!}g''(\alpha)\epsilon_{y_n}^2 + \mathcal{O}(\epsilon_{y_n}^3). \tag{17}$$

Similarly, using the Taylor's series for $\dfrac{1}{g'(y_n)}$ around α, we obtain

$$\frac{1}{g'(y_n)} = \frac{1}{g'(\alpha)} - \frac{g''(\alpha)}{g'^2(\alpha)}\epsilon_{y_n} + \frac{1}{g'(\alpha)}\Big(\frac{g''^2(\alpha)}{g'^2(\alpha)} - \frac{g'''(\alpha)}{2g'(\alpha)}\Big)\epsilon_{y_n}^2 + \mathcal{O}(\epsilon_{y_n}^3). \tag{18}$$

Multiplying (17) and (18), we obtain

$$\frac{g(y_n)}{g'(y_n)} = \epsilon_{y_n} - \frac{g''(\alpha)}{2g'(\alpha)}\epsilon_{y_n}^2 + \Big(\frac{g''^2(\alpha)}{2g'^2(\alpha)} - \frac{g'''(\alpha)}{2g'(\alpha)}\Big)\epsilon_{y_n}^3 + \mathcal{O}(\epsilon_{y_n}^4). \tag{19}$$

Now, substituting (19) in the second step of (8), we obtain

$$\epsilon_{z_n} = \frac{g''(\alpha)}{2g'(\alpha)}\epsilon_{y_n}^2 + \frac{1}{2g'^2(\alpha)}\Big(g'(\alpha)g'''(\alpha) - g''^2(\alpha)\Big)\epsilon_{y_n}^3 + \mathcal{O}(\epsilon_{y_n}^4), \tag{20}$$

where $\epsilon_{z_n} = z_n - \alpha$. Using the Taylor's series (1) for $g(z_n)$ around α, we obtain

$$g(z_n) = g'(\alpha)\epsilon_{z_n} + \frac{1}{2}g''(\alpha)\epsilon_{z_n}^2 + \mathcal{O}(\epsilon_{z_n}^3). \tag{21}$$

Using the Taylor's series (1) for $g'(y_n)$ around α, we obtain

$$g'(y_n) = g'(\alpha) + g''(\alpha)\epsilon_{y_n} + \frac{1}{2}g'''(\alpha)\epsilon_{y_n}^2 + \mathcal{O}(\epsilon_{y_n}^3). \tag{22}$$

Using the Taylor's series (1) for $g'(z_n)$ around α, we obtain

$$g'(z_n) = g'(\alpha) + g''(\alpha)\epsilon_{z_n} + \frac{1}{2}g'''(\alpha)\epsilon_{z_n}^2 + +\mathcal{O}(\epsilon_{z_n}^3). \tag{23}$$

Expanding the Taylor series $\dfrac{1}{5g'(z_n) - g'(y_n)}$ and using Equations (22) and (23), we obtain

$$\begin{aligned}\frac{g'(z_n) + 3g'(y_n)}{5g'(z_n) - g'(y_n)} = 1 &+ \frac{g''(\alpha)}{g'(\alpha)}\Big(\epsilon_{y_n} - \epsilon_{z_n}\Big) + \left(\frac{g'''(\alpha)}{2g'(\alpha)} + \frac{g''^2(\alpha)}{4g'^2(\alpha)}\right)\epsilon_{y_n}^2 - \left(\frac{g'''(\alpha)}{2g'(\alpha)} - \frac{5g''^2(\alpha)}{4g'^2(\alpha)}\right)\epsilon_{z_n}^2 \\ &- \frac{3g''^2(\alpha)}{2g'^2(\alpha)}\epsilon_{y_n}\epsilon_{z_n} + \mathcal{O}(\epsilon_{z_n}^3).\end{aligned} \tag{24}$$

Finally, substituting (24) in the third step of (8) and using Equations (16), (18), (20) and (21), we obtain

$$\epsilon_{x_{n+1}} = \frac{g''^7(\alpha)}{4096g'^9(\alpha)}\Big(52g'(\alpha)g'''(\alpha) - 35g''^2(\alpha)\Big)\epsilon_{x_n}^{10} + \mathcal{O}(\epsilon_{x_n}^{11}). \tag{25}$$

Hence, the tenth-order convergence of the proposed method PM_{10} given by (8) for $g(x) = 0$ is proved. □

3.2. Convergence under Vector Case

This subsection extends the proof for the tenth-order convergence of the proposed method PM_{10} given in (8) regarding solving the system of nonlinear functions $\boldsymbol{G}(\boldsymbol{x}) = \boldsymbol{0}$, where $\boldsymbol{G} = [g_1(\boldsymbol{x}), g_2(\boldsymbol{x}), \ldots, g_n(\boldsymbol{x})]'$ from $\boldsymbol{R}^n$ to $\boldsymbol{R}^n$. For the system of nonlinear functions, PM_{10} can be described as follows:

$$
\begin{aligned}
y_n &= x_n - G'(x_n)^{-1}G(x_n),\\
z_n &= y_n - G'(y_n)^{-1}G(y_n), \qquad\qquad\qquad\qquad\qquad\qquad\qquad n = 0,1,2,\ldots.\\
x_{n+1} &= z_n - \left[5G'(z_n) - G'(y_n)\right]^{-1}\left[G'(z_n) + 3G'(y_n)\right]\left[G'(y_n)^{-1}G(z_n)\right].
\end{aligned}
\tag{26}
$$

We present the following theorem to obtain the asymptotic error term and the order of convergence for PM_{10}.

Theorem 4. *Let the function $G : P^n \subset R^n \rightarrow R^n$ be sufficiently differentiable in a convex set P^n containing the zero α of $G(x)$. Let us consider that $G'(x)$ is continuous and nonsingular in α. Then, the solution x obtained by using proposed three-step method PM_{10} converging to α has tenth-order convergence, if an initial guess x_0 is chosen close to α.*

Proof. Suppose that $\epsilon_{xn} = ||x_n - \alpha||$. Now, using the Taylor series described in the Theorem (2) for $G(x_n)$, we obtain

$$G(x_n) = G'(\alpha)\epsilon_{xn} + \frac{1}{2!}G''(\alpha)\epsilon_{xn}^2 + \mathcal{O}(\epsilon_{xn}^3). \tag{27}$$

Similarly, using the Taylor's series for $G'(x_n)$ around α, we obtain

$$G'(x_n) = G'(\alpha) + G''(\alpha)\epsilon_{xn} + \frac{1}{2!}G'''(\alpha)\epsilon_{xn}^2 + \mathcal{O}(\epsilon_{xn}^3). \tag{28}$$

Employing the Taylor's series for the inverted Jacobian matrix $G'(x_n)^{-1}$ around α, we obtain

$$
\begin{aligned}
G'(x_n)^{-1} &= G'(\alpha)^{-1} - G'^2(\alpha)^{-1}G''(\alpha)\epsilon_{xn} +\\
G'(\alpha)^{-1}&\left[G'^2(\alpha)^{-1}G''^2(\alpha) - 2G'(\alpha)^{-1}G'''(\alpha)\right]\epsilon_{xn}^2 + \mathcal{O}(\epsilon_{xn}^3).
\end{aligned}
\tag{29}
$$

Multiplying (27) and (29), we obtain

$$G'(x_n)^{-1}G(x_n) = \epsilon_{xn} - 2G'(\alpha)^{-1}G''(\alpha)\epsilon_{xn}^2 + \left[2G'^2(\alpha)^{-1}G''^2(\alpha) - 2G'(\alpha)^{-1}G'''(\alpha)\right]\epsilon_{xn}^3 + \mathcal{O}(\epsilon_{xn}^4). \tag{30}$$

Now, substituting (30) in the first step of (26), we obtain

$$\epsilon_{y_n} = 2G'(\alpha)^{-1}G''(\alpha)\epsilon_{xn}^2 + 2G'^2(\alpha)^{-1}\left[G'(\alpha)G'''(\alpha) - G''^2(\alpha)\right]\epsilon_{xn}^3 + \mathcal{O}(\epsilon_{xn}^4). \tag{31}$$

Similarly, employing the Taylor series for $G(y_n)$ and $G'(y_n)$ about α, and also for the inverted Jacobian matrix $G'(y_n)^{-1}$ in the second step of (26), we obtain

$$\epsilon_{zn} = 2G(\alpha)^{-1}G''(\alpha)\epsilon_{y_n}^2 + 2G'^2(\alpha)^{-1}\left[G'(\alpha)G'''(\alpha) - G''^2(\alpha)\right]\epsilon_{y_n}^3 + \mathcal{O}(\epsilon_{y_n}^4). \tag{32}$$

Using the Taylor series for $G'(z_n)$ and $G'(y_n)$ around α, we obtain the following:

$$G'(z_n) = G'(\alpha) + G''(\alpha)\epsilon_{zn} + \frac{1}{2}G'''(\alpha)\epsilon_{zn}^2 + \mathcal{O}(\epsilon_{zn}^3), \tag{33}$$

$$G'(y_n) = G'(\alpha) + G''(\alpha)\epsilon_{y_n} + \frac{1}{2}G'''(\alpha)\epsilon_{y_n}^2 + \mathcal{O}(\epsilon_{y_n}^3). \tag{34}$$

Using the above two equations and the inverted Jacobian matrix for $[5G'(z_n) - G'(y_n)]^{-1}$, we obtain

$$
\begin{aligned}
&[5G'(z_n) - G'(y_n)]^{-1}[G'(z_n) + 3G'(y_n)] = 1 - G'(\alpha)^{-1}G''(\alpha)\epsilon_{zn} - 2G'(\alpha)^{-1}G'''(\alpha)\epsilon_{zn}^2 +\\
&G'(\alpha)^{-1}G''(\alpha)\epsilon_{y_n} + \left[2G'(\alpha)^{-1}G'''(\alpha) + 4G'^2(\alpha)^{-1}G''^2(\alpha)\right]\epsilon_{y_n}^2 + 4G'(\alpha)^{-1}5G''^2(\alpha)\epsilon_{zn}^2\\
&- 2G'^2(\alpha)^{-1}3G''^2(\alpha)\epsilon_{y_n}\epsilon_{zn} + \mathcal{O}(\epsilon_{zn}^3).
\end{aligned}
\tag{35}
$$

Finally, substituting (35) and other required expansions in the third step of (26), we obtain

$$\epsilon_{z_{n+1}} = \frac{1}{4096} G'^{9}(\alpha)^{-1} G''^{7}(\alpha)\left[52G'(\alpha)G'''(\alpha) - 35G''^{2}(\alpha)\right]\epsilon_{z_n}^{10} + \mathcal{O}(\epsilon_{z_n}^{11}). \quad (36)$$

Hence, the tenth-order convergence of the proposed method PM_{10} given by (26) for a non-linear system of equations $\boldsymbol{G}(\boldsymbol{x}) = \boldsymbol{0}$ is proved. □

3.3. Computational Estimation of the Convergence Order

When a new iterative method is proposed to compute an approximate solution for $g(x) = 0$, in order to verify its theoretical local order of convergence, one needs to use a parameter called the Computational Order of Convergence (COC). However, this is possible only when we have information about the exact root α for $g(x) = 0$. Thus, the following parameters can alternatively be employed under some constraints as described below.

Approximated Computational Order of Convergence [27]:

$$ACOC = \frac{log|\epsilon_n/\epsilon_{n-1}|}{log|\epsilon_{n-1}/\epsilon_{n-2}|}, \quad \epsilon_n = x_n - x_{n-1}, \quad n \geq 4. \quad (37)$$

Computational Order of Convergence [28]:

$$COC = \frac{log|\bar{\epsilon}_n/\bar{\epsilon}_{n-1}|}{log|\bar{\epsilon}_{n-1}/\bar{\epsilon}_{n-2}|}, \quad \bar{\epsilon} = x_n - \alpha, \quad n \geq 3. \quad (38)$$

Extrapolated Computational Order of Convergence [29]:

$$ECOC = \frac{log|\hat{\epsilon}_n/\hat{\epsilon}_{n-1}|}{log|\hat{\epsilon}_{n-1}/\hat{\epsilon}_{n-2}|}, \quad \hat{\epsilon} = x_n - \alpha_n, \quad n \geq 5, \quad (39)$$

where

$$\alpha_n = x_n - \frac{(\rho x_{n-1})^2}{\rho^2 x_{n-1}}, \quad \rho x_n = x_{n+1} - x_n.$$

Petkovic Computational Order of Convergence [30]:

$$PCOC = \frac{log|\widehat{\epsilon}_n|}{log|\widehat{\epsilon}_{n-1}|}, \quad \widehat{\epsilon}_n = \frac{f(x_n)}{f(x_{n-1})}, \quad n \geq 2. \quad (40)$$

All of the above formulas can be used to test the convergence order, but in the present study, ACOC as given by (37) is used since the number of iterations taken by the method PM_{10} (8) is at least four in various numerically tested problems, as discussed in Section 5. Additionally, this is the best-known approach employed in most of the recently conducted research studies to verify the theoretical order of convergence.

4. Basins of Attraction

The stability of solutions (roots) for the nonlinear function $g(z) = 0$ using an iterative method can be analyzed with the help of a concept called the basins of attraction. Basins of attractions are phase-planes that demonstrate iterations employed by the iterative method, which can assume different choices for the initial guess z_0. Such 2D regions are esthetically so beautiful that their applications are not only found in applied mathematics, but also people working in fields such as architecture, arts, and design also use the concept to obtain pleasing designs. Many other fields of applications for these basins can be seen in [31–36] and most of the references cited therein. It may be noted that linear models do not depict such dynamically eye-catching behavior, whereas the non-linearity results in features such as those seen herein under the proposed iterative method PM_{10}. MATLAB's built-in routines, including contour, colormap, and color bar, are utilized to obtain the basins of attraction in the present study. In this connection, a squared region R on $[-4,4] \times [-4,4]$ containing 2000 by 2000 mesh points is selected for the selected functions in complex form. Some of these complex-valued functions are taken as follows for the illustration of the regions by the proposed method.

Example 1.

$$P_1(z) = z^8 + z^5 - 4, \quad P_2(z) = z^7 - 1, \quad P_3(z) = z^3 - 1,$$
$$P_4(z) = cos(z) + cos(2z) + z, \quad P_5(z) = \exp(z) - z, \quad P_6(z) = \cosh(z) - 1. \tag{41}$$

To maintain diversity, different kinds of functions including polynomials and transcendental functions are used. The regions are achieved with a tolerance of $\epsilon = 10^{-2}$ and a maximum number of iterations allowed of $n = 12$. As can be seen in Figures 3–8, for the Example (1), the maximum number of iterations is needed by initial guesses that reside near boundaries of the regions, whereas if z_0 lies within the neighborhood of the required solution, the proposed method PM_{10} does not require as many iterations as those needed by most of the methods found in the literature. The average time required to produce the dynamical planes shown in Figures 3–8 is stored in the Table 2. As expected, the complex functions that have exponential and hyperbolic components are computationally expensive.

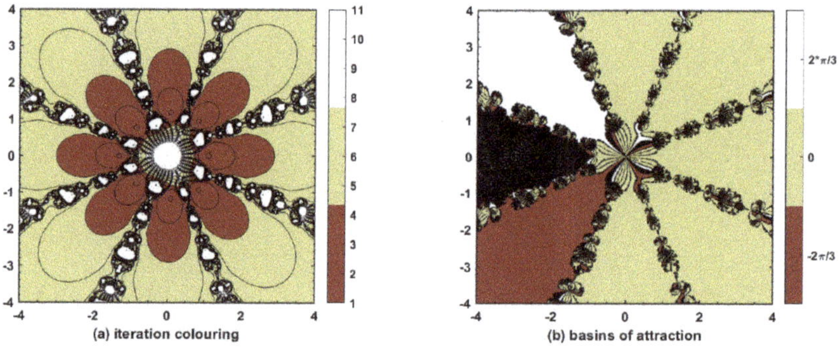

Figure 3. The polynomiographs by the proposed method PM_{10} for $P_1(z)$.

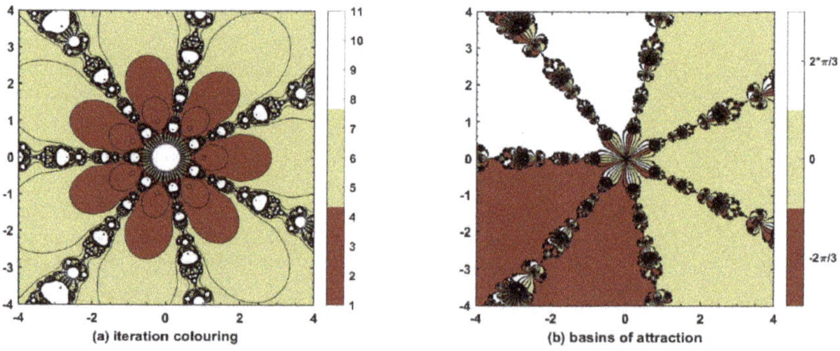

Figure 4. The polynomiographs by the proposed method PM_{10} for $P_2(z)$.

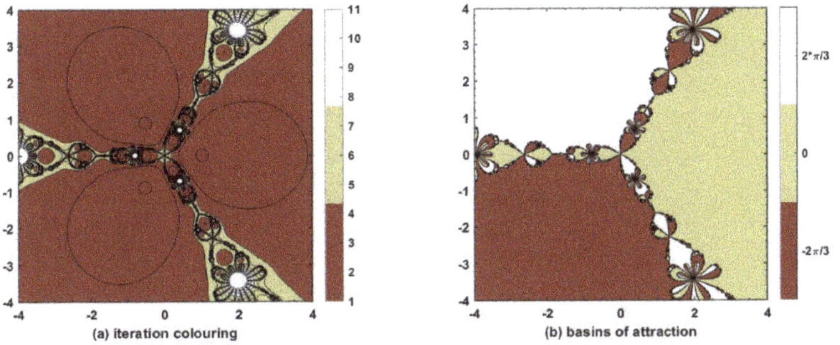

Figure 5. The polynomiographs by the proposed method PM_{10} for $P_3(z)$.

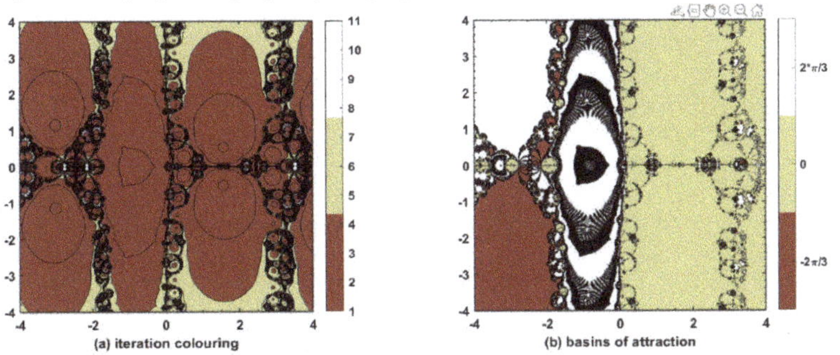

Figure 6. The basins of attractions by the proposed method PM_{10} for $P_4(z)$.

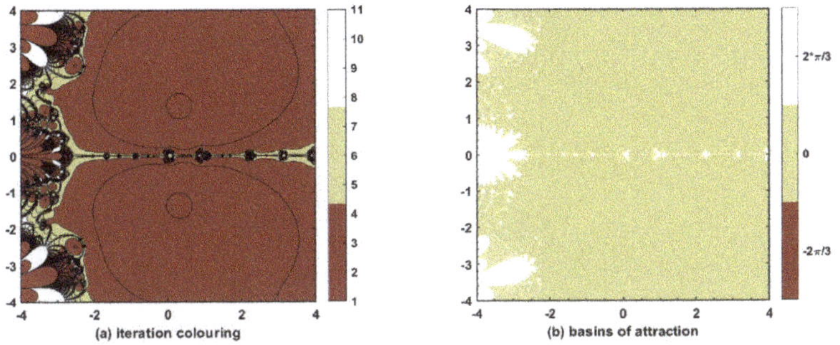

Figure 7. The basins of attractions by the proposed method PM_{10} for $P_5(z)$.

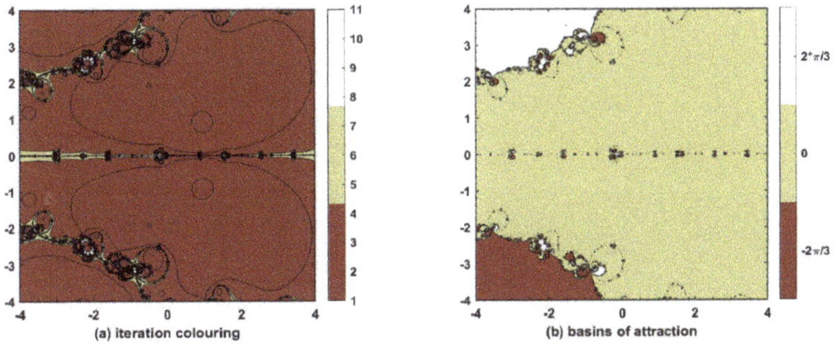

Figure 8. The basins of attractions by the proposed method PM_{10} for $P_6(z)$.

Table 2. Execution time in seconds required by PM_{10} for each $P_i(z)$, $i = 1, 2, \ldots, 6$.

$P_1(z)$	$P_2(z)$	$P_3(z)$	$P_4(z)$	$P_5(z)$	$P_6(z)$
767.413 s	570.103 s	289.626 s	537.531 s	782.588 s	730.300 s

5. Numerical Experiments with Discussion

Various types of test problems are considered from different sources including [3,19,22]. The approximate solutions $x^\star$ up to 50 decimal places are shown against each test function. The error tolerance $|\epsilon_N| = |x_N - x^\star|$ to stop the number of iterations is set to 10^{-200}, whereas the precision is chosen to be as large as 4000, and N shows the total number of iterations taken by the method to achieve the required tolerance. In addition, physically applicable nonlinear models such as the Van der Waals equation, the Shockley ideally diode electric circuit model, the conversion of substances in a batch reactor, and the Lorenz equations in meteorology are taken into consideration to demonstrate the applicability of the proposed method PM_{10}. Obtained numerical results are tabulated for further analysis. Each table contains different initial guesses, numbers of iterations required by a method to achieve the preset error tolerance, function evaluations needed for each method, approximated computational orders of convergence $\left(ACOC = \dfrac{log|\epsilon_n/\epsilon_{n-1}|}{log|\epsilon_{n-1}/\epsilon_{n-2}|}\right)$ where $\epsilon_n = x_n - x_{n-1}$, absolute errors, absolute values of functions at the last iteration, and the execution (CPU) time in seconds.

For the test problem 2 ($g_1(x)$), two initial guesses are chosen for simulations, as can be seen in Table 3. For the initial guess $x_0 = 1.5$, it is observed that the minimum number of iterations is taken by the methods PM_{10} and WO_8 to achieve the error tolerance; however, the smallest error is given by PM_{10} while consuming an equivalent amount of CPU time. The method NR_2 takes the maximum number of iterations at $x_0 = 1.5$. Under the second initial guess $x_0 = 2.0$, although many methods including PM_{10} take the equal number of iterations to achieve $\epsilon = 10^{-200}$, the smallest absolute error and thus smallest absolute functional value is achieved by PM_{10}. This shows that if an initial guess lying near to the solution of $g(x) = 0$ is passed to PM_{10}, then the method yields the smallest error.

For the test problem 2 ($g_2(x)$), two initial guesses are chosen for simulations as can be seen in Table 4. One of them is taken far away from the approximate solution of the quintic equation $g_2(x)$. For $x_0 = -3.8$, the smallest possible absolute error is yielded by WO_8, but at the cost of the maximum number of iterations and largest amount of CPU time. Next comes HM_6 with an absolute error of 9.2097×10^{-670} and greater time efficiency, but it requires one more iteration when compared with PM_{10}, which achieves an absolute error of 1.9515×10^{-572} with only four iterations while consuming a reasonable amount of CPU time. When an initial guess lying near to the root is chosen, the method NM_9 achieves the smallest error, but it requires one extra iteration when compared with PM_{10} and WO_8. The most expensive methods (in terms of N, FV, and CPU time) for this particular test problem seem to be WO_8 and NR_2.

For the transcendental problem 2 ($g_3(x)$), two initial guesses are chosen for simulations, as can be seen in Table 5. Under both of the initial guesses, the maximum numbers of function evaluations are taken by HM_6 followed by PM_{10} to achieve the desired error tolerance. The smallest absolute functional values are obtained with HM_6 and PM_{10}, where NR_2 seems to be the most expensive method in terms of the number of iterations. Although the method MH_{10} consumes the fewest number of CPU seconds, its ACOC is only eight, contradicting the theoretical order of convergence found in [21].

For the transcendental problem 2 ($g_4(x)$), two initial guesses are chosen for simulations as can be seen in Table 6. One of the guesses is taken far away from the approximate solution of $g_4(x)$. Under both initial guesses, it is observed that the method MHM_5 takes the fewest iterations, fewest function evaluations, and fewest CPU seconds with unsatisfactory absolute errors under $x_0 = -9.5$. The method WO_8 diverges for the second initial guess $x_0 = -9.5$, whereas NR_2 is the most expensive method concerning N and

FV, in particular. The proposed method PM_{10} performs reasonably well under the initial guesses and does not diverge under any situation.

For the test problem 2 ($g_5(x)$), three initial guesses are chosen for simulations as can be seen in Table 7. Two of the guesses lie far away from the approximate root of $g_5(x)$. It is easy to observe that the method PM_{10} performs better than other methods, even when the initial guesses are not near to the approximate root, since the number of iterations to attain the required accuracy is the smallest with PM_{10}. Once again, the most expensive method regarding N and FV proves to be NR_2, whereas WO_8 does not succeed towards the desired root when the initial guesses are assumed to be away from the root. The absolute error achieved by PM_{10} with $x_0 = 2.9$ is the smallest when compared with the results of other methods.

Example 2.

$$
\begin{aligned}
&g_1(x) = x^3 - 10,\\
&x^\star \approx 2.1544346900318837217592935665193504952593449421921 0,\\
&g_2(x) = x^5 + x - 10000,\\
&x^\star \approx 6.3087771299726890947675717717830591133775580582111 0,\\
&g_3(x) = \frac{x}{2} - \sin(x),\\
&x^\star \approx 1.8954942670339809471440357380936016917513466273854 0,\\
&g_4(x) = x\exp(x^2) - \sin^2(x) + 3\cos(x) + 5,\\
&x^\star \approx -1.2076478271309189270094167583560840977602358189495 0,\\
&g_5(x) = \exp^{\sin(x)} - x + 1,\\
&x^\star \approx 2.6306641479279036339753270523505985685847319547332 0.
\end{aligned}
\tag{42}
$$

Table 3. Numerical results for problem 2: $g_1(x)$.

Method	IG	N	FV	ACOC	$\lvert\epsilon\rvert$	$\lvert f(x_N)\rvert$	CPU
PM_{10}	1.5	4	24	10	4.3384×10^{-427}	1×10^{-3998}	1.6×10^{-2}
	2.0	4	24	10	7.3775×10^{-1117}	$7.7598 \times 10^{-11,164}$	2.1720×10^{0}
MH_{10}	1.5	5	25	8	6.1001×10^{-1501}	6×10^{-3999}	3.2×10^{-2}
	2.0	4	20	8	8.7875×10^{-538}	7.6607×10^{-4298}	1.3391×10^{1}
NM_9	1.5	5	25	9	1.3799×10^{-1487}	6×10^{-3999}	1.6×10^{-2}
	2.0	4	20	9	2.5853×10^{-772}	6×10^{-3999}	1.6×10^{-2}
WO_8	1.5	4	16	7.9999×10^{0}	3.7895×10^{-250}	5.4086×10^{-1999}	1.6×10^{-2}
	2.0	4	16	8	2.2967×10^{-676}	1×10^{-3998}	1.6×10^{-2}
HM_6	1.5	5	25	6	4.6527×10^{-496}	6.0868×10^{-2973}	3.1×10^{-2}
	2.0	4	20	6	2.7077×10^{-230}	2.3643×10^{-1378}	2.66×10^{-1}
MHM_5	1.5	5	20	5	2.5498×10^{-291}	3.4832×10^{-1454}	3.1×10^{-2}
	2.0	5	20	5	2.7042×10^{-743}	4.6736×10^{-3714}	1.2180×10^{0}
NR_2	1.5	10	20	2	4.7719×10^{-221}	1.4717×10^{-440}	3.1×10^{-2}
	2.0	9	18	2	4.5282×10^{-288}	1.3253×10^{-574}	3.1×10^{-2}

Table 4. Data using fixed step-size problem 2: $g_2(x)$.

Method	IG	N	FV	ACOC	$\|\epsilon\|$	$\|f(x_N)\|$	CPU
PM_{10}	−3.8	4	24	10	1.9515×10^{-572}	0	3.1×10^{-2}
	8.8	4	24	9.9999×10^{0}	1.7260×10^{-277}	6.0030×10^{-2769}	3.1×10^{-2}
MH_{10}	−3.8	4	20	8	4.4858×10^{-276}	1.2527×10^{-2202}	3.2×10^{-2}
	8.8	5	25	8	1.1988×10^{-910}	0	3.1×10^{-2}
NM_9	−3.8	4	20	9	1.3626×10^{-350}	5.8823×10^{-3149}	3.1×10^{-2}
	8.8	5	25	9	3.4812×10^{-1362}	0	1.5×10^{-2}
WO_8	−3.8	11	44	8	2.4211×10^{-933}	0	9.4×10^{-2}
	8.8	4	16	8.0239×10^{0}	1.7219×10^{-242}	1.1652×10^{-1938}	3.2×10^{-2}
HM_6	−3.8	5	25	6	9.2097×10^{-670}	0	3.1×10^{-2}
	8.8	5	25	6	5.0406×10^{-303}	8.3149×10^{-1813}	3.1×10^{-2}
MHM_5	−3.8	5	20	5	8.1401×10^{-297}	2.1439×10^{-1479}	3.2×10^{-2}
	8.8	5	20	5	1.3486×10^{-221}	2.6761×10^{-1103}	3.1×10^{-2}
NR_2	−3.8	10	20	2	2.1403×10^{-292}	1.1503×10^{-580}	3.2×10^{-2}
	8.8	11	22	2	6.8822×10^{-280}	1.1893×10^{-555}	1.6×10^{-2}

Table 5. Numerical results for problem 2: $g_3(x)$.

Method	IG	N	FV	ACOC	$\|\epsilon\|$	$\|f(x_N)\|$	CPU
PM_{10}	3.5	4	24	10	1.3985×10^{-540}	3×10^{-4000}	1.41×10^{-1}
	2.5	4	24	10	9.4449×10^{-603}	3×10^{-4000}	1.41×10^{-1}
MH_{10}	3.5	4	20	8	4.4067×10^{-244}	1.1093×10^{-1948}	9.4×10^{-2}
	2.5	4	20	8	7.5796×10^{-277}	8.4971×10^{-2211}	9.4×10^{-2}
NM_9	3.5	4	20	9	1.9811×10^{-308}	4.6535×10^{-2771}	2.81×10^{-1}
	2.5	4	20	9	3.3116×10^{-366}	4.7431×10^{-3291}	1.4×10^{-1}
WO_8	3.5	5	20	8	5.0436×10^{-1292}	3×10^{-4000}	2.66×10^{-1}
	2.5	4	16	8	6.6531×10^{-270}	1.4143×10^{-2155}	2.18×10^{-1}
HM_6	3.5	5	25	6	3.9257×10^{-663}	3.8877×10^{-3976}	1.41×10^{-1}
	2.5	5	25	6	7.7054×10^{-742}	5×10^{-4000}	1.56×10^{-1}
MHM_5	3.5	5	20	5	9.1683×10^{-320}	1.7296×10^{-1597}	1.25×10^{-1}
	2.5	5	20	5	1.4187×10^{-412}	1.5343×10^{-2061}	1.25×10^{-1}
NR_2	3.5	10	20	2	5.1202×10^{-289}	1.2423×10^{-577}	9.3×10^{-2}
	2.5	10	20	2	4.5680×10^{-321}	9.8883×10^{-642}	1.09×10^{-1}

Table 6. Numerical results for problem 2: $g_4(x)$ with * showing the divergence of the method.

Method	IG	N	FV	ACOC	$\|\epsilon\|$	$\|f(x_N)\|$	CPU
PM_{10}	−4.5	10	60	10	4.1220×10^{-954}	8×10^{-3999}	4.38×10^{-1}
	−9.5	30	180	10	1.0834×10^{-353}	1.0796×10^{-3527}	1.39×10^{0}
MH_{10}	−4.5	12	60	8	5.5553×10^{-552}	8×10^{-3999}	5.62×10^{-1}
	−9.5	39	195	8	1.0488×10^{-636}	8×10^{-3999}	1.8590×10^{0}
NM_9	−4.5	12	60	9	4.4565×10^{-1467}	8×10^{-3999}	9.69×10^{-1}
	−9.5	37	185	9.0001×10^{0}	6.8×10^{-223}	4.0517×10^{-1997}	3.3590×10^{0}
WO_8	−4.5	17	68	8	2.3107×10^{-1507}	8×10^{-3999}	1.5940×10^{0}
	−9.5	200 *	800	9.5356×10^{-1}	2.0883×10^{-2}	$9.0918 \times 10^{+72}$	1.9266×10^{1}
HM_6	−4.5	13	65	6	1.0498×10^{-243}	4.1588×10^{-1456}	5.47×10^{-1}
	−9.5	43	215	6	3.9815×10^{-227}	1.2374×10^{-1356}	1.8910×10^{0}
MHM_5	−4.5	7	28	5	2.1606×10^{-978}	8×10^{-3999}	2.34×10^{-1}
	−9.5	17	68	5	3.1372×10^{-238}	4.1824×10^{-1186}	5.62×10^{-1}
NR_2	−4.5	30	60	2	1.4030×10^{-243}	6.0043×10^{-485}	4.84×10^{-1}
	−9.5	101	202	2	6.9221×10^{-226}	1.4616×10^{-449}	1.5620×10^{0}

Table 7. Numerical results for problem 2: $g_5(x)$.

Method	IG	N	FV	ACOC	$\lvert\epsilon\rvert$	$\lvert f(x_N)\rvert$	CPU
PM_{10}	2.9	4	24	10	3.0523×10^{-1325}	0	1.57×10^{-1}
	−3.7	5	30	10	3.1294×10^{-1405}	0	2.19×10^{-1}
	7.4	6	36	10	1.2713×10^{-454}	0	2.5×10^{-1}
MH_{10}	2.9	4	20	8	3.3060×10^{-632}	0	9.4×10^{-2}
	−3.7	9	45	8	7.1797×10^{-378}	4.0342×10^{-3023}	2.97×10^{-1}
	7.4	22	110	8	5.3293×10^{-202}	3.7179×10^{-1616}	7.81×10^{-1}
NM_9	2.9	4	20	9	3.6544×10^{-815}	0	2.5×10^{-1}
	−3.7	5	25	9	2.2115×10^{-563}	0	3.13×10^{-1}
	7.4		failed	–	–	–	–
WO_8	2.9	4	16	8	8.7376×10^{-540}	0	1.87×10^{-1}
	−3.7	failed	–	–	–	–	–
	7.4	failed	–	–	–	–	–
HM_6	2.9	4	20	6	2.1574×10^{-305}	3.1362×10^{-1833}	1.41×10^{-1}
	−3.7	6	30	6	3.8865×10^{-904}	0	1.72×10^{-1}
	7.4	12	60	6	3.3038×10^{-311}	4.0442×10^{-1868}	4.22×10^{-1}
MHM_5	2.9	5	20	5	4.0088×10^{-716}	5.2603×10^{-3580}	1.1×10^{-1}
	−3.7	8	32	5	7.1662×10^{-833}	0	2.97×10^{-1}
	7.4	8	32	5	2.16×10^{-326}	2.3890×10^{-1631}	2.19×10^{-1}
NR_2	2.9	9	18	2	4.2361×10^{-377}	3.9782×10^{-754}	9.4×10^{-2}
	−3.7	11	22	2	4.9340×10^{-262}	5.3970×10^{-524}	3.13×10^{-1}
	7.4	16	32	2	1.0369×10^{-370}	2.3835×10^{-741}	2.66×10^{-1}

Example 3. *Volume from Van der Waals' Equation [37].*
The Van der Waals equation is represented by the following model:

$$(P + \frac{an^2}{V^2})(V - bn) = nRT. \quad (43)$$

After some simplifications, one obtains the following polynomial of nonlinear form:

$$g(V) = PV^3 - n(RT + bP)V^2 + n^2aV - n^3ab. \quad (44)$$

The Van der Waals equation of state was formulated in 1873, with two constants a and b (Van der Waals constants) determined from the behavior of a substance at the critical point. The equation is based on two effects, which are the molecular size and attractive force between the molecules. The model (43) is a modified version of the ideal gas equation $V = RT/nP$, where n shows the number of moles, R stands for the universal gas constant (0.0820578), T is the absolute temperature, V shows the volume, and P denotes the absolute pressure. If $V = 1.4$ moles of benzene vapor form under $P = 40$ atm with $a = 18$ and $b = 0.1154$, then one has

$$g_1(V) = 40V^3 - 95.26535116V^2 + 35.28V - 5.6998368. \quad (45)$$

The approximate solution up to 50 dp is given as:

$$V^* = 1.9707842194070294114471303720868563598618121603538.$$

Being cubic, the Equation (45) certainly possesses one real root. Here, we aim to show the performance of PM_{10} on this model. Therefore, the model is numerically solved by PM_{10} and the other six methods chosen for comparison. It can be observed in Table 8 that PM_{10} achieves the smallest possible error ϵ along with functional values nearest to 0 in a reasonable amount of time, irrespective of the initial guesses.

Table 8. Numerical results for problem 3.

Method	IG	N	FV	ACOC	$\|\epsilon\|$	$\|f(x_N)\|$	CPU
PM_{10}	2.0	4	24	10	4.6111×10^{-1485}	1.3×10^{-3997}	4.7×10^{-2}
	10.3	6	36	10	1.6261×10^{-1641}	1.1×10^{-3997}	6.2×10^{-2}
MH_{10}	2.0	4	20	8	1.5202×10^{-726}	1.1×10^{-3997}	1.6×10^{-2}
	10.3	6	30	8	6.2361×10^{-305}	2.2532×10^{-2431}	1.6×10^{-2}
NM_9	2.0	4	20	9	1.0867×10^{-1015}	1.3×10^{-3997}	1.6×10^{-2}
	10.3	6	30	9	9.6289×10^{-479}	1.3×10^{-3997}	3.2×10^{-2}
WO_8	2	4	16	8	3.9765×10^{-833}	1.1×10^{-3997}	1.6×10^{-2}
	10.3	6	24	8	4.8401×10^{-690}	1.1×10^{-3997}	4.7×10^{-2}
HM_6	2.0	4	20	6	9.3310×10^{-311}	2.9554×10^{-1858}	1.6×10^{-2}
	10.3	7	35	6	1.3806×10^{-532}	3.1008×10^{-3189}	1.6×10^{-2}
MHM_5	2.0	4	16	5	1.6172×10^{-201}	8.3557×10^{-1003}	0
	10.3	7	28	5	7.9826×10^{-804}	1.1×10^{-3997}	0
NR_2	2.0	9	18	2	1.7818×10^{-383}	4.4839×10^{-764}	0
	10.3	15	30	2	1.1910×10^{-271}	2.0034×10^{-540}	1.6×10^{-2}

Example 4. *The Shockley Ideally Diode Electric Circuit Model.*

The Shockley diode model giving the voltage going through the diode V_D is represented by the following equation:

$$J = J_S\Big(\exp(V_D/nV_T) - 1\Big),$$

where J_S stands for the saturation current, n is the emission coefficient, V_T is the thermal voltage, and J is the diode current. Using the Kirchhoff's second law ($V_R + V_D = V_S$) and Ohm's law ($V = JR$), a root-finding model can be found. The final structure for the model would be as follows:

$$V_S = JR + nV_T ln\Big(\frac{J}{J_S} + 1\Big). \tag{46}$$

Assuming values of parameters V_S, R, n, V_T, J_S from [38], we obtain the following equation that is nonlinear in the variable J:

$$g(J) = 1.4ln(J+1) + 0.1J - 0.5. \tag{47}$$

The approximate solution of the above equation correct to 50 dp is as follows:

$$J^\star = 0.38997719839007758658645353264634118996836946243662.$$

Table 9 presents numerical simulations for (47) with two different initial conditions of 0.5 and 1.8, under which the smallest absolute error seems to lie under the column of the proposed method PM_{10}. Further analysis can easily be conducted by the careful examination of the results tabulated therein.

Example 5. *Conversion of Species within a Chemical Reactor [39].*

The following nonlinear equation arises in chemical engineering during the conversion of species in a chemical reactor:

$$g(x) = \frac{x}{1-x} - 5\ln\left(\frac{0.4(1-x)}{0.4-0.5x}\right) + 4.45977, \tag{48}$$

where x stands for the fractional conversion of the species; thus, it must lie in (0,1). The approximate solution of the above equation correct to 50 dp is as follows:

$$x^\star \approx 0.75739624625375387945964129792914529342795578042081.$$

Table 9. Numerical results for problem 4.

Method	IG	N	FV	ACOC	$\|\epsilon\|$	$\|f(x_N)\|$	CPU
PM_{10}	0.5	4	24	10	7.2498×10^{-1614}	2×10^{-4000}	7.8×10^{-2}
	1.8	4	24	10	1.1065×10^{-661}	2×10^{-4000}	7.8×10^{-2}
MH_{10}	0.5	4	20	8	5.7732×10^{-741}	2×10^{-4000}	3.1×10^{-2}
	1.8	4	20	8.0001×10^{0}	1.3446×10^{-210}	5.0780×10^{-1683}	3.1×10^{-2}
NM_9	0.5	4	20	9	1.7171×10^{-1088}	2×10^{-4000}	4.7×10^{-2}
	1.8	4	20	9	5.9947×10^{-330}	8.3091×10^{-2968}	4.7×10^{-2}
WO_8	0.5	4	16	8	3.2491×10^{-712}	2×10^{-4000}	6.2×10^{-2}
	1.8	4	16	8.0001×10^{0}	8.3705×10^{-212}	3.1312×10^{-1692}	6.3×10^{-2}
HM_6	0.5	4	20	6	1.1747×10^{-300}	2.1830×10^{-1802}	3.1×10^{-2}
	1.8	5	25	6	6.2470×10^{-444}	4.9378×10^{-2662}	4.7×10^{-2}
MHM_5	0.5	5	20	5	2.9184×10^{-855}	2×10^{-4000}	1.6×10^{-2}
	1.8	5	20	5	3.7051×10^{-207}	1.9490×10^{-1034}	1.6×10^{-2}
NR_2	0.5	9	18	2	1.3×10^{-371}	6.1228×10^{-743}	1.5×10^{-2}
	1.8	10	20	2×10^{0}	1.1395×10^{-201}	4.7044×10^{-403}	3.2×10^{-2}

Numerical results can be found in Table 10, where it is shown that PM_{10} has outperformed all other methods in terms of the absolute errors under consideration under two different initial conditions.

Table 10. Numerical results for problem 5.

Method	IG	N	FV	ACOC	$\|\epsilon\|$	$\|f(x_N)\|$	CPU
PM_{10}	0.71	5	30	10	2.5434×10^{-1635}	4×10^{-3999}	1.25×10^{-1}
	0.76	4	24	10	5.0617×10^{-1424}	1×10^{-3998}	9.4×10^{-2}
MH_{10}	0.71	5	25	8	5.1147×10^{-640}	4×10^{-3999}	6.2×10^{-2}
	0.76	4	20	8	6.3350×10^{-690}	4×10^{-3999}	4.7×10^{-2}
NM_9	0.71	5	25	9	3.8480×10^{-559}	1×10^{-3998}	7.8×10^{-2}
	0.76	4	20	9	4.0774×10^{-981}	4×10^{-3999}	4.7×10^{-2}
WO_8	0.71	5	20	8	3.3478×10^{-584}	1.3×10^{-3998}	1.56×10^{-1}
	0.76	4	16	8	1.3297×10^{-696}	2×10^{-3999}	6.3×10^{-2}
HM_6	0.71	6	30	6	5.4957×10^{-618}	4.4077×10^{-3696}	6.3×10^{-2}
	0.76	4	20	6	9.3504×10^{-288}	1.0692×10^{-1714}	4.7×10^{-2}
MHM_5	0.71	6	24	5	2.0592×10^{-216}	2.9538×10^{-1072}	4.7×10^{-2}
	0.76	5	20	5	4.2638×10^{-834}	4×10^{-3999}	3.1×10^{-2}
NR_2	0.71	12	24	2	5.5571×10^{-216}	3.9060×10^{-428}	4.7×10^{-2}
	0.76	9	18	2	5.3583×10^{-356}	3.6316×10^{-708}	3.2×10^{-2}

Example 6. *The Two-Dimensional Bratu Model [40].*

The two-dimension Bratu system is given by the following partial differential equation:

$$\frac{\partial^2 U}{\partial x^2} + \frac{\partial^2 U}{\partial y^2} + \lambda \exp(U) = 0, \; x, y \in D = [0,1] \times [0,1], \tag{49}$$

subject to the following boundary conditions

$$U(x,y) = 0 \; x, y \in D. \tag{50}$$

The two-dimensional Bratu system has two bifurcated exact solutions for $\lambda < \lambda_c$, a unique solution for $\lambda = \lambda_c$, and no solutions for $\lambda > \lambda_c$. The exact solution to (49) is determined as follows:

$$U(x,y) = 2\ln\left[\frac{cosh(\frac{\theta}{4})\cosh((x-\frac{1}{2})(y-\frac{1}{2})\theta)}{\cosh((x-\frac{1}{2})\frac{\theta}{2})\cosh((y-\frac{1}{2})\frac{\theta}{2})}\right], \quad (51)$$

where θ is an undetermined constant satisfying the boundary conditions and assumed to be the solution of (49). Using the procedure described in [41], one obtains

$$\theta^2 = \lambda\cosh^2\left(\frac{\theta}{4}\right). \quad (52)$$

Differentiating (52) with respect to θ and setting $\frac{d\lambda}{d\theta} = 0$, the critical value λ_c satisfies

$$\theta = \frac{1}{4}\lambda_c\cosh\left(\frac{\theta}{4}\right)\sinh\left(\frac{\theta}{4}\right). \quad (53)$$

By eliminating λ from (52) and (53), we have the value of θ_c for the critical λ_c satisfying

$$\frac{\theta_c}{4} = \coth\left(\frac{\theta_c}{4}\right), \quad (54)$$

and $\theta_c = 4.798714561$. We then obtain $\lambda_c = 7.027661438$ from (53). Numerical simulations performed in Table 11 show that the proposed three-step method takes a smaller number of iterations and produces considerably smaller absolute errors with a reasonable amount of CPU time.

Table 11. Numerical results for problem 6.

Method	IG	N	FV	ACOC	$\lvert\epsilon\rvert$	$\lvert f(x_N)\rvert$	CPU
PM_{10}	4.0	4	24	10	3.9×10^{-1086}	0	1.41×10^{-1}
	15.5	4	24	10	1.9074×10^{-553}	0	2.19×10^{-1}
MH_{10}	4.0	4	20	8	5.2357×10^{-556}	0	1.09×10^{-1}
	15.5	4	20	8	3.6499×10^{-277}	6.3171×10^{-2220}	1.09×10^{-1}
NM_9	4.0	4	20	9	4.3486×10^{-847}	0	6.3×10^{-2}
	15.5	4	20	9	2.3344×10^{-332}	1.9566×10^{-2994}	6.3×10^{-2}
WO_8	4.0	4	16	7.9999×10^{0}	1.4375×10^{-487}	1.6570×10^{-3902}	1.09×10^{-1}
	15.5	5	20	8	5.7605×10^{-1311}	0	9.4×10^{-2}
HM_6	4.0	4	20	6	1.4310×10^{-221}	3.9699×10^{-1331}	1.25×10^{-1}
	15.5	5	25	6	2.0883×10^{-638}	3.8348×10^{-3832}	1.25×10^{-1}
MHM_5	4.0	5	20	5	8.6954×10^{-585}	4.7414×10^{-2925}	9.4×10^{-2}
	15.5	5	20	5	5.0264×10^{-265}	3.06×10^{-1326}	9.4×10^{-2}
NR_2	4.0	9	18	2	5.7111×10^{-278}	1.0742×10^{-556}	7.8×10^{-2}
	15.5	10	20	2	1.2645×10^{-281}	5.2661×10^{-564}	1.09×10^{-1}

Now, we consider five different kinds of nonlinear multidimensional equations and numerically solve them with PM_{10}, HM_6, MH_5, and NR_2 since the methods MH_{10}, NM_9, and WO_8 were either divergent or not applicable on systems of nonlinear equations. For the systems considered, various types of initial guesses are used, and for comparison purposes, the approximate solution and the normed error $\epsilon = ||x_{n+1} - x_n||_\infty$ having the same tolerance 10^{-200} and CPU time are taken into consideration. It can be observed in Tables 12–16 that the smallest possible absolute error is achieved only with the proposed method—that is, PM_{10}—in a reasonably affordable time period.

Example 7. *The nonlinear system of two equations from [3,41] is given as:*

$$\begin{aligned} x_1 + exp(x_2) - cos(x_2) &= 0, \\ 3x_1 - x_2 - sin(x_1) &= 0. \end{aligned} \tag{55}$$

The exact solution of the system (55) is $\mathbf{x} = [0,0]'$. *The numerical results for this system are shown in Table 12 under the proposed* PM_{10} *and other three methods.*

Table 12. Numerical results for problem 7.

N	$[x_{1,0}, x_{2,0}]^T$	$[x_1, x_2]^T$	$\lvert\epsilon\rvert$	CPU
PM_{10}	$-1.0, 1.0$	$5.0151 \times 10^{-4003}, 1.0030 \times 10^{-4002}$	7.7834×10^{-3318}	0
HM_6	–	$1.2648 \times 10^{-2287}, 2.5296 \times 10^{-2287}$	9.9330×10^{-382}	1.4×10^{-1}
MH_5	–	$4.6720 \times 10^{-968}, 9.8372 \times 10^{-968}$	5.2359×10^{-194}	1.5×10^{-2}
NR_2	–	$1.2203 \times 10^{-11}, 2.4406 \times 10^{-11}$	6.0505×10^{-06}	4.7×10^{-2}
PM_{10}	$-1.9, 1.8$	$5.0151 \times 10^{-4003}, 1.0030 \times 10^{-4002}$	7.0614×10^{-1865}	3.2×10^{-2}
HM_6	–	$5.4937 \times 10^{-1225}, 1.0987 \times 10^{-1224}$	1.2688×10^{-204}	6.2×10^{-2}
MH_5	–	$2.1088 \times 10^{-559}, 4.4402 \times 10^{-559}$	2.8177×10^{-112}	6.2×10^{-2}
NR_2	–	$6.9588 \times 10^{-07}, 1.3916 \times 10^{-06}$	1.4445×10^{-03}	4.6×10^{-2}
PM_{10}	$2.5, -2.3$	$1.9407 \times 10^{-4002}, 3.8815 \times 10^{-4002}$	4.3934×10^{-3428}	1.6×10^{-2}
HM_6	–	$1.2918 \times 10^{-4000}, 2.5836 \times 10^{-4000}$	1.8909×10^{-691}	6.3×10^{-2}
MH_5	–	$2.8427 \times 10^{-695}, 5.9858 \times 10^{-695}$	1.8873×10^{-139}	6.2×10^{-2}
NR_2	–	$8.2443 \times 10^{-15}, 1.6489 \times 10^{-14}$	1.5727×10^{-07}	4.7×10^{-2}
PM_{10}	$8.9, 5.5$	$3.9204 \times 10^{-1989}, 8.3019 \times 10^{-1989}$	2.1886×10^{-199}	7.8×10^{-2}
HM_6	–	$3.6432 \times 10^{-93}, 7.2864 \times 10^{-93}$	5.4995×10^{-16}	9.3×10^{-2}
MH_5	–	$5.6511 \times 10^{-70}, 1.1901 \times 10^{-69}$	2.1654×10^{-14}	1.41×10^{-1}
NR_2	–	$2.7716 \times 10^{-1}, 5.1674 \times 10^{-1}$	7.9930×10^{-1}	4.7×10^{-2}
PM_{10}	$1.9, 6.5$	$1.5819 \times 10^{-1006}, 3.3499 \times 10^{-1006}$	3.9880×10^{-101}	7.8×10^{-2}
HM_6	–	$5.9575 \times 10^{-44}, 1.1915 \times 10^{-43}$	8.7617×10^{-08}	9.4×10^{-2}
MH_5	–	$9.8552 \times 10^{-36}, 2.0743 \times 10^{-35}$	1.5268×10^{-07}	1.8700×10^{-1}
NR_2	–	$6.7641 \times 10^{-1}, 1.3228 \times 10^{0}$	1.0621×10^{0}	4.6×10^{-2}
PM_{10}	$0.1, 0.1$	$9.2473 \times 10^{-4002}, 1.8495 \times 10^{-4001}$	0	0
HM_6	–	$4.9611 \times 10^{-4002}, 9.9222 \times 10^{-4002}$	2.4509×10^{-1479}	7.8×10^{-2}
MH_5	–	$1.8099 \times 10^{-3702}, 3.8109 \times 10^{-3702}$	6.8648×10^{-741}	4.7×10^{-2}
NR_2	–	$2.8142 \times 10^{-39}, 5.6283 \times 10^{-39}$	9.1883×10^{-20}	4.7×10^{-2}

Example 8. *Another nonlinear system of three equations taken from [42] is shown below:*

$$\begin{aligned} 3x_1 - \cos(x_2 x_3) - 1/2 &= 0, \\ x_1^2 - 81(x_2 + 0.1)^2 + \sin(x_3) + 1.06 &= 0, \\ \exp(-x_1 x_2) + 20x_3 + (10\pi/3 - 1) &= 0, \end{aligned} \tag{56}$$

where its approximate solution up to 50 dp is shown below:

$$x \approx \begin{bmatrix} 0.49814468458949119126228211413809456132099782481239 \\ 0 \\ -0.52882597757338745562224205210357569604547206124467 \end{bmatrix}. \tag{57}$$

The numerical results for the system (56) are shown in Table 13.

Table 13. Numerical results for problem 8.

Method	$[x_{1,0}, x_{2,0}, x_{3,0}]^T$	$[x_1, x_2, x_3]^T$	$\lvert\epsilon\rvert$	CPU
PM_{10}	1.1, 1.1, −1.1	$5 \times 10^{-1}, -1.4199 \times 10^{-4001}, -5.2360 \times 10^{-1}$	3.6961×10^{-541}	4.6×10^{-2}
HM_6	–	$5 \times 10^{-1}, 5.9801 \times 10^{-445}, -5.2360 \times 10^{-1}$	2.1352×10^{-75}	4.7×10^{-2}
MH_5	–	$5 \times 10^{-1}, 3.9207 \times 10^{-162}, -5.2360 \times 10^{-1}$	2.4120×10^{-40}	7.8×10^{-2}
NR_2	–	$5.0001 \times 10^{-1}, 9.5530 \times 10^{-04}, -5.2357 \times 10^{-1}$	1.3842×10^{-2}	1.5×10^{-2}
PM_{10}	3.3, 2.1, −2.1	$5 \times 10^{-1}, -4.5590 \times 10^{-2904}, -5.2360 \times 10^{-1}$	2.9346×10^{-362}	6.3×10^{-2}
HM_6	–	$5 \times 10^{-1}, 2.1651 \times 10^{-257}, -5.2360 \times 10^{-1}$	3.8835×10^{-44}	4.7×10^{-2}
MH_5	–	$5 \times 10^{-1}, -7.2199 \times 10^{-131}, -5.2360 \times 10^{-1}$	5.4152×10^{-33}	9.3×10^{-2}
NR_2	–	$5.0007 \times 10^{-1}, 7.6756 \times 10^{-03}, -5.2340 \times 10^{-1}$	3.9888×10^{-2}	3.1×10^{-2}
PM_{10}	−1.3, 1.1, −0.1	$5 \times 10^{-1}, -8.2172 \times 10^{-4001}, -5.2360 \times 10^{-1}$	1.1159×10^{-551}	6.3×10^{-2}
HM_6	–	$5 \times 10^{-1}, 1.2373 \times 10^{-456}, -5.2360 \times 10^{-1}$	2.4102×10^{-77}	4.6×10^{-2}
MH_5	–	$5 \times 10^{-1}, 3.9887 \times 10^{-168}, -5.2360 \times 10^{-1}$	7.6082×10^{-42}	4.7×10^{-2}
NR_2	–	$5.0001 \times 10^{-1}, 8.3811 \times 10^{-04}, -5.2358 \times 10^{-1}$	1.2961×10^{-2}	1.6×10^{-2}
PM_{10}	1.9, 4.1, 0.1	$5 \times 10^{-1}, -4.6794 \times 10^{-1318}, -5.2360 \times 10^{-1}$	4.8484×10^{-164}	6.3×10^{-2}
HM_6	–	$5 \times 10^{-1}, 3.4949 \times 10^{-117}, -5.2360 \times 10^{-1}$	9.0619×10^{-21}	4.7×10^{-2}
MH_5	–	$5 \times 10^{-1}, -5.3226\text{e}{-56}, -5.2360 \times 10^{-1}$	1.0093×10^{-12}	4.7×10^{-2}
NR_2	–	$5.0048 \times 10^{-1}, 5.5684 \times 10^{-2}, -5.2215 \times 10^{-1}$	1.1923×10^{-1}	3.1×10^{-2}
PM_{10}	6.5, 2.2, −3.3	$5 \times 10^{-1}, -7.0328\text{e-}2659, -5.2360 \times 10^{-1}$	1.3210×10^{-331}	6.3×10^{-2}
HM_6	–	$5 \times 10^{-1}, 6.7653 \times 10^{-236}, -5.2360 \times 10^{-1}$	1.4849×10^{-40}	4.7×10^{-2}
MH_5	–	$5 \times 10^{-1}, -3.4190 \times 10^{-98}, -5.2360 \times 10^{-1}$	3.6270×10^{-24}	1.1×10^{-1}
NR_2	–	$5.0009 \times 10^{-1}, 1.0418 \times 10^{-2}, -5.2333 \times 10^{-1}$	4.6778×10^{-2}	3.2×10^{-2}
PM_{10}	3.5, 3.7, −2.3	$5 \times 10^{-1}, -9.2310 \times 10^{-1487}, -5.2360 \times 10^{-1}$	3.8829×10^{-185}	6.3×10^{-2}
HM_6	–	$5 \times 10^{-1}, 2.4238 \times 10^{-131}, -5.2360 \times 10^{-1}$	3.9573×10^{-23}	6.3×10^{-2}
MH_5	–	$5 \times 10^{-1}, -1.2128 \times 10^{-61}, -5.2360 \times 10^{-1}$	3.6152×10^{-14}	1.41×10^{-1}
NR_2	–	$5.0039 \times 10^{-1}, 4.4303 \times 10^{-2}, -5.2244 \times 10^{-1}$	1.0395×10^{-1}	3.1×10^{-2}

Example 9. *A three-dimensional nonlinear system is taken from [3] as given below:*

$$\begin{aligned} x_1^2 + x_2^2 + x_3^2 - 1 &= 0, \\ 2x_1^2 + x_2^2 - 4x_3 &= 0, \\ 3x_1^2 - 4x_2^2 + x_3^2 &= 0, \end{aligned} \tag{58}$$

where its approximate solution up to 50 dp is as follows:

$$x \approx \begin{bmatrix} 0.69828860997151390091867421225192307770469334334732 \\ 0.62852429796021380638277617781675123954652671431496 \\ 0.34256418968956943776230136116401106884202074401616 \end{bmatrix}. \tag{59}$$

The numerical results for the system (58) are shown in Table 14.

Table 14. Numerical results for problem 9.

Method	$[x_{1,0}, x_{2,0}, x_{3,0}]^T$	$[x_1, x_2, x_3]^T$	$\|\epsilon\|$	CPU
PM_{10}	0.5, 0.5, 0.5	$6.9829 \times 10^{-1}, 6.2852 \times 10^{-1}, 3.4256 \times 10^{-1}$	1×10^{-4000}	0
HM_6	–	$6.9829 \times 10^{-1}, 6.2852 \times 10^{-1}, 3.4256 \times 10^{-1}$	3.8739×10^{-864}	0
MH_5	–	$6.9829 \times 10^{-1}, 6.2852 \times 10^{-1}, 3.4256 \times 10^{-1}$	6.0191×10^{-527}	1.5×10^{-2}
NR_2	–	$6.9829 \times 10^{-1}, 6.2852 \times 10^{-1}, 3.4256 \times 10^{-1}$	3.8598×10^{-12}	1.5×10^{-2}
PM_{10}	1.0, 1.0, 1.0	$6.9829 \times 10^{-1}, 6.2852 \times 10^{-1}, 3.4256 \times 10^{-1}$	1×10^{-4000}	0
HM_6	–	$6.9829 \times 10^{-1}, 6.2852 \times 10^{-1}, 3.4256 \times 10^{-1},$	4.1436×10^{-596}	1.6×10^{-2}
MH_5	–	$6.9829 \times 10^{-1}, 6.2852 \times 10^{-1}, 3.4256 \times 10^{-1},$	3.5513×10^{-269}	0
NR_2	–	$6.9829 \times 10^{-1}, 6.2852 \times 10^{-1}, 3.4256 \times 10^{-1},$	1.1136×10^{-08}	0
PM_{10}	2.8, 3.2, 6.1	$6.9829 \times 10^{-1}, 6.2852 \times 10^{-1}, 3.4256 \times 10^{-1}$	7.0153×10^{-894}	0
HM_6	–	$6.9829 \times 10^{-1}, 6.2852 \times 10^{-1}, 3.4256 \times 10^{-1}$	1.4280×10^{-103}	0
MH_5	–	$6.9829 \times 10^{-1}, 6.2852 \times 10^{-1}, 3.4256 \times 10^{-1}$	6.9270×10^{-53}	0
NR_2	–	$6.9929 \times 10^{-1}, 6.2876 \times 10^{-1}, 3.4257 \times 10^{-1}$	3.7312×10^{-2}	0
PM_{10}	5.1, 4.2, 1.1	$6.9829 \times 10^{-1}, 6.2852 \times 10^{-1}, 3.4256 \times 10^{-1}$	4.8091×10^{-1119}	0
HM_6	–	$6.9829 \times 10^{-1}, 6.2852 \times 10^{-1}, 3.4256 \times 10^{-1}$	8.9729×10^{-126}	0
MH_5	–	$6.9829 \times 10^{-1}, 6.2852 \times 10^{-1}, 3.4256 \times 10^{-1}$	6.3425×10^{-68}	0
NR_2	–	$6.9851 \times 10^{-1}, 6.2861 \times 10^{-1}, 3.4256 \times 10^{-1}$	1.7731×10^{-2}	0
PM_{10}	5.1, 4.2, −1.1	$6.9829 \times 10^{-1}, 6.2852 \times 10^{-1}, 3.4256 \times 10^{-1}$	1.1175×10^{-174}	0
HM_6	–	$6.9829 \times 10^{-1}, 6.2852 \times 10^{-1}, 3.4256 \times 10^{-1}$	2.7993×10^{-19}	0
MH_5	–	$4.3060 \times 10^{0}, -9.1902 \times 10^{-1}, -2.7461 \times 10^{0}$	2.4533×10^{0}	0
NR_2	–	$1.0879 \times 10^{0}, 8.0413 \times 10^{-1}, 5.3209 \times 10^{-1}$	7.7974×10^{-1}	0
PM_{10}	10.2, 14.7, 11.1	$6.9829 \times 10^{-1}, 6.2852 \times 10^{-1}, 3.4256 \times 10^{-1}$	2.9425×10^{-319}	0
HM_6	–	$6.9829 \times 10^{-1}, 6.2852 \times 10^{-1}, 3.4256 \times 10^{-1}$	2.8736×10^{-34}	0
MH_5	–	$6.9829 \times 10^{-1}, 6.2852 \times 10^{-1}, 3.4256 \times 10^{-1}$	1.1693×10^{-19}	1.6×10^{-2}
NR_2	–	$7.5545 \times 10^{-1}, 7.3493 \times 10^{-1}, 3.4367 \times 10^{-1}$	3.7993×10^{-1}	0

Example 10. *(Catenary curve and the ellipse ([43], p. 83)):*

Given below is a nonlinear system of two equations that describe trajectories for the catenary and the ellipse, respectively. We are interested in finding their intersection point that lies in the first quadrant of the cartesian plane. The system has been solved under the proposed PM_{10} method and other methods under consideration. The performance of each method is shown in Table 15, whereas an approximate solution up to 50 dp of the system (60) is shown in comparison to the system.

$$\begin{aligned} &x_2 - \frac{1}{2}\Big(\exp(x_1/2) + \exp(-x_1/2)\Big) = 0, \\ &9x_1^2 + 25x_2^2 - 225 = 0. \end{aligned} \tag{60}$$

Approximate solution:

$$x \approx \begin{bmatrix} 3.03115539171898395365249644784606508519370920650 81 \\ 2.38586565356288572812288096276522630814193233451 76 \end{bmatrix}. \tag{61}$$

Table 15. Numerical results for problem 10 with * showing the divergence of the method.

Method	$[x_{1,0}, x_{2,0}]^T$	$[x_1, x_2]^T$	$\|\epsilon\|$	CPU
PM_{10}	9.3, 8.6	$3.0312 \times 10^0, 2.3859 \times 10^0$	8.4435×10^{-523}	3.1×10^{-2}
HM_6	–	–	3.1353×10^{-78}	4.6×10^{-2}
MH_5	–	–	1.4228×10^{-40}	3.1×10^{-2}
NR_2	–	–	2.1386×10^{-1}	1.6×10^{-2}
PM_{10}	11.6, 13.1	$3.0312 \times 10^0, 2.3859 \times 10^0$	1.4738×10^{-275}	4.7×10^{-2}
HM_6	–	–	8.2367×10^{-31}	1.6×10^{-2}
MH_5	–	–	1.8199×10^{-17}	3.1×10^{-2}
NR_2	–	–	1.0020×10^0	1.6×10^{-2}
PM_{10}	4.6, 3.6	$3.0312 \times 10^0, 2.3859 \times 10^0$	1.7497×10^{-2436}	3.1×10^{-2}
HM_6	–	–	2.4449×10^{-528}	3.1×10^{-2}
MH_5	–	–	1.3516×10^{-167}	1.6×10^{-2}
NR_2	–	–	2.0398×10^{-07}	1.6×10^{-2}
PM_{10}	16.6, 14.5	$3.0312 \times 10^0, 2.3859 \times 10^0$	8.0467×10^{-58}	3.1×10^{-2}
HM_6	–	–	3.8892×10^{-04}	3.2×10^{-2}
MH_5	–	–	1.1246×10^{-05}	3.1×10^{-2}
NR_2	–	$6.6073 \times 10^0, -1.1486 \times 10^0$	2.5508×10^0 *	1.6×10^{-2}
PM_{10}	2.9, 1.9	$3.0312 \times 10^0, 2.3859 \times 10^0$	0	3.1×10^{-2}
HM_6	–	–	1.1004×10^{-1156}	3.1×10^{-2}
MH_5	–	–	5.7581×10^{-307}	3.1×10^{-2}
NR_2	–	–	3.1421×10^{-15}	0
PM_{10}	10.3, 11.7	$3.0312 \times 10^0, 2.3859 \times 10^0$	5.3262×10^{-397}	9.4×10^{-2}
HM_6	–	–	7.9396×10^{-54}	9.3×10^{-2}
MH_5	–	–	1.6306×10^{-28}	6.2×10^{-2}
NR_2	–	–	5.0395×10^{-1}	3.1×10^{-2}

Example 11. *Steady-State Lorenz Equations ([44], p. 816).*

In this problem, we consider a system developed by Edward Lorenz, who was an American meteorologist studying atmospheric convection around the Earth's surface. Lorenz's nonlinear system is a set of three ordinary differential equations, as given below:

$$\begin{aligned} \dot{x_1}(t) &= a(x_2 - x_1), \\ \dot{x_2}(t) &= x_1(b - x_3) - x_2, \\ \dot{x_3}(t) &= x_1 x_2 - c x_3. \end{aligned} \tag{62}$$

In order to study the steady-state behavior of the system (62), we take $\dot{x_1}(t) = \dot{x_2}(t) = \dot{x_3}(t) = 0$ and $a = -1, b = 2, c = 3$ to obtain the following nonlinear algebraic system:

$$\begin{aligned} & x_1 - x_2 = 0, \\ & 2x_1 - x_1 x_3 - x_2 = 0, \\ & x_1 x_2 - 3x_3 = 0. \end{aligned} \tag{63}$$

The approximate solution for the system (63) correct to 50 dp is given as

$$x \approx \begin{bmatrix} 1.73205080756887729352744634150587236694280525381 04 \\ 1.7320508075688772935274463415058723669428052538104 \\ 1 \end{bmatrix}. \tag{64}$$

The nonlinear steady-state system (63) has been numerically solved in Table 16.

Table 16. Numerical results for problem 11 with * showing the divergence of the method.

Method	$[x_{1,0}, x_{2,0}, x_{3,0}]^T$	$[x_1, x_2, x_3]^T$	$\|\epsilon\|$	CPU
PM_{10}	$-2.5, -3.5, -1.5$	1.7321, 1.7321, 1	3.1099×10^{-36}	1.5×10^{-2}
HM_6	–	1.7321, 1.7321, 1	3.8304×10^{-2}	1.6×10^{-2}
MH_5	–	$-1.7321, -1.7321, 1$	8.8956×10^{-08} *	0
NR_2	–	–	1.1129×10^{-1} *	0
PM_{10}	$-1.0, -1.0, 2.0$	$-1.7321, -1.7321, 1$	2.2477×10^{-1164}	0
HM_6	–	–	2.5855×10^{-138}	1.6×10^{-2}
MH_5	–	–	2.6805×10^{-77}	0
NR_2	–	–	2.8563×10^{-04}	0
PM_{10}	$-3.9, -3.3, -6.2$	1.7321, 1.7321, 1	5.2666×10^{-84}	1.6×10^{-2}
HM_6	–	–	5.8363×10^{-07}	0
MH_5	–	–	1.0692×10^{-05}	1.5×10^{-2}
NR_2	–	1.9410, 1.9410, 1.1534	5.5426×10^{-1} *	0
PM_{10}	1, 1, 2	1.7321, 1.7321, 1	2.2477×10^{-1164}	1.6×10^{-2}
HM_6	–	–	2.5855×10^{-138}	0
MH_5	–	–	2.6805×10^{-77}	1.6×10^{-2}
NR_2	–	–	2.8563×10^{-04}	0
PM_{10}	5.9, 3.3, 6.2	1.7321, 1.7321, 1	5.9770×10^{-811}	1.6×10^{-2}
HM_6	–	–	4.0516×10^{-136}	1.5×10^{-2}
MH_5	–	–	5.5239×10^{-62}	0
NR_2	–	–	1.4635×10^{-2}	0
PM_{10}	2.4, 3.0, 1.0	1.7321, 1.7321, 1	0	0
HM_6	–	–	1.8162×10^{-811}	0
MH_5	–	–	5.2175×10^{-433}	1.6×10^{-2}
NR_2	–	–	4.2917×10^{-11}	0

6. Concluding Remarks

This research study is based on devising a new, highly effective three-step iterative method with tenth-order convergence. The convergence is proved theoretically via Taylor's series expansion for single and multi-variable nonlinear equations, and the approximate computational order of convergence confirms such findings. Thus, the proposed method PM_{10} is applicable not only to single nonlinear equations but also to nonlinear systems. Moreover, dynamical aspects of PM_{10} are also explored with basins of attraction that show quite esthetic phase plane diagrams when applied to complex-valued functions, thereby proving the stability of the method when initial guesses are taken within the vicinity of the underlying nonlinear model. Finally, different types of nonlinear equations and systems, including those used in physical and natural sciences, are chosen to be tested with PM_{10} and with various well-known optimal and non-optimal methods in the sense of King–Traub. In most of the cases, PM_{10} is found to have better results, particularly when it comes to the number of iterations N to achieve required accuracy, ACOC, absolute error, and absolute functional value. It is also worthwhile to note that the proposed method always converges, irrespective of whether the initial guess passed to it lies near to or away from the approximate solution. Hence, PM_{10} is a competitive iterative method with tenth-order convergence for solving nonlinear equations and systems.

We understand that methods of very high order are only of academic interest since approximations to the solutions of very high accuracy are not needed in practice. On the other hand, such methods are, to some extent, complicated and do not offer much of an increase in computational efficiency. Moreover, the method proposed in this article lies in the family of methods without memory, requiring the evaluation of three Jacobian matrices, and thereby becomes computationally expensive. To avoid computational complexity, we will propose, in future studies, a modification of the proposed method by replacing the first-order derivative with a suitable finite-difference approximation. In addition, the proposed method will also be analyzed for its semi-local convergence.

Author Contributions: Conceptualization, A.T., S.Q. and A.S.; methodology, S.Q. and D.B.; validation, A.S. and E.H.; formal analysis, S.Q., A.T. and A.A.S.; investigation, S.Q., D.B. and E.H.; data curation, S.Q. and A.A.S.; writing—original draft preparation, A.T., S.Q., A.S. and A.A.S.; writing—review and editing, A.T., S.Q., A.S., E.H. and D.B. All authors have read and agreed to the published version of the manuscript.

Funding: This research did not receive any specific external funding.

Institutional Review Board Statement: Not applicable.

Informed Consent Statement: Not applicable.

Data Availability Statement: Not applicable.

Acknowledgments: The authors extend their appreciation to the deputyship for Research & Innovation, Ministry of Education in Saudi Arabia for funding this research work through the project number (IFP-2020-64).

Conflicts of Interest: The authors declare no conflict of interest.

References

1. Ortega, J.M. *Numerical Analysis: A Second Course*; Society for Industrial and Applied Mathematics: Philadelphia, PA, USA, 1990.
2. Ham, Y.; Chun, C. A fifth-order iterative method for solving nonlinear equations. *Appl. Math. Comput.* **2007**, *194*, 287–290. [CrossRef]
3. Abro, H.A.; Shaikh, M.M. A new time-efficient and convergent nonlinear solver. *Appl. Math. Comput.* **2019**, *355*, 516–536. [CrossRef]
4. Cordero, A.; Torregrosa, J.R.; Vassileva, M.P. Design, Analysis, and Applications of Iterative Methods for Solving Nonlinear Systems. In *Nonlinear Systems-Design, Analysis, Estimation and Control*; Lee, D., Burg, T., Volos, C., Eds.; IntechOpen: Rijeka, Croatia, 2016. [CrossRef]
5. Hafiz, M.A.; Bahgat, M.S. An efficient two-step iterative method for solving system of nonlinear equations. *J. Math. Res.* **2012**, *4*, 28.
6. Noor, M.A.; Khan, W.A.; Hussain, A. A new modified Halley method without second derivatives for nonlinear equation. *Appl. Math. Comput.* **2007**, *189*, 1268–1273. [CrossRef]
7. Sharifi, M.; Babajee, D.K.R.; Soleymani, F. Finding the solution of nonlinear equations by a class of optimal methods. *Comput. Math. Appl.* **2012**, *63*, 764–774. [CrossRef]
8. Noor, M.A. Some iterative methods for solving nonlinear equations using homotopy perturbation method. *Int. J. Comput. Math.* **2010**, *87*, 141–149. [CrossRef]
9. Kung, H.T.; Traub, J.F. Optimal order of one-point and multipoint iteration. *J. ACM (JACM)* **1974**, *21*, 643–651. [CrossRef]
10. Householder, A.S. *The Numerical Treatment of a Single Nonlinear Equation*; McGraw-Hill: New York, NY, USA, 1970.
11. Bahgat, M.S.; Hafiz, M.A. Three-step iterative method with eighteenth order convergence for solving nonlinear equations. *Int. J. Pure Appl. Math.* **2014**, *93*, 85–94. [CrossRef]
12. Chun, C. Some fourth-order iterative methods for solving nonlinear equations. *Appl. Math. Comput.* **2008**, *195*, 454–459. [CrossRef]
13. Sharma, R.; Bahl, A. An optimal fourth order iterative method for solving nonlinear equations and its dynamics. *J. Complex Anal.* **2015**, *2015*, 259167. [CrossRef]
14. Geum, Y.H.; Kim, Y.I.; Neta, B. Constructing a family of optimal eighth-order modified Newton-type multiple-zero finders along with the dynamics behind their purely imaginary extraneous fixed points. *J. Comput. Appl. Math.* **2018**, *333*, 131–156. [CrossRef]
15. Qureshi, S.; Ramos, H.; Soomro, A.K. A New Nonlinear Ninth-Order Root-Finding Method with Error Analysis and Basins of Attraction. *Mathematics* **2021**, *9*, 1996. [CrossRef]
16. Ramos, H.; Monteiro, M.T.T. A new approach based on the Newton's method to solve systems of nonlinear equations. *J. Comput. Appl. Math.* **2017**, *318*, 3–13. [CrossRef]
17. Ramos, H.; Vigo-Aguiar, J. The application of Newton's method in vector form for solving nonlinear scalar equations where the classical Newton method fails. *J. Comput. Appl. Math.* **2015**, *275*, 228–237. [CrossRef]
18. Darvishi, M.T.; Barati, A. A third-order Newton-type method to solve systems of nonlinear equations. *Appl. Math. Comput.* **2007**, *187*, 630–635. [CrossRef]
19. Wang, X.; Liu, L. New eighth-order iterative methods for solving nonlinear equations. *J. Comput. Appl. Math.* **2010**, *234*, 1611–1620. [CrossRef]
20. Hu, Z.; Guocai, L.; Tian, L. An iterative method with ninth-order convergence for solving nonlinear equations. *Int. J. Contemp. Math. Sci.* **2011**, *6*, 17–23.

21. Hafiz, M.A.; Al-Goria, S.M. Solving nonlinear equations using a new tenth-and seventh-order methods free from second derivative. *Int. J. Differ. Equ. Appl.* **2013**, *12(4)*.
22. Cordero, A.; Hueso, J.L.; Martínez, E.; Torregrosa, J.R. New modifications of Potra–Pták's method with optimal fourth and eighth orders of convergence. *J. Comput. Appl. Math.* **2010**, *234*, 2969–2976. [CrossRef]
23. Lotfi, T.; Bakhtiari, P.; Cordero, A.; Mahdiani, K.; Torregrosa, J.R. Some new efficient multipoint iterative methods for solving nonlinear systems of equations. *Int. J. Comput. Math.* **2015**, 92, 1921–1934. [CrossRef]
24. Cordero, A.; Hueso, J.L.; Martínez, E.; Torregrosa, J.R. A modified Newton-Jarratt's composition. *Numer. Algorithms* **2010**, *55*, 87–99. [CrossRef]
25. Waseem, M.; Noor, M.A.; Noor, K.I. Efficient method for solving a system of nonlinear equations. *Appl. Math. Comput.* **2016**, *275*, 134–146. [CrossRef]
26. Noor, M.A.; Noor, K.I.; Al-Said, E.; Waseem, M. Some new iterative methods for nonlinear equations. *Math. Probl. Eng.* **2010**, *2010*, 198943. [CrossRef]
27. Hueso, J.L.; Martínez, E.; Torregrosa, J.R. Third and fourth order iterative methods free from second derivative for nonlinear systems. *Appl. Math. Comput.* **2009**, *211*, 190–197. [CrossRef]
28. Weerakoon, S.; Fernando, T.G.I. A variant of Newton's method with accelerated third-order convergence. *Appl. Math. Lett.* **2000**, *13*, 87–93. [CrossRef]
29. Grau-Sánchez, M.; Gutiérrez, J.M. Zero-finder methods derived from Obreshkov's techniques. *Appl. Math. Comput.* **2009**, *215*, 2992–3001. [CrossRef]
30. Petkovi ´c, M.S. Remarks on "On a general class of multipoint root-finding methods of high computational efficiency". *SIAM J. Numer. Anal.* **2011**, *49*, 1317–1319. [CrossRef]
31. Scott, M.; Neta, B.; Chun, C. Basin attractors for various methods. *Appl. Math. Comput.* **2011**, *218*, 2584–2599. [CrossRef]
32. Stewart, B.D. *Attractor Basins of Various Root-Finding Methods*; Naval Postgraduate School: Monterey CA, USA, 2001.
33. Halley, E. A new, exact, and easy method of finding the roots of any equations generally, and that without any previous reduction. *Philos. Trans. R. Soc. Lond.* **1694**, *18*, 136–145.
34. Chen, H.; Zheng, X. Improved Newton Iterative Algorithm for Fractal Art Graphic Design. *Complexity* **2020**, *2020*, 6623049. [CrossRef]
35. Susanto, H.; Karjanto, N. Newton's method's basins of attraction revisited. *Appl. Math. Comput.* **2009**, *215*, 1084–1090. [CrossRef]
36. Tao, Y.; Madhu, K. Optimal Fourth, Eighth and Sixteenth Order Methods by Using Divided Difference Techniques and Their Basins of Attraction and Its Application. *Mathematics* **2019**, *7*, 322. [CrossRef]
37. Said Solaiman, O.; Hashim, I. Efficacy of optimal methods for nonlinear equations with chemical engineering applications. *Math. Probl. Eng.* **2019**, *2019*, 1728965. [CrossRef]
38. Khoury, R.; Harder, D.W. *Numerical Methods and Modelling for Engineering*; Springer: Berlin/Heidelberg, Germany, 2016; pp. 120–124.
39. Shacham, M. Numerical solution of constrained non-linear algebraic equations. *Int. J. Numer. Methods Eng.* **1986**, *23*, 1455–1481. [CrossRef]
40. Madhu, K.; Babajee, D.K.R.; Jayaraman, J. An improvement to double-step Newton method and its multi-step version for solving system of nonlinear equations and its applications. *Numer. Algorithms* **2017**, *74*, 593–607. [CrossRef]
41. Madhu, K.; Elango, A.; Landry, R., Jr.; Al-arydah, M.T. New multi-step iterative methods for solving systems of nonlinear equations and their application on GNSS pseudorange equations. *Sensors* **2020**, *20*, 5976. [CrossRef] [PubMed]
42. Burden, R.L.; Faires, J.D. *Numerical Analysis*; Brooks/Cole: Boston, MA, USA, 2010; Volume 7.
43. Amos, G.; Subramaniam, V. *Numerical Methods for Engineers and Scientists: An Introduction with Applications Using MATLAB*; Department of Mechanical Engineering, The Ohio State University: Columbus, OH, USA, 2014.
44. Chapra, S.C.; Canale, R.P. *Numerical Methods for Engineers*; McGraw-Hill Higher Education: Boston, MA, USA, 2010.

fractal and fractional

MDPI

Article

Existence Results for the Solution of the Hybrid Caputo–Hadamard Fractional Differential Problems Using Dhage's Approach

Muhammad Yaseen [1], Sadia Mumtaz [1], Reny George [2,*] and Azhar Hussain [1]

1 Department of Mathematics, University of Sargodha, Sargodha 40100, Pakistan; yaseen.yaqoob@uos.edu.pk (M.Y.); mumtazsadia6@gmail.com (S.M.); azhar.hussain@uos.edu.pk (A.H.)
2 Department of Mathematics, College of Science and Humanities in Al-Kharj, Prince Sattam bin Abdulaziz University, Al-Kharj 11942, Saudi Arabia
* Correspondence: renygeorge02@yahoo.com

Abstract: In this work, we explore the existence results for the hybrid Caputo–Hadamard fractional boundary value problem (CH-FBVP). The inclusion version of the proposed BVP with a three-point hybrid Caputo–Hadamard terminal conditions is also considered and the related existence results are provided. To achieve these goals, we utilize the well-known fixed point theorems attributed to Dhage for both BVPs. Moreover, we present two numerical examples to validate our analytical findings.

Keywords: Caputo–Hadamard fractional derivative; thermostat modeling; Caputo–Hadamard fractional integral; hybrid Caputo–Hadamard fractional differential equation and inclusion

Citation: Yaseen, M.; Mumtaz, S.; George, R.; Hussain, A. Existence Results for the Solution of the Hybrid Caputo–Hadamard Fractional Differential Problems Using Dhage's Approach. *Fractal Fract.* **2022**, *6*, 17. https://doi.org/10.3390/fractalfract6010017

Academic Editor: Mark Edelman

Received: 6 November 2021
Accepted: 24 December 2021
Published: 30 December 2021

Publisher's Note: MDPI stays neutral with regard to jurisdictional claims in published maps and institutional affiliations.

1. Introduction

Fractional differential equations are utilized for mathematical modeling of real life problems. Scientists working in various fields of science are encouraged to improve the explanation of their findings by including more accurate knowledge into their problems. In this regard, they are employing a variety of mathematical methods in their models in which fractional order derivatives are very beneficial. Differential equations of a fractional order provide more accurate information than standard differential equations in mathematical modeling of many scientific situations. In these days, differential equations of a fractional order have been constantly utilized in chemistry, biophysics, control theory, mechanics, image processing, polymer rheology, aerodynamics, etc. [1,2].

In recent years, many researchers have been attracted by fractional hybrid differential equations and inclusions with terminal conditions [3–6]. Moreover, in various fields, there are several efforts on the Caputo–Hadamard derivative of fractional order and its implementations [7–15].

In 2010, a new class of differential models named hybrid differential equations (HDEs) was formulated by Dhage and Lakshmikantham [16]. Moreover, they investigated properties of this type of a differential equation's solution. Zhao et al. [17] generalized Dhage's effort and investigated the related HDEs of fractional order. In 2012, a fractional hybrid problem with two-point terminal conditions was presented by Sun et al. [18]

$$\begin{cases} \mathfrak{D}_0^w\left[\frac{s(\ell)}{g(\ell,s(\ell))}\right] + w(\ell,s(\ell)) = 0, & \ell \in \mathcal{E} = [0,1], w \in (1,2], \\ s(0) = s(1) = 0. \end{cases}$$

They obtained some existence results by using a fixed point theorem presented by Dhage in Banach algebra with Lipschitz and mixed Caratheodory conditions. In 2015, the existence criteria for the solutions of the hybrid Caputo problem with terminal conditions was studied by Hilal and Kajouni [19]

$$\begin{cases} {}^{C}\mathfrak{D}_0^{w}\left[\frac{s(\ell)}{g(\ell,s(\ell))}\right] + w(\ell,s(\ell)) = 0, & \ell \in \mathcal{E} = [0,L], \\ y\frac{s(0)}{g(0,s(0))} + z\frac{s(L)}{g(L,s(L))} = c, \end{cases}$$

where, $w \in (0,1)$, $y,z,c \in \Re$ with $y + z \neq 0$ and $g\colon \mathcal{E} \times \Re \longrightarrow \Re \setminus 0$ and $w\colon \mathcal{E} \times \Re \longrightarrow \Re$ are continuous functions. Baleanu et al. [20] studied the existence criteria and significance of the solution for a new kind of hybrid inclusion problem of fractional order,

$${}^{C}\mathfrak{D}_0^{w}\left(\frac{s(\ell)}{g(\ell,s(\ell)),\Im_0^{p_1}s(\ell),\ldots,\Im_0^{p_t}s(\ell))}\right) \in \mathcal{Q}(\ell,s(\ell),\Im_0^{q_1}s(\ell),\ldots,\Im_0^{q_u}s(\ell)), \quad \ell \in [0,1]$$

equipped with boundary conditions $s(0) = s_0^*$ and $s(1) = s_1^*$, where, $w \in (1,2]$, $\Im_0^{\beta}$ and ${}^{C}\mathfrak{D}_0^{w}$ symbolize the Riemann–Liouville fractional integral operator of order $\beta \in \{p_k, q_j\} \subset (0,\infty)$ for $k = 1,\ldots,t$ and $j = 1,2,\ldots,u$ and a fractional Caputo derivative of order w, respectively. In 2006, a thermostat model enclosed at $\ell = 0$ with a restrainer at $\ell = 1$ was studied by Infante and Webb [21],

$$\begin{cases} s''(\ell) + \nu(\ell,s(\ell)) = 0, & 0 \leq \ell \leq 1, \\ s'(0) = 0, & \tau s'(1) + s(\zeta) = 0, \end{cases}$$

where, $\zeta \in [0,1]$ is a real constant and $\tau > 0$ is a positive number. The thermostat includes or excludes heat based on the temperature exposed by the sensor at $\ell = \zeta$ by using this second order approach. They applied a fixed point criteria on Hammerstein integral equations to obtain the existence results for the above BVP. A fractional order problem was presented by Nieto and Pimentel [22],

$$\begin{cases} {}^{C}\mathfrak{D}_0^{w}s(\ell) + \nu(\ell,s(\ell)) = 0, & \ell \in [0,1], \\ s'(0) = 0, \quad \tau\,{}^{C}\mathfrak{D}_0^{w-1}s(1) + s(\zeta) = 0, \end{cases}$$

where, $\zeta \in [0,1]$, $\tau > 0$ is any positive real number and ${}^{C}\mathfrak{D}_0^{w}$ represents the fractional Caputo derivative of order $w \in (1,2]$. It is noticeable that a thermostat is a main component which plays an important role in physical systems to maintain its temperature near a required set-point, which motivated many researchers to study the various models of thermostat systems. In 2020, Baleanu et al. [4] constructed the following Caputo fractional hybrid problem for thermostat model,

$${}^{C}\mathfrak{D}_0^{w}\left(\frac{s(\ell)}{g(\ell,s(\ell))}\right) + \nu(\ell,s(\ell)) = 0, \qquad \ell \in [0,1],$$

supplemented with the hybrid terminal conditions,

$$\begin{cases} \mathfrak{D}\left(\frac{s(\ell)}{g(\ell,s(\ell))}\right)\Big|_{\ell=0} = 0, \\ \tau\,{}^{C}\mathfrak{D}_0^{w-1}\left(\frac{s(\ell)}{g(\ell,s(\ell))}\right)\Big|_{\ell=1} + \left(\frac{s(\ell)}{g(\ell,s(\ell))}\right)\Big|_{\ell=\zeta} = 0, \end{cases}$$

where, $w \in (1,2]$, $w - 1 \in (0,1]$, $\mathfrak{D} = \frac{d}{d\ell}$, τ is any positive real number, $\zeta \in [0,1]$ and ${}^{C}\mathfrak{D}_0^{\beta}$ is the fractional Caputo derivative of order $\beta \in \{w, w-1\}$. Furthermore, $\nu\colon \mathcal{E} \times \Re \to \Re$ and $g\colon \mathcal{E} \times \Re \to \Re \setminus \{0\}$ are continuous functions. Moreover, they studied the related hybrid Caputo inclusion model of a fractional order for a thermostat system, as given below:

$${}^{C}\mathfrak{D}_0^{w}\left(\frac{s(\ell)}{g(\ell,s(\ell))}\right) \in \varpi(\ell,s(\ell)), \quad \ell \in \mathcal{E} = [0,1],$$

supplemented with three-point hybrid Caputo terminal conditions,

$$\begin{cases} \mathfrak{D}\left(\frac{s(\ell)}{g(\ell,s(\ell))}\right)\Big|_{\ell=0} = 0, \\ \tau^{C}\mathfrak{D}_0^{w-1}\left(\frac{s(\ell)}{g(\ell,s(\ell))}\right)\Big|_{\ell=1} + \left(\frac{s(\ell)}{g(\ell,s(\ell))}\right)\Big|_{\ell=\zeta} = 0, \end{cases}$$

where, $\varpi\colon \mathcal{E}\times\Re \to \mathcal{I}(\Re)$ is a multi-valued map. Motivated by the previous studies, we construct the following hybrid CH-FBVP for thermostat model

$$^{CH}\mathfrak{D}_{1^+}^{w}\left(\frac{s(\ell)}{g(\ell,s((\ell))}\right) + \eta(\ell,s(\ell)) = 0, \qquad \ell\in[1,e], \tag{1}$$

supplemented with the three-point hybrid terminal conditions,

$$\begin{cases} \mathfrak{D}\left(\frac{s(\ell)}{g(\ell,s(\ell))}\right)\Big|_{\ell=1} = 0, \\ \tau^{CH}\mathfrak{D}_{1^+}^{w-1}\left(\frac{s(\ell)}{g(\ell,s(\ell))}\right)\Big|_{\ell=e} + \left(\frac{s(\ell)}{g(\ell,s(\ell))}\right)\Big|_{\ell=\zeta} = 0, \end{cases} \tag{2}$$

where, $w\in(1,2], w-1\in(0,1], \mathfrak{D} = \frac{d}{d\ell}, \tau\in\Re^+, \zeta\in[1,e]$ and $^{CH}\mathfrak{D}_{1^+}^{\alpha}$ is the fractional Caputo–Hadamard derivative of order $\alpha\in\{w,w-1\}$. Moreover, $\eta\colon \mathcal{E}\times\Re\to\Re$ and $g\colon \mathcal{E}\times\Re\to\Re\setminus\{0\}$ are continuous functions. Moreover, we study the related hybrid Caputo–Hadamard fractional inclusion boundary value problem (CH–FIBVP) for thermostat system as given below:

$$^{CH}\mathfrak{D}_{1^+}^{w}\left(\frac{s(\ell)}{g(\ell,s(\ell))}\right) \in \varpi(\ell,s(\ell)), \quad \ell\in\mathcal{E}=[1,e] \tag{3}$$

supplemented with three-point hybrid Caputo–Hadamard boundary conditions

$$\begin{cases} \mathfrak{D}\left(\frac{s(\ell)}{g(\ell,s(\ell))}\right)\Big|_{\ell=1} = 0, \\ \tau^{CH}\mathfrak{D}_{1^+}^{w-1}\left(\frac{s(\ell)}{g(\ell,s(\ell))}\right)\Big|_{\ell=e} + \left(\frac{s(\ell)}{g(\ell,s(\ell))}\right)\Big|_{\ell=\zeta} = 0, \end{cases} \tag{4}$$

where, $\varpi\colon [1,e]\times\Re\to\mathcal{I}(\Re)$ is a multi-valued map.

The main motivation behind this work is that there are no research manuscripts based on the authors' knowledge on the problems involving Caputo–Hadamard hybrid fractional boundary conditions. Furthermore, the proposed structure is expressed in a unique and broad manner, allowing us to explore certain specific cases previously addressed (see for example [4]). Here, we establish certain analytical criteria to validate the suggested novel existence results of hybrid Caputo–Hadamard fractional differential problems. The method used to accomplish the goals is based on Dhage's fixed point result.

The following structure is used to arrange the current manuscript. In Section 2, we collected the basic concepts regarding Caputo–Hadamard fractional operators and some requisite notions which are related to multi-valued mappings. In Section 3, we present the existence of a solution to both problems utilizing Dhage's analytical criteria. Section 4 presents two numerical examples to demonstrate the applicability of our analytical conclusions. The concluding remarks are addressed in Section 5.

2. Preliminaries

In this section, we present some basic definitions and notations utilized in the proof of the main results. Let $w\geq 0$ and suppose that the function $s\colon (a,b)\to\Re$ is integrable. Fractional Caputo–Hadamard integral of a function $s\in C((a,b),\Re)$ of order w is presented by $^{CH}\Im_{a^+}^{0}(s(\ell)) = s(\ell)$ and

$$^{CH}\Im_{a^+}^{w}(s(\ell)) = \frac{1}{\Gamma(w)}\int_a^{\ell}\left(\ln\frac{\ell}{x}\right)^{w-1} s(x)\,\frac{dx}{x}$$

whenever the RHS integral exists. Keep in mind that for each $w_1, w_2 \in \Re^+$, the following equality

$$^{CH}\mathfrak{I}^{w_1}_{a^+}\, ^{CH}\mathfrak{I}^{w_2}_{a^+}(s(\ell)) = {}^{CH}\mathfrak{I}^{w_1+w_2}_{a^+}(s(\ell))$$

holds true and

$$^{CH}\mathfrak{I}^{w_1}_{a^+}\left(\ln\frac{\ell}{a}\right)^{w_2} = \frac{\Gamma(w_2+1)}{\Gamma(w_1+w_2+1)}\left(\ln\frac{\ell}{a}\right)^{w_1+w_2}$$

for $\ell > a$. It is noticeable that for $w_2 = 0$ [2], the above equation reduces to

$$^{CH}\mathfrak{I}^{w_1}_{a^+}1 = \frac{1}{\Gamma(w_1+1)}\left(\ln\frac{\ell}{a}\right)^{w_1}, \forall\, \ell > a.$$

Now, assume $\beta = [w] + 1$ or $w \in [\beta - 1, \beta)$. For a real-valued continuous function s defined on (a, b), the fractional Caputo–Hadamard derivative of order w is defined as follows:

$$^{CH}\mathfrak{D}^{w}_{a^+}(s(\ell)) = \frac{1}{\Gamma(\beta - w)}\left(\ell\frac{d}{d\ell}\right)^{\beta}\int_a^{\ell}\left(\ln\frac{\ell}{x}\right)^{(\beta-w-1)} s(x)\,\frac{dx}{x}$$

whenever the RHS integral exists [17]. Assuming $s \in AC^m_{\Re}([a,b])$ and $\beta - 1 < w \leq \beta$, a general solution for the Caputo–Hadamard differential equation $^{CH}\mathfrak{D}^{w}_{a^+}(s(\ell)) = 0$ is of the form $s(\ell) = \sum_{k=0}^{\beta-1} d_k\left(\ln\frac{\ell}{a}\right)^k$, and we have

$$^{CH}\mathfrak{I}^{w}_{a^+}\, ^{CH}\mathfrak{D}^{w}_{a^+}(s(\ell)) = s(\ell) + d_0 + d_1\left(\ln\frac{\ell}{a}\right) + d_2\left(\ln\frac{\ell}{a}\right)^2 + \ldots + d_{\beta-1}\left(\ln\frac{\ell}{a}\right)^{\beta-1},$$

where, $d_0, d_1, \ldots, d_{\beta-1}$ are real constants and $\beta = [w] + 1$ [23].

In the sequel, we assume $(\mathcal{N}, \|.\|_{\mathcal{N}})$ a normed space, collection of all subsets of $\mathcal{N}$, all compact subsets of $\mathcal{N}$, all convex subsets of $\mathcal{N}$, all bounded subsets of $\mathcal{N}$, all closed subsets of $\mathcal{N}$ by $\mathcal{I}(\mathcal{N}), \mathcal{I}_{cmp}(\mathcal{N}), \mathcal{I}_{cvx}(\mathcal{N}), \mathcal{I}_{bnd}(\mathcal{N}), \mathcal{I}_{cls}(\mathcal{N})$, respectively. Moreover, the following notions from [24,25] are essential:

- If for each $\sigma \in \mathcal{N}$, the set $\varpi(\sigma)$ has convex values, then we say that the set-valued map ϖ is convex.
- The set-valued map ϖ is said to be an upper semi-continuous map if for every $\sigma^* \in \mathcal{N}$, $\varpi(\sigma^*)$ belongs to $\mathcal{I}_{cls}(\mathcal{N})$ and for every open set $\mathcal{O}$ with $\varpi(\sigma^*) \subset \mathcal{O}$, there is a neighborhood $\mathcal{U}_0^*$ of σ^* such that $\varpi(\mathcal{U}_0^*) \subset \mathcal{O}$.
- The set-valued map $\varpi\colon \mathcal{N} \to \mathcal{I}(\mathcal{N})$ has a fixed point $\sigma^* \in \mathcal{N}$ if $\sigma^* \in \varpi(\sigma^*)$. The collection of all fixed points of ϖ is represented by $\Upsilon(\varpi)$.
- Assume $(\mathcal{N}, d_{\mathcal{N}})$ to be a metric space. For each $F_1, F_2 \in \mathcal{I}(\mathcal{N})$, the Pompeiu–Hausdorff metric $PH_d\colon \mathcal{I}(\mathcal{N}) \times \mathcal{I}(\mathcal{N}) \to \Re \cup \{\infty\}$ is defined as

$$PH_d(F_1, F_2) = \max\{\sup_{a_1\in F_1} d_N(a_1, F_2), \sup_{a_2\in F_2} d_N(F_1, a_2)\},$$

  where, $d_{\mathcal{N}}(F_1, a_2) = \inf_{a_1\in F_1} d_{\mathcal{N}}(a_1, a_2)$ and $d_{\mathcal{N}}(a_1, F_2) = \inf_{a_2\in F_2} d_{\mathcal{N}}(a_1, a_2)$.
- The set-valued map $\varpi\colon \mathcal{N} \to \mathcal{I}_{cls}(\mathcal{N})$ is said to be Lipschitzian if $PH_{d_N}(\varpi(\sigma_1), \varpi(\sigma_2)) \leq m^* d_{\mathcal{N}}(\sigma_1, \sigma_2)$ holds for every $\sigma_1, \sigma_2 \in \mathcal{N}$, where $m^* > 0$ is a Lipschitz constant. If $0 < m^* < 1$, then we say that the Lipschitz map is a contractive map.
- We say that $\varpi\colon [1, e] \to \mathcal{I}_{cls}(\Re)$ is measurable if the function $\ell \to d_{\mathcal{N}}(r, \varpi(\ell))$ is measurable $\forall\, r \in \Re$.
- The graph of $\varpi\colon \mathcal{N} \to \mathcal{I}_{cls}(\mathcal{T})$ is defined by $Graph(\varpi) = \{(\sigma_1, \sigma_2) \in \mathcal{N} \times \mathcal{T}\colon s^* \in \varpi(\sigma)\}$. It is noticeable that the graph of ϖ is said to be closed if for every arbitrary sequence $\{s_n\}_{n\geq 1} \in \mathcal{N}$ and $\{\sigma_n\}_{n\geq 1} \in \mathcal{T}$ with $s_n \to s_0, \sigma_n \to \sigma_0$ and $\sigma_n \in \varpi(s_n)$, we obtain $\sigma_0 \in \varpi(s_0)$. If $\varpi\colon \mathcal{N} \to \mathcal{I}_{cls}(\mathcal{T})$ is an upper semi-continuous map, then $Graph(\varpi) \subseteq \mathcal{N} \times \mathcal{T}$ is a closed set.

- A set-valued map ϖ is completely continuous operator if the $\varpi(\mathcal{M})$ is relatively compact $\forall \mathcal{M} \in \mathcal{I}_{bnd}(\mathcal{N})$. Furthermore, we assume that by the complete continuity assumption, the map ϖ has a closed graph. Then, the multi-valued map ϖ is upper semi-continuous.
- The set-valued map $\varpi\colon [1,e] \times \Re \to \mathcal{I}(\Re)$ has a Caratheodory property if the function $\sigma \to \varpi(\ell,\sigma)$ is upper semi-continuous $\forall \ell \in [1,e]$ and the function $\ell \to \varpi(\ell,\sigma)$ is measurable for every $\sigma \in \Re$. Furthermore, A Caratheodory multi-valued mapping $\varpi\colon [1,e] \times \Re \to \mathcal{I}(\Re)$ has $\mathcal{L}^1$-Caratheodory property if for every $s > 0$ there exists $\theta_s \in \mathcal{L}^1_{\Re+}([1,e])$ provided that

$$\|\varpi(\ell,\sigma)\| = \sup_{\ell \in [1,e]} \{|p|\colon p \in \varpi(\ell,\sigma)\} \leq \theta_s(\ell)$$

  for all $\ell \in [1,e]$ and for every $|\sigma| \leq s$.
- The selections of ϖ at $\sigma \in C_{\Re}([1,e])$ are represented by

$$(\mathfrak{SEL})_{\varpi,\sigma} := \{v \in \mathcal{L}^1_{\Re+}([1,e])\colon v(\ell) \in \varpi(\ell,\sigma(\ell)),\ \forall \ell \in [1,e]\}.$$

  It is known that $(\mathfrak{SEL})_{\varpi,\sigma} \neq \varphi$ for all $\sigma \in C_{\mathcal{N}}([1,e])$ whenever $dim\, \mathcal{N} < \infty$.

We now state fixed point results due to Dhage and a closed graph theorem, which will be used to prove the existence results of our proposed problems.

Theorem 1 ([26]). *Assume a Banach space $\mathcal{N}$. For almost all $\lambda \in \Re^+$, assume an open ball $v_\lambda(0)$ and its closure $\overline{v}_\lambda(0)$. Suppose that $\theta_1\colon \mathcal{N} \to \mathcal{N}$ and $\theta_2\colon \overline{v}_\lambda(0) \to \mathcal{N}$ are two operators that meet the properties listed*

1. *θ_1 is a Lipschitzian map so that m^* is a Lipschitz constant;*
2. *θ_2 is completely continuous;*
3. *$m^*\triangle^* < 1$, where $\triangle^* = \|\theta_2(\overline{v}_\lambda(0))\|_{\mathcal{N}} = \sup\{\|\theta_2 l\|_{\mathcal{N}}\colon l \in \overline{v}_\lambda(0)\}$.*

Then either

(i) *The operator equation $\theta_1\ell\theta_2\ell = \ell$ has a solution contained in $\overline{v}_\lambda(0)$ or;*
(ii) *There exists $\vartheta^* \in \mathcal{N}$ with $\|\vartheta^*\|_{\mathcal{N}} = \lambda$ so that $w_0\theta_1\vartheta^*\theta_2\vartheta^* = \vartheta^*$ for some $w_0 \in (0,1)$.*

Theorem 2 ([27]). *Assume a separable Banach space $\mathcal{N}$, an $\mathcal{L}^1$-Caratheodory multi-valued function $\varpi\colon [1,e] \times \mathcal{N} \to \mathcal{I}_{cmp,cvx}(\mathcal{N})$ and a linear continuous function $\Lambda\colon \mathcal{L}^1_{\mathcal{N}}([1,e]) \to C_{\mathcal{N}}([1,e])$. Then, $\Lambda\ o\ ((\mathfrak{SEL})_\varpi\colon C_{\mathcal{N}}([1,e]) \to \mathcal{I}_{cmp,cvx}(C_{\mathcal{N}}([1,e]))$ is an operator belonging to $C_{\mathcal{N}}([1,e]) \times C_{\mathcal{N}}([1,e])$ defined by $\sigma \to (\lambda\ o\ (\mathfrak{SEL})_\varpi)(\sigma) = \lambda((\mathfrak{SEL})_{\varpi,\sigma})$ and has a closed graph property.*

Theorem 3 ([28]). *Assume the Banach space $\mathcal{N}$. Consider that there is a single-valued map $\theta_1\colon \mathcal{N} \to \mathcal{N}$ and a multi-valued map $\theta_2\colon \mathcal{N} \to \mathcal{I}_{cmp,cvx}(\mathcal{N})$ satisfying the following properties:*

1. *θ_1 is a Lipschitzian map so that m^* is a Lipschitz constant;*
2. *θ_2 has compactness and an upper-semi continuity property;*
3. *$2m^*\triangle^* < 1$ with $\triangle^* = \|\theta_2(\mathcal{N})\|$.*

Then either

(i) *there exists a solution contained in $\mathcal{N}$ for the inclusion $\ell \in \theta_1\ell\theta_2 l$ or;*
(ii) *the set $O = \{\vartheta^* \in \mathcal{N} | w_0\vartheta^* \in \theta_1\vartheta^*\theta_2\vartheta^*, w_0 > 1\}$ is an unbounded set.*

3. Main Results

Here, we assume $\mathcal{N} = C_{\Re}([1,e])$ to be a Banach space with the standard norm, $\|s\|_{\mathcal{N}} = \sup\{|s(\ell)|\colon \ell \in [1,e]\}$.

Lemma 1. *Consider $\gamma \in \mathcal{N}$. The hybrid CH-FBVP,*

$$ {}^{CH}\mathfrak{D}^{w}_{1^+}\left(\frac{s(\ell)}{g(\ell,s(\ell))}\right) + \gamma(\ell) = 0, \quad \ell \in [1,e], w \in (1,2], \tag{5} $$

supplemented with the three-point hybrid Caputo–Hadamard terminal conditions

$$ \begin{aligned} &\mathfrak{D}\left(\frac{s(\ell)}{g(\ell,s(\ell))}\right)\Big|_{\ell=1} = 0, \\ &\tau\,{}^{CH}\mathfrak{D}^{w-1}_{1^+}\left(\frac{s(\ell)}{g(\ell,s(\ell))}\right)\Big|_{\ell=e} + \left(\frac{s(\ell)}{g(\ell,s(\ell))}\right)\Big|_{\ell=\zeta} = 0, \end{aligned} \tag{6} $$

has a solution s_0, iff s_0 is a solution for the Caputo–Hadamard integral equation,

$$ \begin{aligned} s(\ell) = g(\ell,s(\ell))\Bigg[&-\frac{1}{\Gamma(w)}\int_1^{\ell}\left(\ln\frac{\ell}{x}\right)^{w-1}\gamma(x)\,\frac{dx}{x} + \tau\int_1^{e}\gamma(x)\,\frac{dx}{x} \\ &+\frac{1}{\Gamma(w)}\int_1^{\zeta}\left(\ln\frac{\zeta}{x}\right)^{w-1}\gamma(x)\,\frac{dx}{x}\Bigg]. \end{aligned} \tag{7} $$

Proof. Assume that hybrid Equation (5) has a solution s_0 . Then, by utilizing the equality, $\frac{s_0(\ell)}{g(\ell,s_0(\ell))} = -{}^{CH}\mathfrak{I}^{w}_{1^+}\gamma(\ell) + c_0 + c_1\ln(\ell)$, where c_0 and $c_1 \in \Re$, the homogeneous Equation (5) has general solution given below:

$$ s_0(\ell) = g(\ell,s_0(\ell))\left[-\frac{1}{\Gamma(w)}\int_1^{\ell}\left(\ln\frac{\ell}{x}\right)^{w-1}\gamma(x)\,\frac{dx}{x} + c_0 + c_1\ln(\ell)\right]. \tag{8} $$

Applying ordinary derivative $\mathfrak{D} = \frac{d}{d\ell}$ on (8), we get

$$ \begin{aligned} \mathfrak{D}\left(\frac{s_0}{g(\ell,s_0(\ell))}\right) &= -\frac{1}{\Gamma(w)}\int_1^{\ell}(w-1)\left(\ln\frac{\ell}{x}\right)^{w-2}\frac{\gamma(x)}{\ell}\,dx + \frac{1}{\ell}c_1 \\ &= -\frac{1}{\Gamma(w-1)}\int_1^{\ell}\left(\ln\frac{\ell}{x}\right)^{w-2}\frac{\gamma(x)}{\ell}\,dx + \frac{1}{\ell}c_1. \end{aligned} $$

Using the first condition given in (6), we get $c_1 = 0$. Now, by applying ${}^{CH}\mathfrak{D}^{w-1}_{1^+}$ on both sides of (8), we have

$$ {}^{CH}\mathfrak{D}^{w-1}_{1^+}\left(\frac{s_0(\ell)}{g(\ell,s_0(\ell))}\right) = -\int_1^{\ell}\gamma(x)\,\frac{dx}{x}. $$

$$ \Rightarrow \tau\,{}^{CH}\mathfrak{D}^{w-1}_{1^+}\left(\frac{s_0(\ell)}{g(\ell,s_0(\ell))}\right)\Big|_{\ell=e} = -\tau\int_1^{e}\gamma(x)\,\frac{dx}{x}. $$

$$ \Rightarrow \left(\frac{s_0(\ell)}{g(\ell,s_0(\ell))}\right)\Big|_{\ell=\zeta} = -\frac{1}{\Gamma(w)}\int_1^{\zeta}\left(\ln\frac{\zeta}{x}\right)^{w-1}\gamma(x)\,\frac{dx}{x} + c_0. $$

Using the second condition given in (6), we obtain

$$c_0 = \tau \int_1^e \gamma(x) \frac{dx}{x} + \frac{1}{\Gamma(w)} \int_1^\zeta \left(\ln \frac{\zeta}{x}\right)^{w-1} \gamma(x) \frac{dx}{x}.$$

Now, by using the values of c_0 and c_1 in (8), we obtain

$$s_0(\ell) = g(\ell, s_0(\ell)) \left[-\frac{1}{\Gamma(w)} \int_1^\ell \left(\ln \frac{\ell}{x}\right)^{w-1} \gamma(x) \frac{dx}{x} + \tau \int_1^e \gamma(x) \frac{dx}{x} \right.$$
$$\left. + \frac{1}{\Gamma(w)} \int_1^\zeta \left(\ln \frac{\zeta}{x}\right)^{w-1} \gamma(x) \frac{dx}{x} \right].$$

This implies that the fractional integral Equation (7) has a unique solution s_0. In the reverse order, it is easy to see that if the fractional integral Equation (7) has a solution s_0, then s_0 satisfies the fractional hybrid CH-FBVP (5) and (6). □

Now, we provide the existence result for the solution of problem (1) and (2).

Theorem 4. *Suppose that $g \in C([1,e] \times \Re, \Re \setminus \{0\})$ and $\eta \in C([1,e] \times \Re, \Re)$ and*
$(\mathcal{S}_1)$ there exists bounded function $k\colon [1,e] \to \Re^+$ such that $\forall s_1, s_2 \in \Re$, we have

$$|g(\ell, s_1) - g(\ell, s_2)| \leq k(\ell)|s_1(\ell) - s_2(\ell)|$$

for all $s_1, s_2 \in \Re$ and $\ell \in [1,e]$,
$(\mathcal{S}_2)$ there exists a continuous increasing map $\mathcal{M}\colon [0,\infty) \to (0,\infty)$ and a continuous map $h\colon [1,e] \to \Re^+$ such that

$$|\eta(\ell, s)| \leq h(\ell)\mathcal{M}(\|s\|), \ \forall\, \ell \in [1,e] \text{ and } s \in \Re, \tag{9}$$

$(\mathcal{S}_3)$ there exits a number $\epsilon \in \Re^+$ such that

$$\epsilon > \frac{G^* \Delta^* H^* \mathcal{M}(\|s\|)}{1 - k^* \Delta^* H^* \mathcal{M}(\|s\|)}, \tag{10}$$

where, $G^* = \sup_{\ell \in [1,e]} |g(\ell, 0)|$, $H^* = \sup_{\ell \in [1,e]} |h(\ell)|$, $k^* = \sup_{\ell \in [1,e]} |k(\ell)|$ and

$$\Delta^* = \frac{1}{\Gamma(w+1)} + \frac{(\ln \zeta)^w}{\Gamma(w+1)} + \tau. \tag{11}$$

If $k^* \Delta^* H^* \mathcal{M}(\|s\|) < 1$, then the hybrid CH–FBVP, (1)–(2) has unique solution.

Proof. Assume a closed ball $\overline{\mu}_\epsilon(0)\colon = \{s \in \mathcal{N}\colon \|s\|_{\mathcal{N}} \leq \epsilon\}$ in the Banach space $\mathcal{N}$, where ϵ meets the inequality (10). By utilizing fractional integral Equation (7) and Lemma 1, we define two operators $\mathcal{A}_1, \mathcal{A}_2\colon \overline{\mu}_\epsilon(0) \to \mathcal{N}$ by $(\mathcal{A}_1 s)(\ell) = g(\ell, s(\ell))$ and

$$(\mathcal{A}_2 s)(\ell) = -\int_1^\ell \frac{1}{\Gamma(w)} \left(\ln \frac{\ell}{x}\right)^{w-1} \eta(x, s(x)) \frac{dx}{x} + \tau \int_1^e \eta(x, s(x)) \frac{dx}{x}$$
$$+ \int_1^\zeta \frac{1}{\Gamma(w)} \left(\ln \frac{\zeta}{x}\right)^{w-1} \eta(x, s(x)) \frac{dx}{x}.$$

Obviously, $s \in \mathcal{N}$ is a solution for the fractional hybrid BVP (1) and (2) and satisfies the operator equation $\mathcal{A}_1 s \mathcal{A}_2 s = s$. By utilizing the conditions of Theorem 1, we show that

a solution exists. First, we show that the operator $\mathcal{A}_1$ is a Lipschitzian map on normed algebra $\mathcal{N}$ having constant $k^* = \sup_{[1,e]} |k(\ell))|$.
Let $s_1, s_2 \in \mathcal{N}$. By using assumption $(\mathcal{S}_1)$, we get

$$|(\mathcal{A}_1 s_1)(\ell) - (\mathcal{A}_1 s_2)(\ell)| = |g(\ell, s_1(\ell)) - g(\ell, s_2(\ell))| \leq k(\ell)|s_1(\ell) - s_2(\ell)|$$

for all $s_1, s_2 \in \overline{\mu}_\epsilon(0)$. This means that $\mathcal{A}_1$ is a Lipschitzian map on $\overline{\mu}_\epsilon(0)$ having a Lipschitz constant k^*. Now, we show the complete continuity of the operator $\mathcal{A}_2$ on $\overline{\mu}_\epsilon(0)$. Firstly, it is required to check that the map $\mathcal{A}_2$ is continuous on $\overline{\mu}_\epsilon(0)$. Assume $\{s_n\}$ is a sequence in the closed ball $\overline{\mu}_\epsilon(0)$ with $s_n \to s$, where $s \in \overline{\mu}_\epsilon(0)$.

As we know η is continuous on $[1, e] \times \Re$, we conclude that $\lim_{n\to\infty} \eta(\ell, s_n(\ell)) = \eta(\ell, s(\ell))$. With the help of the Lebesgue dominated convergence theorem, we get

$$\begin{aligned}
\lim_{n\to\infty}(\mathcal{A}_2 s_n)(\ell) = &-\int_1^\ell \frac{1}{\Gamma(w)}\left(\ln\frac{\ell}{x}\right)^{w-1} \lim_{n\to\infty} \eta(x, s_n(x))\,\frac{dx}{x} \\
&+ \tau\int_1^e \lim_{n\to\infty} \eta(x, s_n(x))\,\frac{dx}{x} \\
&+ \int_1^\zeta \frac{1}{\Gamma(w)}\left(\ln\frac{\zeta}{x}\right)^{w-1} \lim_{n\to\infty} \eta(x, s_n(x))\,\frac{dx}{x} \\
= &-\int_1^\ell \frac{1}{\Gamma(w)}\left(\ln\frac{\ell}{x}\right)^{w-1} \lim_{n\to\infty} \eta(x, s(x))\,\frac{dx}{x} \\
&+ \tau\int_1^e \lim_{n\to\infty} \eta(x, s(x))\,\frac{dx}{x} \\
&+ \int_1^\zeta \frac{1}{\Gamma(w)}\left(\ln\frac{\zeta}{x}\right)^{w-1} \lim_{n\to\infty} \eta(x, s(x))\,\frac{dx}{x} \\
= &\ (\mathcal{A}_2 s)(\ell), \ \forall\, \ell \in [1, e].
\end{aligned}$$

Thus, $\mathcal{A}_2 s_n \to \mathcal{A}_2 s$ and so $\mathcal{A}_2$ is a continuous operator on $\overline{\mu}_\epsilon(0)$. Now, we check the uniform boundedness of the operator $\mathcal{A}_2$ on $\overline{\mu}_\epsilon(0)$. By assumption $(\mathcal{S}_1)$, we get

$$\begin{aligned}
|(\mathcal{A}_2 s)(\ell)| = &\int_1^\ell \frac{1}{\Gamma(w)}\left(\ln\frac{\ell}{x}\right)^{w-1} \lim_{n\to\infty} |\eta(x, s(x))|\,\frac{dx}{x} + \tau\int_1^e \lim_{n\to\infty} |\eta(x, s(x))|\,\frac{dx}{x} \\
&+ \int_1^\zeta \frac{1}{\Gamma(w)}\left(\ln\frac{\zeta}{x}\right)^{w-1} \lim_{n\to\infty} |\eta(x, s(x))|\,\frac{dx}{x} \\
\leq &\ \frac{(\ln\ell)^w}{\Gamma(w+1)} h(x)\mathcal{M}(\|s\|) + \tau h(x)\mathcal{M}(\|s\|) + \frac{(\ln\zeta)^w}{\Gamma(w+1)} h(x)\mathcal{M}(\|s\|) \\
= &\ h(x)\mathcal{M}(\|s\|)\left[\frac{(\ln\ell)^w}{\Gamma(w+1)} + \tau + \frac{(\ln\zeta)^w}{\Gamma(w+1)}\right],
\end{aligned}$$

$\forall\, \ell \in [1, e]$ and $s \in \overline{\mu}_\epsilon(0)$. By taking supremum over $[1, e]$, we have

$$\|\mathcal{A}_2 s\| \leq H^* \mathcal{M}(\|s\|)\triangle^*,$$

where, $\triangle^*$ is given in (11). This implies that in normed algebra $\mathcal{N}$, the set $\mathcal{A}_2(\overline{\mu}_\epsilon(0))$ is uniformly bounded. The equi-continuity of the operator $\mathcal{A}_2$ is now being explored. For this purpose, we suppose $\ell_1, \ell_2 \in [1, e]$ with $\ell_1 < \ell_2$. Then, we have

$$\begin{aligned}
|(\mathcal{A}_2 s)(\ell_2) - (\mathcal{A}_2 s)(\ell_1)| &= \left| \int_1^{\ell_2} \frac{1}{\Gamma(w)} \left(\ln \frac{\ell_2}{x} \right)^{w-1} \eta(x, s(x)) \frac{dx}{x} \right. \\
&\left. - \int_1^{\ell_1} \frac{1}{\Gamma(w)} \left(\ln \frac{\ell_1}{x} \right)^{w-1} \eta(x, s(x)) \frac{dx}{x} \right| \\
&\leq H^* \mathcal{M}(\|s\|) \left[\int_1^{\ell_1} \left(\frac{1}{\Gamma(w)} (\ln \frac{\ell_2}{x})^{w-1} - \frac{1}{\Gamma(w)} (\ln \frac{\ell_1}{x})^{w-1} \right) \frac{dx}{x} \right. \\
&\left. + \int_{\ell_1}^{\ell_2} \frac{1}{\Gamma(w)} (\ln \frac{\ell_2}{x})^{w-1} \frac{dx}{x} \right].
\end{aligned}$$

It is noticeable that the RHS of the above inequality approaches zero independent of $s \in \overline{\mu}_\epsilon(0)$ as $\ell_1 \to \ell_2$. Thus, the operator $\mathcal{A}_2$ is equi-continuous. By utilizing the Arzela–Ascoli theorem, it is inferred that $\mathcal{A}_2$ is completely continuous on $s \in \overline{\mu}_\epsilon(0)$.
Now, by utilizing $(\mathcal{S}_3)$, we have

$$\begin{aligned}
M_0^* &= \|\mathcal{A}_2(\overline{\mu}_\epsilon(0))\|_{\mathcal{N}} = \sup\{|\mathcal{A}_2 s| : s \in \overline{\mu}_\epsilon(0)\} \\
&= H^* \mathcal{M}(\|s\|) \left[\frac{1}{\Gamma(w+1)} + \tau + \frac{(\ln \zeta)^w}{\Gamma(w+1)} \right] \\
&= H^* \mathcal{M}(\|s\|) \triangle^*.
\end{aligned}$$

Setting $l^* = k^*$, we get $M_0^* l^* < 1$. So, one of the condition (i) or (ii) in Theorem 1 is satisfied. For any $\nu \in (0,1)$, assume that s satisfies the operator equation, $s = \nu \mathcal{A}_1 \mathcal{A}_2 s$ so that $\|s\| = \epsilon$ and we have

$$\begin{aligned}
|s(\ell)| &= \nu |(\mathcal{A}_1 s)(\ell)||(\mathcal{A}_2 s)(\ell)| = \nu |g(\ell, s(\ell))| \\
&\times \left| - \int_1^{\ell} \frac{1}{\Gamma(w)} (\ln \frac{\ell}{x})^{w-1} \eta(x, s(x)) \frac{dx}{x} \right. \\
&\left. + \tau \int_1^{e} \eta(x, s(x)) \frac{dx}{x} \int_1^{\zeta} \frac{1}{\Gamma(w)} (\ln \frac{\zeta}{x})^{w-1} \eta(x, s(x)) \frac{dx}{x} \right| \\
&\leq (|g(\ell, s(\ell)) - g(\ell, 0))| + |g(\ell, 0)|) \\
&\times \left(\int_1^{\ell} \frac{1}{\Gamma(w)} (\ln \frac{\ell}{x})^{w-1} |\eta(x, s(x))| \frac{dx}{x} \right. \\
&\left. + \tau \int_1^{e} |\eta(x, s(x))| \frac{dx}{x} \int_1^{\zeta} \frac{1}{\Gamma(w)} (\ln \frac{\zeta}{x})^{w-1} |\eta(x, s(x))| \frac{dx}{x} \right) \\
&\leq (k(\ell)|s(\ell)| + G^*) \triangle^* H^* \mathcal{M}(\|s\|) \\
&\leq (k^* \|s\| + G^*) \triangle^* H^* \mathcal{M}(\|s\|).
\end{aligned}$$

So $\epsilon \leq \frac{G^* \triangle^* H^* \mathcal{M}(\|s\|)}{1 - k^* \triangle^* H^* \mathcal{M}(\|s\|)}$ which is inconsistent with (10). This implies that the condition (ii) of Theorem 1 is not possible. Hence, the condition (i) in Theorem 1 is satisfied and the fractional hybrid problem (1) and (2) has a solution. □

From the above theorem, we conclude that the solution of the hybrid Caputo–Hadamard fractional differential equation exists provided that the stated conditions hold.

Now, we provide our main results related to hybrid CH–FIBVP for the thermostat model (3) and (4).

Definition 1. *We call the function $s \in AC_{\Re}([1,e])$ a solution set for hybrid CH–FIBVP, (3) and (4) whenever there exists an integrable function $v \in \mathcal{L}^1([1,e],\Re)$ with $v(\ell) \in \varpi(\ell, s(\ell))$ $\forall\, \ell \in [1,e]$, such that*

$$\mathfrak{D}\left(\frac{s(\ell)}{g(\ell,s(\ell))}\right)\Big|_{\ell=1} = 0,\quad \tau\, {}^{CH}\mathfrak{D}_{1^+}^{w-1}\left(\frac{s(\ell)}{g(\ell,s(\ell))}\right)\Big|_{\ell=e} + \left(\frac{s(\ell)}{g(\ell,s(\ell))}\right)\Big|_{\ell=\zeta} = 0$$

and

$$s(\ell) = g(\ell,s(\ell))\left[-\frac{1}{\Gamma(w)}\int_1^{\ell}\left(\ln\frac{\ell}{x}\right)^{w-1} v(x)\frac{dx}{x} + \tau\int_1^{e} v(x)\frac{dx}{x} + \frac{1}{\Gamma(w)}\int_1^{\zeta}\left(\ln\frac{\zeta}{x}\right)^{w-1} v(x)\frac{dx}{x}\right].$$

Theorem 5. *Assume that*

$(\mathcal{S}_4)$ there is a bounded function $k\colon [1,e] \to \Re^+$ such that $\forall\, s_1, s_2 \in \Re$ and $\ell \in [1,e]$, we get

$$|g(\ell,s_1(\ell)) - g(\ell,s_2(\ell))| \le k(\ell)|s_1(\ell) - s_2(\ell)|,$$

$(\mathcal{S}_5)$ the compact and convex-valued multi-function $\varpi\colon [1,e]\times\Re \to \mathcal{I}_{cp,cv}(\Re)$ is $\mathcal{L}^1$-Caratheodary,

$(\mathcal{S}_6)$ there exits a positive function $p \in \mathcal{L}^1([1,e],\Re^+)$ such that

$$\|\varpi(\ell,s)\| = \sup\{|v|\colon v \in \varpi(\ell,s(\ell))\} \le p(\ell)$$

$\forall\, s \in \Re$, almost $\forall\, \ell \in [1,e]$ and $\|p\|_{\mathcal{L}^1} = \int_1^e |p(x)|dx$,

$(\mathcal{S}_7)$ there is a number $\tilde{\varepsilon} \in \Re^+$ such that

$$\tilde{\varepsilon} > \frac{G^*\Delta^*\|p\|_{\mathcal{L}^1}}{1 - k^*\Delta^*\|p\|_{\mathcal{L}^1}}, \tag{12}$$

where $G^* = \sup_{\ell\in[1,e]}|g(\ell,0)|$, $k^* = \sup_{\ell\in[1,e]}|k(\ell)|$ and Δ^* is as given in (11). Then, hybrid CH–FIBVP (3) and (4) has at least one solution whenever $k^*\Delta^*\|p\|_{\mathcal{L}^1} < \frac{1}{2}$.

Proof. Consider an operator $\mathcal{K}\colon \mathcal{N} \to \mathcal{I}(\mathcal{N})$ defined by

$$\mathcal{K}(s) = \{w \in \mathcal{N}\colon w(\ell) = k_1(\ell) \text{ for all } 1 \le \ell \le e\},$$

where

$$k_1(\ell) = g(\ell,s(\ell))\left(-\frac{1}{\Gamma(w)}\int_1^{\ell}\left(\ln\frac{\ell}{x}\right)^{w-1} v(x)\frac{dx}{x} + \tau\int_1^{e} v(x)\frac{dx}{x} + \frac{1}{\Gamma(w)}\int_1^{\zeta}\left(\ln\frac{\zeta}{x}\right)^{w-1} v(x)\frac{dx}{x}\right), v \in (\mathfrak{SEL})_{\varpi,s}.$$

Note that the solution for hybrid CH–FIBVP (3) and (4) is a fixed point of the map $\mathcal{K}$. Define a single-valued function, $\mathcal{A}_1 : \mathcal{N} \to \mathcal{N}$ by $(\mathcal{A}_1 s)(\ell) = g(\ell, s(\ell))$ and the set-valued map, $\mathcal{A}_2 : \mathcal{N} \to \mathcal{I}(\mathcal{N})$ by

$$(\mathcal{A}_2 s)(\ell) = \{\psi \in \mathcal{N} : \psi(\ell) = k_2(\ell) \; for \; all \; \ell \in [1, e]\},$$

where

$$k_2(\ell) = \left(-\frac{1}{\Gamma(w)} \int_1^{\ell} \left(\ln \frac{\ell}{x} \right)^{w-1} v(x) \frac{dx}{x} + \tau \int_1^{e} v(x) \frac{dx}{x} + \frac{1}{\Gamma(w)} \int_1^{\zeta} \left(\ln \frac{\zeta}{x} \right)^{w-1} v(x) \frac{dx}{x} \right), \quad v \in (\mathfrak{S}\mathfrak{C}\mathfrak{L})_{\varpi,s}.$$

Note that $\mathcal{K}(s) = \mathcal{A}_1 s \mathcal{A}_2 s$. We prove that $\mathcal{A}_1$ and $\mathcal{A}_2$ satisfy the assumptions of Theorem 3. Using $(\mathcal{S}_4)$ and by a comparable conclusion in Theorem 4, $\mathcal{A}_1$ is Lipschitz on $\mathcal{N}$. We can now see that the multi-valued function $\mathcal{A}_2$ has convex values. Assume $s_1, s_2 \in \mathcal{A}_2 s$. Then, select $v_1, v_2 \in (\mathfrak{S}\mathfrak{C}\mathfrak{L})_{\varpi,s}$ such that for $i = 1, 2$, we have

$$s_i(\ell) = -\frac{1}{\Gamma(w)} \int_1^{\ell} \left(\ln \frac{\ell}{x} \right)^{w-1} v_i(x) \frac{dx}{x} + \tau \int_1^{e} v_i(x) \frac{dx}{x} + \frac{1}{\Gamma(w)} \int_1^{\zeta} \left(\ln \frac{\zeta}{x} \right)^{w-1} v_i(x) \frac{dx}{x}, \; \forall \, \ell \in [1, e].$$

For every constant $q \in (0, 1)$, we get

$$\begin{aligned} q s_1(\ell) + (1-q) s_2(\ell) = & -\frac{1}{\Gamma(w)} \int_1^{\ell} \left(\ln \frac{\ell}{x} \right)^{w-1} [q v_1(x) + (1-q) v_2(x)] \frac{dx}{x} \\ & + \tau \int_1^{e} [q v_1(x) + (1-q) v_2(x)] \frac{dx}{x} \\ & + \frac{1}{\Gamma(w)} \int_1^{\zeta} \left(\ln \frac{\zeta}{x} \right)^{w-1} [q v_1(x) + (1-q) v_2(x)] \frac{dx}{x} \end{aligned}$$

$\forall \, \ell \in [1, e]$. As we know that ϖ is convex-valued, $(\mathfrak{S}\mathfrak{C}\mathfrak{L})_{\varpi,s}$ has convex values and so $q v_1(\ell) + (1-q) v_2(\ell) \in (\mathfrak{S}\mathfrak{C}\mathfrak{L})_{\varpi,s} \; \forall \, \ell \in [1, e]$. So that $\mathcal{A}_2 s$ is a convex set for all $s \in \mathcal{N}$. To check the complete continuity of the operator $\mathcal{A}_2$, we must verify that $\mathcal{A}_2(\mathcal{N})$ is uniformly bounded and an equi-continuous set. For this reason, we prove that $\mathcal{A}_2$ mapped all bounded sets into bounded subsets of the space $\mathcal{N}$. For a number $\epsilon^* \in \Re^+$, assume a bounded ball $v_\epsilon^* = \{s \in \mathcal{N} : \|s\|_{\mathcal{N}} \leq \epsilon^*\}$. For every $s \in v_\epsilon^*$ and $\psi \in \mathcal{A}_2 s$, a function $v \in (\mathfrak{S}\mathfrak{C}\mathfrak{L})_{\varpi,s}$ exists such that

$$\psi(\ell) = -\frac{1}{\Gamma(w)} \int_1^{\ell} \left(\ln \frac{\ell}{x} \right)^{w-1} v(x) \frac{dx}{x} + \tau \int_1^{e} v(x) \frac{dx}{x} + \frac{1}{\Gamma(w)} \int_1^{\zeta} \left(\ln \frac{\zeta}{x} \right)^{w-1} v(x) \frac{dx}{x}, \forall \, \ell \in [1, e].$$

Then, we get

$$\begin{aligned}
|\psi(\ell)| &\leq \frac{1}{\Gamma(w)} \int_1^{\ell} \left(\ln \frac{\ell}{x}\right)^{w-1} |v(x)| \frac{dx}{x} + \tau \int_1^{e} |v(x)| \frac{dx}{x} \\
&+ \frac{1}{\Gamma(w)} \int_1^{\zeta} \left(\ln \frac{\zeta}{x}\right)^{w-1} |v(x)| \frac{dx}{x} \\
&\leq \frac{1}{\Gamma(w)} \int_1^{\ell} \left(\ln \frac{\ell}{x}\right)^{w-1} p(x) \frac{dx}{x} + \tau \int_1^{e} p(x) \frac{dx}{x} \\
&+ \frac{1}{\Gamma(w)} \int_1^{\zeta} \left(\ln \frac{\zeta}{x}\right)^{w-1} p(x) \frac{dx}{x} \\
&\leq \left[\frac{1}{\Gamma(w+1)} + \frac{(\ln \zeta)^{w}}{\Gamma(w+1)} + \tau\right] \|p\|_{\mathcal{L}^1} \\
&= \triangle^* \|p\|_{\mathcal{L}^1}
\end{aligned}$$

where, $\triangle^*$ is as given in (11). Thus, $|\psi(\ell)| \leq \triangle^* \|p\|_{\mathcal{L}^1}$ and this implies that $\mathcal{A}_2(\mathcal{N})$ is a uniformly bounded set. We now see how the operator $\mathcal{A}_2$ mapped bounded sets onto equi-continuous sets. Let $s \in v_{\epsilon}^*$ and $\psi \in \mathcal{A}_2 s$. We select $v \in (\mathfrak{S}\mathfrak{C}\mathfrak{L})_{\varpi,s}$ so that

$$\begin{aligned}
\psi(\ell) &= -\frac{1}{\Gamma(w)} \int_1^{\ell} \left(\ln \frac{\ell}{x}\right)^{w-1} v(x) \frac{dx}{x} + \tau \int_1^{e} v(x) \frac{dx}{x} \\
&+ \frac{1}{\Gamma(w)} \int_1^{\zeta} \left(\ln \frac{\zeta}{x}\right)^{w-1} v(x) \frac{dx}{x}, \ \forall \ell \in [1, e].
\end{aligned}$$

For each $l_1, l_2 \in [1, e]$ with $\ell_1 < \ell_2$, we obtain

$$\begin{aligned}
|\psi(\ell_2) - \psi(\ell_1)| &\leq \left| \frac{1}{\Gamma(w)} \int_1^{\ell_2} \left(\ln \frac{\ell_2}{x}\right)^{w-1} v(x) \frac{dx}{x} - \frac{1}{\Gamma(w)} \int_1^{\ell_1} \left(\ln \frac{\ell_1}{x}\right)^{w-1} v(x) \frac{dx}{x} \right| \\
&= \left| \frac{1}{\Gamma(w)} \int_1^{\ell_1} \left(\ln \frac{\ell_2}{x}\right)^{w-1} v(x) \frac{dx}{x} + \frac{1}{\Gamma(w)} \int_{\ell_1}^{\ell_2} \left(\ln \frac{\ell_2}{x}\right)^{w-1} v(x) \frac{dx}{x} \right. \\
&\left. - \frac{1}{\Gamma(w)} \int_1^{\ell_1} \left(\ln \frac{\ell_1}{x}\right)^{w-1} v(x) \frac{dx}{x} \right| \\
&\leq \int_1^{\ell_1} \left(\frac{[(\ln \frac{\ell_2}{x})^{w-1} - (\ln \frac{\ell_2}{x})^{w-1}]}{\Gamma(w)} \right) p(x) \frac{dx}{x} \\
&+ \frac{1}{\Gamma(w)} \int_{\ell_1}^{\ell_2} \left(\ln \frac{l_2}{x}\right)^{w-1} p(x) \frac{dx}{x}.
\end{aligned}$$

It is noticeable that the RHS of the above inequalities goes to 0 independent of $s \in v_{\epsilon}^*$ as $\ell_1 \to \ell_2$. By utilizing the Arzela–Ascoli theorem, the operator $\mathcal{A}_2 \colon C([1, e], \Re) \to \mathcal{I}(C([1, e], \Re))$ has the complete continuity property. We can now check that $\mathcal{A}_2$ has a closed graph which means that due to the complete continuity of $\mathcal{A}_2$, the operator $\mathcal{A}_2$ is upper semi-continuous. In this way, we assume that $s_n \in v_{\epsilon}^*$ and $\psi_n \in \mathcal{A}_2 s_n$ are such that

$s_n \to s^*$ and $\psi_n \to \psi^*$. We claim that $\psi^* \in \mathcal{A}_2 s^*$. For each $n \geq 1$ and $\psi_n \in \mathcal{A}_2 s_n$, choose $v_n \in (\mathfrak{SEL})_{\varpi,s_n}$ such that

$$\psi_n(\ell) = -\frac{1}{\Gamma(w)}\int_1^{\ell}\left(\ln\frac{\ell}{x}\right)^{w-1} v_n(x)\,\frac{dx}{x} + \tau\int_1^{e} v_n(x)\,\frac{dx}{x} + \frac{1}{\Gamma(w)}\int_1^{\zeta}\left(\ln\frac{\zeta}{x}\right)^{w-1} v_n(x)\,\frac{dx}{x}, \quad \forall \ell \in [1,e].$$

It is enough to check that a function $v^* \in (\mathfrak{SEL})_{\varpi,s^*}$ exists such that

$$\psi^*(\ell) = -\frac{1}{\Gamma(w)}\int_1^{\ell}\left(\ln\frac{\ell}{x}\right)^{w-1} v^*(x)\,\frac{dx}{x} + \tau\int_1^{e} v^*(x)\,\frac{dx}{x} + \frac{1}{\Gamma(w)}\int_1^{\zeta}\left(\ln\frac{\zeta}{x}\right)^{w-1} v^*(x)\,\frac{dx}{x}, \forall \ell \in [1,e].$$

Assume that the continuous linear operator,

$$\mathrm{Y}\colon \mathcal{L}^1([1,e],\mathfrak{R}) \to \mathcal{N} = C([1,e],\mathfrak{R})$$

is defined by

$$\mathrm{Y}(v)(\ell) = s(\ell) = -\frac{1}{\Gamma(w)}\int_1^{\ell}\left(\ln\frac{\ell}{x}\right)^{w-1} v(x)\,\frac{dx}{x} + \tau\int_1^{e} v(x)\,\frac{dx}{x} + \frac{1}{\Gamma(w)}\int_1^{\zeta}\left(\ln\frac{\zeta}{x}\right)^{w-1} v(x)\,\frac{dx}{x}, \quad \forall \ell \in [1,e]$$

so that

$$\|\psi_n(\ell) - \psi^*(\ell)\| = \left\| -\frac{1}{\Gamma(w)}\int_1^{\ell}\left(\ln\frac{\ell}{x}\right)^{w-1} (v_n(x) - v^*(x))\,\frac{dx}{x} + \tau\int_1^{e} (v_n(x) - v^*(x))\,\frac{dx}{x} + \frac{1}{\Gamma(w)}\int_1^{\zeta}\left(\ln\frac{\zeta}{x}\right)^{w-1} (v_n(x) - v^*(x))\,\frac{dx}{x} \right\| \to 0$$

as $n \to \infty$. Thus, by utilizing Theorem 2, we go to the conclusion that the operator $\mathrm{Y} \circ (\mathfrak{SEL})_{\varpi,s_n}$ has a closed graph. As we know that $\psi_n \in ((\mathfrak{SEL})_{\varpi,s_n})$ and $s_n \to s^*$, there exists $v^* \in (\mathfrak{SEL})_{\varpi,s^*}$ such that

$$\psi^*(\ell) = -\frac{1}{\Gamma(w)}\int_1^{\ell}\left(\ln\frac{\ell}{x}\right)^{w-1} v^*(x)\,\frac{dx}{x} + \tau\int_1^{e} v^*(x)\,\frac{dx}{x} + \frac{1}{\Gamma(w)}\int_1^{\zeta}\left(\ln\frac{\zeta}{x}\right)^{w-1} v^*(x)\,\frac{dx}{x}, \quad \forall \ell \in [1,e].$$

Thus, $\psi^* \in \mathcal{A}_2 s^*$ and so $\mathcal{A}_2$ has a closed graph. Thus, $\mathcal{A}_2$ is upper semi-continuous. Moreover, by using the hypothesis, the operator $\mathcal{A}_2$ has compact values. This implies that

$\mathcal{A}_2$ is an upper semi-continuous and compact operator. Now, by using condition $(\mathcal{S}_6)$, we obtain

$$\begin{aligned}M_0^* &= \|\mathcal{A}_2(\mathcal{N})\| = \sup\{|\mathcal{A}_2 s| : s \in \mathcal{N}\} \\ &= \left[\frac{1}{\Gamma(w+1)} + \tau + \frac{(\ln \zeta)^w}{\Gamma(w+1)}\right]\|p\|_{\mathcal{L}^1} \\ &= \triangle^* \|p\|_{\mathcal{L}^1}.\end{aligned}$$

Setting $\ell^* = k^*$, we get $M_0^* \ell^* < \frac{1}{2}$, and thus the assumptions of Theorem 3 are satisfied for $\mathcal{A}_1$ and $\mathcal{A}_2$. Thus, one of the conditions (i) or (ii) holds. We prove that the condition (ii) is not possible. By using Theorem 3 and the assumption $(\mathcal{S}_7)$, consider that $s \in O$ with $\|s\| = \tilde{e}$. Then, $\nu s(\ell) \in \mathcal{A}_1 s(\ell)\mathcal{A}_2 s(\ell)$ for each $\nu > 1$. Choose a related function $v \in (\mathfrak{SEL})_{\varpi,s}$. Then, $\forall\, \nu > 1$, we get

$$\begin{aligned}s(\ell) = \frac{1}{\nu} g(\ell, s(\ell)) \Bigg[&-\frac{1}{\Gamma(w)} \int_1^\ell \left(\ln \frac{\ell}{x}\right)^{w-1} v(x) \,\frac{dx}{x} + \tau \int_1^e v(x)\,\frac{dx}{x} \\ &+ \frac{1}{\Gamma(w)} \int_1^\zeta \left(\ln \frac{\zeta}{x}\right)^{w-1} v(x)\,\frac{dx}{x}\Bigg], \quad \forall \ell \in [1, e].\end{aligned}$$

Thus, we obtain

$$\begin{aligned}|s(\ell)| = \frac{1}{\nu} g(\ell, s(\ell)) \Bigg[&-\frac{1}{\Gamma(w)} \int_1^\ell \left(\ln \frac{\ell}{x}\right)^{w-1} |v(x)| \,\frac{dx}{x} + \tau \int_1^e |v(x)|\,\frac{dx}{x} \\ &+ \frac{1}{\Gamma(w)} \int_1^\zeta \left(\ln \frac{\zeta}{x}\right)^{w-1} |v(x)|\,\frac{dx}{x}\Bigg] \\ \leq [|g(\ell, s(\ell)) - g(\ell, 0)| + |g(\ell, 0)|] \Bigg[&\frac{1}{\Gamma(w)} \int_1^\ell \left(\ln \frac{\ell}{x}\right)^{w-1} |v(x)| \,\frac{dx}{x} + \tau \int_1^e |v(x)|\,\frac{dx}{x} \\ &+ \frac{1}{\Gamma(w)} \int_1^\zeta \left(\ln \frac{\zeta}{x}\right)^{w-1} |v(x)|\,\frac{dx}{x}\Bigg] \\ \leq (k^* \|s\| + G^*) \Bigg[&\frac{1}{\Gamma(w)} \int_1^\ell \left(\ln \frac{\ell}{x}\right)^{w-1} p(x) \,\frac{dx}{x} + \tau \int_1^e p(x)\,\frac{dx}{x} \\ &+ \frac{1}{\Gamma(w)} \int_1^\zeta \left(\ln \frac{\zeta}{x}\right)^{w-1} p(x)\,\frac{dx}{x}\Bigg] \\ \leq (k^* \tilde{e} + G^*) \triangle^* \|p\|_{\mathcal{L}^1}, &\forall \ell \in [1, e].\end{aligned}$$

So

$$\tilde{e} \leq \frac{G^* \triangle^* \|p\|_{\mathcal{L}^1}}{1 - k^* \triangle^* \|p\|_{\mathcal{L}^1}}.$$

According to condition (11), we can see that the condition (ii) of Theorem 3 is not possible. Thus, $s \in \mathcal{A}_1 s \mathcal{A}_2 s$. Hence, it is satisfied that ϖ has a fixed point and so that the hybrid CH–FIBVP (3) and (4) has a solution. □

We conclude from the above result that the solution for the hybrid Caputo–Hadamard fractional differential inclusion exists provided that the stated conditions hold.

4. Examples

Now, we provide numerical examples to demonstrate our theoretical findings.

Example 1. *Consider hybrid CH-FBVP (1) and (2). Choose $w = 1.5$, $w - 1 = 0.5$, $\zeta = 1.8$ and $\tau = 0.1$. Assume that the continuous maps $g\colon [1,e] \times \Re \to \Re/\{0\}$ and $\eta\colon [1,e] \times \Re \to \Re$ are defined by*

$$g(\ell, s(\ell)) = \frac{\ell |s(\ell)|^2}{6 + |s(\ell)|} + 5 \ \textit{with}\ G^* = \sup_{\ell \in [1,e]} |g(\ell, 0)| = 5$$

and

$$\eta(\ell, s(\ell)) = \frac{\ell \sin^2(\frac{\pi\ell}{2}) \sin(s(\ell))}{1000}.$$

Now, put $k(\ell) = \ell$ and $h(\ell) = \frac{\ell \sin^2(\frac{\pi}{2}(\ell))}{1000}$. Then, hybrid CH-FBVP takes the form

$$^{CH}\mathfrak{D}^{1.5}_{1^+}\left(\frac{s(\ell)}{\frac{\ell|s(\ell)|^2}{6+|s(\ell)|} + 5}\right) + \frac{\ell \sin^2(\frac{\pi}{2}(\ell)) \sin(s(\ell))}{1000} = 0,\ (\ell \in [1,e]) \tag{13}$$

supplemented with the three-point hybrid terminal conditions

$$\begin{aligned} &\mathfrak{D}\left(\frac{s(\ell)}{\frac{\ell|s(\ell)|^2}{6+|s(\ell)|} + 5}\right)\Bigg|_{\ell=1} = 0, \\ &0.1\,^{CH}\mathfrak{D}^{0.5}_{1^+}\left(\frac{s(\ell)}{\frac{\ell|s(\ell)|^2}{6+|s(\ell)|} + 5}\right)\Bigg|_{\ell=e} + \left(\frac{s(\ell)}{\frac{\ell|s(\ell)|^2}{6+|s(\ell)|} + 5}\right)\Bigg|_{\ell=1.8} = 0. \end{aligned} \tag{14}$$

Then, we get $k^* = \sup_{\ell\in[1,e]} |k(\ell)| \simeq 2.7$, $H^* = \sup_{\ell\in[1,e]} |h(\ell)| = \frac{e \sin^2(\frac{\pi}{2}(e))}{1000} \simeq 0.0021$, $\mathcal{M}(\|s\|) = 1$ and $\triangle^* \simeq 1.1915$. Then, $k^*\triangle^* H^* \mathcal{M}(\|s\|) \simeq 0.0068 < 1$. Choose $\varsigma > 0.0126$. The hybrid CH-FBVP (13) and (14) has a solution according to Theorem 4. The graphical illustration of the inequality (9) in assumption S_2 of Theorem 4 is given in Figure 1. Figures 2 and 3 depict the graphs of the functions η and g, respectively.

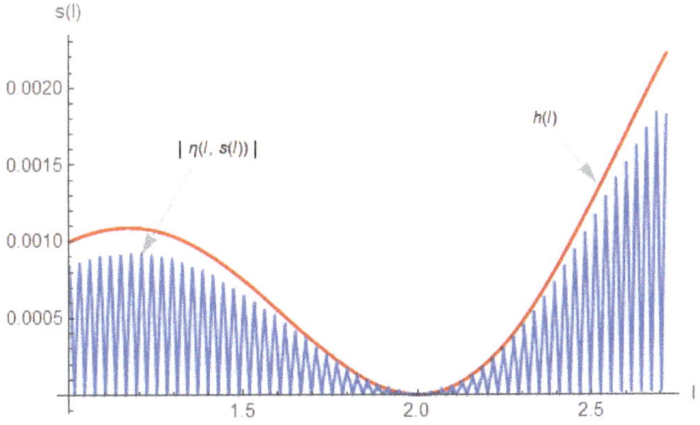

Figure 1. Graphical illustration of inequality (9).

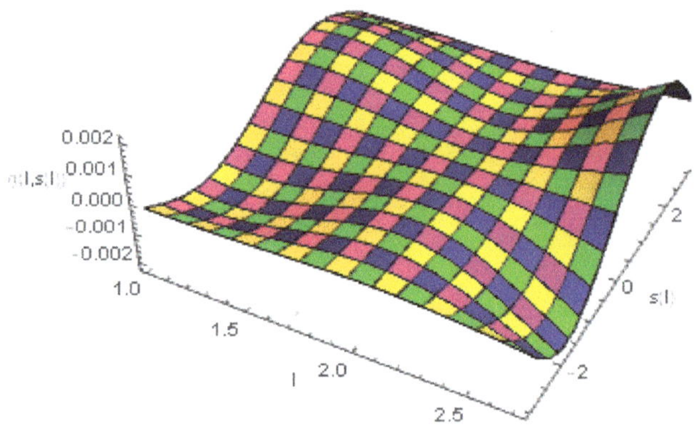

Figure 2. The graph of $\eta(l, s(l))$.

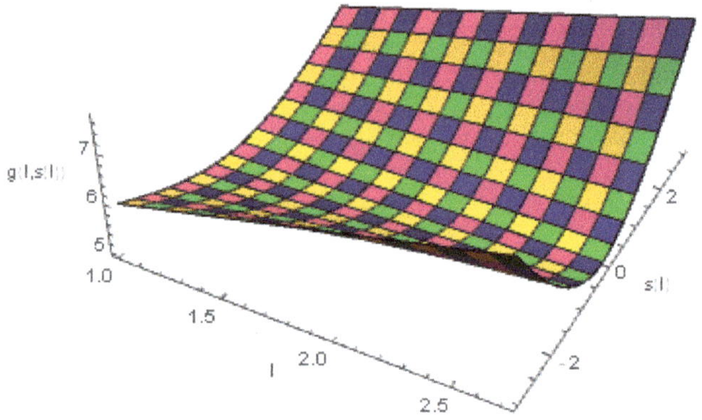

Figure 3. The graph of $g(l, s(l))$.

Example 2. *Let* $w = 1.6$, $w - 1 = 0.6$, $\zeta = 1.89$, $\tau = 1.75$ *and a continuous map* $g\colon [1,e] \times \Re\backslash\{0\}$ *defined by* $g(\ell, s(\ell)) = \ell \sin \frac{s(\ell)}{120} + 0.008$ *with* $G^* = \sup_{\ell\in[1,e]} |g(\ell, 0)| = 0.008$. *Define a multi-valued map* $\varpi\colon [1,e] \times \Re \to \mathcal{I}(\Re)$ *by*

$$\varpi(\ell, s(\ell)) = \left[\frac{|s(\ell)|}{4(|s(\ell)|+1)} + 0.5, \frac{|\cos\, s(\ell)|^4}{5(1+|cos\, s(\ell)|^4)} + 1.5\right]$$

in the proposed hybrid CH-FBVP (3) and (4), then the hybrid CH–FIBVP takes the form,

$$^{CH}\mathfrak{D}_{1^+}^{1.6}\left(\frac{s(\ell)}{\ell \sin \frac{s(\ell)}{120} + 0.008}\right) \in \left[\frac{|s(\ell)|}{4(|s(\ell)|+1)} + 0.5, \frac{|\cos\, s(\ell)|^4}{5(1+|cos\, s(\ell)|^4)} + 1.5\right], \quad (\ell \in [1,e]) \tag{15}$$

supplemented with three-point hybrid terminal conditions,

$$\begin{aligned} &\mathfrak{D}\left(\frac{s(\ell)}{\ell \sin \frac{s(\ell)}{120} + 0.008}\right)\Bigg|_{\ell=1} = 0 \\ &(1.75)^{CH}\mathfrak{D}_{1^+}^{0.6}\left(\frac{s(\ell)}{\ell \sin \frac{s(\ell)}{120} + 0.008}\right)\Bigg|_{\ell=e} + \left(\frac{s(\ell)}{\ell \sin \frac{s(\ell)}{120} + 0.008}\right)\Bigg|_{\ell=1.89} = 0. \end{aligned} \tag{16}$$

If $k(\ell) = \frac{\ell}{120}$, *then* $k^* = \sup_{\ell \in [1,e]} |k(\ell)| = \frac{e}{120} \simeq 0.0225$. *Since*

$$|\psi| \leq \max\left(\frac{|s(\ell)|}{4(|s(\ell)|+1)} + 0.5, \frac{|\cos\ s(\ell)|^4}{5(1+|cos\ s(\ell)|^4)} + 1.5\right) \leq 2$$

$\forall\ \psi \in \varpi(\ell, s(\ell))$. *We obtain* $\|\varpi(\ell, s(\ell))\| = \sup\{|v| : v \in \varpi(\ell, s(\ell))\} \leq 2$. *Put* $p(\ell) = 2$ *for any* $(\ell \in [1, e])$. *Then,* $\|p\|_{\mathcal{L}^1} = \int_1^e |p(s)|ds = 2(e-1) \simeq 3.42$. *Hence we get* $\triangle^* \simeq 2.7891$. *Then* $k^* \triangle^* \|p\|_{\mathcal{L}^1} \simeq 0.2146 < \frac{1}{2}$. *So we have* $\tilde{\epsilon} > 0.0971$. *Hence, hybrid CH–FIBVP (15) and (16) has at least one solution according to Theorem 5.*

5. Concluding Remarks

Fractional differential equations and inclusions can be used to model the many real-world problems. In this paper, we look at two new hybrid Caputo–Hadamard FBVP classes with three-point hybrid Caputo–Hadamard terminal conditions. Using Dhage's fixed point theorems, we investigated the essential requirements for the existence and uniqueness of solutions to both problems. Our results are natural extensions of the hybrid Caputo fractional model of a thermostat along with hybrid boundary conditions due to Baleanu et al. [4]. Furthermore, we provided numerical examples to support the validity of our findings. The Caputo–Hadamard fractional derivative can be used to prove the existence of solutions for more sophisticated FDEs and FDIs utilizing the fixed point theory and functional analysis.

Author Contributions: Conceptualization, M.Y. and A.H.; methodology, S.M. and A.H.; validation, R.G., A.H.; formal analysis, M.Y., R.G. and A.H.; investigation, A.H.; writing—original draft preparation, S.M.; writing—review and editing, M.Y., R.G. and A.H.; supervision, R.G. and A.H.; project administration, R.G.; funding acquisition, R.G. All authors have read and agreed to the published version of the manuscript.

Funding: This research received no external funding.

Institutional Review Board Statement: Not applicable.

Informed Consent Statement: Not applicable.

Data Availability Statement: No data were used.

Conflicts of Interest: The authors declare no conflict of interest.

Abbreviations

The following abbreviations are used in this manuscript:

${}^{CH}\mathfrak{D}$	Caputo–Hadamard Fractional Derivative
${}^{CH}\mathfrak{J}$	Caputo–Hadamard Fractional Integral
CH-FBVP	Caputo–Hadamard Fractional Boundary Value Problem
CH–FIBVP	Caputo–Hadamard Fractional Inclusion Boundary Value Problem
$\mathcal{N}$	Normed Space
$\mathcal{I}(\mathcal{N})$	Set of all Subsets of $\mathcal{N}$
PH_d	Pompeiu–Hausdorff metric
$(\mathfrak{SEL})_{\omega,\sigma}$	Selections of ω at σ

References

1. Hilfer, R. *Application of Fractional Calculus in Physics;* World Scientific: Singapore, 2000.
2. Kilbas, A.A.; Srivastava, H.M.; Rujillo, J.J. *Theory and Applications of Fractional Diferential Equations;* Elsevier: Amsterdam, The Netherlands, 2006.
3. Mohammadi, H.; Rezapour, S. Two existence results for nonlinear fractional diferential equations by using fixed point theory on ordered gauge spaces. *J. Adv. Math. Stud.* **2013**, *6*, 154–158.

4. Baleanu, D.; Etemad, S.; Rezapour, S. A hybrid Caputo fractional modeling for thermostat with hybrid boundary value conditions. *Bound. Value Probl.* **2020**, *2020*, 64. [CrossRef]
5. Baleanu, D.; Nazemi, S.Z.; Rezapour, S. Attractivity for a k-dimensional system of fractional functional diferential equations and global attractivity for a k-dimensional system of nonlinear fractional diferential equations. *J. Inequal. Appl.* **2014**, *2014*, 31. [CrossRef]
6. Baleanu, D.; Nazemi, S.Z.; Rezapour, S. The existence of solution for a k-dimensional system of multi-term fractional integro-diferential equations with anti-periodic boundary value problems. *Abstr. Appl. Anal.* **2014**, *2014*, 896871. [CrossRef]
7. Wang, G.; Ren, X.; Zhang, L.; Ahmad, B. Explicit iteration and unique positive solution for a Caputo–Caputo–Hadamard fractional turbulent fow model. *IEEE Access* **2019**, *7*, 109833–109839. [CrossRef]
8. Wang, G.; Pei, K.; Chen, Y.Q. Stability analysis of nonlinear Caputo–Hadamard fractional diferential system. *J. Franklin Inst.* **2019**, *356*, 6538–6546. [CrossRef]
9. Wang, G.; Pei, K.; Agarwal, R.P.; Zhang, L.; Ahmad, B. Nonlocal Caputo–Hadamard fractional boundary value problem with Caputo–Hadamard integral and discrete boundary conditions on a half-line. *J. Comput. Appl. Math.* **2018**, *343*, 230–239. [CrossRef]
10. Pei, K.; Wang, G.; Sun, Y. Successive iterations and positive extremal solutions for a Caputo–Hadamard type fractional integro-diferential equations on infnite domain. *Appl. Math. Comput.* **2017**, *312*, 158–168.
11. Abbas, S.; Benchohra, M.; Hamani, S.; Henderson, J. Upper and lower solutions method for Caputo–Hadamard fractional differential inclusions. *Math. Moravica* **2019**, *23*, 107–118. [CrossRef]
12. Benchohra, M.; Graef, J.R.; Guerraiche, N.; Hamani, S. Nonlinear boundary value problems for fractional differential inclusions with Caputo-Hadamard derivatives on the half line. *AIMS Math.* **2016**, *6*, 6278–6292. [CrossRef]
13. Hamani, S.; Henderson, J. Boundary value problems for fractional differential inclusions with nonlocal conditions. *Mediterr. J. Math.* **2016**, *13*, 967–979. [CrossRef]
14. Benhamidaa, W.; Hamania, S.; Hendersonb, J. Boundary Value Problems For Caputo-Hadamard fractional differential equations. *Adv. Theory Nonlinear Anal. Its Appl.* **2018**, *2*, 138–145.
15. Benchohra, M.; Hamani, S.; Zhou, Y. Oscillation and nonoscillation for Caputo–Hadamard impulsive fractional differential inclusions. *Adv. Differ. Equations* **2019**, *2019*, 74. [CrossRef]
16. Dhage, B.C.; Lakshmikantham, V. Basic results on hybrid diferential equation. *Nonlinear Anal. Hybrid Syst.* **2010**, *4*, 414–424 [CrossRef]
17. Zhao, Y.; Sun, S.; Han, Z.; Li, Q. Theory of fractional hybrid diferential equations. *Comput. Math. Appl.* **2011**, *62*, 1312–1324. [CrossRef]
18. Sun, S.; Zhao, Y.; Han, Z.; Li, Y. The existence of solutions for boundary value problem of fractional hybrid differential equations. *Commun. Nonlinear Sci. Numer. Simul.* **2012**, *17*, 4961–4967. [CrossRef]
19. Hilal, K.; Kajouni, A. Boundary value problems for hybrid diferential equations with fractional order. *Adv. Differ. Equations* **2015**, *2015*, 183. [CrossRef]
20. Baleanu, D.; Hedayati, V.; Rezapour, S.; Al Qurashi, M.M. On two fractional diferential inclusions. *SpringerPlus* **2016**, *5*, 882. [CrossRef] [PubMed]
21. Infante, G.; Webb, J. Loss of positivity in a nonlinear scalar heat equation. *Nonlinear Differ. Equ. Appl.* **2006**, *13*, 249–261. [CrossRef]
22. Nieto, J.J.; Pimentel, J. Positive solutions of a fractional thermostat model. *Bound. Value Probl.* **2013**, *2013*, 5. [CrossRef]
23. Miller, K.S.; Ross, B. *An Introduction to Fractional Calculus and Fractional Diferential Equations*; Wiley: New York, NY, USA, 1993.
24. Deimling, K. *Multi-Valued Diferential Equations*; de Gruyter: Berlin, Germany, 1992.
25. Aubin, J.; Cellina, A. *Diferential Inclusions: Set-Valued Maps and Viability Theory*; Springer: Berlin, Germany, 1984.
26. Dhage, B.C. Nonlinear functional boundary value problems involving Carathedory. *Kyungpook Math. J.* **2006**, *46*, 427–441.
27. Lasota, A.; Opial, Z. An application of the Kakutani–Ky Fan theorem in the theory of ordinary diferential equations. *Bull. Acad. Pol. Sci. Set. Sci. Math. Astronom. Phys.* **1965**, *13*, 781–786.
28. Dhage, B.C. Existence results for neutral functional diferential inclusions in Banach algebras. *Nonlinear Anal.* **2006**, *64*, 1290–1306. [CrossRef]

fractal and fractional

MDPI

Article

Discretization, Bifurcation, and Control for a Class of Predator-Prey Interactions

Asifa Tassaddiq [1,*], Muhammad Sajjad Shabbir [2], Qamar Din [2] and Humera Naaz [1]

1 Department of Basic Sciences and Humanities, College of Computer and Information Sciences, Majmaah University, Al-Majmaah 11952, Saudi Arabia; h.naaz@mu.edu.sa
2 Department of Mathematics, University of the Poonch Rawalakot, Azad Kashmir 12350, Pakistan; sajjadmust@gmail.com (M.S.S.); qamar.sms@gmail.com (Q.D.)
* Correspondence: a.tassaddiq@mu.edu.sa

Abstract: The present study focuses on the dynamical aspects of a discrete-time Leslie-Gower predator-prey model accompanied by a Holling type III functional response. Discretization is conducted by applying a piecewise constant argument method of differential equations. Moreover, boundedness, existence, uniqueness, and a local stability analysis of biologically feasible equilibria were investigated. By implementing the center manifold theorem and bifurcation theory, our study reveals that the given system undergoes period-doubling and Neimark-Sacker bifurcation around the interior equilibrium point. By contrast, chaotic attractors ensure chaos. To avoid these unpredictable situations, we establish a feedback-control strategy to control the chaos created under the influence of bifurcation. The fractal dimensions of the proposed model are calculated. The maximum Lyapunov exponents and phase portraits are depicted to further confirm the complexity and chaotic behavior. Finally, numerical simulations are presented to confirm the theoretical and analytical findings.

Keywords: prey-predator model; boundedness; period-doubling bifurcation; Neimark-Sacker bifurcation; hybrid control; fractal dimensions

Citation: Tassaddiq, A.; Shabbir, M.S.; Din, Q.; Naaz, H. Discretization, Bifurcation, and Control for a Class of Predator-Prey Interactions. *Fractal Fract.* **2022**, *6*, 31. https://doi.org/10.3390/fractalfract6010031

Academic Editor: Tassos C. Bountis

Received: 30 November 2021
Accepted: 27 December 2021
Published: 6 January 2022

1. Introduction

predator-prey models have a wide range of applications in ecological and biological fields. Although various fundamental aspects of the nonlinear dynamics of predator-prey population models related to continuous dynamical systems have been studied, the characteristics of discrete dynamical systems remain comparatively unknown. A discrete dynamical structure possesses a solitary dynamical nature as compared to a continuous system. There are several critical and practical problems in daily life that can be characterized with the help of a discrete dynamical system. To consider the analytical aspects of a solution that is difficult to calculate, various schemes can be implemented to discretize a continuous system and discuss the numerical solution. Therefore, detailed critical inspections of discrete-time dynamical systems have contributed immensely to various fields such as engineering, physics, chemistry, and mathematics. There have been numerous studies conducted that are related to the dynamics of predator-prey models.

Chen et al. [1] applied the Euler scheme and center manifold theorem to a ratio-dependent prey-predator model and scrutinized the dynamic characteristics of the model. Ghaziani et al. [2] studied a prey-predator system with a Holling functional response and discussed the resonance and bifurcation analyses. Jana [3] found extremely powerful dynamical conditions through numerical and theoretical investigations of discrete-time prey-predator models, such as stability conditions, flip, and hopf-bifurcation. Misra et al. [4] studied a predator-prey model based on age predation and discussed the dynamic behavior of the models. Zhang et al. [5] presented a biological economic system related to the predator-prey model of a differential algebraic system by applying a new normal form. Hu and Cao [6] investigated the Holling and Leslie type predator-prey model and discussed a chaos and bifurcation analysis. Wang and Li [7] proposed a lemma that is extremely meaningful for discussing the stability and bifurcation of the systems. The fundamental finding in the dynamics of prey-predator species is the classical Lotka-Volterra prey-predator model, which exhibits unrealistic behavior (see, Murdoch et al. [8]). To remove such unrealistic behavior, Holling introduced three different types of functional responses (see, Holling [9]). Rosenzweig and MacArthur [10] implemented a functional response to modify the predator-prey model. An investigation into population interaction focused on the continuous dynamical system of two species [11–13]. By contrast, a recent study led to the discrete dynamical system becoming more suitable than a continuous version when the population is non-overlapping (e.g., see, Jing et al. [14], Liu et al. [15], Lopez-Ruiz and Fournier-Prunaret [16], Neubert and Kot [17]). Furthermore, multiple existing studies related to the dynamics of predator-prey models are described in [18–26]. In [27], the Holling type-III functional response was introduced in both populations (prey and predator). The stability conditions around biologically suitable equilibria were further discussed. Diagrams of the phase portraits, bifurcation, and time series were plotted. It was shown that the system is sensitive to the initial conditions, which means that the system is chaotic. A two-dimensional continuous model with a Holling-III functional response in both prey and predator was presented [28]. Furthermore, Euler's scheme was used to discretize the model and study the complex behavior of the system. Elettreby et al. [29] discussed a discrete-time prey-predator model with predator and prey populations having Holling type I and III functional responses, respectively. Moreover, they described a fascinating dynamical nature of the model, including stability, bifurcation, and chaos, which ensure the rich dynamics of discrete-time models.

In this study, we evaluate the specific prey –predator model discussed by Murray [30]:

$$\begin{aligned} \frac{dx}{dt} &= x\left[r\left(1-\frac{x}{k}\right)-\frac{axy}{b^2+x^2}\right], \\ \frac{dy}{dt} &= ys\left(1-\frac{hy}{x}\right), \end{aligned} \tag{1}$$

where $x(t)$ and $y(t)$ denote the densities of prey and predator species at any time t, respectively; the carrying capacity of prey in the absence of predator is k, and $r,\ b,\ a,\ s,$ and h are positive constants. Moreover, the carrying capacity is proportional to the prey population size and population of prey attacked by predators, as specified by the Holling type III functional response. He and Lai [31] examined the bifurcation and chaos control of the discrete-time version of model (1) by implementing Euler's forward scheme with step size h as the bifurcation parameter. The numerical results in [31] show that period-doubling bifurcation occurs when a large step size is considered in Euler's method; this fact contravenes the precision of the numerical method for discretization. To overcome this deficiency, the following discretization method was implemented. Considering the regular time interval for the average growth rate in both populations, by resorting to piecewise constant arguments for solving nonlinear differential equations, system (1) can then be rewritten as follows:

$$\begin{aligned} \frac{1}{x(t)}\frac{dx(t)}{dt} &= \left[r\left(1-\frac{x[t]}{k}\right)-\frac{ax[t]y[t]}{b^2+x[t]^2}\right], \\ \frac{1}{y(t)}\frac{dy(t)}{dt} &= s\left(1-\frac{hy[t]}{x[t]}\right), \end{aligned} \tag{2}$$

where the integer part of t is given by $[t]$ within the interval $0 < t < 1$. In addition, by integrating system (1) for $t \in [n : n+1]$, $(n = 0, 1, 2, \ldots)$, we have the following system:

$$\begin{aligned} x(t) =& x_n \exp\left(\left[r\left(1 - \frac{x_n}{k}\right) - \frac{a x_n y_n}{b^2 + x_n{}^2}\right](t - n)\right), \\ y(t) =& y_n \exp\left(\left[s\left(1 - \frac{h y_n}{x_n}\right)\right](t - n)\right). \end{aligned} \tag{3}$$

Applying $t \to n+1$, we obtain the following prey-predator system:

$$\begin{aligned} x_{n+1} =& x_n \exp\left(r\left(1 - \frac{x_n}{k}\right) - \frac{a x_n y_n}{b^2 + x_n{}^2}\right), \\ y_{n+1} =& y_n \exp\left(s\left(1 - \frac{h y_n}{x_n}\right)\right). \end{aligned} \tag{4}$$

The key contributions and findings of the current study are as follows for model (4):

- The existence and uniqueness of biologically feasible equilibria and their stability analysis are discussed.
- Our findings indicate that model (4) undergoes periodic doubling as well as a Neimark-Sacker bifurcation at its unique positive equilibrium.
- The direction and existance criteria for both types of bifurcation are examined under interior equilibrium.
- A hybrid control strategy is applied to control the chaos in model (4).

The remainder of this paper is organized as follows. After presenting some related preliminaries in Section 2, the boundedness of the steady state is analyzed in Section 3. In Section 4, the dynamics of system (4), including the existence of equilibria and local stability, are presented. Section 5 describes an investigation of the birfurcation analysis at the interior fixed point of system (4). In Section 6, we study a hybrid control method to control the chaos. The fractal dimensions are calculated in Section 7. Finally, numerical simulations are provided in Section 8 to verify our analytical approach. Conclusions related to these results are presented in Section 9 and the future directions are providing in Section 10.

Furthermore, a detailed investigation of some charismatic population models and their qualitative behavior are provided (see, Din and Din et al. [18–26] and the references therein).

2. Preliminaries

Definition 1. *([32]) A point $x^*$is said to be a fixed point of the map for an equilibrium point if $f(x^*) = x^*$.*

Theorem 1. *([32]) Let $f : I \to I$ be a continuous map, where $I = [a, b]$ is a closed interval in$R$. Then, f has a fixed point.*

Theorem 2. *([32]) Let $f : I = [a, b] \to R$ be a continuous map such that $f(I) \supset I$. Then, f has a fixed point in I.*

Definition 2. *([32]) Let $f : I \to I$ be a map and x^* be a fixed point of f, where I is an interval in the set of real numbers R. Then, the following conditions hold true:*

1. *x^* is said to be stable if for any $\varepsilon > 0$, there exists $\delta > 0$ such that for all $x_0 \in I$ with $|x_0 - x^*| < \delta$ we have $|f^n(x_0) - x^*| < \varepsilon$ for all $n \in Z^+$. Otherwise, the fixed point x^* is unstable.*
2. *x^* is said to be attractive if $\eta > 0$ exists, such that $|x_0 - x^*| < \eta$ implies $\lim_{n\to\infty} f^n(x_0) = x^*$.*
3. *x^* is asymptotically stable if it is both stable and attractive. If in (2), $\eta = \infty$, then x^* is said to be globally asymptotically stable.*

Definition 3. ([32]) *A fixed point x^* of a map f is said to be hyperbolic if $|f'(x^*)| = 1$. Otherwise, it is non-hyperbolic.*

Theorem 3. ([32]) *Let x^* be a hyperbolic fixed point of a map f, where f is continuously differentiable at x^*. The following statements hold true:*

1. *If $|f'(x^*)| < 1$, then x^* is asymptotically stable.*
2. *If $|f'(x^*)| > 1$, then x^* is unstable.*

3. Boundedness

The boundedness of system (4) is based on the following Remark.

Remark 1. ([25]) *Assuming that $x_0 > 0$ for every x_t and $x_{t+1} \leq x_t exp(A[1 - Bx_t])$ for every $t \in [t_1, \infty]$, where $B > 0$ is constant. Then,*

$$\lim_{n\to\infty} \mathrm{Sup} x_t \leq \frac{1}{AB} \exp(A-1)$$

Using Remark 1, we state the following theorem for the uniform boundedness of system (4).

Theorem 4. *Any positive solution (x_n, y_n) of model (4) is uniformly bounded.*

Proof. Assuming that (x_n, y_n) is any positive solution of system (4), we then have

$$x_{n+1} \leq x_n \exp\left(r\left(1 - \frac{x_n}{k}\right)\right), \text{ for all } n = 0, 1, 2, \ldots.$$

Let $x_0 > 0$. Using Remark 1, we obtain the following result.

$$\lim_{n\to\infty} \mathrm{Sup} x_n \leq \frac{k}{r} \exp(r-1) = l_1. \tag{5}$$

Furthermore, from the second part of system (4), we obtain the following:

$$y_{n+1} \leq y_n \exp\left(s\left(1 - \frac{hy_n}{l_1}\right)\right).$$

Let $y_0 > 0$. Applying Remark 1, we obtain the following result:

$$\lim_{n\to\infty} \mathrm{Sup} y_n \leq \frac{l_1}{sh}(s-1) = l_2 \tag{6}$$

Thus, it follows that $\lim_{n\to\infty} \mathrm{Sup}(x_n, y_n) \leq l$, where $l = \max\{l_1, l_2\}$. The proof is completed. □

4. Existence of a Positive Fixed Point and Local Stability

To explore the existence of a fixed point of model (4), suppose that (x, y) is any arbitrary fixed point of (4). Then, (x, y) must satisfy the following algebraic system of equations:

$$\begin{aligned} x =& x \exp\left(r\left(1 - \frac{x}{k}\right) - \frac{axy}{b^2 + x^2}\right), \\ y =& y \exp\left(s\left(1 - \frac{hy}{x}\right)\right). \end{aligned} \tag{7}$$

Then, (7) has a boundary equilibrium point $(k, 0)$. In addition, we also explore the existence and uniqueness of the solution of system (4) because the positive fixed points are not in a closed form. For this purpose, the following computation, using Theorem 4, exhibits the existence and uniqueness of the solution to model (4).

Theorem 5. *There exists a unique positive steady-state* $(x_*,\ y_*) \in [0,\ l_1] \times [0,\ l_2]$ *of system (4).*

Proof. To attain the fixed point by solving system (7), we have

$$\begin{aligned} r(1-\tfrac{x}{k}) &= \tfrac{axy}{x^2+b^2}, \\ x &= yh. \end{aligned} \tag{8}$$

Suppose that

$$F(x) = r\left(1-\frac{x}{k}\right) - \frac{\frac{ax^2}{h}}{x^2+b^2}$$

for all $x \in [0,\ l_1]$. Then, we can see that $F(0) = r > 0$ and

$$F(l_1) = -\frac{a\exp(2(r-1))k^2}{h\left(b^2+\frac{\exp(2(r-1))k^2}{r^2}\right)r^2} + r\left(1-\frac{\exp(r-1)}{r}\right) < 0$$

for all $a,\ b,\ s,\ r,\ k,$ and $h > 0$. Hence, there exists at least one root of $F(x) = 0$, for $x \in [0,\ l_1]$. In addition,

$$F'(x) = -\frac{r}{k} - \frac{2ab^2x}{h(b^2+x^2)^2} < 0$$

for all $x \in [0,\ l_1]$. Therefore, the system (4) has a unique positive fixed point $(x_*,\ y_*) \in [0,\ l_1] \times [0,\ l_2]$.

Initially, we explored the stability analysis of the boundary equilibrium $(k, 0)$. The Jacobian matrix F_J evaluated at $(k, 0)$, is expressed as

$$F_J(k,0) = \begin{bmatrix} 1-r & -\frac{k^2a}{b^2+k^2} \\ 0 & \exp(s) \end{bmatrix},$$

and the characteristic equation computed at $(k, 0)$ is given by

$$\mathbb{F}(\eta) = \eta^2 - (1-r+\exp(s))\eta + (1-r)\exp(s)$$

Hence, $\mathbb{F}(\eta) = 0$ has two roots, namely, $\eta_1 = \exp(s)$ and $\eta_2 = 1-r$. In addition, $(k, 0)$ is the source if $r > 2$, and is the saddle point if $0 < r < 2$. Next, we explored the stability analysis of the fixed points. To investigate the stability of the equilibria, we calculated the Jacobian F_J of system (4) at any point (x, y) as follows:

$$F_J(x,y) = \begin{bmatrix} b_{11} & b_{12} \\ b_{21} & b_{22} \end{bmatrix}$$

The characteristic polynomial of F_J at (x, y) is given by

$$\mathbb{R}(\eta) = \eta^2 - T_1\eta + D_1, \tag{9}$$

where

$$T_1 = (b_{11} + b_{22}),$$

and

$$D_1 = b_{11}b_{22} - b_{12}b_{21}$$

The following Lemma is extremely useful to examine the stability of the equilibria. □

Lemma 1. *Let $\Re(\eta) = \eta^2 - T_1\eta + D_1$ and $\Re(1) > 0$. Moreover, η_1,η_2 are the roots of equation $\Re(\eta) = 0$, and thus*

(i) $|\eta_1| < 1$ & $|\eta_2| < 1 \Leftrightarrow \Re(-1) > 0$ *and* $D_1 < 1$;
(ii) $|\eta_1| < 1$ & $|\eta_2| > 1$ or ($|\eta_1| > 1$ *and* $|\eta_2| < 1$) $\Leftrightarrow \Re(-1) < 0$;
(iii) $|\eta_1| > 1$ & $|\eta_2| > 1 \Leftrightarrow \Re(-1) > 0$ *and* $D_1 > 1$;
(iv) $\eta_1 = -1$ & $|\eta_2| \neq 1 \Leftrightarrow \Re(-1) = 0$ *and* $T_1 \neq 0, 2$;
(v) η_1 *and* η_2 *are complex and* $|\eta_1| = 1$ *and* $|\eta_2| = 1 \Leftrightarrow {T_1}^2 - 4D_1 < 0$ *and* $D_1 = 1$.

Because η_1 and η_2 are the eigenvalues of (9), the following topological results are obtained.

The equilibrium (x, y) is known as a sink if $|\eta_1| < 1$ and $|\eta_2| < 1$, which is locally asymptotically stable, and as a source if $|\eta_1| > 1$ and $|\eta_2| > 1$; thus, the nature of the source is always unstable. Moreover, the equilibrium point (x, y) is always known as the saddle point if $|\eta_1| < 1$ and $|\eta_2| > 1$ or ($|\eta_1| > 1$ and $|\eta_2| < 1$). In the case of a non-hyperbolic equilibrium (x, y), either $|\eta_1| = 1$ or $|\eta_2| = 1$.

Our next aim is to discuss the local stability of the unique positive equilibrium (x_*, y_*) of system (4). Let (9) be the characteristic polynomial of the variational matrix evaluated at (x_*, y_*), such that

$$T_1 = \left(2 - \frac{rx_*}{k} - \Omega - s)\right) \text{ and } D_1 = \left(1 - \frac{rx_*}{k} - \Omega\right)(1 - s) + \frac{s\Phi}{h}$$

where $\Omega = \frac{ax_*y_*(b^2 - x_*^2)}{(b^2 + x_*^2)^2}$ and $\Phi = \frac{ax_*^2}{b^2 + x_*^2}$. Thus, by applying Lemma 1, we discuss the local stability of system (4) around (x_*, y_*) by stating the following proposition.

Proposition 1. *The interior equilibrium point (x_*, y_*) of system (4) satisfies the following results:*

(i) *The interior equilibrium(x_*, y_*) is stable iff:*

$$\left|2 - \frac{rx_*}{k} - \Omega - s\right| < \left|1 + \left(1 - \frac{rx_*}{k} - \Omega\right)(1 - s) + \frac{s\Phi}{h}\right|,$$

and

$$\left|\frac{s\Phi}{h} + (1 - s)\left(1 - \frac{rx_*}{k} - \Omega\right)\right| > 1$$

(ii) *The positive fixed point (x_*, y_*) is a saddle point if and only if*

$$\left[2 - \frac{rx_*}{k} - \Omega - s\right]^2 > 4\left[(1 - s)\left(1 - \frac{rx_*}{k} - \Omega\right) + \frac{s\Phi}{h}\right],$$

and

(iii) *The interior fixed point (x_*, y_*) is non-hyperbolic if and only if*

$$\left|2 - \frac{rx_*}{k} - \Omega - s\right| = \left|1 + \left(1 - \frac{rx_*}{k} - \Omega\right)(1 - s) + \frac{s\Phi}{h}\right| \tag{10}$$

or

$$\left(1 - \frac{rx_*}{k} - \Omega\right)(1 - s) + \frac{s\Phi}{h} = 1 \text{ and } \left|2 - \frac{rx_*}{k} - \Omega - s\right| < 2. \tag{11}$$

To explore the local stability criteria for (x_*, y_*) of model (4), we have the following theorem:

Theorem 6. *If neither (10) nor (11) is satisfied, then the positive steady-state (x_*, y_*) of system (4) is locally asymptotically stable if and only if the following condition is satisfied.*

$$\left|2-\frac{rx_*}{k}-\Omega-s\right|<1+\left(1-\frac{rx_*}{k}-\Omega\right)(1-s)+\frac{s\Phi}{h}<2$$

5. Bifurcation Analysis

In this section, we discuss the period-doubling and Neimark-Sacker bifurcations of system (4) around the interior equilibrium. Initially, we explored the period-doubling bifurcation at a positive fixed point (x_*, y_*) of system (4). To study the period-doubling bifurcation, assume that $T_1^2 > 4D_1$, that is,

$$\left(2-\frac{rx_*}{k}-\Omega-s\right)>4\left[\left(1-\frac{rx_*}{k}-\Omega\right)(1-s)+\frac{s\Phi}{h}\right] \tag{12}$$

and $T_1 + D_1 + 1 = 0$. It then follows that

$$s := \frac{2h(rx_* - (2-\Omega)k)}{h(\Omega\, k + rx_*) + k(\Phi - 2\, h)} \tag{13}$$

Then, $\eta_1 = -1$ and $\eta_2 \neq 1$ if

$$\left(1-\frac{rx_*}{k}-\Omega\right)(1-s)+\frac{s\Phi}{h}\neq \pm 1. \tag{14}$$

Consider the map $T_{PB} = \{(a,b,k,r,s) \in R_+^5$ for which (12)–(14) are thus satisfied. Then, the equilibrium (x_*, y_*) of system (4) sustains period-doubling bifurcation whenever the parameters deviate within the small neighborhood of T_{PB}. Thus, system (4) along with parameters $(a,b,k,r,s_1) \in T_{PB}$, can be written as follows:

$$\begin{pmatrix} x \\ y \end{pmatrix} \rightarrow \begin{pmatrix} xe^{r(1-\frac{x}{k})-\frac{axy}{b^2+x^2}} \\ ye^{s_1(1-\frac{hy}{x})} \end{pmatrix} \tag{15}$$

The following perturbation of system (15) can be obtained by taking $\bar{s}$ as a bifurcation parameter:

$$\begin{pmatrix} x \\ y \end{pmatrix} \rightarrow \begin{pmatrix} xe^{r(1-\frac{x}{k})-\frac{axy}{b^2+x^2}} \\ ye^{(s_1+\bar{s})(1-\frac{hy}{x})} \end{pmatrix} \tag{16}$$

where $|\bar{s}| << 1$ denotes the least perturbation parameter. Assuming that $N = x - x_*$, $P = y - y_*$, system (16) is reduced to the following form:

$$\begin{pmatrix} N \\ P \end{pmatrix} \rightarrow \begin{pmatrix} b_{11} & b_{12} \\ b_{21} & b_{22} \end{pmatrix}\begin{pmatrix} N \\ P \end{pmatrix} + \begin{pmatrix} f_1(N,P,\bar{s}) \\ f_2(N,P,\bar{s}) \end{pmatrix} \tag{17}$$

Here,

$$f_1(N,P,\bar{s}) = b_{13}N^2 + b_{14}NP + b_{15}P^2 + a_1N^3 + a_2N^2P + a_3NP^2 + a_4P^3 + O\Big((|N|+|P|+|\bar{s}|)^4\Big),$$
$$f_2(N,P,\bar{s}) = b_{23}N^2 + b_{24}NP + b_{25}P^2 + d_1N^3 + d_2N^2P + d_3NP^2 + d_4P^3$$
$$+c_1\bar{s}N + c_2\bar{r}P + c_3\bar{r}^2 + c_4\bar{r}NP + c_5\bar{r}N^2 + c_6\bar{r}P^2 + c_7\bar{r}^2N + c_8\bar{s}^2P + c_9\bar{r}^3 + O\Big((|N|+|P|+|\bar{s}|)^4\Big)$$

Whereas the descriptions and computations of the involved coefficients are given in Appendix A.

The canonical form of (17) at $s_1 = 0$ can be obtained by assuming the following map:

$$\begin{pmatrix} N \\ P \end{pmatrix} = \begin{pmatrix} b_{12} & b_{12} \\ -1-b_{11} & \eta_2 - b_{11} \end{pmatrix}\begin{pmatrix} u \\ v \end{pmatrix}. \tag{18}$$

The normal form of system (17) under translation (18) can be expressed as

$$\begin{pmatrix} u \\ v \end{pmatrix} \to \begin{pmatrix} -1 & 0 \\ 0 & \eta_2 \end{pmatrix} \begin{pmatrix} u \\ v \end{pmatrix} + \begin{pmatrix} \tilde{f}(u, v, \bar{s}) \\ \tilde{g}(u, v, \bar{s}) \end{pmatrix}, \tag{19}$$

the descriptions and computations of the involved functions and parameters leading to the following expression are provided in Appendix B.

$$F_1 := \left(\frac{\partial^2 \tilde{f}}{\partial u \partial \bar{s}} + \frac{1}{2} \frac{\partial F}{\partial \bar{r}} \frac{\partial^2 F}{\partial u^2} \right)_{(0,0)} = \frac{c_2(1+b_{11})}{\eta_2+1} - \frac{c_1 b_{12}}{\eta_2+1},$$

$$F_2 := \left(\frac{1}{6} \frac{\partial^3 F}{\partial u^3} + \left(\frac{1}{2} \frac{\partial^2 F}{\partial u^2} \right)^2 \right)_{(0,0)} = t_1^2 + t_5$$

Hence, we arrive at the following conclusions based on the aforementioned calculations.

Theorem 7. *There exists a period-doubling bifurcation at* (x_*, y_*) *of system (4), whenever* $F_2 \neq 0$ *and* r *deviates within a small neighboring point of* s_1*. In addition, if* $F_2 > 0$, $(F_2 < 0)$, *the orbit is period-2 stable (unstable).*

Next, we investigated the Neimark-Sacker bifurcation around (x_*, y_*) of system (4). For identical results, we referred to the studies by Din [24,25], Shabbir et al. [20], and Jing et al. [14]. Furthermore, the equilibrium point moves around the close invariant curve, owing to the Neimark-Sacker bifurcation. To explore the Neimark-Sacker bifurcation, we find the conditions for which (x_*, y_*) is a non-hyperbolic point with a complex conjugate root of the characteristic equation of the unit modulus. Thus, if the following results hold true, then $\Re(\eta) = 0$ has two complex conjugate roots with a unit modulus.

$$s := \frac{(\Omega k + r x_*) h}{h(\Omega k + r x_*) + k(\Phi - h)},$$

and

$$\left| 2 - \frac{r x_*}{k} - \Omega - s \right| < 2.$$

Consider

$$T_{NS} = \left\{ (a, b, k, r, s) \in R_+^5 : s = \frac{(\Omega k + r x_*) h}{h(\Omega k + r x_*) + k(\Phi - h)} \text{ and } \left| 2 - \frac{r x_*}{k} - \Omega - s \right| < 2 \right\}.$$

Assuming that $s_2 = \frac{(\Omega k + r x_*) h}{h(\Omega k + r x_*) + k(\Phi - h)}$, the fixed point (x_*, y_*) ensures the Neimark-Sacker bifurcation when the parameters fluctuate in the least neighborhood of T_{NS}. Thus, system (4) along with parameters (a, b, k, r_2, s) can be expressed as follows:

$$\begin{pmatrix} x \\ y \end{pmatrix} \to \begin{pmatrix} x e^{r(1-\frac{x}{k}) - \frac{axy}{b^2+x^2}} \\ y e^{s_2(1-\frac{hy}{x})} \end{pmatrix}. \tag{20}$$

The following perturbation of system (20) can be obtained by taking $\tilde{s}$ as the bifurcation parameter, i.e.,

$$\begin{pmatrix} x \\ y \end{pmatrix} \to \begin{pmatrix} x e^{r(1-\frac{x}{k}) - \frac{axy}{b^2+x^2}} \\ y e^{(s_2+\tilde{s})(1-\frac{hy}{x})} \end{pmatrix} \tag{21}$$

where $|\tilde{s}| << 1$ denotes the least perturbation. Assuming that $N = x - x_*,\ P = y - y_*$, then system (21) takes the following modified form:

$$\begin{pmatrix} N \\ P \end{pmatrix} \to \begin{pmatrix} a_{11} & a_{12} \\ a_{21} & a_{22} \end{pmatrix} \begin{pmatrix} N \\ P \end{pmatrix} + \begin{pmatrix} g_1(N, P) \\ g_2(N, P) \end{pmatrix}, \tag{22}$$

where

$$g_1(N,P) = b_{13}N^2 + b_{14}NP + b_{15}P^2 + a_1N^3 + a_2N^2P + a_3NP^2 + a_4P^3 + O\Big((|N| + |P| + |\bar{r}|)^4\Big),$$
$$g_2(N,P) = b_{23}N^2 + b_{24}NP + b_{25}P^2 + d_1N^3 + d_2N^2P + d_3NP^2 + d_4P^3 + O\Big((|N| + |P| + |\bar{r}|)^4\Big).$$

Here, b_{11}, b_{12}, b_{21}, b_{22}, b_{13}, b_{14}, b_{15}, a_1, a_2, a_3, a_4, b_{23}, b_{24}, b_{25}, d_1, d_2, d_3, and d_4, are defined in (17) by replacing s_1 by $s_2 + \widetilde{s}$. Let

$$\eta^2 - T_1(\widetilde{s})\eta + D_1(\widetilde{s}) = 0, \tag{23}$$

be the characteristic equation of the variational matrix of (22) evaluated at $(0,0)$, where

$$T_1(\widetilde{s}) = \Big(2 - \frac{rx_*}{k} - \Omega - (s_2 + \widetilde{s})\Big) \text{ and } D_1(\widetilde{s}) = \Big(1 - \frac{rx_*}{k} - \Omega\Big)(1 - (s_2 + \widetilde{s})) + \frac{(s_2 + \widetilde{s})\Phi}{h}$$

where $\Omega = \frac{ax_*y_*(b^2-x_*^2)}{(b^2+x_*^2)^2}$ and $\Phi = \frac{ax_*^2}{b^2+x_*^2}$. Because $(a,b,k,r,s_2) \in T_{NS}$, $|\eta_1| = |\eta_2|$ such that η_1 and η_2 are the complex conjugate roots of (23), it follows that

$$\eta_1, \eta_2 = \frac{T_1(\widetilde{s})}{2} \pm \frac{i}{2}\sqrt{4D_1(\widetilde{s}) - T_1^2(\widetilde{s})}$$

We then obtain

$$|\eta_1| = |\eta_2| = \sqrt{D_1(\widetilde{s})}, \left(\frac{d\sqrt{D_1(\widetilde{s})}}{d\widetilde{s}}\right)_{\widetilde{s}=0} = \frac{(\Omega\, h + \Phi - h)k + hrx}{2\sqrt{((\Omega-1)k + rx)(s-1)h + \Phi\, ks}} \neq 0$$

Moreover, $T_1(0) = (2 - \frac{rx_*}{k} - \Omega - s_2) \neq 0, -1$. Because $(a,b,k,r,s_2) \in T_{NS}$, it follows that $-2 < T_1(0) = (2 - \frac{rx_*}{k} - \Omega - s_2) < 2$. Thus, we have $\eta_1^m, \eta_2^m \neq 1$ for all $m = 1, 2, 3, 4$ at $\widetilde{s} = 0$, for $T_1(0) \neq 0, -1, \pm 2$. Hence, for $\widetilde{s} = 0$, zeros of (23) do not belong to the intersection of the unit circle with coordinate axes if the following condition is satisfied:

$$2 - \Omega - s_2 \neq \frac{rx_*}{k}, \; 3 - \Omega - s_2 \neq \frac{rx_*}{k} \tag{24}$$

The canonical form of (22) at $\widetilde{s} = 0$ can be obtained by taking $\gamma = \frac{T_1(0)}{2}$, $\delta = \frac{1}{2}\sqrt{4D_1(0) - T_1^2(0)}$ and assuming

$$\begin{pmatrix} N \\ P \end{pmatrix} = \begin{pmatrix} b_{12} & 0 \\ \gamma - b_{11} & -\delta \end{pmatrix}\begin{pmatrix} u \\ v \end{pmatrix} \tag{25}$$

Using transformation (25), we obtain the following canonical form of system (22):

$$\begin{pmatrix} u \\ v \end{pmatrix} \to \begin{pmatrix} \gamma & -\delta \\ \delta & \gamma \end{pmatrix}\begin{pmatrix} u \\ v \end{pmatrix} + \begin{pmatrix} \widetilde{f}(u,v) \\ \widetilde{g}(u,v) \end{pmatrix} \tag{26}$$

where

$$\widetilde{f}(u,v) = \frac{a_1N^3}{b_{12}} + \frac{a_2N^2P}{b_{12}} + \frac{b_{13}N^2}{b_{12}} + \frac{a_3NP^2}{b_{12}} + \frac{b_{14}NP}{b_{12}} + \frac{a_4P^3}{b_{12}} + \frac{b_{15}P^2}{b_{12}} + O\Big((|u| + |v|)^4\Big)$$

$$\widetilde{g}(u,v) = \left(\frac{(\gamma - b_{11})a_1}{b_{12}\delta} - \frac{d_1}{\delta}\right)N^3 + \left(\frac{(\gamma - b_{11})a_2}{b_{12}\delta} - \frac{d_2}{\delta}\right)N^2P$$
$$+ \left(\frac{(\gamma - b_{11})b_{13}}{b_{12}\delta} - \frac{b_{23}}{\delta}\right)N^2 + \left(\frac{(\gamma - b_{11})a_3}{b_{12}\delta} - \frac{d_3}{\delta}\right)NP^2$$

$$+\left(\frac{(\gamma-b_{11})b_{14}}{b_{12}\delta}-\frac{b_{24}}{\delta}\right)NP+\left(\frac{(\gamma-b_{11})a_4}{b_{12}\delta}-\frac{d_4}{\delta}\right)P^3$$
$$+\left(\frac{(\gamma-b_{11})b_{15}}{b_{12}\delta}-\frac{b_{25}}{\delta}\right)P^2+O\Big((|u|+|v|)^4\Big)$$
$$N=a_{12}u \text{ and } < P=(\gamma-b_{11})u-\delta v$$

Owing to the aforementioned computation, we state a nonzero real number

$$L := \left(\left[-Re\left(\frac{(1-2\eta_1)\eta_2^2}{1-\eta_1}\xi_{20}\xi_{11}\right)-\frac{1}{2}|\xi_{11}|2-|\xi_{02}|^2+Re(\eta_2\xi_{21})\right]\right)_{\tilde{c}=0}$$

where,

$$\xi_{11}=\frac{1}{4}\left[\tilde{f}_{uu}+\tilde{f}_{vv}+i(\tilde{g}_{uu}+\tilde{g}_{vv})\right],$$
$$\xi_{02}=\frac{1}{8}\left[\tilde{f}_{uu}-\tilde{f}_{vv}-2\tilde{g}_{uv}+i\Big(\tilde{g}_{uu}-\tilde{g}_{vv}+2\tilde{f}_{uv}\Big)\right]$$
$$\xi_{20}=\frac{1}{8}\left[\tilde{f}_{uu}-\tilde{f}_{vv}+2\tilde{g}_{uv}+i\Big(\tilde{g}_{uu}-\tilde{g}_{vv}-2\tilde{f}_{uv}\Big)\right],$$
$$\xi_{21}=\frac{1}{16}\left[\tilde{f}_{uuu}+\tilde{f}_{uvv}+\tilde{g}_{uuv}+\tilde{g}_{vvv}+i\Big(\tilde{g}_{uuu}+\tilde{g}_{uvv}-\tilde{f}_{uuv}-\tilde{f}_{vvv}\Big)\right].$$

Ultimately, we deduced the following conclusions for the direction and existence of the Neimark-Sacker bifurcation, based on the aforementioned calculation:

Theorem 8. *There exists a Neimark-Sacker bifurcation around (x_*, y_*) whenever s deviates wtihin the neighborhood of $s_2 = \frac{(\Omega\, k+rx_*)h}{h(\Omega\, k+rx_*)+k(\Phi-h)}$. In addition, if $L<0$ $(L>0)$, then an attracting (or repelling) invariant closed curve fluctuates in the range (x_*, y_*) for $s > s_2$ (or $s < s_2$).*

6. Chaos Control

In this section, we implement the hybrid control method for controlling the chaos caused by the period-doubling bifurcation and for controlling the Neimark-Sacker bifurcation in (4). Such strategies have been discussed elsewhere in [21,33–37]. We assume the following controlled system corresponding to model (4):

$$\begin{aligned} x_{n+1} =& \epsilon x_n \exp\left(r\left(1-\frac{x_n}{k}\right)-\frac{a x_n y_n}{b^2+x_n{}^2}\right)+(1-\epsilon)x_n, \\ y_{n+1} =& \epsilon y_n \exp\left(s\left(1-\frac{h y_n}{x_n}\right)\right)+(1-\epsilon)y_n, \end{aligned} \tag{27}$$

where $0<\epsilon<1$. Furthermore, both types of bifurcations can be controlled by choosing an appropriate value of parameter ϵ. The controlled system (27) and the original system (4) have the same equilibrium point; the Jacobian matrix of the controlled system (27) at (x_*, y_*) is expressed by

$$\begin{bmatrix} 1-\frac{\epsilon\, xr}{k}-\epsilon\,\Omega & -\epsilon\,\Phi \\ \frac{\epsilon\, s}{h} & 1-\epsilon\, s \end{bmatrix}$$

Consequently, the necessary and sufficient condition for local stability around (x_*, y_*) of the controlled system (27) yields the following result.

Theorem 9. *The interior fixed point (x_*, y_*) of (27) is locally asymptotically stable if*

$$\left|2-\frac{\varepsilon r x_*}{k}-\varepsilon\Omega-\varepsilon s\right| < 1+\left(1-\frac{\varepsilon r x_*}{k}-\varepsilon\Omega\right)(1-\varepsilon s)+\frac{s\Phi}{h} < 2.$$

7. Fractal Dimension

The fractal dimension that describes the strange attractors of discrete-time models is defined as follows [38,39]:

$$\mathfrak{D}_\ell = m + \frac{\sum_{j=1}^{m} k_m}{|k_m|},$$

where $k_1, k_2, \ldots, k_m$ are Lyapunov exponents, and m is the largest integer such that $\sum_{j=1}^{m} k_m \geq 0$ and $\sum_{j=1}^{m+1} k_m < 0$. For the discrete-time model (4), the fractal dimension takes the following form:

$$\mathfrak{D}_\ell = 1 + \frac{k_1}{|k_2|}, \quad k_1 > 0 > k_2$$

Furthermore, for the values of parameters a, b, r, k, h, and s, the two Lyapunov exponents F_1 and F_2 are computed numerically. If $b = 3.3$, $a = 0.8$, $r = 1.3$, $h = 2.7$, and $k = 1.8$, then F_1 and F_2 corresponding to the values of the bifurcation (period-doubling) parameter s from the chaotic region, with the help of Mathematica software, are shown in Table 1.

Table 1. Fractal dimension of model (4).

Values of s	1st Lyapunov Exponents F_1	2nd Lyapunov Exponents F_2	Fractal Dimension $\mathfrak{D}_\ell$
2.85	0.08462596943938297	−1.1368798813345231	1.0744370366903186
2.90	0.22741225178613755	−1.2160641186798002	1.187006793714976
3.0	0.31895100399320747	−1.0790255038850196	1.2955917194216706
3.1	0.22493177216760443	−1.244489487648406	1.1807422034497348
3.2	0.4025124673527987	−1.1944261491614452	1.3369923436751492
3.3	0.3894200849244259	−1.2328810276916689	1.3158618521801246
3.4	0.47745428811163265	−1.2528556268034083	1.3810928233844715
3.5	0.47582043180971084	−1.2925566246939908	1.3681234715131803

The strange attractors for fixed parametric values illustrate that the discrete model (4) has a complex dynamical behavior as parameter s increases by $s > 2.1894756175566834$. Similarly, for the Neimark-Sacker bifurcation, the Lyapunov exponents and fractal dimension can be calculated for the values of parameter s from the chaotic region, that is, $s = 1.66$, 1.68, 1.89, and so on. The strange attractors corresponding to these values are also shown in Figure 1. In particular, Figure 1g,h,k below, demonstrate that the discrete time model (4) has a complex dynamical nature when parameter $s > 1.3874082082631611$.

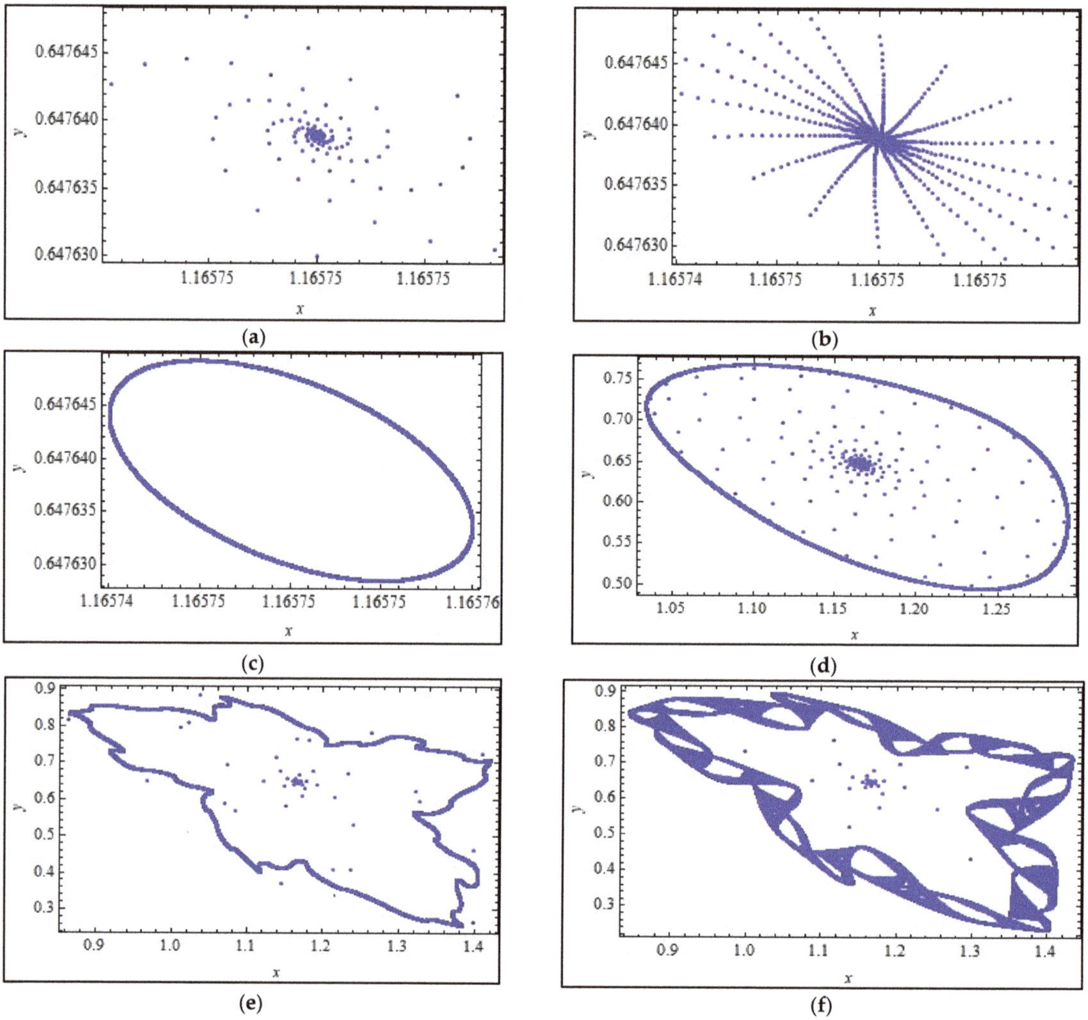

Figure 1. *Cont.*

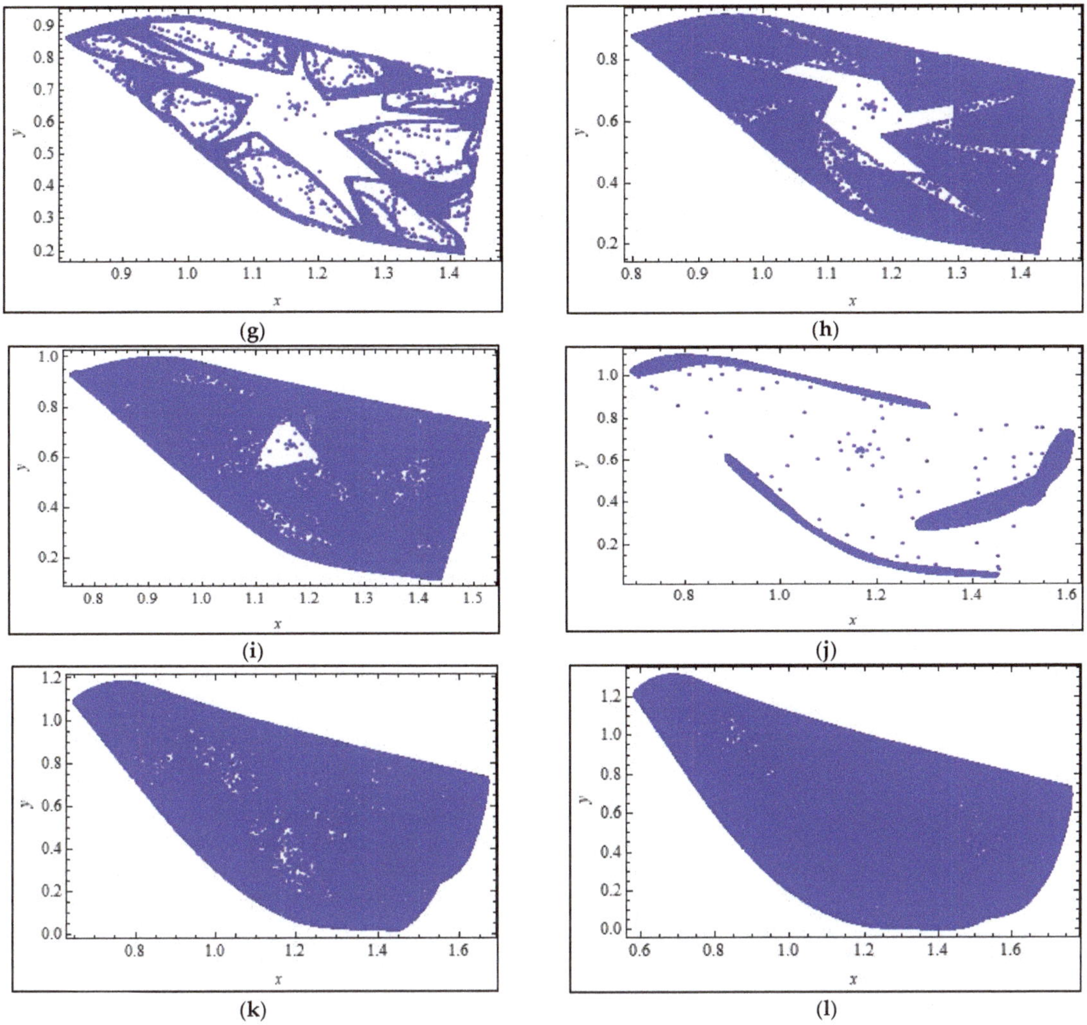

Figure 1. (**a**) $s = 1.34$, (**b**) $s = 1.3811$, (**c**) $s = 1.3874082082631611$, (**d**) $s = 1.43$, (**e**) $s = 1.607$, (**f**) $s = 1.625$, (**g**) $s = 1.66$, (**h**) $s = 1.68$, (**i**) $s = 1.735$, (**j**) $s = 1.83$, (**k**) $s = 1.89$, (**l**) $s = 1.99$. (**a**)–(**l**) Phase portraits of system (4) for different values of $s \in [1,2]$ with $b = 1.7, a = 2.26, r = 2.4, h = 1.8$, and $k = 1.4$ under the initial conditions $x_0 = 1.0219, y_0 = 0.567721$.

8. Numerical Simulation

This section verifies the aforementioned theoretical discussion. The first example is related to the existence and direction of the Neimark-Sacker bifurcation. The second example shows that for a suitable choice of parameters, system (4) undergoes period-doubling bifurcation. Moreover, to confirm the control of flip and Neimark-Sacker bifurcation, we provide two examples for different choices of parameters defined in T_{PB} and T_{NS}.

Example 1. *Let $b = 1.7$, $a = 2.26$, $r = 2.4$, $h = 1.8$, $k = 1.4$, $s \in [1, 1.8]$, and initial conditions $(x_0, y_0) = (1.0219, 0.567721)$. Then, both species undergo a Neimark-Sacker bifurcation, as shown in Figure 2. To confirm the chaotic behavior of model (4) MLE is shown in Figure 2c.*

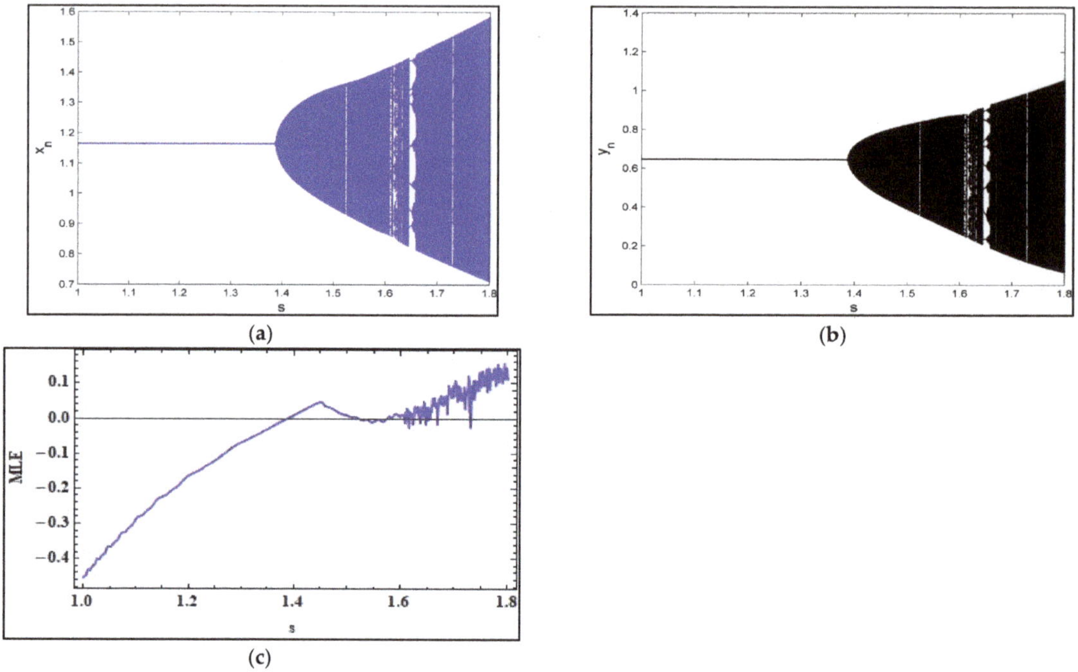

Figure 2. Bifurcation diagrams and maximum Lypunov exponents (MLE) for system (4) with parametric values $b = 1.7$, $a = 2.26$, $r = 2.4$, $h = 1.8$, $k = 1.4$, and $s \in [1, 1.8]$ and initial conditions $(x_0, y_0) = (1.0219, 0.567721)$: (**a**) bifurcation for x_n, (**b**) bifurcation for y_n, and (**c**) MLE.

Furthermore, Figure 1a–l shows the interesting behavior of system (4). Figure 1f–l shows the chaotic behavior of system (4). Assuming that $b = 1.7$, we have a positive fixed point $(x_*, y_*) = (1.165750001, 0.6476388892)$, which loses stability and undergoes a Neimark-Sacker bifurcation. Thus, for the aforementioned parameters, we have the following control system:

$$\begin{aligned} x_{n+1} &= \epsilon\, x_n e^{2.4 - 1.714285714\, x_n - 2.26\, \frac{x_n y_n}{x_n{}^2 + 2.89}} + (1-\epsilon) x_n, \\ y_{n+1} &= \epsilon\, y_n e^{1.5 - 2.70\, \frac{y_n}{x_n}} + (1-\epsilon) y_n. \end{aligned} \tag{28}$$

It can be clearly observed that the controlled system (28) has a unique positive equilibrium point $(x_*, y_*) = (1.165750001,\ 0.6476388892)$, which is similar to the original system (4). In addition, the Jacobian at equilibrium $(x_*, y_*) = (1.165750001, 0.6476388892)$ has the following form:

$$\begin{bmatrix} -2.143126283\,\epsilon + 1 & -0.7228285706\,\epsilon \\ 0.8333333325\,\epsilon & -1.500000000\,\epsilon + 1 \end{bmatrix}.$$

Example 2. *Assuming the parameters $b = 3.3, a = 0.8, r = 1.3, h = 2.7, k = 1.8$, and $s \in [2,\ 3.5]$, and the initial conditions $(x_0, y_0) = (1.74693, 0.647013)$, both species then undergo period-doubling bifurcation when the bifurcation parameter passes through $s = 2.1894756175566834$, as shown in Figure 3. In particular this fact is obvious in Figure 3a,b. Moreover, to confirm the chaotic behavior of model (4) MLE is shown in Figure 3c.*

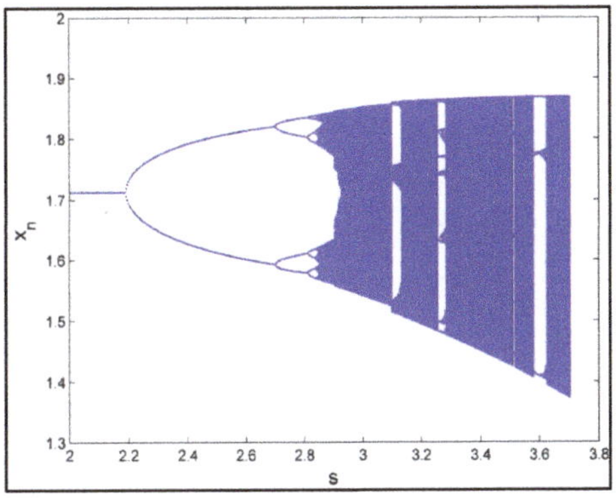

(**a**)

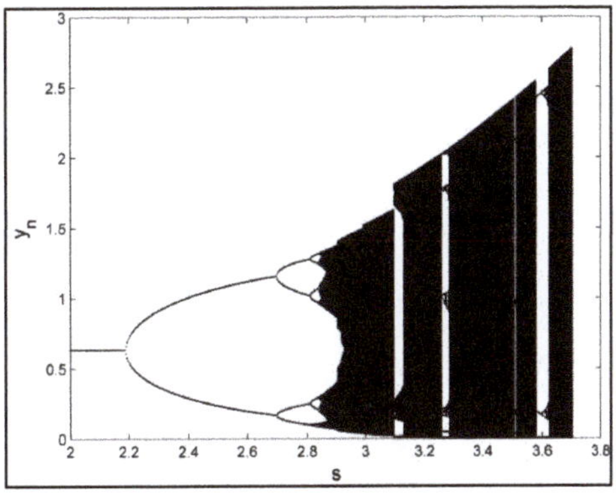

(**b**)

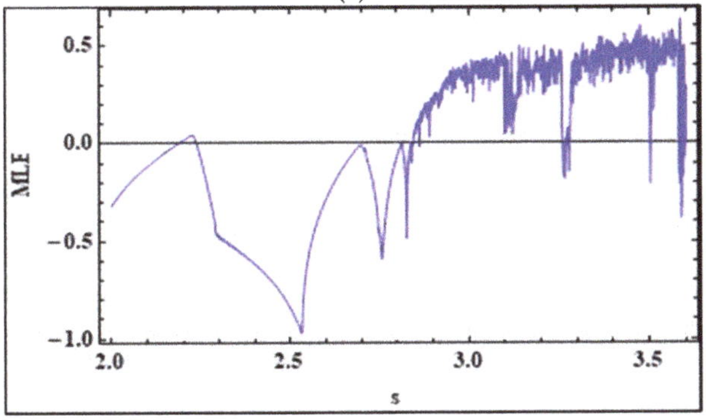

(**c**)

Figure 3. Bifurcation diagrams and MLE for system (4) with parameters $b = 3.3$, $a = 0.8$, $r = 1.3$, $h = 2.7$, $k = 1.8$, $s \in [2, 3]$, and $(x_0, y_0) = (1.74693, 0.647013)$: (**a**) period-doubling bifurcation for x_n, (**b**) period-doubling bifurcation for y_n, and (**c**) MLE.

Furthermore, if $s = 2.4$, then the equilibrium point $(x_*, y_*) = (1.712924751, 0.6344165743)$ loses its stability and undergoes periodic doubling (see Figure 4).

Thus, for the aforementioned parameters, we present the following control system:

$$\begin{aligned} x_{n+1} &= \epsilon\, x_n e^{1.3 - 0.7222222223\, x_n - 0.8\, \frac{x_n y_n}{x_n{}^2 + 10.89}} + (1-\epsilon) x_n, \\ y_{n+1} &= \epsilon\, y_n e^{2.4 - 6.48\, \frac{y_n}{x_n}} + (1-\epsilon) y_n, \end{aligned} \tag{29}$$

The fixed point $(x_*, y_*) = (1.712924751,\ 0.6344165743)$ was preserved in the case of a controlled system (29). Furthermore, the variational matrix of the aforementioned controlled system computed at a fixed point $(x_*, y_*) = (1.712924751,\ 0.6344165743)$ is given by

$$\begin{bmatrix} -1.273304693\,\epsilon + 1 & -0.1697967362\,\epsilon \\ 0.8888888882\,\epsilon & -2.399999999\,\epsilon + 1 \end{bmatrix}$$

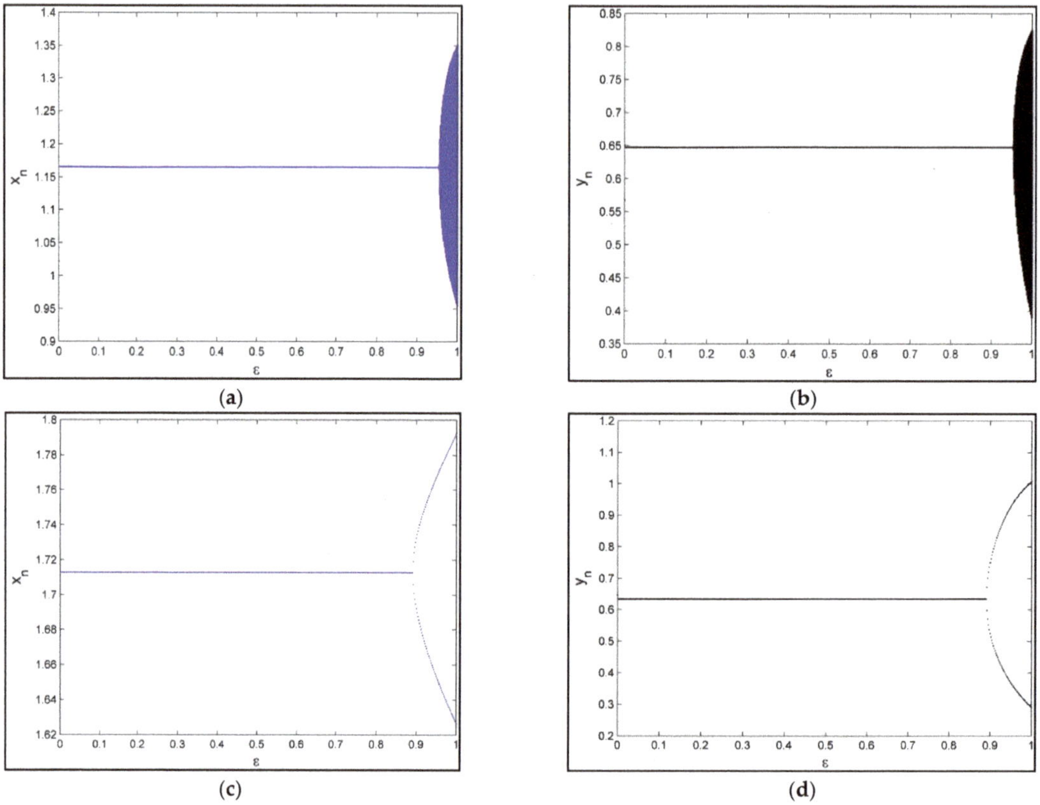

Figure 4. (**a**) Bifurcation diagrams of x_n for controlled system (28), (**b**) bifurcation diagrams of y_n for controlled system (28), (**c**) bifurcation diagrams of x_n for controlled system (30), and (**d**) bifurcation diagrams of y_n for controlled system (30).

The characteristic polynomial of the aforementioned Jacobian matrix is given by

$$\eta^2 + (3.673304692\,\epsilon - 2)\eta + 3.206861694\,\epsilon^2 - 3.673304692\,\epsilon + 1 = 0.$$

According to Lemma 1, the control system is locally asymptotically stable, if $0 < \epsilon < 0.8910230450195268$ and bifurcation is controlled for $0 < \epsilon < 0.8910230450195268$ (see Figure 4c,d).

Finally, some local implications of the MLE diagrams, shown in Figures 1c and 2c for the Neimark-Sacker bifurcation and period-doubling bifurcation, respectively, are plotted in Figure 5a,b, respectively. It has also been verified that the system undergoes Neimark-Sacker bifurcation at $s = 1.3874082082631611$, where the phase portrait at this point shows a closed invariant curve, as already shown in Figure 4c.

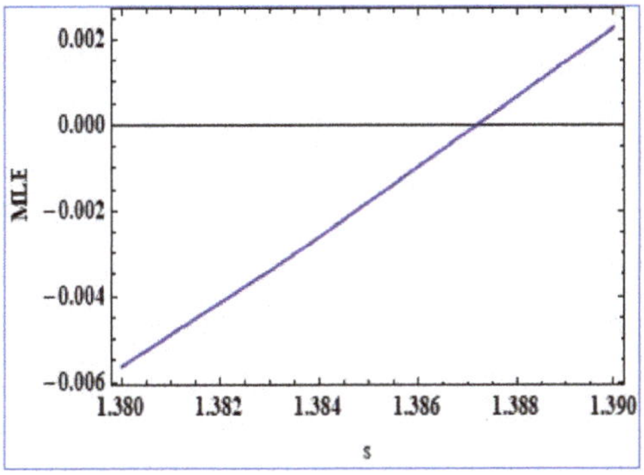

(a)

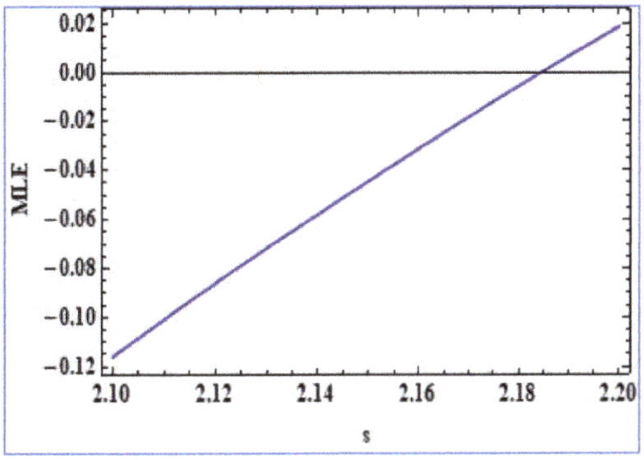

(b)

Figure 5. (**a**) Local implication for the Neimark-Sacker bifurcation and (**b**) local implication for period-doubling bifurcation.

9. Concluding Remarks

In this study, we examined the qualitative and dynamical analyses of a discrete-time predator-prey model. Piecewise constant arguments have been applied to achieve the discrete-time counterpart of a continuous model. Thus, a comprehensive analysis of model (4) was presented. In particular, we investigated the boundedness, local stability of the boundary, and positive equilibrium points, which seem to present more challenging cases of Euler's discretization scheme in [31]. Moreover, it was proved that the population sustains both period-doubling bifurcation and Neimark-Sacker bifurcation near the interior equilibrium. The parametric conditions were obtained for the direction and existence of both types of bifurcations using the theory of bifurcation and center manifold theorem. Moreover, the chaotic attractors shown in Figure 5 ensures chaos in the system. To control the chaotic behavior of system (4), a hybrid control method was implemented. Hence, by applying a control strategy, both types of bifurcations can be controlled for a maximum range of control parameters. We also presented the fractal dimension of model (4), which characterizes the strange attractors provided in Figure 5 thereby illustrating the complexity and rich dynamics of discrete model (4). Finally, numerical simulations were conducted to verify the analytical and theoretical approaches.

10. Future Direction

Our future research will include the Leslie–Gower predator-prey model with the functional response of Holling type-II. In this case, we aim to conduct stability, bifurcation, and chaos-control analyses of the model. A comparison of both functional responses will be conducted in a future study.

Author Contributions: Conceptualization, A.T., M.S.S. and Q.D.; methodology, A.T., M.S.S. and Q.D.; validation, A.T., M.S.S., Q.D. and H.N.; formal analysis, A.T., M.S.S., Q.D. and H.N.; investigation, A.T., M.S.S., Q.D. and H.N.; data curation, A.T., M.S.S., Q.D. and H.N.; writing—original draft preparation, A.T., M.S.S., Q.D. and H.N.; writing—review and editing, A.T., M.S.S., Q.D. and H.N. All authors have read and agreed to the published version of the manuscript.

Funding: This research did not receive any specific external funding.

Institutional Review Board Statement: Not applicable.

Informed Consent Statement: Not applicable.

Data Availability Statement: Not applicable.

Acknowledgments: Asifa Tassaddiq would like to thank the Deanship of Scientific Research at Majmaah University for supporting this work under Project No. R-2021-274. The authors are also thankful to the worthy reviewers and editors for their useful and valuable suggestions for the improvement of this paper which led to a better presentation.

Conflicts of Interest: None of the authors have any competing interests in this manuscript.

Appendix A

$$b_{11} = 1 + x_* B_1,\ b_{12} = -\frac{a x_*^2}{b^2 + x_*^2},\ b_{21} = \frac{s_1}{h}, b_{22} = 1 - s_1,\ b_{15} = \frac{a^2 x_*^3}{2(b^2+x_*^2)^2}$$

$$b_{13} = B_1 + \frac{1}{2} x_* y_* B_3 + \frac{1}{2} x_* B_1^2, b_{14} = -\frac{x_*\left(a x_* B_1 - b^2 B_2 - x_*^2 B_2 + a\right)}{b^2 + x_*^2},$$

$$a_1 = y_* B_3 + \frac{1}{2} B_1^2 + \frac{x_* a y_* \left(b^4 - 6b^2 x_*^2 + x_*^4\right)}{\left(b^2+x_*^2\right)^4} + \frac{1}{2} x_* y_* B_3 B_1 + \frac{1}{6} x_* B_1^3,$$

$$a_2 = B_2 - \frac{B_1 a x_*}{b^2+x_*^2} + \frac{1}{2} x_* B_3 - \frac{1}{2}\frac{x_*^2 y_* B_3 a}{b^2+x_*^2} + x_* B_1 B_2 - \frac{1}{2}\frac{x_*^2 B_1^2 a}{b^2+x_*^2},$$

$$a_3 = \frac{1}{2}\frac{a^2 x_*^2}{\left(b^2+x_*^2\right)^2} - \frac{x_*^2 B_2 a}{b^2+x_*^2} + \frac{1}{2}\frac{x_*^3 B_1 a^2}{\left(b^2+x_*^2\right)^2}, a_4 = -\frac{1}{6}\frac{x_*^4 a^3}{\left(b^2+x_*^2\right)^3},$$

$$b_{23} = \frac{h s_1 y_*^2 (h s_1 y_* - 2x_*)}{2x_*^4},\quad b_{24} = \frac{h s_1 y_* (2x_* - h s_1 y_*)}{x_*^3},$$

$$b_{25} = \frac{h s_1 (h s_1 y_* - 2x_*)}{2x_*^2},\quad d_1 = \frac{h s_1 y_*^2 \left(6x_*^2 - 6h s_1 x_* y_* + h^2 s_1^2 y_*^2\right)}{6x_*^6},$$

$$d_2 = -\frac{h s_1 y_* (h s_1 y_* - 4x_*)(h s_1 y_* - x_*)}{2x_*^5}, d_3 = \frac{h s_1 \left(2x_*^2 - 4h s_1 x_* y_* + h^2 s_1^2 y_*^2\right)}{2x_*^4},$$

$$d_4 = -\frac{h^2 s_1^2 (h s_1 y_* - 3x_*)}{6x_*^3},\quad c_1 = \frac{h y_*^2 ((1+s_1)x_* - h s_1 y_*)}{x_*^3},$$

$$c_2 = \frac{\left(x_*^2 - h(2+s_1)x_* y_* + h^2 s_1 y_*^2\right)}{x_*^2},$$

$$c_3 = \frac{h y_* \left(2(1+s_1)x_*^2 - h s_1 (4+s_1) x_* y_* + h^2 s_1^2 y_*^2\right)}{x_*^4},$$

$$c_4 = -\frac{h y_*^2 \left(2(1+s_1)x_*^2 - h s_1 (4+s_1) x_* y_* + h^2 s_1^2 y_*^2\right)}{2x_*^5},$$

$$c_5 = -\frac{h\left(2(1+s_1)x_*^2 - h s_1 (4+s_1) x_* y_* + h^2 s_1^2 y_*^2\right)}{2x_*^3}.$$

where $B_1 = -\left(\frac{r_1}{k} + \frac{a y_* \left(b^2 - x_*^2\right)}{\left(b^2+x_*^2\right)^2}\right)$, $B_2 = -\frac{a\left(b^2 - x_*^2\right)}{\left(b^2+x_*^2\right)^2}$, $B_3 = \frac{2a x_* \left(3b^2 - x_*^2\right)}{\left(b^2+x_*^2\right)^3}$, $B_4 = y_* B_3$.

Appendix B

$$\begin{aligned}
\widetilde{f}(u,v,\bar{s}) = {} & \left(\frac{b_{13}(\eta_2 - b_{11})}{b_{12}(1+\eta_2)} - \frac{b_{23}}{1+\eta_2}\right)N^2 - \left(\frac{b_{24}}{1+\eta_2} + \frac{b_{14}(b_{11}-\eta_2)}{b_{12}(1+\eta_2)}\right)NP \\
& - \left(\frac{b_{25}}{1+\lambda_2} + \frac{b_{15}(b_{11}-\eta_2)}{b_{12}(1+\eta_2)}\right)P^2 - \left(\frac{d_1}{1+\lambda_2} + \frac{a_1(b_{11}-\eta_2)}{b_{12}(1+\eta_2)}\right)N^3 \\
& - \left(\frac{d_2}{1+\lambda_2} + \frac{a_2(b_{11}-\eta_2)}{b_{12}(1+\eta_2)}\right)N^2 P - \left(\frac{d_3}{1+\eta_2} + \frac{a_3(b_{11}-\eta_2)}{b_{12}(1+\eta_2)}\right)NP^2 \\
& - \left(\frac{d_4}{1+\eta_2} + \frac{a_4(b_{11}-\eta_2)}{b_{12}(1+\eta_2)}\right)P^3 - \frac{c_1}{1+\eta_2}\bar{s}N - \frac{c_2}{1+\eta_2}\bar{r}P - \frac{c_3}{1+\eta_2}NP - \frac{c_4}{1+\eta_2}\bar{r}N^2 - \frac{c_5}{1+\eta_2}\bar{r}P^2 + O\Big((|u| + |v| + |\bar{s}|)^4\Big), \\
\widetilde{g}(u,v,\bar{s}) = {} & \left(\frac{b_{23}}{1+\eta_2} + \frac{(1+b_{11})b_{13}}{b_{12}(1+\eta_2)}\right)N^2 + \left(\frac{b_{24}}{1+\eta_2} + \frac{(1+b_{11})b_{14}}{b_{12}(1+\eta_2)}\right)NP \\
& + \left(\frac{b_{25}}{1+\eta_2} + \frac{(1+b_{11})b_{15}}{b_{12}(1+\eta_2)}\right)P^2 + \left(\frac{a_1(1+b_{11})}{b_{12}(1+\eta_2)} + \frac{d_1}{1+\eta_2}\right)N^3 \\
& + \left(\frac{a_2(1+b_{11})}{b_{12}(1+\eta_2)} + \frac{d_2}{1+\eta_2}\right)N^2 P + \left(\frac{a_3(1+b_{11})}{b_{12}(1+\eta_2)} + \frac{d_3}{1+\eta_2}\right)NP^2 \\
& + \left(\frac{a_4(1+b_{11})}{b_{12}(1+\eta_2)} + \frac{d_4}{1+\eta_2}\right)P^3 + \frac{c_1}{1+\eta_2}\bar{s}N + \frac{c_2}{1+\eta_2}\bar{s}P + \frac{c_3}{1+\eta_2}\bar{s}NP + \frac{c_4}{1+\eta_2}\bar{s}N^2 + \frac{c_5}{1+\eta_2}\bar{r}P^2 + O\Big((|u| + |v| + |\bar{s}|)^4\Big),
\end{aligned}$$

where $N = b_{12}(u+v)$ and $P = (-1 - b_{11})u + (\eta_2 - b_{11})v$.

Thus, the approximation of the center manifold $W^c(0,0,0)$ of (19) within the neighborhood of $\bar{s}=0$ evaluated at the origin can be expressed as

$$W^c(0,0,0) = \left\{ \left(u, v, s_1 \in R^3\right) = M_3 s^2 + M_2 su + m_1 u^2 + (O|u|, |s_1|)^4 \right\},$$

where

$$M_1 = \frac{b_{12}^2\left(\frac{b_{23}}{1+\eta_2}+\frac{(1+b_{11})b_{13}}{b_{12}(1+\eta_2)}\right)}{1-\eta_2} - \frac{(1+b_{11})b_{12}\left(\frac{b_{24}}{1+\eta_2}+\frac{(1+b_{11})b_{14}}{b_{12}(1+\eta_2)}\right)}{1-\eta_2} + \frac{(1+b_{11})^2\left(\frac{b_{25}}{1+\eta_2}+\frac{(1+b_{11})b_{15}}{b_{12}(1+\eta_2)}\right)}{1-\eta_2},$$
$$M_2 = \frac{c_2(1+b_{11})-c_1 b_{12}}{\eta_2{}^2-1}, \quad M_3 = 0.$$

Consequently, the restricted map to center manifold $W^c(0,0,0)$ is expressed as follows:

$$F : u \to -u + t_1 u^2 + t_2 us + t_3 u^2 s + t_4 us^2 + t_5 u^3 + (O|u|, |s_1|)^4$$

where

$$\begin{aligned}
t_1 = \;& -b_{12}^2\left(\frac{b_{23}}{1+\eta_2} + \frac{b_{13}(b_{11}-\eta_2)}{b_{12}(1+\eta_2)}\right) - (1+b_{11})^2\left(\frac{b_{25}}{1+\eta_2} + \frac{b_{15}(b_{11}-\eta_2)}{b_{12}(1+\eta_2)}\right) \\
& +(1+b_{11})b_{12}\left(\frac{b_{24}}{1+\eta_2} + \frac{b_{14}(b_{11}-\eta_2)}{b_{12}(1+\eta_2)}\right), \\
t_2 = \;& \frac{c_2(1+b_{11})}{\eta_2+1} - \frac{c_1 b_{12}}{\eta_2+1}, \\
t_3 = \;& 2\left(\frac{(\eta_2-b_{11})b_{13}}{b_{12}(\eta_2+1)} - \frac{b_{23}}{\eta_2+1}\right)b_{12}^2 M_2 + \frac{c_3 b_{12}(1+b_{11})}{\eta_2+1} - \frac{c_4 b_{12}{}^2}{\eta_2+1} + \left(\frac{(\eta_2-b_{11})b_{14}}{b_{12}(\eta_2+1)} - \frac{b_{24}}{\eta_2+1}\right)b_{12}(\eta_2-b_{11})M_2 - \frac{c_2(\eta_2-b_{11})M_1}{\eta_2+1} \\
& +\left(\frac{(b_{11}-\eta_2)b_{14}}{b_{12}(\eta_2+1)} + \frac{b_{24}}{\eta_2+1}\right)b_{12}M_2(1+b_{11}) - \frac{c_5(1+b_{11})^2}{\eta_2+1} + 2\left(\frac{(b_{11}-\eta_2)b_{15}}{b_{12}(\eta_2+1)} + \frac{b_{25}}{\eta_2+1}\right)(1+b_{11})(\eta_2-b_{11})M_2 - \frac{c_1 b_{12} M_1}{\eta_2+1}, \\
t_4 = \;& \left(2\frac{(\eta_2-b_{11})b_{13}}{b_{12}(\eta_2+1)} - \frac{b_{23}}{\eta_2+1}\right)b_{12}{}^2 M_3 - \frac{c_2(\eta_2-b_{11})M_2}{\eta_2+1} \\
& +\left(\frac{(\eta_2-b_{11})b_{14}}{b_{12}(\eta_2+1)} - \frac{b_{24}}{\eta_2+1}\right)b_{12}(\eta_2-b_{11})M_3 - \frac{c_1 b_{12} M_2}{\eta_2+1} \\
& +2\left(\frac{(b_{11}-\eta_2)b_{15}}{b_{12}(\eta_2+1)} + \frac{b_{25}}{\eta_2+1}\right)(1+b_{11})(\eta_2-b_{11})M_3 \\
& +\left(\frac{(b_{11}-\eta_2)b_{14}}{b_{12}(\eta_2+1)} + \frac{b_{24}}{\eta_2+1}\right)b_{12}M_3(1,+,b_{11}) \\
t_5 = \;& \left(\frac{(\eta_2-b_{11})b_{14}}{b_{12}(\eta_2+1)} - \frac{b_{24}}{\eta_2+1}\right)b_{12}M_1(\eta_2-b_{11}) \\
& +\left(\frac{(b_{11}b_{11}-\eta_2)b_{14}}{b_{12}(\eta_2+1)} + \frac{b_{24}}{\eta_2+1}\right)b_{12}M_1(1+b_{11}) \\
& +\left(\frac{(b_{11}-\eta_2)a_4}{b_{12}(\eta_2+1)} + \frac{d_4}{\eta_2+1}\right)(1+b_{11})^3 \\
& +2\left(\frac{(b_{11}-\eta_2)b_{15}}{b_{12}(\eta_2+1)} + \frac{b_{25}}{\eta_2+1}\right)(1+b_{11})(\eta_2-b_{11})M_1 \\
& +2\left(\frac{(\eta_2-b_{11})b_{13}}{b_{12}(\eta_2+1)} - \frac{b_{23}}{\eta_2+1}\right)b_{12}{}^2 M_1 \\
& \left(\frac{+(\eta_2-b_{11})a_3}{b_{12}(\eta_2+1)} - \frac{d_3}{\eta_2+1}\right)b_{12}(1+b_{11})^2 + \left(\frac{(\eta_2-b_{11})a_1}{b_{12}(\eta_2+1)} - \frac{d_1}{\eta_2+1}\right)b_{12}{}^3 \\
& +\left(\frac{(b_{11}-\eta_2)a_2}{b_{12}(\eta_2+1)} + \frac{d_2}{\eta_2+1}\right)b_{12}{}^2(1+b_{11}).
\end{aligned}$$

References

1. Chen, B.S.; Chen, J.J. Bifurcation and chaotic behavior of a discrete singular biological economic system. *Appl. Math. Comput.* **2012**, *219*, 2371–2386. [CrossRef]
2. Ghaziani, R.K.; Govaerts, W.; Sonck, C. Resonance and bifurcation in a discrete-time predator-prey system with Holling functional response. *Nonlinear Anal. RWA* **2012**, *13*, 1451–1465. [CrossRef]
3. Jana, D. Chaotic dynamics of a discrete predator-prey system with prey refuge. *Appl. Math. Comput.* **2013**, *224*, 848–865. [CrossRef]
4. Misra, O.P.; Sinha, P.; Singh, C. Stability and bifurcation analysis of a prey-predator model with age based predation. *Appl. Math. Model.* **2013**, *37*, 6519–6529. [CrossRef]
5. Zhang, G.D.; Shen, Y.; Chen, B.S. Bifurcation analysis in a discrete differential-algebraic predator-prey system. *Appl. Math. Model.* **2014**, *38*, 4835–4848. [CrossRef]

6. Hu, D.P.; Cao, H.J. Bifurcation and chaos in a discrete-time predator-prey system of Holling and Leslie type. *Commun. Nonlinear Sci. Numer. Simulat.* **2015**, *22*, 702–715. [CrossRef]
7. Wang, C.; Li, X.Y. Further investigations into the stability and bifurcation of a discrete predator-prey model. *J. Math. Anal. Appl.* **2015**, *422*, 920–939. [CrossRef]
8. Murdoch, W.; Briggs, C.; Nisbet, R. *Consumer-Resource Dynamics*; Princeton University Press: New York, NY, USA, 2003.
9. Holling, C.S. The functional response of predator to prey density and its role in mimicry and population regulation. *Mem. Entomol. Soc. Can.* **1965**, *97*, 5–60. [CrossRef]
10. Rosenzweig, M.L.; MacArthur, R.H. Graphical representation and stability conditions of predator-prey interactions. *Am. Nat.* **1963**, *97*, 209–223. [CrossRef]
11. Freedman, H.I.; Mathsen, R.M. Persistence in predator-prey systems with ratio-dependent predator influence. *Bull. Math. Biol.* **1993**, *55*, 817–827. [CrossRef]
12. Hastings, A. Multiple limit cycles in predator-prey models. *J. Math. Biol.* **1981**, *11*, 51–63. [CrossRef]
13. Lindstrom, T. Qualitative analysis of a predator-prey systems with limit cycles. *J. Math. Biol.* **1993**, *31*, 541–561. [CrossRef]
14. Jing, Z.J.; Yang, J. Bifurcation and chaos in discrete-time predator-prey system. *Chaos Solitons Fractals* **2006**, *27*, 259–277. [CrossRef]
15. Liu, X.; Xiao, D. Complex dynamic behaviors of a discrete-time predator-prey system. *Chaos Solitons Fractals* **2007**, *32*, 80–94. [CrossRef]
16. Lopez-Ruiz, R.; Fournier-Prunaret, D. Indirect Allee effect, bistability and chaotic oscillations in a predator-prey discrete model of logistic type. *Chaos Solitons Fractals* **2005**, *24*, 85–101. [CrossRef]
17. Neubert, M.G.; Kot, M. The subcritical collapse of predator populations in discrete-time predator-prey models. *Math. Biosci.* **1992**, *110*, 45–66. [CrossRef]
18. Shabbir, M.S.; Din, Q.; Ahmad, K.; Tassaddiq, A.; Soori, A.H.; Khan, M.A. Stability, bifurcation and chaos control of a novel discrete-time model involving Allee effect and cannibalism. *Adv. Differ. Equ.* **2020**, *2020*, 379. [CrossRef]
19. Din, Q.; Shabbir, M.S.; Khan, M.A.; Ahmad, K. Bifurcation analysis and chaos control for a plant-herbivore model with weak predator functional response. *J. Biol. Dyn.* **2019**, *13*, 481–501. [CrossRef]
20. Shabbir, M.S.; Din, Q.; Safeer, M.; Khan, M.A.; Ahmad, K.A. A dynamically consistent nonstandard finite difference scheme for a predator-prey model. *Adv. Differ. Equ.* **2019**, *2019*, 381. [CrossRef]
21. Samaddar, S.; Dhar, M.; Bhattacharya, P. Effect of fear on prey-predator dynamics: Exploring the role of prey refuge and additional food. *Chaos Interdiscip. J. Nonlinear Sci.* **2020**, *30*, 63129. [CrossRef]
22. Anacleto, M.; Vidal, C. Dynamics of a delayed predator-prey model with Allee effect and Holling type II functional response. *Math. Methods Appl. Sci.* **2020**, *43*, 5708–5728. [CrossRef]
23. Tang, B. Dynamics for a fractional-order predator-prey model with group defense. *Sci. Rep.* **2020**, *10*, 4906. [CrossRef]
24. Sarwardi, S.; Haque, M.M.; Hossain, S. Analysis of Bogdanov-Takens bifurcations in a spatiotemporal harvested-predator and prey system with Beddington–DeAngelis-type response function. *Nonlinear Dyn.* **2020**, *100*, 1755–1778. [CrossRef]
25. Din, Q. Complexity and chaos control in a discrete-time prey-predator model. *Commun. Nonlinear. Sci. Numer. Simulat.* **2017**, *49*, 113–134. [CrossRef]
26. Shabbir, M.S.; Din, Q.; Alabdan, R.; Tassaddiq, A.; Ahmad, K. Dynamical complexity in a class of novel discrete-time predator-prey interaction with cannibalism. *IEEE Access* **2020**, *8*, 100226–100240. [CrossRef]
27. Selvam, A.G.; Vianny, D.A.; Jacob, S.B. Dynamical behaviour of discrete time prey-predator model with Holling type III functional response. *Cikitusi J. Multidiscip. Res.* **2019**, *6*, 75–81.
28. Jiangang, Z.; Tian, D.; Yandong, C.; Shuang, Q.; Wenju, D.; Hongwei, L. Stability and bifurcation analysis of a discrete predator-prey model with Holling type III functional response. *J. Nonlinear Sci. Appl.* **2016**, *2016*, 6228–6243.
29. Elettreby, M.F.; Nabil, T.; Khawagi, A. Stability and bifurcation analysis of a discrete predator-prey model with mixed Holling interaction. *Comput. Modeling Eng. Sci.* **2020**, *122*, 907–921. [CrossRef]
30. Murray, J.D. *Mathematical Biology*, 2nd ed.; Springer: Berlin, Germany, 1993.
31. He, Z.; Lai, X. Bifurcation and chaotic behavior of a discrete-time predator-prey system. *Nonlinear Anal. Real World Appl.* **2011**, *12*, 403–417. [CrossRef]
32. Saber, N. *Elaydi, Discrete Chaos*; Chapman & Hall/CRC: Boca Raton, FL, USA, 2007.
33. Chen, Y.U. Controlling and anti-controlling Hopf bifurcations in discrete maps using polynomial functions. *Chaos Solitons Fractals* **2005**, *26*, 1231–1248. [CrossRef]
34. ELabbasy, E.M.; Agiza, H.N.; Metwally, H.E.L.; Elsadany, A.A. Bifurcation analysis, chaos and control in the Burgers mapping. *Int. J. Nonlinear Sci.* **2007**, *4*, 171–185.
35. Tassaddiq, A.; Shabbir, M.S.; Din, Q.; Ahmad, K. A ratio-dependent nonlinear predator-prey model with certain dynamical results. *IEEE Access* **2020**, *8*, 195074–195088. [CrossRef]
36. Din, Q.; Saleem, N.; Shabbir, M.S. A class of discrete predator-prey interaction with bifurcation analysis and chaos control. *Math. Model. Nat. Phenom.* **2020**, *15*, 1–27. [CrossRef]
37. Din, Q.; Shabbir, M.S. A Cubic autocatalator chemical reaction model with limit cycle analysis and consistency preserving discretization. *MATCH Commun. Math. Comput. Chem.* **2022**, *87*, 441–462. [CrossRef]

38. Cartwright, J.H.E. Nonlinear stiffness Lyapunov exponents and attractor dimension. *Phys. Lett. A* **1999**, *264*, 298–304. [CrossRef]
39. Kaplan, J.L.; Yorke, J.A. Preturbulence: A regime observed in a fluid flow model of Lorenz. *Commun. Math. Phys.* **1979**, *67*, 93–108. [CrossRef]

fractal and fractional

Article

Comparative Numerical Study of Spline-Based Numerical Techniques for Time Fractional Cattaneo Equation in the Sense of Caputo–Fabrizio

Muhammad Yaseen [1,*], Qamar Un Nisa Arif [1], Reny George [2] and Sana Khan [1]

[1] Department of Mathematics, University of Sargodha, Sargodha 40100, Pakistan; qamararif449@gmail.com (Q.U.N.A.); sanakhanpm28@gmail.com (S.K.)
[2] Department of Mathematics, College of Science and Humanities in Al-Kharj, Prince Sattam bin Abdulaziz University, Al-Kharj 11942, Saudi Arabia; r.kunnelchacko@psau.edu.sa
* Correspondence: yaseen.yaqoob@uos.edu.pk

Abstract: This study focuses on numerically addressing the time fractional Cattaneo equation involving Caputo–Fabrizio derivative using spline-based numerical techniques. The splines used are the cubic B-splines, trigonometric cubic B-splines and extended cubic B-splines. The space derivative is approximated using B-splines basis functions, Caputo–Fabrizio derivative is discretized, using a finite difference approach. The techniques are also put through a stability analysis to verify that the errors do not pile up. The proposed scheme's convergence analysis is also explored. The key advantage of the schemes is that the approximation solution is produced as a smooth piecewise continuous function, allowing us to approximate a solution at any place in the domain of interest. A numerical study is performed using various splines, and the outcomes are compared to demonstrate the efficiency of the proposed schemes.

Keywords: cubic B-splines; trigonometric cubic B-splines; extended cubic B-splines; Caputo–Fabrizio derivative; Cattaneo equation

Citation: Yaseen, M.; Nisa Arif, Q.U.; George, R.; Khan, S. Comparative Numerical Study of Spline-Based Numerical Techniques for Time Fractional Cattaneo Equation in the Sense of Caputo–Fabrizio. *Fractal Fract.* **2022**, *6*, 50. https://doi.org/10.3390/fractalfract6020050

Academic Editors: Amar Debbouche and Hari Mohan Srivastava

Received: 15 December 2021
Accepted: 11 January 2022
Published: 18 January 2022

1. Introduction

The time fractional Cattaneo differential equation (TFCDE) under consideration is [1]

$$\frac{\partial v(s,t)}{\partial t} + {}^{CF}_{a}\mathfrak{D}^{\alpha}_{t} v(s,t) = \frac{\partial^2 v(s,t)}{\partial^2 s} + g(s,t), \tag{1}$$

with initial conditions

$$\begin{cases} v(s,0) = \phi(s), \\ v_t(s,0) = \psi(s), \end{cases} \qquad 0 \le s \le L, \tag{2}$$

and the boundary conditions,

$$\begin{cases} v(0,t) = f_1(t), \\ v(L,t) = f_2(t), \end{cases} \qquad t \ge 0, \tag{3}$$

where $(s,t) \in \Delta = [0,L] \times [0,T]$, $1 < \alpha < 2$, $g \in C[0,T]$, and $f_1(t), f_2(t), \phi(s), \psi(s)$ are known functions. Moreover, ${}^{CF}_{a}\mathfrak{D}^{\alpha}_{t} v(s,t)$ is the Caputo-Fabrizio derivative given by

$${}^{CF}_{a}\mathfrak{D}^{\alpha}_{t} v(s,t) = \frac{M(\alpha)}{2-\alpha} \int_a^t v''(s,x) exp[\sigma(t-x)]dx,$$

where $M(0) = M(1) = 1$ and $\sigma = \frac{1-\alpha}{2-\alpha}$.

For mathematical modeling of real-world problems, fractional differential equations are often used. Scientists in a variety of fields are pushed to improve the interpretations of their findings by utilizing the fractional order derivatives, which are particularly useful. In mathematical modeling of many scientific situations, fractional order differential equations provide more accurate information than regular differential equations. Fractional derivatives are used to describe a variety of physical phenomena [2]. This is owing to the fact that fractional operators assess both global and local properties when analyzing system evolution. In addition, integer-order calculus can sometimes contradict the experimental results; therefore, non-integer order derivatives may be preferable [3]. It is difficult to determine the solution to fractional differential equations (FDEs). As a result, a numerical method must be used to obtain the solution to these partial differential equations. To tackle these problems numerically, many approaches have been developed and extended. The existence of solution of FDEs can be seen in [4]. Diethelm et al. presented the predictor-corrector method [5] for the numerical solution of FDEs. Meerschaert and Tadjern [6] developed a finite difference method for a fractional advection–dispersion equation. The homotopy analysis method [7] for the fractional initial value problem was developed by Hashim et al. An eigenvector expansion method for motion containing fractional derivatives was presented by Suarez and Shokooh [8].

When compared to the finite difference approach, other spectral methods, such as the operational matrix method, are particularly popular since they provide good accuracy and take less time to compute. This method works well with fractional ordinary differential equations (ODEs), fractional partial differential equations (PDEs), and variable order PDEs. Jafari et al. [9] gave applications of Legendre wavelets in solving FDEs numerically. The Haar wavelet operational matrix of fractional order integration and its applications in solving fractional order differential equations can be seen in [10]. Chebyshev wavelets [11] were used by Yuanlu for solving a nonlinear fractional order differential equation. Li and Sun [12] developed a generalized block pulse operational matrix method for the solution of FDEs. Obidat [13] used Legendre polynomials to approximate the solution of nonlinear FDEs. Genocchi polynomials [14] were used by Araci to find numerical solutions of FDEs. Grbz and Sezer [15] solved a class of initial and boundary value problems arising in science and engineering using Laguerre polynomials. Caputo and Fabrizio proposed one of the most recent fractional order derivatives. For more applications of this new derivative and the related work, the reader is referred to [1,16–29].

In comparison to polynomials, the B-splines based collocation methods provide a good approximation rate, are computationally quick, numerically consistent, and have second-order continuity. To obtain numerical solutions to differential equations, multiple numerical approaches based on various forms of B-splines functions were recently utilized. Inspired by the popularity of spline approaches in finding numerical solutions of fractional partial differential equations, various splines-based numerical techniques have been developed for the numerical solution of the Cattaneo equation involving the Caputo–Fabrizio derivative. The main motivation behind this work is that to the authors' knowledge, this equation has not been solved using the B-splines basis functions. In the current work, B-splines are used to approximate the space derivative, while the Caputo–Fabrizio derivative is approximated using finite differences. Moreover, the presented schemes are tested for stability and convergence analysis.

2. Numerical Schemes

In this section, the cubic B-splines, extended cubic B-splines and the trigonometric cubic B-splines are used to develop numerical techniques for the numerical solution of time fractional Cattaneo equation (TFCE) (1).

2.1. Numerical Scheme Based on Cubic B-Splines

Let $\tau = \frac{T}{N}$ and $h = \frac{L}{M}$ be the step length in space and time direction, respectively. Set $t_m = m\tau, s_j = jh$, where the positive integers, N and M, are used. The knots s_j divide the solution domain Δ equally into M equal subintervals $[s_j, s_{j+1}]$, $j = 0, 1, \ldots, M-1$, where $a = s_0 < s_1 < \cdots < s_M = b$. The approximate solution $V(s,t)$ to the exact solution $v(s,t)$ in the following form is acquired by our scheme for solving (1)

$$V(s,t) = \sum_{j=-1}^{M+1} C_j(t) B_j(s), \tag{4}$$

where $C_j(t)$ are unknowns to be found, and $B_j(s)$ [30] are cubic B-splines basis (CuBS) functions given by

$$B_j(s) = \frac{1}{6h^3}\begin{cases} (s-s_j)^3, & s \in [s_j, s_{j+1}] \\ h^3 + 3h^2(s - s_{j+1}) & \\ +3h(s-s_{j+1})^2 - 3(s-s_{j+1})^3, & s \in [s_{j+1}, s_{j+2}] \\ h^3 + 3h^2(s_{j+3} - s) & \\ +3h(s_{j+3}-s)^2 - 3(s_{j+1}-s)^3, & s \in [s_{j+2}, s_{j+3}] \\ (s_{j+4}-s)^3, & s \in [s_{j+3}, s_{j+4}] \\ 0, & \text{otherwise.} \end{cases} \tag{5}$$

Here, $B_{j-1}(s), B_j(s)$ and $B_{j+1}(s)$ are survived due to the local support characteristic of the cubic B-splines so that the approximation v_j^m at the grid point (s_j, t_m) at the mth time level is given as

$$v(s_j, t^m) = v_j^m = \sum_{w=j-1}^{j+1} C_w^m(t) B_w(s). \tag{6}$$

The time-dependent unknowns $C_j^m(t)$ are found using the specified initial and boundary conditions as well as the collocation conditions on $B_j(s)$. As a result, the approximation v_j^m and its required derivatives are

$$\begin{cases} v_j^m = a_1 C_{j-1}^m + a_2 C_j^m + a_1 C_{j+1}^m, \\ (v_j^m)_s = -b_1 C_{j-1}^m + b_1 C_{j+1}^m, \\ (v_j^m)_{ss} = c_1 C_{j-1}^m + c_2 C_j^m + c_1 C_{j+1}^m, \end{cases} \tag{7}$$

where $a_1 = \frac{1}{6}$, $a_2 = \frac{4}{6}$, $b_1 = \frac{1}{2h}$, $c_1 = \frac{1}{h^2}$, and $c_2 = -\frac{2}{h^2}$. Let $g = \{g^m : 0 \le m \le N\}$ be the collection of grid functions on a uniform mesh of the interval $[0,T]$ such that $\delta_t g^m = \frac{g^m - g^{m-1}}{\tau}$. A discrete approximation to ${}_0^{CF}\mathfrak{D}_t^\alpha v(s,t)$ at $(s_j, t_{m+\frac{1}{2}})$ can be obtained as [1]

$${}_0^{CF}\mathfrak{D}_t^\alpha v(s_j, t_{m+\frac{1}{2}}) = \frac{1}{(1-\alpha)\tau}\Big(M_0 \delta_t v_j^{m+1} - \sum_{l=1}^{m}(M_{m-l} - M_{m-l+1})\delta_t v_j^l - M_m \psi_j\Big) + R_j^{m+\frac{1}{2}}, \tag{8}$$

where,

$$M_j = \exp\Big(\frac{1-\alpha}{2-\alpha}\tau j\Big) - \exp\Big(\frac{1-\alpha}{2-\alpha}\tau(j+1)\Big), \tag{9}$$

and

$$|R_i^{m+\frac{1}{2}}| = O(\tau^2).$$

Lemma 1 ([1]). *From the definition of M_j in (9), we have $M_j > 0$ and $M_{j+1} < M_j$, $\forall j \le m$.*

Lemma 2 ([1]). *Suppose that $v(t) \in C_{s,t}^{4,4}([0,L] \times [0,T])$, then*

$$0 \le M_j \le C\tau$$

and

$$0 \leq M_j - M_{j+1} \leq C\tau M_j.$$

Now, we employ the Caputo–Fabrizio fractional derivative and CuBS to establish the numerical scheme for solving (1). Using CuBS and the approximation given in (8), we obtain

$$(v_j^m)_t + \frac{1}{(\alpha-1)\tau}(M_0\delta_t v_j^{m+1} - \sum_{l=1}^{m}(M_{m-l} - M_{m-l+1})\delta_t v_j^l - M_m\psi_j) = (v_j^{m+1})_{ss} + g_j^{m+1} + R^{m+1}, \tag{10}$$

where $R^{m+1} = O(\tau^2 + h^2)$. Thus, by ignoring R^{m+1} and using the discretization $(v_j^m)_t = \frac{v_j^{m+1} - v_j^m}{\tau}$, we have

$$\begin{aligned}&(\alpha-1)(v_j^{m+1} - v_j^m) + \frac{M_0}{\tau}(v_j^{m+1} - v_j^m) - \frac{1}{\tau}\sum_{l=1}^{m}(M_{m-l} - M_{m-l+1})(v_j^l - v_j^{l-1})\\&- M_m\psi_j) = (\alpha-1)(v_j^{m+1})_{ss} + (\alpha-1)\tau g_j^{m+1}.\end{aligned}$$

Rearranging the above equation, we obtain

$$\sigma v_j^{m+1} - \mu(v_j^{m+1})_{ss} = \sigma v_j^m + \frac{1}{\tau}\sum_{l=1}^{m}(M_{m-l} - M_{m-l+1})(v_j^l - v_j^{l-1}) + M_m\psi_j + \mu g_j^{m+1}, \tag{11}$$

where $\sigma = (\alpha - 1 + \frac{M_0}{\tau})$ and $\mu = (\alpha-1)\tau$. Using the CuBS approximation (7) in (11), we obtain

$$\begin{aligned}\eta_1 C_{j-1}^{m+1} + \eta_2 C_j^{m+1} + \eta_1 C_{j+1}^{m+1} &= \eta_3 C_{j-1}^m + \eta_4 C_j^m + \eta_3 C_{j+1}^m\\&+ \frac{1}{\tau}\sum_{l=1}^{m}(M_{m-l} - M_{m-l+1})[(a_1 C_{j-1}^l + a_2 C_j^l + a_1 C_{j+1}^l)\\&- (a_1 C_{j-1}^{l-1} + a_2 C_j^{l-1} + a_1 C_{j+1}^{l-1})] + M_m\psi_j + \mu g_j^{m+1},\end{aligned} \tag{12}$$

where $\eta_1 = \sigma a_1 - \mu c_1$, $\eta_2 = \sigma a_2 - \mu c_2$, $\eta_3 = \sigma a_1$,and $\eta_4 = \sigma a_2$. In matrix notation, the above equation is expressed as

$$A_1 C^{m+1} = A_2 C^m + B_1(\frac{1}{\tau}\sum_{l=1}^{m}(M_{m-l} - M_{m-l+1})(C^l - C^{l-1})) + M_m\Psi{+}\mu G,$$

where the matrices A_1, A_2, B_1, Ψ and G are

$$A_1 = \begin{bmatrix}\eta_1 & \eta_2 & \eta_1 & 0 & \dots & 0\\ 0 & \eta_1 & \eta_2 & \eta_1 & \ddots & \vdots\\ \vdots & \ddots & \ddots & \ddots & \ddots & 0\\ 0 & \dots & \eta_1 & \eta_2 & \eta_1 & 0\\ 0 & \dots & 0 & \eta_1 & \eta_2 & \eta_1\end{bmatrix},$$

$$A_2 = \begin{bmatrix}\eta_3 & \eta_4 & \eta_3 & 0 & \dots & 0\\ 0 & \eta_3 & \eta_4 & \eta_3 & \ddots & \vdots\\ \vdots & \ddots & \ddots & \ddots & \ddots & 0\\ 0 & \dots & \eta_3 & \eta_4 & \eta_3 & 0\\ 0 & \dots & 0 & \eta_3 & \eta_4 & \eta_3\end{bmatrix},$$

$$B_1 = \begin{bmatrix} a_1 & a_2 & a_1 & 0 & \dots & 0 \\ 0 & a_1 & a_2 & a_1 & \ddots & \vdots \\ \vdots & \ddots & \ddots & \ddots & \ddots & \vdots \\ 0 & \dots & a_1 & a_2 & a_1 & 0 \\ 0 & \dots & 0 & a_1 & a_2 & a_1 \end{bmatrix},$$

$$\Psi = \left[\psi_0^{m+1}, \quad \psi_1^{m+1} \quad ,\dots, \quad \psi_M^{m+1}\right]^T,$$

and

$$G = \left[g_0^{m+1}, \quad g_1^{m+1} \quad ,\dots, \quad g_M^{m+1}\right]^T.$$

The above system gives $(M+1)$ equations in $(M+3)$ unknowns. For a unique solution, two additional linear equations are necessary. For this purpose, the boundary conditions are utilized as

$$\begin{cases} a_1 C_{-1}^{m+1} + a_2 C_0^{m+1} + a_1 C_1^{m+1} = f_1(t_{m+1}), \\ a_1 C_{M-1}^{m+1} + a_2 C_M^{m+1} + a_1 C_{M+1}^{m+1} = f_2(t_{m+1}). \end{cases} \tag{13}$$

By combining Equations (12) and (13), we have $(M+3) \times (M+3)$, a system of linear equations which can be solved uniquely.

2.2. Initial State

First of all, it is essential to find the initial vector $C^0 = \left[C_{-1}^0, \quad C_0^0, \quad \dots \quad ,C_M^0, \quad C_{M+1}^0\right]^T$ to initiate the iteration procedure. This vector is obtained from initial conditions as

$$\begin{cases} v_0' &= \phi'(s_0), \\ v_j^0 &= \phi(s_j), \quad j = 0,1,2,3,\dots,M, \\ v_M' &= \phi'(s_M). \end{cases}$$

Thus, $(M+3) \times (M+3)$ a system of linear equations results, and this system can be written in matrix notation as

$$A_3 C^0 = B_2,$$

where the matrices A_3, C^0 and B_2 are

$$A_3 = \begin{bmatrix} -b_1 & 0 & b_1 & 0 & \dots & 0 \\ a_1 & a_2 & a_1 & 0 & \dots & 0 \\ 0 & a_1 & a_2 & a_1 & \ddots & \vdots \\ \vdots & \ddots & \ddots & \ddots & \ddots & 0 \\ 0 & \dots & 0 & a_1 & a_2 & a_1 \\ 0 & \dots & 0 & -b_1 & 0 & b_1 \end{bmatrix},$$

$$C^0 = \left[C_{-1}^0, \quad C_0^0 \quad ,\dots, \quad C_M^0, \quad C_{M+1}^0\right]^T,$$

and

$$B_2 = \left[\phi'(s_0), \quad \phi(s_0) \quad ,\dots, \quad \phi(s_M), \quad \phi'(s_M)\right]^T.$$

2.3. Numerical Scheme Based on Extended Cubic B-Splines

A cubic B-spline of degree four with a free parameter η is called an extended cubic B-spline. This kind of cubic B-spline was introduced by Han and Liu in 2003. We follow the same notations for the time and space discretizations that we used before. The extended cubic B-spline (ECuBS) basis functions, $B_j^4(s,\eta)$ are given by

$$B_j^4(s,\eta) = \frac{1}{24h^4}\begin{cases} 4h(1-\eta)(s-s_j)^3 + 3\eta(s-s_j)^4, & s \in [s_j, s_{j+1}] \\ (4-\eta)h^4 + 12h^3(s-s_{j+1}) + 6h^2(2+\eta)(s-s_{j+1})^2 & \\ -12h(s-s_{j+1})^3 - 3\eta(s-s_{j+1})^4, & s \in [s_{j+1}, s_{j+2}] \\ (4-\eta)h^4 + 12h^3(s_{j+3}-s) + 6h^2(2+\eta)(s_{j+3}-s)^2 & \\ -12h(s_{j+3}-s)^3 - 3\eta(s_{j+1}-s)^4, & s \in [s_{j+2}, s_{j+3}] \\ 4h(1-\eta)(s_{j+4}-s)^3 + 3\eta(s_{j+4}-s)^4, & s \in [s_{j+3}, s_{j+4}] \\ 0, & \text{otherwise,} \end{cases} \tag{14}$$

where $\eta \in [-8, 1]$. Here, $B_{j-1}^4(s), B_j^4(s)$ and $B_{j+1}^4(s)$ are survived due to local support characteristic of the cubic B-splines so that the approximation v_j^m at the grid point (s_j, t_m) at mth time level is given as

$$v(s_j, t^m) = v_j^m = \sum_{w=j-1}^{j+1} C_w^m(t) B_w^4(s,\eta). \tag{15}$$

The time-dependent unknowns $C_j^m(t)$ are found using the specified initial and boundary conditions as well as the collocation conditions on $B_j(s)$. As a result, the approximation v_j^m and its required derivatives are

$$\begin{cases} v_j^m = \omega_1 C_{j-1}^m + \omega_2 C_j^m + \omega_1 C_{j+1}^m, \\ (v_j^m)_s = -\omega_3 C_{j-1}^m + \omega_4 C_j^m + \omega_3 C_{j+1}^m, \\ (v_j^m)_{ss} = \omega_5 C_{j-1}^m + \omega_6 C_j^m + \omega_5 C_{j+1}^m, \end{cases} \tag{16}$$

where $\omega_1 = \frac{4-\eta}{24}$, $\omega_2 = \frac{8+\eta}{12}$, $\omega_3 = \frac{1}{2h}$, $\omega_4 = 0$, $\omega_5 = \frac{2+\eta}{2h^2}$ and $\omega_6 = -\frac{2+\eta}{2h^2}$. By following the same procedure as was done for cubic B-splines and using the ECuBS approximation given in (16), we obtain the following approximation to the solution of (1)

$$\begin{aligned} \eta_5 C_{j-1}^{m+1} + \eta_6 C_j^{m+1} + \eta_5 C_{j+1}^{m+1} &= \eta_7 C_{j-1}^m + \eta_8 C_j^m + \eta_7 C_{j+1}^m \\ &+ \frac{1}{\tau}\sum_{l=1}^{m}(M_{m-l} - M_{m-l+1})[(\omega_1 C_{j-1}^l + \omega_2 C_j^l + \omega_1 C_{j+1}^l) \\ &- (\omega_1 C_{j-1}^{l-1} + \omega_2 C_j^{l-1} + \omega_1 C_{j+1}^{l-1})] + M_m \psi_j + \mu g_j^{m+1}, \end{aligned} \tag{17}$$

where, $\eta_5 = \sigma\omega_1 - \mu\omega_5$, $\eta_6 = \sigma\omega_2 - \mu\omega_5$, $\eta_7 = \sigma\omega_1$ and $\eta_8 = \sigma\omega_2$. In matrix notation, the above Equation (17) is expressed as

$$A_4 C^{m+1} = A_5 C^m + B_3\left(\frac{1}{\tau}\sum_{l=1}^{m}(M_{m-l} - M_{m-l+1})(C^l - C^{l-1})\right) + M_m \Psi + \mu G,$$

where the matrices A_4, A_5 and B_3 are

$$A_4 = \begin{bmatrix} \eta_5 & \eta_6 & \eta_5 & 0 & \dots & 0 \\ 0 & \eta_5 & \eta_6 & \eta_5 & \ddots & \vdots \\ \vdots & \ddots & \ddots & \ddots & \ddots & 0 \\ 0 & \dots & \eta_5 & \eta_6 & \eta_5 & 0 \\ 0 & \dots & 0 & \eta_5 & \eta_6 & \eta_5 \end{bmatrix},$$

$$A_5 = \begin{bmatrix} \eta_7 & \eta_8 & \eta_7 & 0 & \dots & 0 \\ 0 & \eta_7 & \eta_8 & \eta_7 & \ddots & \vdots \\ \vdots & \ddots & \ddots & \ddots & \ddots & 0 \\ 0 & \dots & \eta_7 & \eta_8 & \eta_7 & 0 \\ 0 & \dots & 0 & \eta_7 & \eta_8 & \eta_7 \end{bmatrix},$$

and

$$B_3 = \begin{bmatrix} \omega_1 & \omega_2 & \omega_1 & 0 & \dots & 0 \\ 0 & \omega_1 & \omega_2 & \omega_1 & \ddots & \vdots \\ \vdots & \ddots & \ddots & \ddots & \ddots & \vdots \\ 0 & \dots & \omega_1 & \omega_2 & \omega_1 & 0 \\ 0 & \dots & 0 & \omega_1 & \omega_2 & \omega_1 \end{bmatrix}.$$

The above system gives $(M+1)$ equations in $(M+3)$ unknowns. For a unique solution, two additional linear equations are necessary. From the boundary conditions, we obtain the required equations as follows

$$\begin{cases} \omega_1 C_{-1}^{m+1} + \omega_2 C_0^{m+1} + \omega_1 C_1^{m+1} = f_1(t_{m+1}), \\ \omega_1 C_{M-1}^{m+1} + \omega_2 C_M^{m+1} + \omega_1 C_{M+1}^{m+1} = f_2(t_{m+1}). \end{cases} \tag{18}$$

By combining Equations (17) and (18), we have $(M+3) \times (M+3)$, a system of linear equations, which can be solved uniquely.

2.4. Numerical Scheme Based on Trigonometric Cubic B-Splines

We follow the same notations for the time and space discretizations used before. The trigonometric cubic B-spline (TCuBS) basis functions are given by [31]

$$TB_j^4(s) = \frac{1}{p} \begin{cases} l^3(s_j), & s \in [s_j, s_{j+1}) \\ l(s_j)(l(s_j)m(s_{j+2}) + m(s_{j+3})l(s_{j+1})) + m(s_{j+4})l^2(s_{j+1}), & s \in [s_{j+1}, s_{j+2}) \\ m(s_{j+4})(l(s_{j+1})m(s_{j+3}) + m(s_{j+4})l(s_{j+2})) + l(s_j)m^2(s_{j+3}), & s \in [s_{j+2}, s_{j+3}) \\ m^3(s_{j+4}), & s \in [s_{j+3}, s_{j+4}), \end{cases} \tag{19}$$

where $l(s_j) = \sin(\frac{s-s_j}{2}), m(s_j) = \sin(\frac{s_j-s}{2})$ and $p = \sin(\frac{h}{2})\sin(h)\sin(\frac{3h}{2})$.

Here, $TB_{j-1}^4(s), TB_j^4(s)$ and $TB_{j+1}^4(s)$ are survived due to the local support characteristic of the trigonometric cubic B-splines so that the approximation v_j^m at the grid point (s_j, t_m) at mth time level is given as

$$v(s_j, t^m) = v_j^m = \sum_{w=j-1}^{j+1} C_w^m(t) TB_w^4(s). \tag{20}$$

The time-dependent unknowns $C_j^m(t)$ are found using the specified initial and boundary conditions as well as the collocation conditions on $B_j(s)$. As a result, the approximation v_j^m and its required derivatives are

$$\begin{cases} v_j^m = \zeta_1 C_{j-1}^m + \zeta_2 C_j^m + \zeta_1 C_{j+1}^m, \\ (v_j^m)_s = -\zeta_3 C_{j-1}^m + \zeta_4 C_{j+1}^m + \zeta_3 C_{j+1}^m, \\ (v_j^m)_{ss} = \zeta_5 C_{j-1}^m + \zeta_6 C_j^m + \zeta_5 C_{j+1}^m, \end{cases} \tag{21}$$

where

$$\begin{cases} \zeta_1 = \csc(h)\csc(\frac{3h}{2})\sin^2(\frac{h}{2}), \\ \zeta_2 = \frac{2}{1+2\cos(h)}, \\ \zeta_3 = \frac{3}{4}\csc(\frac{3h}{2}), \\ \zeta_4 = 0, \\ \zeta_5 = \frac{3+9\cos(h)}{4\cos(\frac{h}{2})} - 4\cos(\frac{5h}{2}), \\ \zeta_6 = -\frac{3\cot^2(\frac{h}{2})}{2+4\cos(h)}. \end{cases}$$

By following the same procedure as was done for cubic B-splines and using the approximation (8) in (1), we obtain

$$\sigma v_j^{m+1} - \mu(v_j^{m+1})_{ss} = \sigma v_j^m + \frac{1}{\tau}\sum_{l=1}^{m}(M_{m-l} - M_{m-l+1})(v_j^l - v_j^{l-1}) + M_m\psi_j + \mu g_j^{m+1}, \tag{22}$$

where $\sigma = (\alpha - 1 + \frac{M_0}{\tau})$ and $\mu = (\alpha - 1)\tau$. Using the CuTBS approximation given in (21), we obtain the following approximation to the solution of (1)

$$\begin{aligned} \eta_9 C_{j-1}^{m+1} + \eta_{10} C_j^{m+1} + \eta_9 C_{j+1}^{m+1} &= \eta_{11} C_{j-1}^m + \eta_{12} C_j^m + \eta_{11} C_{j+1}^m \\ &+ \frac{1}{\tau}\sum_{l=1}^{m}(M_{m-l} - M_{m-l+1})[(\zeta_1 C_{j-1}^l + \zeta_2 C_j^l + \zeta_1 C_{j+1}^l) \\ &- (\zeta_1 C_{j-1}^{l-1} + \zeta_2 C_j^{l-1} + \zeta_1 C_{j+1}^{l-1})] + M_m\psi_j + \mu g_j^{m+1}, \end{aligned} \tag{23}$$

where $\eta_9 = \sigma\zeta_1 - \mu\zeta_5$, $\eta_{10} = \sigma\zeta_2 - \mu\zeta_6$, $\eta_{11} = \sigma\zeta_1$, and $\eta_{12} = \sigma\zeta_2$. In matrix notation, (23) is expressed as

$$A_7 C^{m+1} = A_8 C^m + B_5(\frac{1}{\tau}\sum_{l=1}^{m}(M_{m-l} - M_{m-l+1})(C^l - C^{l-1})) + M_m\Psi + \mu G,$$

where the matrices A_7, A_8 and B_5 are

$$A_7 = \begin{bmatrix} \eta_9 & \eta_{10} & \eta_9 & 0 & \dots & 0 \\ 0 & \eta_9 & \eta_{10} & \eta_9 & \ddots & \vdots \\ \vdots & \ddots & \ddots & \ddots & \ddots & 0 \\ 0 & \dots & \eta_9 & \eta_{10} & \eta_9 & 0 \\ 0 & \dots & 0 & \eta_9 & \eta_{10} & \eta_9 \end{bmatrix},$$

$$A_8 = \begin{bmatrix} \eta_{11} & \eta_{12} & \eta_{11} & 0 & \dots & 0 \\ 0 & \eta_{11} & \eta_{12} & \eta_{11} & \ddots & \vdots \\ \vdots & \ddots & \ddots & \ddots & \ddots & 0 \\ 0 & \dots & \eta_{11} & \eta_{12} & \eta_{11} & 0 \\ 0 & \dots & 0 & \eta_{11} & \eta_{12} & \eta_{11} \end{bmatrix},$$

and

$$B_5 = \begin{bmatrix} \zeta_1 & \zeta_2 & \zeta_1 & 0 & \dots & 0 \\ 0 & \zeta_1 & \zeta_2 & \zeta_1 & \ddots & \vdots \\ \vdots & \ddots & \ddots & \ddots & \ddots & 0 \\ 0 & \dots & \zeta_1 & \zeta_2 & \zeta_1 & 0 \\ 0 & \dots & 0 & \zeta_1 & \zeta_2 & \zeta_1 \end{bmatrix}.$$

The above system gives $(M+1)$ equations in $(M+3)$ unknowns. For a unique solution, two additional linear equations are necessary. From the boundary conditions, we obtained these equations as follows

$$\begin{cases} \zeta_1 C_{-1}^{m+1} + \zeta_2 C_0^{m+1} + \zeta_1 C_1^{m+1} = f_1(t_{m+1}), \\ \zeta_1 C_{M-1}^{m+1} + \zeta_2 C_M^{m+1} + \zeta_1 C_{M+1}^{m+1} = f_2(t_{m+1}). \end{cases} \tag{24}$$

By combining Equations (23) and (24), we have $(M+3) \times (M+3)$, a system of linear equations, which can be solved uniquely.

2.5. Stability Analysis

This section deals with stability analysis of the scheme based on cubic B-splines. The stability analysis of the schemes based on extended and cubic trigonometric B-splines can be carried out by a similar argument. We use the Fourier method to study the stability analysis of the scheme. Let $\tilde{V}^0$ be the perturbation vector of initial values V^0 and $\tilde{V}^m$, $1 \le m \le N-1$ be the approximate solution of the scheme (12). The error vector δ^m is defined as

$$\delta^m = V^m - \tilde{V}^m, \quad 0 \le m \le N-1, \tag{25}$$

where,

$$V^m = \left[V_1^m, \quad V_2^m, \quad \dots, \quad V_{M-1}^m\right]^T,$$
$$\tilde{V}^m = \left[\tilde{V}_1^m, \quad \tilde{V}_2^m, \quad \dots, \quad \tilde{V}_{M-1}^m\right]^T,$$

and

$$\delta_j^m = V_j^m - \tilde{V}_j^m = \left[\delta_1^m, \quad \delta_2^m, \quad \dots, \quad \delta_{M-1}^m\right]^T.$$

Define the grid functions as follows:

$$\delta^m(s) = \begin{cases} \delta_j^m, \; s_j - \frac{h}{2} < s < s_j + \frac{h}{2}, \\ 0, \; 0 \le s \le \frac{h}{2} \; or \; L - \frac{h}{2} < s < L. \end{cases}$$

We can expand $\delta^m(s)$ into Fourier series as

$$\delta^m(s) = \sum_{l=-\infty}^{\infty} d_m(l) \exp\left(\frac{I2\pi ls}{L}\right),$$

where,

$$d_m(l) = \frac{1}{L} \int_0^L \delta^m(s) \exp\left(\frac{-I2\pi ls}{L}\right) ds.$$

Denoting

$$||\delta^m||_2 = (\int_0^L ||\delta^m(s)||^2 ds)^{\frac{1}{2}},$$

and using the Parseval's equality,

$$\int_0^L ||\delta^m(s)||^2 ds = \sum_{l=-\infty}^{\infty} ||d_m(l)||^2,$$

we obtain

$$||\delta^m(s)||^2 = \sum_{l=-\infty}^{\infty} ||d_m(l)||^2.$$

We can expand δ_j^m into Fourier series, and because the difference equations are linear, we can analyze the behavior of total error by tracking the behavior of an arbitrary nth component. Based on the above analysis, we can suppose that the solution of (11) has the following form

$$\delta_j^m = d_m \exp(I\sigma_s jh), \tag{26}$$

where $\sigma_s = \frac{2\pi l}{L}, I = \sqrt{-1}$. Substituting the above expression into (11), we obtain

$$\sigma(\delta^{m+1}) - \mu(\delta_{ss}^{m+1}) = \sum_{l=1}^{m}(M_{m-l} - M_{m-l+1})\delta_t(\delta^l) + \sigma(\delta^m). \tag{27}$$

Using the CuBS approximation given in (7) and Equation (26) in the above equation, we obtain

$$\begin{aligned}
&d_{m+1}(\sigma(a_1\exp(-I\sigma_s h) + a_2 + a_1\exp(I\sigma_s h) - \mu(c_1\exp(-I\sigma_s h) + c_2 + c_1\exp(I\sigma_s h)))\\
&= \sum_{l=1}^{m}(M_{m-l} - M_{m-l+1})\delta_t d_l(a_1\exp(-I\sigma_s h) + a_2 + a_1\exp(I\sigma_s h)) + \sigma d_m(a_1\exp(-I\sigma_s h))\\
&+ a_2 + a_1\exp(I\sigma_s h)),
\end{aligned}$$

$$\begin{aligned}
&\Rightarrow\ d_{m+1}(\sigma(a_2 + a_1(2\cos(\sigma_s h)) - \mu(c_2 + c_1(2\cos(\sigma_s h)))\\
&= \sum_{l=1}^{m}(M_{m-l} - M_{m-l+1})\delta_t d_l(a_2 + a_1(2\cos(\sigma_s h)) + \sigma d_m(a_2 + a_1(2\cos(\sigma_s h)),
\end{aligned}$$

which, on further simplification, reduces to

$$d_{m+1} = \frac{1}{\sigma - \mu r}\sum_{l=1}^{m}(M_{m-l} - M_{m-l+1})(\frac{d_l - d_{l-1}}{\tau}) + \frac{\sigma}{\sigma - \mu r}d_m,\ 1 \leq m \leq M-1, \tag{28}$$

where, $r = (\frac{c_2 + 2c_1\cos(\sigma_s h)}{a_2 + 2a_1\cos(\sigma_s h)})$.

Definition 1 ([32,33]). *A scheme is called stable if there exists a positive number C, independent of j and m such that*

$$||V^n - \tilde{V}^n|| \leq C||V^0 - \tilde{V}^0||,$$

where V^n and $\tilde{V}^n$ are the exact solutions of the difference scheme and its perturbed equation, respectively.

Theorem 1. *Suppose that d_m, $(1 \leq m \leq N-1)$ are defined by (28), then for $\alpha \in (1,2)$, we have*

$$|d_m| \leq (1 + 2C\tau)^m |d_0|, \quad m = 1, 2, \dots, M-1.$$

Proof. We use the mathematical induction for proof. For $m = 1$, we have from (28),

$$|d_1| = |\frac{\sigma}{\sigma - \mu r}||d_0| \leq (1 + 2C\tau)|d_0|.$$

Now, suppose that

$$|d_m| \leq (1 + 2C\tau)^m |d_0|, \quad m = 1, 2, \ldots, M - 2.$$

Then, by using Lemmas 1 and 2, we obtain

$$\begin{aligned} |d_{m+1}| &\leq \frac{C\tau}{\sigma - \mu r} \sum_{l=1}^{m} |(d_l - d_{l-1})| + \frac{\sigma}{\sigma - \mu r} |d_m| \\ &= \frac{C\tau}{\sigma - \mu r} |(d_m - d_0)| + \frac{\sigma}{\sigma - \mu r} |d_m| \\ &\leq \frac{2C'\tau + \sigma}{\sigma - \mu r} (1 + 2C\tau)^m |d_0| \leq (1 + 2C\tau)^{m+1} |d_0|. \end{aligned}$$

This completes the proof. □

Theorem 2. *The scheme (12) is unconditionally stable for* $\alpha \in (1, 2)$.

Proof. By using Theorem 1, Parseval's equality and $m\tau \leq T$, we obtain

$$\begin{aligned} ||V^m - \tilde{V}^m||_{l_2}^2 &= \sum_{-\infty}^{\infty} ||d_m(l)||^2 \\ &\leq (1 + 2C\tau)^{2m} \sum_{l=-\infty}^{\infty} ||d_0(l)||^2 \\ &= (1 + 2C\tau)^{2m} ||\delta^0(l)||_{l_2}^2 \\ &\leq \exp(4C\tau m) ||V^0 - \tilde{V}^0||_{l_2}^2. \end{aligned}$$

so that

$$||V^m - \tilde{V}^m||_{l_2} \leq \exp(2\sqrt{C\tau}) ||V^0 - \tilde{V}^0||_{l_2}.$$

which means that the scheme is unconditionally stable. □

3. Convergence Analysis

The convergence of the scheme based on cubic B-splines is presented in this section. The convergence analysis of the extended and cubic trigonometric B-splines based numerical scheme follows accordingly. Let $e_j^m = v_j^m - V_j^m$, $1 \leq j \leq M - 1$, $1 \leq m \leq N - 1$ and

$$\begin{aligned} e^m &= (e_1^m, e_2^m, \ldots, e_{M-1}^m), \\ \mathbf{R}^m &= (R_1^m, R_2^m, \ldots, R_{M-1}^m), \quad 0 \leq m \leq N - 1. \end{aligned}$$

From Equation (11) and $R_j^{m+1} = O(\tau^2 + h^2)$ and noting that $e_j^0 = 0$, we have

$$\sigma e_j^{m+1} - \mu (e_j^{m+1})_{ss} = \sum_{l=1}^{m} (M_{m-l} - M_{m-l+1}) \delta_t e_j^l + \sigma e_j^m + R_j^{m+1}. \tag{29}$$

Define the functions

$$e^m(s) = \begin{cases} e_j^m, \ s_j - \dfrac{h}{2} < s \leq s_j + \dfrac{h}{2}, & 1 \leq j \leq M - 1, \\ 0, \ 0 \leq s \leq \dfrac{h}{2} \ or \ L - \dfrac{h}{2} < s \leq L. \end{cases}$$

and

$$R^m(s) = \begin{cases} R_j^m, \ s_j - \dfrac{h}{2} < s \le s_j + \dfrac{h}{2}, & 1 \le j \le M-1, \\ 0, \ 0 \le s \le \dfrac{h}{2} \ or \ L - \dfrac{h}{2} < s \le L. \end{cases}$$

We expand the above functions into Fourier series expansions as

$$\begin{cases} e^m(s) = \sum_{l=-\infty}^{\infty} \xi_m(l) \exp^{\frac{I2\pi ls}{L}}, \\ R^m(s) = \sum_{l=-\infty}^{\infty} \lambda_m(l) \exp^{\frac{I2\pi ls}{L}}. \end{cases}$$

where,

$$\begin{cases} \xi_m(l) = \dfrac{1}{L} \displaystyle\int_0^L e^m(s) \exp^{\frac{-I2\pi ls}{L}} ds, \\ \lambda_m(l) = \dfrac{1}{L} \displaystyle\int_0^L R^m(s) \exp^{\frac{-I2\pi ls}{L}} ds, \end{cases}$$

Applying Parseval's equalities,

$$\int_0^L \|e^m(s)\|^2 ds = \sum_{j=1}^{M-1} h\|e_j^m\|^2, \text{ and}$$

$$\int_0^L \|R^m(s)\|^2 ds = \sum_{j=1}^{M-1} h\|R_j^m\|^2$$

to the above expression, we have

$$\begin{cases} \|e^m\|_2^2 & = \sum_{l=-\infty}^{\infty} \|\xi_m(l)\|^2, \\ \|R^m\|_2^2 & = \sum_{l=-\infty}^{\infty} \|\lambda_m(l)\|^2. \end{cases} \tag{30}$$

Now, we suppose that

$$\begin{cases} e_j^m = \xi_m \exp^{I\sigma_s jh}, \\ R_j^m = \lambda_m \exp^{I\sigma_s jh}, \end{cases} \tag{31}$$

where $\sigma_s = \frac{2\pi l}{L}$. Substituting relations (31) in Equation (29).

$$\begin{aligned} &\xi_{m+1}[\sigma(a_1 \exp^{I\sigma_s(j-1)h} + a_2 \exp^{I\sigma_s(j)h} + a_1 \exp^{I\sigma_s(j+1)h}) - \mu(c_1 \exp^{I\sigma_s(j-1)h} \\ &+ c_2 \exp^{I\sigma_s(j)h} + c_1 \exp^{I\sigma_s(j+1)h})] = \sum_{l=1}^{m}(M_{m-l} - M_{m-l+1})\delta_t \xi_l [a_1 \exp^{I\sigma_s(j-1)h} \\ &+ a_2 \exp^{I\sigma_s(j)h} + a_1 \exp^{I\sigma_s(j+1)h}] + \sigma\xi_m[a_1 \exp^{I\sigma_s(j-1)h} + a_2 \exp^{I\sigma_s(j)h} \\ &+ a_1 \exp^{I\sigma_s(j+1)h}] + \lambda_{m+1}[a_1 \exp^{I\sigma_s(j-1)h} + a_2 \exp^{I\sigma_s(j)h} + a_1 \exp^{I\sigma_s(j+1)h}]. \end{aligned}$$

$$\begin{aligned} \Rightarrow \ & \xi_{m+1}[\sigma(a_2 + 2a_1 \cos(\sigma_s h)) - \mu(c_2 + 2c_1 \cos(\sigma_s h))] \\ & = \sum_{l=1}^{m}(M_{m-l} - M_{m-l+1})\delta_t \xi_l (a_2 + 2a_1 \cos(\sigma_s h)) + \sigma\xi_m(a_2 + 2a_1 \cos(\sigma_s h)) \\ & + \lambda_{m+1}(a_2 + 2a_1 \cos(\sigma_s h)). \end{aligned}$$

The above expression is further simplified as

$$\xi_{m+1} = \frac{1}{(\sigma - \mu r)} \sum_{l=1}^{m} (M_{m-l} - M_{m-l+1})(\frac{\xi_l - \xi_{l-1}}{\tau}) + \frac{\sigma}{(\sigma - \mu r)} \xi_m$$
$$+ \frac{1}{(\sigma - \mu r)} \lambda_{m+1}, \quad 1 \le m \le M - 1. \tag{32}$$

Theorem 3. *Let ξ_m be the solution of (32), then, there is a positive constant C such that*

$$|\xi_m| \le C(1+\tau)^m |\lambda_1|, \qquad m = 0, \ldots, N-1.$$

Proof. We use the mathematical induction to prove this claim. For $m = 1$, we have from (32)

$$|\xi_1| \le |\frac{\lambda_1}{\sigma - \mu r}| \le C(1+\tau)|\lambda_1|.$$

Assume that

$$|\xi_m| \le C(1+\tau)^m |\lambda_1|, \qquad m = 0, \ldots, N-2.$$

Now by using the convergence of the series on the RHS of (30), we know that there exists a constant C_2 such that

$$|\lambda_m| \le C_2 \tau |\lambda_1|, \qquad m = 1, \ldots, N-1.$$

From (32), we have

$$\begin{aligned}
|\xi_{m+1}| &\le \frac{C\tau}{\sigma - \mu r} \sum_{l=1}^{m} |\xi_l - \xi_{l-1}| + \frac{\sigma}{\sigma - \mu r} |\xi_m| + |\frac{\lambda_{m+1}}{\sigma - \mu r}| \\
&= \frac{C\tau}{\sigma - \mu r} \sum_{l=1}^{m} |\xi_m - \xi_0| + \frac{\sigma}{\sigma - \mu r} |\xi_m| + |\frac{\lambda_{m+1}}{\sigma - \mu r}| \\
&= C_1 \tau (1+\tau)^m |\lambda_1| + C_3 \tau (1+\tau)^m |\lambda_1| + C_2 \tau |\lambda_1| \\
&\qquad \le (1+\tau)^{m+1} C |\lambda_1|.
\end{aligned}$$

□

Theorem 4. *The scheme (12) is convergent, and the order of convergence is $O(\tau^2 + h^2)$.*

Proof. By Theorem 3, Equation (30) and $m\tau \le T$, we have

$$\begin{aligned}
\|e^m\|_{l_2}^2 &= \sum_{l=-\infty}^{\infty} \|\xi_m(l)\|^2 \le \sum_{l=-\infty}^{\infty} C^2 (1+\tau)^{2m} \|\lambda_1(l)\|^2 \\
&= C^2 (1+\tau)^{2m} \|R^1\|_{l_2}^2 \\
&\le C^2 C_1^2 e^{2m\tau} (\tau^2 + h^2)^2 \\
&\le C'^2 (\tau^2 + h^2)^2.
\end{aligned}$$

This completes the proof. □

4. Numerical Findings and Discussion

The efficiency and the validity of the suggested methodologies are confirmed in this part using various test problems by utilizing the L_2 and L_∞ error norms. The numerical results obtained by the proposed schemes are compared. Mathematica 12 was used to obtain the numerical and graphical results.

Example 1. *Consider the time fractional Cattaneo equation,*

$$\frac{\partial v(s,t)}{\partial t} + {}_0^{CF}\mathfrak{D}_t^{\alpha} v(s,t) = \frac{\partial^2 v(s,t)}{\partial^2 s} + g(s,t),\ 1 < \alpha < 2,$$

with initial constraint,

$$v(s,0) = 0,\ \ v_t(s,0) = 0, s > 0,$$

and with boundary constraint,

$$v(0,t) = 0,\ \ v(1,t) = 0,\ \ 0 \leqslant t \leqslant 1.$$

The corresponding source term is

$$g(s,t) = 2(1-s^2)s^{\frac{16}{3}}[t + \frac{1}{\alpha - 1}(1 - \exp(\frac{1-\alpha}{2-\alpha}t))] + t^2(\frac{418}{9}s^{\frac{16}{3}} - \frac{208}{9}s^{\frac{10}{3}}).$$

The analytic solution of the given problem is $v(s,t) = t^2(1-s^2)s^{\frac{16}{3}}$. The suggested schemes are implemented on the aforementioned problem to obtain the numerical results. The errors obtained by the schemes are compared with each other in Tables 1–3. Figure 1 presents an efficient comparison of approximate and exact solutions at various times. Figure 2 exhibits the 2D error profile. The 3D comparison between the exact and approximate solutions is depicted in Figure 3. The approximate solution using the scheme based on cubic B-splines when $\tau = 0.01$ and $M = 20$ at $T = 0.5$ and $T = 1$ for Example 1 are given by

$$V(s,0.5) = \begin{cases} -2.3293 \times 10^{-21} + 1.9849 \times 10^{-4}s - 3.03577 \times 10^{-18}s^2 + 1.5661 \times 10^{-3}s^3, & s \in [0, \frac{1}{20}) \\ -8.7474 \times 10^{-7} + 2.5097 \times 10^{-4}s - 1.0497 \times 10^{-3}s^2 + 8.5639 \times 10^{-3}s^3, & s \in [\frac{1}{20}, \frac{1}{10}) \\ -1.7230 \times 10^{-5} + 7.4162 \times 10^{-4}s - 5.9562 \times 10^{-3}s^2 + 0.0249s^3, & s \in [\frac{1}{10}, \frac{3}{20}) \\ \vdots \\ \vdots \\ 1.5989 - 6.1887s + 8.0414s^2 - 3.4502s^3, & s \in [\frac{17}{20}, \frac{9}{10}) \\ 2.5132 - 9.2363s + 11.4277s^2 - 4.7044s^3, & s \in [\frac{9}{10}, \frac{19}{20}) \\ 3.8346 - 13.4089s + 15.8199s^2 - 6.2455s^3, & s \in [\frac{19}{20}, 1). \end{cases}$$

and

$$V(s,1) = \begin{cases} -8.3009 \times 10^{-20} + 1.4517 \times 10^{-3}s + 2.9490 \times 10^{-17}s^2 + 6.8016 \times 10^{-3}s^3, & s \in [0, \frac{1}{20}) \\ -3.5058 \times 10^{-6} + 1.6620 \times 10^{-3}s - 4.2069 \times 10^{-3}s^2 + 3.485 \times 10^{-2}s^3, & s \in [\frac{1}{20}, \frac{1}{10}) \\ -6.8991 \times 10^{-5} + 3.6266 \times 10^{-3}s - 2.3853 \times 10^{-2}s^2 + 0.1003s^3, & s \in [\frac{1}{10}, \frac{3}{20}) \\ \vdots \\ \vdots \\ 6.3895 - 24.7327s + 32.1435s^2 - 13.7945s^3, & s \in [\frac{17}{20}, \frac{9}{10}) \\ 10.0419 - 36.9074s + 45.6709s^2 - 18.8046s^3, & s \in [\frac{9}{10}, \frac{19}{20}) \\ 15.3213 - 53.5792s + 63.2202s^2 - 24.9623s^3, & s \in [\frac{19}{20}, 1). \end{cases}$$

respectively.

Table 1. Comparison of errors using various B-splines when $\alpha = 1.1$, $dt = 0.001$, $T = 1$ for Example 1.

M	CuBS		TCuBS		ECuBS	
	L_2 Norm	L_∞ Norm	L_2 Norm	L_∞ Norm	L_2 Norm	L_∞ Norm
20	1.63×10^{-3}	2.60×10^{-3}	1.65×10^{-3}	2.64×10^{-3}	9.46×10^{-4}	1.47×10^{-3}
40	4.11×10^{-4}	6.55×10^{-4}	4.16×10^{-4}	6.65×10^{-4}	2.45×10^{-4}	3.94×10^{-4}
80	1.08×10^{-4}	1.71×10^{-4}	1.09×10^{-4}	1.73×10^{-4}	6.61×10^{-5}	1.04×10^{-4}
160	3.20×10^{-5}	5.00×10^{-5}	3.23×10^{-5}	5.06×10^{-5}	2.03×10^{-5}	3.25×10^{-5}

Table 2. Comparison of errors using various B-splines with $\alpha = 1.5$, $dt = 0.001$, $T = 1$ for Example 1.

M	CuBS		TCuBS		ECuBS	
	L_2 Norm	L_∞ Norm	L_2 Norm	L_∞ Norm	L_2 Norm	L_∞ Norm
20	1.54×10^{-3}	2.49×10^{-3}	1.56×10^{-3}	2.53×10^{-3}	1.42×10^{-3}	8.89×10^{-4}
40	3.90×10^{-4}	6.27×10^{-4}	3.94×10^{-4}	6.37×10^{-4}	2.33×10^{-4}	3.77×10^{-4}
80	1.02×10^{-4}	1.64×10^{-4}	1.04×10^{-4}	1.67×10^{-4}	6.42×10^{-5}	9.99×10^{-5}
160	3.08×10^{-5}	4.83×10^{-5}	3.11×10^{-5}	4.89×10^{-5}	2.00×10^{-5}	3.21×10^{-5}

Table 3. Comparison of errors using various B-splines with $\alpha = 1.9$, $dt = 0.001$, $T = 1$ for Example 1.

M	CuBS		TCuBS		ECuBS	
	L_2 Norm	L_∞ Norm	L_2 Norm	L_∞ Norm	L_2 Norm	L_∞ Norm
20	1.39×10^{-3}	2.31×10^{-3}	1.41×10^{-3}	2.35×10^{-3}	8.01×10^{-4}	1.30×10^{-3}
40	3.50×10^{-4}	5.85×10^{-4}	3.56×10^{-4}	5.95×10^{-4}	2.09×10^{-4}	3.44×10^{-4}
80	9.18×10^{-5}	1.51×10^{-4}	9.30×10^{-5}	1.54×10^{-4}	5.70×10^{-5}	9.18×10^{-5}
160	2.73×10^{-5}	4.29×10^{-5}	2.75×10^{-5}	4.35×10^{-5}	1.85×10^{-5}	2.96×10^{-5}

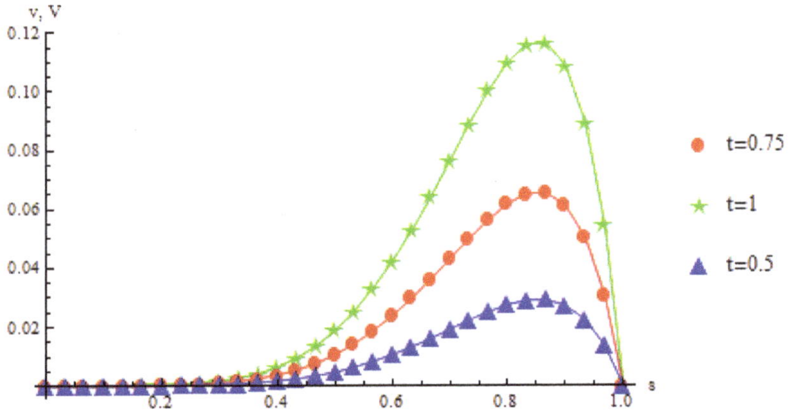

Figure 1. The exact and approximate (triangles, stars, circles) solutions using cubic B-spline-based scheme for Example 1 at various times when $h = \frac{1}{60}$.

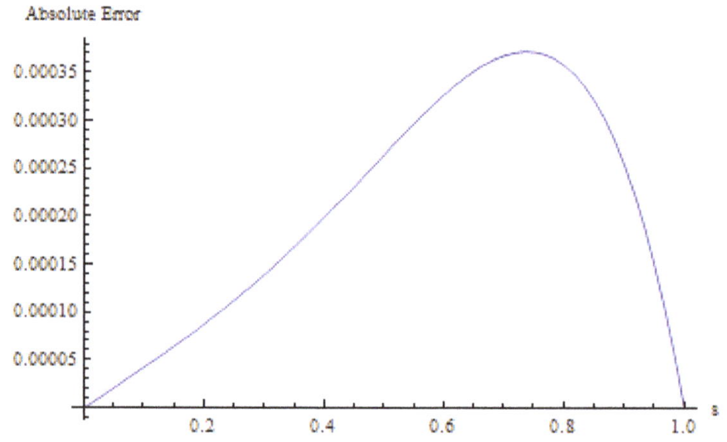

Figure 2. The 2D error profile using cubic B-spline-based scheme for Example 1 when $h = \frac{1}{60}$, $T = 1$, $dt = 0.01$, $\alpha = 1.5$.

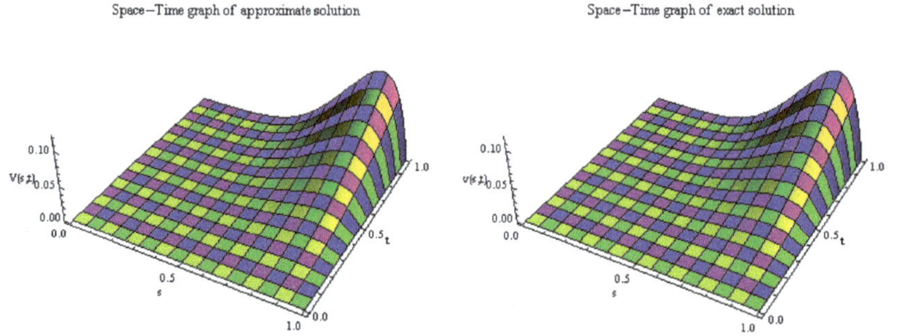

Figure 3. The approximate (**left**) and exact (**right**) solutions using cubic B-spline-based scheme for Example 1 when $h = \frac{1}{60}$, $T = 1$, $dt = 0.01$, $\alpha = 1.5$.

Example 2. *Consider the time fractional Cattaneo equation,*

$$\frac{\partial v(s,t)}{\partial t} + {}_0^{CF}\mathfrak{D}_t^{\alpha} v(s,t) = \frac{\partial^2 v(s,t)}{\partial^2 s} + g(s,t),\ 1 < \alpha < 2,$$

with ICs,

$$v(s,0) = 0,\ \ v_t(s,0) = \sin x,\ \ 0 \leqslant s \leqslant 1,$$

and BCs,

$$v(0,t) = 0,\ \ v(1,t) = t\sin(1),\ \ t > 0.$$

The corresponding source term is $g(s,t) = (1+t)\sin s$. The analytic solution of the given problem is $v(s,t) = t\sin s$. In order to achieve the desired numerical results the presented schemes are applied on Example 2. The errors obtained by the schemes are compared with each other in Tables 4–6. For various time stages, a sharp contrast between the exact and approximate solutions is presented in Figure 4. The 2D absolute error profile is plotted in Figure 5. Figure 6 depicts a 3D comparison between the exact and approximate solutions.

The approximate solution using cubic B-spline-based scheme when $\tau = 0.01$ and $M = 20$ at $T = 0.5$ and $T = 1$ for Example 2 are given by

$$V(s,0.5) = \begin{cases} 3.0358 \times 10^{-17} + 0.499995s + 4.4409 \times 10^{-15}s^2 - 0.08331s^3, & s \in [0, \frac{1}{20}) \\ -2.6046 \times 10^{-8} + 0.499996s - 3.12551 \times 10^{-5}s^2 - 0.08311s^3, & s \in [\frac{1}{20}, \frac{1}{10}) \\ -4.4227 \times 10^{-7} + 0.50001s - 1.56124 \times 10^{-4}s^2 - 0.08267s^3, & s \in [\frac{1}{10}, \frac{3}{20}) \\ \vdots \\ \vdots \\ -7.78029 \times 10^{-3} + 0.5336s - 0.05168s^2 - 0.05339s^3, & s \in [\frac{17}{20}, \frac{9}{10}) \\ -0.01016 + 0.5415s - 0.06050s^2 - 0.05013s^3, & s \in [\frac{9}{10}, \frac{19}{20}) \\ -0.01307 + 0.5507s - 0.07016s^2 - 0.04673s^3, & s \in [\frac{19}{20}, 1). \end{cases}$$

and

$$V(s,1) = \begin{cases} -4.5103 \times 10^{-17} + 0.99997s - 5.32907 \times 10^{-14}s^2 - 0.1666s^3, & s \in [0, \frac{1}{20}) \\ -5.2096 \times 10^{-8} + 0.99998s - 6.25146 \times 10^{-5}s^2 - 0.1662s^3, & s \in [\frac{1}{20}, \frac{1}{10}) \\ -8.8457 \times 10^{-7} + s - 3.12256 \times 10^{-4}s^2 - 0.1654s^3, & s \in [\frac{1}{10}, \frac{3}{20}) \\ \vdots \\ \vdots \\ -1.5548 \times 10^{-2} + 1.0671s - 0.1033s^2 - 0.1068s^3, & s \in [\frac{17}{20}, \frac{9}{10}) \\ -2.0307 \times 10^{-2} + 1.08297s - 0.1209s^2 - 0.1003s^3, & s \in [\frac{9}{10}, \frac{19}{20}) \\ -2.6118 \times 10^{-2} + 1.10132s - 0.1402s^2 - 0.0935s^3, & s \in [\frac{19}{20}, 1). \end{cases}$$

respectively.

Table 4. Comparison of errors using various B-splines with $\alpha = 1.1,\ dt = 0.001,\ T = 1$ for Example 2.

M	CuBS		TCuBS		ECuBS	
	L_2 Norm	L_∞ Norm	L_2 Norm	L_∞ Norm	L_2 Norm	L_∞ Norm
20	7.15×10^{-6}	9.97×10^{-6}	6.71×10^{-6}	9.35×10^{-6}	2.87×10^{-7}	3.99×10^{-7}
40	1.79×10^{-6}	2.49×10^{-6}	1.68×10^{-6}	2.34×10^{-6}	1.43×10^{-8}	1.98×10^{-8}
80	4.47×10^{-7}	6.23×10^{-7}	4.19×10^{-7}	5.84×10^{-7}	6.15×10^{-9}	8.57×10^{-9}
160	1.12×10^{-7}	1.56×10^{-7}	1.05×10^{-7}	1.46×10^{-7}	1.79×10^{-10}	2.50×10^{-10}

Table 5. Comparison of errors using various B-splines with $\alpha = 1.5,\ dt = 0.001,\ T = 1$ for Example 2.

M	CuBS		TCuBS		ECuBS	
	L_2 Norm	L_∞ Norm	L_2 Norm	L_∞ Norm	L_2 Norm	L_∞ Norm
20	7.04×10^{-6}	9.81×10^{-5}	6.60×10^{-6}	9.20×10^{-6}	2.82×10^{-7}	3.93×10^{-7}
40	1.76×10^{-6}	2.45×10^{-6}	1.65×10^{-6}	2.30×10^{-6}	1.40×10^{-8}	1.96×10^{-8}
80	4.40×10^{-7}	6.13×10^{-7}	4.12×10^{-7}	5.75×10^{-7}	6.05×10^{-9}	8.44×10^{-9}
160	1.09×10^{-7}	1.53×10^{-7}	1.03×10^{-7}	1.44×10^{-7}	1.76×10^{-10}	2.46×10^{-10}

Table 6. Comparison of errors using various B-splines with $\alpha = 1.9,\ dt = 0.001,\ T = 1$ for Example 2.

M	CuBS		TCuBS		ECuBS	
	L_2 Norm	L_∞ Norm	L_2 Norm	L_∞ Norm	L_2 Norm	L_∞ Norm
20	7.20×10^{-6}	1.01×10^{-5}	6.75×10^{-6}	9.43×10^{-6}	2.89×10^{-7}	4.03×10^{-7}
40	1.80×10^{-6}	2.51×10^{-6}	1.69×10^{-6}	2.36×10^{-6}	1.44×10^{-8}	2.01×10^{-8}
80	4.50×10^{-7}	6.29×10^{-7}	4.22×10^{-7}	5.90×10^{-7}	6.19×10^{-9}	8.65×10^{-9}
160	1.12×10^{-7}	1.57×10^{-7}	1.05×10^{-7}	1.47×10^{-7}	1.80×10^{-10}	2.52×10^{-10}

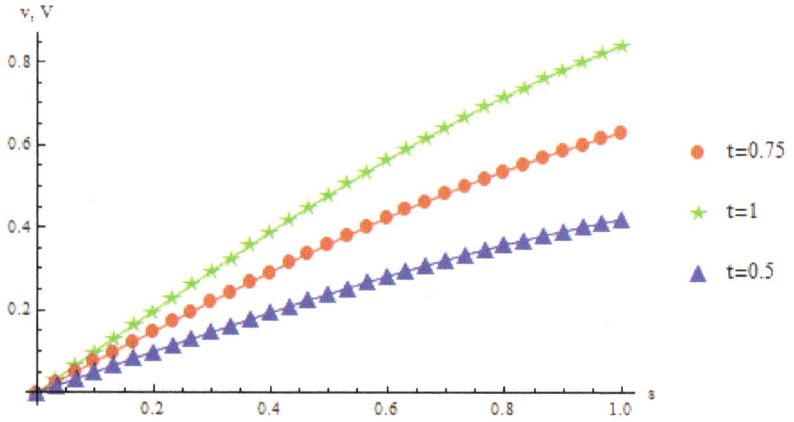

Figure 4. The exact and approximate (triangles, stars, circles) solutions using cubic B-spline-based scheme for Example 2 at various times when $h = \frac{1}{60}$.

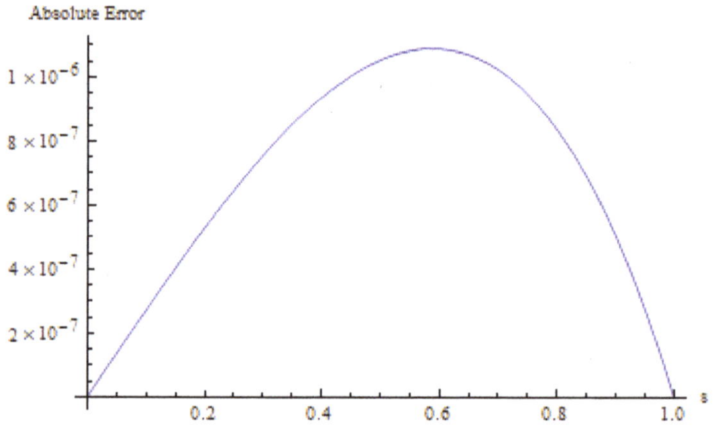

Figure 5. The 2D error profile using cubic B-spline-based scheme for Example 2 when $h = \frac{1}{60}$, $T = 1$, $dt = 0.01$, $\alpha = 1.5$.

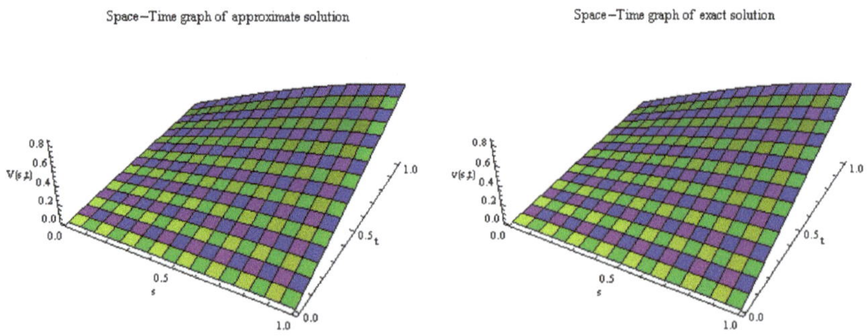

Figure 6. The approximate (**left**) and exact (**right**) solutions using cubic B-spline-based scheme for Example 2 when $h = \frac{1}{60}$, $T = 1$, $dt = 0.01$, $\alpha = 1.5$.

Example 3. *Consider the time fractional Cattaneo equation,*

$$\frac{\partial v(s,t)}{\partial t} + {}_0^{CF}\mathfrak{D}_t^{\alpha} v(s,t) = \frac{\partial^2 v(s,t)}{\partial^2 s} + g(s,t),\ 1 < \alpha < 2,$$

with ICs,

$$v(s,0) = 0,\ \ v_t(s,0) = (1-s)\cos s,\ \ 0 \leqslant s \leqslant 1,$$

and BCs

$$v(0,t) = t,\ \ v(1,t) = 0,\ \ t > 0.$$

The corresponding source term is $g(s,t) = (1+t)(1-s)\cos s - 2t\sin s$. The analytic solution of the given problem is $v(s,t) = t(1-s)\cos s$. The proposed methodologies are utilized to acquire the numerical results for Example 3.

A comparison of computed errors is provided in Tables 7–9. For various time stages, a close comparison between the exact and approximate solutions is displayed in Figure 7. The 2D error function is plotted in Figure 8. Figure 9 depicts a 3D comparison between the exact and approximate solutions.

The approximate solution using cubic B-spline-based scheme when $\tau = 0.01$ and $M = 20$ at $T = 0.5$ and $T = 1$ for Example 3 are given by

$$V(s,0.5) = \begin{cases} 0.5 - 0.5s - 0.25s^2 + 0.2519s^3, & s \in [0, \frac{1}{20}) \\ 0.5 - 0.49997s - 0.2505s^2 + 0.2551s^3, & s \in [\frac{1}{20}, \frac{1}{10}) \\ 0.499998 - 0.4999s - 0.2511s^2 + 0.2571s^3, & s \in [\frac{1}{10}, \frac{3}{20}) \\ \vdots & \\ \vdots & \\ 0.5282 - 0.6185s - 0.07785s^2 + 0.1681s^3, & s \in [\frac{17}{20}, \frac{9}{10}) \\ 0.5375 - 0.6496s - 0.04329s^2 + 0.1553s^3, & s \in [\frac{9}{10}, \frac{19}{20}) \\ 0.5491 - 0.6860s - 4.990 \times 10^{-3}s^2 + 0.1419s^3, & s \in [\frac{19}{20}, 1). \end{cases}$$

and

$$V(s,1) = \begin{cases} 1 - 0.9999s - 0.5s^2 + 0.5039s^3, & s \in [0, \frac{1}{20}) \\ 0.9999 - 0.9999s - 0.5009s^2 + 0.5101s^3, & s \in [\frac{1}{20}, \frac{1}{10}) \\ 0.9999 - 0.9998s - 0.5022s^2 + 0.5142s^3, & s \in [\frac{1}{10}, \frac{3}{20}) \\ \vdots & \\ \vdots & \\ 1.0563 - 1.2367s - 0.1560s^2 + 0.3364s^3, & s \in [\frac{17}{20}, \frac{9}{10}) \\ 1.0750 - 1.2989s - 0.0869s^2 + 0.3108s^3, & s \in [\frac{9}{10}, \frac{19}{20}) \\ 1.0980 - 1.3716s - 0.0103s^2 + 0.2839s^3, & s \in [\frac{19}{20}, 1). \end{cases}$$

respectively.

Table 7. Comparison of errors using various B-splines with $\alpha = 1.1$, $dt = 0.001$, $T = 1$ for Example 3.

M	CuBS		TCuBS		ECuBS	
	L_2 Norm	L_∞ Norm	L_2 Norm	L_∞ Norm	L_2 Norm	L_∞ Norm
20	2.21×10^{-5}	3.19×10^{-5}	2.23×10^{-6}	3.55×10^{-6}	4.04×10^{-6}	1.47×10^{-6}
40	5.53×10^{-6}	7.99×10^{-6}	5.58×10^{-7}	8.89×10^{-7}	1.01×10^{-6}	3.94×10^{-6}
80	1.38×10^{-6}	1.99×10^{-6}	1.40×10^{-7}	2.23×10^{-7}	2.97×10^{-7}	1.04×10^{-7}
160	3.46×10^{-7}	4.99×10^{-7}	3.49×10^{-8}	5.57×10^{-8}	6.71×10^{-8}	3.25×10^{-8}

Table 8. Comparison of errors using various B-splines with $\alpha = 1.5$, $dt = 0.001$, $T = 1$ for Example 3.

M	CuBS		TCuBS		ECuBS	
	L_2 Norm	L_∞ Norm	L_2 Norm	L_∞ Norm	L_2 Norm	L_∞ Norm
20	2.18×10^{-5}	3.14×10^{-5}	2.24×10^{-6}	3.56×10^{-6}	4.06×10^{-6}	6.07×10^{-6}
40	5.44×10^{-6}	7.88×10^{-6}	5.60×10^{-7}	8.90×10^{-7}	1.01×10^{-6}	1.51×10^{-6}
80	1.36×10^{-6}	1.97×10^{-6}	1.40×10^{-7}	2.23×10^{-7}	2.98×10^{-7}	4.58×10^{-7}
160	3.40×10^{-7}	4.92×10^{-7}	3.50×10^{-8}	5.57×10^{-8}	6.74×10^{-8}	9.56×10^{-8}

Table 9. Comparison of errors using various B-splines with $\alpha = 1.9$, $dt = 0.001$, $T = 1$ for Example 3.

M	CuBS		TCuBS		ECuBS	
	L_2 Norm	L_∞ Norm	L_2 Norm	L_∞ Norm	L_2 Norm	L_∞ Norm
20	2.23×10^{-5}	3.23×10^{-5}	2.33×10^{-6}	3.69×10^{-6}	4.22×10^{-6}	6.30×10^{-6}
40	5.57×10^{-6}	8.08×10^{-6}	5.81×10^{-7}	9.24×10^{-7}	1.05×10^{-6}	1.57×10^{-6}
80	1.39×10^{-6}	2.02×10^{-6}	1.45×10^{-7}	2.32×10^{-7}	3.09×10^{-7}	4.76×10^{-7}
160	3.48×10^{-7}	5.05×10^{-7}	3.63×10^{-8}	5.79×10^{-8}	7.20×10^{-8}	9.96×10^{-8}

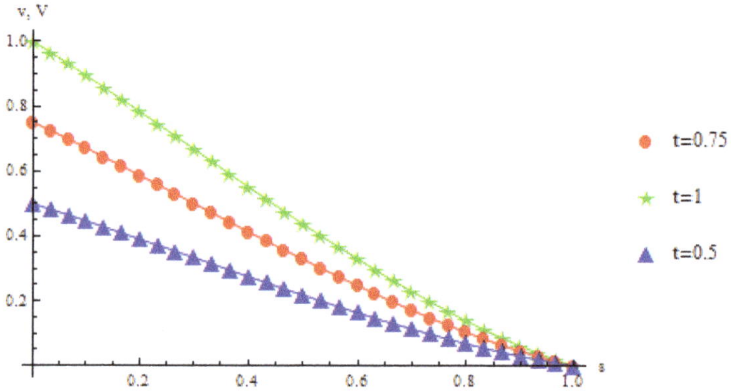

Figure 7. The exact and approximate (triangles, stars, circles) solutions using cubic B-spline-based scheme for Example 3 at various times when $h = \frac{1}{60}$.

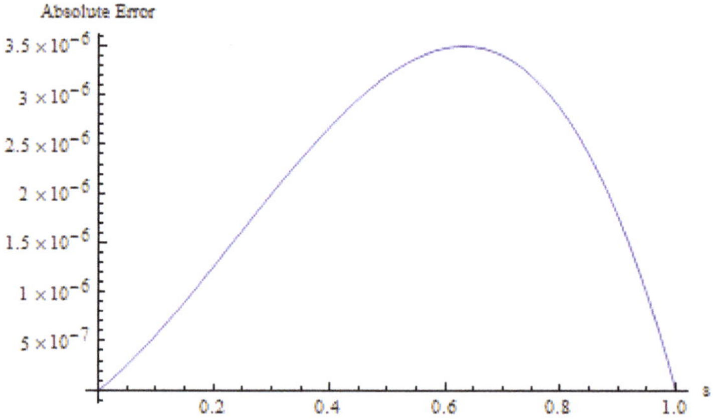

Figure 8. The 2D error profile using cubic B-spline-based scheme when for Example 2 when $h = \frac{1}{60}$, $T = 1, dt = 0.01, \alpha = 1.5$.

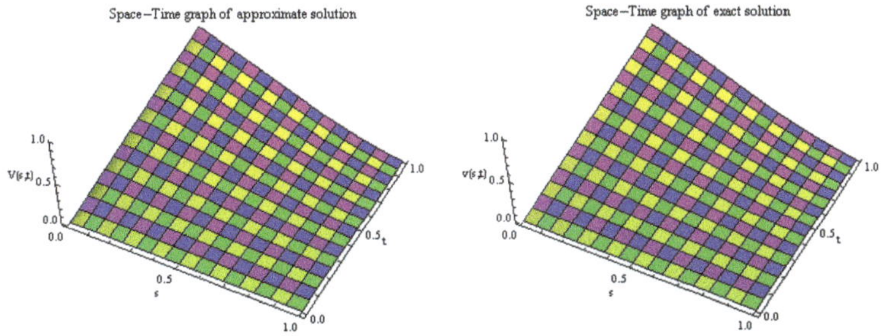

Figure 9. The approximate (**left**) and exact (**right**) solutions using cubic B-spline-based scheme for Example 3 when $h = \frac{1}{60}$, $T = 1$, $dt = 0.01$, $\alpha = 1.5$.

5. Concluding Remarks

The spline-based collocation schemes are developed for the numerical solution of the time fractional Cattaneo differential equation involving the Caputo–Fabrizio time fractional derivative. To begin with, the space derivative involved is approximated using the cubic B-spline. Secondly, using finite differences, the Caputo–Fabrizio derivative is approximated. The stability and convergence analysis of the schemes are also discussed in detail. The splines used are the cubic B-splines, extended cubic B-splines and the trigonometric cubic B-splines. The key advantage is that the approximate solution is obtained as a piecewise continuous function so that approximate solution at any desired position in the domain can be tracked. The efficiency and accuracy of the proposed approaches are confirmed by the experimental findings. The suggested schemes can be applied to a wide range of problems in varied fields of applied sciences.

Author Contributions: Conceptualization, M.Y. , Q.U.N.A. and S.K.; methodology, Q.U.N.A. and R.G.; validation, R.G. and S.K.; formal analysis, M.Y., Q.U.N.A. and S.K.; investigation, S.K.; writing—original draft preparation, Q.U.N.A.; writing—review and editing, M.Y., R.G. and S.K.; supervision, R.G. and M.Y.; project administration, R.G.; funding acquisition, R.G. All authors have read and agreed to the published version of the manuscript.

Funding: This research is supported by Deanship of Scientific Research, Prince Sattam bin Abdulaziz University, Alkharj, Saudi Arabia.

Institutional Review Board Statement: Not applicable:

Informed Consent Statement: Not applicable.

Data Availability Statement: No data were used.

Conflicts of Interest: The authors declare no conflict of interest.

References

1. Liu, Z.; Cheng, A.; Li, X. A second order crank-Nicolson scheme for fractional Cattaneo equation based on new fractional derivative. *Appl. Math. Comput.* **2017**, *311*, 361–374. [CrossRef]
2. Zheng, B. (G'/G)-expansion method for solving fractional partial differential equations in the theory of mathematical physics. *Commun. Theor. Phys.* **2012**, *58*, 623–630. [CrossRef]
3. Tavares, D.; Almeida, R.; Torres, D.F.M. Caputo derivatives of fractional variable order: Numerical approximations. *Commun. Nonlinear Sci. Numer. Simul.* **2016**, *35*, 69–87. [CrossRef]
4. Baleanu, D.; Rezapour, S.; Mohammadi, H. Some existence results on nonlinear fractional differential equations. *Philos. Trans. R. Soc. A Math. Phys. Eng. Sci.* **2013**, *371*, 20120144. [CrossRef] [PubMed]
5. Diethelm, K.; Ford, N.J.; Freed, A.D. A predictor-corrector approach for the numerical solution of fractional differential equations. *Nonlinear Dyn.* **2002**, *29*, 3–22. [CrossRef]
6. Meerschaert, M.M.; Tadjeran, C. Finite difference approximations for fractional advection–dispersion flow equations. *J. Comput. Appl. Math.* **2004**, *172*, 65–77. [CrossRef]
7. Hashim, I.; Abdulaziz, O.; Momani, S. Homotopy analysis method for fractional IVPS. *Commun. Nonlinear Sci. Numer. Simul.* **2009**, *14*, 674–684. [CrossRef]

8. Suarez, L.; Shokooh, A. An eigenvector expansion method for the solution of motion containing fractional derivatives. *J. Appl. Mech.* **1997**, *64*, 629–635. [CrossRef]
9. Jafari, H.; Yousefi, S.; Firoozjaee, M.; Momani, S.; Khalique, C.M. Application of Legendre wavelets for solving fractional differential equations. *Comput. Math. Appl.* **2011**, *62*, 1038–1045. [CrossRef]
10. Li, Y.; Zhao, W. Haar wavelet operational matrix of fractional order integration and its applications in solving the fractional order differential equations. *Appl. Math. Comput.* **2010**, *216*, 2276–2285. [CrossRef]
11. Yuanlu, L. Solving a nonlinear fractional differential equation using Chebyshev wavelets. *Commun. Nonlinear Sci. Numer. Simul.* **2010**, *15*, 2284–2292.
12. Li, Y.; Sun, N. Numerical solution of fractional differential equations using the generalized block pulse operational matrix. *Comput. Math. Appl.* **2011**, *62*, 1046–1054. [CrossRef]
13. Odibat, Z. On Legendre polynomial approximation with the vim or ham for numerical treatment of nonlinear fractional differential eqations. *J. Comput. Appl. Math.* **2011**, *235*, 2956–2968. [CrossRef]
14. Araci, S. Novel identities for q-Genocchi numbers and polynomials. *J. Funct. Spaces Appl.* **2012**, *2012*, 214961. [CrossRef]
15. Gürbüz, B.; Sezer, M. Laguerre polynomial solutions of a class of initial and boundary value problems arising in science and engineering fields. *Acta Phys. Pol. A* **2016**, *130*, 194–197. [CrossRef]
16. Caputo, M.; Fabrizio, M. A new definition of fractional derivative without singular kernel. *Prog. Fract. Differ. Appl.* **2015**, *1*, 73–85.
17. Caputo, M.; Fabrizio, M. Applications of new time and spatial fractional derivatives with exponential kernels. *Prog. Fract. Differ. Appl.* **2016**, *2*, 1–11. [CrossRef]
18. Feulefack, P.A.; Djida, J.D.; Abdon, A. A new model of groundwater flow within an unconfined aquifer: Application of Caputo-Fabrizio fractional derivative. *Discret. Cont. Dyn. Syst.-B* **2019**, *24*, 3227–3247. [CrossRef]
19. Delgado, V.F.M.; Aguilar, J.F.G.; Saad, K.; Jiménez, R.F.E. Application of the Caputo-Fabrizio and Atangana-Baleanu fractional derivatives to mathematical model of cancer chemotherapy effect. *Math. Methods Appl. Sci.* **2019**, *42*, 1167–1193. [CrossRef]
20. Abdon, A. On the new fractional derivative and application to nonlinear fishers reaction-diffusion equation. *Appl. Math. Comput.* **2016**, *273*, 948–956.
21. Liu, Z.; Cheng, A.; Li, X. A fully discrete spectral method for fractional Cattaneo equation based on Caputo-Fabrizo derivative. *Numer. Methods Partial Differ. Equ.* **2019**, *35*, 936–954. [CrossRef]
22. Compte, A.; Metzler, R. The generalized Cattaneo equation for the description of anomalous transport processes. *J. Phys. A* **1997**, *30*, 72–77. [CrossRef]
23. Giusti, A. Why fractional derivatives with nonsingular kernels should not be used. *Nonlinear Dyn.* **2018**, *93*, 1757–1763. [CrossRef]
24. Li, X.; Rui, H.; Liu, Z. A block-centered finite difference method for fractional Cattaneo equaton. *Numer. Methods Partial Differ. Equ.* **2018**, *34*, 296–316. [CrossRef]
25. Dhiman, N.; Huntul, M.J.; Tamsir, M. A modified trigonometric cubic B-spline collocation technique for solving the time-fractional diffusion equation. *Eng. Comput.* **2021**, *38*. [CrossRef]
26. Tamsir, M.; Dhiman, N.; Gill, F.C. Approximation of 3D convection diffusion equation using DQM based on modified cubic trigonometric B-splines. *J. Comput. Methods Sci. Eng.* **2020**, *20*, 1357–1366.
27. Tamsir, M.; Dhiman, N. DQM Based on the Modified Form of CTB Shape Functions for Coupled Burgers' Equation in 2D and 3D. *Int. J. Math. Eng. Manag. Sci.* **2019**, *4*, 1051–1067.
28. Dhiman, N.; Tamsir, M. A collocation technique based on modified form of trigonometric cubic B-spline basis functions for Fisher's reaction-diffusion equation. *Multidiscip. Model. Mater. Struct.* **2018**, *14*, 923–939. [CrossRef]
29. Tamsir, M. Cubic trigonometric B-spline differential quadrature method for numerical treatment of Fisher's reaction-diffusion equations. *Alex. Eng. J.* **2018**, *53*, 2019–2026.
30. Abbas, M.; Majid, A.A.; Ismail, A.I.M.; Rashid, A. Numerical method using cubic b-spline for a strongly coupled reaction-diffusion system. *PLoS ONE* **2014**, *9*, e83265.
31. Abbas, M.; Majid, A.A.; Ismail, A.I.M.; Rashid, A. The application of cubic trigonometric B-spline to the numerical solution of the hyperbolic problems. *Appl. Math. Comput.* **2014**, *239*, 74–88. [CrossRef]
32. Morton, K.W.; Mayers, D.F. *Numerical Solution of Partial Differential Equations: An Introduction*; Cambridge University Press: Cambridge, UK, 2005.
33. Smith, G.D. *Numerical Solution of Partial Differential Equations: Finite Difference Methods*; Oxford University Press: Oxford, UK, 1987.

fractal and fractional

MDPI

Article

Hermite-Hadamard Fractional Inequalities for Differentiable Functions

Muhammad Samraiz [1], Zahida Perveen [1], Gauhar Rahman [2], Muhammad Adil Khan [3] and Kottakkaran Sooppy Nisar [4,*]

[1] Department of Mathematics, University of Sargodha, Sargodha 40100, Pakistan; muhammad.samraiz@uos.edu.pk (M.S.); zahidaperveen1995@gmail.com (Z.P.)
[2] Department of Mathematics and Statistics, Hazara University, Mansehra 21300, Pakistan; gauhar55uom@gmail.com or drgauhar.rahman@hu.edu.pk
[3] Department of Mathematics, University of Peshawar, Peshawar 25000, Pakistan; adilswati@gmail.com or madilkhan@uop.edu.pk
[4] Department of Mathematics, College of Arts and Sciences, Prince Sattam Bin Abdulaziz University, Wadi Aldawser 11991, Saudi Arabia
* Correspondence: n.sooppy@psau.edu.sa or ksnisar1@gmail.com

Abstract: In this article, we look at a variety of mean-type integral inequalities for a well-known Hilfer fractional derivative. We consider twice differentiable convex and s-convex functions for $s \in (0, 1]$ that have applications in optimization theory. In order to infer more interesting mean inequalities, some identities are also established. The consequences for Caputo fractional derivative are presented as special cases to our general conclusions.

Keywords: Hermite-Hadamard-type inequalities; Hilfer fractional derivative; Hölder's inequality

MSC: 26D15; 26D07; 26D10; 26A33; 26A51

Citation: Samraiz, M.; Perveen, Z.; Rahman, G.; Adil Khan, M.; Nisar, K.S. Hermite-Hadamard Fractional Inequalities for Differentiable Functions. *Fractal Fract.* **2022**, *6*, 60. https://doi.org/10.3390/fractalfract6020060

Academic Editors: Mark Edelman and Ricardo Almeida

Received: 29 November 2021
Accepted: 11 January 2022
Published: 25 January 2022

1. Introduction

The subject of fractional calculus has achieved a significant prominence during the most recent couple of years due to its demonstrated applications in the field of science and engineering. This offers useful strategies to solve differential and integral equations, see the books [1,2] and articles [3–5]. Fractional calculus has been applied in different areas of science, engineering, financial mathematics, applied sciences, bio engineering, etc.

Mathematical inequalities are significantly important in the study of mathematics and related fields. Nowadays, fractional integral inequalities are fruitful in generating the uniqueness of solutions for fractional partial differential equations. They also provide boundedness of the solutions of fractional boundary value problems. These recommendations have inspired various researchers in the field of integral inequalities to inquire the extensions by involving fractional calculus operators. Recently, Peter Korus presented a class of Hermite-Hadamard inequalities by considering the class of convex or generalized convex derivative in [6], Farid et al. explored Fejér-Hadamard type inequalities [7] for $(\alpha, h - m) - p$-convex functions by involving the fractional operators. We further refer the reader to, e.g., [8–10].

The convex functions are utilized to create numerous inequalities like Alomari et al. [11] present Ostrowski's inequalities via s convexity in second sense, Dragomir et al. discuss some properties of convex functions in [12] and explored some important quadrature rules in [13]. More applications can be observed from literature [14–16] on convex functions and inequalities. Hermite-Hadamard's inequality [17] is one of the most important classical inequalities, as it has a rich geometrical meaning and applications [18–20]. Hermite-Hadamard's double inequality is one of the most widely studied concerning convex functions. The inequality is defined as follows:

Let $\psi : I \subseteq \mathbb{R} \to \mathbb{R}$ be a convex mapping and $\theta, \zeta \in I$ with $\theta < \zeta$. Then,

$$\psi\left(\frac{\theta+\zeta}{2}\right) \leq \frac{1}{\zeta-\theta}\int_{\theta}^{\zeta}\psi(\nu)d\nu \leq \frac{\psi(\theta)+\psi(\zeta)}{2}. \quad (1)$$

If ψ is concave, then the inequalities (1) hold in reverse direction. For particular choices of function ψ, some classical inequalities for means can be derived from (1) (see [21]). The principle point of this paper is to infer Hermite-Hadamard-type integral inequalities for Hilfer fractional derivative. Such inequalities were proved by many scientists for different convexities and for many fractional operators, but the main results of this paper are more general then the existing literature.

2. Preliminaries

In this section, we recall some basic preliminary results.

Definition 1 ([22])**.** *Let $\psi : [\theta, \zeta] \to \mathbb{R}$ is said to be convex if the inequality*

$$\psi(\nu\gamma + (1-\nu)\beta) \leq \nu\psi(\gamma) + (1-\nu)\psi(\beta),$$

holds for $\gamma, \beta \in [\theta, \zeta]$ and $\nu \in [0,1]$.

The definition of classical Riemann–Liouville fractional derivative (see [23] (Chapter 4)) is given as follows.

Definition 2. *Let $\Phi \in L^1[\theta, \zeta]$, then the right-sided and left-sided Riemann–Liouville fractional derivative of order $\alpha > 0$ are defined by*

$$D^{\gamma}_{\theta+}\psi(\nu) = \frac{1}{\Gamma(n-\gamma)}\left(\frac{d}{d\nu}\right)^{n}\int_{\theta}^{\nu}(\nu-\tau)^{n-\gamma-1}\psi(\tau)d\tau,$$

and

$$D^{\gamma}_{\zeta-}\psi(\nu) = \frac{1}{\Gamma(n-\gamma)}\left(\frac{d}{d\nu}\right)^{n}\int_{\nu}^{\zeta}(\tau-\nu)^{n-\gamma-1}\psi(\tau)d\tau,$$

where $n = [\gamma]+1,\ \nu \in [\theta, \zeta]$.

Let $x > \theta > 0$ and $L^1(\theta, x)$, denote the space of all Lebesgue integrable functions on the interval (θ, x). Then, for any $\psi \in L^1(\theta, x)$ the Riemann–Liouville fractional integral of order γ is defined by

$$(I^{\gamma}_{\theta_+}\psi)(\nu) = \frac{1}{\Gamma(\nu)}\int_{\theta}^{\nu}(x-\tau)^{\gamma-1}\psi(\tau)d\tau = (\psi * K_{\gamma})(\nu), \quad \nu \in [\theta, x], \quad (\gamma > 0), \quad (2)$$

where $K_{\gamma}(\nu) = \frac{\nu^{\gamma-1}}{\Gamma(\gamma)}$. The integral on the right side of (2) exists for almost $\nu \in [\theta, x]$ and $I^{\gamma}_{\theta_+}\psi \in L^1(\theta, x)$.

Throughout this paper, the space of all continuous differentiable functions up to order m, on $[\theta, x]$ is presented by $C^m[\theta, x]$. By $AC[\theta, x]$, we mean the space of all absolutely continuous functions on $[\theta, x]$ and the space $AC^m[\theta, x]$, denote the space of all such functions $\psi \in C^m[\theta, x]$ with $\psi^{(m-1)} \in AC[\theta, x]$. By $L_{\infty}(\theta, x)$, we denote the space of all measurable functions essentially bounded on $[\theta, x]$. Let $\mu > 0, m = [\mu]+1$ and $f \in AC^m[a, b]$. The Caputo derivative of order $\gamma > 0$ is defined as

$$({}^{C}D^{\gamma}_{\theta_+}\psi)(\nu) = \left(I^{m-\gamma}_{\theta_+}\frac{d^m}{d\nu^m}\psi\right)(\nu) = \frac{1}{\Gamma(m-\gamma)}\int_{\theta}^{\nu}(\nu-\tau)^{m-\gamma-1}\frac{d^m}{d\nu^m}\psi(\tau)d\tau.$$

Definition 3 ([24]). *Let* $\psi \in L^1[\theta,\zeta], \psi * K_{(1-\beta)(1-\gamma)} \in AC^1[\theta,\zeta]$. *The fractional derivative operator* $D^{\gamma,\beta}_{\theta^+}$ *of order* $0<\gamma<1$ *and type* $0<\beta\le 1$ *with respect to* $\nu \in [\theta,\zeta]$ *is defined by*

$$\left(D^{\gamma,\beta}_{\theta^+}\psi\right)(\nu) := I^{\beta(1-\gamma)}_{\theta^+}\frac{d}{d\nu}\left(I^{(1-\beta)(1-\gamma)}_{\theta^+}\psi(\nu)\right). \tag{3}$$

The derivative (3) *is usually called Hilfer fractional derivative.*

The more general integral representation of Equation (3) given in [24] is defined as follows:

Let $\psi \in L^1[\theta,\zeta], \psi * K_{(1-\beta)(n-\gamma)} \in AC^n[\theta,\zeta], n-1<\gamma<n, 0<\beta\le 1, n\in\mathbb{N}$. Then,

$$\left(D^{\gamma,\beta}_{\theta^+}\psi\right)(\nu) = \left(I^{\beta(n-\gamma)}_{\theta^+}\frac{d^n}{d\nu^n}\left(I^{(1-\beta)(n-\gamma)}_{\theta^+}\psi(\nu)\right)\right), \tag{4}$$

which coincide with (3) for $n=1$.

Specially for $\beta=0$, $D^{\gamma,0}_{\theta^+}\psi = D^{\gamma}_{\theta^+}\psi$ is Riemann–Liouville fractional derivative of order γ and for $\beta=1$ it is Caputo fractional derivative $D^{\gamma,1}_{\theta^+}\psi = {}^{C}D^{\gamma}_{\theta^+}\psi$ of order γ. Applying the properties of Riemann–Liouville integral the relation (4) can be rewritten in the form

$$\begin{aligned}\left(D^{\gamma,\beta}_{\theta^+}\psi\right)(\nu) &= \left(I^{\beta(n-\gamma)}_{\theta^+}\left(\left(D^{n-(1-\beta)(n-\gamma)}_{\theta^+}\psi\right)(\nu)\right)\right)\\ &= \frac{1}{\Gamma(\beta(n-\gamma))}\int_{\theta}^{\nu}(\nu-\tau)^{\beta(n-\gamma)-1}\left(\left(D^{\gamma+\beta(n-\gamma)}_{\theta^+}\psi\right)(\tau)\right)d\tau. \end{aligned}\tag{5}$$

The geometric arithmetically s-convex function given in [25] presented in the following definition.

Definition 4. *Let* $\psi : I \subset \mathbb{R}^+ \to \mathbb{R}^+$ *and* $s\in(0,1]$. *A function* ψ *is geometric-arithmetically* s*-convex function on* I *if for every* $\gamma,\beta\in I$ *and* $\nu\in[0,1]$, *we have*

$$\psi(\gamma^{\nu}\beta^{1-\nu}) \le \nu^s(\psi(\gamma)) + (1-\nu)^s\psi(\beta).$$

The following lemma was given by Liao et al. [25].

Lemma 1. *For* $\theta\in[0,1]$, γ, $\beta>0$, *we have*

$$\theta\gamma + (1-\theta)\beta \ge \beta^{1-\theta}\gamma^{\theta}.$$

Deng et al. [26] prove the following lemma.

Lemma 2. *For* $\theta\in[0,1]$, *we have*

$$\begin{aligned}(1-\theta)^{\gamma} &\le 2^{1-\gamma}-\theta^{\gamma}, \gamma\in[0,1],\\ (1-\theta)^{\gamma} &\ge 2^{1-\gamma}-\theta^{\gamma}, \gamma\in[1,\infty).\end{aligned}$$

3. Main Results

This section includes several mean-type fractional integral inequalities involving Hilfer fractional derivative. The first main result for the fractional derivative is presented in the following theorem.

Theorem 1. *Let* $\psi \in L^1[\theta,\zeta], \psi * K_{(1-\beta)(n-\gamma)} \in AC^n[\theta,\zeta], n\in N$ *and* $D^{\gamma+\beta(n-\gamma)}_{(\theta,\zeta)}\psi : [\theta,\zeta]\to\mathbb{R}$ *be a positive function with* $0\le\theta<\zeta$, $n-1<\gamma<n$, $0<\beta\le 1$ *and* $D^{\gamma+\beta(n-\gamma)}_{(\theta,\zeta)}\psi \in$

$L^1[\theta,\zeta]$. If $D^{\gamma+\beta(n-\gamma)}_{(\theta,\zeta)}\psi$ is convex function on $[\theta,\zeta]$, then the following inequality for fractional derivative holds.

$$\begin{aligned} &D^{\gamma+\beta(n-\gamma)}_{(\theta,\zeta)}\Phi\left(\frac{\theta+\zeta}{2}\right) \\ &\leq \frac{\Gamma(\beta(n-\gamma)+1)}{(\zeta-\theta)^{\beta(n-\gamma)}}\left[D^{\gamma,\beta}_{\theta^+}\Phi(\zeta)+D^{\gamma,\beta}_{\zeta^-}\Phi(\theta)\right] \\ &\leq D^{\gamma+\beta(n-\gamma)}_{\theta^+}\Phi(\zeta)+D^{\gamma+\beta(n-\gamma)}_{\zeta^-}\Phi(\theta). \end{aligned} \tag{6}$$

Proof. We define functions $\tilde{\psi}(\nu)=\psi(\theta+\zeta-\nu)$, $\nu\in[\theta,\zeta]$ and $\Phi(\nu)=\psi(\nu)+\tilde{\psi}(\nu)$, $\nu\in[\theta,\zeta]$. Since $D^{\gamma+\beta(n-\gamma)}_{\theta^+}\psi$ is convex on $[\theta,\zeta]$, therefore with $\mu=\frac{1}{2}$, we have

$$D^{\gamma+\beta(n-\gamma)}_{\theta^+}\psi\left(\frac{x+y}{2}\right)\leq\frac{D^{\gamma+\beta(n-\gamma)}_{\theta^+}\psi(x)+D^{\gamma+\beta(n-\gamma)}_{\theta^+}\psi(y)}{2}.$$

Choosing $x=\nu\theta+(1-\nu)\zeta$ and $y=(1-\nu)\theta+\nu\zeta$, we get

$$\begin{aligned} &2D^{\gamma+\beta(n-\gamma)}_{\theta^+}\psi\left(\frac{\theta+\zeta}{2}\right) \\ &\leq D^{\gamma+\beta(n-\gamma)}_{\theta^+}\psi(\nu\theta+(1-\nu)\zeta)+D^{\gamma+\beta(n-\gamma)}_{\theta^+}\psi((1-\nu)\theta+\nu\zeta) \\ &= D^{\gamma+\beta(n-\gamma)}_{\theta^+}\Phi(\nu\theta+(1-\nu)\zeta). \end{aligned}$$

Now, we multiply both sides of above inequality by $\nu^{\beta(n-\gamma)-1}$ and then integrating the resulting inequality with respect to ν over $[0,1]$, we have

$$\frac{1}{\beta(n-\gamma)}D^{\gamma+\beta(n-\gamma)}_{\theta^+}\Phi\left(\frac{\theta+\zeta}{2}\right)\leq\int_0^1\nu^{\beta(n-\gamma)-1}D^{\gamma+\beta(n-\gamma)}_{\theta^+}\Phi(\nu\theta+(1-\nu)\zeta)d\nu. \tag{7}$$

By substituting $u=\nu\theta+(1-\nu)\zeta$, the inequality (7) becomes

$$D^{\gamma+\beta(n-\gamma)}_{\theta^+}\Phi\left(\frac{\theta+\zeta}{2}\right)\leq\frac{\Gamma(\beta(n-\gamma)+1)}{(\zeta-\theta)^{\beta(n-\gamma)}}D^{\gamma,\beta}_{\theta^+}\Phi(\zeta). \tag{8}$$

Similarly, for the choice

$$D^{\gamma+\beta(n-\gamma)}_{\zeta^-}\psi\left(\frac{x+y}{2}\right)\leq\frac{D^{\gamma+\beta(n-\gamma)}_{\zeta^-}\psi(x)+D^{\gamma+\beta(n-\gamma)}_{\zeta^-}\psi(y)}{2},$$

we get

$$D^{\gamma+\beta(n-\gamma)}_{\zeta^-}\Phi\left(\frac{\theta+\zeta}{2}\right)\leq\frac{\Gamma(\beta(n-\gamma)+1)}{(\zeta-\theta)^{\beta(n-\gamma)}}D^{\gamma,\beta}_{\zeta^-}\Phi(\theta). \tag{9}$$

By adding (8) and (9), we obtain

$$\begin{aligned} &D^{\gamma+\beta(n-\gamma)}_{\theta^+}\Phi\left(\frac{\theta+\zeta}{2}\right)+D^{\gamma+\beta(n-\gamma)}_{\zeta^-}\Phi\left(\frac{\theta+\zeta}{2}\right) \\ &\leq\frac{\Gamma(\beta(n-\gamma)+1)}{(\zeta-\theta)^{\beta(n-\gamma)}}\left[D^{\gamma,\beta}_{\theta^+}\Phi(\zeta)+D^{\gamma,\beta}_{\zeta^-}\Phi(\theta)\right], \end{aligned} \tag{10}$$

which proves the left half part of inequality (6).

For the proof of the second half, we first note that if $D_{\theta^+}^{\gamma+\beta(n-\gamma)}\psi$ is convex, then for $\nu \in [0,1]$, yields

$$D_{\theta^+}^{\gamma+\beta(n-\gamma)}\psi(\nu\theta+(1-\nu)\zeta) \leq \nu D_{\theta^+}^{\gamma+\beta(n-\gamma)}\tilde{\psi}(\zeta)+(1-\nu)D_{\theta^+}^{\gamma+\beta(n-\gamma)}\psi(\zeta)$$
$$D_{\theta^+}^{\gamma+\beta(n-\gamma)}\psi((1-\nu)\theta+\nu\zeta) \leq (1-\nu)D_{\theta^+}^{\gamma+\beta(n-\gamma)}\tilde{\psi}(\zeta)+\nu D_{\theta^+}^{\gamma+\beta(n-\gamma)}\psi(\zeta).$$

By adding above two inequalities, we have

$$D_{\theta^+}^{\gamma+\beta(n-\gamma)}\Phi(\nu\theta+(1-\nu)\zeta) \leq D_{\theta^+}^{\gamma+\beta(n-\gamma)}\Phi(\zeta). \tag{11}$$

Similarly,

$$D_{\zeta^-}^{\gamma+\beta(n-\gamma)}\Phi((1-\nu)\theta+\nu\zeta) \leq D_{\zeta^-}^{\gamma+\beta(n-\gamma)}\Phi(\theta). \tag{12}$$

From (11) and (12), we get

$$\begin{aligned} D_{\theta^+}^{\gamma+\beta(n-\gamma)}&\Phi(\nu\theta+(1-\nu)\zeta)+D_{\zeta^-}^{\gamma+\beta(n-\gamma)}\Phi((1-\nu)\theta+\nu\zeta) \\ &\leq D_{\theta^+}^{\gamma+\beta(n-\gamma)}\Phi(\zeta)+D_{\zeta^-}^{\gamma+\beta(n-\gamma)}\Phi(\theta). \end{aligned} \tag{13}$$

Now, first, we multiply both sides of (13) by $\nu^{\beta(n-\gamma)-1}$, and then we integrate the resulting inequality with respect to ν over $[0,1]$, we have

$$\begin{aligned} &\int_0^1 \nu^{\beta(n-\gamma)-1}D_{\theta^+}^{\gamma+\beta(n-\gamma)}\Phi(\nu\theta+(1-\nu)\zeta)d\nu \\ &+\int_0^1 \nu^{\beta(n-\gamma)-1}D_{\zeta^-}^{\gamma+\beta(n-\gamma)}\Phi((1-\nu)\theta+\nu\zeta)d\nu \\ &\leq [D_{\theta^+}^{\gamma+\beta(n-\gamma)}\Phi(\zeta)+D_{\zeta^-}^{\gamma+\beta(n-\gamma)}\Phi(\theta)]\int_0^1 \nu^{\beta(n-\gamma)-1}. \end{aligned}$$

By substituting $u=\nu\theta+(1-\nu)\zeta$ and $v=(1-\nu)\theta+\nu\zeta$, the above inequality becomes

$$\frac{\Gamma(\beta(n-\gamma)+1)}{(\zeta-\theta)^{\beta(n-\gamma)}}\left[D_{\theta^+}^{\gamma,\beta}\Phi(\zeta)+D_{\zeta^-}^{\gamma,\beta}\Phi(\theta)\right] \leq D_{\theta^+}^{\gamma+\beta(n-\gamma)}\Phi(\zeta)+D_{\zeta^-}^{\gamma+\beta(n-\gamma)}\Phi(\theta). \tag{14}$$

From (10) and (14), we get inequality (6). □

The special case of Theorem 1 presented in [27] (Theorem 2.3) is given as follows.

Corollary 1. *If we choose $\beta=1$ and ψ is symmetric about $\frac{\theta+\zeta}{2}$ in Theorem 1, we get*

$$\psi^n\left(\frac{\theta+\zeta}{2}\right) \leq \frac{\Gamma(n-\gamma+1)}{2(\zeta-\theta)^{n-\gamma}}\left[{}^C D_{\theta+}^{\gamma}\psi(\zeta)+(-1)^{n C}D_{\zeta-}^{\gamma}\psi(\theta)\right] \leq \frac{\psi^n(\theta)+\psi^n(\zeta)}{2}.$$

Lemma 3. *Let $\psi \in L^1[\theta,\zeta]$, $\psi * K_{(1-\beta)(n-\gamma)} \in AC^n[\theta,\zeta]$, $n \in N$. For the differentiable function $D_{(\theta,\zeta)}^{\gamma+\beta(n-\gamma)}\psi : [\theta,\zeta] \to \mathbb{R}$ with $n-1<\gamma<n$, $0<\beta\leq 1$ and $D_{(\theta,\zeta)}^{\gamma+\beta(n-\gamma)+1}\psi \in L^1[\theta,\zeta]$ the following equality*

$$\begin{aligned} &\frac{D_{\theta^+}^{\gamma+\beta(n-\gamma)}\Phi(\zeta)+D_{\zeta^-}^{\gamma+\beta(n-\gamma)}\Phi(\theta)}{2}-\frac{\Gamma(\beta(n-\gamma)+1)}{2(\zeta-\theta)^{\beta(n-\gamma)}}\left[D_{\theta^+}^{\gamma,\beta}\Phi(\zeta)+D_{\zeta^-}^{\gamma,\beta}\Phi(\theta)\right] \\ &=\frac{\zeta-\theta}{2}\int_0^1\left[(1-\nu)^{\beta(n-\gamma)}-\nu^{\beta(n-\gamma)}\right]D_{(\theta,\zeta)}^{\gamma+\beta(n-\gamma)+1}\psi(\nu\theta+(1-\nu)\zeta)d\nu, \end{aligned}$$

holds.

Proof. Consider

$$
\begin{aligned}
I &= \int_0^1 \left[(1-\nu)^{\beta(n-\gamma)} - \nu^{\beta(n-\gamma)}\right] D_{(\theta,\zeta)}^{\gamma+\beta(n-\gamma)+1}\psi(\nu\theta+(1-\nu)\zeta)d\nu \\
&= \int_0^1 \left[(1-\nu)^{\beta(n-\gamma)} - \nu^{\beta(n-\gamma)}\right] D_{\theta^+}^{\gamma+\beta(n-\gamma)+1}\psi(\nu\theta+(1-\nu)\zeta)d\nu \\
&+ \int_0^1 \left[(1-\nu)^{\beta(n-\gamma)} - \nu^{\beta(n-\gamma)}\right] D_{\zeta^-}^{\gamma+\beta(n-\gamma)+1}\psi(\nu\theta+(1-\nu)\zeta)d\nu \\
&= I_1 + I_2.
\end{aligned} \tag{15}
$$

Integrating I_1 by parts, we get

$$
\begin{aligned}
I_1 &= (1-\nu)^{\beta(n-\gamma)} \frac{D_{\theta^+}^{\gamma+\beta(n-\gamma)}\psi(\nu\theta+(1-\nu)\zeta)}{\theta-\zeta} \Big|_0^1 \\
&+ \int_0^1 \beta(n-\gamma)(1-\nu)^{\beta(n-\gamma)-1} \frac{D_{\theta^+}^{\gamma+\beta(n-\gamma)}\psi(\nu\theta+(1-\nu)\zeta)}{\theta-\zeta} d\nu \\
&- \nu^{\beta(n-\gamma)} \frac{D_{\theta^+}^{\gamma+\beta(n-\gamma)}\psi(\nu\theta+(1-\nu)\zeta)}{\theta-\zeta} \Big|_0^1 \\
&+ \int_0^1 \beta(n-\gamma)\nu^{\beta(n-\gamma)-1} \frac{D_{\theta^+}^{\gamma+\beta(n-\gamma)}\psi(\nu\theta+(1-\nu)\zeta)}{\theta-\zeta} d\nu.
\end{aligned}
$$

By substituting $x = \nu\theta + (1-\nu)\zeta$, we obtain

$$
\begin{aligned}
I_1 &= \frac{D_{\theta^+}^{\gamma+\beta(n-\gamma)}\psi(\zeta) + D_{\theta^+}^{\gamma+\beta(n-\gamma)}\tilde{\psi}(\zeta)}{\zeta-\theta} - \frac{\beta(n-\gamma)}{\zeta-\theta}\left[\int_\zeta^\theta \left(\frac{\theta-x}{\theta-\zeta}\right)^{\beta(n-\gamma)-1}\right. \\
&\times \left.\frac{D_{\theta^+}^{\gamma+\beta(n-\gamma)}\psi(x)}{\theta-\zeta}dx + \int_\zeta^\theta \left(\frac{\zeta-x}{\zeta-\theta}\right)^{\beta(n-\gamma)-1} \frac{D_{\theta^+}^{\gamma+\beta(n-\gamma)}\psi(x)}{\theta-\zeta}dx\right] \\
&= \frac{D_{\theta^+}^{\gamma+\beta(n-\gamma)}\Phi(\zeta)}{\zeta-\theta} - \frac{\beta(n-\gamma)}{(\zeta-\theta)^{\beta(n-\gamma)+1}} \int_\theta^\zeta (\zeta-x)^{\beta(n-\gamma)-1} D_{\theta^+}^{\gamma+\beta(n-\gamma)}\Phi(x)dx \\
&= \frac{D_{\theta^+}^{\gamma+\beta(n-\gamma)}\Phi(\zeta)}{\zeta-\theta} - \frac{\Gamma(\beta(n-\gamma)+1)}{(\zeta-\theta)^{\beta(n-\gamma)+1}} D_{\theta^+}^{\gamma,\beta}\Phi(\zeta).
\end{aligned} \tag{16}
$$

Similarly, integrating I_2 by parts, we get

$$
\begin{aligned}
I_2 &= (1-\nu)^{\beta(n-\gamma)} \frac{D_{\zeta^-}^{\gamma+\beta(n-\gamma)}\psi(\nu\theta+(1-\nu)\zeta)}{\theta-\zeta} \Big|_0^1 \\
&+ \int_0^1 \beta(n-\gamma)(1-\nu)^{\beta(n-\gamma)-1} \frac{D_{\zeta^-}^{\gamma+\beta(n-\gamma)}\psi(\nu\theta+(1-\nu)\zeta)}{\theta-\zeta} d\nu \\
&- \nu^{\beta(n-\gamma)} \frac{D_{\zeta^-}^{\gamma+\beta(n-\gamma)}\psi(\nu\theta+(1-\nu)\zeta)}{\theta-\zeta} \Big|_0^1 \\
&+ \int_0^1 \beta(n-\gamma)\nu^{\beta(n-\gamma)-1} \frac{D_{\zeta^-}^{\gamma+\beta(n-\gamma)}\psi(\nu\theta+(1-\nu)\zeta)}{\theta-\zeta}.
\end{aligned}
$$

By substituting again $x = \nu\theta + (1-\nu)\zeta$, we get

$$
I_2 = \frac{D_{\zeta^-}^{\gamma+\beta(n-\gamma)}\Phi(\theta)}{\zeta-\theta} - \frac{\Gamma(\beta(n-\gamma)+1)}{(\zeta-\theta)^{\beta(n-\gamma)+1}} D_{\zeta^-}^{\gamma,\beta}\Phi(\theta). \tag{17}
$$

Using (16) and (17) in (15), we have

$$I = \frac{D_{\theta^+}^{\gamma+\beta(n-\gamma)}\Phi(\zeta) + D_{\zeta^-}^{\gamma+\beta(n-\gamma)}\Phi(\theta)}{\zeta-\theta} - \frac{\Gamma(\beta(n-\gamma)+1)}{(\zeta-\theta)^{\beta(n-\gamma)+1}}\left[D_{\theta^+}^{\gamma,\beta}\Phi(\zeta) + D_{\zeta^-}^{\gamma,\beta}\Phi(\theta)\right].$$

Thus, by multiplying both sides with $\frac{\zeta-\theta}{2}$, we get the desired result. □

The following special case of Lemma 3 was proved by Farid et al. in [27] (Lemma 2.2).

Corollary 2. *If we take $\beta = 1$ and ψ is symmetric about $\frac{\theta+\zeta}{2}$ in Lemma 3, we obtain*

$$\begin{aligned}
&\frac{\psi^n(\theta)+\psi^n(\zeta)}{2} - \frac{\Gamma(n-\gamma+1)}{2(\zeta-\theta)^{n-\gamma}}\left[{}^C D_{\theta^+}^{\gamma}\psi(\zeta) + (-1)^{n\,C} D_{\zeta^-}^{\gamma}\psi(\theta)\right] \\
&= \frac{\zeta-\theta}{2}\int_0^1\left[(1-\nu)^{n-\gamma} - \nu^{n-\gamma}\right]\psi^{n+1}(\nu\theta+(1-\nu)\zeta)d\nu,
\end{aligned}$$

for Caputo fractional derivatives.

Theorem 2. *Let $\psi \in L^1[\theta,\zeta]$, $\psi * K_{(1-\beta)(n-\gamma)} \in AC^n[\theta,\zeta]$, $n \in N$ and $D_{(\theta,\zeta)}^{\gamma+\beta(n-\gamma)}\psi : [\theta,\zeta] \to \mathbb{R}$ be a differentiable function with $n-1 < \gamma < n$ and $0 < \beta \le 1$. If $D_{(\theta,\zeta)}^{\gamma+\beta(n-\gamma)+1}\psi$ is convex on $[\theta,\zeta]$, then the following inequality is true*

$$\begin{aligned}
&\left|\frac{D_{\theta^+}^{\gamma+\beta(n-\gamma)}\Phi(\zeta) + D_{\zeta^-}^{\gamma+\beta(n-\gamma)}\Phi(\theta)}{2} - \frac{\Gamma(\beta(n-\gamma)+1)}{2(\zeta-\theta)^{\beta(n-\gamma)}}\left[D_{\theta^+}^{\gamma,\beta}\Phi(\zeta) + D_{\zeta^-}^{\gamma,\beta}\Phi(\theta)\right]\right| \\
&\le \frac{\zeta-\theta}{2(\beta(n-\gamma)+1)}\left(1-\frac{1}{2^{\beta(n-\gamma)}}\right)\left(|D_{(\theta,\zeta)}^{\gamma+\beta(n-\gamma)+1}\psi(\zeta)| + |D_{(\theta,\zeta)}^{\gamma+\beta(n-\gamma)+1}\psi(\theta)|\right).
\end{aligned}$$

Proof. By using Lemma 3 and Definition 1, we get

$$\begin{aligned}
&\left|\frac{D_{\theta^+}^{\gamma+\beta(n-\gamma)}\Phi(\zeta) + D_{\zeta^-}^{\gamma+\beta(n-\gamma)}\Phi(\theta)}{2} - \frac{\Gamma(\beta(n-\gamma)+1)}{2(\zeta-\theta)^{\beta(n-\gamma)}}\left[D_{\theta^+}^{\gamma,\beta}\Phi(\zeta) + D_{\zeta^-}^{\gamma,\beta}\Phi(\theta)\right]\right| \\
&\le \frac{\zeta-\theta}{2}\int_0^1 |(1-\nu)^{\beta(n-\gamma)} - \nu^{\beta(n-\gamma)}| \\
&\times\left(\nu|D_{(\theta,\zeta)}^{\gamma+\beta(n-\gamma)+1}\psi(\theta)| + (1-\nu)|D_{(\theta,\zeta)}^{\gamma+\beta(n-\gamma)+1}\psi(\zeta)|\right)d\nu \\
&= \frac{\zeta-\theta}{2}\int_0^{\frac{1}{2}}\left[(1-\nu)^{\beta(n-\gamma)} - \nu^{\beta(n-\gamma)}\right] \\
&\times\left(\nu|D_{(\theta,\zeta)}^{\gamma+\beta(n-\gamma)+1}\psi(\theta)| + (1-\nu)|D_{(\theta,\zeta)}^{\gamma+\beta(n-\gamma)+1}\psi(\zeta)|\right)d\nu \\
&+ \int_{\frac{1}{2}}^1\left[\nu^{\beta(n-\gamma)} - (1-\nu)^{\beta(n-\gamma)}\right]\left(\nu|D_{(\theta,\zeta)}^{\gamma+\beta(n-\gamma)+1}\psi(\theta)| + (1-\nu)|D_{(\theta,\zeta)}^{\gamma+\beta(n-\gamma)+1}\psi(\zeta)|\right)d\nu \\
&= \frac{\zeta-\theta}{2}\left[|D_{(\theta,\zeta)}^{\gamma+\beta(n-\gamma)+1}\psi(\zeta)|\int_0^{\frac{1}{2}}\left[(1-\nu)^{\beta(n-\gamma)+1} - (1-\nu)\nu^{\beta(n-\gamma)}\right]d\nu\right. \\
&+ |D_{(\theta,\zeta)}^{\gamma+\beta(n-\gamma)+1}\psi(\theta)|\int_0^{\frac{1}{2}}\left[\nu(1-\nu)^{\beta(n-\gamma)} - \nu^{\beta(n-\gamma)+1}\right]d\nu \\
&+ |D_{(\theta,\zeta)}^{\gamma+\beta(n-\gamma)+1}\psi(\zeta)|\int_{\frac{1}{2}}^1\left[(1-\nu)\nu^{\beta(n-\gamma)} - (1-\nu)^{\beta(n-\gamma)+1}\right]d\nu \\
&\left.+ |D_{(\theta,\zeta)}^{\gamma+\beta(n-\gamma)+1}\psi(\theta)|\int_{\frac{1}{2}}^1\left[\nu^{\beta(n-\gamma)+1} - \nu(1-\nu)^{\beta(n-\gamma)}\right]d\nu\right]
\end{aligned}$$

$$= \frac{\zeta-\theta}{2}\Big[|D_{(\theta,\zeta)}^{\gamma+\beta(n-\gamma)+1}\psi(\zeta)|\Big(\frac{1}{\beta(n-\gamma)+1}-\frac{1}{(\beta(n-\gamma)+1)2^{\beta(n-\gamma)}}\Big)$$
$$+|D_{(\theta,\zeta)}^{\gamma+\beta(n-\gamma)+1}\psi(\theta)|\Big(\frac{1}{\beta(n-\gamma)+1}-\frac{1}{(\beta(n-\gamma)+1)2^{\beta(n-\gamma)}}\Big)\Big]$$
$$= \frac{\zeta-\theta}{2(\beta(n-\gamma)+1)}\Big(1-\frac{1}{2^{\beta(n-\gamma)}}\Big)\Big(|D_{(\theta,\zeta)}^{\gamma+\beta(n-\gamma)+1}\psi(\zeta)|+|D_{(\theta,\zeta)}^{\gamma+\beta(n-\gamma)+1}\psi(\theta)|\Big).$$

Hence, the proof is complete. □

The corollary given below presented in [27] (Theorem 2.4) is a special case of Theorem 2.

Corollary 3. *If we choose $\beta = 1$ and ψ is symmetric about $\frac{\theta+\zeta}{2}$ in Theorem 2, we get*

$$\frac{\psi^n(\zeta)+\psi^n(\theta)}{2}-\frac{\Gamma(n-\gamma+1)}{2(\zeta-\theta)^{n-\gamma}}\Big[{}^C D_{\theta+}^{\gamma}\psi(\zeta)+(-1)^{n C}D_{\zeta-}^{\gamma}\psi(\theta)\Big]$$
$$\leq \frac{\zeta-\theta}{2(n-\gamma+1)}\Big(1-\frac{1}{2^{n-\gamma}}\Big)\Big(|\psi^{n+1}(\zeta)|+|\psi^{n+1}(\theta)|\Big).$$

Lemma 4. *Let $\psi \in L^1[\theta,\zeta]$, $\psi * K_{(1-\beta)(n-\gamma)} \in AC^n[\theta,\zeta]$, $n \in N$ and $D_{(\theta,\zeta)}^{\gamma+\beta(n-\gamma)}\psi : [\theta,\zeta] \to \mathbb{R}$ be twice differential mapping on (θ,ζ) with $n-1<\gamma<n$ and $0<\beta\leq 1$. If $D_{(\theta,\zeta)}^{\gamma+\beta(n-\gamma)+2}$ $\psi \in L^1[\theta,\zeta]$, then we have the following equality.*

$$\frac{D_{\theta+}^{\gamma+\beta(n-\gamma)}\Phi(\zeta)+D_{\zeta-}^{\gamma+\beta(n-\gamma)}\Phi(\theta)}{2}-\frac{\Gamma(\beta(n-\gamma)+1)}{2(\zeta-\theta)^{\beta(n-\gamma)}}\Big[D_{\theta+}^{\gamma,\beta}\Phi(\zeta)+D_{\zeta-}^{\gamma,\beta}\Phi(\theta)\Big]$$
$$= \frac{(\zeta-\theta)^2}{2}\int_0^1 \frac{1-(1-\nu)^{\beta(n-\gamma)+1}-\nu^{\beta(n-\gamma)+1}}{\beta(n-\gamma)+1}D_{(\theta,\zeta)}^{\gamma+\beta(n-\gamma)+2}\psi(\nu\theta+(1-\nu)\zeta)d\nu.$$

Proof. By using Lemma 3, we get

$$\frac{D_{\theta+}^{\gamma+\beta(n-\gamma)}\Phi(\zeta)+D_{\zeta-}^{\gamma+\beta(n-\gamma)}\Phi(\theta)}{2}-\frac{\Gamma(\beta(n-\gamma)+1)}{2(\zeta-\theta)^{\beta(n-\gamma)}}[D_{\theta+}^{\gamma,\beta}\Phi(\zeta)+D_{\zeta-}^{\gamma,\beta}\Phi(\theta)]$$
$$= \frac{\zeta-\theta}{2}\Big[\int_0^1\Big[(1-\nu)^{\beta(n-\gamma)}-\nu^{\beta(n-\gamma)}\Big]D_{\theta+}^{\gamma+\beta(n-\gamma)+1}\psi(\nu\theta+(1-\nu)\zeta)d\nu$$
$$+\int_0^1\Big[(1-\nu)^{\beta(n-\gamma)}-\nu^{\beta(n-\gamma)}\Big]D_{\zeta-}^{\gamma+\beta(n-\gamma)+1}\psi(\nu\theta+(1-\nu)\zeta)d\nu\Big].$$

Integrating by parts, we get

$$= \frac{\zeta-\theta}{2}\Big[\frac{D_{\theta+}^{\gamma+\beta(n-\gamma)+1}\psi(\zeta)-D_{\theta+}^{\gamma+\beta(n-\gamma)+1}\psi(\theta)+D_{\zeta-}^{\gamma+\beta(n-\gamma)+1}\psi(\zeta)-D_{\zeta-}^{\gamma+\beta(n-\gamma)+1}\psi(\theta)}{\beta(n-\gamma)+1}$$
$$-\frac{\zeta-\theta}{\beta(n-\gamma)+1}\int_0^1\Big[(1-\nu)^{\beta(n-\gamma)+1}+\nu^{\beta(n-\gamma)+1}\Big]D_{(\theta,\zeta)}^{\gamma+\beta(n-\gamma)+2}\psi(\nu\theta+(1-\nu)\zeta)d\nu\Big]. \quad (18)$$

Since

$$D_{\theta+}^{\gamma+\beta(n-\gamma)+1}\psi(\zeta)-D_{\theta+}^{\gamma+\beta(n-\gamma)+1}\psi(\theta)=\int_\theta^\zeta D_{\theta+}^{\gamma+\beta(n-\gamma)+2}\psi(u)du.$$

By substituting $u = \nu\theta + (1-\nu)\zeta$, we get

$$\begin{aligned} &D_{\theta+}^{\gamma+\beta(n-\gamma)+1}\psi(\zeta) - D_{\theta+}^{\gamma+\beta(n-\gamma)+1}\psi(\theta) \\ &= (\zeta-\theta)\int_0^1 D_{\theta+}^{\gamma+\beta(n-\gamma)+2}\psi(\nu\theta+(1-\nu)\zeta)d\nu, \end{aligned} \tag{19}$$

and

$$D_{\zeta-}^{\gamma+\beta(n-\gamma)+1}\psi(\zeta) - D_{\zeta-}^{\gamma+\beta(n-\gamma)+1}\psi(\theta) = \int_\theta^\zeta D_{\zeta-}^{\gamma+\beta(n-\gamma)+2}\psi(u)du.$$

By substituting again $u = \nu\theta + (1-\nu)\zeta$, we get

$$D_{\zeta-}^{\gamma+\beta(n-\gamma)+1}\psi(\zeta) - D_{\zeta-}^{\gamma+\beta(n-\gamma)+1}\psi(\theta) = (\zeta-\theta)\int_0^1 D_{\zeta-}^{\gamma+\beta(n-\gamma)+2}\psi(\nu\theta+(1-\nu)\zeta)d\nu. \tag{20}$$

By adding (19) and (20), we obtain

$$\begin{aligned} &D_{\theta+}^{\gamma+\beta(n-\gamma)+1}\psi(\zeta) - D_{\theta+}^{\gamma+\beta(n-\gamma)+1}\psi(\theta) + D_{\zeta-}^{\gamma+\beta(n-\gamma)+1}\psi(\zeta) - D_{\zeta-}^{\gamma+\beta(n-\gamma)+1}\psi(\theta) \\ &= (\zeta-\theta)\int_0^1 D_{(\theta,\zeta)}^{\gamma+\beta(n-\gamma)+2}\psi(\nu\theta+(1-\nu)\zeta)d\nu. \end{aligned} \tag{21}$$

Using Equation (21) into (18), we get the required result. □

Corollary 4. *If we take $\beta = 1$ and ψ is symmetric about $\frac{\theta+\zeta}{2}$ in Lemma 4, we get the following equality for Caputo fractional derivatives:*

$$\begin{aligned} &\frac{\psi^n(\theta)+\psi^n(\zeta)}{2} - \frac{\Gamma(n-\gamma+1)}{2(\zeta-\theta)^{n-\gamma}}\left[{}^C D_{\theta+}^{\gamma}\psi(\zeta) + (-1)^{n\,C} D_{\zeta-}^{\gamma}\psi(\theta)\right] \\ &= \frac{(\zeta-\theta)^2}{2}\int_0^1 \frac{1-(1-\nu)^{n-\gamma+1}-\nu^{n-\gamma+1}}{n-\gamma+1}\psi^{n+2}(\nu\theta+(1-\nu)\zeta)d\nu. \end{aligned}$$

Lemma 5. *Let $\psi \in L^1[\theta,\zeta]$, $\psi * K_{(1-\beta)(n-\gamma)} \in AC^n[\theta,\zeta]$, $n \in N$ and $D_{(\theta,\zeta)}^{\gamma+\beta(n-\gamma)}\psi : [\theta,\zeta] \to \mathbb{R}$ is twice differentiable and measurable on $[\theta,\zeta]$, $n-1 < \gamma < n$ and $0 < \beta \le 1$, then the equation*

$$\begin{aligned} &\frac{\Gamma(\beta(n-\gamma)+1)}{2(\zeta-\theta)^{\beta(n-\gamma)}}\left[D_{\theta+}^{\gamma,\beta}\Phi(\zeta) + D_{\zeta-}^{\gamma,\beta}\Phi(\theta)\right] - D_{(\theta,\zeta)}^{\gamma+\beta(n-\gamma)}\psi\left(\frac{\theta+\zeta}{2}\right) \\ &= \frac{(\zeta-\theta)^2}{2}\int_0^1 m(\nu)D_{(\theta,\zeta)}^{\gamma+\beta(n-\gamma)+2}\psi(\nu\theta+(1-\nu)\zeta)d\nu, \end{aligned}$$

holds for $m(\nu) = \begin{cases} \nu - \frac{1-(1-\nu)^{\beta(n-\gamma)+1}-\nu^{\beta(n-\gamma)+1}}{\beta(n-\gamma)+1}, & \nu \in [0,\frac{1}{2}); \\ 1-\nu - \frac{1-(1-\nu)^{\beta(n-\gamma)+1}-\nu^{\beta(n-\gamma)+1}}{\beta(n-\gamma)+1}, & \nu \in [\frac{1}{2},1). \end{cases}$

Proof. Consider

$$\begin{aligned}
&\frac{(\zeta-\theta)^2}{2}\int_0^1 m(\nu)D_{(\theta,\zeta)}^{\gamma+\beta(n-\gamma)+2}\psi(\nu\theta+(1-\nu)\zeta)d\nu \\
&=\frac{(\zeta-\theta)^2}{2}\Big[\int_0^{\frac{1}{2}}\nu\Big(D_{\theta^+}^{\gamma+\beta(n-\gamma)+2}\psi(\nu\theta+(1-\nu)\zeta) \\
&+D_{\zeta^-}^{\gamma+\beta(n-\gamma)+2}\psi(\nu\theta+(1-\nu)\zeta)\Big)d\nu \\
&+\int_{\frac{1}{2}}^1(1-\nu)\Big(D_{\theta^+}^{\gamma+\beta(n-\gamma)+2}\psi(\nu\theta+(1-\nu)\zeta)+D_{\zeta^-}^{\gamma+\beta(n-\gamma)+2}\psi(\nu\theta+(1-\nu)\zeta)\Big)d\nu \\
&-\int_0^1\left(\frac{1-(1-\nu)^{\beta(n-\gamma)+1}-\nu^{\beta(n-\gamma)+1}}{\beta(n-\gamma)+1}\right)D_{(\theta,\zeta)}^{\gamma+\beta(n-\gamma)+2}\psi(\nu\theta+(1-\nu)\zeta)d\nu\Big].
\end{aligned}$$

Let

$$\begin{aligned}
I&=\int_0^{\frac{1}{2}}\nu\Big(D_{\theta^+}^{\gamma+\beta(n-\gamma)+2}\psi(\nu\theta+(1-\nu)\zeta)+D_{\zeta^-}^{\gamma+\beta(n-\gamma)+2}\psi(\nu\theta+(1-\nu)\zeta)\Big)d\nu \\
&+\int_{\frac{1}{2}}^1(1-\nu)\Big(D_{\theta^+}^{\gamma+\beta(n-\gamma)+2}\psi(\nu\theta+(1-\nu)\zeta)+D_{\zeta^-}^{\gamma+\beta(n-\gamma)+2}\psi(\nu\theta+(1-\nu)\zeta)\Big)d\nu \\
&=I_1+I_2.
\end{aligned}\tag{22}$$

Integrating I_1 by parts, we get

$$\begin{aligned}
I_1&=\frac{D_{\theta^+}^{\gamma+\beta(n-\gamma)+1}\psi\left(\frac{\theta+\zeta}{2}\right)+D_{\zeta^-}^{\gamma+\beta(n-\gamma)+1}\psi\left(\frac{\theta+\zeta}{2}\right)}{2(\theta-\zeta)} \\
&-\frac{\Big[D_{\theta^+}^{\gamma+\beta(n-\gamma)}\psi\left(\frac{\theta+\zeta}{2}\right)+D_{\zeta^-}^{\gamma+\beta(n-\gamma)}\psi\left(\frac{\theta+\zeta}{2}\right)-D_{\theta^+}^{\gamma+\beta(n-\gamma)}\psi(\zeta)-D_{\zeta^-}^{\gamma+\beta(n-\gamma)}\tilde{\psi}(\theta)\Big]}{(\theta-\zeta)^2}.
\end{aligned}\tag{23}$$

Now integrating I_2 by parts, we get

$$\begin{aligned}
I_2&=-\frac{D_{\theta^+}^{\gamma+\beta(n-\gamma)+1}\psi\left(\frac{\theta+\zeta}{2}\right)+D_{\zeta^-}^{\gamma+\beta(n-\gamma)+1}\psi\left(\frac{\theta+\zeta}{2}\right)}{2(\theta-\zeta)} \\
&-\frac{\Big[D_{\theta^+}^{\gamma+\beta(n-\gamma)}\psi\left(\frac{\theta+\zeta}{2}\right)+D_{\zeta^-}^{\gamma+\beta(n-\gamma)}\psi\left(\frac{\theta+\zeta}{2}\right)--D_{\theta^+}^{\gamma+\beta(n-\gamma)}\tilde{\psi}(\zeta)-D_{\zeta^-}^{\gamma+\beta(n-\gamma)}\psi(\theta)\Big]}{(\theta-\zeta)^2}.
\end{aligned}\tag{24}$$

By substituting (23) and (24) to (22), we get

$$I=\frac{D_{\theta^+}^{\gamma+\beta(n-\gamma)}\Phi(\zeta)+D_{\zeta^-}^{\gamma+\beta(n-\gamma)}\Phi(\theta)}{(\zeta-\theta)^2}-\frac{2D_{\theta^+}^{\gamma+\beta(n-\gamma)}\psi\left(\frac{\theta+\zeta}{2}\right)+2D_{\zeta^-}^{\gamma+\beta(n-\gamma)}\psi\left(\frac{\theta+\zeta}{2}\right)}{(\zeta-\theta)^2}.$$

Thus

$$\begin{aligned}
&\frac{(\zeta-\theta)^2}{2}\int_0^1 m(\nu)D_{(\theta,\zeta)}^{\gamma+\beta(n-\gamma)+2}\psi(\nu\theta+(1-\nu)\zeta)d\nu \\
&=\frac{D_{\theta^+}^{\gamma+\beta(n-\gamma)}\Phi(\zeta)+D_{\zeta^-}^{\gamma+\beta(n-\gamma)}\Phi(\theta)}{2}-D_{(\theta,\zeta)}^{\gamma+\beta(n-\gamma)}\psi\left(\frac{\theta+\zeta}{2}\right)
\end{aligned}$$

$$-\frac{(\zeta-\theta)^2}{2}\int_0^1\left(\frac{1-(1-\nu)^{\beta(n-\gamma)+1}-\nu^{\beta(n-\gamma)+1}}{\beta(n-\gamma)+1}\right)$$
$$\times D^{\gamma+\beta(n-\gamma)+2}_{(\theta,\zeta)}\psi(\nu\theta+(1-\nu)\zeta)d\nu.$$

By using Lemma (4), we arrive at the desired result. □

Corollary 5. *If we take $\beta=1$ and ψ is symmetric about $\frac{\theta+\zeta}{2}$ in Lemma 5, then the following equality for Caputo fractional derivatives*

$$\frac{\Gamma(n-\gamma+1)}{2(\zeta-\theta)^{n-\gamma}}\left[{}^{C}D^{\gamma}_{\theta+}\psi(\zeta)+(-1)^{n\,C}D^{\gamma}_{\zeta-}\psi(\theta)\right]-\psi^{n}\left(\frac{\theta+\zeta}{2}\right)$$
$$=\frac{(\zeta-\theta)^2}{2}\int_0^1 m(\nu)\psi^{n+2}(\nu\theta+(1-\nu)\zeta)d\nu,$$

holds, where $m(\nu)=\begin{cases}\nu-\frac{1-(1-\nu)^{n-\gamma+1}-\nu^{n-\gamma+1}}{n-\gamma+1}, & \nu\in[0,\frac{1}{2});\\ 1-\nu-\frac{1-(1-\nu)^{n-\gamma+1}-\nu^{n-\gamma+1}}{n-\gamma+1}, & \nu\in[\frac{1}{2},1).\end{cases}$

Theorem 3. *Let $\psi\in L^1[\theta,\zeta]$, $\psi * K_{(1-\beta)(n-\gamma)}\in AC^n[\theta,\zeta]$, $n\in N$ and $D^{\gamma+\beta(n-\gamma)}_{(\theta,\zeta)}\psi:[\theta,\zeta]\to\mathbb{R}$ be a twice differentiable function with $n-1<\gamma<n$ and $0<\beta\le 1$. If $|D^{\gamma+\beta(n-\gamma)+2}_{(\theta,\zeta)}\psi|$ is measurable, decreasing and geometric-arithmetically s-convex on $[\theta,\zeta]$ for some fixed $\gamma\in(0,\infty)$, $s\in(0,1]$, $0\le\theta<\zeta$, then the inequality*

$$\left|\frac{D^{\gamma+\beta(n-\gamma)}_{\theta+}\Phi(\zeta)+D^{\gamma+\beta(n-\gamma)}_{\zeta-}\Phi(\theta)}{2}-\frac{\Gamma(\beta(n-\gamma)+1)}{2(\zeta-\theta)^{\beta(n-\gamma)}}\left[D^{\gamma,\beta}_{\theta+}\Phi(\zeta)+D^{\gamma,\beta}_{\zeta-}\Phi(\theta)\right]\right|$$
$$\le\frac{(\zeta-\theta)^2\left(|D^{\gamma+\beta(n-\gamma)+2}_{(\theta,\zeta)}\psi(\theta)|+|D^{\gamma+\beta(n-\gamma)+2}_{(\theta,\zeta)}\psi(\zeta)|\right)}{2(\beta(n-\gamma)+1)}$$
$$\times\left(\frac{1}{s+1}-\frac{1}{\beta(n-\gamma)+s+2}\right),$$

holds.

Proof. By using Lemmas 1, 4 and Definition 4, we have

$$\left|\frac{D^{\gamma+\beta(n-\gamma)}_{\theta+}\Phi(\zeta)+D^{\gamma+\beta(n-\gamma)}_{\zeta-}\Phi(\theta)}{2}-\frac{\Gamma(\beta(n-\gamma)+1)}{2(\zeta-\theta)^{\beta(n-\gamma)}}\left[D^{\gamma,\beta}_{\theta+}\Phi(\zeta)+D^{\gamma,\beta}_{\zeta-}\Phi(\theta)\right]\right|$$
$$\le\frac{(\zeta-\theta)^2}{2(\beta(n-\gamma)+1)}\int_0^1|1-(1-\nu)^{\beta(n-\gamma)+1}-\nu^{\beta(n-\gamma)+1}|$$
$$\times|D^{\gamma+\beta(n-\gamma)+2}_{(\theta,\zeta)}\psi(\nu\theta+(1-\nu)\zeta)|d\nu$$
$$\le\frac{(\zeta-\theta)^2}{2(\beta(n-\gamma)+1)}\int_0^1|1-(1-\nu)^{\beta(n-\gamma)+1}-\nu^{\beta(n-\gamma)+1}||D^{\gamma+\beta(n-\gamma)+2}_{(\theta,\zeta)}\psi(\theta^{\nu}\zeta^{1-\nu})|d\nu$$

$$\leq \frac{(\zeta-\theta)^2}{2(\beta(n-\gamma)+1)}\int_0^1\left(1-(1-\nu)^{\beta(n-\gamma)+1}-\nu^{\beta(n-\gamma)+1}\right)\left[\nu^s|D_{(\theta,\zeta)}^{\gamma+\beta(n-\gamma)+2}\psi(\theta)|\right.$$
$$\left.+(1-\nu)^s|D_{(\theta,\zeta)}^{\gamma+\beta(n-\gamma)+2}\psi(\zeta)|\right]d\nu$$
$$=\frac{(\zeta-\theta)^2}{2(\beta(n-\gamma)+1)}\left[\int_0^1\left[\nu^s|D_{(\theta,\zeta)}^{\gamma+\beta(n-\gamma)+2}\psi(\theta)|+(1-\nu)^s|D_{(\theta,\zeta)}^{\gamma+\beta(n-\gamma)+2}\psi(\zeta)|\right]d\nu\right.$$
$$-\int_0^1\left[\nu^s(1-\nu)^{\beta(n-\gamma)+1}|D_{(\theta,\zeta)}^{\gamma+\beta(n-\gamma)+2}\psi(\theta)|+(1-\nu)^{\beta(n-\gamma)+s+1}|D_{(\theta,\zeta)}^{\gamma+\beta(n-\gamma)+2}\psi(\zeta)|\right]d\nu$$
$$\left.-\int_0^1\left[\nu^{\beta(n-\gamma)+s+1}|D_{(\theta,\zeta)}^{\gamma+\beta(n-\gamma)+2}\psi(\theta)|+\nu^{\beta(n-\gamma)+1}(1-\nu)^s|D_{(\theta,\zeta)}^{\gamma+\beta(n-\gamma)+2}\psi(\zeta)|\right]d\nu\right].$$

By using the definition of the beta function, we get

$$\left|\frac{D_{\theta^+}^{\gamma+\beta(n-\gamma)}\Phi(\zeta)+D_{\zeta^-}^{\gamma+\beta(n-\gamma)}\Phi(\theta)}{2}-\frac{\Gamma(\beta(n-\gamma)+1)}{2(\zeta-\theta)^{\beta(n-\gamma)}}\left[D_{\theta^+}^{\gamma,\beta}\Phi(\zeta)+D_{\zeta^-}^{\gamma,\beta}\Phi(\theta)\right]\right|$$
$$\leq\frac{(\zeta-\theta)^2}{2(\beta(n-\gamma)+1)}\left[\frac{|D_{(\theta,\zeta)}^{\gamma+\beta(n-\gamma)+2}\psi(\theta)|}{s+1}+\frac{|D_{(\theta,\zeta)}^{\gamma+\beta(n-\gamma)+2}\psi(\zeta)|}{s+1}\right.$$
$$-\frac{|D_{(\theta,\zeta)}^{\gamma+\beta(n-\gamma)+2}\psi(\theta)|}{\beta(n-\gamma)+s+2}-|D_{(\theta,\zeta)}^{\gamma+\beta(n-\gamma)+2}\psi(\zeta)|B(s+1,\beta(n-\gamma)+2)$$
$$\left.-|D_{(\theta,\zeta)}^{\gamma+\beta(n-\gamma)+2}\psi(\theta)|B(s+1,\beta(n-\gamma)+2)-\frac{|D_{(\theta,\zeta)}^{\gamma+\beta(n-\gamma)+2}\psi(\zeta)|}{\beta(n-\gamma)+s+2}\right]$$
$$\leq\frac{(\zeta-\theta)^2\left(|D_{(\theta,\zeta)}^{\gamma+\beta(n-\gamma)+2}\psi(\zeta)|+|D_{(\theta,\zeta)}^{\gamma+\beta(n-\gamma)+2}\psi(\theta)|\right)}{2(\beta(n-\gamma)+1)}$$
$$\times\left(\frac{1}{s+1}-\frac{1}{\beta(n-\gamma)+s+2}\right).$$

Which completes the proof of the result. □

Corollary 6. *If we take $\beta=1$ and ψ is symmetric about $\frac{\theta+\zeta}{2}$ in Theorem 3, then the following result for Caputo fractional derivatives holds.*

$$\left|\frac{\psi^n(\theta)+\psi^n(\zeta)}{2}-\frac{\Gamma(n-\gamma+1)}{2(\zeta-\theta)^{n-\gamma}}\left[{}^CD_{\theta^+}^{\gamma}\psi(\zeta)+{}^CD_{\zeta^-}^{\gamma}\psi(\theta)\right]\right|$$
$$\leq\frac{(\zeta-\theta)^2\left(|\psi^{n+2}(\theta)|+|\psi^{n+2}(\zeta)|\right)}{2(n-\gamma+1)}\left(\frac{1}{s+1}-\frac{1}{n-\gamma+s+2}\right).$$

Theorem 4. *Let $\psi\in L^1[\theta,\zeta]$, $\psi*K_{(1-\beta)(n-\gamma)}\in AC^n[\theta,\zeta]$, $n\in N$. Consider $D_{(\theta,\zeta)}^{\gamma+\beta(n-\gamma)}\psi:[\theta,\zeta]\to\mathbb{R}$ to be a twice differentiable function with $n-1<\gamma<n$ and $0<\beta\leq1$. If $|D_{(\theta,\zeta)}^{\gamma+\beta(n-\gamma)+2}\psi|^q$ is measurable, decreasing and geometric arithmetically s-convex on $[\theta,\zeta]$ for some fixed $\gamma\in(0,\infty)$, $s\in(0,1]$, $0\leq\theta<\zeta$, then the inequality*

$$\left|\frac{D_{\theta^+}^{\gamma+\beta(n-\gamma)}\Phi(\zeta)+D_{\zeta^-}^{\gamma+\beta(n-\gamma)}\Phi(\theta)}{2}-\frac{\Gamma(\beta(n-\gamma)+1)}{2(\zeta-\theta)^{\beta(n-\gamma)}}[D_{\theta^+}^{\gamma,\beta}\Phi(\zeta)+D_{\zeta^-}^{\gamma,\beta}\Phi(\theta)]\right|$$
$$\leq\frac{(\zeta-\theta)^2\max\left(1-2^{1-\beta(n-\gamma)},2^{1-\beta(n-\gamma)}-1\right)}{2(\beta(n-\gamma)+1)}$$
$$\times\left(\frac{|D_{(\theta,\zeta)}^{\gamma+\beta(n-\gamma)+2}\psi(\zeta)|^q+|D_{(\theta,\zeta)}^{\gamma+\beta(n-\gamma)+2}\psi(\theta)|^q}{s+1}\right)^{\frac{1}{q}},$$

is true.

Proof. We shall prove this theorem in two cases:

Case 1: Let $\gamma\in(0,1)$ and $\beta(n-\gamma)\in[0,1]$, then by using Lemma 4, Holder's inequality, Lemma 1, Definition 4 and Lemma 2, we obtain

$$\left|\frac{D_{\theta^+}^{\gamma+\beta(n-\gamma)}\Phi(\zeta)+D_{\zeta^-}^{\gamma+\beta(n-\gamma)}\Phi(\theta)}{2}-\frac{\Gamma(\beta(n-\gamma)+1)}{2(\zeta-\theta)^{\beta(n-\gamma)}}[D_{\theta^+}^{\gamma,\beta}\Phi(\zeta)+D_{\zeta^-}^{\gamma,\beta}\Phi(\theta)]\right|$$
$$\leq\frac{(\zeta-\theta)^2}{2(\beta(n-\gamma)+1)}\left(\int_0^1|1-(1-\nu)^{\beta(n-\gamma)+1}-\nu^{\beta(n-\gamma)+1}|^p d\nu\right)^{\frac{1}{p}}$$
$$\times\left(\int_0^1|D_{(\theta,\zeta)}^{\gamma+\beta(n-\gamma)+2}\psi(\nu\theta+(1-\nu)\zeta)|^q d\nu\right)^{\frac{1}{q}}$$
$$\leq\frac{(\zeta-\theta)^2}{2(\beta(n-\gamma)+1)}\left(\int_0^1|1-(1-\nu)^{\beta(n-\gamma)+1}-\nu^{\beta(n-\gamma)+1}|^p d\nu\right)^{\frac{1}{p}}$$
$$\times\left(\int_0^1|D_{(\theta,\zeta)}^{\gamma+\beta(n-\gamma)+2}\psi(\theta^\nu\zeta^{1-\nu})|^q d\nu\right)^{\frac{1}{q}}$$
$$\leq\frac{(\zeta-\theta)^2}{2(\beta(n-\gamma)+1)}\left(\int_0^1|1-(1-\nu)^{\beta(n-\gamma)+1}-\nu^{\beta(n-\gamma)+1}|^p d\nu\right)^{\frac{1}{p}}$$
$$\times\left(\int_0^1\left[\nu^s|D_{(\theta,\zeta)}^{\gamma+\beta(n-\gamma)+2}\psi(\theta)|^q+(1-\nu)^s|D_{(\theta,\zeta)}^{\gamma+\beta(n-\gamma)+2}\psi(\zeta)|^q\right]d\nu\right)^{\frac{1}{q}}$$
$$\leq\frac{(\zeta-\theta)^2}{2(\beta(n-\gamma)+1)}\left(\frac{|D_{(\theta,\zeta)}^{\gamma+\beta(n-\gamma)+2}\psi(\theta)|^q+|D_{(\theta,\zeta)}^{\gamma+\beta(n-\gamma)+2}\psi(\zeta)|^q}{s+1}\right)^{\frac{1}{q}}$$
$$\times\left(\int_0^1\left[(1-\nu)^{\beta(n-\gamma)}+\nu^{\beta(n-\gamma)}-1\right]^p d\nu\right)^{\frac{1}{p}}$$
$$\leq\frac{(\zeta-\theta)^2}{2(\beta(n-\gamma)+1)}\left(\frac{|D_{(\theta,\zeta)}^{\gamma+\beta(n-\gamma)+2}\psi(\theta)|^q+|D_{(\theta,\zeta)}^{\gamma+\beta(n-\gamma)+2}\psi(\zeta)|^q}{s+1}\right)^{\frac{1}{q}}$$
$$\times\left(\int_0^1(2^{1-\beta(n-\gamma)}-1)^p d\nu\right)^{\frac{1}{p}}$$
$$=\frac{(\zeta-\theta)^2}{2(\beta(n-\gamma)+1)}\left(\frac{|D_{(\theta,\zeta)}^{\gamma+\beta(n-\gamma)+2}\psi(\theta)|^q+|D_{(\theta,\zeta)}^{\gamma+\beta(n-\gamma)+2}\psi(\zeta)|^q}{s+1}\right)^{\frac{1}{q}}$$
$$\times\left(2^{1-\beta(n-\gamma)}-1\right).\tag{25}$$

Case 2: Let $\gamma \in [1,\infty)$ and $\beta(n-\gamma) \in [1,\infty)$. By using Lemma 4, Holder's inequality, Lemma 1, Definition 4 and Lemma 2, we obtain

$$\begin{aligned}
&\left|\frac{D_{\theta^+}^{\gamma+\beta(n-\gamma)}\Phi(\zeta) + D_{\zeta^-}^{\gamma+\beta(n-\gamma)}\Phi(\theta)}{2} - \frac{\Gamma(\beta(n-\gamma)+1)}{2(\zeta-\theta)^{\beta(n-\gamma)}}\left[D_{\theta^+}^{\gamma,\beta}\Phi(\zeta) + D_{\zeta^-}^{\gamma,\beta}\Phi(\theta)\right]\right| \\
&\leq \frac{(\zeta-\theta)^2}{2(\beta(n-\gamma)+1)}\left(\frac{|D_{(\theta,\zeta)}^{\gamma+\beta(n-\gamma)+2}\psi(\theta)|^q + |D_{(\theta,\zeta)}^{\gamma+\beta(n-\gamma)+2}\psi(\zeta)|^q}{s+1}\right)^{\frac{1}{q}} \\
&\times\left(\int_0^1 (1-2^{1-\beta(n-\gamma)})^p dv\right)^{\frac{1}{p}} \\
&= \frac{(\zeta-\theta)^2}{2(\beta(n-\gamma)+1)}\left(\frac{|D_{(\theta,\zeta)}^{\gamma+\beta(n-\gamma)+2}\psi(\theta)|^q + |D_{(\theta,\zeta)}^{\gamma+\beta(n-\gamma)+2}\psi(\zeta)|^q}{s+1}\right)^{\frac{1}{q}} \\
&\times\left(1-2^{1-\beta(n-\gamma)}\right).
\end{aligned} \tag{26}$$

Now, from (25) and (26), we obtain the required result. □

Corollary 7. *If we take $\beta = 1$ and ψ is symmetric about $\frac{\theta+\zeta}{2}$ in Theorem 4, we get the following inequality for Caputo fractional derivatives:*

$$\begin{aligned}
&\left|\frac{\psi^n(\theta)+\psi^n(\zeta)}{2} - \frac{\Gamma(n-\gamma+1)}{2(\zeta-\theta)^{n-\gamma}}[{}^C D_{\theta+}^{\gamma}\psi(\zeta) + (-1)^n {}^C D_{\zeta-}^{\gamma}\psi(\theta)]\right| \\
&\leq \frac{(\zeta-\theta)^2 \max(1-2^{1-n+\gamma}, 2^{1-n+\gamma}-1)}{2(n-\gamma+1)}\left(\frac{|\psi^{n+2}(\theta)|^q + |\psi^{n+2}(\zeta)|^q}{s+1}\right)^{\frac{1}{q}}.
\end{aligned}$$

Theorem 5. *Let $\psi \in L^1[\theta,\zeta]$, $\psi * K_{(1-\beta)(n-\gamma)} \in AC^n[\theta,\zeta]$, $n \in N$ and $D_{(\theta,\zeta)}^{\gamma+\beta(n-\gamma)}\psi : [0,\zeta] \to \mathbb{R}$ be differentiable function with $n-1<\gamma<n$ and $0<\beta\leq 1$. If $|D_{(\theta,\zeta)}^{\gamma+\beta(n-\gamma)+2}\psi|^q$ is measurable for $1<q<\infty$, decreasing and geometric arithmetically s-convex on $[0,\zeta]$ for some fixed $\gamma \in (0,\infty)$, $s \in (0,1]$, $0 \leq \theta < \zeta$, then the following fractional inequality holds.*

$$\begin{aligned}
&\frac{\Gamma(\beta(n-\gamma)+1)}{2(\zeta-\theta)^{\beta(n-\gamma)}}\left[D_{\theta^+}^{\gamma,\beta}\Phi(\zeta) + D_{\zeta^-}^{\gamma,\beta}\Phi(\theta)\right] - D_{(\theta,\zeta)}^{\gamma+\beta(n-\gamma)}\psi\left(\frac{\theta+\zeta}{2}\right) \\
&\leq \frac{(\zeta-\theta)^2}{2(\beta(n-\gamma)+1)}\left(\frac{|D_{(\theta,\zeta)}^{\gamma+\beta(n-\gamma)+2}\psi(\theta)|^q + |D_{(\theta,\zeta)}^{\gamma+\beta(n-\gamma)+2}\psi(\zeta)|^q}{s+1}\right)^{\frac{1}{q}} \\
&\times\left(\frac{(\beta(n-\gamma)+1)2^{-p-1} + (\beta(n-\gamma)+0.5)^{p+1} - (\beta(n-\gamma))^{p+1}}{p+1}\right)^{\frac{1}{p}},
\end{aligned}$$

where $\frac{1}{p}+\frac{1}{q}=1$.

Proof. By using Lemmas 1 and 5, Holder's inequality and Definition 4, we get

$$
\begin{aligned}
&\left|\frac{\Gamma(\beta(n-\gamma)+1)}{2(\zeta-\theta)^{\beta(n-\gamma)}}\left[D_{\theta^+}^{\gamma,\beta}\Phi(\zeta)+D_{\zeta^-}^{\gamma,\beta}\Phi(\theta)\right]-D_{(\theta,\zeta)}^{\gamma+\beta(n-\gamma)}\psi\left(\frac{\theta+\zeta}{2}\right)\right| \\
&\leq \frac{(\zeta-\theta)^2}{2}\int_0^1 |m(v)||D_{(\theta,\zeta)}^{\gamma+\beta(n-\gamma)+2}\psi(v\theta+(1-v)\zeta)|dv \\
&\leq \frac{(\zeta-\theta)^2}{2}\int_0^1 |m(v)||D_{(\theta,\zeta)}^{\gamma+\beta(n-\gamma)+2}\psi(\theta^v\zeta^{1-v})|dv \\
&\leq \frac{(\zeta-\theta)^2}{2}\left(\int_0^1 |m(v)|^p dv\right)^{\frac{1}{p}}\left(\int_0^1 \left|D_{(\theta,\zeta)}^{\gamma+\beta(n-\gamma)+2}\psi(\theta^v\zeta^{1-v})\right|^q dv\right)^{\frac{1}{q}} \\
&\leq \frac{(\zeta-\theta)^2}{2}\left(\int_0^1 |m(v)|^p dv\right)^{\frac{1}{p}}\left(\int_0^1 \left[v^s|D_{(\theta,\zeta)}^{\gamma+\beta(n-\gamma)+2}\psi(\theta)|^q\right.\right. \\
&\left.\left.+(1-v)^s|D_{(\theta,\zeta)}^{\gamma+\beta(n-\gamma)+2}\psi(\zeta)|^q\right]dv\right)^{\frac{1}{q}} \\
&= \frac{(\zeta-\theta)^2}{2}\left(\frac{|D_{(\theta,\zeta)}^{\gamma+\beta(n-\gamma)+2}\psi(\theta)|^q+|D_{(\theta,\zeta)}^{\gamma+\beta(n-\gamma)+2}\psi(\zeta)|^q}{s+1}\right)^{\frac{1}{q}} \\
&\times\left[\int_0^{\frac{1}{2}}\left|v-\frac{1-(1-v)^{\beta(n-\gamma)+1}-v^{\beta(n-\gamma)+1}}{\beta(n-\gamma)+1}\right|^p dv\right. \\
&\left.+\int_{\frac{1}{2}}^1\left|(1-v)-\frac{1-(1-v)^{\beta(n-\gamma)+1}-v^{\beta(n-\gamma)+1}}{\beta(n-\gamma)+1}\right|^p dv\right]^{\frac{1}{p}} \\
&= \frac{(\zeta-\theta)^2}{2(\beta(n-\gamma)+1)}\left(\frac{|D_{(\theta,\zeta)}^{\gamma+\beta(n-\gamma)+2}\psi(\theta)|^q+|D_{(\theta,\zeta)}^{\gamma+\beta(n-\gamma)+2}\psi(\zeta)|^q}{s+1}\right)^{\frac{1}{q}} \\
&\times\left[\int_0^{\frac{1}{2}}\left|\beta(n-\gamma)v-1+(1-v)^{\beta(n-\gamma)+1}+v^{\beta(n-\gamma)+1}\right|^p dv\right. \\
&\left.+\int_{\frac{1}{2}}^1\left|\beta(n-\gamma)+1-\beta(n-\gamma)v-v-1+(1-v)^{\beta(n-\gamma)+1}+v^{\beta(n-\gamma)+1}\right|^p dv\right]^{\frac{1}{p}} \\
&\leq \frac{(\zeta-\theta)^2}{2(\beta(n-\gamma)+1)}\left(\frac{|D_{(\theta,\zeta)}^{\gamma+\beta(n-\gamma)+2}\psi(\theta)|^q+|D_{(\theta,\zeta)}^{\gamma+\beta(n-\gamma)+2}\psi(\zeta)|^q}{s+1}\right)^{\frac{1}{q}} \\
&\times\left(\int_0^{\frac{1}{2}}((\beta(n-\gamma)+1)v)^p dv+\int_{\frac{1}{2}}^1(\beta(n-\gamma)-v+1)^p dv\right)^{\frac{1}{p}} \\
&\leq \frac{(\zeta-\theta)^2}{2(\beta(n-\gamma)+1)}\left(\frac{|D_{(\theta,\zeta)}^{\gamma+\beta(n-\gamma)+2}\psi(\theta)|^q+|D_{(\theta,\zeta)}^{\gamma+\beta(n-\gamma)+2}\psi(\zeta)|^q}{s+1}\right)^{\frac{1}{q}} \\
&\times\left((\beta(n-\gamma)+1)\int_0^{\frac{1}{2}}v^p dv+\int_{\frac{1}{2}}^1(\beta(n-\gamma)-v+1)^p dv\right)^{\frac{1}{p}} \\
&= \frac{(\zeta-\theta)^2}{2(\beta(n-\gamma)+1)}\left(\frac{|D_{(\theta,\zeta)}^{\gamma+\beta(n-\gamma)+2}\psi(\theta)|^q+|D_{(\theta,\zeta)}^{\gamma+\beta(n-\gamma)+2}\psi(\zeta)|^q}{s+1}\right)^{\frac{1}{q}} \\
&\times\left(\frac{(\beta(n-\gamma)+1)2^{-p-1}+(\beta(n-\gamma)+0.5)^{p+1}-(\beta(n-\gamma))^{p+1}}{p+1}\right)^{\frac{1}{p}}.
\end{aligned}
$$

Which completes the proof of the result. □

Corollary 8. *If we take $\beta = 1$ and ψ is symmetric about $\frac{\theta+\zeta}{2}$ in Theorem 5, then the inequality*

$$\frac{\Gamma(n-\gamma+1)}{2(\zeta-\theta)^{n-\gamma}}\left[{}^C D_{\theta^+}^{\gamma}\psi(\zeta) + (-1)^{nC} D_{\zeta^-}^{\gamma}\psi(\theta)\right] - \psi^n\left(\frac{\theta+\zeta}{2}\right)$$

$$\leq \frac{(\zeta-\theta)^2}{2(n-\gamma+1)}\left(\frac{|\psi^{n+2}(\theta)|^q + |\psi^{n+2}(\zeta)|^q}{s+1}\right)^{\frac{1}{q}}$$

$$\times \left(\frac{(n-\gamma+1)2^{-p-1} + (n-\gamma+0.5)^{p+1} - (n-\gamma)^{p+1}}{p+1}\right)^{\frac{1}{p}},$$

where $\frac{1}{p} + \frac{1}{q} = 1$, holds for Caputo fractional derivatives.

4. Concluding Remarks

The Hilfer fractional derivative has been used to set up a class of Hermite-Hadamard-type inequalities by involving convexity theory. Our results present many of the earlier inequalities that exist in the literature. The methodology used to generate the new inequalities is based on Hilfer's fractional derivative and skillful use of Hölder's inequality that has a wide range of applications in optimization theory. The findings of this work may stimulate the interest of researchers working in this field can pursue further investigation.

Author Contributions: Conceptualization, M.S., G.R., Z.P.; methodology, K.S.N.; software, Z.P., M.A.K., K.S.N.; validation, M.A.K., G.R.; formal analysis, K.S.N.; investigation, M.S., Z.P., G.R., M.A.K.; writing—original draft preparation, M.S., Z.P., G.R., M.A.K., K.S.N.; writing—review and editing, M.S., G.R., K.S.N.; funding acquisition, K.S.N. All authors have read and agreed to the published version of the manuscript.

Funding: This research received no external funding.

Institutional Review Board Statement: Not applicable.

Informed Consent Statement: Not applicable.

Data Availability Statement: Not applicable.

Conflicts of Interest: The authors declare that there is no conflict of interests regarding the publication of this paper.

References

1. Guo, B.; Pu, X.; Huange, F. *Fractional Partial Differential Equations and Thier Numerical Solution*; World Scientific: Hong Kong, China, 2015.
2. Kilbas, A.A.; Srivastava, H.M.; Trujillo, J.J. *Theory and Applications of Fractional Differential Equations*; North-Holland Mathematics Studies; Elsevier Science: Amsterdam, The Netherlands, 2006.
3. Dokuyucua, M.A. Caputo and Atangana-Baleanu-Caputo Fractional Derivative Applied to Garden Equation. *Turk. J. Sci.* **2020**, *5*, 1–7.
4. Samraiz, M.; Perveen, Z.; Abdeljawad, T.; Iqbal, S.; Naheed, S. On certain fractional calculus operators and applications in mathematical physics. *Phys. Scr.* **2020**, *95*, 115210.
5. Samraiz, M.; Perveen, Z.; Rahman, G.; Nisar, K.S.; Kumar, D. On the (k, s)-Hilfer-Prabhakar Fractional Derivative with Applications to Mathematical Physics. *Front. Phys.* **2020**, *8*, 309. [CrossRef]
6. Korus, P. Some Hermite-Hadamard type inequalities for functions of generalized convex derivative. *Acta Math. Hung.* **2021**, *165*, 463–473. [CrossRef]
7. Farid, G.; Yousaf, M.; Nonlaopon, K. Fejér-Hadamard type inequalities for $(a, h - m) - p$-convex functions via extended generalized fractional integrals. *Fractal Fract.* **2021**, *5*, 253. [CrossRef]
8. Ekinci, A.; Özdemir, M.E. Some new integral inequalities via Riemann-Liouville integral operators. *Appl. Comput. Math.* **2019**, *18*, 288–295. [CrossRef]
9. Butt, S.I.; Nadeem, M.; Farid, G. On Caputo fractional derivatives via exponential s-convex functions. *Turk. J. Sci.* **2020**, *5*, 140–146.
10. Zhou, S.-S.; Rashid, S.; Parveen, S.; Akdemir, A.O.; Hammouch, Z. New computations for extended weighted functionals within the Hilfer generalized proportional fractional integral operators. *AIMS Math.* **2021**, *6*, 507–4525.
11. Alomari, M.; Darus, M.; Dragomir, S.S.; Seron, P. Ostrowski's inequalities for functions whose derivatives are s-convex in the second sense. *Appl. Math. Lett.* **2010**, *23*, 1071–1076. [CrossRef]

12. Dragomir, S.S.; Ionescu, N.M. Some remarks on convex functions. *Rev. d'analyse Numer. Theor. Approx. Tome* **1992**, *21*, 31–36. [CrossRef]
13. Dragomir, S.S.; Wang, S. A new inequality of Ostrowski's type in L_p norm and applications to some special means and to some numerical quadrature rules. *Indian J. Math.* **2008**, *40*, 299–304.
14. McShane, E.J. Jensen's inequality. *Bull. Am. Math. Soc.* **1937**, *43*, 521–527.
15. Samraiz, M.; Nawaz, F.; Iqbal, S.; Abdeljawad, T.; Rahman, G.; Nisar, K.S. Certain mean-type fractional integral inequalities via different convexities with applications. *J. Inequal. Appl.* **2020**, *2020*, 208. [CrossRef]
16. Sarikaya, M.Z.; Set, E.; Yaldiz, H.; Basak, B. Hermite-Hadamard's inequalities for fractional integrals and related fractional inequalities. *Math. Comp. Model.* **2013**, *57*, 2403–2407. [CrossRef]
17. Hadamard, J. Etude sur les proprietes des fonctions entieres et en particulier d'une fonction consideree par Riemann. *J. Math. Pures Appl.* **1893**, *58*, 171–216. [CrossRef]
18. Lyu, S.L. On the Hermite-Hadamard inequality for convex functions of two variables. *Numer. Algebra Control Optim.* **2014**, *4*, 1–8.
19. Niculescu, C.P.; Persson, L.E. Old and new on the Hermite Hadamard inequality. *Real Anal. Exch.* **2003**, *29*, 663–683. [CrossRef]
20. Wang, J.; Li, X.; Feckan, M.; Zhou, Y. Hermite-Hadamard-type inequalities for Riemann-Liouville fractional integrals via two kinds of convexity. *Appl. Anal.* **2013**, *92*, 2241–2253. [CrossRef]
21. Mihai, M.V.; Awan, M.U.; Noor, M.A.; Kim, J.K. Hermite-Hadamard inequalities and their applications. *J. Inequal. Appl.* **2018**, *2018*, 309. [CrossRef]
22. Okur, N.; Yalcin, F.B.; Karahan, V. Some Hermite-Hadamard Type Integral Inequalities for Multidimentional Preinvex Functions. *Turk. J. Ineq.* **2019**, *3*, 54–63. [CrossRef] [PubMed]
23. Anastassiou, G.A. *Fractional Differentiation Inequalities*; Springer: New York, NY, USA, 2009.
24. Hilfer, R.; Luchko, Y.; Tomovski, Z. Operational method for solution of fractional differential equations with generalized Riemann-Liouville fractional derivative. *Fract. Calc. Appl. Anal.* **2009**, *12*, 299–318.
25. Liao, Y.; Deng, J.; Wang, J. Riemann-Liouville fractional Hermite-Hadamard inequalities, part I, for twice differentiable geometric-arithmetically s-convex functions. *J. Inequal. Appl.* **2013**, *2013*, 517.
26. Deng, J.; Wang, J. Fractional Hermite-Hadamard's inequalities for (α, m)-logrithmically convex functions. *J. Inequal. Appl.* **2013**, *2013*, 364. [CrossRef]
27. Farid, G.; Naqvi, S.; Javed, A. Hadamard and Fejer-Hadamard inequalities and related results via Caputo fractional derivative. *Bull. Math. Anal. Appl.* **2017**, *9*, 16–30. [CrossRef]

fractal and fractional

Article

Numerical Study of Caputo Fractional-Order Differential Equations by Developing New Operational Matrices of Vieta–Lucas Polynomials

Zulfiqar Ahmad Noor [1,2], Imran Talib [1], Thabet Abdeljawad [3,4,*] and Manar A. Alqudah [5]

1 Nonlinear Analysis Group (NAG), Mathematics Department, Virtual University of Pakistan, Lahore 54000, Pakistan; zulfiqar@vu.edu.pk (Z.A.N.); imrantalib@vu.edu.pk (I.T.)
2 Department of Mathematics, University of Management and Technology, Lahore 54770, Pakistan
3 Department of Mathematics and Sciences, Prince Sultan University, Riyadh 11586, Saudi Arabia
4 Department of Medical Research, China Medical University, Taichung 40402, Taiwan
5 Department of Mathematical Sciences, Faculty of Sciences, Princess Nourah Bint Abdulrahman University, Riyadh 11671, Saudi Arabia; maalqudah@pnu.edu.sa
* Correspondence: tabdeljawad@psu.edu.sa

Abstract: In this article, we develop a numerical method based on the operational matrices of shifted Vieta–Lucas polynomials (VLPs) for solving Caputo fractional-order differential equations (FDEs). We derive a new operational matrix of the fractional-order derivatives in the Caputo sense, which is then used with spectral tau and spectral collocation methods to reduce the FDEs to a system of algebraic equations. Several numerical examples are given to show the accuracy of this method. These examples show that the obtained results have good agreement with the analytical solutions in both linear and non-linear FDEs. In addition to this, the numerical results obtained by using our method are compared with the numerical results obtained otherwise in the literature.

Keywords: fractional-order differential equations; operational matrices; shifted Vieta–Lucas polynomials; Caputo derivative

Citation: Noor, Z.A.; Talib, I.; Abdeljawad, T.; Alqudah, M.A. Numerical Study of Caputo Fractional-Order Differential Equations by Developing New Operational Matrices of Vieta–Lucas Polynomials. *Fractal Fract.* **2022**, *6*, 79. https://doi.org/10.3390/fractalfract6020079

Academic Editor: Paul Eloe

Received: 22 November 2021
Accepted: 25 January 2022
Published: 31 January 2022

1. Introduction

Fractional calculus has been playing a very important role in scientific computations. Scientists are able to describe and model many physical phenomena with fractional-order differential equations. As a result, fractional-order differential operators are widely used to solve systems by developing more accurate models [1–4]. The nonlocal property of the fractional-order operators makes them more efficient for modeling the various problems of physics, fluid dynamics and their related disciplines [1,5–9]. For example, consider a thin rigid plate of mass a_1 and area R immersed in a Newtonian fluid of infinite extent and connected by a massless spring of stiffness K to a fixed point. A force $g(z)$ is applied to the plate. Assume that the spring has no effect on the fluid and that the area of the plate is large enough to produce the fluid adjacent to the plate, whereas stresses $\sigma(z,x)$ can be defined by the following relation:

$$\sigma(z,x) = \sqrt{\mu\rho}D^{0.5}v(z,x); \tag{1}$$

where x is the distance of a point in the fluid from the spring to the submerged plate. By some assumptions discussed in [10], the dynamics of the system are given by

$$a_1D^2v(z) = g(z) - Kv(z) - 2R\sigma(z,0) \tag{2}$$

where $\sigma(z,0) = Dv(z)$. Equation (2), with some assumptions considered in [10], takes the following form of a Bagley–Torvik-type problem solved in (Section 6, Example 1):

$$a_1\mathcal{D}^2 v(z) + a_2\mathcal{D}^{3/2}v(z) + a_3 v(z) = g(z),\ z \in [0,1]. \tag{3}$$

The existence and uniqueness results of fractional-order differential equations (FDEs) have been investigated extensively in the literature. Some of them are presented as follows: Fazli and Nieto [11] investigated the existence and uniqueness of the solution of FDEs of Bagley–Torvik type by considering the existence of coupled lower and upper solutions. Pang et al. [12] investigated the existence and uniqueness of the solution of the generalized FDEs with initial conditions by proposing a novel max-metric containing a Caputo derivative. Abbas [13] studied the existence and uniqueness of the solution of FDEs by using Banach's contraction principle together with Krasnoselskii fixed point theorem. For more works on existence and uniqueness results, we refer the reader to [14–17].

As most FDEs do not have closed-form solutions, different numerical techniques, including the finite difference method, variational iteration method and spectral methods, are preferably used. Among them, spectral methods have received considerable attention from the fractional community for solving FDEs, both ordinary and partial. Spectral methods are classified into three types, known as the collocation, tau and Galerkin methods. The basic idea of the spectral methods is to write the solution as a linear combination of basis vectors of global polynomials, typically Legendre, Jacobi and Chebyshev. The speed of convergence is considered the best advantage of the spectral methods, as the rate of convergence is exponential in these methods, which gives a high level of accuracy. Many efficient spectral techniques are obtained in the literature using the various global polynomials [18–21].

The construction of the operational matrices of fractional derivative operators defined with singular or nonsingular kernels has played a key role in the development of spectral methods. Many researchers have worked on the construction of the operational matrices of fractional derivatives using different types of global polynomials. For example, Benattia et al. [22] introduced the operational matrix of the fractional derivatives to develop a numerical method that is based on the Chebyshev wavelet for solving FDEs. Saadatmandi et al. [23] derived an operational matrix of derivatives of fractional order using the fractional-order Chebyshev functions. They also extended the results of [23] for solving the coupled system of FDEs with variable coefficients [24]. Additionally, Bharway et al. [25] introduced a new shifted Chebyshev operational matrix of fractional integration for solving linear FDEs. Moreover, Talib et al. [26] developed a new operational matrix based on the orthogonal shifted Legendre polynomials to numerically solve the fractional partial differential equations. Meanwhile, Rahimkhani et al. [27] introduced a Bernoulli wavelet operational matrix of fractional integration for obtaining the approximate solution of a fractional delay differential equation. Kazem et al. [18] derived an operational matrix that generalized the results presented in [19]. Recently, Dehastani et al. [28] calculated modified operational matrices of integration and pseudo-operational of fractional derivatives for the Lucas wavelet functions to compute the numerical solution of fractional Fredholm–Volterra integro-differential equations. Moreover, Dehastani et al. [29,30] also derived operational matrices of fractional-order derivatives and integration for fractional-order Bessel functions and fractional-order hybrid Bessel functions. In the derived numerical techniques, the operational matrices are applied to reduce the FDEs to a system of algebraic equations.

Dehestani et al. [31] also presented a novel collocation method based on the Genocchi wavelet for the numerical solution of FDEs and time-fractional partial differential equations with delay.

Motivated by the aforementioned works, we extend the study of the spectral methods by constructing a numerical algorithm that is based on the fractional-order derivative operational matrix of VLPs in Caputo sense, together with the spectral tau method and spectral collocation method. The basis vectors of VLPs are used to approximate the solution of the problems. The derivative terms are approximated by using the fractional-order derivative operational matrix of VLPs. It is important to mention that the proposed algorithm is computer-oriented and is capable of reducing the FDEs to a system of algebraic equations, which greatly simplifies the problems. Subsequently, we use the operational matrices approach together with the spectral tau method in the case of linear FDEs, and the operational matrices approach together with the spectral collocation method in the case of nonlinear FDEs. Our proposed numerical algorithm produces highly efficient numerical results as obtained otherwise in the literature [32–34].

The novel aspects of our proposed study are the development of the new fractional-order derivative operational matrix of VLPs in Caputo sense and the construction of the numerical algorithm that is based on this newly developed operational matrix. To the best of our knowledge, this is the first result where the numerical algorithm is presented by using the operational matrix of VLPs. Moreover, the proposed numerical algorithm is fit to solve both linear and nonlinear FDEs with initial conditions. In addition, our proposed method has advantages over other methods, such as the Homotopy perturbation method, because, in our case, the perturbation, linearization or discretization are not necessary to be implemented.

The structure of this paper is set in the following way. In Section 2, we discuss the VLPs along with their properties. In Section 3 , the Vieta–Lucas operational matrix of fractional-order derivatives is derived. In Section 4, the numerical method is developed by using the operational matrices of VLPs. In Section 5, the error bound is determined. In Section 6, the accuracy and the stability of the proposed method are analyzed by taking some numerical examples. In Section 6, we conclude and give the summary of this paper.

2. Preliminaries

In this section, we summarize some definitions, properties and results of fractional calculus that are essential to construct the numerical algorithm to solve the linear and nonlinear FDEs.

Definition 1. *The Riemann–Liouville fractional integral operator of order $\alpha > 0$, of a function v, is defined as:*

$$J^{\alpha}v(z) = \frac{1}{\Gamma(\alpha)}\int_0^z (z-s)^{\alpha-1}v(s)ds, \quad \alpha > 0,$$
$$J^{0}v(z) = v(z).$$

Definition 2. *The Caputo operator of the fractional derivative is defined as follows:*

$$D^{\alpha}v(z) = \frac{1}{\Gamma(n-\alpha)}\int_0^z \frac{v^{(n)}(s)}{(z-s)^{\alpha+1-n}}ds, \quad \alpha > 0, \ z > 0, \tag{4}$$

where $n-1 < \alpha \leq n$, $n \in \mathbb{N}$, and $v \in C^n[0,1]$.

Hence, the Caputo operator follows:

$$\mathcal{D}^{\alpha}z^k = \begin{cases} 0, & k \in 0,1,2,\ldots,\lceil\alpha\rceil-1, \\ \frac{\Gamma(1+k)}{\Gamma(1+k-\alpha)}z^{k-\alpha}, & k \in \mathbb{N} \wedge k \geq \lceil\alpha\rceil. \end{cases} \tag{5}$$

2.1. Vieta–Lucas Polynomials

Vieta–Lucas polynomials belong to the class of orthogonal polynomials and can be created by using the recurrence relation [35]. Consider $|z| \leq 2$; then, the Vieta–Lucas polynomials of degree $n \in \mathbb{N}_0$ in the variable z can be defined as

$$\mathrm{VL}_n(z) = 2\cos(n\theta), \quad \theta = \cos^{-1}\left(\frac{z}{2}\right), \ \theta \in [0,\pi]. \tag{6}$$

The VLPs can be created by using the following recurrence relation:

$$\begin{gathered}\mathrm{VL}_n(z) = z\,\mathrm{VL}_{n-1}(z) - \mathrm{VL}_{n-2}(z), \quad n = 2,3,\ldots, \\ \mathrm{VL}_0(z) = 2, \quad \mathrm{VL}_1(z) = z.\end{gathered}$$

Moreover, $\mathrm{VL}_n(z)$ can be expressed using the following power series formula:

$$\mathrm{VL}_n(z) = \sum_{j=0}^{\lceil \frac{n}{2} \rceil} (-1)^j \frac{n\Gamma(n-j)}{\Gamma(j+1)\Gamma(n+1-2j)} z^{n-2j}, \quad n = \{2,3,\ldots\}, \tag{7}$$

where $\lceil \frac{n}{2} \rceil$ is the ceiling function.

Moreover, the orthogonality of $\mathrm{VL}_n(z)$ can be expressed as:

$$\langle \mathrm{VL}_m(z), \mathrm{VL}_n(z) \rangle = \int_{-2}^{2} \frac{1}{\sqrt{4-z^2}} \mathrm{VL}_m(z)\,\mathrm{VL}_n(z)\,dz = \begin{cases} 0, & m \neq n \neq 0, \\ 4\pi, & m = n = 0, \\ 2\pi, & m = n \neq 0, \end{cases} \tag{8}$$

where $\frac{1}{\sqrt{4-z^2}}$ is the weight function.

2.2. Shifted VLPs

As a new class of orthogonal polynomials, the shifted VLPs, $\mathrm{VL}_n^*(z)$ of degree n, defined on the closed interval $[0,1]$, can be obtained as follows:

$$\mathrm{VL}_n^*(z) = \mathrm{VL}_n(4z-2) = \mathrm{VL}_{2n}(2\sqrt{z}). \tag{9}$$

Moreover, $\mathrm{VL}_n^*(z)$ are created by the following formula:

$$\mathrm{VL}_{n+1}^*(z) = (4z-2)\,\mathrm{VL}_n^*(z) - \mathrm{VL}_{n-1}^*(z), \quad n = 1,2,\ldots, \tag{10}$$

with the starting values

$$\mathrm{VL}_0^*(z) = 2, \qquad \mathrm{VL}_1^*(z) = 4z-2. \tag{11}$$

Moreover, analytically, $\mathrm{VL}_n^*(z)$ can be expressed as:

$$\mathrm{VL}_n^*(z) = 2n \sum_{j=0}^{n} (-1)^j \frac{4^{n-j}\Gamma(2n-j)}{\Gamma(j+1)\Gamma(2n-2j+1)} z^{n-j}, \quad n = \{2,3,\ldots\}. \tag{12}$$

Let the function $u(z)$ be Lebesgue-square-integrable on the interval $[0,1]$, which can be expressed in terms of $\mathrm{VL}_n(z)$ as follows:

$$u(z) = \sum_{j=0}^{\infty} c_j\,\mathrm{VL}_j^*(z), \tag{13}$$

where the undetermined coefficients, $c_j, j = 0,1,2,\ldots,n,$ can be determined through the following expression:

$$c_j = \frac{1}{\delta_j \pi} \int_{-2}^{2} \frac{u(\frac{z+2}{4})\,\mathrm{VL}_j(z)}{\sqrt{4-z^2}} dz, \tag{14}$$

or

$$c_j = \frac{1}{\delta_j \pi} \int_{0}^{1} \frac{u(z)\,\mathrm{VL}_j^*(z)}{\sqrt{z-z^2}} dz, \tag{15}$$

where

$$\delta_j = \begin{cases} 4, & j = 0, \\ 2, & j = \{1, 2, \ldots, n\}. \end{cases} \tag{16}$$

For approximation, we can take the first $n+1$ terms of the series; therefore, $u(z)$ can be expanded in the form

$$u_n(z) \simeq \sum_{j=0}^{n} c_j \, \mathrm{VL}_j^*(z) = C^T \Psi(z), \tag{17}$$

where the shifted VLP coefficient vector C and the shifted VLP vector $\Psi(z)$ are given by

$$C^T = [c_0, c_1, c_2, \ldots, c_n],$$

$$\Psi(z) = [\mathrm{VL}_0^*(z), \mathrm{VL}_1^*(z), \ldots, \mathrm{VL}_n^*(z)]^T. \tag{18}$$

3. Operational Matrices of Differentiation

Theorem 1. *Let $\Psi(z)$ be the shifted VLP vector defined in (18) and also suppose that $\alpha > 0$; then,*

$$D^\alpha(\Psi(z)) \simeq P^\alpha \Psi(z),$$

where P^α is the $(m+1) \times (m+1)$ operational matrix of the fractional derivative of order α in the Caputo sense and is defined as follows:

$$P^\alpha = \begin{pmatrix} 0 & 0 & \cdots & 0 \\ \vdots & \vdots & \cdots & \vdots \\ 0 & 0 & \cdots & 0 \\ \sum_{k=0}^{i-\lceil\alpha\rceil} \xi_{i,0,k} & \sum_{k=0}^{i-\lceil\alpha\rceil} \xi_{i,1,k} & \cdots & \sum_{k=0}^{i-\lceil\alpha\rceil} \xi_{i,m,k} \\ \vdots & \vdots & \cdots & \vdots \\ \sum_{k=0}^{m-\lceil\alpha\rceil} \xi_{m,0,k} & \sum_{k=0}^{m-\lceil\alpha\rceil} \xi_{m,1,k} & \cdots & \sum_{k=0}^{m-\lceil\alpha\rceil} \xi_{m,m,k} \end{pmatrix}$$ *and $\xi_{i,j,k}$ is given by*

$$\xi_{i,j,k} = \begin{cases} i \sum_{k=0}^{i-\lceil\alpha\rceil} (-1)^k \frac{4^{i-k}\Gamma(2i-k)\Gamma(i-k+1)\Gamma(i-k-\alpha+1/2)}{\sqrt{\pi}\Gamma(k+1)\Gamma(2i-2k+1)\Gamma(i-k-\alpha+1)^2}, & j = 0, \\ 2i \sum_{k=0}^{i-\lceil\alpha\rceil} \sum_{r=0}^{j} \frac{(-1)^{k+r}}{\sqrt{\pi}} \frac{4^{i-k}\Gamma(2i-k)\Gamma(i-k+1)}{\Gamma(k+1)\Gamma(2i-2k+1)\Gamma(i-k-\alpha+1)} \\ \qquad \times \frac{4^{j-r}\Gamma(2j-r)\Gamma(i+j-k-r-\alpha+1/2)}{\Gamma(r+1)\Gamma(2j-2r+1)\Gamma(i+j+k-r-\alpha+1)}, & j = 1, 2, 3, \ldots. \end{cases} \tag{19}$$

Proof. Applying the Caputo derivative to (12), we have

$$D^\alpha(\mathrm{VL}_i^*(z)) = D^\alpha\left(\sum_{k=0}^{i} (-1)^k \frac{4^{i-k} 2i \Gamma(2i-k)}{\Gamma(k+1)\Gamma(2i-2k+1)} z^{i-k}\right). \tag{20}$$

Applying the linearity of the Caputo derivative, and using (5), we have

$$D^\alpha(\mathrm{VL}_i^*(z)) = \sum_{k=0}^{i-\lceil\alpha\rceil} (-1)^k \frac{4^{i-k} 2i \Gamma(2i-k)\Gamma(i-k+1)}{\Gamma(k+1)\Gamma(2i-2k+1)\Gamma(i-k+1-\alpha)} z^{i-k-\alpha}, \; i = \lceil\alpha\rceil, \ldots, n, \tag{21}$$

and

$$D^\alpha(\mathrm{VL}_i^*(z)) = 0, \quad i = 0, 1, \ldots, \lceil\alpha\rceil - 1. \tag{22}$$

Now, approximate $z^{i-k-\alpha}$ by $(m+1)$ terms of the series, as

$$z^{i-k-\alpha} \simeq \sum_{j=0}^{m} c_{kj}\, \mathrm{VL}_j^*(z), \tag{23}$$

where

$$c_{kj} = \frac{1}{\delta_j \pi} \int_0^1 \frac{u(z)\, \mathrm{VL}_j^*(z)}{\sqrt{z-z^2}}\, dz, \tag{24}$$

and $u(z) = z^{i-k-\alpha}$.

Now, inserting the value of $u(z) and \mathrm{VL}_j^*(z)$ into Equation (24), we obtain

$$c_{kj} = \begin{cases} \frac{1}{4\pi} \int_0^1 \frac{2z^{i-k-\alpha}}{\sqrt{z-z^2}} dz, & j=0, \\ \frac{1}{2\pi} \int_0^1 \frac{z^{i-k-\alpha}}{\sqrt{z-z^2}} \sum_{r=0}^{j} \frac{(-1)^r 2j 4^{j-r} \Gamma(2j-r)}{\Gamma(r+1)\Gamma(2j-2r+1)} z^{j-r} dx, & j=1,2,3,\ldots, \end{cases}$$

$$= \begin{cases} \frac{1}{2\pi} \int_0^1 \frac{z^{i-k-\alpha}}{\sqrt{z-z^2}} dz, & j=0, \\ \frac{1}{\pi} \int_0^1 \sum_{r=0}^{j} \frac{(-1)^r j 4^{j-r} \Gamma(2j-r)}{\Gamma(r+1)\Gamma(2j-2r+1)} \frac{z^{i+j-k-r-\alpha}}{\sqrt{z-z^2}} dz, & j=1,2,3,\ldots, \end{cases}$$

$$= \begin{cases} \frac{1}{2\sqrt{\pi}} \frac{\Gamma(i-k-\alpha+1/2)}{\Gamma(i-k-\alpha+1)}, & j=0, \\ \frac{1}{\sqrt{\pi}} \sum_{r=0}^{j} \frac{(-1)^r j 4^{j-r} \Gamma(2j-r)\Gamma(i+j-k-r-\alpha+1/2)}{\Gamma(r+1)\Gamma(2j-2r+1)\Gamma(i+j-k-r-\alpha+1)}, & j=1,2,3,\ldots, \end{cases} \tag{25}$$

by inserting the value of $z^{i-k-\alpha}$ into Equation (21), we obtain

$$D^{\alpha}(\mathrm{VL}_i^*(z)) \simeq \sum_{j=0}^{m} S_v(i,j)\, \mathrm{VL}_j^*(z)), \tag{26}$$

where $S_v(i,j) = \sum_{k=0}^{i-\lceil \alpha \rceil} \xi_{i,j,k}$, and

$$\xi_{i,j,k} = \begin{cases} (-1)^k \frac{4^{i-k} i \Gamma(2i-k)\Gamma(i-k+1)\Gamma(i-k-\alpha+1/2)}{\sqrt{\pi}\Gamma(k+1)\Gamma(2i-2k+1)\Gamma(i-k-\alpha+1)^2}, & j=0, \\ \sum_{r=0}^{j} \frac{(-1)^{k+r}}{\sqrt{\pi}} \frac{4^{i-k} 2i \Gamma(2i-k)\Gamma(i-k+1)}{\Gamma(k+1)\Gamma(2i-2k+1)\Gamma(i-k-\alpha+1)} \\ \qquad \times \frac{4^{j-r} j \Gamma(2j-r)\Gamma(i+j-k-r-\alpha+1/2)}{\Gamma(r+1)\Gamma(2j-2r+1)\Gamma(i+j+k-r-\alpha+1)}, & j=1,2,3,\ldots. \end{cases} \tag{27}$$

Rewriting Equation (26) in vector form, we obtain

$$D^{\alpha}(\mathrm{VL}_i^*(z)) \simeq \left(\sum_{k=0}^{i-\lceil \alpha \rceil} \xi_{i,0,k}, \sum_{k=0}^{i-\lceil \alpha \rceil} \xi_{i,1,k}, \ldots, \sum_{k=0}^{i-\lceil \alpha \rceil} \xi_{i,m,k} \right) \Psi(z). \tag{28}$$

For simplicity, we can write Equation (28) as:

$$D^{\alpha}(\mathrm{VL}_i^*(z)) = P^{(\alpha)} \Psi(z), \tag{29}$$

where

$$P^{(\alpha)} = \left(\sum_{k=0}^{i-\lceil \alpha \rceil} \xi_{i,0,k}, \sum_{k=0}^{i-\lceil \alpha \rceil} \xi_{i,1,k}, \ldots, \sum_{k=0}^{i-\lceil \alpha \rceil} \xi_{i,m,k} \right). \tag{30}$$

Equations (22) and (29) prove the required result. □

4. Application of Operational Matrices Method

In this section, we apply the Vieta–Lucas operational matrix method to find the analytical-approximate solution of linear and nonlinear FDEs.

4.1. Linear FDEs

Consider the linear FDE

$$D^{\alpha}v(z) = b_1 D^{\vartheta_1}v(z) + b_2 D^{\vartheta_2}v(z) + \ldots + b_k D^{\vartheta_k}v(z) + b_{k+1}v(z) + b_{k+2}g(z), \qquad (31)$$

with initial conditions

$$v^{(i)}(0) = d_i, \qquad i = 0, \ldots, n, \qquad (32)$$

where b_l, for $l = 1, \ldots, k+2$, are real constant coefficients and also $n < \alpha \leq n+1$, and $0 < \vartheta_1 < \vartheta_2 < \ldots < \vartheta_k < \alpha$.

The unknown function $v(z)$ and the source term $g(z)$ can be approximated as:

$$v(z) \simeq \sum_{i=0}^{M} c_i \, \mathrm{VL}_i^*(z) = C^T \Psi(z), \qquad (33)$$

$$g(z) \simeq \sum_{i=0}^{M} h_i \, \mathrm{VL}_i^*(z) = H^T \Psi(z), \qquad (34)$$

where $H = [h_0, \ldots, h_M]^T$ is known, and $C = [c_0, \ldots, c_M]^T$ is an unknown to be determined.

Now, using Equations (29) and (33), we have

$$D^{\alpha}v(z) \simeq C^T D^{\alpha}\Psi(z) \simeq C^T P^{\alpha}\Psi(z), \qquad (35)$$

$$D^{\vartheta_j}v(z) \simeq C^T D^{\vartheta_j}\Psi(z) \simeq C^T P^{\vartheta_j}\Psi(z), \qquad j = 1, \ldots, k. \qquad (36)$$

Using Equations (33)–(36), the residual $R(x)$ for Equation (31) can be written as

$$R_M(z) \simeq (C^T P^{\alpha} - b_1 C^T P^{\vartheta_1} - \ldots - b_k C^T P^{\vartheta_k} - b_{k+1} C^T - b_{k+2} G^T)\Psi(z). \qquad (37)$$

Using the spectral tau method [36], a system of linear equations is generated by applying

$$\langle R_M(z), \Psi(z) \rangle = \int_0^1 R_M(z)\Psi(z)dz, \qquad j = 0, 1, \ldots, M-n-1. \qquad (38)$$

Moreover, by substituting Equation (33) in the initial conditions given in Equation (32), we obtain

$$v^{(i)}(0) = C^T P^{(i)}\Psi(0) = d_i, \qquad i = 0, 1, \ldots, n. \qquad (39)$$

Equations (38) and (39) generate the $(M-n)$ and $(n+1)$ set of linear equations, respectively. This system of linear equations can then be solved easily for the unknown coefficients. Consequently, we can approximate $v(z)$ given in Equation (33).

4.2. Nonlinear FDEs

Consider the nonlinear fractional-order differential equation

$$F(z, v(z), D^{\vartheta_1}v(z), D^{\vartheta_2}v(z), \ldots, D^{\vartheta_k}v(z)) = 0, \qquad (40)$$

with boundary conditions

$$H_j(v(\varsigma_j), v'(\varsigma_j), \ldots, v^{(s)}(\varsigma_j)) = d_j, \qquad (41)$$

where $0 \leq s < max\{\vartheta_j, j = 1, \ldots, k\} \leq s+1, \varsigma_j \in [0,1], j = 0, \ldots, s$ and H_j are linear combinations of $v(\varsigma_j), v'(\varsigma_j), \ldots, v^{(s)}(\varsigma_j)$. Now, using Theorem 1 and Equation (33), the terms of Equation (40) can be approximated as

$$\begin{cases} v(z) = C^T\Psi(z), \\ D^{\vartheta_1}v(z) = C^T P^{\vartheta_1}\Psi(z), \\ D^{\vartheta_2}v(z) = C^T P^{\vartheta_2}\Psi(z), \\ \vdots \qquad\qquad \vdots \\ D^{\vartheta_k}v(z) = C^T P^{\vartheta_k}\Psi(z). \end{cases} \tag{42}$$

Similarly, the terms of Equation (41) can be approximated as

$$\begin{cases} v(\varsigma_j) = C^T\Psi(\varsigma_j), \\ v'(\varsigma_j) = C^T P^{(1)}\Psi(\varsigma_j), \\ v''(\varsigma_j) = C^T P^{(2)}\Psi(\varsigma_j), \\ \vdots \qquad\qquad \vdots \\ v^{(s)}(\varsigma_j) = C^T P^{(s)}\Psi(\varsigma_j). \end{cases} \tag{43}$$

In light of (42) and (43), we may write Equations (40) and (41), respectively, as

$$F(z, C^T\Psi(z), C^T P^{\vartheta_1}\Psi(z), \ldots, C^T P^{\vartheta_k}\Psi(z)) = 0, \tag{44}$$

$$H_j(C^T\Psi(\varsigma_j), C^T P^{(1)}\Psi(\varsigma_j), \ldots, C^T P^{(s)}\Psi(\varsigma_j)) = d_j. \tag{45}$$

Now, to find the solution $v(z)$, we first collocate Equation (44) at $(M-s)$ points. These equations, along with Equation (45), generate a system of algebraic equations, which can be solved to find $c_i, \quad i = 0, \ldots, M$. Consequently, the function $v(z)$ can be approximated.

5. Error Estimate

Lemma 1. ([37]) *The following assumptions for the function $g(z)$, such that $g(k) = b_k$, must hold true:*

1. *The function $g(z)$ is positive, decreasing and continuous for $z \geq m$.*
2. *$\sum b_m$ is convergent, and $P_m = \sum_{k=m+1}^{\infty} b_k$.*

Then,

$$P_m \leq \int_m^{\infty} g(z)dz.$$

Theorem 2. *If $v(z) \in L_w^2(\Delta)$, $\Delta = [0,1]$, $v(z) = \sum_{k=0}^{\infty} b_k \, \mathrm{VL}_k^*(z)$, b_k is introduced in Equation (15), and $v''(z) \leq N$, then we have*

$$\|v(z) - v_m(z)\|_w \leq \frac{N}{\sqrt{96m^3\pi}}. \tag{46}$$

Proof. It is evident that the shifted VLPs are orthogonal on the interval $[0,1]$ with respect to the weight function, $w(z) = \frac{1}{\sqrt{z-z^2}}$. Hence, these polynomials form a complete $L_w^2(\Delta)$ orthogonal set, where $L_w^2(\Delta)$ represents the space of functions defined as $v: \Delta \to \mathbb{R}$. Thus, the error in space $L_w^2(\Delta)$ is determined as

$$\|v(z) - v_m(z)\|_w^2 = \left(\int_0^1 |v(z) - v_m(z)|^2 w(z)dz\right)^{\frac{1}{2}}. \tag{47}$$

Using Equations (13) and (17) in Equation (47), we have

$$\|v(z) - v_m(z)\|_w^2 = \left(\int_0^1 \left| \sum_{k=m+1}^{\infty} b_k \, \mathrm{VL}_k^*(z) \right|^2 w(z) dz \right)^{\frac{1}{2}}. \tag{48}$$

Now, by applying the orthogonality property of shifted VLPs to Equation (48), we have

$$\|v(z) - v_m(z)\|_w^2 = \frac{1}{\delta_k \pi} \sum_{k=m+1}^{\infty} |b_k|^2. \tag{49}$$

Now, by using the substitution $4x - 2 = 2\cos(\theta)$ in Equation (15), the coefficients, b_k, $k = 0, 1, \cdots, m$, can be determined as

$$b_k = \frac{1}{4\delta_k \pi} \int_0^{\pi} v''\left(\frac{1+\cos(\theta)}{2}\right) \sin(\theta) \left(\frac{\sin(k-1)\theta}{k-1} - \frac{\sin(k+1)\theta}{k+1} \right) d\theta. \tag{50}$$

Equation (50) can also be expressed as

$$|b_k| = \left| \frac{1}{4\delta_k \pi} \int_0^{\pi} v''\left(\frac{1+\cos(\theta)}{2}\right) \sin(\theta) \left(\frac{\sin(k-1)\theta}{k-1} - \frac{\sin(k+1)\theta}{k+1} \right) \right| d\theta. \tag{51}$$

Using $v''(z) \leq N$, and the properties of trigonometric functions, we may express Equation (51) as

$$|b_k| \leq \frac{N}{4k(k-1)(k+1)}, \quad k > 2. \tag{52}$$

Now, using Equation (52) in Equation (49), we have

$$\|v(z) - v_m(z)\|_w^2 \leq \frac{N^2}{32\pi} \sum_{k=m+1}^{\infty} \frac{1}{k^4}. \tag{53}$$

Now, by using the Lemma 1, we have

$$\|v(z) - v_m(z)\|_w^2 \leq \frac{N^2}{32\pi} \int_m^{\infty} z^{-4} dz \tag{54}$$

$$= \frac{N^2}{32\pi} \times \frac{1}{3m^3} = \frac{N^2}{96m^3\pi}. \tag{55}$$

Finally, we have

$$\|v(z) - v_m(z)\|_w \leq \frac{N}{\sqrt{96m^3\pi}}. \tag{56}$$

□

6. Illustrative Examples

In this section, we give some numerical examples to show the accuracy of our proposed method.

Example 1. *Consider the following fractional Bagley–Torvik equation*

$$a_1 \mathcal{D}^2 v(z) + a_2 \mathcal{D}^{3/2} v(z) + a_3 v(z) = g(z), \; z \in [0, 1], \tag{57}$$

subject to the initial conditions with integer order

$$v(0) = 1 = v(0)'. \tag{58}$$

The source term $g(z)$ is as follows:

$$g(z) = 1 + z. \tag{59}$$

The exact solution of the problem in Example 1 is:

$$v(z) = 1 + z. \tag{60}$$

Now, we apply the technique that is described in Section 4.1 by choosing the first three terms of VLPs. We may write the approximation solution as

$$\begin{aligned} v(z) = C^T\Psi(z) &\Longleftrightarrow y(0) = C^T\Psi(0) = d_0 = 1, \\ v^{'}(z) = C^T H^{v}\Psi(z) &\Longleftrightarrow y^{'}(0) = C^T H^{v}\Psi(0) = d_1 = 1. \end{aligned} \tag{61}$$

Now, we have Vieta–Lucas polynomials

$$\begin{aligned} \Psi(z) &= (VL_0^*(z), VL_1^*(z), VL_2^*(z)), \\ &= (2,\ 4z-2,\ 16z^2-16z+2), \end{aligned} \tag{62}$$

and

$$\mathbf{G}^T = \begin{pmatrix} 0.75 \\ 0.25 \\ 0 \end{pmatrix}. \tag{63}$$

The Vieta–Lucas operational matrices can be expressed as

$$\begin{aligned} \mathbf{D}^2 &= \begin{pmatrix} 0 & 0 & 0 \\ 0 & 0 & 0 \\ 16 & 0 & 0 \end{pmatrix}, \\ \mathbf{D}^{3/2} &= \begin{pmatrix} 0 & 0 & 0 \\ 0 & 0 & 0 \\ 11.4936 & 7.6624 & -1.5325 \end{pmatrix}, \\ \mathbf{D}^1 &= \begin{pmatrix} 0 & 0 & 0 \\ 2 & 0 & 0 \\ 0 & 8 & 0 \end{pmatrix}. \end{aligned} \tag{64}$$

The residuals can be evaluated as

$$R(z) = \Big(\mathbf{C}\mathbf{D}^2 + a_2\mathbf{C}\mathbf{D}^{3/2} + a_2 C - a_3 G\Big)\Psi(z),$$

where $C = [c_0, c_1, c_2]$. Now, using initial conditions, we have:

$$\begin{aligned} 2c_0 - 2c_1 + 2c_2 &= 1, \\ 4c_1 - 16c_2 &= 1. \end{aligned} \tag{65}$$

Moreover, using the inner product of the residual with the Vieta–Lucas polynomials, we obtain a system of equations. If we take one equation from this system and two equations from Equation (65), then, by simultaneously solving these equations, we obtain $c_0 = 0.75, c_1 = 0.25, c_2 = 0$, hence

$$v(z) = (\frac{3}{4}, \frac{1}{4}, 0)\begin{pmatrix} 2 \\ 4z-2 \\ 16z^2-16z+2 \end{pmatrix} = 1 + z, \tag{66}$$

which is the exact solution.

Remark 1. *The numerical results computed using our method are compared with the method of [33] by choosing various n. We observe that our proposed method produces efficient numerical results as compared to the numerical results obtained by using the method of [33] (see Tables 1 and 2 and Figure 1). Moreover, for a small value of $n = 2$, the exact solution and the approximate solution computed by using our method coincide (see Table 1 and Figure 1).*

Table 1. Comparison of approximate solution of Example 1.

z	$v(z)$	**Our Method at $n = 2$**	**The Method of [33] at $n = 10$**
0	1.00	1.00	1.024862
0.1	1.10	1.10	1.121206
0.2	1.20	1.20	1.220821
0.3	1.30	1.30	1.323041
0.4	1.40	1.40	1.426952
0.5	1.50	1.50	1.531330
0.6	1.60	1.60	1.634569
0.7	1.70	1.70	1.734591
0.8	1.80	1.80	1.828738
0.9	1.90	1.90	1.913640
1.0	2.00	2.00	1.985057

Table 2. Comparison of absolute errors of Example 1.

z	**Absolute Errors at $n = 10$ Using [33]**	**Absolute Errors at $n = 2, 8, 10$ Using Our Method**
0	2.30×10^{-2}	0
0.1	2.69×10^{-2}	0
0.2	3.13×10^{-2}	0
0.3	3.45×10^{-2}	0
0.4	3.45×10^{-2}	0
0.5	2.87×10^{-2}	0
0.6	1.36×10^{-2}	0
0.7	1.49×10^{-2}	0
0.8	2.30×10^{-2}	0
0.9	2.69×10^{-2}	0
1.0	3.13×10^{-2}	0

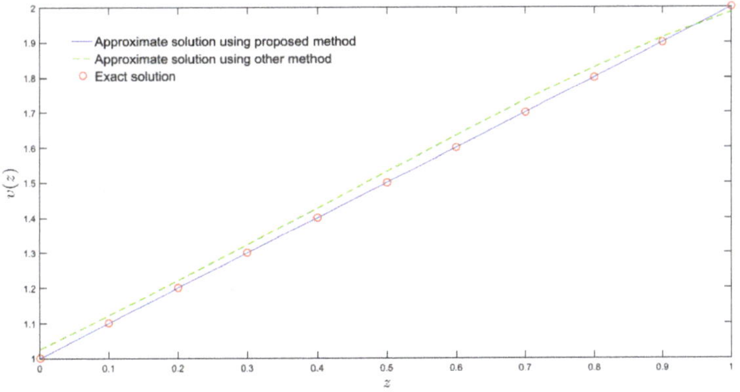

Figure 1. Approximate solutions of Example 1 computed by our method at $n = 2$ is compared with the method of [33] computed at $n = 10$.

Remark 2. *The numerical results of Example 1 computed at $n = 2$ by using our method are compared with the results obtained by using the methods of [32,34] at $n = 6$. We observe that, for a small value of $n = 2$, the approximate solution obtained using our method coincides with the exact* *solution of Example 1 (see Tables 3 and 4). However, the exact solution and the approximate solution computed by using the methods of [32,34] coincide at $n = 6$. This shows that our proposed method is numerically more efficient.*

Table 3. Comparison of approximate solution of Example 1.

z	$v(z)$	Our Method at at $n = 2$	The Method of [32] at $n = 6$
0	1.00	1.00	1.00
0.1	1.10	1.10	1.10
0.2	1.20	1.20	1.20
0.3	1.30	1.30	1.30
0.4	1.40	1.40	1.40
0.5	1.50	1.50	1.50
0.6	1.60	1.60	1.60
0.7	1.70	1.70	1.70
0.8	1.80	1.80	1.80
0.9	1.90	1.90	1.90
1.0	2.00	2.00	2.00

Table 4. Comparison of approximate solution of Example 1.

z	$v(z)$	Our Method at at $n = 2$	The Method of [34] at $n = 6$
0	1.00	1.00	1.00
0.1	1.10	1.10	1.10
0.2	1.20	1.20	1.20
0.3	1.30	1.30	1.30
0.4	1.40	1.40	1.40
0.5	1.50	1.50	1.50
0.6	1.60	1.60	1.60
0.7	1.70	1.70	1.70
0.8	1.80	1.80	1.80
0.9	1.90	1.90	1.90
1.0	2.00	2.00	2.00

Example 2. *Consider the following linear initial value problem [38]:*

$$\begin{aligned} D^{\alpha} v(z) + v(z) = 0, \ \ 0 < \alpha < 2, \\ v(0) = 1, \ \ v'(0) = 0. \end{aligned} \tag{67}$$

The exact solution of the problem is $v(z) = \sum_{k=0}^{\infty} \frac{(-z^{\alpha})^k}{\Gamma(\alpha k+1)}$ [39].

To solve the problem, we use the technique described in Section 4.1. The absolute error for $\alpha = 0.85$ and $n = 2$,5 and 8 is shown in Table 5. An error plot is also shown in Figure 2 for these values.

We can see in Table 5 that a good approximation has been achieved by using some initial terms of VLPs. Moreover, the numerical results for $v(z)$ when $n = 10$ and $\alpha = 0.5$, 0.65, 0.8, 0.95 and 1 are plotted in Figure 3. The exact solution for $\alpha = 1$ is $v(z) = exp(-z)$. It can be noted that the numerical solution converges to the analytical solution when α approaches 1. We also analyze the nonlocal behavior of the fractional derivative by computing the results at various non-integer values of α, which highlights the advantage of using the fractional derivatives, as the next state of the system depends not only upon its current state by also upon all of its historical states.

Table 5. Absolute error for $\alpha = 0.85$ and n = 2, 5 and 8 in Example 2.

z	$n = 2$	$n = 5$	$n = 8$
0.0	0	0	0
0.1	2.04×10^{-2}	7.16×10^{-3}	1.12×10^{-3}
0.2	7.61×10^{-3}	5.08×10^{-3}	1.38×10^{-3}
0.3	1.44×10^{-2}	1.86×10^{-3}	2.73×10^{-3}
0.4	4.06×10^{-2}	4.55×10^{-3}	1.56×10^{-3}
0.5	6.85×10^{-2}	1.55×10^{-3}	1.78×10^{-3}
0.6	9.64×10^{-2}	3.39×10^{-3}	3.82×10^{-4}
0.7	1.23×10^{-1}	4.98×10^{-3}	2.52×10^{-3}
0.8	1.48×10^{-2}	3.24×10^{-3}	1.52×10^{-4}
0.9	1.71×10^{-2}	7.04×10^{-3}	1.16×10^{-3}
1.0	1.91×10^{-2}	3.07×10^{-3}	4.68×10^{-4}

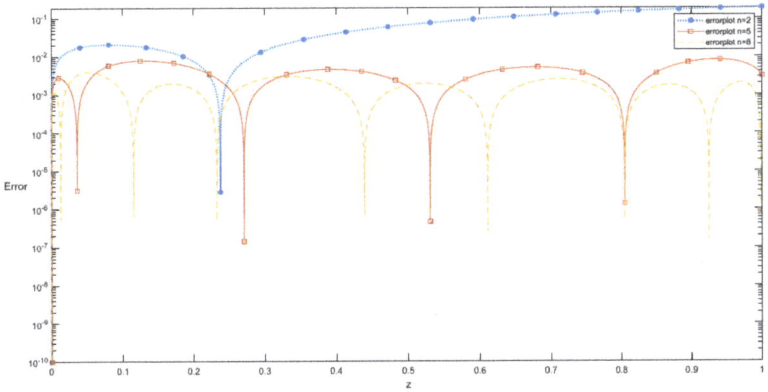

Figure 2. Error plots of Example 2 at different scale levels.

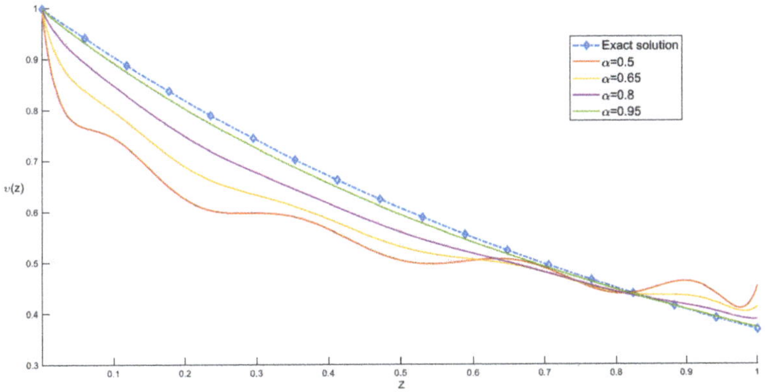

Figure 3. Exact and approximate solutions of Example 2 are compared at different scale levels.

Example 3. *Consider the following initial value problem [10]:*

$$ {}_C\mathcal{D}v(z) = a_1 {}_C\mathcal{D}^{\frac{1}{4}} v(z) - v(z) + g(z), \; z \in [0, 1], \tag{68} $$

subject to the initial condition with integer order

$$v(0) = 0. \tag{69}$$

The source term is given as

$$g(z) = \frac{5z^{\frac{3}{2}}}{2} + z^{\frac{5}{2}} + \frac{15\sqrt{\pi}\,z^{\frac{9}{4}}}{8\Gamma(\frac{13}{4})}.$$

The exact solution at $a_1 = -1$ is given below:

$$v(z) = z^2\sqrt{z}.$$

We test the behavior of our proposed method by solving Example (3) at various values of n. In Table 6, we list the L^∞ and L^2 errors for different values of n. We compare the numerical results obtained by using our proposed method with the numerical results obtained in [10]. It can be observed that the errors computed by using our method are much smaller than those computed by using the method presented in [10]; see Table 6. This highlights the efficiency of our method for this problem. Note that the symbol " – " means that the result for n is unavailable for the method [10].

Table 6. Approximate results of Example 3 at various values of n.

Our Method		**Method in ([10], Example 3)**		
n	L^∞	L^2	L^∞	L^2
3	1.1×10^{-3}	2.2×10^{-3}	–	–
4	2.29×10^{-4}	3.45×10^{-4}	1.21×10^{-3}	5.92×10^{-4}
6	2.11×10^{-5}	3.56×10^{-5}	–	–
8	7.52×10^{-6}	1.85×10^{-5}	5.80×10^{-5}	2.50×10^{-5}
16	4.85×10^{-9}	7.35×10^{-8}	2.45×10^{-6}	9.89×10^{-7}

Example 4. *Consider the following nonlinear initial value problem [40]:*

$$D^\alpha v(z) = \frac{40320}{\Gamma(9-\alpha)} z^{8-\alpha} - 3\frac{\Gamma(5+\alpha/2)}{\Gamma(5-\alpha/2)} z^{4-\alpha/2} + \frac{9}{4}\Gamma(\alpha+1) + \left(\frac{3}{2}z^{\alpha/2} - z^4\right)^3 - v(z)^{\frac{3}{2}},$$

$$v(0) = 0, \quad z'(0) = 0, \quad 0 < \alpha < 2. \tag{70}$$

The exact solution of the problem is $v(z) = z^8 - 3z^{(4+\alpha/2)} + \frac{9}{4}z^\alpha$ [39].

We have solved the problem using the technique described in Section 4.2. The absolute error for $\alpha = 0.85$ and $n = 2, 5$ and 8 is shown in Table 5. An error plot is also shown in Figures 4 and 5 for these values. We can see in Table 7 that a good approximation has been achieved. Numerical results for $v(z)$ when $n = 6$ and $\alpha = 0.6, 0.7, 0.8, 0.9$ and 1 are plotted in Figure 6, along with the exact solutions at the given values of α. It can be noted that, as α approaches 1, the solution of the FDEs approaches that of the integer-order differential equations. We also analyze the nonlocal behavior of the fractional derivative by computing the results at various non-integer values of α, which highlights the advantage of using the fractional derivatives, as the next state of the system depends not only upon its current state but also upon all of its historical states (Table 8).

Table 7. Absolute error for $\alpha = 0.85$ and n = 2, 5 and 8 in Example 4.

z	$n = 2$	$n = 5$	$n = 8$
0.0	0.00×10^{00}	4.02×10^{-41}	3.30×10^{-41}
0.1	3.88×10^{-2}	2.68×10^{-2}	2.60×10^{-3}
0.2	6.48×10^{-2}	1.36×10^{-2}	5.08×10^{-3}
0.3	4.21×10^{-2}	6.53×10^{-3}	9.28×10^{-3}
0.4	2.20×10^{-2}	1.14×10^{-2}	6.92×10^{-3}
0.5	1.08×10^{-1}	2.53×10^{-3}	5.38×10^{-3}
0.6	1.89×10^{-1}	8.36×10^{-3}	4.18×10^{-3}
0.7	2.38×10^{-1}	1.14×10^{-2}	6.41×10^{-3}
0.8	2.38×10^{-1}	5.73×10^{-3}	4.10×10^{-3}
0.9	1.98×10^{-1}	2.61×10^{-4}	4.16×10^{-3}
1.0	1.86×10^{-1}	8.01×10^{-4}	1.59×10^{-3}

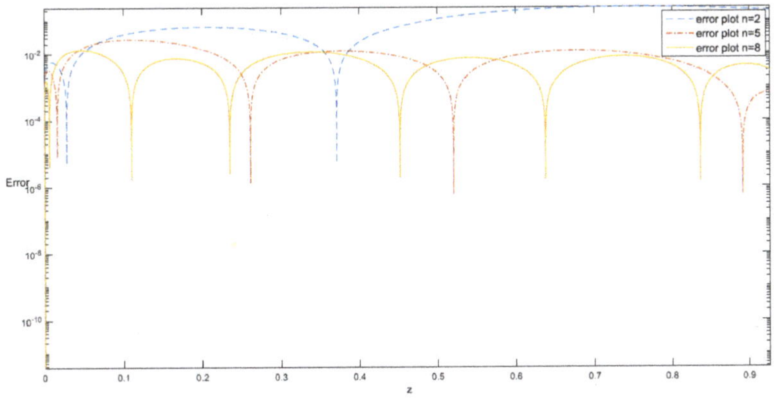

Figure 4. Error plots of Example 4 at different scale levels.

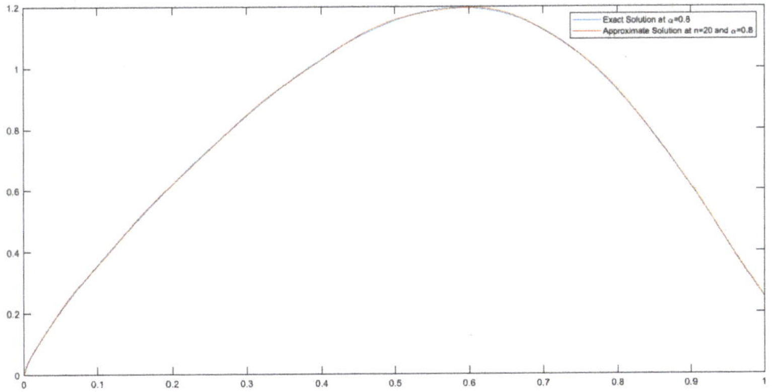

Figure 5. Exact and approximate solutions of Example 4 are compared at n = 20.

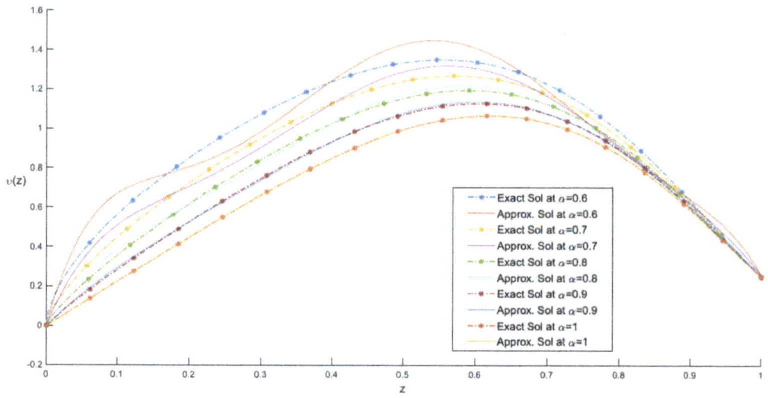

Figure 6. Exact and approximate solutions of Example 4 are compared at different scale levels.

Table 8. Approximate results of Example 4 at various values of α.

Our Method			Method in ([19], Example 3)	
α	$z = 0.5$	$z = 0.9$	$z = 0.5$	$z = 0.9$
0.2	6.94×10^{-1}	6.2×10^{-1}	3.6×10^{-2}	1.7×10^{0}
0.4	1.97×10^{-1}	1.39×10^{-1}	2.4×10^{-2}	3.0×10^{-1}
0.6	4.81×10^{-2}	2.18×10^{-2}	9.6×10^{-3}	3.7×10^{-2}
0.8	8.87×10^{-3}	1.36×10^{-3}	2.3×10^{-3}	2.1×10^{-3}

Example 5. *Consider the following nonlinear initial value problem:*

$$\mathcal{D}^3 v(z) + \mathcal{D}^{5/2} v(z) + v^2(z) = z^4, \;\; v(0) = v'(0) = 0, \;\; v''(0) = 2. \tag{71}$$

We solved this problem by using the same technique as described in Section 4.2 with $n = 3$.

The exact solution of the problem is $v(z) = z^2$, whereas the source term is $g(z) = z^4$. The operational matrices can be expressed as

$$\begin{aligned}
\mathbf{D}^3 &= \begin{pmatrix} 0 & 0 & 0 & 0 \\ 0 & 0 & 0 & 0 \\ 0 & 0 & 0 & 0 \\ 192 & 0 & 0 & 0 \end{pmatrix}, \\
\mathbf{D}^{5/2} &= \begin{pmatrix} 0 & 0 & 0 & 0 \\ 0 & 0 & 0 & 0 \\ 0 & 0 & 0 & 0 \\ 137.9229 & 91.9486 & -18.3897 & 7.8813 \end{pmatrix}, \\
\mathbf{D}^1 &= \begin{pmatrix} 0 & 0 & 0 & 0 \\ 2 & 0 & 0 & 0 \\ 0 & 8 & 0 & 0 \\ 6 & 0 & 12 & 0 \end{pmatrix}, \\
\mathbf{D}^2 &= \begin{pmatrix} 0 & 0 & 0 & 0 \\ 0 & 0 & 0 & 0 \\ 16 & 0 & 0 & 0 \\ 0 & 96 & 0 & 0 \end{pmatrix},
\end{aligned} \tag{72}$$

where $C = [c_0, c_1, c_2, c_3]$.

Now, using the initial condition, we obtain three equations:

$$\begin{aligned} 2c_0 - 2c_1 + 2c_2 - 2c_3 &= 0, \\ 4c_1 - 16c_2 + 36c_3 &= 0, \\ 32c_2 - 192c_3 &= 2. \end{aligned} \tag{73}$$

Meanwhile, using the technique in Section 4.2, we obtain the following equation:

$$C^T\mathbf{D}^{(3)}\Psi(z) + C^T\mathbf{D}^{(\frac{5}{2})}\Psi(z) + [C^T\Psi(z)]^2 - z^4 = 0. \tag{74}$$

Now, we collocate Equation (74) at the first root of $P_4(z)$, and we obtain $z_0 = 0.06698$. Now, solving Equations (73) and (74), we obtain

$$v(z) = (\frac{3}{16}, \frac{1}{4}, \frac{1}{16}, 0)\begin{pmatrix} 2 \\ 4z-2 \\ 16z^2-16z+2 \\ 64z^3-96z^2+36z-2 \end{pmatrix} \approx z^2, \tag{75}$$

which is the exact solution.

Example 6. *Consider the following non-homogenous multi-order fractional problem:*

$$\begin{aligned} {}_C\mathcal{D}^{\alpha}v(z) &= a_C\mathcal{D}^{\beta_0}v(z) + b_C\mathcal{D}^{\beta_1}v(z) + c_C\mathcal{D}^{\beta_2}v(z) \\ &+ d_C\mathcal{D}^{\beta_3}v(z) + g(z), \ z \in [0,1], \ 0 < \alpha < 2, \end{aligned} \tag{76}$$

subject to the following initial conditions

$$v(0) = 0, \ v'(0) = 0.$$

The source term is as below:

$$g(z) = 4z - z^2 - \frac{6776}{4503}z^{\frac{3}{2}} + 42z^5 - 14z^6 + z^7 + \frac{1516}{5629}z^{\frac{13}{2}} - 2. \tag{77}$$

The exact solution corresponding to $\alpha = 2$,$a = c = -1, b = 2, d = 0, \beta_0 = 0, \beta_1 = 1,$ $\beta_2 = \frac{1}{2}$ is given below:

$$v(z) = z^7 - z^2.$$

We can observe in Example 6 that a good approximation of the function has been achieved while using $n = 7$ as a scale level. The absolute error Table 9 at different scale levels is given below. (see Figure 7).

Table 9. Absolute error for n = 2, 5 and 8 in Example 6.

z	$n = 2$	$n = 5$	$n = 8$
0.0	0.00×10^{00}	0.00×10^{00}	0.00×10^{00}
0.1	3.26×10^{-3}	7.98×10^{-06}	1.40×10^{-7}
0.2	5.13×10^{-3}	6.95×10^{-6}	5.90×10^{-8}
0.3	1.98×10^{-3}	1.49×10^{-5}	1.61×10^{-6}
0.4	3.39×10^{-3}	1.39×10^{-5}	1.79×10^{-5}
0.5	6.59×10^{-3}	9.92×10^{-5}	1.18×10^{-4}
0.6	4.97×10^{-3}	5.49×10^{-4}	5.59×10^{-4}
0.7	1.14×10^{-3}	2.11×10^{-3}	2.10×10^{-3}
0.8	9.60×10^{-3}	6.63×10^{-3}	6.63×10^{-3}
0.9	2.08×10^{-2}	1.83×10^{-2}	1.83×10^{-2}
1.0	4.52×10^{-2}	4.56×10^{-2}	4.56×10^{-2}

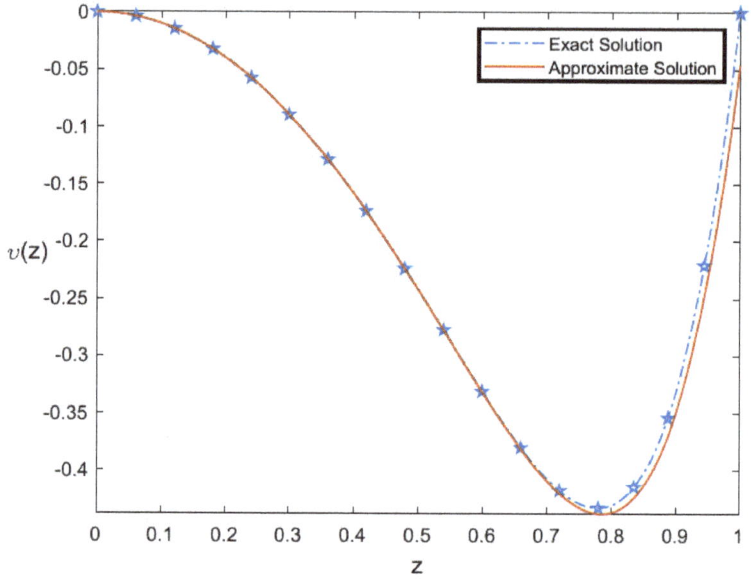

Figure 7. Graphs of exact solution and approximate solution of Example 6 with $n = 7$.

Moreover, a comparison between the exact solution and approximate solution at different scale levels for the values of z is given in Table 10.

Table 10. Comparison of exact solution with approximate solution (AS) at $m = 2, 5$ and 8 in Example 6.

z	$v(z)$	**AS at $n = 2$**	**AS at $n = 5$**	**AS at $n = 8$**
0.0	0.0000	0.0000	0.0000	0.0000
0.1	0.0100	0.0133	0.0100	0.0100
0.2	0.0400	0.0451	0.0400	0.0400
0.3	0.0898	0.0918	0.0898	0.0898
0.4	0.1584	0.1550	0.1584	0.1584
0.5	0.2422	0.2356	0.2423	0.2423
0.6	0.3320	0.3270	0.3326	0.3326
0.7	0.4076	0.4088	0.4098	0.4097
0.8	0.4303	0.4399	0.4369	0.4369
0.9	0.3317	0.3525	0.3500	0.3500
1.0	0.0000	0.0452	0.0456	0.0456

7. Conclusions

In the present study, we introduce a new fractional-order derivative operational matrix of VLPs in Caputo sense. The newly derived operational matrix is used to develop a computer-oriented numerical algorithm to solve the linear and nonlinear FDEs that include the Caputo fractional-order derivative. The proposed numerical algorithm has the advantage of transforming the problems into a system of algebraic equations that are easy to solve using any computational software. To the best of our knowledge, this is the first result where the numerical algorithm is presented using the operational matrix of VLPs and the solution of the problems is approximated using its basis vectors.

We tested the accuracy and efficiency of the algorithm by solving various linear and nonlinear FDEs with initial conditions. We found that with an increase in the values of n, the approximate solutions were in good agreement with the exact solutions. We also demonstrated the high efficiency of the method by determining the amount of absolute error and observed that as we increased n, this amount was decreased significantly. In addition to this, the numerical efficiency was also demonstrated by comparing the results obtained by using our method with results obtained otherwise in the literature [10,32–34]. We observed that our method produced more efficient results.

Author Contributions: Conceptualization, I.T. and T.A.; Formal analysis, Z.A.N., I.T. and M.A.A.; Funding acquisition, T.A. and M.A.A.; Investigation, Z.A.N. and T.A.; Methodology, Z.A.N. and I.T.; Resources, T.A.; Software, Z.A.N. and I.T.; Supervision, I.T. and T.A.; Validation, I.T., T.A. and M.A.A.; Visualization, Z.A.N.; Writing—original draft, Z.A.N.; Writing—review and editing, I.T., T.A. and M.A.A. All authors contributed equally to this article. All authors read and approved the final manuscript.

Funding: This research received no external funding.

Institutional Review Board Statement: Not applicable.

Informed Consent Statement: Not applicable.

Acknowledgments: Prince Nourah bint Abdulrahman University Researchers supporting project number (PNURSP2022R14), Princess Nourah bint Abdulrahman University, Riyadh, Saudi Arabia. The author T. Abdeljawad would like to thank Prince Sultan University for paying the APC and for the support through the TAS research lab.

Conflicts of Interest: The authors declare no conflict of interest.

References

1. Diethelm, K. *The Analysis of Fractional Differential Equations: An Application-Oriented Exposition Using Differential Operators of Caputo Type*; Springer Science & Business Media: New York, NY, USA, 2010.
2. Eckert, M.; Kupper, M.; Hohmann, S. Functional fractional calculus for system identification of battery cells. *AT-Automatisierungstechnik* **2014**, *62*, 272–281. [CrossRef]
3. Kilbas, A.A.; Srivastava, H.M.; Trujillo, J.J. *Theory and Applications of Fractional Differential Equations*; Elsevier: Amsterdam, The Netherlands, 2006; Volume 204.
4. Miller, K.S.; Ross, B. *An Introduction to the Fractional Calculus and Fractional Differential Equations*; Wiley: Hoboken, NJ, USA, 1993.
5. Alam, M.; Talib, I.; Bazighifan, O.; Chalishajar, D.; Almarri, B. An analytical technique implemented in the fractional Clannish Random Walker's Parabolic equation with nonlinear physical phenomena. *Mathematics* **2021**, *9*, 801. [CrossRef]
6. Oldham, K.B.; Spanier, J. *The Fractional Calculus*; Academic Press: New York, NY, USA, 1974.
7. Zhang, H.; Jiang, X.; Yang, X. A time-space spectral method for the time-space fractional Fokker-Planck equation and its inverse problem. *Appl. Math. Comput.* **2018**, *320*, 302–318. [CrossRef]
8. Zhang, H.; Jiang, X.; Zeng, F.; Karniadakis, G.E. A stabilized semi-implicit Fourier spectral method for nonlinear space-fractional reaction-diffusion equations. *J. Comput. Phys.* **2020**, *405*, 109141. [CrossRef]
9. Zheng, X.; Wang, H. Optimal-order error estimates of finite element approximations to variable-order time-fractional diffusion equations without regularity assumptions of the true solutions. *IMA J. Numer. Anal.* **2021**, *41*, 1522–1545. [CrossRef]
10. Talaei, Y.; Asgari, M. An operational matrix based on Chelyshkov polynomials for solving multi-order fractional differential equations. *Neural Comput. Appl.* **2018**, *30*, 1369–1376. [CrossRef]
11. Fazli, H.; Nieto, J.J. An investigation of fractional Bagley-Torvik equation. *Open Math.* **2019**, *17*, 499–512. [CrossRef]
12. Pang, D.; Jiang, W.; Du, J.; Niazi, A. Analytical solution of the generalized Bagley-Torvik equation. *Adv. Differ. Equ.* **2019**, *2019*, 1–13. [CrossRef]
13. Abbas, M.I. Existence and uniqueness results for fractional order differential equations with Riemann-Liouville fractional integral boundary conditions. *Abstr. Appl. Anal.* **2015**, *2015*, 1–7.
14. Deng, J.; Ma, L. Existence and uniqueness of solutions of initial value problems for nonlinear fractional differential equations. *Appl. Math. Lett.* **2010**, *23*, 676–680. [CrossRef]
15. Khan, R.A.; Rehman, M.; Henderson, J. Existence and uniqueness of solution for nonlinear fractional differential equations with integral boundary conditions. *Fract. Differ. Calc.* **2011**, *1*, 29–43. [CrossRef]
16. Liu, X.; Jia, M.; Wu, B. Existence and uniqueness of solution for fractional differential equations with integral boundary conditions. *Electron. J. Qual. Differ. Equ.* **2009**, *2009*, 69. [CrossRef]
17. Zhou, Y. Existence and uniqueness of solutions for a system of fractional differential equations. *Fract. Calc. Appl. Anal.* **2009**, *12*, 195–204.
18. Kazem, S.; Abbasbandy, S.; Kumar, S. Fractional-order Legendre functions for solving fractional-order differential equations. *Appl. Math. Model.* **2013**, *37*, 5498–5510. [CrossRef]

19. Saadatmandi, A.; Dehghan, M. A new operational matrix for solving fractional-order differential equations. *Comput. Math. Appl.* **2010**, *59*, 1326–1336. [CrossRef]
20. Song, L.; Wang, W. A new improved Adomian decomposition method and its application to fractional differential equations. *Appl. Math. Model.* **2013**, *37*, 1590–1598. [CrossRef]
21. Talib, I.; Alam, M.; Baleanu, D.; Zaidi, D.; Marriyam, A. A new integral operational matrix with applications to multi-order fractional differential equations. *AIMS Math.* **2021**, *6*, 8742–8771. [CrossRef]
22. Benattia, M.E.; Belghaba, K. Numerical Solution for Solving Fractional Differential Equations using Shifted Chebyshev Wavelet. *Gen. Lett. Math.* **2017**, *3*, 101–110.
23. Darani, M.A.; Saadatmandi, A. The operational matrix of fractional derivative of the fractional-order Chebyshev functions and its applications. *Comput. Methods Differ. Equ.* **2017**, *5*, 67–87.
24. Khalil, H.; Khan, R.A.; Al-Smadi, M.H.; Freihat, A.A.; Shawagfeh, N. New *Operational Matrix for Shifted Legendre Polynomials and Fractional Differential Equations with Variable Coeffcients*; Punjab University Journal of Mathematics: Lahore, Pakistan, 2020; Volume 47.
25. Bhrawy, A.H.; Alofi, A.S. The operational matrix of fractional integration for shifted Chebyshev polynomials. *Appl. Math. Lett.* **2013**, *26*, 25–31. [CrossRef]
26. Talib, I.; Tunc, C.; Noor, Z.A. New operational matrices of orthogonal Legendre polynomials and their operational. *J. Taibah Univ. Sci.* **2019**, *13*, 377–389. [CrossRef]
27. Rahimkhani, P.; Ordokhani, Y.; Babolian, E. A new operational matrix based on Bernoulli wavelets for solving fractional delay differential equations. *Numer. Algorithms* **2017**, *74*, 223–245. [CrossRef]
28. Dehestani, H.; Ordokhani, Y.; Razzaghi, M. Combination of Lucas wavelets with Legendre-Gauss quadrature for fractional Fredholm-Volterra integro-differential equations. *J. Comput. Appl. Math.* **2021**, *382*, 113070. [CrossRef]
29. Dehestani, H.; Ordokhani, Y.; Razzaghi, M. Fractional-order Bessel functions with various applications. *Appl. Math.* **2019**, *64*, 637–662. [CrossRef]
30. Dehestani, H.; Ordokhani, Y.; Razzaghi, M. Numerical technique for solving fractional generalized pantograph-delay differential equations by using fractional-order hybrid bessel functions. *Int. J. Appl. Comput. Math.* **2020**, *6*, 1–27. [CrossRef]
31. Dehestani, H.; Ordokhani, Y.; Razzaghi, M. On the applicability of Genocchi wavelet method for different kinds of fractional-order differential equations with delay. *Numer. Linear Algebra Appl.* **2019**, *26*, 2259. [CrossRef]
32. Gulsu, M.; Ozturk, Y.; Ayse, A. Numerical solution of the fractional Bagley-Torvik equation arising in fluid mechanics. *Int. J. Comput. Math.* **2017**, *94*, 173–184. [CrossRef]
33. Raja, M.; Khan, J.A.; Qureshi, I. Solution of fractional order system of Bagley-Torvik equation using evolutionary computational intelligence. *Math. Probl. Eng.* **2011**, *2011*, 675075. [CrossRef]
34. Yuzbasi, S. Numerical solution of the Bagley-Torvik equation by the Bessel collocation method. *Math. Methods Appl. Sci.* **2013**, *36*, 300–312. [CrossRef]
35. Agarwal, P.; El-Sayed, A.A. Vieta-Lucas polynomials for solving a fractional-order mathematical physics model. *Adv. Differ. Equ.* **2020**, *2020*, 1–18. [CrossRef]
36. Canuto, C.; Hussaini, M.Y.; Quarteroni, A.; Zang, T.A. *Spectral Methods: Fundamentals in Single Domains*; Springer Science & Business Media: New York, NY, USA, 2007.
37. Stewart, J. *Single Variable Essential Calculus: Early Transcendentals*; Cengage Learning: Belmont, CA, USA, 2012.
38. Hashim, I.; Abdulaziz, O.; Momani, S. Homotopy analysis method for fractional IVPs. *Commun. Nonlinear Sci. Numer. Simul.* **2009**, *14*, 674–684. [CrossRef]
39. Diethelm, K.; Ford, N.J.; Freed, A.D. A predictor-corrector approach for the numerical solution of fractional differential equations. *Nonlinear Dyn.* **2002**, *29*, 3–22. [CrossRef]
40. Kumar, P.; Agrawal, O.P. An approximate method for numerical solution of fractional differential equations. *Signal Process.* **2006**, *86*, 2602–2610. [CrossRef]

fractal and fractional

Article

Numerical Analysis of Time-Fractional Whitham-Broer-Kaup Equations with Exponential-Decay Kernel

Humaira Yasmin

Department of Basic Sciences, Preparatory Year Deanship, King Faisal University, Al-Ahsa 31982, Saudi Arabia; hhassain@kfu.edu.sa

Abstract: This paper presents the semi-analytical analysis of the fractional-order non-linear coupled system of Whitham-Broer-Kaup equations. An iterative process is designed to analyze analytical findings to the specified non-linear partial fractional derivatives scheme utilizing the Yang transformation coupled with the Adomian technique. The fractional derivative is considered in the sense of Caputo-Fabrizio. Two numerical problems show the suggested method. Moreover, the results of the suggested technique are compared with the solution of other well-known numerical techniques such as the Homotopy perturbation technique, Adomian decomposition technique, and the Variation iteration technique. Numerical simulation has been carried out to verify that the suggested methodologies are accurate and reliable, and the results are revealed using graphs and tables. Comparing the analytical and actual solutions demonstrates that the proposed approaches effectively solve complicated non-linear problems. Furthermore, the proposed methodologies control and manipulate the achieved numerical solutions in a vast acceptable region in an extreme manner. It will provide us with a simple process to control and adjust the convergence region of the series solution.

Keywords: Adomian decomposition method; system of Whitham-Broer-Kaup equations; Caputo-Fabrizio derivative; Yang transform

Citation: Yasmin, H. Numerical Analysis of Time-Fractional Whitham-Broer-Kaup Equations with Exponential-Decay Kernel. *Fractal Fract.* **2022**, *6*, 142. https://doi.org/10.3390/fractalfract6030142

Academic Editors: Asifa Tassaddiq and Muhammad Yaseen

Received: 10 February 2022
Accepted: 27 February 2022
Published: 2 March 2022

1. Introduction

Fractional calculus (FC) was invented by Newton, but it has recently piqued the interest of many academics. Fascinating breakthroughs in science and engineering applications have been found within the framework of FC over the last 30 years. Due to the complications involved with a heterogeneity issue, the notion of the fractional derivative has been industrialized. The behaviour of complex media with a diffusion mechanism may be captured using non-integer order differential operators [1–4]. It has proven a handy tool, and differential equations of any order may demonstrate various situations more efficiently and precisely. Numerous scholars began to work on calculus and its generalization to express their viewpoints while investigating many complicated events due to the rapid development of mathematical approaches using computer software [5–8].

Differential equations featuring non-linearities are used in science, technology, and engineering to explain a variety of phenomena, ranging from gravity to dynamical systems [9–11]. Non-linear partial differential equations (PDEs) are significant techniques for modeling non-linear dynamical events in a variety of domains, including mathematical biology, fluid mechanics, material science, and fluid dynamics, as shown in [12]. A sufficient set of partial differential equations can represent the bulk of dynamical systems. PDEs are also well-known for being utilized to solve mathematical difficulties like the Poincare and the Calabi conjectures.

It has already been demonstrated that the non-linear development of shallow-water waves may be represented using the technique of the Whitham-Broer-Kaup equation in fluid mechanics (WBKEs) [13]. Whitham, Broer, and Kaup [14–16] developed the integrated framework of the equations as mentioned earlier. The mentioned equations can be written as the shallow water acoustic waves with various diversity connections, as shown [17]. In the classical order, the governing equations for the phenomenon mentioned above are represented by

$$\begin{cases} \mathbb{U}_{\Im} + \mathbb{U}\mathbb{U}_{\varepsilon} + \mathbb{V}_{\varepsilon} + q\mathbb{U}_{\varepsilon\varepsilon} = 0 \\ \mathbb{V}_{\Im} + \mathbb{V}\mathbb{U}_{\varepsilon} + \mathbb{U}\mathbb{V}_{\varepsilon} - q\mathbb{V}_{\varepsilon\varepsilon} + p\mathbb{U}_{\varepsilon\varepsilon\varepsilon} = 0, \end{cases} \tag{1}$$

where $\mathbb{U} = \mathbb{U}(\varepsilon, \Im), \mathbb{V} = \mathbb{V}(\varepsilon, \Im)$ indicates the horizontal velocity and height of the fluids, respectively, which differ greatly from the equilibrium, and q, p are the constants that are composed of various diffusion powers. For the past few decades, investigating the results of non-linear PDEs has been a major focus of research. Several authors have devised numerous mathematical methods to examine approximate results of non-linear PDEs. Mohyud Din et al. [18] investigated the analysis of many integer order PDEs using homotopy perturbation techniques. To solve the coupled set of Burgers and Brusselator equations, Biazar and Aminikhah [19] used the perturbation technique. For the numerical result of many traditional order differential equations by applying other techniques, interested readers can refer to Refs. [20–25]. Numerous strategies have been used to study the solution to the given non-linear coupled scheme (1) of PDEs. To address the classical order coupled systems of the WBK problem, Mohyud-Din et al. [26] employed perturbation methods. As a result, researchers like Xie et al. 2002 (who studied the solution using the hyperbolic technique) have used several powerful and efficient methods to investigate the problem of the WBK coupled equation of classical order PDEs. In the same way, El-Sayed and Kaya used the Adomian decomposition approach to investigate the scheme (1). Moreover, Ahmad et al. [12] used the Adomian decomposition method and He's polynomial to solve the coupled system (1).

Adomian proposed the Adomian decomposition method in 1980, which is a helpful technique for obtaining an explicit and numerical solution to a system of differential equations that represents a physical problem [27–29]. The Laplace transform technique is a vital technique in technology and applied mathematics. Combining the Adomian decomposition method and Yang transformation leads to a well-known technique named the Yang decomposition method. In this study, we convert differential equations to algebraic equations using the Laplace transform, and the non-linear terms are decomposed using Adomain polynomials. This numerical approach is effective for both deterministic and stochastic differential equation systems. It can be applied to a classical and fractional-order ordinary and a PDEs system, both linear and non-linear. There is no need for perturbation or liberalization in this procedure. Furthermore, unlike RK4, it does not require a pre-defined step size. In addition, this technique does not depend upon a parameter, as required for homotopy analysis and homotopy perturbation methods. However, the solutions achieved via this technique are the same as gained by the Adomian decomposition method (for detail, see [30–33]). It must be mentioned that the Yang decomposition method is more effective than the basic Adomian decomposition method.

The rest of this article is organized as follows. In Section 2, we present some basic definitions and properties. Section 3 describes the Yang decomposition method for solving fractional partial differential equations. The conclusion is presented at the end of the article.

2. Preliminaries Concepts

In this section, we provide the fundamental definitions that will be used throughout the article. For the purpose of simplification, we write the exponential decay kernel as, $K(\Im, \varrho) = e^{[-\wp(\Im-\varrho/1-\wp)]}$.

Definition 1. *If the Caputo-Fabrizio derivative is given as follows [34]:*

$${}^{CF}D_{\Im}^{\wp}[\mathbb{P}(\Im)] = \frac{N(\wp)}{1-\wp}\int_0^{\Im} \mathbb{P}'(\varrho)K(\Im, \varrho)d\varrho, \ \ n-1 < \wp \leq n \tag{2}$$

$N(\wp)$ is the normalization function with $N(0) = N(1) = 1$.

$$^{CF}D^{\wp}_{\Im}[\mathbb{P}(\Im)] = \frac{N(\wp)}{1-\wp}\int_0^{\Im}[\mathbb{P}(\Im) - \mathbb{P}(\varrho)]K(\Im,\varrho)d\varrho. \tag{3}$$

Definition 2. *The fractional integral Caputo-Fabrizio is given as [34]*

$$^{CF}I^{\wp}_{\Im}[\mathbb{P}(\Im)] = \frac{1-\wp}{N(\wp)}\mathbb{P}(\Im) + \frac{\wp}{N(\wp)}\int_0^{\Im}\mathbb{P}(\varrho)d\varrho,\ \Im \geq 0, \wp \in (0,1]. \tag{4}$$

Definition 3. *For $N(\wp) = 1$, the following result shows the Caputo-Fabrizio derivative of Laplace transformation [34]:*

$$L\left[^{CF}D^{\wp}_{\Im}[\mathbb{P}(\Im)]\right] = \frac{vL[\mathbb{P}(\Im) - \mathbb{P}(0)]}{v + \wp(1-v)}. \tag{5}$$

Definition 4. *The Yang transformation of $\mathbb{P}(\Im)$ is expressed as [35]*

$$\mathbb{Y}[\mathbb{P}(\Im)] = \chi(v) = \int_0^{\infty}\mathbb{P}(\Im)e^{-\frac{\Im}{v}}d\Im.\ \Im > 0, \tag{6}$$

Remark 1. *The Yang transformation of a few useful functions is defined as:*

$$\begin{aligned}\mathbb{Y}[1] &= v,\\ \mathbb{Y}[\Im] &= v^2,\\ \mathbb{Y}[\Im^i] &= \Gamma(i+1)v^{i+1}.\end{aligned} \tag{7}$$

Lemma 1. *Let the Laplace transformation of $\mathbb{P}(\Im)$ is $F(v)$, then $\chi(v) = F(1/v)$ [36].*

Proof. From Equation (6), we can achieve another type of the Yang transformation by putting $\Im/v = \zeta$ as

$$L[\mathbb{P}(\Im)] = \chi(v) = v\int_0^{\infty}\mathbb{P}(v\zeta)e^{\zeta}d\zeta.\ \zeta > 0, \tag{8}$$

Since $L[\mathbb{P}(\Im)] = F(v)$, this implies that

$$F(v) = L[\mathbb{P}(\Im)] = \int_0^{\infty}\mathbb{P}(\Im)e^{-v\Im}d\Im. \tag{9}$$

Put $\Im = \zeta/v$ in (9), we have

$$F(v) = \frac{1}{v}\int_0^{\infty}\mathbb{P}\left(\frac{\zeta}{v}\right)e^{\zeta}d\zeta. \tag{10}$$

Thus, from Equation (8), we achieve

$$F(v) = \chi\left(\frac{1}{v}\right). \tag{11}$$

Additionally, from Equations (6) and (9), we achieve

$$F\left(\frac{1}{v}\right) = \chi(v). \tag{12}$$

The connections Equations between (11) and (12) represent the duality link between the Laplace and Yang transformation. □

Lemma 2. *Let $\mathbb{P}(\Im)$ be a continuous function; then, the Caputo-Fabrizio derivative Yang transformation of $\mathbb{P}(\Im)$ is defined by [36]*

$$\mathbb{Y}[\mathbb{P}(\Im)] = \frac{\mathbb{Y}[\mathbb{P}(\Im) - v\mathbb{P}(0)]}{1 + \wp(v-1)}. \tag{13}$$

Proof. The Caputo-Fabrizio fractional Laplace transformation is given by

$$L[\mathbb{P}(\Im)] = \frac{L[v\mathbb{P}(\Im) - \mathbb{P}(0)]}{v + \wp(1-v)}, \tag{14}$$

In addition, we have that the connection among Laplace and Yang property, i.e., $\chi(v) = F(1/v)$. To achieve the necessary result, we substitute v by $1/v$ in Equation (14), and get

$$\begin{aligned}\mathbb{Y}[\mathbb{P}(\Im)] =& \frac{\frac{1}{v}\mathbb{Y}[\mathbb{P}(\Im) - \mathbb{P}(0)]}{\frac{1}{v} + \wp(1-\frac{1}{v})},\\ \mathbb{Y}[\mathbb{P}(\Im)] =& \frac{\mathbb{Y}[\mathbb{P}(\Im) - v\mathbb{P}(0)]}{1 + \wp(v-1)}.\end{aligned} \tag{15}$$

The proof is completed. □

3. The Producer of YDM

In this portion, we discuses the YDM producer for fractional partial differential equations.

$$\begin{aligned}&{}^{CF}D_{\Im}^{\wp}\mathbb{U}(\varepsilon,\Im) + \mathcal{G}_1(\mathbb{U},\mathbb{V}) + \mathcal{L}_1(\mathbb{U},\mathbb{V}) - \mathcal{P}_1(\varepsilon,\Im) = 0,\\ &{}^{CF}D_{\Im}^{\wp}\mathbb{V}(\varepsilon,\Im) + \mathcal{G}_2(\mathbb{U},\mathbb{V}) + \mathcal{L}_2(\mathbb{U},\mathbb{V}) - \mathcal{P}_2(\varepsilon,\Im) = 0, \quad 0 < \wp \leq 1,\end{aligned} \tag{16}$$

with initial condition

$$\mathbb{U}(\varepsilon,0) = g_1(\varepsilon), \quad \mathbb{V}(\varepsilon,0) = g_2(\varepsilon). \tag{17}$$

where $D_{\Im}^{\wp} = \frac{\partial^{\wp}}{\partial\Im^{\wp}}$ is the Caputo fractional derivative of order $\wp$, $\mathcal{G}_1$, $\mathcal{G}_2$ and $\mathcal{L}_1$, $\mathcal{L}_2$ are the linear and non-linear functions, respectively, and $\mathcal{P}_1$,$\mathcal{P}_2$ are the source functions.

Using the Yang transformation to Equation (16),

$$\begin{aligned}&\mathbb{Y}[D_{\Im}^{\wp}\mathbb{U}(\varepsilon,\Im)] + \mathbb{Y}[\mathcal{G}_1(\mathbb{U},\mathbb{V}) + \mathcal{L}_1(\mathbb{U},\mathbb{V}) - \mathcal{P}_1(\varepsilon,\Im)] = 0,\\ &\mathbb{Y}[D_{\Im}^{\wp}\mathbb{V}(\varepsilon,\Im)] + \mathbb{Y}[\mathcal{G}_2(\mathbb{U},\mathbb{V}) + \mathcal{L}_2(\mathbb{U},\mathbb{V}) - \mathcal{P}_2(\varepsilon,\Im)] = 0.\end{aligned} \tag{18}$$

Using the Yang transformation differentiation property, we have

$$\begin{aligned}\mathbb{Y}[\mathbb{U}(\varepsilon,\Im)] &= v\mathbb{U}(\varepsilon,0) + (1+\wp(v-1))\mathbb{Y}[\mathcal{P}_1(\varepsilon,\Im)] - (1+\wp(v-1))\mathbb{Y}\{\mathcal{G}_1(\mathbb{U},\mathbb{V}) + \mathcal{L}_1(\mathbb{U},\mathbb{V})\}],\\ \mathbb{Y}[\mathbb{V}(\varepsilon,\Im)] &= v\mathbb{V}(\varepsilon,0) + (1+\wp(v-1))\mathbb{Y}[\mathcal{P}_2(\varepsilon,\Im)] - (1+\wp(v-1))\mathbb{Y}\{\mathcal{G}_2(\mathbb{U},\mathbb{V}) + \mathcal{L}_2(\mathbb{U},\mathbb{V})\}].\end{aligned} \tag{19}$$

YDM describes the solution of infinite series $\mathbb{U}(\varepsilon,\Im)$ and $\mathbb{V}(\varepsilon,\Im)$,

$$\mathbb{U}(\varepsilon,\Im) = \sum_{m=0}^{\infty}\mathbb{U}_m(\varepsilon,\Im), \quad \mathbb{V}(\varepsilon,\Im) = \sum_{m=0}^{\infty}\mathbb{V}_m(\varepsilon,\Im). \tag{20}$$

Adomian polynomials of non-linear terms of $\mathcal{L}_1$ and $\mathcal{L}_2$ are represented as

$$\mathcal{L}_1(\mathbb{U},\mathbb{V}) = \sum_{m=0}^{\infty}\mathcal{A}_m, \quad \mathcal{L}_2(\mathbb{U},\mathbb{V}) = \sum_{m=0}^{\infty}\mathcal{B}_m. \tag{21}$$

The expression for Adomian polynomials is

$$\mathcal{A}_m = \frac{1}{m!}\left[\frac{\partial^m}{\partial\lambda^m}\left\{\sum_{m=0}^{\infty}\lambda^m\mathbb{U}_m, \sum_{m=0}^{\infty}\lambda^m\mathbb{V}_m\right\}\right]_{\lambda=0},$$
$$\mathcal{B}_m = \frac{1}{m!}\left[\frac{\partial^m}{\partial\lambda^m}\left\{\sum_{m=0}^{\infty}\lambda^m\mathbb{U}_m, \sum_{m=0}^{\infty}\lambda^m\mathbb{V}_m\right\}\right]_{\lambda=0}. \tag{22}$$

Putting Equations (20) and (22) into Equation (19),

$$\begin{aligned}\mathbb{Y}\left[\sum_{m=1}^{\infty}\mathbb{U}_m(\varepsilon,\Im)\right] &= s\mathbb{U}(\varepsilon,0) + (1+\wp(v-1))\mathbb{Y}\{\mathcal{P}_1(\varepsilon,\Im)\}\\ &\quad - (1+\wp(v-1))\mathbb{Y}\left\{\mathcal{G}_1(\sum_{m=0}^{\infty}\mathbb{U}_m, \sum_{m=0}^{\infty}\mathbb{V}_m) + \sum_{m=0}^{\infty}\mathcal{A}_m\right\},\\ \mathbb{Y}[\sum_{m=1}^{\infty}\mathbb{V}_m(\varepsilon,\Im)] &= v\mathbb{V}(\varepsilon,0) + (1+\wp(v-1))\mathbb{Y}\{\mathcal{P}_2(\varepsilon,\Im)\}\\ &\quad - (1+\wp(v-1))\mathbb{Y}\left\{\mathcal{G}_2(\sum_{m=0}^{\infty}\mathbb{U}_m, \sum_{m=0}^{\infty}\mathbb{V}_m) + \sum_{m=0}^{\infty}\mathcal{B}_m\right\}.\end{aligned} \tag{23}$$

The inverse Yang transformation is implemented on Equation (23),

$$\begin{aligned}\sum_{m=1}^{\infty}\mathbb{U}_m(\varepsilon,\Im) &= \mathbb{Y}^{-1}[v\mathbb{U}(\varepsilon,0) + (1+\wp(v-1))\mathbb{Y}\{\mathcal{P}_1(\varepsilon,\Im)\}]\\ &\quad - \mathbb{Y}^{-1}\left[(1+\wp(v-1))\mathbb{Y}\left\{\mathcal{G}_1(\sum_{m=0}^{\infty}\mathbb{U}_m, \sum_{m=0}^{\infty}\mathbb{V}_m) + \sum_{m=0}^{\infty}\mathcal{A}_m\right\}\right],\\ \sum_{m=1}^{\infty}\mathbb{V}_m(\varepsilon,\Im) &= \mathbb{Y}^{-1}[v\mathbb{V}(\varepsilon,0) + (1+\wp(v-1))\mathbb{Y}\{\mathcal{P}_2(\varepsilon,\Im)\}]\\ &\quad - \mathbb{Y}^{-1}\left[(1+\wp(v-1))\mathbb{Y}\left\{\mathcal{G}_2(\sum_{m=0}^{\infty}\mathbb{U}_m, \sum_{m=0}^{\infty}\mathbb{V}_m) + \sum_{m=0}^{\infty}\mathcal{B}_m\right\}\right].\end{aligned} \tag{24}$$

Find the $\mathbb{U}_0$ and $\mathbb{V}_0$ using the initial conditions and sources functions. The following terms are expressed:

$$\begin{aligned}\mathbb{U}_0(\varepsilon,\Im) &= \mathbb{Y}^{-1}[v\mathbb{U}(\varepsilon,0) + (1+\wp(v-1))\mathbb{Y}\{\mathcal{P}_1(\varepsilon,\Im)\}],\\ \mathbb{V}_0(\varepsilon,\Im) &= \mathbb{Y}^{-1}[v\mathbb{V}(\varepsilon,0) + (1+\wp(v-1))\mathbb{Y}\{\mathcal{P}_2(\varepsilon,\Im)\}].\end{aligned} \tag{25}$$

For $m = 1$

$$\begin{aligned}\mathbb{U}_1(\varepsilon,\Im) &= -\mathbb{Y}^{-1}[(1+\wp(v-1))\mathbb{Y}\{\mathcal{G}_1(\mathbb{U}_0,\mathbb{V}_0) + \mathcal{A}_0\}],\\ \mathbb{V}_1(\varepsilon,\Im) &= -\mathbb{Y}^{-1}[(1+\wp(v-1))\mathbb{Y}\{\mathcal{G}_2(\mathbb{U}_0,\mathbb{V}_0) + \mathcal{B}_0\}],\end{aligned}$$

the general for $m \geq 1$, is given by

$$\begin{aligned}\mathbb{U}_{m+1}(\varepsilon,\Im) &= -\mathbb{Y}^{-1}[(1+\wp(v-1))\mathbb{Y}\{\mathcal{G}_1(\mathbb{U}_m,\mathbb{V}_m) + \mathcal{A}_m\}],\\ \mathbb{V}_{m+1}(\varepsilon,\Im) &= -\mathbb{Y}^{-1}[(1+\wp(v-1))\mathbb{Y}\{\mathcal{G}_2(\mathbb{U}_m,\mathbb{V}_m) + \mathcal{B}_m\}],\end{aligned}$$

4. Numerical Results

Example 1. *Consider the fractional-order system of WBKEs [11]*

$$\begin{aligned}
&{}^{CF}D^{\wp}_{\Im}\mathbb{U}(\varepsilon,\Im)+\mathbb{U}(\varepsilon,\Im)\frac{\partial\mathbb{U}(\varepsilon,\Im)}{\partial\varepsilon}+\frac{\partial\mathbb{U}(\varepsilon,\Im)}{\partial\varepsilon}+\frac{\partial\mathbb{V}(\varepsilon,\Im)}{\partial\varepsilon}=0,\\
&{}^{CF}D^{\wp}_{\Im}\mathbb{V}(\varepsilon,\Im)+\mathbb{U}(\varepsilon,\Im)\frac{\partial\mathbb{V}(\varepsilon,\Im)}{\partial\varepsilon}+\mathbb{V}(\varepsilon,\Im)\frac{\partial\mathbb{U}(\varepsilon,\Im)}{\partial\varepsilon}+3\frac{\partial^3\mathbb{U}(\varepsilon,\Im)}{\partial\varepsilon^3}-\frac{\partial^2\mathbb{V}(\varepsilon,\Im)}{\partial\varepsilon^2}=0,\\
&0<\wp\le 1,\quad -1<\Im\le 1,\quad -10\le\varepsilon\le 10,
\end{aligned}\tag{26}$$

with the initial conditions

$$\begin{aligned}
\mathbb{U}(\varepsilon,0)&=\frac{1}{2}-8\tanh(-2\varepsilon),\\
\mathbb{V}(\varepsilon,0)&=16-16\tanh^2(-2\varepsilon).
\end{aligned}\tag{27}$$

Applying the Yang transformation of Equation (26), we have

$$\begin{aligned}
\mathbb{Y}\left\{\frac{\partial^{\wp}\mathbb{U}(\varepsilon,\Im)}{\partial\Im^{\wp}}\right\}&=-\mathbb{Y}\left[\mathbb{U}(\varepsilon,\Im)\frac{\partial\mathbb{U}(\varepsilon,\Im)}{\partial\varepsilon}+\frac{\partial\mathbb{U}(\varepsilon,\Im)}{\partial\varepsilon}+\frac{\partial\mathbb{V}(\varepsilon,\Im)}{\partial\varepsilon}\right],\\
\mathbb{Y}\left\{\frac{\partial^{\wp}\mathbb{V}(\varepsilon,\Im)}{\partial\Im^{\wp}}\right\}&=-\mathbb{Y}\left[\mathbb{U}(\varepsilon,\Im)\frac{\partial\mathbb{V}(\varepsilon,\Im)}{\partial\varepsilon}+\mathbb{V}(\varepsilon,\Im)\frac{\partial\mathbb{U}(\varepsilon,\Im)}{\partial\varepsilon}+3\frac{\partial^3\mathbb{U}(\varepsilon,\Im)}{\partial\varepsilon^3}-\frac{\partial^2\mathbb{V}(\varepsilon,\Im)}{\partial\varepsilon^2}\right],
\end{aligned}$$

$$\begin{aligned}
\frac{1}{(1+\wp(v-1))}\mathbb{Y}\{\mathbb{U}(\varepsilon,\Im)\}-v\mathbb{U}(\varepsilon,0)&=-\mathbb{Y}\left[\mathbb{U}(\varepsilon,\Im)\frac{\partial\mathbb{U}(\varepsilon,\Im)}{\partial\varepsilon}+\frac{\partial\mathbb{U}(\varepsilon,\Im)}{\partial\varepsilon}+\frac{\partial\mathbb{V}(\varepsilon,\Im)}{\partial\varepsilon}\right]\\
\frac{1}{(1+\wp(v-1))}\mathbb{Y}\{\mathbb{V}(\varepsilon,\Im)\}-v\mathbb{V}(\varepsilon,0)&=-\mathbb{Y}\left[\mathbb{U}(\varepsilon,\Im)\frac{\partial\mathbb{V}(\varepsilon,\Im)}{\partial\varepsilon}+\mathbb{V}(\varepsilon,\Im)\frac{\partial\mathbb{U}(\varepsilon,\Im)}{\partial\varepsilon}+3\frac{\partial^3\mathbb{U}(\varepsilon,\Im)}{\partial\varepsilon^3}-\frac{\partial^2\mathbb{V}(\varepsilon,\Im)}{\partial\varepsilon^2}\right].
\end{aligned}$$

The above equation is simplified

$$\begin{aligned}
\mathbb{Y}\{\mathbb{U}(\varepsilon,\Im)\}&=v\{\mathbb{U}(\varepsilon,0)\}-(1+\wp(v-1))\mathbb{Y}\left[\mathbb{U}(\varepsilon,\Im)\frac{\partial\mathbb{U}(\varepsilon,\Im)}{\partial\varepsilon}+\frac{\partial\mathbb{U}(\varepsilon,\Im)}{\partial\varepsilon}+\frac{\partial\mathbb{V}(\varepsilon,\Im)}{\partial\varepsilon}\right],\\
\mathbb{Y}\{\mathbb{V}(\varepsilon,\Im)\}&=v\{\mathbb{V}(\varepsilon,0)\}-(1+\wp(v-1))\mathbb{Y}\left[\mathbb{U}(\varepsilon,\Im)\frac{\partial\mathbb{V}(\varepsilon,\Im)}{\partial\varepsilon}+\mathbb{V}(\varepsilon,\Im)\frac{\partial\mathbb{U}(\varepsilon,\Im)}{\partial\varepsilon}+3\frac{\partial^3\mathbb{U}(\varepsilon,\Im)}{\partial\varepsilon^3}-\frac{\partial^2\mathbb{V}(\varepsilon,\Im)}{\partial\varepsilon^2}\right].
\end{aligned}\tag{28}$$

Using inverse Yang transform, we have

$$\begin{aligned}
\mathbb{U}(\varepsilon,\Im)&=\mathbb{U}(\varepsilon,0)-\mathbb{Y}^{-1}\left[(1+\wp(v-1))\mathbb{Y}\left\{\mathbb{U}(\varepsilon,\Im)\frac{\partial\mathbb{U}(\varepsilon,\Im)}{\partial\varepsilon}+\frac{\partial\mathbb{U}(\varepsilon,\Im)}{\partial\varepsilon}+\frac{\partial\mathbb{V}(\varepsilon,\Im)}{\partial\varepsilon}\right\}\right],\\
\mathbb{V}(\varepsilon,\Im)&=\mathbb{V}(\varepsilon,0)-\mathbb{Y}^{-1}\left[(1+\wp(v-1))\mathbb{Y}\left\{\mathbb{U}(\varepsilon,\Im)\frac{\partial\mathbb{V}(\varepsilon,\Im)}{\partial\varepsilon}+\mathbb{V}(\varepsilon,\Im)\frac{\partial\mathbb{U}(\varepsilon,\Im)}{\partial\varepsilon}+3\frac{\partial^3\mathbb{U}(\varepsilon,\Im)}{\partial\varepsilon^3}-\frac{\partial^2\mathbb{V}(\varepsilon,\Im)}{\partial\varepsilon^2}\right\}\right].
\end{aligned}\tag{29}$$

Assume that the $\mathbb{U}(\varepsilon,\Im)$ and the $\mathbb{V}(\varepsilon,\Im)$ infinite series solution functions as follows:

$$\mathbb{U}(\varepsilon,\Im)=\sum_{m=0}^{\infty}\mathbb{U}_m(\varepsilon,\Im)\quad and\quad \mathbb{V}(\varepsilon,\Im)=\sum_{m=0}^{\infty}\mathbb{V}_m(\varepsilon,\Im).$$

Remember that the Adomian polynomials are given as $\mathbb{U}\mathbb{U}_{\varepsilon}=\sum_{m=0}^{\infty}\mathcal{A}_m$, $\mathbb{U}\mathbb{V}_{\varepsilon}=\sum_{m=0}^{\infty}\mathcal{B}_m$ and $\mathbb{V}\mathbb{U}_{\varepsilon}=\sum_{m=0}^{\infty}\mathcal{C}_m$

$$\begin{aligned}
\sum_{m=0}^{\infty}\mathbb{U}_m(\varepsilon,\Im)&=\mathbb{U}(\varepsilon,0)-\mathbb{Y}^{-1}\left[(1+\wp(v-1))\mathbb{Y}\left\{\sum_{m=0}^{\infty}\mathcal{A}_m+\frac{\partial\mathbb{U}(\varepsilon,\Im)}{\partial\varepsilon}+\frac{\partial\mathbb{V}(\varepsilon,\Im)}{\partial\varepsilon}\right\}\right],\\
\sum_{m=0}^{\infty}\mathbb{V}_m(\varepsilon,\Im)&=\mathbb{V}(\varepsilon,0)-\mathbb{Y}^{-1}\left[(1+\wp(v-1))\mathbb{Y}\left\{\sum_{m=0}^{\infty}\mathcal{B}_m+\sum_{m=0}^{\infty}\mathcal{C}_m+3\frac{\partial^3\mathbb{U}(\varepsilon,\Im)}{\partial\varepsilon^3}-\frac{\partial^2\mathbb{V}(\varepsilon,\Im)}{\partial\varepsilon^2}\right\}\right],
\end{aligned}$$

$$\sum_{m=0}^{\infty} \mathbb{U}_m(\varepsilon,\Im) = \frac{1}{2} - 8\tanh(-2\varepsilon) - \Upsilon^{-1}\left[(1+\wp(v-1))\Upsilon\left\{\sum_{m=0}^{\infty}\mathcal{A}_m + \frac{\partial \mathbb{U}(\varepsilon,\Im)}{\partial \varepsilon} + \frac{\partial \mathbb{V}(\varepsilon,\Im)}{\partial \varepsilon}\right\}\right],$$
$$\sum_{m=0}^{\infty} \mathbb{V}_m(\varepsilon,\Im) = 16 - 16\tanh^2(-2\varepsilon) - \Upsilon^{-1}\left[(1+\wp(v-1))\Upsilon\left\{\sum_{m=0}^{\infty}\mathcal{B}_m + \sum_{m=0}^{\infty}\mathcal{C}_m + 3\frac{\partial^3 \mathbb{U}(\varepsilon,\Im)}{\partial \varepsilon^3} - \frac{\partial^2 \mathbb{V}(\varepsilon,\Im)}{\partial \varepsilon^2}\right\}\right]. \quad (30)$$

With the aid of the Adomian polynomial, according to Equation (22), all forms of non-linear may be stated as

$$\mathcal{A}_0 = \mathbb{U}_0\frac{\partial \mathbb{U}_0}{\partial \varepsilon}, \quad \mathcal{A}_1 = \mathbb{U}_0\frac{\partial \mathbb{U}_1}{\partial \varepsilon} + \mathbb{U}_1\frac{\partial \mathbb{U}_0}{\partial \varepsilon}, \quad \mathcal{B}_0 = \mathbb{U}_0\frac{\partial \mathbb{V}_0}{\partial \beta}, \quad \mathcal{B}_1 = \mathbb{U}_0\frac{\partial \mathbb{V}_1}{\partial \beta} + \mathbb{U}_1\frac{\partial \mathbb{V}_0}{\partial \beta},$$
$$\mathcal{C}_0 = \mathbb{V}_0\frac{\partial \mathbb{U}_0}{\partial \varepsilon}, \quad \mathcal{C}_1 = \mathbb{V}_0\frac{\partial \mathbb{U}_1}{\partial \varepsilon} + \mathbb{V}_1\frac{\partial \mathbb{U}_0}{\partial \varepsilon}.$$

Therefore we can easily obtain

$$\mathbb{U}_0(\varepsilon,\Im) = \frac{1}{2} - 8\tanh(-2\varepsilon), \quad \mathbb{V}_0(\varepsilon,\Im) = 16 - 16\tanh^2(-2\varepsilon).$$

For $m = 0$

$$\mathbb{U}_1(\varepsilon,\Im) = -8\sec h^2(-2\varepsilon)\Big\{1 + \wp\Im - \wp\Big\}, \quad \mathbb{V}_1(\varepsilon,\Im) = -32\sec h^2(-2\varepsilon)\tanh(-2\varepsilon)\Big\{1 + \wp\Im - \wp\Big\}.$$

For $m = 1$

$$\mathbb{U}_2(\varepsilon,\Im) = -16\sec h^2(-2\varepsilon)\Big(4\sec h^2(-2\varepsilon) - 8\tanh^2(-2\varepsilon) + 3\tanh(-2\varepsilon)\Big)\Big\{(1-\wp)2\wp\Im + (1-\wp)^2 + \frac{\wp^2\Im^2}{2}\Big\},$$
$$\mathbb{V}_2(\varepsilon,\Im) = -32\sec h^2(-2\varepsilon)\{40\sec h^2(-2\varepsilon)\tanh(-2\varepsilon) + 96\tanh(-2\varepsilon) - 2\tanh^2(-2\varepsilon)$$
$$- 32\tanh^3(-2\varepsilon) - 25\sec h^2(-2\varepsilon)\}\Big\{(1-\wp)2\wp\Im + (1-\wp)^2 + \frac{\wp^2\Im^2}{2}\Big\}.$$

The remaining steps of the YDM outcomes may be conveniently gathered from $\mathbb{U}m$ *and* $\mathbb{V}m$ $(m \geq 2)$ *using the same methods. Then, we assess the sequence of possibilities as follows:*

$$\mathbb{U}(\varepsilon,\Im) = \sum_{m=0}^{\infty}\mathbb{U}_m(\varepsilon,\Im) = \mathbb{U}_0(\varepsilon,\Im) + \mathbb{U}_1(\varepsilon,\Im) + \mathbb{U}_2(\varepsilon,\Im) + \mathbb{U}_3(\varepsilon,\Im) + \cdots.$$
$$\mathbb{V}(\varepsilon,\Im) = \sum_{m=0}^{\infty}\mathbb{V}_m(\varepsilon,\Im) = \mathbb{V}_0(\varepsilon,\Im) + \mathbb{V}_1(\varepsilon,\Im) + \mathbb{V}_2(\varepsilon,\Im) + \mathbb{V}_3(\varepsilon,\Im) + \cdots.$$

$$\mathbb{U}(\varepsilon,\Im) = \frac{1}{2} - 8\tanh(-2\varepsilon) - 8\sec h^2(-2\varepsilon)\Big\{1 + \wp\Im - \wp\Big\} - 16\sec h^2(-2\varepsilon)\Big(4\sec h^2(-2\varepsilon)$$
$$-8\tanh^2(-2\varepsilon) + 3\tanh(-2\varepsilon)\Big)\Big\{(1-\wp)2\wp\Im + (1-\wp)^2 + \frac{\wp^2\Im^2}{2}\Big\} - \cdots.$$
$$\mathbb{V}(\varepsilon,\Im) = 16 - 16\tanh^2(-2\varepsilon) - 32\sec h^2(-2\varepsilon)\tanh(-2\varepsilon)\Big\{1 + \wp\Im - \wp\Big\} - 32\sec h^2(-2\varepsilon)$$
$$\{40\sec h^2(-2\varepsilon)\tanh(-2\varepsilon) + 96\tanh(-2\varepsilon) - 2\tanh^2(-2\varepsilon) - 32\tanh^3(-2\varepsilon)$$
$$- 25\sec h^2(-2\varepsilon)\}\Big\{(1-\wp)2\wp\Im + (1-\wp)^2 + \frac{\wp^2\Im^2}{2}\Big\} - \cdots.$$

At integer order $\wp = 1$*, the following series form of solution is achieved:*

$$
\begin{aligned}
\mathbb{U}(\varepsilon,\Im) =&\frac{1}{2}-8\tanh(-2\varepsilon)-8\operatorname{sech}(-2\varepsilon)^2\Im+8\operatorname{sech}^2(-2\varepsilon)\\
&\times\left\{3\tanh(-2\varepsilon)+8\tanh(-2\varepsilon)^2+4\operatorname{sech}^2(-2\varepsilon)\right\}\Im^2+\cdots.\\
\mathbb{V}(\varepsilon,\Im) =&16-16\tanh^2(-2\varepsilon)-32\operatorname{sech}^2(-2\varepsilon)\tanh(-2\varepsilon)\Im\\
&-16\sec h^2(-2\varepsilon)\left\{96\tanh(-2\varepsilon)-32\tanh^3(-2\varepsilon)\right.\\
&\left.+40\sec h^2(-2\varepsilon)\tanh(-2\varepsilon)-2\tanh^2(-2\varepsilon)-25\sec h^2(-2\varepsilon)\right\}\Im^2+\cdots.
\end{aligned}
$$

The exact solution of Equation (26) at $\wp=1$,

$$
\begin{aligned}
\mathbb{U}(\varepsilon,\Im) &= \frac{1}{2}-8\tanh\left\{-2\left(\varepsilon-\frac{\Im}{2}\right)\right\},\\
\mathbb{V}(\varepsilon,\Im) &= 16-16\tanh^2\left\{-2\left(\varepsilon-\frac{\Im}{2}\right)\right\}.
\end{aligned}
\tag{31}
$$

In Figures 1 and 2, the actual and Yang decomposition method solutions at an integer-order $\wp=1$ *are represented for both* $\mathbb{U}(\varepsilon,\Im)$ *and* $\mathbb{V}(\varepsilon,\Im)$ *of Example 1. It is observed that Yang decomposition method results are in good contact with the actual result of the models. In Figures 3 and 4, various fractional-order solutions of Example 1, at different fractional-orders,* $\wp=1,0.8,0.6,0.4$ *are plotted. It is investigated that for Example 1, the fractional-order solutions are convergent to an integer-order solution for both* $\mathbb{U}(\varepsilon,\Im)$ *and* $\mathbb{V}(\varepsilon,\Im)$*. In Tables 1 and 2 show that yang decomposition method of different fractional order* $\wp$ *of Example 1. In Tables 3 and 4 compassion of different analytical and numerical methods of Example 1.*

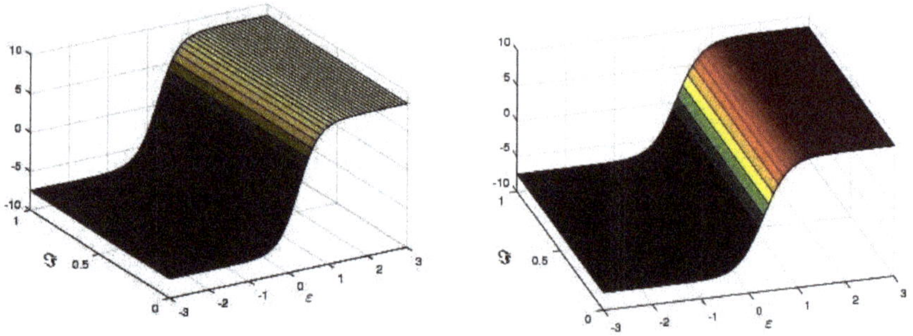

Figure 1. The actual and YDM solution of $\mathbb{U}(\varepsilon,\Im)$ at $\wp=1$ of Example 1.

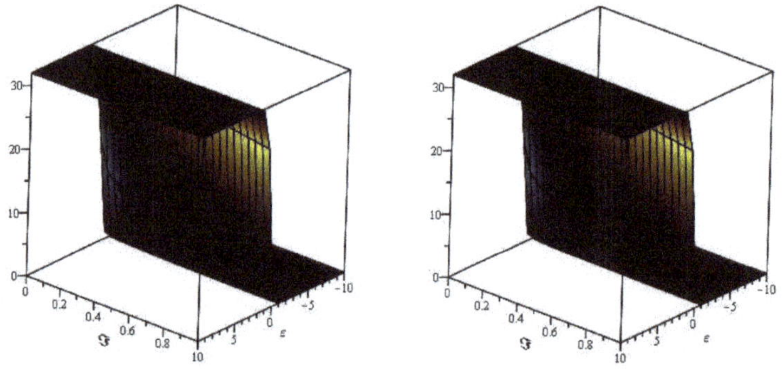

Figure 2. The actual and YDM solution of $\mathbb{V}(\varepsilon,\Im)$ at $\wp=1$ of Example 1.

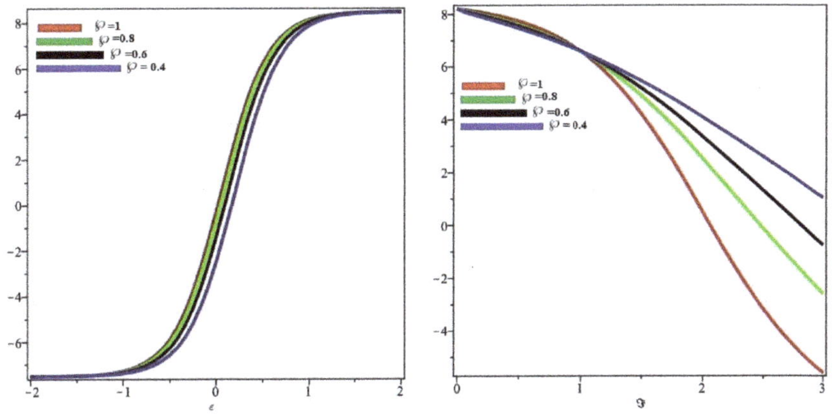

Figure 3. The fractional-order solutions of $\mathbb{V}(\varepsilon, \Im)$ at $\wp$ of Example 1.

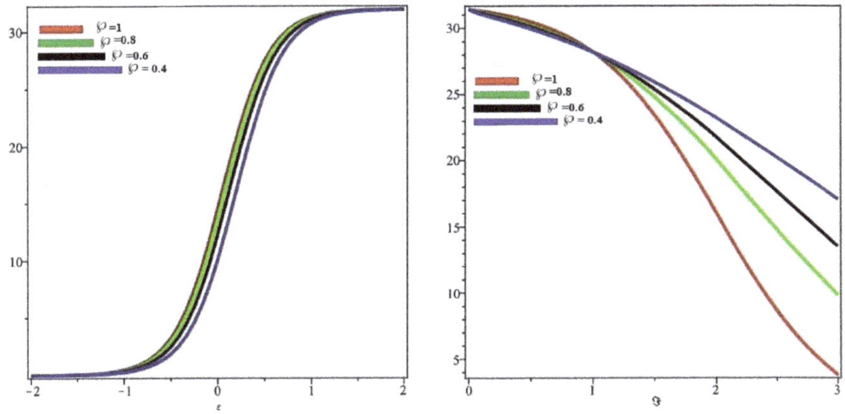

Figure 4. The fractional-order solutions of $\mathbb{V}(\varepsilon, \Im)$ at $\wp$ of Example 1.

Table 1. YDM solution of $\mathbb{U}(\varepsilon, \Im)$ at various fractional-order $\wp$ of Example 1.

$(\varepsilon; \Im)$	$\mathbb{U}(\varepsilon, \Im)$ at $\wp = 0.5$	$\mathbb{U}(\varepsilon, \Im)$ at $\wp = 0.75$	$\mathbb{U}(\varepsilon, \Im)$ at $\wp = 1$	**Exact Result**
(0.1, 0.2)	0.501928	0.501886	0.501893	0.501893
(0.1, 0.4)	0.501964	0.501938	0.501920	0.501920
(0.1, 0.6)	0.501989	0.501968	0.501858	0.501948
(0.2, 0.2)	0.499230	0.497189	0.499196	0.498090
(0.2, 0.4)	0.499265	0.497242	0.499223	0.498223
(0.2, 0.6)	0.499389	0.499269	0.499248	0.499148
(0.3, 0.2)	0.496582	0.496570	0.496569	0.494569
(0.3, 0.4)	0.496636	0.496413	0.496595	0.496595
(0.3, 0.6)	0.496659	0.496638	0.496620	0.496620
(0.4, 0.2)	0.49384	0.493818	0.493988	0.493988
(0.4, 0.4)	0.493874	0.493830	0.493833	0.493833
(0.4, 0.6)	0.493896	0.493877	0.493859	0.493859
(0.5, 0.2)	0.491544	0.491324	0.491512	0.491512
(0.5, 0.4)	0.491576	0.491354	0.491537	0.491327
(0.5, 0.6)	0.491598	0.491578	0.491562	0.491442

Table 2. YDM solution of $\mathbb{V}(\varepsilon, \Im)$ at various fractional-order $\wp$ of Example 1.

$(\varepsilon; \Im)$	$\mathbb{V}(\varepsilon, \Im)$ at $\wp = 0.5$	$\mathbb{V}(\varepsilon, \Im)$ at $\wp = 0.75$	$\mathbb{V}(\varepsilon, \Im)$ at $\wp = 1$	Exact Result
(0.1, 0.2)	0.0828104	0.0828124	0.0827800	0.0828900
(0.1, 0.4)	0.0828425	0.0828208	0.0828235	0.0839235
(0.1, 0.6)	0.0828646	0.0828460	0.0828280	0.0828391
(0.2, 0.2)	0.0804153	0.0803760	0.0803648	0.0803648
(0.2, 0.4)	0.0804264	0.0804054	0.0803886	0.0803886
(0.2, 0.6)	0.0804478	0.0804318	0.0804124	0.0804124
(0.3, 0.2)	0.0780546	0.0782358	0.0780250	0.0782472
(0.3, 0.4)	0.0780847	0.0780843	0.0780481	0.0782481
(0.3, 0.6)	0.0781055	0.0780881	0.0780711	0.0782711
(0.4, 0.2)	0.0757854	0.0757671	0.0757567	0.0757567
(0.4, 0.4)	0.0758148	0.0758148	0.0757810	0.0757780
(0.4, 0.6)	0.0758347	0.0758178	0.0758014	0.0758014
(0.5, 0.2)	0.0735850	0.0735673	0.0735572	0.0735578
(0.5, 0.4)	0.0736133	0.0736141	0.0735788	0.0735788
(0.5, 0.6)	0.0736328	0.0736164	0.0736225	0.0738005

Table 3. Comparison of absolute error (AE) of $\mathbb{U}(\varepsilon, \Im)$ at $\wp = 1$ obtained by various methods.

$(\varepsilon, \Im)$	AE Of ADM [37]	AE Of VIM [38]	AE Of OHAM [39]	AE of YDM
(0.1, 0.2)	1.05983×10^{-5}	1.34144×10^{-6}	1.18169×10^{-7}	1.56432×10^{-11}
(0.1, 0.4)	9.75585×10^{-6}	3.78688×10^{-6}	3.15656×10^{-7}	4.42375×10^{-10}
(0.1, 0.6)	8.77423×10^{-6}	6.27984×10^{-6}	4.92412×10^{-7}	2.18645×10^{-9}
(0.2, 0.2)	4.37319×10^{-5}	1.38978×10^{-6}	1.12395×10^{-7}	1.46768×10^{-11}
(0.2, 0.4)	3.82189×10^{-5}	3.51189×10^{-6}	2.86457×10^{-6}	4.35336×10^{-10}
(0.2, 0.6)	3.51272×10^{-5}	6.10117×10^{-5}	4.51245×10^{-6}	1.86439×10^{-09}
(0.3, 0.2)	9.62833×10^{-5}	1.25698×10^{-5}	1.13664×10^{-6}	1.38262×10^{-11}
(0.3, 0.4)	8.84418×10^{-5}	3.61977×10^{-5}	2.62353×10^{-6}	4.13675×10^{-10}
(0.3, 0.6)	8.33563×10^{-5}	5.96721×10^{-5}	4.46642×10^{-6}	1.46354×10^{-09}
(0.4, 0.2)	1.86687×10^{-4}	1.24938×10^{-5}	9.24537×10^{-5}	1.84245×10^{-11}
(0.4, 0.4)	1.72542×10^{-4}	3.52859×10^{-5}	2.63564×10^{-5}	3.60624×10^{-10}
(0.4, 0.6)	1.58687×10^{-4}	5.81821×10^{-5}	4.65446×10^{-5}	1.56784×10^{-09}
(0.5, 0.2)	2.88628×10^{-4}	1.21847×10^{-5}	9.72736×10^{-5}	1.42355×10^{-11}
(0.5, 0.4)	2.47825×10^{-4}	3.44373×10^{-5}	2.33457×10^{-5}	3.52237×10^{-10}
(0.5, 0.6)	2.47295×10^{-4}	5.47346×10^{-5}	4.38895×10^{-5}	1.66734×10^{-09}

Table 4. Comparison of absolute error of $\mathbb{V}(\varepsilon, \Im)$ at $\wp = 1$ obtained by various methods.

$(\zeta, \Im)$	AE Of ADM [37]	AE Of VIM [38]	AE Of OHAM [39]	AE of YDM
(0.1, 0.2)	6.52318×10^{-4}	1.23581×10^{-5}	5.72451×10^{-6}	3.262182×10^{-11}
(0.1, 0.4)	5.87694×10^{-4}	3.53456×10^{-5}	3.24632×10^{-6}	8.94623×10^{-10}
(0.1, 0.6)	5.72618×10^{-4}	5.63261×10^{-5}	3.38923×10^{-6}	4.23455×10^{-09}
(0.2, 0.2)	1.44292×10^{-3}	1.18127×10^{-5}	5.45771×10^{-6}	3.18974×10^{-11}
(0.2, 0.4)	1.33452×10^{-3}	3.34512×10^{-5}	2.86341×10^{-5}	8.21855×10^{-10}
(0.2, 0.6)	1.25527×10^{-3}	5.47838×10^{-5}	2.82545×10^{-5}	3.72424×10^{-09}
(0.3, 0.2)	2.14563×10^{-3}	1.14848×10^{-5}	5.36746×10^{-5}	2.45694×10^{-11}
(0.3, 0.4)	1.84963×10^{-3}	3.22828×10^{-5}	2.74231×10^{-5}	7.67817×10^{-10}
(0.3, 0.6)	1.72318×10^{-3}	5.32558×10^{-4}	2.66463×10^{-5}	3.4356×10^{-11}
(0.4, 0.2)	2.98211×10^{-3}	1.11468×10^{-4}	5.23838×10^{-5}	2.71232×10^{-11}
(0.4, 0.4)	2.59845×10^{-3}	3.13456×10^{-4}	2.72338×10^{-3}	7.24545×10^{-10}
(0.4, 0.6)	2.61896×10^{-3}	5.15382×10^{-3}	2.54328×10^{-3}	3.25166×10^{-09}
(0.5, 0.2)	3.84384×10^{-3}	9.86396×10^{-3}	4.83832×10^{-3}	2.13536×10^{-11}
(0.5, 0.4)	3.58728×10^{-3}	2.84228×10^{-3}	2.84563×10^{-3}	6.19148×10^{-10}
(0.5, 0.6)	3.35348×10^{-3}	4.72446×10^{-3}	2.52741×10^{-3}	3.24436×10^{-09}

Example 2. *Consider the fractional-order system of WBKEs [11]*

$$
\begin{aligned}
&{}^{CF}D_{\Im}^{\wp}\mathbb{U}(\varepsilon,\Im) + \mathbb{U}(\varepsilon,\Im)\frac{\partial \mathbb{U}(\varepsilon,\Im)}{\partial \varepsilon} + \frac{1}{2}\frac{\partial \mathbb{U}(\varepsilon,\Im)}{\partial \varepsilon} + \frac{\partial \mathbb{V}(\varepsilon,\Im)}{\partial \varepsilon} = 0,\\
&{}^{CF}D_{\Im}^{\wp}\mathbb{V}(\varepsilon,\Im) + \mathbb{U}(\varepsilon,\Im)\frac{\partial \mathbb{V}(\varepsilon,\Im)}{\partial \varepsilon} + \mathbb{V}(\varepsilon,\Im)\frac{\partial \mathbb{U}(\varepsilon,\Im)}{\partial \varepsilon} - \frac{1}{2}\frac{\partial^2 \mathbb{V}(\varepsilon,\Im)}{\partial \varepsilon^2} = 0,\\
&0 < \wp \leq 1, \quad 0 < \Im \leq 1, \quad -100 \leq \varepsilon \leq 100,
\end{aligned} \tag{32}
$$

with the initial conditions

$$
\begin{aligned}
\mathbb{U}(\varepsilon,0) &= \xi - \kappa \coth[\kappa(\varepsilon+\theta)],\\
\mathbb{V}(\varepsilon,0) &= -\kappa^2 \mathrm{cosech}^2[\kappa(\varepsilon+\theta)].
\end{aligned} \tag{33}
$$

Applying the Yang transformation of Equation (32), we have

$$
\begin{aligned}
\mathbb{Y}\left\{\frac{\partial^{\wp}\mathbb{U}(\varepsilon,\Im)}{\partial \Im^{\wp}}\right\} &= -\mathbb{Y}\left[\mathbb{U}(\varepsilon,\Im)\frac{\partial \mathbb{U}(\varepsilon,\Im)}{\partial \varepsilon} + \frac{1}{2}\frac{\partial \mathbb{U}(\varepsilon,\Im)}{\partial \varepsilon} + \frac{\partial \mathbb{V}(\varepsilon,\Im)}{\partial \varepsilon}\right],\\
\mathbb{Y}\left\{\frac{\partial^{\wp}\mathbb{V}(\varepsilon,\Im)}{\partial \Im^{\wp}}\right\} &= -\mathbb{Y}\left[\mathbb{U}(\varepsilon,\Im)\frac{\partial \mathbb{V}(\varepsilon,\Im)}{\partial \varepsilon} + \mathbb{V}(\varepsilon,\Im)\frac{\partial \mathbb{U}(\varepsilon,\Im)}{\partial \varepsilon} - \frac{1}{2}\frac{\partial^2 \mathbb{V}(\varepsilon,\Im)}{\partial \varepsilon^2}\right],
\end{aligned}
$$

$$
\begin{aligned}
\frac{1}{(1+\wp(v-1))}\mathbb{Y}\{\mathbb{U}(\varepsilon,\Im)\} - v\mathbb{U}(\varepsilon,0) &= -\mathbb{Y}\left[\mathbb{U}(\varepsilon,\Im)\frac{\partial \mathbb{U}(\varepsilon,\Im)}{\partial \varepsilon} + \frac{1}{2}\frac{\partial \mathbb{U}(\varepsilon,\Im)}{\partial \varepsilon} + \frac{\partial \mathbb{V}(\varepsilon,\Im)}{\partial \varepsilon}\right],\\
\frac{1}{(1+\wp(v-1))}\mathbb{Y}\{\mathbb{V}(\varepsilon,\Im)\} - v\mathbb{V}(\varepsilon,0) &= -\mathbb{Y}\left[\mathbb{U}(\varepsilon,\Im)\frac{\partial \mathbb{V}(\varepsilon,\Im)}{\partial \varepsilon} + \mathbb{V}(\varepsilon,\Im)\frac{\partial \mathbb{U}(\varepsilon,\Im)}{\partial \varepsilon} - \frac{1}{2}\frac{\partial^2 \mathbb{V}(\varepsilon,\Im)}{\partial \varepsilon^2}\right].
\end{aligned}
$$

The above equation is simplified

$$
\begin{aligned}
\mathbb{Y}\{\mathbb{U}(\varepsilon,\Im)\} &= v\{\mathbb{U}(\varepsilon,0)\} - (1+\wp(v-1))\mathbb{Y}\left[\mathbb{U}(\varepsilon,\Im)\frac{\partial \mathbb{U}(\varepsilon,\Im)}{\partial \varepsilon} + \frac{1}{2}\frac{\partial \mathbb{U}(\varepsilon,\Im)}{\partial \varepsilon} + \frac{\partial \mathbb{V}(\varepsilon,\Im)}{\partial \varepsilon}\right],\\
\mathbb{Y}\{\mathbb{V}(\varepsilon,\Im)\} &= v\{\mathbb{V}(\varepsilon,0)\} - (1+\wp(v-1))\mathbb{Y}\left[\mathbb{U}(\varepsilon,\Im)\frac{\partial \mathbb{V}(\varepsilon,\Im)}{\partial \varepsilon} + \mathbb{V}(\varepsilon,\Im)\frac{\partial \mathbb{U}(\varepsilon,\Im)}{\partial \varepsilon} - \frac{1}{2}\frac{\partial^2 \mathbb{V}(\varepsilon,\Im)}{\partial \varepsilon^2}\right].
\end{aligned} \tag{34}
$$

Using inverse Yang transform, we have

$$
\begin{aligned}
\mathbb{U}(\varepsilon,\Im) &= \mathbb{U}(\varepsilon,0) - \mathbb{Y}^{-1}\left[(1+\wp(v-1))\mathbb{Y}\left\{\mathbb{U}(\varepsilon,\Im)\frac{\partial \mathbb{U}(\varepsilon,\Im)}{\partial \varepsilon} + \frac{1}{2}\frac{\partial \mathbb{U}(\varepsilon,\Im)}{\partial \varepsilon} + \frac{\partial \mathbb{V}(\varepsilon,\Im)}{\partial \varepsilon}\right\}\right],\\
\mathbb{V}(\varepsilon,\Im) &= \mathbb{V}(\varepsilon,0) - \mathbb{Y}^{-1}\left[(1+\wp(v-1))\mathbb{Y}\left\{\mathbb{U}(\varepsilon,\Im)\frac{\partial \mathbb{V}(\varepsilon,\Im)}{\partial \varepsilon} + \mathbb{V}(\varepsilon,\Im)\frac{\partial \mathbb{U}(\varepsilon,\Im)}{\partial \varepsilon} - \frac{1}{2}\frac{\partial^2 \mathbb{V}(\varepsilon,\Im)}{\partial \varepsilon^2}\right\}\right].
\end{aligned} \tag{35}
$$

Assume that the infinite series solution functions $\mathbb{U}(\varepsilon,\Im)$ and $\mathbb{V}(\varepsilon,\Im)$ are as follows:

$$
\mathbb{U}(\varepsilon,\Im) = \sum_{m=0}^{\infty}\mathbb{U}_m(\varepsilon,\Im), \quad \text{and} \quad \mathbb{V}(\varepsilon,\Im) = \sum_{m=0}^{\infty}\mathbb{V}_m(\varepsilon,\Im).
$$

Remember that $\mathbb{U}\mathbb{U}_{\varepsilon} = \sum_{m=0}^{\infty}\mathcal{A}_m$, $\mathbb{U}\mathbb{V}_{\varepsilon} = \sum_{m=0}^{\infty}\mathcal{B}_m$ and $\mathbb{V}\mathbb{U}_{\varepsilon} = \sum_{m=0}^{\infty}\mathcal{C}_m$ are the Adomian polynomials

$$
\begin{aligned}
\sum_{m=0}^{\infty}\mathbb{U}_m(\varepsilon,\Im) &= \mathbb{U}(\varepsilon,0) - \mathbb{Y}^{-1}\left[(1+\wp(v-1))\mathbb{Y}\left\{\sum_{m=0}^{\infty}\mathcal{A}_m + \frac{1}{2}\frac{\partial \mathbb{U}(\varepsilon,\Im)}{\partial \varepsilon} + \frac{\partial \mathbb{V}(\varepsilon,\Im)}{\partial \varepsilon}\right\}\right],\\
\sum_{m=0}^{\infty}\mathbb{V}_m(\varepsilon,\Im) &= \mathbb{V}(\varepsilon,0) - \mathbb{Y}^{-1}\left[(1+\wp(v-1))\mathbb{Y}\left\{\sum_{m=0}^{\infty}\mathcal{B}_m + \sum_{m=0}^{\infty}\mathcal{C}_m - \frac{1}{2}\frac{\partial^2 \mathbb{V}(\varepsilon,\Im)}{\partial \varepsilon^2}\right\}\right],
\end{aligned}
$$

$$\sum_{m=0}^{\infty}\mathbb{U}_m(\varepsilon,\Im)=\xi-\kappa\coth[\kappa(\varepsilon+\theta)]-\mathbb{Y}^{-1}\left[(1+\wp(v-1))\mathbb{Y}\left\{\sum_{m=0}^{\infty}\mathcal{A}_m+\frac{1}{2}\frac{\partial\mathbb{U}(\varepsilon,\Im)}{\partial\varepsilon}+\frac{\partial\mathbb{V}(\varepsilon,\Im)}{\partial\varepsilon}\right\}\right],$$
$$\sum_{m=0}^{\infty}\mathbb{V}_m(\varepsilon,\Im)=-\kappa^2\mathrm{cosech}^2[\kappa(\varepsilon+\theta)]-\mathbb{Y}^{-1}\left[(1+\wp(v-1))\mathbb{Y}\left\{\sum_{m=0}^{\infty}\mathcal{B}_m+\sum_{m=0}^{\infty}\mathcal{C}_m-\frac{1}{2}\frac{\partial^2\mathbb{V}(\varepsilon,\Im)}{\partial\varepsilon^2}\right\}\right]. \quad (36)$$

With the aid of the Adomian polynomial according to Equation (22), all forms of non-linear may be stated as

$$\mathcal{A}_0=\mathbb{U}_0\frac{\partial\mathbb{U}_0}{\partial\varepsilon},\quad \mathcal{A}_1=\mathbb{U}_0\frac{\partial\mathbb{U}_1}{\partial\varepsilon}+\mathbb{U}_1\frac{\partial\mathbb{U}_0}{\partial\varepsilon},\quad \mathcal{B}_0=\mathbb{U}_0\frac{\partial\mathbb{V}_0}{\partial\beta},\quad \mathcal{B}_1=\mathbb{U}_0\frac{\partial\mathbb{V}_1}{\partial\beta}+\mathbb{U}_1\frac{\partial\mathbb{V}_0}{\partial\beta},$$
$$\mathcal{C}_0=\mathbb{V}_0\frac{\partial\mathbb{U}_0}{\partial\varepsilon},\quad \mathcal{C}_1=\mathbb{V}_0\frac{\partial\mathbb{U}_1}{\partial\varepsilon}+\mathbb{V}_1\frac{\partial\mathbb{U}_0}{\partial\varepsilon}.$$

Hence, one can easily obtain

$$\mathbb{U}_0(\varepsilon,\Im)=\xi-\kappa\coth[\kappa(\varepsilon+\theta)],\quad \mathbb{V}_0(\varepsilon,\Im)=-\kappa^2\mathrm{cosech}^2[\kappa(\varepsilon+\theta)].$$

For $m=0$

$$\mathbb{U}_1(\varepsilon,\Im)=-\xi\kappa^2\mathrm{cosech}^2[\kappa(\varepsilon+\theta)]\Big\{1+\wp\Im-\wp\Big\},$$
$$\mathbb{V}_1(\varepsilon,\Im)=-\xi\kappa^2\mathrm{cosech}^2[\kappa(\varepsilon+\theta)]\coth[\kappa(\varepsilon+\theta)]\Big\{1+\wp\Im-\wp\Big\}.$$

For $m=1$

$$\mathbb{U}_2(\varepsilon,\Im)=\xi\kappa^4\mathrm{cosech}^2[\kappa(\varepsilon+\theta)]\Big\{2\xi\kappa\Big\{(1-\wp)^2 3\wp\Im+(1-\wp)^3+\frac{3\wp^2(1-\wp)\Im^2}{2}+\frac{\wp^3\Im^3}{3!}\Big\}$$
$$-(3\coth^2([\kappa(\varepsilon+\theta)]-1))\Big\{(1-\wp)2\wp\Im+(1-\wp)^2+\frac{\wp^2\Im^2}{2}\Big\}\Big\},$$
$$\mathbb{V}_2(\varepsilon,\Im)=[2\xi\kappa^5\mathrm{cosech}^2[\kappa(\varepsilon+\theta)]]\Big[\xi\kappa\mathrm{cosech}^2(3\coth^2([\kappa(\varepsilon+\theta)]-1))\Big\{(1-\wp)^2 3\wp\Im+(1-\wp)^3+\frac{3\wp^2(1-\wp)\Im^2}{2}$$
$$+\frac{\wp^3\Im^3}{3!}\Big\}+\frac{2\xi\kappa\mathrm{cosech}^2\coth^2([\kappa(\varepsilon+\theta)])\Im^{3\wp}}{\Gamma(\wp+1)\Gamma(3\wp+1)}-2\xi\coth(3\mathrm{cosech}^2([\kappa(\varepsilon+\theta)]-1))\Big\{(1-\wp)2\wp\Im+(1-\wp)^2+\frac{\wp^2\Im^2}{2}\Big\}\Big].$$

The remaining steps of the YDM results may be conveniently gathered from $\mathbb{U}_m$ and $\mathbb{V}_m$ ($m\geq 2$) using the same procedure. The alternative series can then be assessed as follows:

$$\mathbb{U}(\varepsilon,\Im)=\sum_{m=0}^{\infty}\mathbb{U}_m(\varepsilon,\Im)=\mathbb{U}_0(\varepsilon,\Im)+\mathbb{U}_1(\varepsilon,\Im)+\mathbb{U}_2(\varepsilon,\Im)+\mathbb{U}_3(\varepsilon,\Im)+\cdots.$$
$$\mathbb{V}(\varepsilon,\Im)=\sum_{m=0}^{\infty}\mathbb{V}_m(\varepsilon,\Im)=\mathbb{V}_0(\varepsilon,\Im)+\mathbb{V}_1(\varepsilon,\Im)+\mathbb{V}_2(\varepsilon,\Im)+\mathbb{V}_3(\varepsilon,\Im)+\cdots.$$

$$\mathbb{U}(\varepsilon,\Im)=\xi-\kappa\coth[\kappa(\varepsilon+\theta)]-\xi\kappa^2\mathrm{cosech}^2[\kappa(\varepsilon+\theta)]\Big\{1+\wp\Im-\wp\Big\}$$
$$+\xi\kappa^4\mathrm{cosech}^2[\kappa(\varepsilon+\theta)]\Big\{2\xi\kappa\Gamma(2\wp+1)\Big\{(1-\wp)^2 3\wp\Im+(1-\wp)^3+\frac{3\wp^2(1-\wp)\Im^2}{2}+\frac{\wp^3\Im^3}{3!}\Big\}$$
$$-(3\coth^2([\kappa(\varepsilon+\theta)]-1))\Big\{(1-\wp)2\wp\Im+(1-\wp)^2+\frac{\wp^2\Im^2}{2}\Big\}\Big\}-\cdots.$$
$$\mathbb{V}(\varepsilon,\Im)=-\kappa^2\mathrm{cosech}^2[\kappa(\varepsilon+\theta)]-\xi\kappa^2\mathrm{cosech}^2[\kappa(\varepsilon+\theta)]\coth[\kappa(\varepsilon+\theta)]\Big\{1+\wp\Im-\wp\Big\}$$
$$+[2\xi\kappa^5\mathrm{cosech}^2[\kappa(\varepsilon+\theta)]]\Big[\xi\kappa\mathrm{cosech}^2(3\coth^2([\kappa(\varepsilon+\theta)]-1))\Big\{(1-\wp)^2 3\wp\Im+(1-\wp)^3+\frac{3\wp^2(1-\wp)\Im^2}{2}+\frac{\wp^3\Im^3}{3!}\Big\}$$
$$+\frac{2\xi\kappa\mathrm{cosech}^2\coth^2([\kappa(\varepsilon+\theta)])\Im^{3\wp}}{\Gamma(\wp+1)\Gamma(3\wp+1)}-2\xi\coth(3\mathrm{cosech}^2([\kappa(\varepsilon+\theta)]-1))\Big\{(1-\wp)2\wp\Im+(1-\wp)^2+\frac{\wp^2\Im^2}{2}\Big\}\Big]-\cdots.$$

We achieve the following series solution at integer order $\wp = 1, \kappa = 0.1, \zeta = 0.005\,, \theta = 10,$ defined by

$$\begin{aligned}\mathbb{U}(\varepsilon,\Im) =& 0.005 - 0.1\coth(0.1\varepsilon+10) - 0.0005\mathrm{cosech}^2(0.1\varepsilon+10)\Im + 5\times10^{-7}\mathrm{cosech}^2(0.1\varepsilon+10)0.003\Im^3\\ &-0.5\Big(3\coth^2(0.1\varepsilon+10)-1.\Big)\Big)\Im^2,\\ \mathbb{V}(\varepsilon,\Im) =& -0.01\mathrm{cosech}^2(0.1\varepsilon+10) - 0.000010\mathrm{cosech}^2(0.1\varepsilon+10)\times\coth(0.1\varepsilon+10)\Im + 1.0\times10^{-7}\mathrm{cosech}^2(0.1\varepsilon+10)\\ &\times\Big[8.3\times10^{-5}\Im^3\mathrm{cosech}^2(0.1\varepsilon+10)(3\coth(0.1\varepsilon+10)-1) - \Im^2\coth(0.1\varepsilon+10)\Big(3\mathrm{cosech}^2(0.1\varepsilon+10)-1\Big)\\ &+1.6\times10^{-4}\Im^3\mathrm{cosech}^2(0.1\varepsilon+10)\coth(0.1\varepsilon+10)\Big].\end{aligned}$$

The exact result of Equation (32) at $\wp = 1$ and taking $\xi = 0.005$, $\theta = 10$ and $\kappa = 0.1$.

$$\begin{aligned}\mathbb{U}(\varepsilon,\Im) &== \xi - \kappa\coth[\kappa(\varepsilon+\theta-\xi\Im)],\\ \mathbb{V}(\varepsilon,\Im) &= -\kappa^2\mathrm{cosech}^2[\kappa(\varepsilon+\theta-\xi\Im)].\end{aligned} \tag{37}$$

In Figures 5 and 6, the actual and Yang decomposition method solutions at an integer-order $\wp = 1$ are represented for both $\mathbb{U}(\varepsilon,\Im)$ and $\mathbb{V}(\varepsilon,\Im)$ of Example 1. It is observed that Yang decomposition method results are in good contact with the actual result of the models. In Figures 7 and 8, various fractional-order solutions of Example 2, at different fractional-orders, $\wp = 1, 0.8, 0.6, 0.4$ are plotted. It is investigated that for Example 2, the fractional-order solutions are convergent to an integer-order solution for both $\mathbb{U}(\varepsilon,\Im)$ and $\mathbb{V}(\varepsilon,\Im)$. In Tables 5 and 6 show that yang decomposition method of different fractional order $\wp$ of Example 2.

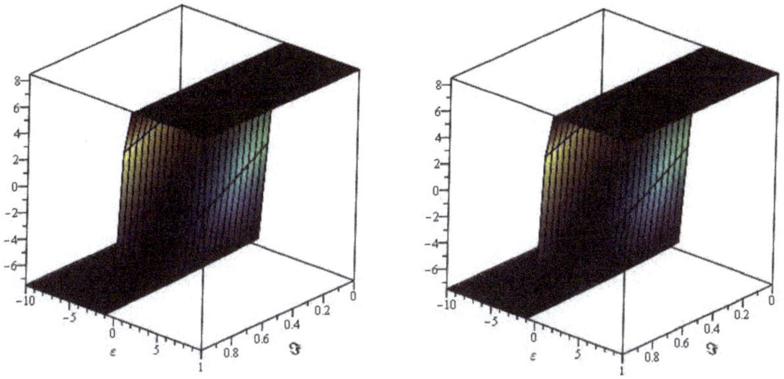

Figure 5. The actual and YDM solution of $\mathbb{U}(\varepsilon,\Im)$ at $\wp = 1$ of Example 2.

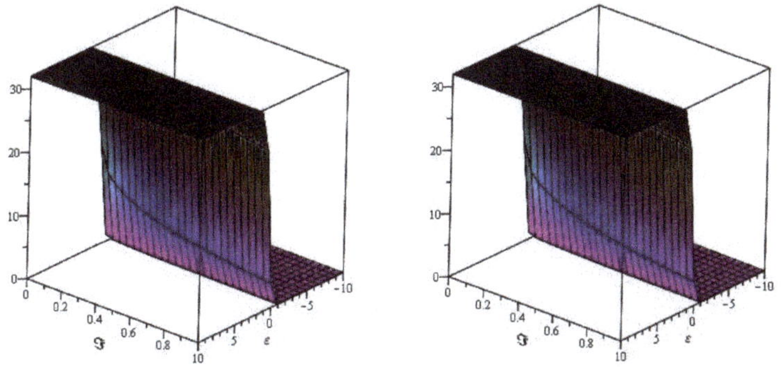

Figure 6. The actual and YDM solution of $\mathbb{V}(\varepsilon,\Im)$ at $\wp = 1$ of Example 2.

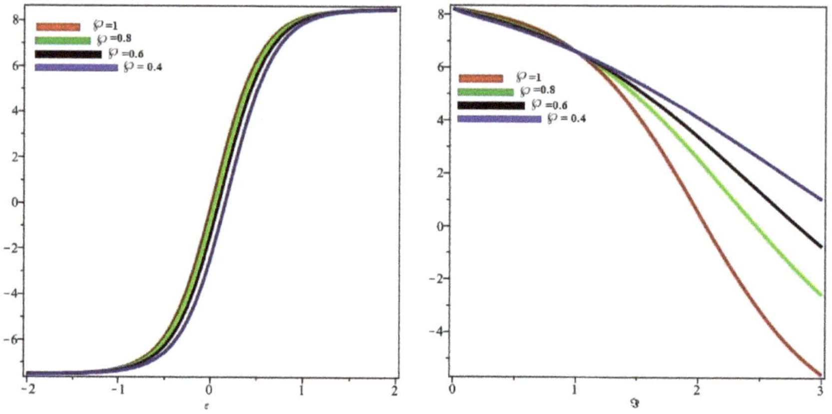

Figure 7. The fractional-order solutions of $\mathbb{V}(\varepsilon, \Im)$ at $\wp$ of Example 2.

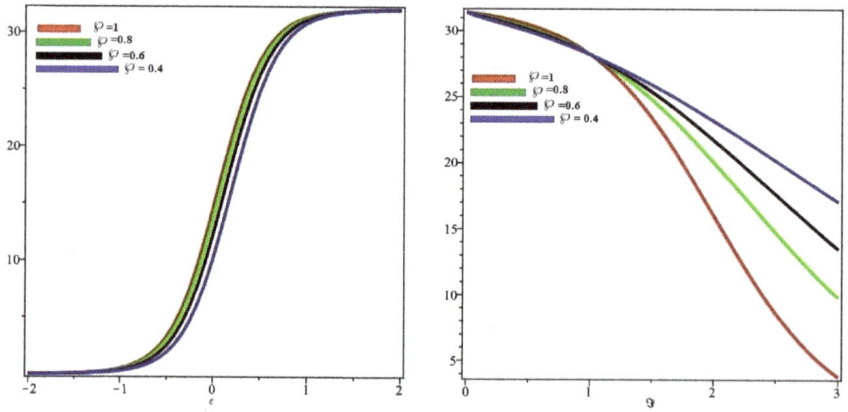

Figure 8. The fractional-order solutions of $\mathbb{V}(\varepsilon, \Im)$ at $\wp$ of Example 2.

Table 5. YDM solution of $\mathbb{U}(\varepsilon, \Im)$ at different fractional-order $\wp$ of Example 2.

$(\varepsilon; \Im)$	$\mathbb{U}(\varepsilon, \Im)$ at $\wp = 0.5$	$\mathbb{U}(\varepsilon, \Im)$ at $\wp = 0.75$	$\mathbb{U}(\varepsilon, \Im)$ at $\wp = 1$	**Exact Result**
(0.1, 0.2)	0.500726	0.500684	0.500671	0.500761
(0.1, 0.4)	0.500742	0.500738	0.500720	0.500720
(0.1, 0.6)	0.500767	0.500746	0.500726	0.500826
(0.2, 0.2)	0.497230	0.497187	0.498174	0.498074
(0.2, 0.4)	0.497243	0.497221	0.497453	0.498121
(0.2, 0.6)	0.496267	0.497047	0.497248	0.498128
(0.3, 0.2)	0.494382	0.494360	0.494347	0.495447
(0.3, 0.4)	0.494414	0.494411	0.494373	0.495473
(0.3, 0.6)	0.494437	0.494418	0.494400	0.49540
(0.4, 0.2)	0.491920	0.491818	0.492786	0.492886
(0.4, 0.4)	0.491852	0.491831	0.492810	0.492911
(0.4, 0.6)	0.491874	0.491855	0.491993	0.492937
(0.5, 0.2)	0.491322	0.491322	0.490312	0.490410
(0.5, 0.4)	0.491354	0.491332	0.490315	0.490415
(0.5, 0.6)	0.491278	0.491358	0.490342	0.490440

Table 6. YDM solution of $\mathbb{V}(\varepsilon, \Im)$ at different fractional-order $\wp$ of Example 2.

$(\varepsilon; \Im)$	$\mathbb{V}(\varepsilon, \Im)$ at $\wp = 0.5$	$\mathbb{V}(\varepsilon, \Im)$ at $\wp = 0.75$	$\mathbb{V}(\varepsilon, \Im)$ at $\wp = 1$	**Exact Result**
(0.1, 0.2)	0.0939215	0.0939015	0.09389	0.09389
(0.1, 0.4)	0.0939536	0.0939319	0.0939146	0.0939146
(0.1, 0.6)	0.0939757	0.0939571	0.0939391	0.0939391
(0.2, 0.2)	0.0915064	0.091487	0.0914759	0.0914759
(0.2, 0.4)	0.0915375	0.0915165	0.0914997	0.0914997
(0.2, 0.6)	0.0915589	0.0915409	0.0915235	0.0915235
(0.3, 0.2)	0.0891657	0.0891469	0.0891361	0.0891361
(0.3, 0.4)	0.0891958	0.0891754	0.0891592	0.0891592
(0.3, 0.6)	0.0892166	0.0891992	0.0891822	0.0891822
(0.4, 0.2)	0.0868965	0.0868782	0.0868678	0.0868678
(0.4, 0.4)	0.0869257	0.0869059	0.0868901	0.08688901
(0.4, 0.6)	0.0869458	0.0869289	0.0869125	0.0869125
(0.5, 0.2)	0.0846961	0.0846784	0.0846683	0.0846683
(0.5, 0.4)	0.0847244	0.0847052	0.0846899	0.0846899
(0.5, 0.6)	0.0847439	0.0847275	0.0847116	0.0847116

5. Conclusions

This research applies the Yang decomposition method to a fractional-order non-linear Whitham-Broer-Kaup equations system. The suggested approach has been thoroughly researched for fractional-order systems of linear and non-linear differential equations. The numerical results show that the approach is accurate and effective in achieving numerical solutions for non-linear fractional partial differential equations. The proposed methodology is an effective and convenient tool for evaluating numerical solutions to non-linear coupled systems of fractional PDEs compared to previous analytical techniques. Furthermore, the proposed scheme is easy and intuitive, requiring less computing time to solve additional fractional-order partial differential equations.

Funding: This work is supported by the Deanship of Scientific Research, King Faisal University, Saudi Arabia (Nasher (third edition) Grant Number: NA000190).

Institutional Review Board Statement: Not applicable.

Informed Consent Statement: Not applicable.

Data Availability Statement: No data were used to support this study.

Acknowledgments: This work is supported by the Deanship of Scientific Research, King Faisal University, Saudi Arabia (Nasher (third edition) Grant Number: NA000190).

Conflicts of Interest: The author declare that I have no conflict of interest.

References

1. Kavitha, K.; Vijayakumar, V.; Udhayakumar, R.; Sakthivel, N.; Sooppy Nisar, K. A note on approximate controllability of the Hilfer fractional neutral differential inclusions with infinite delay. *Math. Methods Appl. Sci.* **2021**, *44*, 4428–4447. [CrossRef]
2. Shah, R.; Farooq, U.; Khan, H.; Baleanu, D.; Kumam, P.; Arif, M. Fractional View Analysis of Third Order Kortewege-De Vries Equations, Using a New Analytical Technique. *Front. Phys.* **2020**, *7*, 244. [CrossRef]
3. Vijayakumar, V.; Nisar, K.S.; Chalishajar, D.; Shukla, A.; Malik, M.; Alsaadi, A.; Aldosary, S.F. A Note on Approximate Controllability of Fractional Semilinear Integrodifferential Control Systems via Resolvent Operators. *Fractal Fract.* **2022**, *6*, 73. [CrossRef]
4. Shah, R.; Khan, H.; Farooq, U.; Baleanu, D.; Kumam, P.; Arif, M. A New Analytical Technique to Solve System of Fractional-Order Partial Differential Equations. *IEEE Access* **2019**, *7*, 150037–150050. [CrossRef]
5. Hammouch, Z.; Mekkaoui, T.; Agarwal, P. Optical solitons for the Calogero-Bogoyavlenskii-Schiff equation in (2 + 1) dimensions with time-fractional conformable derivative. *Eur. Phys. J. Plus* **2018**, *133*, 248. [CrossRef]
6. Shah, R.; Khan, H.; Baleanu, D.; Kumam, P.; Arif, M. The analytical investigation of time-fractional multi-dimensional Navier–Stokes equation. *Alex. Eng. J.* **2020**, *59*, 2941–2956. [CrossRef]

7. Ruzhansky, M.; Cho, Y.J.; Agarwal, P.; Area, I. (Eds.) *Advances in Real and Complex Analysis with Applications*; Springer: Singapore, 2017.
8. Alesemi, M.; Iqbal, N.; Botmart, T. Novel Analysis of the Fractional-Order System of Non-Linear Partial Differential Equations with the Exponential-Decay Kernel. *Mathematics* **2022**, *10*, 615. [CrossRef]
9. Xie, F.; Yan, Z.; Zhang, H. Explicit and exact traveling wave solutions of Whitham-Broer-Kaup shallow water equations. *Phys. Lett. A* **2001**, *285*, 76–80. [CrossRef]
10. Wang, L.; Chen, X. Approximate analytical solutions of time fractional Whitham-Broer-Kaup equations by a residual power series method. *Entropy* **2015**, *17*, 6519–6533. [CrossRef]
11. Ali, A.; Shah, K.; Khan, R.A. Numerical treatment for traveling wave solutions of fractional Whitham-Broer-Kaup equations. *Alex. Eng. J.* **2018**, *57*, 1991–1998. [CrossRef]
12. Ahmad, J.; Mushtaq, M.; Sajjad, N. Exact Solution of Whitham Broer-Kaup Shallow Water Wave Equations. *J. Sci. Arts* **2015**, *15*, 5.
13. Kupershmidt, B. Mathematics of dispersive water waves. *Commun. Math. Phys.* **1985**, *99*, 51–73. [CrossRef]
14. Whitham, G.B. Variational methods and applications to water waves. Proceedings of the Royal Society of London. *Ser. A Math. Phys. Sci.* **1967**, *299*, 6–25.
15. Broer, L.J.F. Approximate equations for long water waves. *Appl. Sci. Res.* **1975**, *31*, 377–395. [CrossRef]
16. Kaup, D. A higher-order water-wave equation and the method for solving it. *Prog. Theor. Phys.* **1975**, *54*, 396–408. [CrossRef]
17. Rashidi, M.M.; Ganji, D.D.; Dinarvand, S. Approximate traveling wave solutions of coupled Whitham-Broer-Kaup shallow water equations by homotopy analysis method. *Differ. Equ. Nonlinear Mech.* **2008**, *2008*, 243459. [CrossRef]
18. Mohyud-Din, S.T.; Noor, M.A. Homotopy perturbation method for solving partial differential equations. *Z. Fur Nat. A* **2009**, *64*, 157–170. [CrossRef]
19. Biazar, J.; Aminikhah, H. Study of convergence of homotopy perturbation method for systems of partial differential equations. *Comput. Math. Appl.* **2009**, *58*, 2221–2230.
20. Yuzbasi, S.; Sahin, N. Numerical solutions of singularly perturbed one-dimensional parabolic convection-diffusion problems by the Bessel collocation method. *Appl. Math. Comput.* **2013**, *220*, 305–315.
21. Shah, R.; Khan, H.; Baleanu, D. Fractional Whitham-Broer-Kaup equations within modified analytical approaches. *Axioms* **2019**, *8*, 125. [CrossRef]
22. Ali, I.; Khan, H.; Shah, R.; Baleanu, D.; Kumam, P.; Arif, M. Fractional view analysis of acoustic wave equations, using fractional-order differential equations. *Appl. Sci.* **2020**, *10*, 610. [CrossRef]
23. Nonlaopon, K.; Alsharif, A.M.; Zidan, A.M.; Khan, A.; Hamed, Y.S.; Shah, R. Numerical investigation of fractional-order Swift-Hohenberg equations via a Novel transform. *Symmetry* **2021**, *13*, 1263. [CrossRef]
24. Xu, J.; Khan, H.; Shah, R.; Alderremy, A.A.; Aly, S.; Baleanu, D. The analytical analysis of nonlinear fractional-order dynamical models. *AIMS Math.* **2021**, *6*, 6201–6219. [CrossRef]
25. Iqbal, N.; Yasmin, H.; Ali, A.; Bariq, A.; Al-Sawalha, M.M.; Mohammed, W.W. Numerical Methods for Fractional-Order Fornberg-Whitham Equations in the Sense of Atangana-Baleanu Derivative. *J. Funct. Spaces* **2021**, *2021*, 2197247. [CrossRef]
26. Mohyud-Din, S.T.; Yildirim, A.; Demirli, G. Traveling wave solutions of Whitham-Broer-Kaup equations by homotopy perturbation method. *J. King Saud Univ. Sci.* **2010**, *22*, 173–176. [CrossRef]
27. Iqbal, N.; Yasmin, H.; Rezaiguia, A.; Kafle, J.; Almatroud, A.O.; Hassan, T.S. Analysis of the Fractional-Order Kaup-Kupershmidt Equation via Novel Transforms. *J. Math.* **2021**, *2021*, 2567927. [CrossRef]
28. Rach, R. On the Adomian (decomposition) method and comparisons with Picard's method. *J. Math. Anal. Appl.* **1987**, *128*, 480–483. [CrossRef]
29. Wazwaz, A.M. A reliable modification of Adomian decomposition method. *Appl. Math. Comput.* **1999**, *102*, 77–86. [CrossRef]
30. Alesemi, M.; Iqbal, N.; Hamoud, A.A. The Analysis of Fractional-Order Proportional Delay Physical Models via a Novel Transform. *Complexity* **2022**, *2022*, 2431533. [CrossRef]
31. Alesemi, M.; Iqbal, N.; Abdo, M.S. Novel Investigation of Fractional-Order Cauchy-Reaction Diffusion Equation Involving Caputo-Fabrizio Operator. *J. Funct. Spaces* **2022**, *2022*, 4284060. [CrossRef]
32. Kumar, M. Numerical solution of singular boundary value problems using advanced Adomian decomposition method. *Eng. Comput.* **2021**, *37*, 2853–2863.
33. Agarwal, R.; Mofarreh, F.; Shah, R.; Luangboon, W.; Nonlaopon, K. An Analytical Technique, Based on Natural Transform to Solve Fractional-Order Parabolic Equations. *Entropy* **2021**, *23*, 1086. [CrossRef] [PubMed]
34. Caputo, M.; Fabrizio, M. On the singular kernels for fractional derivatives. some applications to partial differential equations. *Progr. Fract. Differ. Appl.* **2021**, *7*, 1–4.
35. Yang, X.J. A new integral transform method for solving steady heat-transfer problem. *Therm. Sci.* **2016**, *20* (Suppl. S3), 639–642. [CrossRef]
36. Ahmad, S.; Ullah, A.; Akgul, A.; De la Sen, M. A Novel Homotopy Perturbation Method with Applications to Nonlinear Fractional Order KdV and Burger Equation with Exponential-Decay Kernel. *J. Funct. Spaces* **2021**, *2021*, 8770488. [CrossRef]
37. El-Sayed, S.M.; Kaya, D. Exact and numerical traveling wave solutions of Whitham Broer Kaup equations. *Appl. Math. Comput.* **2005**, *167*, 1339–1349. [CrossRef]

38. Rafei, M.; Daniali, H. Application of the variational iteration method to the Whitham-Broer-Kaup equations. *Comput. Math. Appl.* **2007**, *54*, 1079–1085. [CrossRef]
39. Sirajul, H.; Ishaq, M. Solution of coupled Whitham-Broer-Kaup equations using optimal homotopy asymptotic method. *Ocean. Eng.* **2014**, *84*, 81–88.

fractal and fractional

MDPI

Article

Extremal Solutions of Generalized Caputo-Type Fractional-Order Boundary Value Problems Using Monotone Iterative Method

Choukri Derbazi [1], Zidane Baitiche [1], Mohammed S. Abdo [2], Kamal Shah [3,4], Bahaaeldin Abdalla [3] and Thabet Abdeljawad [3,5,*]

1 Laboratoire Equations Différentielles, Department of Mathematics, Faculty of Exact Sciences, Frères Mentouri University Constantine 1, Ain El Bey Way, P.O. Box 325, Constantine 25017, Algeria; choukri.derbazi@umc.edu.dz (C.D.); zidane.baitiche@umc.edu.dz (Z.B.)
2 Department of Mathematics, College of Education, Hodeidah University, P.O. Box 3114, Al-Hudaydah 207416, Yemen; msabdo@hoduniv.net.ye
3 Department of Mathematics and Sciences, Prince Sultan University, P.O. Box 66833, Riyadh 11586, Saudi Arabia; kshah@psu.edu.sa (K.S.); babdallah@psu.edu.sa (B.A.)
4 Department of Mathematics, University of Malakand, P.O. Box 18000, Chakdara 18800, Dir (L), Khyber Pakhtankhawa, Pakistan
5 Department of Medical Research, China Medical University, Taichung 40402, Taiwan
* Correspondence: tabdeljawad@psu.edu.sa

Abstract: The aim of this research work is to derive some appropriate results for extremal solutions to a class of generalized Caputo-type nonlinear fractional differential equations (FDEs) under nonlinear boundary conditions (NBCs). The aforesaid results are derived by using the monotone iterative method, which exercises the procedure of upper and lower solutions. Two sequences of extremal solutions are generated in which one converges to the upper and the other to the corresponding lower solution. The method does not need any prior discretization or collocation for generating the aforesaid two sequences for upper and lower solutions. Further, the aforesaid techniques produce a fruitful combination of upper and lower solutions. To demonstrate our results, we provide some pertinent examples.

Keywords: ϑ-Caputo derivative; extremal solutions; monotone iterative method; sequences

Citation: Derbazi, C.; Baitiche, Z.; Abdo, M.S.; Shah, K.; Abdalla, B.; Abdeljawad, T. Extremal Solutions of Generalized Caputo-Type Fractional-Order Boundary Value Problems Using Monotone Iterative Method. *Fractal Fract.* **2022**, *6*, 146. https://doi.org/10.3390/fractalfract6030146

Academic Editors: Asifa Tassaddiq and Muhammad Yaseen

Received: 6 February 2022
Accepted: 3 March 2022
Published: 7 March 2022

1. Introduction

Over the last few decades, fractional calculus has attracted the attention of many researchers in the community of science and technology. This is because of its significant applications in different fields of science and engineering such as mathematics, physics, chemistry, biology, economics, finance, rheology, etc. (for more details, see [1–3]). Further, the most important applications of fractional calculus can be found in the description of memory and hereditary processes. The mentioned processes can be more nicely explained by the concept of fractional-order derivatives as compared to traditional ones. Keeping their importance in mind, researchers have given much attention to the use of fractional-order derivatives and integrals in the mathematical modeling of real-world processes instead of classical derivatives and integrals. In this regard, several monographs, and plenty of papers and books have been published, in which various kinds of important results and applications have been reported. Some of these can be found in [4–7]. Nevertheless, the application of the aforesaid area has been traced out just in the last two decades. This is due to the progress in the area of chaos that revealed refined relationships with the concepts of fractional calculus. In addition, in recent times, the application of the theory of fractional calculus to robotics has opened promising aspects for future developments, where in these robots, joint-level control is usually planted by using PID-like schemes with position feedback. For instance, one famous machine using the fractional PD^{α} controller is known as the Hexapod robot (see [8]). Further, some important applications of fractional calculus can be found about the dynamics in the trajectory control of redundant manipulators, where fractional-order derivatives have a more precise appearance than classical ones (see [9]). Further, the fractional derivative is a global operator that preserves greater degrees of freedom as compared to integer-order derivative, because a classical derivative with integer-order is a local operator. It is estimated that the fractional-order derivative operation contains some types of boundary conditions that involve information on the function further out. Some researchers have proved that fractional-order derivatives play a significant role in electrochemical analysis to elucidate the mechanistic behavior of the concentration of a substrate at the electrode surface to the current. Some researchers have proved that fractional-order modelers of contaminant flow in heterogeneous porous media involving fractional derivatives are more powerful than classical ones (see [10]). It should be kept in mind that fractional-order derivative operation of a function produces a complete spectrum or accumulation, which preserves the corresponding integer-order counterpart as a special case.

It is interesting that fractional-order derivatives have not yet been uniquely defined. Various renowned mathematicians have given their own definitions. Among the said definitions, some of them have gained much more popularity and proper attention from researchers, such as Riemann–Liouville (RL), Caputo, Hilfer, Caputo–Hadamard, Caputo–Katugampola, etc. It is interesting that aside from the aforesaid operators various other variants that contain singular kernels have been introduced recently. Hence the fractional differential operators have been divided into two classes including singular and non-singular. Here we state that this partition is not bad but provides a great degree of freedom in the choice of operator for the description of a particular phenomenon. It is remarkable that nearly all mentioned operators preserve memory in their respective kernels. Further, the two partitions of singular and non-singular kernels have their own benefits and drawbacks.

We remark that Hilfer-type fractional calculus unifies the aforesaid definitions. It is important that ϑ-Hilfer operators constitute a wide class of fractional derivatives (FDs). In this respect, some frequent results involving ϑ-FDs have been reported in [11–13]. However, various strategies exist in the literature to handle such types of problems of FDEs for computation of their solutions. Numerous tools and theories have been established so far. Iterative techniques of various kinds have key importance to investigate the aforementioned area. Among them are the monotone iterative algorithm, along with the upper and lower solutions method [14,15], fixed-point technique [16–18], and coincidence degree theory [19,20]. In particular, the monotone iterative method together with the technique of upper and lower solutions is an advantageous and effective tool for the existence as well as the approximation of solutions for nonlinear problems. In this regard, very useful results have been published so far. Among the iterative techniques, those introduced by Ladde, Lakshmikantham, and Vatsala [21] in 1985 for nonlinear differential equations have gained proper attention. Therefore, monotone iterative techniques associated with upper and lower solutions have been extensively used for nonlinear partial differential equations in the last few decades. In this regard, plenty of work has been published to date; a few can be found in [22–25]. Proposals have been made for classical differential equations for the first time [14,15,26–29]. In addition, the aforesaid techniques have been widely used to deal with FDEs subject to initial and boundary conditions. Some significant results can be found in [25,30–37]. We demonstrate that the said method is well known because it not only produces constructive proof for existence theorems but it also yields various comparison results, which are powerful tools to investigate the qualitative properties of solutions. Further, the sequence of iterations has useful behavior in the computation of numerical solutions to various boundary value and initial boundary value problems of classical as well FDEs. In addition, the method of upper and lower solutions is very useful for the construction of Lyapunov functions. Construction of such functions is increasingly used to derive various stability theories in dynamical systems.

Motivated by the work cited above, and particularly by the work in [31], we determine the existence criteria of extremal solutions for the following ϑ-Caputo-type FDE in a Caputo sense with NBCs

$$\begin{cases} {}^{c}\mathbb{D}_{a^+}^{\nu;\vartheta}\phi(\varsigma) = \mathbb{F}(\varsigma,\phi(\varsigma)), \ \varsigma \in \mathrm{J} := [a,b], \\ \mathbb{G}(\phi(a),\phi(b)) = 0, \end{cases} \tag{1}$$

where $0 < \nu \leq 1$, ${}^{c}\mathbb{D}_{a^+}^{\nu;\vartheta}$ is the ϑ-fractional operator of order ν in the Caputo sense and this is investigated. Further, $\mathbb{F} \in C(\mathrm{J} \times \mathbb{R}, \mathbb{R})$, $\mathbb{G} \in C(\mathbb{R}^2, \mathbb{R})$.

The considered problem (1) in the current article includes a wide range of nonlinear FDEs involving the standard Caputo operator (for $\vartheta(\varsigma) = \varsigma$), Caputo–Hadamard (for $\vartheta(\varsigma) = \log\varsigma$), and Caputo–Katugampola (for $\vartheta(\varsigma) = \varsigma^{\mathrm{p}}$, $\mathrm{p} > 0$). Further, fractional operators have been listed in Almeida [11] for further applications. Further, results acquired in the current article include the results of Franco et al. [26] if $a \to 0$, $\vartheta(\varsigma) \to \varsigma$, and $\nu = 1$ as a special case.

In this regard, we also point out some recent and similar findings that used operators on many fractional problems; see [38–40]. To the best of our knowledge in this regard, no one has considered the monotone iterative procedure to obtain the existence of extremal solutions involving a ϑ-Caputo derivative subject to NBCs. Therefore, motivated by the aforesaid gap, we have conducted this study.

The rest of this article is organized as follows. In Section 2, we insert some basic definitions and important results. Section 3 is devoted to studying the existence of extremal solutions for (1). In Section 4, we give two appropriate examples to highlight the feasibility of our abstract results.

2. Basic Results

Some fundamental results about the ϑ-Caputo derivative and integral that are needed throughout this work are given below.

The function $\vartheta \in C(\mathrm{J}, \mathbb{R})$ is non-decreasing differentiable with argument $0 < \vartheta'(\varsigma)$, at every point of J.

Definition 1 ([6,11]). *The ϑ-RL fractional integral of order $\nu > 0$ for an integrable function $\phi\colon \mathrm{J} \longrightarrow \mathbb{R}$ is given by*

$$\mathbb{I}_{a^+}^{\nu;\vartheta}\phi(\varsigma) = \frac{1}{\Gamma(\nu)}\int_a^{\varsigma} \vartheta'(s)(\vartheta(\varsigma) - \vartheta(s))^{\nu-1}\phi(s)\mathrm{d}s, \ \varsigma > a.$$

Definition 2 ([11]). *Let $\vartheta, \phi \in C^n(\mathrm{J}, \mathbb{R})$. The ϑ-RL derivative of fractional order of a function ϕ with $n-1 < \nu < n$ is given by*

$$\begin{aligned} \mathbb{D}_{a^+}^{\nu;\vartheta}\phi(\varsigma) &= \left(\frac{D_t}{\vartheta'(\varsigma)}\right)^n \mathbb{I}_{a^+}^{n-\nu;\vartheta}\phi(\varsigma) \\ &= \frac{1}{\Gamma(n-\nu)}\left(\frac{D_t}{\vartheta'(\varsigma)}\right)^n \int_a^{\varsigma} \vartheta'(s)(\vartheta(\varsigma) - \vartheta(s))^{n-\nu-1}\phi(s)\mathrm{d}s, \end{aligned}$$

where $n = [\nu] + 1$ ($n \in \mathbb{N}$), and $D_t = \frac{d}{dt}$.

Definition 3 ([11]). *Let $\vartheta, \phi \in C^n(\mathrm{J}, \mathbb{R})$. The ϑ-Caputo derivative of fractional order of function ϕ with $n-1 < \nu < n$ is defined by*

$${}^{c}\mathbb{D}_{a^+}^{\nu;\vartheta}\phi(\varsigma) = \mathbb{I}_{a^+}^{n-\nu;\vartheta}\phi_{\vartheta}^{[n]}(\varsigma),$$

where $\phi_{\vartheta}^{[n]}(\varsigma) = \left(\frac{D_{\varsigma}}{\vartheta'(\varsigma)}\right)^n \phi(\varsigma)$, $n = [\nu] + 1$ for $\nu \notin \mathbb{N}$, and $n = \nu$ for $\nu \in \mathbb{N}$.

One has

$$ {}^{c}\mathbb{D}^{\nu;\vartheta}_{a^+}\phi(\varsigma) = \begin{cases} \int_a^{\varsigma} \frac{\vartheta'(s)(\vartheta(\varsigma)-\vartheta(s))^{n-\nu-1}}{\Gamma(n-\nu)}\phi^{[n]}_{\vartheta}(s)ds &, \quad \text{if } \nu \notin \mathbb{N}, \\ \phi^{[n]}_{\vartheta}(\varsigma) &, \quad \text{if } \nu \in \mathbb{N}. \end{cases} $$

Lemma 1 ([11])**.** *Let $\nu, \mu > 0$, and $\phi \in C(J, \mathbb{R})$, for every $\varsigma \in J$*

1. ${}^{c}\mathbb{D}^{\nu;\vartheta}_{a^+}\mathbb{I}^{\nu;\vartheta}_{a^+}\phi(\varsigma) = \phi(\varsigma)$,
2. $\mathbb{I}^{\nu;\vartheta}_{a^+}{}^{c}\mathbb{D}^{\nu;\vartheta}_{a^+}\phi(\varsigma) = \phi(\varsigma) - \phi(a), \quad 0 < \nu \leq 1$,
3. $\mathbb{I}^{\nu;\vartheta}_{a^+}(\vartheta(\varsigma) - \vartheta(a))^{\mu-1} = \frac{\Gamma(\mu)}{\Gamma(\mu+\nu)}(\vartheta(\varsigma) - \vartheta(a))^{\mu+\nu-1}$,
4. ${}^{c}\mathbb{D}^{\nu;\vartheta}_{a^+}(\vartheta(\varsigma) - \vartheta(a))^{\mu-1} = \frac{\Gamma(\mu)}{\Gamma(\mu-\nu)}(\vartheta(\varsigma) - \vartheta(a))^{\mu-\nu-1}$,
5. ${}^{c}\mathbb{D}^{\nu;\vartheta}_{a^+}(\vartheta(\varsigma) - \vartheta(a))^{k} = 0, \ \forall k < n \in \mathbb{N}$.

Definition 4 ([7])**.** *One- and two-parameter Mittag–Leffler functions (MLFs) are recalled as*

$$ \mathbb{E}_{\nu}(z) = \sum_{k=0}^{\infty} \frac{z^k}{\Gamma(\nu k + 1)}, \quad (z \in \mathbb{R}, \ \nu > 0), $$

and

$$ \mathbb{E}_{\nu,\mu}(z) = \sum_{k=0}^{\infty} \frac{z^k}{\Gamma(\nu k + \mu)}, \ \nu, \mu > 0 \text{ and } z \in \mathbb{R}, $$

respectively. Clearly $\mathbb{E}_{1,1}(z) = \mathbb{E}_1(z) = \exp(z)$.

Further properties of MLFs are given below.

Lemma 2 ([41])**.** *Let $\nu \in (0,1)$ and $z \in \mathbb{R}$, one has*

1. *$\mathbb{E}_{\nu,1}$ and $\mathbb{E}_{\nu,\nu}$ are non-negative.*
2. *$\mathbb{E}_{\nu,1}(z) \leq 1, \mathbb{E}_{\nu,\nu}(z) \leq \frac{1}{\Gamma(\nu)}$, for any $z < 0$.*

For further analysis, we recall the following Lemma [31] as:

Lemma 3 ([31] (Lemma 4))**.** *Let $\nu \in (0,1]$, $\lambda \in \mathbb{R}$ and $h \in C(J, \mathbb{R})$, then, the linear version*

$$ \begin{cases} {}^{c}\mathbb{D}^{\nu;\vartheta}_{a^+}\phi(\varsigma) + \lambda\phi(\varsigma) = h(\varsigma), \ \varsigma \in J. \\ \phi(a) = \phi_a, \end{cases} $$

has a unique solution that is described as

$$ \begin{aligned} \phi(\varsigma) &= \theta_a \mathbb{E}_{\nu,1}(-\lambda(\vartheta(\varsigma) - \vartheta(a))^{\nu}) \\ &\quad + \int_a^{\varsigma} \vartheta'(s)(\vartheta(\varsigma) - \vartheta(s))^{\nu-1}\mathbb{E}_{\nu,\nu}(-\lambda(\vartheta(\varsigma) - \vartheta(s))^{\nu})h(s)ds, \end{aligned} $$

where $\mathbb{E}_{\nu,\mu}(\cdot)$ is the two-parametric MLF defined earlier.

The given comparison results comprise a central rule in the following analysis.

Lemma 4 ([31] (Lemma 5))**.** *Let $\nu \in (0,1]$, and $\lambda \in \mathbb{R}$, if $\gamma \in C(J, \mathbb{R})$ obey the given relation*

$$ \begin{cases} {}^{c}\mathbb{D}^{\nu;\vartheta}_{a^+}\gamma(\varsigma) \geq -\lambda\gamma(\varsigma), \quad \varsigma \in (a,b], \\ \gamma(a) \geq 0, \end{cases} $$

then $\gamma(\varsigma) \geq 0$, for all $\varsigma \in J$.

3. Main Results

Here, key findings are established about the ϑ-Caputo FDEs (1). We develop two monotone iterative sequences for upper and lower solutions, respectively.

Definition 5. *The function $\phi \in C(J, \mathbb{R})$ such that ${}^{c}D_{a^+}^{\nu;\vartheta}\phi$ exists and is continuous on J and is known to be a solution of the problem (1). Further, ϕ gives the statistics of the equation ${}^{c}\mathbb{D}_{a^+}^{\nu;\vartheta}\phi(\varsigma) = \mathbb{F}(\varsigma, \phi(\varsigma))$, for each $\varsigma \in J$ and the NBCs*

$$\mathbb{G}(\phi(a), \phi(b)) = 0.$$

Subsequently, we mention the definitions of extremal solutions of ϑ-Caputo FDEs (1).

Definition 6. *The mapping $\phi_0 \in C(J, \mathbb{R})$ is known as lower solution (1), if it satisfies*

$$\begin{cases} {}^{c}\mathbb{D}_{a^+}^{\nu;\vartheta}\phi_0(\varsigma) \leq \mathbb{F}(\varsigma, \phi_0(\varsigma)), & t \in (a, b], \\ \mathbb{G}(\phi_0(a), \phi_0(b)) \leq 0. \end{cases}$$

An upper solution $\varpi_0 \in C(J, \mathbb{R})$ of the problem (1) can be defined in a similar way by reversing the above inequality.

Now to move forward, we will introduce the following conditions:

Hypothesis 1. *There exist ϕ_0 and ϖ_0 as lower and upper solutions of problem (1) in $C(J, \mathbb{R})$ respectively, with $\phi_0(\varsigma) \leq \varpi_0(\varsigma), \varsigma \in J$.*

Hypothesis 2. *There exists a constant $\lambda > 0$ with*

$$\mathbb{F}(\varsigma, y) - \mathbb{F}(\varsigma, x) \geq -\lambda(y - x) \quad \text{for } \phi_0(\varsigma) \leq x \leq y \leq \varpi_0(\varsigma), \varsigma \in J.$$

Hypothesis 3. *There exist constants $c > 0$ and $d \geq 0$ with $\phi_0(a) \leq u_1 \leq u_2 \leq \varpi_0(a)$, $\phi_0(b) \leq v_1 \leq v_2 \leq \varpi_0(b)$, such that*

$$\mathbb{G}(u_2, v_2) - \mathbb{G}(u_1, v_1) \leq c(u_2 - u_1) - d(v_2 - v_1),$$

Now, we shall apply the monotonous method to prove our key findings.

Theorem 1. *Let $\mathbb{F} : J \times \mathbb{R} \longrightarrow \mathbb{R}$ be continuous. Assume that Hypotheses 1–3 hold. Then there exist two monotone iterative sequences $\{\phi_n\}$ and $\{\varpi_n\}$, which are converging uniformly on J to the extremal solutions of (1) in the sector $[\phi_0, \varpi_0]$, where*

$$[\phi_0, \varpi_0] = \{z \in C(J, \mathbb{R}) : \phi_0(\varsigma) \leq z(\varsigma) \leq \varpi_0(\varsigma), \quad \varsigma \in J\}.$$

Proof. First, for any $\phi_0(\varsigma), \varpi_0(\varsigma) \in C(J, \mathbb{R})$ and $\lambda > 0$, we consider the following FDEs

$$\begin{cases} {}^{c}\mathbb{D}_{a^+}^{\nu;\vartheta}\phi_{n+1}(\varsigma) = \mathbb{F}(\varsigma, \phi_n(\varsigma)) - \lambda(\phi_{n+1}(\varsigma) - \phi_n(\varsigma)), & \varsigma \in J, \\ \phi_{n+1}(a) = \phi_n(a) - \frac{1}{c}\mathbb{G}(\phi_n(a), \phi_n(b)), \end{cases} \tag{2}$$

and

$$\begin{cases} {}^{c}\mathbb{D}_{a^+}^{\nu;\vartheta}\varpi_{n+1}(\varsigma) = \mathbb{F}(\varsigma, \varpi_n(\varsigma)) - \lambda(\varpi_{n+1}(\varsigma) - \varpi_n(\varsigma)), & \varsigma \in J, \\ \varpi_{n+1}(a) = \varpi_n(a) - \frac{1}{c}\mathbb{G}(\varpi_n(a), \varpi_n(b)), \end{cases} \tag{3}$$

According to Lemma 3, one can deduce that (2) and (3) preserve at most one solution in $C(J, \mathbb{R})$. Thus we have

$$\phi_{n+1}(\varsigma) = \left(\phi_n(a) - \frac{1}{c}\mathbb{G}(\phi_n(a), \phi_n(b))\right)\mathbb{E}_{\nu,1}(-\lambda(\vartheta(\varsigma) - \vartheta(a))^{\nu})$$
$$+ \int_a^{\varsigma} \vartheta'(s)(\vartheta(\varsigma) - \vartheta(s))^{\nu-1}\mathbb{E}_{\nu,\nu}(-\lambda(\vartheta(\varsigma) - \vartheta(s))^{\nu})(\mathbb{F}(s, \phi_n(s)) + \lambda\phi_n(s))ds, \ \varsigma \in J,$$

$$\varpi_{n+1}(\varsigma) = \left(\varpi_n(a) - \frac{1}{c}\mathbb{G}(\varpi_n(a), \varpi_n(b))\right)\mathbb{E}_{\nu,\nu}(-\lambda(\vartheta(\varsigma) - \vartheta(a))^{\nu})$$
$$+ \int_a^{\varsigma} \vartheta'(s)(\vartheta(\varsigma) - \vartheta(s))^{\nu-1}\mathbb{E}_{\nu,\nu}(-\lambda(\vartheta(\varsigma) - \vartheta(s))^{\nu})(\mathbb{F}(s, \varpi_n(s)) + \lambda\varpi_n(s))ds, \ \varsigma \in J.$$

For appropriateness, the proof will be divided into a number of steps.

Step 1: The sequences $\phi_n(\varsigma), \varpi_n(\varsigma) (n \geq 1)$ are lower and upper solutions of (1), correspondingly. Moreover, we assume

$$\phi_0(\varsigma) \leq \phi_1(\varsigma) \leq \cdots \leq \phi_n(\varsigma) \leq \cdots \leq \varpi_n(\varsigma) \leq \cdots \leq \varpi_1(\varsigma) \leq \varpi_0(\varsigma), \quad \varsigma \in J. \tag{4}$$

Firstly, we show that

$$\phi_0(\varsigma) \leq \phi_1(\varsigma) \leq \varpi_1(\varsigma) \leq \varpi_0(\varsigma), \quad \varsigma \in J.$$

Set $\gamma(\varsigma) = \phi_1(\varsigma) - \phi_0(\varsigma)$. From (2) and Definition 6, we obtain

$$\begin{aligned} {}^c\mathbb{D}_{a^+}^{\nu;\vartheta}\gamma(\varsigma) &= {}^c\mathbb{D}_{a^+}^{\nu;\vartheta}\phi_1(\varsigma) - {}^c\mathbb{D}_{a^+}^{\nu;\vartheta}\phi_0(\varsigma) \\ &\geq \mathbb{F}(\varsigma, \phi_0(\varsigma)) - \lambda(\phi_1(\varsigma) - \phi_0(\varsigma)) - \mathbb{F}(\varsigma, \phi_0(\varsigma)) \\ &= -\lambda\gamma(\varsigma). \end{aligned}$$

Again, since $\gamma(a) = -\frac{1}{c}\mathbb{G}(\phi_0(a), \phi_0(b)) \geq 0$, $\gamma(\varsigma) \geq 0$, for $\varsigma \in J$ due to Lemma 4. Thus, $\phi_0(\varsigma) \leq \phi_1(\varsigma)$.

Similarly, we can find that $\varpi_1(\varsigma) \leq \varpi_0(\varsigma)$, $\varsigma \in J$.

Now, let $\gamma(\varsigma) = \varpi_1(\varsigma) - \phi_1(\varsigma)$. Using (2) and (3) together with Hypotheses 2 and 3, we obtain

$$\begin{aligned} {}^c\mathbb{D}_{a^+}^{\nu;\vartheta}\gamma(\varsigma) &= \mathbb{F}(\varsigma, \varpi_0(\varsigma)) - \mathbb{F}(\varsigma, \phi_0(\varsigma)) - \lambda(\varpi_1(\varsigma) - \varpi_0(\varsigma)) + \lambda(\phi_1(\varsigma) - \phi_0(\varsigma)) \\ &\geq -\lambda(\varpi_0(\varsigma) - \phi_0(\varsigma)) - \lambda(\varpi_1(\varsigma) - \varpi_0(\varsigma)) + \lambda(\phi_1(\varsigma) - \phi_0(\varsigma)) \\ &= -\lambda\gamma(\varsigma). \end{aligned}$$

Since

$$\begin{aligned} \gamma(a) &= (\varpi_0(a) - \phi_0(a)) - \frac{1}{c}(\mathbb{G}(\varpi_0(a), \varpi_0(b)) - \mathbb{G}(\phi_0(a), \phi_0(b))) \\ &\geq \frac{d}{c}(\varpi_0(b) - \phi_0(b)) \geq 0, \end{aligned}$$

we obtain $\phi_1(\varsigma) \leq \varpi_1(\varsigma)$, $\varsigma \in J$ due to Lemma 4.

Secondly, we show that $\phi_1(\varsigma), \varpi_1(\varsigma)$ are extremum solutions of (1). Since ϕ_0 and ϖ_0 are lower and upper solutions of (1), by Hypotheses 2 and 3, we obtain

$${}^c\mathbb{D}_{a^+}^{\nu;\vartheta}\phi_1(\varsigma) = \mathbb{F}(\varsigma, \phi_0(\varsigma)) - \lambda(\phi_1(\varsigma) - \phi_0(\varsigma)) \leq \mathbb{F}(\varsigma, \phi_1(\varsigma)),$$

and

$$\begin{aligned} \mathbb{G}(\phi_1(a), \phi_1(b)) &\leq \mathbb{G}(\phi_0(a), \phi_0(b)) + c(\phi_1(a) - \phi_0(a)) - d(\phi_1(b) - \phi_0(b)) \\ &= -d(\phi_1(b) - \phi_0(b)) \\ &\leq 0. \end{aligned}$$

Therefore, $\phi_1(\varsigma)$ is a lower solution of (1). Analogously, it is obvious that $\varpi_1(\varsigma)$ is an upper solution of (1).

Through the above debates and induction, we can show that $\phi_n(\varsigma), \varpi_n(\varsigma), (n \geq 1)$ are lower and upper solutions of (1), respectively, and the assumption (4) is true.

Step 2: $\phi_n \to \phi$ and $\varpi_n \to \varpi$.

First, we prove that $\{\phi_n\}$ is uniformly bounded. By considering supposition Hypothesis 2, we may write

$$\mathbb{F}(\varsigma, \phi_0(\varsigma)) + \lambda\phi_0(\varsigma) \leq \mathbb{F}(\varsigma, \phi_n(\varsigma)) + \lambda\phi_n(\varsigma) \leq \mathbb{F}(\varsigma, \varpi_0(\varsigma)) + \lambda\varpi_0(\varsigma), \quad \varsigma \in \mathrm{J}$$

i.e.,

$$\begin{aligned} 0 &\leq \mathbb{F}(\varsigma, \phi_n(\varsigma)) - \mathbb{F}(\varsigma, \phi_0(\varsigma)) + \lambda(\phi_n(\varsigma) - \phi_0(\varsigma)) \\ &\leq \mathbb{F}(\varsigma, \varpi_0(\varsigma)) - \mathbb{F}(\varsigma, \phi_0(\varsigma)) + \lambda(\varpi_0(\varsigma) - \phi_0(\varsigma)). \end{aligned}$$

Hence, we obtain

$$\begin{aligned} |\mathbb{F}(\varsigma, \phi_n(\varsigma)) - \mathbb{F}(\varsigma, \phi_0(\varsigma)) + \lambda(\phi_n(\varsigma) - \phi_0(\varsigma))| \leq |\mathbb{F}(\varsigma, \varpi_0(\varsigma)) - \mathbb{F}(\varsigma, \phi_0(\varsigma)) \\ + \lambda(\varpi_0(\varsigma) - \phi_0(\varsigma))|. \end{aligned}$$

Thus

$$\begin{aligned} |\mathbb{F}(\varsigma, \phi_n(\varsigma)) + \lambda\phi_n(\varsigma)| &\leq |\mathbb{F}(\varsigma, \phi_n(\varsigma)) - \mathbb{F}(\varsigma, \phi_0(\varsigma)) + \lambda(\phi_n(\varsigma) - \phi_0(\varsigma))| \\ &\quad + |\mathbb{F}(\varsigma, \phi_0(\varsigma)) + \lambda\phi_0(\varsigma)| \\ &\leq |\mathbb{F}(\varsigma, \varpi_0(\varsigma)) - \mathbb{F}(\varsigma, \phi_0(\varsigma)) + \lambda(\varpi_0(\varsigma) - \phi_0(\varsigma))| \\ &\quad + |\mathbb{F}(\varsigma, \phi_0(\varsigma)) + \lambda\phi_0(\varsigma)| \\ &\leq 2|\mathbb{F}(\varsigma, \phi_0(\varsigma)) + \lambda\phi_0(\varsigma)| + |\mathbb{F}(\varsigma, \varpi_0(\varsigma)) + \lambda\varpi_0(\varsigma)|. \end{aligned}$$

Since $\phi_0, \mathbb{F}$ are continuous on J, we can see a constant $\mathbb{M}$ independent of n with

$$|\mathbb{F}(\varsigma, \phi_n(\varsigma)) + \lambda\phi_n(\varsigma)| \leq \mathbb{M}. \tag{5}$$

Furthermore, from Hypothesis 3, we can obtain

$$\phi_0(a) - \frac{1}{c}\mathbb{G}(\phi_0(a), \phi_0(b)) \leq \phi_n(a) - \frac{1}{c}\mathbb{G}(\phi_n(a), \phi_n(b)) \leq \varpi_0(a) - \frac{1}{c}\mathbb{G}(\varpi_0(a), \varpi_0(b)),$$

i.e.,

$$\begin{aligned} 0 &\leq \phi_n(a) - \phi_0(a) - \frac{1}{c}(\mathbb{G}(\phi_n(a), \phi_n(b)) - \mathbb{G}(\phi_0(a), \phi_0(b))) \\ &\leq \varpi_0(a) - \phi_0(a) - \frac{1}{c}(\mathbb{G}(\varpi_0(a), \varpi_0(b)) - \mathbb{G}(\phi_0(a), \phi_0(b))). \end{aligned}$$

Hence, we obtain

$$\begin{aligned} &\left|\phi_n(a) - \phi_0(a) - \frac{1}{c}(\mathbb{G}(\phi_n(a), \phi_n(b)) - \mathbb{G}(\phi_0(a), \phi_0(b)))\right| \leq \\ &\left|\varpi_0(a) - \phi_0(a) - \frac{1}{c}(\mathbb{G}(\varpi_0(a), \varpi_0(b)) - \mathbb{G}(\phi_0(a), \phi_0(b)))\right| \\ &\leq \left|\phi_0(a) - \frac{1}{c}\mathbb{G}(\phi_0(a), \phi_0(b))\right| + \left|\varpi_0(a) - \frac{1}{c}\mathbb{G}(\varpi_0(a), \varpi_0(b))\right|. \end{aligned}$$

Thus

$$\begin{aligned}\left|\phi_n(a)-\frac{1}{c}\mathbb{G}(\phi_n(a),\phi_n(b))\right| &\le \left|\phi_n(a)-\phi_0(a)-\frac{1}{c}(\mathbb{G}(\phi_n(a),\phi_n(b))-\mathbb{G}(\phi_0(a),\phi_0(b)))\right| \\ &+\left|\phi_0(a)-\frac{1}{c}\mathbb{G}(\phi_0(a),\phi_0(b))\right| \\ &\le 2\left|\phi_0(a)-\frac{1}{c}\mathbb{G}(\phi_0(a),\phi_0(b))\right|+\left|\varpi_0(a)-\frac{1}{c}\mathbb{G}(\varpi_0(a),\varpi_0(b))\right|.\end{aligned}$$

Since ϕ_0, ϖ_0 and $\mathbb{G}$ are continuous functions, we can find a constant $\mathbb{K}$ independent of n, such that

$$\left|\phi_n(a)-\frac{1}{c}\mathbb{G}(\phi_n(a),\phi_n(b))\right| \le \mathbb{K}. \tag{6}$$

Moreover, by (2) and (3) we have

$$\begin{aligned}|\phi_{n+1}(\varsigma)| &= \left|\phi_n(a)-\frac{1}{c}\mathbb{G}(\phi_n(a),\phi_n(b))\right|\mathbb{E}_{\nu,1}(-\lambda(\vartheta(\varsigma)-\vartheta(a))^{\nu}) \\ &+\int_a^{\varsigma}\vartheta'(s)(\vartheta(\varsigma)-\vartheta(s))^{\nu-1}\mathbb{E}_{\nu,\nu}(-\lambda(\vartheta(\varsigma)-\vartheta(s))^{\nu})|\mathbb{F}(s,\phi_n(s))+\lambda\phi_n(s)|\mathrm{ds},\ \varsigma\in \mathrm{J}.\end{aligned}$$

Using Lemma 2 along with (5) and (6), we obtain

$$\begin{aligned}|\phi_{n+1}(\varsigma)| &= \mathbb{K}+\frac{\mathbb{M}}{\Gamma(\nu)}\int_a^{\varsigma}\vartheta'(s)(\vartheta(\varsigma)-\vartheta(s))^{\nu-1}\mathrm{ds} \\ &\le \mathbb{K}+\frac{\mathbb{M}(\vartheta(b)-\vartheta(a))^{\nu}}{\Gamma(\nu+1)}.\end{aligned}$$

Hence, $\{\phi_n\}$ is uniformly bounded in $C(\mathrm{J},\mathbb{R})$. With same argument one can deduce that $\{\varpi_n\}$ is uniformly bounded.

It is necessary to derive that the sequences $\{\phi_n\}$ and $\{\varpi_n\}$ are equi-continuous on J. To do this, choosing $\varsigma_1, \varsigma_2 \in \mathrm{J}$, with $\varsigma_1 \le \varsigma_2$. By (5) and (6) and Lemma 2, we have

$$\begin{aligned}&|\phi_{n+1}(\varsigma_2)-\phi_{n+1}(\varsigma_1)| \le \left|\phi_n(a)-\frac{1}{c}\mathbb{G}(\phi_n(a),\phi_n(b))\right||\mathbb{E}_{\nu,1}(-\lambda(\vartheta(\varsigma_2)-\vartheta(a))^{\nu})- \\ &\mathbb{E}_{\nu,1}(-\lambda(\vartheta(\varsigma_1)-\vartheta(a))^{\nu})| \\ &+\int_a^{\varsigma_1}\frac{\vartheta'(s)\left[(\vartheta(\varsigma_1)-\vartheta(s))^{\nu-1}-(\vartheta(\varsigma_2)-\vartheta(s))^{\nu-1}\right]}{\Gamma(\nu)}|\mathbb{F}(s,\phi_n(s))+\lambda\phi_n(s)|\mathrm{ds} \\ &+\int_{\varsigma_1}^{\varsigma_2}\frac{\vartheta'(s)(\vartheta(\varsigma_2)-\vartheta(s))^{\nu-1}}{\Gamma(\nu)}|\mathbb{F}(s,\phi_n(s))+\lambda\phi_n(s)|\,\mathrm{ds} \\ &\le \mathbb{K}|\mathbb{E}_{\nu,1}(-\lambda(\vartheta(\varsigma_2)-\vartheta(a))^{\nu})-\mathbb{E}_{\nu,1}(-\lambda(\vartheta(\varsigma_1)-\vartheta(a))^{\nu})| \\ &+\frac{2\mathbb{M}}{\Gamma(\nu+1)}(\vartheta(\varsigma_2)-\vartheta(\varsigma_1))^{\nu}.\end{aligned}$$

By the continuity of $\mathbb{E}_{\nu,1}(-\lambda(\vartheta(\varsigma)-\vartheta(a))^{\nu})$ on J, the right-hand-side of the preceding inequality approaches zero, when $\varsigma_1 \to \varsigma_2$. This implies that $\{\phi_n(\varsigma)\}$ is equi-continuous on J. Likewise, we can demonstrate that $\{\varpi_n(\varsigma)\}$ is equi-continuous. Hence, by using Ascoli-Arzelás theorem, the sequence $\phi_{n_k}(\varsigma) \to \phi^*(\varsigma)$ and $\varpi_{n_k}(\varsigma) \to \varpi^*(\varsigma)$ as $k \to \infty$. Hence the aforesaid relation combined under the monotonicity of sequences $\{\phi_n(\varsigma)\}$ and $\{\varpi_n(\varsigma)\}$ yields

$$\lim_{n\to\infty}\phi_n(\varsigma)=\phi^*(\varsigma) \quad \text{and} \quad \lim_{n\to\infty}\varpi_n(t)=\varpi^*(\varsigma),$$

uniformly on $\varsigma \in \mathrm{J}$ and the limit functions ϕ^*, ϖ^* satisfy (1).

Step 3: ϕ^* and ϖ^* are maximal and minimal solutions of (1) in $[\phi_0, \varpi_0]$.

Let $z \in [\phi_0, \varpi_0]$ be any solution of (1). Suppose that

$$\phi_n(\varsigma) \leq z(\varsigma) \leq \varpi_n(\varsigma), \quad \varsigma \in \mathrm{J}, \text{ for some } n \in \mathbb{N}^*. \tag{7}$$

Setting $\gamma(\varsigma) = z(\varsigma) - \phi_{n+1}(\varsigma)$. It follows that

$$\begin{aligned}
{}^{c}\mathbb{D}_{a^+}^{\nu;\vartheta}\gamma(\varsigma) &= \mathbb{F}(\varsigma, z(\varsigma)) - \mathbb{F}(\varsigma, \phi_n(\varsigma)) + \lambda(\phi_{n+1}(\varsigma) - \phi_n(\varsigma)) \\
&\geq -\lambda(z(\varsigma) - \phi_n(\varsigma)) + \lambda(\phi_{n+1}(\varsigma) - \phi_n(\varsigma)) \\
&= -\lambda\gamma(\varsigma).
\end{aligned}$$

Furthermore

$$\begin{aligned}
\phi_{n+1}(a) &= \phi_n(a) - \frac{1}{c}\mathbb{G}(\phi_n(a), \phi_n(b)) \\
&= \phi_n(a) + \frac{1}{c}\mathbb{G}(z(a), z(b)) - \frac{1}{c}\mathbb{G}(\phi_n(a), \phi_n(b)) \\
&\leq z(a) - \frac{d}{c}(z(b) - \phi_n(b)) \\
&\leq z(a),
\end{aligned}$$

that is

$$\gamma(a) \geq 0.$$

In view of Lemma 4, we obtain $\gamma(\varsigma) \geq 0,\ \varsigma \in \mathrm{J}$, which implies

$$\phi_{n+1}(\varsigma) \leq z(\varsigma),\ \varsigma \in \mathrm{J}.$$

Utilizing the same procedure, we can derive that

$$z(\varsigma) \leq \varpi_{n+1}(\varsigma),\ \varsigma \in \mathrm{J}.$$

Hence,

$$\phi_{n+1}(\varsigma) \leq z(\varsigma) \leq \varpi_{n+1}(\varsigma),\ \varsigma \in \mathrm{J}.$$

Hence (7) is satisfied on J for all $n \in \mathbb{N}^*$. Employing $n \to \infty$ on (7) from either side, one has

$$\phi^* \leq z \leq \varpi^*.$$

This confirms that ϕ^*, ϖ^* are the extremal solutions of (1) in $[\phi_0, \varpi_0]$. □

4. Illustrative Problems

This section includes some test problems for the illustration of our main results.

Example 1. *Consider the ϑ-Caputo FDE (1) with*

$$\nu = 0.5,\ a = 1,\ b = e,\ \vartheta(\varsigma) = \ln\varsigma. \tag{8}$$

In order to justify that Theorem 1 is valid, we take

$$\begin{cases} \mathbb{F}(\varsigma, \phi(\varsigma)) = 1 - \phi^2(\varsigma) + 2\sqrt{\ln\varsigma}, & \text{for } \varsigma \in [1, e], \\ \mathbb{G}(\phi(1), \phi(e)) = \phi(1) - 1. \end{cases} \tag{9}$$

Without difficulty, we can infer that

$$\phi_0(\varsigma) = 1, \quad \varpi_0(\varsigma) = 1 + \sqrt{\ln\varsigma},$$

are lower and upper solutions of (1), *respectively. It is apparent that* $\phi_0(\varsigma) \leq \varpi_0(\varsigma)$, *for* $\varsigma \in [1, e]$. *In addition, for* $\phi_0(\varsigma) \leq \mathrm{x} \leq \mathrm{y} \leq \varpi_0(\varsigma)$, *we have*

$$\mathbb{F}(\varsigma, \mathrm{y}) - \mathbb{F}(\varsigma, \mathrm{x}) \geq -4(\mathrm{y} - \mathrm{x}), \quad \varsigma \in [1, e].$$

Hence the Hypothesis 2 holds with $\lambda = 4$. *Further, if* $\phi_0(1) \leq u_1 \leq u_2 \leq \varpi_0(1)$, $\phi_0(e) \leq v_1 \leq v_2 \leq \varpi_0(e)$, *we have*

$$\mathbb{G}(u_2, v_2) - \mathbb{G}(u_1, v_1) \leq (u_2 - u_1).$$

Therefore, Hypothesis 3 holds with $c = 1$ *and* $d = 0$. *An application of Theorem 1 shows that the problem* (1) *with the data* (8) *and* (9) *has extremal solutions in the region* $[\phi_0, \varpi_0]$. *Moreover, the monotone iterative sequences* $\{\phi_n\}_{n\in\mathbb{N}}, \{\varpi_n\}_{n\in\mathbb{N}}$ *can be acquired by*

$$\begin{aligned} \phi_{n+1}(\varsigma) = {} & \mathbb{E}_{0.5,1}\left(-4\sqrt{\ln\frac{\varsigma}{s}}\right) + \int_1^{\varsigma}\left(\ln\frac{\varsigma}{s}\right)^{-0.5}\mathbb{E}_{0.5,0.5}\left(-4\sqrt{\ln\frac{\varsigma}{s}}\right) \\ & \times\left(1 - \phi_n^2(s) + 2\sqrt{\ln s} + 4\phi_n(s)\right)\frac{\mathrm{ds}}{\mathrm{s}}, \ n \geq 0, \end{aligned} \tag{10}$$

$$\begin{aligned} \varpi_{n+1}(\varsigma) = {} & \mathbb{E}_{0.5,1}\left(-4\sqrt{\ln\frac{\varsigma}{s}}\right) + \int_1^{\varsigma}\left(\ln\frac{\varsigma}{s}\right)^{-0.5}\mathbb{E}_{0.5,0.5}\left(-4\sqrt{\ln\frac{\varsigma}{s}}\right) \\ & \times\left(1 - \varpi_n^2(s) + 2\sqrt{\ln s} + 4\varpi_n(s)\right)\frac{\mathrm{ds}}{\mathrm{s}}, \ n \geq 0. \end{aligned} \tag{11}$$

Example 2. *Consider the following Caputo FDE*

$$\begin{cases} {}^{c}\mathbb{D}_{0^+}^{\nu}\phi(\varsigma) = \sin(\phi(\varsigma)) - \phi(\varsigma), \ \varsigma \in [0, 1], \\ 0.5\phi(0) - 3\phi(0)\phi(1) = 0. \end{cases} \tag{12}$$

Note that problem (12) *is a particular case of problem* (1), *where*

$$\nu = 0.5, \ a = 0, \ b = 1, \ \vartheta(\varsigma) = \varsigma,$$

and

$$\begin{cases} \mathbb{F}(\varsigma, \phi(\varsigma)) = \sin(\phi(\varsigma)) - \phi(\varsigma), \quad \varsigma \in \mathrm{J}, \\ \mathbb{G}(\phi(0), \phi(1)) = 0.5\phi(0) - 3\phi(0)\phi(1) \end{cases}$$

Taking $\phi_0(\varsigma) = 0$ *and* $\varpi_0(\varsigma) = \sqrt{\varsigma}$, *it is easy to verify that* ϕ_0, ϖ_0 *are lower and upper solutions of* (12), *respectively, and* $\phi_0(\varsigma) \leq \varpi_0(\varsigma)$, *for* $\varsigma \in [0, 1]$. *Therefore, Hypothesis 1 of Theorem 1 holds. However, if* $\phi_0(\varsigma) \leq \mathrm{x} \leq \mathrm{y} \leq \varpi_0(\varsigma)$ *we have*

$$\mathbb{F}(\varsigma, \mathrm{y}) - \mathbb{F}(\varsigma, \mathrm{x}) \geq -2(\mathrm{y} - \mathrm{x}), \quad \varsigma \in [0, 1].$$

Hence Hypothesis 2 holds with $\lambda = 2$, *and if* $\phi_0(0) \leq u_1 \leq u_2 \leq \varpi_0(0)$, $\phi_0(1) \leq v_1 \leq v_2 \leq \varpi_0(1)$, *we have*

$$\mathbb{G}(u_2, v_2) - \mathbb{G}(u_1, v_1) \leq (u_2 - u_1).$$

Therefore, Hypothesis 3 holds with $c = 1$ *and* $d = 0$. *According to Theorem 1, there exist monotone iterative sequences* $\{\phi_n\}$ *and* $\{\varpi_n\}$ *that are uniformly converging to* ϕ^* *and* ϖ^*, *respectively, and* ϕ^*, ϖ^* *are the extremal solutions in* $[\phi_0, \varpi_0]$ *of problem* (12).

Example 3. *Consider the following* ϑ*-Caputo FDE*

$$\begin{cases} {}^{c}\mathbb{D}_{0^+}^{0.5;e^{2\varsigma}}\phi(\varsigma) = \frac{2}{\sqrt{\pi}}\sqrt{e^{2\varsigma} - 1} + e^{2\varsigma} - 1 - \sin(e^{2\varsigma} - 1) + \sin(\phi(\varsigma)) - \phi(\varsigma), \ \varsigma \in [0, 1], \\ \phi(0) = \frac{2\pi}{e^2 - 1 + 2\pi}\phi(1). \end{cases} \tag{13}$$

Comparing the above problem with problem (1), we obtain

$$\nu = 0.5,\ a = 0,\ b = 1,\ \vartheta(\varsigma) = e^{2\varsigma},$$

and

$$\begin{cases} \mathbb{F}(\varsigma, \phi(\varsigma)) = \frac{2}{\sqrt{\pi}}\sqrt{e^{2\varsigma} - 1} + e^{2\varsigma} - 1 - \sin(e^{2\varsigma} - 1) + \sin(\phi(\varsigma)) - \phi(\varsigma), & \varsigma \in \mathrm{J}, \\ \mathbb{G}(\phi(0), \phi(1)) = \phi(0) - \frac{2\pi}{e^2 - 1 + 2\pi}\phi(1). \end{cases}$$

One can verify that $\phi(\varsigma) = e^{2\varsigma} - 1$ is an exact solution of the problem (1). Moreover, taking $\phi_0(\varsigma) = 0$ and $\varpi_0(\varsigma) = e^{2\varsigma} - 1 + 2\pi$, it is easy to verify that ϕ_0, ϖ_0 are lower and upper solutions of (13), respectively, and $\phi_0(\varsigma) \leq \varpi_0(\varsigma)$, for $\varsigma \in [0,1]$. Therefore, Hypothesis 1 of Theorem 1 holds. However, if $\phi_0(\varsigma) \leq \mathrm{x} \leq \mathrm{y} \leq \varpi_0(\varsigma)$ we have

$$\mathbb{F}(\varsigma, \mathrm{y}) - \mathbb{F}(\varsigma, \mathrm{x}) \geq -2(\mathrm{y} - \mathrm{x}), \quad \varsigma \in [0,1].$$

Hence Hypothesis 2 holds with $\lambda = 2$, and if $\phi_0(0) \leq u_1 \leq u_2 \leq \varpi_0(0)$, $\phi_0(1) \leq v_1 \leq v_2 \leq \varpi_0(1)$, we have

$$\mathbb{G}(u_2, v_2) - \mathbb{G}(u_1, v_1) \leq (u_2 - u_1) - \frac{2\pi}{e^2 - 1 + 2\pi}(v_2 - v_1).$$

Therefore, Hypothesis 3 holds with $c = 1$ and $d = \frac{2\pi}{e^2 - 1 + 2\pi}$. According to Theorem 1, there exist monotone iterative sequences $\{\phi_n\}$ and $\{\varpi_n\}$ that are uniformly converging to ϕ^* and ϖ^*, respectively, and ϕ^*, ϖ^* are the extremal solutions in $[\phi_0, \varpi_0]$ of problem (13).

5. Conclusions

We have established sufficient results by using monotone iterative techniques together with upper and lower solutions for a class of boundary value problem involving a generalized form of Caputo derivative of fractional order. By using the mentioned tool, we have established fruitful combinations between lower and upper solutions. Further, the said method has the ability to produce two sequences of upper and lower solutions, respectively. For the construction of the aforesaid sequences this method does not need any kind of discretization or collocation like other methods usually require. The two sequences we have generated are of a monotonic type with increasing and decreasing behaviors, respectively. Moreover, the sequence that is monotonically decreasing converges to its lower bound. However, the other one that is monotonically increasing is converging to its upper bound. The bounds for upper and lower solutions have also been investigated for their uniqueness using Banach contraction theorem. For the justification of our results, we have provided some examples. Overall we have concluded that the proposed procedure is a powerful and efficient tool to study various classes of FDEs for their extremal solutions. In future this technique can be applied to investigate those classes of FDEs involving non-singular-type derivatives under boundary conditions for upper and lower solution. In addition, the mentioned tool can be applied to investigate fractal-fractional-type problems corresponding to boundary conditions. Overall we have concluded that the monotone iterative technique of applied analysis is a powerful technique for dealing with various kinds of problems involving different types of fractional-order operators.

Author Contributions: Conceptualization, C.D., Z.B.; writing—original draft preparation, C.D., Z.B. and M.S.A.; methodology, C.D. and Z.B.; software, K.S. and B.A.; validation, C.D., Z.B., M.S.A. and T.A.; formal analysis, C.D., Z.B. and M.S.A.; writing—review and editing, C.D., Z.B., M.S.A., K.S. and T.A.; investigation, M.S.A., K.S., B.A. and T.A. All authors have read and agreed to the published version of the manuscript.

Funding: This work was funded by the Prince Sultan University, Riyadh P.O. Box 11586, Saudi Arabia.

Institutional Review Board Statement: Not applicable.

Informed Consent Statement: Not applicable.

Data Availability Statement: Not applicable.

Acknowledgments: Authors Kamal Shah, Bahaaeldin Abdalla and Thabet Abdeljawad would like to thank Prince Sultan University for paying the APC and support through TAS research lab.

Conflicts of Interest: The authors declare no conflicts of interest.

References

1. Hilfer, R. *Applications of Fractional Calculus in Physics*; World Scientific: Singapore, 2000.
2. Mainardi, F. *Fractional Calculus and Waves in Linear Viscoelasticity*; Imperial College Press: London, UK, 2010.
3. Tarasov, V.E. *Fractional Dynamics*; Nonlinear Physical Science; Springer: Heidelberg, Germany, 2010.
4. Abbas, S.; Benchohra, M.; Graef, J.R.; Henderson, J. *Implicit Fractional Differential and Integral Equations: Existence and Stability*; De Gruyter: Berlin, Germany, 2018.
5. Diethelm, K. *The Analysis of Fractional Differential Equations: An Application-Oriented Exposition Using Differential Operators of Caputo Type*; Springer Science & Business Media: New York, NY, USA, 2010.
6. Kilbas, A.A.; Srivastava, H.M.; Trujillo, J.J. *Theory and Applications of Fractional Differential Equations*; North-Holland Mathematics Sudies; Elsevier: Amsterdam, The Netherlands, 2006; Volume 204.
7. Gorenflo, R.; Kilbas, A.A.; Mainardi, F.; Rogosin, S.V. *Mittag–Leffler Functions, Related Topics and Applications*; Springer: New York, NY, USA, 2014.
8. Tenreiro Machado, J.A.; Silva, M.F.; Barbosa, R.S.; Jesus, I.S.; Reis, C.M.; Marcos, M.G.; Galhano, A.F. Some applications of fractional calculus in engineering. *Math. Probl. Eng.* **2010**, *34*, 639801. [CrossRef]
9. da Graça Marcos, M.; Duarte, F.B.M.; Machado, J.A.T. Complex dynamics in the trajectory control of redundant manipulators. *Nonlinear Sci. Complex.* **2007**, *2007*, 134143.
10. Atangana, A.; Kilicman, A. On the generalized mass transport equation to the concept of variable fractional derivative. *Math. Probl. Eng.* **2014**, *2014*, 542809. [CrossRef]
11. Almeida, R. A Caputo fractional derivative of a function with respect to another function. *Commun. Nonlinear Sci. Numer. Simul.* **2017**, *44*, 460–481. [CrossRef]
12. Vanterler da C. Sousa, J.; Capelas de Oliveira, E. On the ψ-Hilfer fractional derivative. *Commun. Nonlinear Sci. Numer. Simul.* **2018**, *60*, 72–91. [CrossRef]
13. Sousa, J.V.D.C.; de Oliveira, E.C. On the Stability of a Hyperbolic Fractional Partial Differential Equation. *Differ. Equ. Dyn. Syst.* **2019**, *2019*, 730465. [CrossRef]
14. Du, S.W.; Lakshmikantham, V. Monotone iterative technique for differential equations in a Banach space. *J. Math. Anal. Appl.* **1982**, *87*, 454–459. [CrossRef]
15. Guo, D.; Lakshmikantham, V. *Nonlinear Problems in Abstract Spaces*; Academic Press: New York, NY, USA, 1988.
16. Derbazi, C.; Baitiche, Z. Coupled systems of ψ-Caputo differential equations with initial conditions in Banach spaces. *Mediterr. J. Math.* **2020**, *17*, 169. [CrossRef]
17. Derbazi, C.; Baitiche, Z.; Fečkan, M. Some new uniqueness and Ulam stability results for a class of multiterms fractional differential equations in the framework of generalized Caputo fractional derivative using the Φ-fractional Bielecki-type norm. *Turkish J. Math.* **2021**, *45*, 2307–2322. [CrossRef]
18. Derbazi, C.; Baitiche, Z.; Benchohra, M.; N'Guérékata, G. Existence, uniqueness, and Mittag-Leffler-Ulam stability results for Cauchy problem involving ψ-Caputo derivative in Banach and Fréchet spaces. *Int. J. Differ. Equ.* **2020**, *2020*, 6383916. [CrossRef]
19. Baitiche, Z.; Guerbati, K.; Benchohra, M.; Zhou, Y. Solvability of Fractional Multi-Point BVP with Nonlinear Growth at Resonance. *J. Contemp. Math. Anal.* **2020**, *55*, 126–142. [CrossRef]
20. Benchohra, M.; Bouriah, S.; Nieto, J.J. Existence of periodic solutions for nonlinear implicit Hadamard's fractional differential equations. *Rev. R. Acad. Cienc. Exactas Fís. Nat. Ser. A Mat.* **2018**, *112*, 25–35. [CrossRef]
21. Ladde, G.S.; Lakshmikantham, V.; Vatsala, A.S. *Monotone Iterative Techniques for Nonlinear Differential Equations*; Pitman Publishing: Bay City, MI, USA, 1985; Volume 27.
22. Heikkilä, S.; Lakshmikantham, V. *Monotone Iterative Techniques for Discontinuous Nonlinear Differential Equations*; Taylor & Francis: New York, NY, USA, 2017.

23. Hristova, S.S.; Bainov, D.D. Monotone-iterative techniques of V. Lakshmikantham for a boundary value problem for systems of impulsive differential-difference equations. *J. Nath. Anal. Appl.* **1996**, *197*, 1–13. [CrossRef]
24. Bhaskar, T.G.; Farzana, A.M. Monotone iterative techniques for nonlinear problems involving the difference of two monotone functions. *Appl. Math. Comput.* **2002**, *133*, 187–192. [CrossRef]
25. Lakshmikantham, V.; Vatsala, A.S. General uniqueness and monotone iterative technique for fractional differential equations. *Appl. Math. Lett.* **2008**, *21*, 828–834. [CrossRef]
26. Franco, D.; Nieto, J.J.; O'Regan, D. Existence of solutions for first order ordinary differential equations with nonlinear boundary conditions. *Appl. Math. Comput.* **2004**, *153*, 793–802. [CrossRef]
27. Lin, X.; Zhao, Z. Iterative technique for a third-order differential equation with three-point nonlinear boundary value conditions. *Electron. J. Qual. Theory Differ. Equ.* **2016**, *2016*, 1–10. [CrossRef]
28. West, I.H.; Vatsala, A.S. Generalized monotone iterative method for initial value problems. *Appl. Math. Lett.* **2004**, *17*, 1231–1237. [CrossRef]
29. Xu, H.K.; Nieto, J.J. Extremal solutions of a class of nonlinear integro-differential equations in Banach spaces. *Proc. Amer. Math. Soc.* **1997**, *125*, 2605–2614. [CrossRef]
30. Al-Refai, M.; Ali Hajji, M. Monotone iterative sequences for nonlinear boundary value problems of fractional order. *Nonlinear Anal.* **2011**, *74*, 3531–3539. [CrossRef]
31. Derbazi, C.; Baitiche, Z.; Benchohra, M.; Cabada, A. Initial value problem for nonlinear fractional differential equations with ψ-Caputo derivative via monotone iterative technique. *Axioms* **2020**, 9, 57. [CrossRef]
32. Derbazi, C.; Baitiche, Z.; Benchohra, M.; N'Guérékata, G. Existence, uniqueness, approximation of solutions and E_α-Ulam stability results for a class of nonlinear fractional differential equations involving ψ-Caputo derivative with initial conditions. *Math. Morav.* **2021**, *25*, 1–30. [CrossRef]
33. Baitiche, Z.; Derbazi, C.; Alzabut, J.; Samei, M.E.; Kaabar, M.K.A.; Siri, Z. Monotone Iterative Method for ϑ-Caputo Fractional Differential Equation with Nonlinear Boundary Conditions. *Fractal Fract.* **2021**, *5*, 81. [CrossRef]
34. Baitiche, Z.; Derbazi, C.; Wang, G. Monotone iterative method for nonlinear fractional p-Laplacian differential equation in terms of ψ-Caputo fractional derivative equipped with a new class of nonlinear boundary conditions. *Math. Meth. Appl. Sci.* **2021**, *45*, 967–976. [CrossRef]
35. Wang, G. Monotone iterative technique for boundary value problems of a nonlinear fractional differential equation with deviating arguments. *J. Comput. Appl. Math.* **2012**, *236*, 2425–2430. [CrossRef]
36. Wang, G.; Sudsutad, W.; Zhang, L.; Tariboon, J. Monotone iterative technique for a nonlinear fractional q-difference equation of Caputo type. *Adv. Difference Equ.* **2016**, *211*, 11. [CrossRef]
37. Zhang, S. Monotone iterative method for initial value problem involving Riemann-Liouville fractional derivatives. *Nonlinear Anal.* **2009**, *71*, 2087–2093. [CrossRef]
38. Abdo, M.S.; Ibrahim, A.G.; Panchal, S.K. Nonlinear implicit fractional differential equation involving-Caputo fractional derivative, *Proc. Jangjeon Math. Soc.* **2019**, *22*, 387–400.
39. Patil, J.; Chaudhari, A.; Abdo, M.S.; Hardan, B. Upper and lower solution method for positive solution of generalized Caputo fractional differential equations. *Adv. Theo. Nonl. Anal. Appl.* **2020**, *4*, 279–291. [CrossRef]
40. Wahash, H.A.; Panchal, S.K.; Abdo, M.S. Positive solutions for generalized Caputo fractional differential equations with integral boundary conditions. *J. Math. Model.* **2020**, *8*, 393–414.
41. Wei, Z.; Li, Q.; Che, J. Initial value problems for fractional differential equations involving Riemann-Liouville sequential fractional derivative. *J. Math. Anal. Appl.* **2010**, *367*, 260–272. [CrossRef]

fractal and fractional

MDPI

Article

Some q-Fractional Estimates of Trapezoid like Inequalities Involving Raina's Function

Kamsing Nonlaopon [1], Muhammad Uzair Awan [2,*], Muhammad Zakria Javed [2], Hüseyin Budak [3] and Muhammad Aslam Noor [4]

1 Department of Mathematics, Faculty of Science, Khon Kaen University, Khon Kaen 40002, Thailand; nkamsi@kku.ac.th
2 Department of Mathematics, Government College University, Faisalabad 38000, Pakistan; zakriajaved071@gmail.com
3 Department of Mathematics, Faculty of Science and Arts, Düzce University, Düzce 81620, Turkey; huseyinbudak@duzce.edu.tr
4 Department of Mathematics, COMSATS University Islamabad, Islamabad 45550, Pakistan; noormaslam@gmail.com
* Correspondence: muawan@gcuf.edu.pk

Abstract: In this paper, we derive two new identities involving q-Riemann-Liouville fractional integrals. Using these identities, as auxiliary results, we derive some new q-fractional estimates of trapezoidal-like inequalities, essentially using the class of generalized exponential convex functions.

Keywords: convex; exponential convex; fractional; quantum; inequalities

MSC: 05A30; 26A33; 26A51; 34A08; 26D07; 26D10; 26D15

Citation: Nonlaopon, K.; Awan, M.U.; Javed, M.Z.; Budak, H.; Noor, A.M. Some q-Fractional Estimates of Trapezoid like Inequalities Involving Raina's Function. *Fractal Fract.* **2022**, *6*, 185. https://doi.org/10.3390/fractalfract6040185

Academic Editors: Asifa Tassaddiq and Muhammad Yaseen

Received: 24 February 2022
Accepted: 22 March 2022
Published: 25 March 2022

1. Introduction

A set $\mathcal{C} \subseteq \mathbb{R}$ is said to be convex, if

$$(1-\nu)x + \nu y \in \mathcal{C}$$

for all $x, y \in \mathcal{C}$ and $\nu \in [0,1]$.

A function $\mathscr{F} : \mathcal{C} \to \mathbb{R}$ is said to be convex, if

$$\mathscr{F}((1-\nu)x + \nu y) \leqslant (1-\nu)\mathscr{F}(x) + \nu\mathscr{F}(y)$$

for all $x, y \in \mathcal{C}$ and $\nu \in [0,1]$.

The classical concepts of convexity are simple but have many applications in different fields of pure and applied sciences. For example, they play a significant role in the theory of optimization, mathematical economics, operations research, etc. In recent years, the classical concepts of convexity have been extended and generalized in different directions using novel and innovative ideas. It has been observed that these new generalizations of classical convexity enjoy some nice properties which classical convexity has. Recently, Cortez et al. [1] presented a new generalization of convexity class as follows:

Definition 1 ([1]). *Let $\rho, \lambda > 0$ and $\sigma = (\sigma(0), \ldots, \sigma(k), \ldots)$ be a bounded sequence of positive real numbers. A non-empty set $\mathfrak{I}$ is said to be generalized convex, if*

$$\varpi_1 + \tau\mathcal{R}_{\rho,\lambda,\sigma}(\varpi_2 - \varpi_1) \in \mathfrak{I}$$

for all $\varpi_1, \varpi_2 \in \mathfrak{I}$ and $\tau \in [0,1]$.

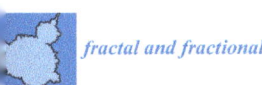

Here, $\mathcal{R}_{\rho,\lambda,\sigma}(z)$ is the Raina's function and is defined as:

$$\mathcal{R}_{\rho,\lambda,\sigma}(z) = \mathcal{R}^{\sigma(0),\sigma(1),\dots}_{\rho,\lambda}(z) = \sum_{k=0}^{\infty} \frac{\sigma(k)}{\Gamma(\rho k + \lambda)} z^k,$$

where $\rho, \lambda > 0, |z| < R$, $\sigma = \{\sigma(0), \sigma(1), \dots, \sigma(k), \dots\}$ is a bounded sequence of positive real numbers and $\Gamma(\eta) = \int\limits_0^{\infty} x^{\eta-1} e^{-x} dx$ is the gamma function. For details, see [2].

Cortez et al. [1] also defined the class of generalized convex functions as:

Definition 2 ([1]). *Let $\rho, \lambda > 0$ and $\sigma = (\sigma(0), \dots, \sigma(k), \dots)$ be a bounded sequence of positive real numbers. A function $\mathscr{F} : \mathcal{J} \to \mathbb{R}$ is said to be generalized convex, if*

$$\mathscr{F}(\varpi_1 + \tau \mathcal{R}_{\rho,\lambda,\sigma}(\varpi_2 - \varpi_1)) \leqslant (1 - \tau)\mathscr{F}(\varpi_1) + \tau \mathscr{F}(\varpi_2)$$

for all $\varpi_1, \varpi_2 \in \mathcal{J}$ and $\tau \in [0, 1]$.

Awan et al. [3] introduced the class of exponential convex functions as:

Definition 3 ([3]). *A function $\mathscr{F} : \mathcal{C} \to \mathbb{R}$ is said to be exponentially convex, if*

$$\mathscr{F}((1 - \nu)x + \nu y) \leqslant (1 - \nu) \frac{\mathscr{F}(x)}{\exp(\alpha x)} + \nu \frac{\mathscr{F}(y)}{\exp(\alpha y)},$$

for all $x, y \in \mathcal{C}$ and $\nu \in [0, 1]$.

Besides its applications, the theory of convexity has also played a dynamic role in developing the theory of inequalities. A wide class of inequalities is just a direct consequence of the applications of the convexity property of the functions. Hermite–Hadamard's inequality, also known as trapezium-like inequality, is one of the most studied results. It reads as:

Let $\mathscr{F} : I \subseteq \mathbb{R} \to \mathbb{R}$ be a convex function, then

$$\mathscr{F}\left(\frac{\varpi_1 + \varpi_2}{2}\right) \leqslant \frac{1}{\varpi_2 - \varpi_1} \int\limits_{\varpi_1}^{\varpi_2} \mathscr{F}(x) dx \leqslant \frac{\mathscr{F}(\varpi_1) + \mathscr{F}(\varpi_2)}{2}.$$

For some recent developments related to Hermite–Hadamard's inequality and its applications, see [4].

In recent years, several new techniques have been used to obtain new versions of Hermite–Hadamard's inequality. For instance, Sarikaya et al. [5] utilized the concepts of fractional calculus and obtained the fractional analogues of Hermite–Hadamard's inequality. Alp et al. [6] obtained quantum analogue of Hermite–Hadamard's inequality. Awan et al. [3] obtained a new refinement of Hermite–Hadamard's inequality using the class of exponentially convex functions. Cortez et al. [1] obtained Hermite–Hadamard's inequality using the class of generalized convex functions. Kunt and Aljasem [7] obtained fractional quantum versions of Hermite–Hadamard type of inequalities. Noor et al. [8] obtained some more quantum estimates for Hermite–Hadamard inequalities using the class of convex functions. Sudsutad [9] obtained various new quantum integral inequalities for convex functions. Zhang et al. [10] obtained a new generalized quantum-integral identity and obtained new q-integral inequalities via (α, m)-convexity property of the functions.

The main motivation of this article is to obtain two new identities involving q-Riemann–Liouville fractional integrals. Using these identities as auxiliary results, we derive some new q-fractional estimates of trapezoidal-like inequalities, essentially using the class of generalized exponential convex functions. We hope that the ideas and techniques of this article will inspire interested readers working in this field.

2. Preliminaries

In this section, we recall some previously known concepts and results.

The following concept of q-derivative was introduced and studied in [11].

Definition 4 ([11]). *For a continuous function $\mathscr{F} : [\varpi_1, \varpi_2] \to \mathbb{R}$ the q-derivative of $\mathscr{F}$ at $x \in [\varpi_1, \varpi_2]$ is defined as:*

$$_{\varpi_1}D_q\mathscr{F}(x) = \frac{\mathscr{F}(x) - \mathscr{F}(qx + (1-q)\varpi_1)}{(1-q)(x-\varpi_1)}, \quad x \neq \varpi_1. \tag{1}$$

The q-definite integral is defined as:

Definition 5 ([11]). *Let $\mathscr{F} : [\varpi_1, \varpi_2] \to \mathbb{R}$ be a continuous function. Then the q-definite integral on $[\varpi_1, \varpi_2]$ is defined as:*

$$\int_{\varpi_1}^{x} \mathscr{F}(\nu)\ {}_{\varpi_1}d_q\nu = (1-q)(x-\varpi_1)\sum_{n=0}^{\infty} q^n \mathscr{F}(q^n x + (1-q^n)\varpi_1), \tag{2}$$

for $x \in [\varpi_1, \varpi_2]$.

Interesting additional details of the following concepts can be found in [9,12].

$$[m]_q = \frac{1-q^m}{1-q}, \quad m \in \mathbb{R}. \tag{3}$$

The q-analogue of power function is defined as, if $\gamma \in \mathbb{R}$, then

$$(r-m)^{(\gamma)} = r^{\gamma}\prod_{n=0}^{\infty}\frac{r-q^n m}{r-q^{\gamma+n}m}, \quad r \neq 0. \tag{4}$$

The q-gamma function is defined as:

$$\Gamma_q(\nu) = \frac{(1-q)^{(\nu-1)}}{(1-q)^{\nu-1}}, \quad \nu \in \mathbb{R}/\{0,-1,-2,\ldots\}. \tag{5}$$

For any $s, \nu > 0$, the q-beta function is defined as:

$$B_q(s,\nu) = \int_0^1 u^{(s-1)}(1-qu)^{(\nu-1)}d_q u, \tag{6}$$

and

$$B_q(s,\nu) = \frac{\Gamma_q(s)\Gamma_q(\nu)}{\Gamma_q(s+\nu)}.$$

The q-Pochhammer symbol is defined as:

$$(m;q)_0 = 1, \quad \text{and} \quad (m;q)_k = \prod_{n=0}^{k-1}(1-q^n m) \tag{7}$$

for $k \in \mathbb{N} \cup \{\infty\}$.

Theorem 1 ([13]). *Suppose $\lambda, \mu \in \mathbb{R}$, then*

$$\lim_{q\to 1^-}\frac{(q^{\lambda}x;q)}{(q^{\mu}x;q)} = (1-x)^{\mu-\lambda}, \tag{8}$$

uniformly on $\{x \in \mathbb{C} : |x| \leqslant 1\}$, *if* $\mu \geqslant \lambda$, $\lambda + \mu \geqslant 1$, *and uniformly on compact subset of* $\{x \in \mathbb{C} : |x| \leqslant 1, x \neq 1\}$ *for other choices of* μ *and* λ.

The q-shifting operator is defined as:

$$_{\varpi_1}\Phi_q(m) = qm + (1-q)\varpi_1. \tag{9}$$

For any positive integer k, one has:

$$_{\varpi_1}\Phi_q^k(m) = {}_{\varpi_1}\Phi_q^{k-1}({}_{\varpi_1}\Phi_q(m)), \quad {}_{\varpi_1}\Phi_q^0(m) = m. \tag{10}$$

The following properties for q-shifting operator hold:

Theorem 2 ([9,12]). *For any* $r, m \in \mathbb{R}$ *and for all positive integers* k, j, *one has:*

1. $_{\varpi_1}\Phi_q^k(m) = {}_{\varpi_1}\Phi_{q^k}(m)$;
2. $_{\varpi_1}\Phi_q^k({}_{\varpi_1}\Phi_q^j(m)) = {}_{\varpi_1}\Phi_q^j({}_{\varpi_1}\Phi_q^k(m)) = {}_{\varpi_1}\Phi_q^{j+k}(m)$;
3. $_{\varpi_1}\Phi_q(\varpi_1) = \varpi_1$;
4. $_{\varpi_1}\Phi_q^k(m) - \varpi_1 = q^k(m - \varpi_1)$;
5. $m - {}_{\varpi_1}\Phi_q^k(m) = (1-q^k)(m-\varpi_1)$;
6. $_{\varpi_1}\Phi_q^k(m) = m\, {}_{\frac{\varpi_1}{m}}\Phi_q^k(1)$, *for* $m \neq 0$;
7. $_{\varpi_1}\Phi_q(m) - {}_{\varpi_1}\Phi_q^k(r) = q(m - {}_{\varpi_1}\Phi_q^{k-1}(r))$.

The power of q-shifting operator is defined as:

$$_{\varpi_1}(r-m)_q^{(\gamma)} = (r-\varpi_1)^\gamma \prod_{n=0}^{\infty} \frac{r - {}_{\varpi_1}\Phi_q^n(m)}{r - {}_{\varpi_1}\Phi_q^{\gamma+n}(m)}, \quad \gamma \in \mathbb{R}. \tag{11}$$

Theorem 3 ([9,12]). *For any* $\gamma, r, m \in \mathbb{R}$, $r \neq \varpi_1$ *and* $k \in \mathbb{N}$, *one has:*

1. $_{\varpi_1}(r-m)_q^{(k)} = (r-\varpi_1)^k \left(\frac{m-\varpi_1}{r-\varpi_1}; q\right)_k$;
2. $_{\varpi_1}(r-m)_q^{(\gamma)} = (r-\varpi_1)^\gamma \prod_{n=0}^{\infty} \frac{1-\frac{m-\varpi_1}{r-\varpi_1}q^n}{1-\frac{m-\varpi_1}{r-\varpi_1}q^{n+\gamma}} = (r-\varpi_1)^\gamma \frac{\left(\frac{m-\varpi_1}{r-\varpi_1}; q\right)_\infty}{\left(\frac{m-\varpi_1}{r-\varpi_1}q^\gamma; q\right)_\infty}$;
3. $_{\varpi_1}(r - {}_{\varpi_1}\Phi_q^k(r))_q^{(\gamma)} = (r-\varpi_1)^\gamma \frac{(q^k;q)_\infty}{(q^{\gamma+k};q)_\infty}$.

Definition 6 ([9,12]). *Let* $\alpha \geqslant 0$ *and* $\mathscr{F}$ *be a continuous function on* $[\varpi_1, \varpi_2]$. *Then the Riemann–Liouville-type fractional quantum integral is given by* $({}_{\varpi_1}J_q^0\mathscr{F})(v) = \mathscr{F}(v)$ *and*

$$\begin{aligned}({}_{\varpi_1}J_q^\alpha\mathscr{F})(x) = ({}_{\varpi_1}J_q^\alpha\mathscr{F}(v))(x) &= \frac{1}{\Gamma_q(\alpha)} \int_{\varpi_1}^{x} {}_{\varpi_1}(x - {}_{\varpi_1}\Phi_q(v))_q^{(\alpha-1)} \mathscr{F}(v)\, {}_{\varpi_1}d_q v \\ &= \frac{(1-q)(x-\varpi_1)}{\Gamma_q(\alpha)} \sum_{n=0}^{\infty} q^n\, {}_{\varpi_1}(x - {}_{\varpi_1}\Phi_q^{n+1}(x))_q^{(\alpha-1)} \mathscr{F}({}_{\varpi_1}\Phi_q^n(x)),\end{aligned} \tag{12}$$

where $\alpha > 0$ *and* $x \in [\varpi_1, \varpi_2]$.

3. Results and Discussions

In this section, we will discuss our main results. First of all we define the class of generalized exponential convex functions.

Definition 7. *Let $\rho, \lambda > 0$ and $\sigma = (\sigma(0), \dots, \sigma(k), \dots)$ be a bounded sequence of positive real numbers. A function $\mathscr{F} : \mathcal{I} \to \mathbb{R}$ is said to be generalized exponential convex, if*

$$\mathscr{F}(\varpi_1 + \tau \mathcal{R}_{\rho,\lambda,\sigma}(\varpi_2 - \varpi_1)) \leqslant (1-\tau)\frac{\mathscr{F}(\varpi_1)}{\chi^{\alpha\varpi_1}} + \tau \frac{\mathscr{F}(\varpi_2)}{\chi^{\alpha\varpi_2}}$$

for all $\varpi_1, \varpi_2 \in \mathcal{I}$, $\tau \in [0,1]$ and $\chi \geqslant 1$.

Note that if we take $\alpha = 0$ or $\chi = 1$, then the class of generalized exponential convex functions reduces to the class of generalized convex functions introduced and studied in [1]. If we take $\chi = \exp$, then we have the class of exponentially convex functions involving Raina's function. This class is defined as:

Definition 8. *Let $\rho, \lambda > 0$ and $\sigma = (\sigma(0), \dots, \sigma(k), \dots)$ be a bounded sequence of positive real numbers. A function $\mathscr{F} : \mathcal{I} \to \mathbb{R}$ is said to be generalized exponential convex, if*

$$\mathscr{F}(\varpi_1 + \tau \mathcal{R}_{\rho,\lambda,\sigma}(\varpi_2 - \varpi_1)) \leqslant (1-\tau)\frac{\mathscr{F}(\varpi_1)}{\exp(\alpha\varpi_1)} + \tau \frac{\mathscr{F}(\varpi_2)}{\exp(\alpha\varpi_2)}$$

for all $\varpi_1, \varpi_2 \in \mathcal{I}$ and $\tau \in [0,1]$.

Now, we derive our auxiliary results. Before we proceed, for the sake of simplicity, we consider $\Omega = [\varpi_1, \varpi_1 + \mathcal{R}_{\rho,\lambda,\sigma}(\varpi_2 - \varpi_1)]$ and $\Omega^\circ = (\varpi_1, \varpi_1 + \mathcal{R}_{\rho,\lambda,\sigma}(\varpi_2 - \varpi_1))$.

Lemma 1. *Let $\mathscr{F} : \Omega \to \mathbb{R}$ be a continuous function and $\alpha > 0$. If ${}_{\varpi_1}D_q\mathscr{F}$ is q-integrable on Ω°, then*

$$\begin{aligned}
&\frac{\Gamma_q(\alpha+1)}{\mathcal{R}^{\alpha}_{\rho,\lambda,\sigma}(\varpi_2 - \varpi_1)}\left({}_{\varpi_1}J^{\alpha}_{q}\mathscr{F}\right)(\varpi_1 + \mathcal{R}_{\rho,\lambda,\sigma}(\varpi_2 - \varpi_1)) - \frac{([\alpha+1]_q - 1)\mathscr{F}(\varpi_1) + \mathscr{F}(\varpi_1 + \mathcal{R}_{\rho,\lambda,\sigma}(\varpi_2 - \varpi_1))}{[\alpha+1]_q} \\
&= \frac{\mathcal{R}_{\rho,\lambda,\sigma}(\varpi_2 - \varpi_1)}{[\alpha+1]_q}\int_0^1 \left([\alpha+1]_q\, {}_0(1 - {}_0\Phi_q(\nu))^{(\alpha)}_q - 1\right) {}_{\varpi_1}D_q\mathscr{F}(\varpi_1 + \nu\mathcal{R}_{\rho,\lambda,\sigma}(\varpi_2 - \varpi_1))\, {}_0d_q\nu. \qquad (13)
\end{aligned}$$

Proof. It suffices to show that

$$\begin{aligned}
&\frac{\mathcal{R}_{\rho,\lambda,\sigma}(\varpi_2 - \varpi_1)}{[\alpha+1]_q}\int_0^1 \left([\alpha+1]_q\, {}_0(1 - {}_0\Phi_q(\nu))^{(\alpha)}_q - 1\right) {}_{\varpi_1}D_q\mathscr{F}(\varpi_1 + \nu\mathcal{R}_{\rho,\lambda,\sigma}(\varpi_2 - \varpi_1))\, {}_0d_q\nu \\
&= \mathcal{R}_{\rho,\lambda,\sigma}(\varpi_2 - \varpi_1)\int_0^1 {}_0(1 - {}_0\Phi_q(\nu))^{(\alpha)}_q\, {}_{\varpi_1}D_q\mathscr{F}(\varpi_1 + \nu\mathcal{R}_{\rho,\lambda,\sigma}(\varpi_2 - \varpi_1))\, {}_0d_q\nu \\
&- \frac{\mathcal{R}_{\rho,\lambda,\sigma}(\varpi_2 - \varpi_1)}{[\alpha+1]_q}\int_0^1 {}_{\varpi_1}D_q\mathscr{F}(\varpi_1 + \nu\mathcal{R}_{\rho,\lambda,\sigma}(\varpi_2 - \varpi_1))\, {}_0d_q\nu \\
&= \mathcal{S}_1 - \mathcal{S}_2. \qquad (14)
\end{aligned}$$

Now,

$$
\begin{aligned}
\mathcal{S}_1 &= \mathcal{R}_{\rho,\lambda,\sigma}(\varpi_2-\varpi_1)\int_0^1 {}_0(1-{}_0\Phi_q(\nu))_q^{(\alpha)}\;{}_{\varpi_1}D_q\mathscr{F}(\varpi_1+\nu\mathcal{R}_{\rho,\lambda,\sigma}(\varpi_2-\varpi_1))\,{}_0d_q\nu \\
&= \mathcal{R}_{\rho,\lambda,\sigma}(\varpi_2-\varpi_1)\int_0^1 {}_0(1-{}_0\Phi_q(\nu))_q^{(\alpha)}\frac{\mathscr{F}(\varpi_1+\nu\mathcal{R}_{\rho,\lambda,\sigma}(\varpi_2-\varpi_1))-\mathscr{F}(\varpi_1+q\nu\mathcal{R}_{\rho,\lambda,\sigma}(\varpi_2-\varpi_1))}{(1-q)\mathcal{R}_{\rho,\lambda,\sigma}(\varpi_2-\varpi_1)\nu}\,{}_0d_q\nu \\
&= \frac{1}{1-q}\int_0^1 {}_0(1-{}_0\Phi_q(\nu))_q^{(\alpha)}\frac{\mathscr{F}(\varpi_1+\nu\mathcal{R}_{\rho,\lambda,\sigma}(\varpi_2-\varpi_1))}{\nu}\,{}_0d_q\nu \\
&\quad -\frac{1}{1-q}\int_0^1 {}_0(1-{}_0\Phi_q(\nu))_q^{(\alpha)}\frac{\mathscr{F}(\varpi_1+q\nu\mathcal{R}_{\rho,\lambda,\sigma}(\varpi_2-\varpi_1))}{\nu}\,{}_0d_q\nu \\
&= \sum_{n=0}^{\infty} q^n\,{}_0(1-{}_0\Phi_q^{n+1}(1))_q^{(\alpha)}\frac{\mathscr{F}\left(\varpi_1+{}_0\Phi_q^n(1)\mathcal{R}_{\rho,\lambda,\sigma}(\varpi_2-\varpi_1)\right)}{{}_0\Phi_q^n(1)} \\
&\quad -\sum_{n=0}^{\infty} q^n\,{}_0(1-{}_0\Phi_q^{n+1}(1))_q^{(\alpha)}\frac{\mathscr{F}\left(\varpi_1+q\,{}_0\Phi_q^n(1)\mathcal{R}_{\rho,\lambda,\sigma}(\varpi_2-\varpi_1)\right)}{{}_0\Phi_q^n(1)} \\
&= \left[\begin{array}{l}\sum\limits_{n=0}^{\infty}\frac{(q^{n+1};q)_\infty}{(q^{\alpha+n+1};q)_\infty}\mathscr{F}(\varpi_1+q^n\mathcal{R}_{\rho,\lambda,\sigma}(\varpi_2-\varpi_1)) \\ -\sum\limits_{n=0}^{\infty}\frac{(q^{n+1};q)_\infty}{(q^{\alpha+n+1};q)_\infty}\mathscr{F}(\varpi_1+q^{n+1}\mathcal{R}_{\rho,\lambda,\sigma}(\varpi_2-\varpi_1))\end{array}\right] \\
&= \left[\begin{array}{l}\sum\limits_{n=0}^{\infty}(1-q^{\alpha+n})\frac{(q^{n+1};q)_\infty}{(q^{\alpha+n};q)_\infty}\mathscr{F}(\varpi_1+q^n\mathcal{R}_{\rho,\lambda,\sigma}(\varpi_2-\varpi_1)) \\ -\sum\limits_{n=0}^{\infty}(1-q^{n+1})\frac{(q^{n+2};q)_\infty}{(q^{\alpha+n+1};q)_\infty}\mathscr{F}(\varpi_1+q^{n+1}\mathcal{R}_{\rho,\lambda,\sigma}(\varpi_2-\varpi_1))\end{array}\right] \\
&= \left[\begin{array}{l}\sum\limits_{n=0}^{\infty}\frac{(q^{n+1};q)_\infty}{(q^{\alpha+n};q)_\infty}\mathscr{F}(\varpi_1+q^n\mathcal{R}_{\rho,\lambda,\sigma}(\varpi_2-\varpi_1)) \\ -\sum\limits_{n=0}^{\infty}\frac{(q^{n+2};q)_\infty}{(q^{\alpha+n+1};q)_\infty}\mathscr{F}(\varpi_1+q^{n+1}\mathcal{R}_{\rho,\lambda,\sigma}(\varpi_2-\varpi_1))\end{array}\right] \\
&= \left[\begin{array}{l}\sum\limits_{n=0}^{\infty}q^{\alpha+n}\frac{(q^{n+1};q)_\infty}{(q^{\alpha+n};q)_\infty}\mathscr{F}(\varpi_1+q^n\mathcal{R}_{\rho,\lambda,\sigma}(\varpi_2-\varpi_1)) \\ -\sum\limits_{n=0}^{\infty}q^{n+2}\frac{(q^{n+2};q)_\infty}{(q^{\alpha+n+1};q)_\infty}\mathscr{F}(\varpi_1+q^{n+1}\mathcal{R}_{\rho,\lambda,\sigma}(\varpi_2-\varpi_1))\end{array}\right] \\
&= \frac{(q^1;q)_\infty}{(q^\alpha;q)_\infty}\mathscr{F}(\varpi_1+\mathcal{R}_{\rho,\lambda,\sigma}(\varpi_2-\varpi_1))-\mathscr{F}(\varpi_1) \\
&\quad -\left[\begin{array}{l}\sum\limits_{n=0}^{\infty}q^{\alpha+n}\frac{(q^{n+1};q)_\infty}{(q^{\alpha+n};q)_\infty}\mathscr{F}(\varpi_1+q^n\mathcal{R}_{\rho,\lambda,\sigma}(\varpi_2-\varpi_1)) \\ -\sum\limits_{n=1}^{\infty}q^{n}\frac{(q^{n+1};q)_\infty}{(q^{\alpha+n};q)_\infty}\mathscr{F}(\varpi_1+q^{n}\mathcal{R}_{\rho,\lambda,\sigma}(\varpi_2-\varpi_1))\end{array}\right] \\
&= \frac{(q^1;q)_\infty}{(q^\alpha;q)_\infty}\mathscr{F}(\varpi_1+\mathcal{R}_{\rho,\lambda,\sigma}(\varpi_2-\varpi_1))-\mathscr{F}(\varpi_1) \\
&\quad -\left[\begin{array}{l}\sum\limits_{n=0}^{\infty}q^{\alpha+n}\frac{(q^{n+1};q)_\infty}{(q^{\alpha+n};q)_\infty}\mathscr{F}(\varpi_1+q^n\mathcal{R}_{\rho,\lambda,\sigma}(\varpi_2-\varpi_1)) \\ -\sum\limits_{n=0}^{\infty}q^{n}\frac{(q^{n+1};q)_\infty}{(q^{\alpha+n};q)_\infty}\mathscr{F}(\varpi_1+q^{n}\mathcal{R}_{\rho,\lambda,\sigma}(\varpi_2-\varpi_1)) \\ \quad +\frac{(q^1;q)_\infty}{(q^\alpha;q)_\infty}\mathscr{F}(\varpi_1+\mathcal{R}_{\rho,\lambda,\sigma}(\varpi_2-\varpi_1))\end{array}\right]
\end{aligned}
$$

$$
\begin{aligned}
&= -\mathscr{F}(\varpi_1) + (1-q^n)\sum_{n=0}^{\infty} q^n \frac{(q^{n+1};q)_\infty}{(q^{\alpha+n};q)_\infty}\mathscr{F}(\varpi_1 + q^n \mathcal{R}_{\rho,\lambda,\sigma}(\varpi_2-\varpi_1)) \\
&= -\mathscr{F}(\varpi_1) + [\alpha]_q(1-q)\sum_{n=0}^{\infty} q^n \frac{(q^{n+1};q)_\infty}{(q^{\alpha+n};q)_\infty}\mathscr{F}(\varpi_1 + q^n \mathcal{R}_{\rho,\lambda,\sigma}(\varpi_2-\varpi_1)) \\
&= -\mathscr{F}(\varpi_1) + \frac{[\alpha]_q\Gamma_q(\alpha)}{\mathcal{R}^{\alpha}_{\rho,\lambda,\sigma}(\varpi_2-\varpi_1)} \\
&\quad \times \left(\frac{(1-q)\mathcal{R}_{\rho,\lambda,\sigma}(\varpi_2-\varpi_1)}{\Gamma_q(\alpha)} \sum_{n=0}^{\infty} q^n \zeta^{\alpha-1}(\varpi_2,\varpi_1)\frac{(q^{n+1};q)_\infty}{(q^{\alpha+n};q)_\infty}\mathscr{F}(\varpi_1 + q^n \mathcal{R}_{\rho,\lambda,\sigma}(\varpi_2-\varpi_1))\right) \\
&= -\mathscr{F}(\varpi_1) + \frac{\Gamma_q(\alpha+1)}{\mathcal{R}^{\alpha}_{\rho,\lambda,\sigma}(\varpi_2-\varpi_1)} \\
&\quad \times \left(\frac{(1-q)\mathcal{R}_{\rho,\lambda,\sigma}(\varpi_2-\varpi_1)}{\Gamma_q(\alpha)} \sum_{n=0}^{\infty} q^n \zeta^{\alpha-1}(\varpi_2,\varpi_1)\frac{(q^{n+1};q)_\infty}{(q^{(\alpha+1)+(n+1)};q)_\infty}\mathscr{F}(\varpi_1 + q^n \mathcal{R}_{\rho,\lambda,\sigma}(\varpi_2-\varpi_1))\right) \\
&= -\mathscr{F}(\varpi_1) + \frac{\Gamma_q(\alpha+1)}{\mathcal{R}^{\alpha}_{\rho,\lambda,\sigma}(\varpi_2-\varpi_1)} \Bigg(\frac{(1-q)\mathcal{R}_{\rho,\lambda,\sigma}(\varpi_2-\varpi_1)}{\Gamma_q(\alpha)} \sum_{n=0}^{\infty} q^n \,{}_{\varpi_1}(\varpi_1 + \mathcal{R}_{\rho,\lambda,\sigma}(\varpi_2-\varpi_1)) \\
&\quad - {}_{\varpi_1}\Phi_q^{n+1}(\varpi_1 + \mathcal{R}_{\rho,\lambda,\sigma}(\varpi_2-\varpi_1))_q^{(\alpha-1)}\mathscr{F}\left({}_{\varpi_1}\Phi_q^{n}(\varpi_1 + \mathcal{R}_{\rho,\lambda,\sigma}(\varpi_2-\varpi_1))\right)\Bigg) \\
&= -\mathscr{F}(\varpi_1) + \frac{\Gamma_q(\alpha+1)}{\mathcal{R}^{\alpha}_{\rho,\lambda,\sigma}(\varpi_2-\varpi_1)} \\
&\quad \times \left(\frac{1}{\Gamma_q(\alpha)} \int_{\varpi_1}^{\varpi_1+\mathcal{R}_{\rho,\lambda,\sigma}(\varpi_2-\varpi_1)} {}_{\varpi_1}(\varpi_1 + \mathcal{R}_{\rho,\lambda,\sigma}(\varpi_2-\varpi_1) - {}_{\varpi_1}\Phi_q(\nu))_q^{(\alpha-1)}\mathscr{F}(\nu)\,{}_{\varpi_1}d_q\nu \right) \\
&= -\mathscr{F}(\varpi_1) + \frac{\Gamma_q(\alpha+1)}{\mathcal{R}^{\alpha}_{\rho,\lambda,\sigma}(\varpi_2-\varpi_1)}({}_{\varpi_1}J_q^{\alpha}\mathscr{F})(\varpi_1 + \mathcal{R}_{\rho,\lambda,\sigma}(\varpi_2-\varpi_1)).
\end{aligned} \tag{15}
$$

Similarly, we have

$$
\begin{aligned}
\mathcal{S}_2 &= \frac{\mathcal{R}_{\rho,\lambda,\sigma}(\varpi_2-\varpi_1)}{[\alpha+1]_q}\int_0^1 {}_{\varpi_1}D_q\mathscr{F}(\varpi_1 + \nu\mathcal{R}_{\rho,\lambda,\sigma}(\varpi_2-\varpi_1))\,{}_0d_q\nu \\
&= \frac{\mathcal{R}_{\rho,\lambda,\sigma}(\varpi_2-\varpi_1)}{[\alpha+1]_q}\int_0^1 \frac{\mathscr{F}(\varpi_1 + \nu\mathcal{R}_{\rho,\lambda,\sigma}(\varpi_2-\varpi_1)) - \mathscr{F}(\varpi_1 + q\nu\mathcal{R}_{\rho,\lambda,\sigma}(\varpi_2-\varpi_1))}{(1-q)\mathcal{R}_{\rho,\lambda,\sigma}(\varpi_2-\varpi_1)\nu}\,{}_0d_q\nu \\
&= \frac{1}{(1-q)[\alpha+1]_q}\int_0^1 \frac{\mathscr{F}(\varpi_1 + \nu\mathcal{R}_{\rho,\lambda,\sigma}(\varpi_2-\varpi_1))}{\nu}\,{}_0d_q\nu \\
&\quad - \frac{1}{(1-q)[\alpha+1]_q}\int_0^1 \frac{\mathscr{F}(\varpi_1 + q\nu\mathcal{R}_{\rho,\lambda,\sigma}(\varpi_2-\varpi_1))}{\nu}\,{}_0d_q\nu \\
&= \frac{1}{[\alpha+1]_q}\left[\sum_{n=0}^{\infty}\mathscr{F}(\varpi_1 + q^n\mathcal{R}_{\rho,\lambda,\sigma}(\varpi_2-\varpi_1)) - \sum_{n=0}^{\infty}\mathscr{F}(\varpi_1 + q^{n+1}\mathcal{R}_{\rho,\lambda,\sigma}(\varpi_2-\varpi_1))\right] \\
&= \frac{\mathscr{F}(\varpi_1 + \mathcal{R}_{\rho,\lambda,\sigma}(\varpi_2-\varpi_1)) - \mathscr{F}(\varpi_1)}{[\alpha+1]_q}.
\end{aligned} \tag{16}
$$

Using the equalities (15) and (16) in (14), we obtain the desired result. □

Corollary 1. *Under the assumptions of Lemma 1, if we choose $\alpha = 1$, then*

$$\frac{1}{\mathcal{R}_{\rho,\lambda,\sigma}(\varpi_2-\varpi_1)}\int_{\varpi_1}^{\varpi_1+\mathcal{R}_{\rho,\lambda,\sigma}(\varpi_2-\varpi_1)}\mathscr{F}(\nu)\,_{\varpi_1}d_q\nu-\frac{q\mathscr{F}(\varpi_1)+\mathscr{F}(\varpi_1+\mathcal{R}_{\rho,\lambda,\sigma}(\varpi_2-\varpi_1))}{1+q}$$
$$=\frac{q\mathcal{R}_{\rho,\lambda,\sigma}(\varpi_2-\varpi_1)}{1+q}\int_0^1(1-(1+q)\nu)\,_{\varpi_1}D_q\mathscr{F}(\varpi_1+\nu\mathcal{R}_{\rho,\lambda,\sigma}(\varpi_2-\varpi_1))\,_0d_q\nu. \tag{17}$$

Lemma 2. *Let $\mathscr{F}:\Omega\to\mathbb{R}$ be a continuous function and $\alpha>0$. If $_{\varpi_1}D_q\mathscr{F}$ is q-integrable on Ω°, then the following equality holds:*

$$\mathscr{F}\left(\frac{([\alpha+1]_q-1)\varpi_1+(\varpi_1+\mathcal{R}_{\rho,\lambda,\sigma}(\varpi_2-\varpi_1))}{[\alpha+1]_q}\right)-\frac{\Gamma_q(\alpha+1)}{\mathcal{R}^{\alpha}_{\rho,\lambda,\sigma}(\varpi_2-\varpi_1)}(\,_{\varpi_1}J_q^{\alpha}\mathscr{F})(\varpi_1+\mathcal{R}_{\rho,\lambda,\sigma}(\varpi_2-\varpi_1))$$
$$=\mathcal{R}_{\rho,\lambda,\sigma}(\varpi_2-\varpi_1)\left[\begin{array}{l}\int_0^{\frac{1}{[\alpha+1]_q}}\left(1-\,_0(1-\Phi_q(\nu))_q^{(\alpha)}\right)\,_{\varpi_1}D_q\mathscr{F}(\varpi_1+\nu\mathcal{R}_{\rho,\lambda,\sigma}(\varpi_2-\varpi_1))\,_0d_q\nu\\+\int_{\frac{1}{[\alpha+1]_q}}^{1}\left(-\,_0(1-\Phi_q(\nu))_q^{(\alpha)}\right)\,_{\varpi_1}D_q\mathscr{F}(\varpi_1+\nu\mathcal{R}_{\rho,\lambda,\sigma}(\varpi_2-\varpi_1))\,_0d_q\nu\end{array}\right]. \tag{18}$$

Proof. Let $S_3=\int_0^{\frac{1}{[\alpha+1]_q}}\,_{\varpi_1}D_q\mathscr{F}(\varpi_1+\nu\mathcal{R}_{\rho,\lambda,\sigma}(\varpi_2-\varpi_1))\,_0d_q\nu$. Then

$$S_3=\int_0^{\frac{1}{[\alpha+1]_q}}\,_{\varpi_1}D_q\mathscr{F}(\varpi_1+\nu\mathcal{R}_{\rho,\lambda,\sigma}(\varpi_2-\varpi_1))\,_0d_q\nu$$
$$=\int_0^{\frac{1}{[\alpha+1]_q}}\frac{\mathscr{F}(\varpi_1+\nu\mathcal{R}_{\rho,\lambda,\sigma}(\varpi_2-\varpi_1))-\mathscr{F}(\varpi_1+q\nu\mathcal{R}_{\rho,\lambda,\sigma}(\varpi_2-\varpi_1))}{(1-q)\mathcal{R}_{\rho,\lambda,\sigma}(\varpi_2-\varpi_1)\nu}\,_0d_q\nu$$
$$=\frac{1}{(1-q)\mathcal{R}_{\rho,\lambda,\sigma}(\varpi_2-\varpi_1)}\int_0^{\frac{1}{[\alpha+1]_q}}\frac{\mathscr{F}(\varpi_1+\nu\mathcal{R}_{\rho,\lambda,\sigma}(\varpi_2-\varpi_1))}{\nu}\,_0d_q\nu$$
$$-\frac{1}{(1-q)\mathcal{R}_{\rho,\lambda,\sigma}(\varpi_2-\varpi_1)}\int_0^{\frac{1}{[\alpha+1]_q}}\frac{\mathscr{F}(\varpi_1+q\nu\mathcal{R}_{\rho,\lambda,\sigma}(\varpi_2-\varpi_1))}{\nu}\,_0d_q\nu$$
$$=\frac{1}{\mathcal{R}_{\rho,\lambda,\sigma}(\varpi_2-\varpi_1)[\alpha+1]_q}\sum_{n=0}^{\infty}q^n\frac{\mathscr{F}\left(\varpi_1+\frac{q^n}{[\alpha+1]_q}\mathcal{R}_{\rho,\lambda,\sigma}(\varpi_2-\varpi_1)\right)}{\frac{q^n}{[\alpha+1]_q}}$$
$$-\frac{1}{\mathcal{R}_{\rho,\lambda,\sigma}(\varpi_2-\varpi_1)[\alpha+1]_q}\sum_{n=0}^{\infty}q^n\frac{\mathscr{F}\left(\varpi_1+\frac{q^{n+1}}{[\alpha+1]_q}\mathcal{R}_{\rho,\lambda,\sigma}(\varpi_2-\varpi_1)\right)}{\frac{q^n}{[\alpha+1]_q}}$$
$$=\frac{1}{\mathcal{R}_{\rho,\lambda,\sigma}(\varpi_2-\varpi_1)}\left[\sum_{n=0}^{\infty}\mathscr{F}\left(\varpi_1+\frac{q^n}{[\alpha+1]_q}\mathcal{R}_{\rho,\lambda,\sigma}(\varpi_2-\varpi_1)\right)-\sum_{n=0}^{\infty}\mathscr{F}\left(\varpi_1+\frac{q^{n+1}}{[\alpha+1]_q}\mathcal{R}_{\rho,\lambda,\sigma}(\varpi_2-\varpi_1)\right)\right]$$
$$=\frac{1}{\mathcal{R}_{\rho,\lambda,\sigma}(\varpi_2-\varpi_1)}\left[\mathscr{F}\left(\frac{([\alpha+1]_q-1)\varpi_1+(\varpi_1+\mathcal{R}_{\rho,\lambda,\sigma}(\varpi_2-\varpi_1))}{[\alpha+1]_q}\right)-\mathscr{F}(\varpi_1)\right]. \tag{19}$$

By (15), we have

$$
\begin{aligned}
\mathcal{S}_1 &= \mathcal{R}_{\rho,\lambda,\sigma}(\varpi_2-\varpi_1)\int_0^1 {}_0(1-{}_0\Phi_q(\nu))_q^{(\alpha)}\ {}_{\varpi_1}D_q\mathscr{F}(\varpi_1+\nu\mathcal{R}_{\rho,\lambda,\sigma}(\varpi_2-\varpi_1))\,{}_0d_q\nu \\
&= -\mathscr{F}(\varpi_1)+\frac{\Gamma_q(\alpha+1)}{\mathcal{R}^{\alpha}_{\rho,\lambda,\sigma}(\varpi_2-\varpi_1)}\left({}_{\varpi_1}J_q^{\alpha}\mathscr{F}\right)(\varpi_1+\mathcal{R}_{\rho,\lambda,\sigma}(\varpi_2-\varpi_1)).
\end{aligned}
\tag{20}
$$

Using the equalities (19) and (20), we have

$$
\begin{aligned}
&\mathcal{R}_{\rho,\lambda,\sigma}(\varpi_2-\varpi_1)\left[\begin{array}{l}\int_0^{\frac{1}{[\alpha+1]_q}}\left(1-{}_0(1-\Phi_q(\nu))_q^{(\alpha)}\right){}_{\varpi_1}D_q\mathscr{F}(\varpi_1+\nu\mathcal{R}_{\rho,\lambda,\sigma}(\varpi_2-\varpi_1))\,{}_0d_q\nu \\ +\int_{\frac{1}{[\alpha+1]_q}}^{1} -{}_0(1-\Phi_q(\nu))_q^{(\alpha)}\ {}_{\varpi_1}D_q\mathscr{F}(\varpi_1+\nu\mathcal{R}_{\rho,\lambda,\sigma}(\varpi_2-\varpi_1))\,{}_0d_q\nu\end{array}\right] \\
&= \mathcal{R}_{\rho,\lambda,\sigma}(\varpi_2-\varpi_1)\left[\begin{array}{l}\int_0^{\frac{1}{[\alpha+1]_q}} {}_{\varpi_1}D_q\mathscr{F}(\varpi_1+\nu\mathcal{R}_{\rho,\lambda,\sigma}(\varpi_2-\varpi_1))\,{}_0d_q\nu \\ -\int_0^1 {}_0(1-\Phi_q(\nu))_q^{(\alpha)}\ {}_{\varpi_1}D_q\mathscr{F}(\varpi_1+\nu\mathcal{R}_{\rho,\lambda,\sigma}(\varpi_2-\varpi_1))\,{}_0d_q\nu\end{array}\right] \\
&= \mathcal{R}_{\rho,\lambda,\sigma}(\varpi_2-\varpi_1)\left[\begin{array}{l}\frac{1}{\mathcal{R}_{\rho,\lambda,\sigma}(\varpi_2-\varpi_1)}\left[\mathscr{F}\left(\frac{([\alpha+1]_q-1)\varpi_1+(\varpi_1+\mathcal{R}_{\rho,\lambda,\sigma}(\varpi_2-\varpi_1))}{[\alpha+1]_q}\right)-\mathscr{F}(\varpi_1)\right] \\ -\frac{1}{\mathcal{R}_{\rho,\lambda,\sigma}(\varpi_2-\varpi_1)}\left[-\mathscr{F}(\varpi_1)+\frac{\Gamma_q(\alpha+1)}{\mathcal{R}^{\alpha}_{\rho,\lambda,\sigma}(\varpi_2-\varpi_1)}\left({}_{\varpi_1}J_q^{\alpha}\mathscr{F}\right)(\varpi_1+\mathcal{R}_{\rho,\lambda,\sigma}(\varpi_2-\varpi_1))\right]\end{array}\right] \\
&= \mathscr{F}\left(\frac{([\alpha+1]_q-1)\varpi_1+(\varpi_1+\mathcal{R}_{\rho,\lambda,\sigma}(\varpi_2-\varpi_1))}{[\alpha+1]_q}\right)-\frac{\Gamma_q(\alpha+1)}{\mathcal{R}^{\alpha}_{\rho,\lambda,\sigma}(\varpi_2-\varpi_1)}\left({}_{\varpi_1}J_q^{\alpha}\mathscr{F}\right)(\varpi_1+\mathcal{R}_{\rho,\lambda,\sigma}(\varpi_2-\varpi_1)).
\end{aligned}
$$

This completes the proof. □

Corollary 2. *Under the assumptions of Lemma 2, if we take $\alpha = 1$, then the following result holds:*

$$
\begin{aligned}
&\mathscr{F}\left(\frac{(1+q)\varpi_1+\mathcal{R}_{\rho,\lambda,\sigma}(\varpi_2-\varpi_1)}{1+q}\right)-\frac{1}{\mathcal{R}_{\rho,\lambda,\sigma}(\varpi_2-\varpi_1)}\int_{\varpi_1}^{\varpi_1+\mathcal{R}_{\rho,\lambda,\sigma}(\varpi_2-\varpi_1)}\mathscr{F}(\nu)\,{}_{\varpi_1}d_q\nu \\
&= q\mathcal{R}_{\rho,\lambda,\sigma}(\varpi_2-\varpi_1)\left[\begin{array}{l}\int_0^{\frac{1}{1+q}}\nu\ {}_{\varpi_1}D_q\mathscr{F}(\varpi_1+\nu\mathcal{R}_{\rho,\lambda,\sigma}(\varpi_2-\varpi_1))\,{}_0d_q\nu \\ +\int_{\frac{1}{1+q}}^{1}\left(\nu-\frac{1}{q}\right){}_{\varpi_1}D_q\mathscr{F}(\varpi_1+\nu\mathcal{R}_{\rho,\lambda,\sigma}(\varpi_2-\varpi_1))\,{}_0d_q\nu\end{array}\right].
\end{aligned}
\tag{21}
$$

Theorem 4. *Let $\mathscr{F}:\Omega\to\mathbb{R}$ be a continuous function and $\alpha>0$ and ${}_{\varpi_1}D_q\mathscr{F}$ be q-integrable on Ω°. If $|{}_{\varpi_1}D_q\mathscr{F}|$ is generalized exponential convex on Ω, then*

$$
\begin{aligned}
&\left|\frac{\Gamma_q(\alpha+1)}{\mathcal{R}^{\alpha}_{\rho,\lambda,\sigma}(\varpi_2-\varpi_1)}\left({}_{\varpi_1}J_q^{\alpha}\mathscr{F}\right)(\varpi_1+\mathcal{R}_{\rho,\lambda,\sigma}(\varpi_2-\varpi_1))-\frac{([\alpha+1]_q-1)\mathscr{F}(\varpi_1)+\mathscr{F}(\varpi_1+\mathcal{R}_{\rho,\lambda,\sigma}(\varpi_2-\varpi_1))}{[\alpha+1]_q}\right| \\
&\leqslant \frac{\mathcal{R}_{\rho,\lambda,\sigma}(\varpi_2-\varpi_1)}{[\alpha+1]_q}\left(A_1\frac{|{}_{\varpi_1}D_q\mathscr{F}(\varpi_1)|}{\chi^{\alpha\varpi_1}}+A_2\frac{|{}_{\varpi_1}D_q\mathscr{F}(\varpi_2)|}{\chi^{\alpha\varpi_2}}\right),
\end{aligned}
\tag{22}
$$

where

$$A_1 = \int_0^1 \left| [\alpha+1]_q \, {}_0(1 - {}_0\Phi_q(\nu))_q^{(\alpha)} - 1 \right| (1-\nu) \, {}_0d_q\nu,$$

and

$$A_2 = \int_0^1 \left| [\alpha+1]_q \, {}_0(1 - {}_0\Phi_q(\nu))_q^{(\alpha)} - 1 \right| \nu \, {}_0d_q\nu.$$

Proof. Using the Lemma (17), property of modulus and the generalized exponential convexity of $|{}_{\varpi_1}D_q\mathscr{F}|$, we have

$$\begin{aligned}
&\left| \frac{\Gamma_q(\alpha+1)}{\mathcal{R}^{\alpha}_{\rho,\lambda,\sigma}(\varpi_2-\varpi_1)} ({}_{\varpi_1}J_q^{\alpha}\mathscr{F})(\varpi_1 + \mathcal{R}_{\rho,\lambda,\sigma}(\varpi_2-\varpi_1)) - \frac{([\alpha+1]_q - 1)\mathscr{F}(\varpi_1) + \mathscr{F}(\varpi_1 + \mathcal{R}_{\rho,\lambda,\sigma}(\varpi_2-\varpi_1))}{[\alpha+1]_q} \right| \\
&\leqslant \frac{\mathcal{R}_{\rho,\lambda,\sigma}(\varpi_2-\varpi_1)}{[\alpha+1]_q} \int_0^1 \left| \left([\alpha+1]_q \, {}_0(1 - {}_0\Phi_q(\nu))_q^{(\alpha)} - 1 \right) {}_{\varpi_1}D_q\mathscr{F}(\varpi_1 + \nu\mathcal{R}_{\rho,\lambda,\sigma}(\varpi_2-\varpi_1)) \right| {}_0d_q\nu \\
&\leqslant \frac{\mathcal{R}_{\rho,\lambda,\sigma}(\varpi_2-\varpi_1)}{[\alpha+1]_q} \int_0^1 \left| [\alpha+1]_q \, {}_0(1 - {}_0\Phi_q(\nu))_q^{(\alpha)} - 1 \right| \left[(1-\nu) \frac{|{}_{\varpi_1}D_q\mathscr{F}(\varpi_1)|}{\chi^{\alpha\varpi_1}} + \nu \frac{|{}_{\varpi_1}D_q\mathscr{F}(\varpi_2)|}{\chi^{\alpha\varpi_2}} \right] {}_0d_q\nu \\
&= \frac{\mathcal{R}_{\rho,\lambda,\sigma}(\varpi_2-\varpi_1)}{[\alpha+1]_q} \left[\begin{array}{l} \frac{|{}_{\varpi_1}D_q\mathscr{F}(\varpi_1)|}{\chi^{\alpha\varpi_1}} \int_0^1 \left| [\alpha+1]_q \, {}_0(1 - {}_0\Phi_q(\nu))_q^{(\alpha)} - 1 \right| (1-\nu) \, {}_0d_q\nu \\ + \frac{|{}_{\varpi_1}D_q\mathscr{F}(\varpi_2)|}{\chi^{\alpha\varpi_2}} \int_0^1 \left| [\alpha+1]_q \, {}_0(1 - {}_0\Phi_q(\nu))_q^{(\alpha)} - 1 \right| \nu \, {}_0d_q\nu \end{array} \right] \\
&= \frac{\mathcal{R}_{\rho,\lambda,\sigma}(\varpi_2-\varpi_1)}{[\alpha+1]_q} \left(\frac{|{}_{\varpi_1}D_q\mathscr{F}(\varpi_1)|}{\chi^{\alpha\varpi_1}} A_1 + \frac{|{}_{\varpi_1}D_q\mathscr{F}(\varpi_2)|}{\chi^{\alpha\varpi_2}} A_2 \right),
\end{aligned}$$

which completes the proof. □

Corollary 3. *Under the assumptions of Theorem 4, if we choose $\alpha = 1$, then we have*

$$\begin{aligned}
&\left| \frac{1}{\mathcal{R}_{\rho,\lambda,\sigma}(\varpi_2-\varpi_1)} \int_{\varpi_1}^{\varpi_1+\mathcal{R}_{\rho,\lambda,\sigma}(\varpi_2-\varpi_1)} \mathscr{F}(\nu) \, {}_{\varpi_1}d_q\nu - \frac{q\mathscr{F}(\varpi_1) + \mathscr{F}(\varpi_1 + \mathcal{R}_{\rho,\lambda,\sigma}(\varpi_2-\varpi_1))}{1+q} \right| \\
&\leqslant \frac{q^2\mathcal{R}_{\rho,\lambda,\sigma}(\varpi_2-\varpi_1)}{(1+q)^4} \left(A_1^* \frac{|{}_{\varpi_1}D_q\mathscr{F}(\varpi_1)|}{\chi^{\alpha\varpi_1}} + A_2^* \frac{|{}_{\varpi_1}D_q\mathscr{F}(\varpi_2)|}{\chi^{\alpha\varpi_2}} \right),
\end{aligned} \tag{23}$$

where

$$A_1^* = \frac{q + 3q^3 + 2q^4}{1+q+q^2},$$

and

$$A_2^* = \frac{1 + 4q + q^2}{1+q+q^2}.$$

Theorem 5. *Let $\mathscr{F} : \Omega \to \mathbb{R}$ be a continuous function and $\alpha > 0$ and ${}_{\varpi_1}D_q\mathscr{F}$ be q-integrable on Ω°. If $|{}_{\varpi_1}D_q\mathscr{F}|^r$ is generalized exponential convex on Ω for $r > 1$ and $p^{-1} + r^{-1} = 1$, then the following inequality holds:*

$$\left|\frac{\Gamma_q(\alpha+1)}{\mathcal{R}^{\alpha}_{\rho,\lambda,\sigma}(\varpi_2-\varpi_1)}\left({}_{\varpi_1}J^{\alpha}_q\mathscr{F}\right)(\varpi_1+\mathcal{R}_{\rho,\lambda,\sigma}(\varpi_2-\varpi_1))-\frac{([\alpha+1]_q-1)\mathscr{F}(\varpi_1)+\mathscr{F}(\varpi_1+\mathcal{R}_{\rho,\lambda,\sigma}(\varpi_2-\varpi_1))}{[\alpha+1]_q}\right|$$

$$\leqslant\frac{\mathcal{R}_{\rho,\lambda,\sigma}(\varpi_2-\varpi_1)}{[\alpha+1]_q}A_3^{\frac{1}{p}}\left(\frac{q\frac{|{}_{\varpi_1}D_q\mathscr{F}(\varpi_1)|^r}{\chi^{\alpha\varpi_1}}+|{}_{\varpi_1}D_q\mathscr{F}(\varpi_2)^r}{1+q}\right)^{\frac{1}{r}}, \tag{24}$$

where

$$A_3=\int_0^1\left|[\alpha+1]_q\,{}_0(1-{}_0\Phi_q(\nu))_q^{(\alpha)}-1\right|^p{}_0d_q\nu.$$

Proof. Using Lemma (17), Hölder's integral inequality and generalized exponential convexity of $|{}_{\varpi_1}D_q\mathscr{F}|^r$, we have

$$\left|\frac{\Gamma_q(\alpha+1)}{\mathcal{R}^{\alpha}_{\rho,\lambda,\sigma}(\varpi_2-\varpi_1)}\left({}_{\varpi_1}J^{\alpha}_q\mathscr{F}\right)(\varpi_1+\mathcal{R}_{\rho,\lambda,\sigma}(\varpi_2-\varpi_1))-\frac{([\alpha+1]_q-1)\mathscr{F}(\varpi_1)+\mathscr{F}(\varpi_1+\mathcal{R}_{\rho,\lambda,\sigma}(\varpi_2-\varpi_1))}{[\alpha+1]_q}\right|$$

$$\leqslant\frac{\mathcal{R}_{\rho,\lambda,\sigma}(\varpi_2-\varpi_1)}{[\alpha+1]_q}\int_0^1\left|[\alpha+1]_q\,{}_0(1-{}_0\Phi_q(\nu))_q^{(\alpha)}-1\right|\left|{}_{\varpi_1}D_q\mathscr{F}(\varpi_1+\nu\mathcal{R}_{\rho,\lambda,\sigma}(\varpi_2-\varpi_1))\right|{}_0d_q\nu$$

$$\leqslant\frac{\mathcal{R}_{\rho,\lambda,\sigma}(\varpi_2-\varpi_1)}{[\alpha+1]_q}\left(\int_0^1\left|[\alpha+1]_q\,{}_0(1-{}_0\Phi_q(\nu))_q^{(\alpha)}-1\right|^p{}_0d_q\nu\right)^{\frac{1}{p}}\left(\int_0^1\left|{}_{\varpi_1}D_q\mathscr{F}(\varpi_1+\nu\mathcal{R}_{\rho,\lambda,\sigma}(\varpi_2-\varpi_1))\right|^r{}_0d_q\nu\right)^{\frac{1}{r}}$$

$$\leqslant\frac{\mathcal{R}_{\rho,\lambda,\sigma}(\varpi_2-\varpi_1)}{[\alpha+1]_q}\left(\int_0^1\left|[\alpha+1]_q\,{}_0(1-{}_0\Phi_q(\nu))_q^{(\alpha)}-1\right|^p{}_0d_q\nu\right)^{\frac{1}{p}}$$

$$\times\left(\frac{|{}_{\varpi_1}D_q\mathscr{F}(\varpi_1)|^r}{\chi^{\alpha\varpi_1}}\int_0^1(1-\nu)\,{}_0d_q\nu+\frac{|{}_{\varpi_1}D_q\mathscr{F}(\varpi_2)|^r}{\chi^{\alpha\varpi_2}}\int_0^1\nu\,{}_0d_q\nu\right)^{\frac{1}{r}}$$

$$=\frac{\mathcal{R}_{\rho,\lambda,\sigma}(\varpi_2-\varpi_1)}{[\alpha+1]_q}\left(\int_0^1\left|[\alpha+1]_q\,{}_0(1-{}_0\Phi_q(\nu))_q^{(\alpha)}-1\right|^p{}_0d_q\nu\right)^{\frac{1}{p}}\left(\frac{q\frac{|{}_{\varpi_1}D_q\mathscr{F}(\varpi_1)|^r}{\chi^{\alpha\varpi_1}}+\frac{|{}_{\varpi_1}D_q\mathscr{F}(\varpi_2)|^r}{\chi^{\alpha\varpi_2}}}{1+q}\right)^{\frac{1}{r}},$$

which completes the proof. □

Corollary 4. *Under the assumptions of Theorem 5, if we choose $\alpha=1$, then we have*

$$\left|\frac{1}{\mathcal{R}_{\rho,\lambda,\sigma}(\varpi_2-\varpi_1)}\int_{\varpi_1}^{\varpi_1+\mathcal{R}_{\rho,\lambda,\sigma}(\varpi_2-\varpi_1)}\mathscr{F}(\nu)\,{}_{\varpi_1}d_q\nu-\frac{q\mathscr{F}(\varpi_1)+\mathscr{F}(\varpi_1+\mathcal{R}_{\rho,\lambda,\sigma}(\varpi_2-\varpi_1))}{1+q}\right|$$

$$\leqslant\frac{q\mathcal{R}_{\rho,\lambda,\sigma}(\varpi_2-\varpi_1)}{(1+q)}A_3^{*\frac{1}{p}}\left(\frac{q\frac{|{}_{\varpi_1}D_q\mathscr{F}(\varpi_1)|^r}{\chi^{\alpha\varpi_1}}+|{}_{\varpi_1}D_q\mathscr{F}(\varpi_2)^r}{1+q}\right)^{\frac{1}{r}}, \tag{25}$$

where

$$A_3^* = \int_0^1 |1-(1+q)\nu|^p \, {}_0d_q\nu.$$

Theorem 6. *Let $\mathscr{F} : \Omega \to \mathbb{R}$ be a continuous function and $\alpha > 0$ and ${}_{\varpi_1}D_q\mathscr{F}$ be q-integrable on Ω°. If $|{}_{\varpi_1}D_q\mathscr{F}|^r$ is generalized exponential convex on Ω for $r \geqslant 1$, then the following inequality holds:*

$$\left|\frac{\Gamma_q(\alpha+1)}{\mathcal{R}^\alpha_{\rho,\lambda,\sigma}(\varpi_2-\varpi_1)}({}_{\varpi_1}J^\alpha_q\mathscr{F})(\varpi_1+\mathcal{R}_{\rho,\lambda,\sigma}(\varpi_2-\varpi_1)) - \frac{([\alpha+1]_q-1)\mathscr{F}(\varpi_1)+\mathscr{F}(\varpi_1+\mathcal{R}_{\rho,\lambda,\sigma}(\varpi_2-\varpi_1))}{[\alpha+1]_q}\right|$$
$$\leqslant \frac{\mathcal{R}_{\rho,\lambda,\sigma}(\varpi_2-\varpi_1)}{[\alpha+1]_q}A_4^{1-\frac{1}{r}}\left(A_1\frac{|{}_{\varpi_1}D_q\mathscr{F}(\varpi_1)|^r}{\chi^{\alpha\varpi_1}}+A_2\frac{|{}_{\varpi_1}D_q\mathscr{F}(\varpi_2)|^r}{\chi^{\alpha\varpi_2}}\right)^{\frac{1}{r}}, \tag{26}$$

where A_1, A_2 are given in Theorem 4 and A_4 is given as:

$$A_4 = \int_0^1 \left|[\alpha+1]_q\,{}_0(1-{}_0\Phi_q(\nu))^{(\alpha)}_q - 1\right| {}_0d_q\nu.$$

Proof. Using Lemma (17), the power mean integral inequality and generalized exponential convexity of $|{}_{\varpi_1}D_q\mathscr{F}|^r$, we have

$$\left|\frac{\Gamma_q(\alpha+1)}{\mathcal{R}^\alpha_{\rho,\lambda,\sigma}(\varpi_2-\varpi_1)}({}_{\varpi_1}J^\alpha_q\mathscr{F})(\varpi_1+\mathcal{R}_{\rho,\lambda,\sigma}(\varpi_2-\varpi_1)) - \frac{([\alpha+1]_q-1)\mathscr{F}(\varpi_1)+\mathscr{F}(\varpi_1+\mathcal{R}_{\rho,\lambda,\sigma}(\varpi_2-\varpi_1))}{[\alpha+1]_q}\right|$$
$$\leqslant \frac{\mathcal{R}_{\rho,\lambda,\sigma}(\varpi_2-\varpi_1)}{[\alpha+1]_q}\int_0^1\left|[\alpha+1]_q\,{}_0(1-{}_0\Phi_q(\nu))^{(\alpha)}_q-1\right||{}_{\varpi_1}D_q\mathscr{F}(\varpi_1+\nu\mathcal{R}_{\rho,\lambda,\sigma}(\varpi_2-\varpi_1))|\,{}_0d_q\nu$$
$$\leqslant \frac{\mathcal{R}_{\rho,\lambda,\sigma}(\varpi_2-\varpi_1)}{[\alpha+1]_q}\left(\int_0^1\left|[\alpha+1]_q\,{}_0(1-{}_0\Phi_q(\nu))^{(\alpha)}_q-1\right|{}_0d_q\nu\right)^{1-\frac{1}{r}}$$
$$\times\left(\int_0^1\left|[\alpha+1]_q\,{}_0(1-{}_0\Phi_q(\nu))^{(\alpha)}_q-1\right||{}_{\varpi_1}D_q\mathscr{F}(\varpi_1+\nu\mathcal{R}_{\rho,\lambda,\sigma}(\varpi_2-\varpi_1))|^r\,{}_0d_q\nu\right)^{\frac{1}{r}}$$
$$\leqslant \frac{\mathcal{R}_{\rho,\lambda,\sigma}(\varpi_2-\varpi_1)}{[\alpha+1]_q}\left(\int_0^1\left|[\alpha+1]_q\,{}_0(1-{}_0\Phi_q(\nu))^{(\alpha)}_q-1\right|{}_0d_q\nu\right)^{1-\frac{1}{r}}$$
$$\times\left(\int_0^1\left|[\alpha+1]_q\,{}_0(1-{}_0\Phi_q(\nu))^{(\alpha)}_q-1\right|\left[\frac{|{}_{\varpi_1}D_q\mathscr{F}(\varpi_1)|^r}{\chi^{\alpha\varpi_1}}(1-\nu)+\frac{|{}_{\varpi_1}D_q\mathscr{F}(\varpi_2)|^r}{\chi^{\alpha\varpi_2}}\nu\right]{}_0d_q\nu\right)^{\frac{1}{r}}$$
$$\leqslant \frac{\mathcal{R}_{\rho,\lambda,\sigma}(\varpi_2-\varpi_1)}{[\alpha+1]_q}\left(\int_0^1\left|[\alpha+1]_q\,{}_0(1-{}_0\Phi_q(\nu))^{(\alpha)}_q-1\right|{}_0d_q\nu\right)^{1-\frac{1}{r}}$$
$$\times\left[\begin{array}{l}\frac{|{}_{\varpi_1}D_q\mathscr{F}(\varpi_1)|^r}{\chi^{\alpha\varpi_1}}\int_0^1\left|[\alpha+1]_q\,{}_0(1-{}_0\Phi_q(\nu))^{(\alpha)}_q-1\right|(1-\nu)\,{}_0d_q\nu\\ +\frac{|{}_{\varpi_1}D_q\mathscr{F}(\varpi_2)|^r}{\chi^{\alpha\varpi_2}}\left|[\alpha+1]_q\,{}_0(1-{}_0\Phi_q(\nu))^{(\alpha)}_q-1\right|\nu\,{}_0d_q\nu\end{array}\right]^{\frac{1}{r}}$$
$$= \frac{\mathcal{R}_{\rho,\lambda,\sigma}(\varpi_2-\varpi_1)}{[\alpha+1]_q}A_4^{1-\frac{1}{r}}\left(A_1\frac{|{}_{\varpi_1}D_q\mathscr{F}(\varpi_1)|^r}{\chi^{\alpha\varpi_1}}+A_3\frac{|{}_{\varpi_1}D_q\mathscr{F}(\varpi_2)|^r}{\chi^{\alpha\varpi_2}}\right)^{\frac{1}{r}},$$

which completes the proof. □

Corollary 5. *Under the assumptions of Theorem 6, if we choose $\alpha = 1$, then we have*

$$\left| \frac{1}{\mathcal{R}_{\rho,\lambda,\sigma}(\varpi_2 - \varpi_1)} \int_{\varpi_1}^{\varpi_1 + \mathcal{R}_{\rho,\lambda,\sigma}(\varpi_2 - \varpi_1)} \mathscr{F}(\nu)\, {}_{\varpi_1}d_q\nu - \frac{q\mathscr{F}(\varpi_1) + \mathscr{F}(\varpi_1 + \mathcal{R}_{\rho,\lambda,\sigma}(\varpi_2 - \varpi_1))}{1+q} \right|$$
$$\leqslant \frac{q\mathcal{R}_{\rho,\lambda,\sigma}(\varpi_2 - \varpi_1)}{(1+q)} A_4^{*1-\frac{1}{r}} \left(A_1^* \frac{|{}_{\varpi_1}D_q\mathscr{F}(\varpi_1)|^r}{\chi^{\alpha\varpi_1}} + A_2^* \frac{|{}_{\varpi_1}D_q\mathscr{F}(\varpi_2)|^r}{\chi^{\alpha\varpi_2}} \right)^{\frac{1}{r}}, \tag{27}$$

where A_1^*, A_2^* are already defined in Corollary 3 and

$$A_4^* = \frac{2q + q^2 + q^4}{(1+q)^3}.$$

Theorem 7. *Let $\mathscr{F} : \Omega \to \mathbb{R}$ be a continuous function and $\alpha > 0$ and ${}_{\varpi_1}D_q\mathscr{F}$ be q-integrable on Ω°. If $|{}_{\varpi_1}D_q\mathscr{F}|$ is generalized exponential convex on Ω, then*

$$\left| \mathscr{F}\left(\frac{([\alpha+1]_q - 1)\varpi_1 + (\varpi_1 + \mathcal{R}_{\rho,\lambda,\sigma}(\varpi_2 - \varpi_1))}{[\alpha+1]_q} \right) - \frac{\Gamma_q(\alpha+1)}{\mathcal{R}^{\alpha}_{\rho,\lambda,\sigma}(\varpi_2 - \varpi_1)} ({}_{\varpi_1}J_q^{\alpha}\mathscr{F})(\varpi_1 + \mathcal{R}_{\rho,\lambda,\sigma}(\varpi_2 - \varpi_1)) \right|$$
$$\leqslant \mathcal{R}_{\rho,\lambda,\sigma}(\varpi_2 - \varpi_1) \left[(A_5 + A_7) \frac{|{}_{\varpi_1}D_q\mathscr{F}(\varpi_1)|}{\chi^{\alpha\varpi_1}} + (A_6 + A_8) \frac{|{}_{\varpi_1}D_q\mathscr{F}(\varpi_2)|}{\chi^{\alpha\varpi_2}} \right], \tag{28}$$

where

$$A_5 = \int_0^{\frac{1}{[\alpha+1]_q}} \left| 1 - {}_0(1 - \Phi_q(\nu))_q^{(\alpha)} \right| (1-\nu)\, {}_0d_q\nu$$
$$A_6 = \int_0^{\frac{1}{[\alpha+1]_q}} \left| 1 - {}_0(1 - \Phi_q(\nu))_q^{(\alpha)} \right| \nu\, {}_0d_q\nu$$
$$A_7 = \int_{\frac{1}{[\alpha+1]_q}}^{1} |- {}_0(1 - \Phi_q(\nu))_q^{(\alpha)}| (1-\nu)\, {}_0d_q\nu$$
$$A_8 = \int_{\frac{1}{[\alpha+1]_q}}^{1} |- {}_0(1 - \Phi_q(\nu))_q^{(\alpha)}| \nu\, {}_0d_q\nu.$$

Proof. Using Lemma (21) and the generalized exponential convexity of $|{}_{\varpi_1}D_q\mathscr{F}|$, we have

$$\left|\mathscr{F}\left(\frac{([\alpha+1]_q-1)\varpi_1+(\varpi_1+\mathcal{R}_{\rho,\lambda,\sigma}(\varpi_2-\varpi_1))}{[\alpha+1]_q}\right)-\frac{\Gamma_q(\alpha+1)}{\mathcal{R}^{\alpha}_{\rho,\lambda,\sigma}(\varpi_2-\varpi_1)}({}_{\varpi_1}J^{\alpha}_q\mathscr{F})(\varpi_1+\mathcal{R}_{\rho,\lambda,\sigma}(\varpi_2-\varpi_1))\right|$$
$$\leqslant \mathcal{R}_{\rho,\lambda,\sigma}(\varpi_2-\varpi_1)\left[\begin{array}{l}\int_0^{\frac{1}{[\alpha+1]_q}}|1-{}_0(1-\Phi_q(\nu))_q^{(\alpha)}||{}_{\varpi_1}D_q\mathscr{F}(\varpi_1+\nu\mathcal{R}_{\rho,\lambda,\sigma}(\varpi_2-\varpi_1))|{}_0d_q\nu\\+\int_{\frac{1}{[\alpha+1]_q}}^1|-{}_0(1-\Phi_q(\nu))_q^{(\alpha)}||{}_{\varpi_1}D_q\mathscr{F}(\varpi_1+\nu\mathcal{R}_{\rho,\lambda,\sigma}(\varpi_2-\varpi_1))|{}_0d_q\nu\end{array}\right]$$
$$\leqslant \mathcal{R}_{\rho,\lambda,\sigma}(\varpi_2-\varpi_1)\left[\begin{array}{l}\int_0^{\frac{1}{[\alpha+1]_q}}|1-{}_0(1-\Phi_q(\nu))_q^{(\alpha)}|[(1-\nu)\frac{|{}_{\varpi_1}D_q\mathscr{F}(\varpi_1)|}{\chi^{\alpha\varpi_1}}+\nu|{}_{\varpi_1}D_q\mathscr{F}(\varpi_2)]{}_0d_q\nu\\+\int_{\frac{1}{[\alpha+1]_q}}^1|-{}_0(1-\Phi_q(\nu))_q^{(\alpha)}|[(1-\nu)|{}_{\varpi_1}D_q\mathscr{F}(\varpi_1)+\nu\frac{|{}_{\varpi_1}D_q\mathscr{F}(\varpi_2)|}{\chi^{\alpha\varpi_2}}]{}_0d_q\nu\end{array}\right]$$
$$=\mathcal{R}_{\rho,\lambda,\sigma}(\varpi_2-\varpi_1)\left[\begin{array}{l}\frac{|{}_{\varpi_1}D_q\mathscr{F}(\varpi_1)|}{\chi^{\alpha\varpi_1}}\left[\int_0^{\frac{1}{[\alpha+1]_q}}|1-{}_0(1-\Phi_q(\nu))_q^{(\alpha)}|(1-\nu)\,{}_0d_q\nu+\int_{\frac{1}{[\alpha+1]_q}}^1|-{}_0(1-\Phi_q(\nu))_q^{(\alpha)}|(1-\nu)\,{}_0d_q\nu\right]\\+\frac{|{}_{\varpi_1}D_q\mathscr{F}(\varpi_2)|}{\chi^{\alpha\varpi_2}}\left[\int_0^{\frac{1}{[\alpha+1]_q}}|1-{}_0(1-\Phi_q(\nu)_q^{(\alpha)})|\nu\,{}_0d_q\nu+\int_{\frac{1}{[\alpha+1]_q}}^1|-{}_0(1-\Phi_q(\nu))_q^{(\alpha)}|\nu\,{}_0d_q\nu\right]\end{array}\right].$$

This completes the proof. □

Corollary 6. *Under the assumptions of Theorem 7, if we set $\alpha=1$, then we have the following inequality*

$$\left|\mathscr{F}\left(\frac{(1+q)\varpi_1+\mathcal{R}_{\rho,\lambda,\sigma}(\varpi_2-\varpi_1)}{1+q}\right)-\frac{1}{\mathcal{R}_{\rho,\lambda,\sigma}(\varpi_2-\varpi_1)}\int_{\varpi_1}^{\varpi_1+\mathcal{R}_{\rho,\lambda,\sigma}(\varpi_2-\varpi_1)}\mathscr{F}(\nu)\,{}_{\varpi_1}d_q\nu\right|$$
$$\leqslant\frac{q\mathcal{R}_{\rho,\lambda,\sigma}(\varpi_2-\varpi_1)}{(1+q+q^2)(1+q)^4}\left[3\frac{|{}_{\varpi_1}D_q\mathscr{F}(\varpi_1)|}{\chi^{\alpha\varpi_1}}+(2q^2+2q-1)\frac{|{}_{\varpi_1}D_q\mathscr{F}(\varpi_2)|}{\chi^{\alpha\varpi_2}}\right].$$

Theorem 8. *Let $\mathscr{F}:\Omega\to\mathbb{R}$ be a continuous function and $\alpha>0$ and ${}_{\varpi_1}D_q\mathscr{F}$ be q-integrable on Ω°. If $|{}_{\varpi_1}D_q\mathscr{F}|^r$ is generalized exponential convex on Ω, then the following inequality holds for $p^{-1}+r^{-1}=1$:*

$$\left|\mathscr{F}\left(\frac{([\alpha+1]_q-1)\varpi_1+(\varpi_1+\mathcal{R}_{\rho,\lambda,\sigma}(\varpi_2-\varpi_1))}{[\alpha+1]_q}\right)-\frac{\Gamma_q(\alpha+1)}{\mathcal{R}^{\alpha}_{\rho,\lambda,\sigma}(\varpi_2-\varpi_1)}({}_{\varpi_1}J^{\alpha}_q\mathscr{F})(\varpi_1+\mathcal{R}_{\rho,\lambda,\sigma}(\varpi_2-\varpi_1))\right|$$
$$\leqslant\mathcal{R}_{\rho,\lambda,\sigma}(\varpi_2-\varpi_1)\left[\begin{array}{l}A_9^{\frac{1}{p}}\left(\frac{|{}_{\varpi_1}D_q\mathscr{F}(\varpi_1)|^r}{\chi^{\alpha\varpi_1}}\left(\frac{(1+q)[\alpha+1]_q-1}{(1+q)([\alpha+1]_q)^2}\right)+\frac{|{}_{\varpi_1}D_q\mathscr{F}(\varpi_2)|^r}{\chi^{\alpha\varpi_2}}\left(\frac{1}{(1+q)([\alpha+1]_q)^2}\right)\right)^{\frac{1}{r}}\\+A_{10}^{\frac{1}{p}}\left(\frac{|{}_{\varpi_1}D_q\mathscr{F}(\varpi_1)|^r}{\chi^{\alpha\varpi_1}}\left(\frac{q}{1+q}-\frac{(1+q)[\alpha+1]_q-1}{(1+q)([\alpha+1]_q)^2}\right)+\frac{|{}_{\varpi_1}D_q\mathscr{F}(\varpi_2)|^r}{\chi^{\alpha\varpi_2}}\left(\frac{1}{1+q}-\frac{1}{(1+q)([\alpha+1]_q)^2}\right)\right)^{\frac{1}{r}}\end{array}\right], \quad (29)$$

where

$$A_9=\int_0^{\frac{1}{[\alpha+1]_q}}\left|1-{}_0(1-\Phi_q(\nu))_q^{(\alpha)}\right|{}_0d_q\nu,$$

and

$$A_{10} = \int_{\frac{1}{[\alpha+1]_q}}^{1} |-{}_0(1-\Phi_q(\nu))_q^{(\alpha)}|\,{}_0d_q\nu.$$

Proof. Using Lemma (21), Hölder's inequality and the generalized exponential convexity of $|{}_{\varpi_1}D_q\mathscr{F}|^r$, we have

$$\left|\mathscr{F}\left(\frac{([\alpha+1]_q-1)\varpi_1+(\varpi_1+\mathcal{R}_{\rho,\lambda,\sigma}(\varpi_2-\varpi_1))}{[\alpha+1]_q}\right)-\frac{\Gamma_q(\alpha+1)}{\mathcal{R}^{\alpha}_{\rho,\lambda,\sigma}(\varpi_2-\varpi_1)}({}_{\varpi_1}J_q^{\alpha}\mathscr{F})(\varpi_1+\mathcal{R}_{\rho,\lambda,\sigma}(\varpi_2-\varpi_1))\right|$$

$$\leqslant \mathcal{R}_{\rho,\lambda,\sigma}(\varpi_2-\varpi_1)\left[\begin{array}{c}\int_0^{\frac{1}{[\alpha+1]_q}}|1-{}_0(1-\Phi_q(\nu))_q^{(\alpha)}||{}_{\varpi_1}D_q\mathscr{F}(\varpi_1+\nu\mathcal{R}_{\rho,\lambda,\sigma}(\varpi_2-\varpi_1))|\,{}_0d_q\nu \\ +\int_{\frac{1}{[\alpha+1]_q}}^{1}|-{}_0(1-\Phi_q(\nu))_q^{(\alpha)}||{}_{\varpi_1}D_q\mathscr{F}(\varpi_1+\nu\mathcal{R}_{\rho,\lambda,\sigma}(\varpi_2-\varpi_1))|\,{}_0d_q\nu\end{array}\right]$$

$$\leqslant \mathcal{R}_{\rho,\lambda,\sigma}(\varpi_2-\varpi_1)\left[\begin{array}{c}\left(\int_0^{\frac{1}{[\alpha+1]_q}}|1-{}_0(1-\Phi_q(\nu))_q^{(\alpha)}|^p\,{}_0d_q\nu\right)^{\frac{1}{p}}\left(\int_{\frac{1}{[\alpha+1]_q}}^{1}|{}_{\varpi_1}D_q\mathscr{F}(\varpi_1+\nu\mathcal{R}_{\rho,\lambda,\sigma}(\varpi_2-\varpi_1))|^r\,{}_0d_q\nu\right)^{\frac{1}{r}} \\ +\left(\int_{\frac{1}{[\alpha+1]_q}}^{1}|-{}_0(1-\Phi_q(\nu))_q^{(\alpha)}|^p\,{}_0d_q\nu\right)^{\frac{1}{p}}\left(\int_{\frac{1}{[\alpha+1]_q}}^{1}|{}_{\varpi_1}D_q\mathscr{F}(\varpi_1+\nu\mathcal{R}_{\rho,\lambda,\sigma}(\varpi_2-\varpi_1))|^r\,{}_0d_q\nu\right)^{\frac{1}{r}}\end{array}\right]$$

$$\leqslant \mathcal{R}_{\rho,\lambda,\sigma}(\varpi_2-\varpi_1)\left[\begin{array}{c}\left(\int_0^{\frac{1}{[\alpha+1]_q}}|1-{}_0(1-\Phi_q(\nu))_q^{(\alpha)}|^p\,{}_0d_q\nu\right)^{\frac{1}{p}} \\ \times\left[\frac{|{}_{\varpi_1}D_q\mathscr{F}(\varpi_2)|^r}{\chi^{\alpha\varpi_2}}\int_0^{\frac{1}{[\alpha+1]_q}}(1-\nu)\,{}_0d_q\nu+\frac{|{}_{\varpi_1}D_q\mathscr{F}(\varpi_2)|^r}{\chi^{\alpha\varpi_2}}\int_0^{\frac{1}{[\alpha+1]_q}}\nu\,{}_0d_q\nu\right]^{\frac{1}{r}} \\ +\left(\int_{\frac{1}{[\alpha+1]_q}}^{1}|-{}_0(1-\Phi_q(\nu))_q^{(\alpha)}|^p\,{}_0d_q\nu\right)^{\frac{1}{p}} \\ \times\left[\frac{|{}_{\varpi_1}D_q\mathscr{F}(\varpi_1)|^r}{\chi^{\alpha\varpi_1}}\int_{\frac{1}{[\alpha+1]_q}}^{1}(1-\nu)\,{}_0d_q\nu+\frac{|{}_{\varpi_1}D_q\mathscr{F}(\varpi_2)|^r}{\chi^{\alpha\varpi_2}}\int_{\frac{1}{[\alpha+1]_q}}^{1}\nu\,{}_0d_q\nu\right]^{\frac{1}{r}}\end{array}\right]$$

$$= \mathcal{R}_{\rho,\lambda,\sigma}(\varpi_2-\varpi_1)\left[\begin{array}{c}\left(\int_0^{\frac{1}{[\alpha+1]_q}}|1-{}_0(1-\Phi_q(\nu))_q^{(\alpha)}|^p\,{}_0d_q\nu\right)^{\frac{1}{p}} \\ \times\left(\frac{|{}_{\varpi_1}D_q\mathscr{F}(\varpi_1)|^r}{\chi^{\alpha\varpi_1}}\left(\frac{(1+q)[\alpha+1]_q-1}{(1+q)([\alpha+1]_q)^2}\right)+\frac{|{}_{\varpi_1}D_q\mathscr{F}(\varpi_2)|^r}{\chi^{\alpha\varpi_2}}\left(\frac{1}{(1+q)([\alpha+1]_q)^2}\right)\right)^{\frac{1}{r}} \\ +\left(\int_{\frac{1}{[\alpha+1]_q}}^{1}|-{}_0(1-\Phi_q(\nu))_q^{(\alpha)}|^p\,{}_0d_q\nu\right)^{\frac{1}{p}} \\ \times\left(\frac{|{}_{\varpi_1}D_q\mathscr{F}(\varpi_1)|^r}{\chi^{\alpha\varpi_1}}\left(\frac{q}{1+q}-\frac{(1+q)[\alpha+1]_q-1}{(1+q)([\alpha+1]_q)^2}\right)+\frac{|{}_{\varpi_1}D_q\mathscr{F}(\varpi_2)|^r}{\chi^{\alpha\varpi_2}}\left(\frac{1}{1+q}-\frac{1}{(1+q)([\alpha+1]_q)^2}\right)\right)^{\frac{1}{r}}\end{array}\right].$$

This completes the proof. □

Corollary 7. *Under the assumptions of Theorem 8, if we set* $\alpha = 1$*, then*

$$\left| \mathscr{F}\left(\frac{(1+q)\varpi_1 + \mathcal{R}_{\rho,\lambda,\sigma}(\varpi_2 - \varpi_1)}{1+q}\right) - \frac{1}{\mathcal{R}_{\rho,\lambda,\sigma}(\varpi_2 - \varpi_1)} \int_{\varpi_1}^{\varpi_1 + \mathcal{R}_{\rho,\lambda,\sigma}(\varpi_2 - \varpi_1)} \mathscr{F}(\nu)\, {}_{\varpi_1}d_q\nu \right|$$
$$\leqslant q\mathcal{R}_{\rho,\lambda,\sigma}(\varpi_2 - \varpi_1)\left[A_9^{*\frac{1}{p}}\left(\frac{|{}_{\varpi_1}D_q\mathscr{F}(\varpi_1)|^r}{\chi^{\alpha\varpi_1}}\left(\frac{q^2+2q}{(1+q)^3}\right) + \frac{|{}_{\varpi_1}D_q\mathscr{F}(\varpi_2)|^r}{\chi^{\alpha\varpi_2}}\left(\frac{1}{(1+q)^3}\right)\right)^{\frac{1}{r}} \right.$$
$$\left. + A_{10}^{*\frac{1}{p}}\left(\frac{|{}_{\varpi_1}D_q\mathscr{F}(\varpi_1)|^r}{\chi^{\alpha\varpi_1}}\left(\frac{q^3+q^2-q}{(1+q)^3}\right) + \frac{|{}_{\varpi_1}D_q\mathscr{F}(\varpi_2)|^r}{\chi^{\alpha\varpi_2}}\left(\frac{q^2+2q}{(1+q)^3}\right)\right)^{\frac{1}{r}}\right], \tag{30}$$

where

$$A_9^* = \frac{(1-q)}{(1+q)^{p+1}(1-q^{p+1})}$$

and

$$A_{10}^* = \int_{\frac{1}{1+q}}^{1}\left(\frac{1}{q} - \nu\right)^p {}_0d_q\nu.$$

Theorem 9. *Let* $\mathscr{F} : \Omega \to \mathbb{R}$ *be a continuous function and* $\alpha > 0$ *and* ${}_{\varpi_1}D_q\mathscr{F}$ *be q-integrable on* Ω°*. If* $|{}_{\varpi_1}D_q\mathscr{F}|^r, r \geqslant 1$ *is generalized exponential convex on* Ω*, then*

$$\left| \mathscr{F}\left(\frac{([\alpha+1]_q - 1)\varpi_1 + (\varpi_1 + \mathcal{R}_{\rho,\lambda,\sigma}(\varpi_2 - \varpi_1))}{[\alpha+1]_q}\right) - \frac{\Gamma_q(\alpha+1)}{\mathcal{R}^{\alpha}_{\rho,\lambda,\sigma}(\varpi_2 - \varpi_1)}({}_{\varpi_1}J_q^{\alpha}\mathscr{F})(\varpi_1 + \mathcal{R}_{\rho,\lambda,\sigma}(\varpi_2 - \varpi_1)) \right|$$
$$= \mathcal{R}_{\rho,\lambda,\sigma}(\varpi_2 - \varpi_1)\left[\begin{array}{c} A_9^{1-\frac{1}{r}}\left[A_5\frac{|{}_{\varpi_1}D_q\mathscr{F}(\varpi_2)|^r}{\chi^{\alpha\varpi_2}} + A_6\frac{|{}_{\varpi_1}D_q\mathscr{F}(\varpi_2)|^r}{\chi^{\alpha\varpi_2}}\right]^{\frac{1}{r}} \\ + A_{10}^{1-\frac{1}{r}}\left[A_7\frac{|{}_{\varpi_1}D_q\mathscr{F}(\varpi_1)|^r}{\chi^{\alpha\varpi_1}} + A_8\frac{|{}_{\varpi_1}D_q\mathscr{F}(\varpi_2)|^r}{\chi^{\alpha\varpi_2}}\right]^{\frac{1}{r}} \end{array}\right],$$

where

$$A_9 = \int_0^{\frac{1}{[\alpha+1]_q}} \left|1 - {}_0(1 - \Phi_q(\nu))_q^{(\alpha)}\right| {}_0d_q\nu$$

and

$$A_{10} = \int_{\frac{1}{[\alpha+1]_q}}^{1} |1 - {}_0(1 - \Phi_q(\nu))_q^{(\alpha)}| {}_0d_q\nu.$$

Proof. Using Lemma (21), power mean integral inequality and the generalized exponential convexity of $|{}_{\varpi_1}D_q\mathscr{F}|^r$, we have

$$\left|\mathscr{F}\left(\frac{([\alpha+1]_q-1)\varpi_1+(\varpi_1+\mathcal{R}_{\rho,\lambda,\sigma}(\varpi_2-\varpi_1))}{[\alpha+1]_q}\right)-\frac{\Gamma_q(\alpha+1)}{\mathcal{R}^{\alpha}_{\rho,\lambda,\sigma}(\varpi_2-\varpi_1)}({}_{\varpi_1}J_q^{\alpha}\mathscr{F})(\varpi_1+\mathcal{R}_{\rho,\lambda,\sigma}(\varpi_2-\varpi_1))\right|$$

$$\leqslant \mathcal{R}_{\rho,\lambda,\sigma}(\varpi_2-\varpi_1)\left[\begin{array}{l}\int_0^{\frac{1}{[\alpha+1]_q}}|1-{}_0(1-\Phi_q(\nu))_q^{(\alpha)}||{}_{\varpi_1}D_q\mathscr{F}(\varpi_1+\nu\mathcal{R}_{\rho,\lambda,\sigma}(\varpi_2-\varpi_1))|{}_0d_q\nu\\ +\int_{\frac{1}{[\alpha+1]_q}}^1|-{}_0(1-\Phi_q(\nu))_q^{(\alpha)}||{}_{\varpi_1}D_q\mathscr{F}(\varpi_1+\nu\mathcal{R}_{\rho,\lambda,\sigma}(\varpi_2-\varpi_1))|{}_0d_q\nu\end{array}\right]$$

$$\leqslant \mathcal{R}_{\rho,\lambda,\sigma}(\varpi_2-\varpi_1)\left[\begin{array}{c}\left(\int_0^{\frac{1}{[\alpha+1]_q}}|1-{}_0(1-\Phi_q(\nu))_q^{(\alpha)}|{}_0d_q\nu\right)^{1-\frac{1}{r}}\\ \times\left(\int_{\frac{1}{[\alpha+1]_q}}^1|1-{}_0(1-\Phi_q(\nu)_q^{(\alpha)})||{}_{\varpi_1}D_q\mathscr{F}(\varpi_1+\nu\mathcal{R}_{\rho,\lambda,\sigma}(\varpi_2-\varpi_1))|^r{}_0d_q\nu\right)^{\frac{1}{r}}\\ +\left(\int_{\frac{1}{[\alpha+1]_q}}^1|-{}_0(1-\Phi_q(\nu))_q^{(\alpha)}|{}_0d_q\nu\right)^{1-\frac{1}{r}}\\ \times\left(\int_{\frac{1}{[\alpha+1]_q}}^1|-{}_0(1-\Phi_q(\nu))_q^{(\alpha)}||{}_{\varpi_1}D_q\mathscr{F}(\varpi_1+\nu\mathcal{R}_{\rho,\lambda,\sigma}(\varpi_2-\varpi_1))|^r{}_0d_q\nu\right)^{\frac{1}{r}}\end{array}\right]$$

$$\leqslant \mathcal{R}_{\rho,\lambda,\sigma}(\varpi_2-\varpi_1)\left[\begin{array}{c}\left(\int_0^{\frac{1}{[\alpha+1]_q}}|1-{}_0(1-\Phi_q(\nu))_q^{(\alpha)}|{}_0d_q\nu\right)^{1-\frac{1}{r}}\\ \times\left(\int_0^{\frac{1}{[\alpha+1]_q}}|1-{}_0(1-\Phi_q(\nu)_q^{(\alpha)})|\left[(1-\nu)\frac{|{}_{\varpi_1}D_q\mathscr{F}(\varpi_2)|^r}{\chi^{\alpha\varpi_2}}+\nu\frac{|{}_{\varpi_1}D_q\mathscr{F}(\varpi_2)|^r}{\chi^{\alpha\varpi_2}}\right]{}_0d_q\nu\right)^{\frac{1}{r}}\\ +\left(\int_{\frac{1}{[\alpha+1]_q}}^1|-{}_0(1-\Phi_q(\nu))_q^{(\alpha)}|{}_0d_q\nu\right)^{1-\frac{1}{r}}\\ \times\left(\int_{\frac{1}{[\alpha+1]_q}}^1|-{}_0(1-\Phi_q(\nu))_q^{(\alpha)}|\left[(1-\nu)\frac{|{}_{\varpi_1}D_q\mathscr{F}(\varpi_1)|^r}{\chi^{\alpha\varpi_1}}+\nu\frac{|{}_{\varpi_1}D_q\mathscr{F}(\varpi_2)|^r}{\chi^{\alpha\varpi_2}}\right]{}_0d_q\nu\right)^{\frac{1}{r}}\end{array}\right]$$

$$= \mathcal{R}_{\rho,\lambda,\sigma}(\varpi_2-\varpi_1)\left[\begin{array}{c}\left(\int_0^{\frac{1}{[\alpha+1]_q}}|1-{}_0(1-\Phi_q(\nu))_q^{(\alpha)}|{}_0d_q\nu\right)^{1-\frac{1}{r}}\\ \times\left[\frac{|{}_{\varpi_1}D_q\mathscr{F}(\varpi_2)|^r}{\chi^{\alpha\varpi_2}}\int_0^{\frac{1}{[\alpha+1]_q}}|1-{}_0(1-\Phi_q(\nu))_q^{(\alpha)}|(1-\nu){}_0d_q\nu+\frac{|{}_{\varpi_1}D_q\mathscr{F}(\varpi_2)|^r}{\chi^{\alpha\varpi_2}}\int_0^{\frac{1}{[\alpha+1]_q}}|1-{}_0(1-\Phi_q(\nu)_q^{(\alpha)})|\nu{}_0d_q\nu\right]^{\frac{1}{r}}\\ +\left(\int_{\frac{1}{[\alpha+1]_q}}^1|-{}_0(1-\Phi_q(\nu))_q^{(\alpha)}|{}_0d_q\nu\right)^{1-\frac{1}{r}}\\ \times\left[\frac{|{}_{\varpi_1}D_q\mathscr{F}(\varpi_1)|^r}{\chi^{\alpha\varpi_1}}\int_{\frac{1}{[\alpha+1]_q}}^1|-{}_0(1-\Phi_q(\nu))_q^{(\alpha)}|(1-\nu){}_0d_q\nu+\frac{|{}_{\varpi_1}D_q\mathscr{F}(\varpi_2)|^r}{\chi^{\alpha\varpi_2}}\int_{\frac{1}{[\alpha+1]_q}}^1|-{}_0(1-\Phi_q(\nu))_q^{(\alpha)}|\nu{}_0d_q\nu\right]^{\frac{1}{r}}\end{array}\right].$$

This completes the proof. □

Corollary 8. *Under the assumptions of Theorem 9, if we set $\alpha=1$, then*

$$\left|\mathscr{F}\left(\frac{(1+q)\varpi_1+\mathcal{R}_{\rho,\lambda,\sigma}(\varpi_2-\varpi_1)}{1+q}\right)-\frac{1}{\mathcal{R}_{\rho,\lambda,\sigma}(\varpi_2-\varpi_1)}\int_{\varpi_1}^{\varpi_1+\mathcal{R}_{\rho,\lambda,\sigma}(\varpi_2-\varpi_1)}\mathscr{F}(\nu)\ {}_{\varpi_1}d_q\nu\right|$$
$$\leqslant q\mathcal{R}_{\rho,\lambda,\sigma}(\varpi_2-\varpi_1)\left(\frac{1}{(1+q)^3}\right)\Bigg[\left(\frac{|{}_{\varpi_1}D_q\mathscr{F}(\varpi_1)|^{\mathfrak{r}}}{\chi^{\alpha\varpi_1}}\left(\frac{q^2+q}{(1+q+q^2)}\right)+\frac{|{}_{\varpi_1}D_q\mathscr{F}(\varpi_2)|^{\mathfrak{r}}}{\chi^{\alpha\varpi_2}}\left(\frac{1}{(1+q+q^2)}\right)\right)^{\frac{1}{\mathfrak{r}}}$$
$$+\left(\frac{|{}_{\varpi_1}D_q\mathscr{F}(\varpi_1)|^{\mathfrak{r}}}{\chi^{\alpha\varpi_1}}\left(\frac{q^2+q-1}{1+q+q^2}\right)+\frac{|{}_{\varpi_1}D_q\mathscr{F}(\varpi_2)|^{\mathfrak{r}}}{\chi^{\alpha\varpi_2}}\left(\frac{2}{1+q+q^2}\right)\right)^{\frac{1}{\mathfrak{r}}}\Bigg].$$

4. Conclusions

We have introduced the class of generalized exponential convex functions involving Raina's function. We have derived two new identities involving q-Riemann–Liouville fractional integrals. Using these identities, as auxiliary results, we have derived several new q-fractional estimates of trapezoidal-like inequalities, essentially using the class of generalized exponential convex functions. We hope that the ideas within this paper will inspire interested readers. The results of this paper can be extended by using other classes of convexity, for instance by using the exponential preinvexity property of the functions. One can also extend these results using the concepts of post-quantum calculus, which is an interesting problem for future research. It is worth mentioning here that many inequalities e.g., Lipschitz, Hölders, Minkowski, etc., are used to solve the control problems and stability analysis for dynamical systems; for details, see [14–19]. So it can also be an interesting problem for future research to use the inequalities obtained in this paper to solve physical problems.

Author Contributions: Conceptualization, M.U.A.; formal analysis, K.N., M.U.A., M.Z.J. and H.B.; investigation, K.N., M.U.A., M.Z.J., H.B. and M.A.N.; writing—original draft preparation, K.N., M.U.A., M.Z.J., H.B. and M.A.N.; supervision, M.A.N.; All authors have read and agreed to the published version of the manuscript.

Funding: This research received funding support from the National Science, Research and Innovation Fund (NSRF), Thailand.

Institutional Review Board Statement: Not applicable.

Informed Consent Statement: Not applicable.

Data Availability Statement: Not applicable.

Acknowledgments: The authors are grateful to the editor and the anonymous reviewers for their valuable comments and suggestions.

Conflicts of Interest: The authors declare no conflict of interest.

References

1. Cortez, M.V.J.; Liko, R.; Kashuri, A.; Hernández, J.E.H. New quantum estimates of trapezium—Type inequalities for generalized φ-convex functions. *Mathematics* **2019**, *7*, 1047. [CrossRef]
2. Raina, R.K. On generalized Wright's hypergeometric functions and fractional calculus operators. *East Asian Math. J.* **2015**, *21*, 191–203.
3. Awan, M.U.; Noor, M.A.; Noor, K.I. Hermite-Hadamard inequalities for exponentially convex functions. *Appl. Math. Inf. Sci.* **2018**, *12*, 405–409. [CrossRef]
4. Dragomir, S.S.; Pearce, C.E.M. *Selected Topics on Hermite—Hadamard Inequality and Applications*; Victoria University: Melbourne, Australia, 2000.
5. Sarikaya, M.Z.; Set, E.; Yaldiz, H.; Basak, N. Hermite—Hadamard's inequalities for fractional integrals and related fractional inequalities. *Math. Comput. Model.* **2013**, *57*, 2403–2407. [CrossRef]
6. Alp, N.; Sarıkaya, M.Z.; Kunt, M.; İşcan, İ. q-Hermite Hadamard inequalities and quantum estimates for midpoint type inequalities via convex and quasi-convex functions. *J. King Saud Univ. Sci.* **2018**, *30*, 193–203. [CrossRef]

7. Kunt, M.; Aljasem, M. Fractional quantum Hermite-Hadamard type inequalities. *Konuralp J. Math.* **2020**, *8*, 122–136.
8. Noor, M.A.; Noor, K.I.; Awan, M.U. Some quantum estimates for Hermite-Hadamard inequalities. *Appl. Math. Comput.* **2015**, *251*, 675–679. [CrossRef]
9. Sudsutad, W.; Ntouyas, S.K.; Tariboon, J. Quantum integral inequalities for convex functions. *J. Math. Inequal.* **2015**, *9*, 781–793. [CrossRef]
10. Zhang, Y.; Du, T.-S.; Wang, H.; Shen, Y.-J. Different types of quantum integral inequalities via $(\alpha, \mathfrak{m})$-convexity. *J. Inequal. Appl.* **2018**, *2018*, 264. [CrossRef]
11. Tariboon, J.; Ntouyas, S.K. Quantum calculus on finite intervals and applications to impulsive difference equations. *Adv. Diff. Equ.* **2013**, *282*, 1–19. [CrossRef]
12. Tariboon, J.; Ntouyas, S.K.; Agarwal, P. New concepts of fractional quantum calculus and applications to impulsive fractional q-difference equations. *Adv. Diff. Equ.* **2015**, *18*, 1–19. [CrossRef]
13. Annaby, M.H.; Mansour, Z.S. q-*Fractional Calculus and Equations;* Springer: Berlin/Heidelberg, Germany, 2012.
14. Cheng, Y.; Huo, L.; Zhao, L. Stability analysis and optimal control of rumor spreading model under media coverage considering time delay and pulse vaccination. *Chaos Solitons Fractals* **2022**, *157*, 111931. [CrossRef]
15. Mahmudov, N.I. Finite–approximate controllability of Riemann—Liouville fractional evolution systems via resolvent—Like operators. *Fractal Fract.* **2021**, *5*, 199. [CrossRef]
16. Patel, R.; Shukla, A.; Jadon, S.S. Existence and optimal control problem for semilinear fractional order $(1, 2]$ control system. *Math. Meth. Appl. Sci.* **2020**, 43, 1–12. [CrossRef]
17. Shukla, A.; Sukavanam, N.; Pandey, D.N. Complete controllability of semi–linear stochastic system with delay. *Rend. Circ. Mat. Palermo* **2015**, *64*, 209–220. [CrossRef]
18. Shukla, A.; Sukavanam, N.; Pandey, D.N. Approximate controllability of semilinear fractional control systems of order $\alpha \in (1, 2]$ with infinite delay. *Mediterranean J. Math.* **2016**, *13*, 2539–2550. [CrossRef]
19. Singh, A.; Shukla, A.; Vijayakumar, V.; Udhayakumar, R. Asymptotic stability of fractional order $(1, 2]$ stochastic delay differential equations in Banach spaces. *Chaos Solitons Fractals* **2021**, *150*, 111095. [CrossRef]

fractal and fractional

MDPI

Article

Certain Hybrid Matrix Polynomials Related to the Laguerre-Sheffer Family

Tabinda Nahid [1] and Junesang Choi [2,*]

[1] Department of Mathematics, Aligarh Muslim University, Aligarh 202001, India; tabindanahid@gmail.com
[2] Department of Mathematics, Dongguk University, Gyeongju 38066, Korea
* Correspondence: junesang@dongguk.ac.kr; Tel.: +82-010-6525-2262

Abstract: The main goal of this article is to explore a new type of polynomials, specifically the Gould-Hopper-Laguerre-Sheffer matrix polynomials, through operational techniques. The generating function and operational representations for this new family of polynomials will be established. In addition, these specific matrix polynomials are interpreted in terms of quasi-monomiality. The extended versions of the Gould-Hopper-Laguerre-Sheffer matrix polynomials are introduced, and their characteristics are explored using the integral transform. Further, examples of how these results apply to specific members of the matrix polynomial family are shown.

Keywords: Gould-Hopper-Laguerre-Sheffer matrix polynomials; quasi-monomiality; umbral calculus; fractional calculus; Euler's integral of gamma functions; beta function; generalized hypergeometric series; operational methods

MSC: 05A40; 08A40; 26A33; 33B10; 33C45; 33C50; 44A20

Citation: Nahid, T.; Choi, J. Certain Hybrid Matrix Polynomials Related to the Laguerre-Sheffer Family. *Fractal Fract.* **2022**, *6*, 211. http://doi.org/10.3390/fractalfract6040211

Academic Editors: Asifa Tassaddiq and Muhammad Yaseen

Received: 17 February 2022
Accepted: 7 April 2022
Published: 9 April 2022

1. Introduction and Preliminaries

Significant discoveries in the theory of group representation, statistics, quadrature and interpolation, scattering theory, imaging of medicine, and splines have led to the development of matrix polynomials and special matrix functions. Numerous disciplines of mathematics and engineering make use of special matrix polynomials (see, for example, [1,2], and the citations included therein). For instance, many mathematicians investigate and explore special matrix polynomials.

The Sheffer sequences [3] are used extensively in mathematics, theoretical physics, theory of approximation, and various different mathematical disciplines. Roman [4] naturally discusses the Sheffer polynomials' properties in the context of contemporary classical umbral calculus. The Sheffer polynomials are given as follows (see [4], p. 17): Set $p(\tau)$ and $q(\tau)$ power series, which are formally given as follows:

$$p(\tau) = \sum_{\ell=0}^{\infty} p_\ell \frac{\tau^\ell}{\ell!} \quad (p_\ell \in \mathbb{C},\ \ell \in \mathbb{Z}_{\geq 0};\ p_0 = 0,\ p_1 \neq 0), \tag{1a}$$

and

$$q(\tau) = \sum_{\ell=0}^{\infty} q_\ell \frac{\tau^n}{\ell!} \quad (q_\ell \in \mathbb{C},\ \ell \in \mathbb{Z}_{\geq 0};\ q_0 \neq 0), \tag{1b}$$

which are referred to as delta series and invertible series, respectively. Here and elsewhere, let $\mathbb{C}$, $\mathbb{R}$, and $\mathbb{Z}$ be, respectively, the sets of complex numbers, real numbers, and integers. Let

$$\mathbb{E}_{\leq \xi}, \quad \mathbb{E}_{<\xi}, \quad \mathbb{E}_{\geq \xi}, \quad \text{and} \quad \mathbb{E}_{>\xi}$$

be the sets of numbers in $\mathbb{E}$ less than or equal to ξ, less than ξ, greater than or equal to ξ, and greater than ξ, respectively, for some $\xi \in \mathbb{R}$, where $\mathbb{E}$ is either $\mathbb{Z}$ or $\mathbb{R}$.

With each pairing of an invertible series $q(\tau)$ and a delta series $p(\tau)$, there is a unique sequence $s_\ell(x)$ of polynomials that satisfies the conditions of orthogonality (consult [4], p. 17):

$$\left\langle q(\tau)\, p(\tau)^k \middle| s_\ell(x) \right\rangle = \ell!\, \delta_{\ell,k} \quad (\ell,\, k \in \mathbb{Z}_{\geq 0}), \tag{2}$$

where $\delta_{\ell,k}$ is the Kronecker delta function defined by $\delta_{\ell,k} = 1$ $(\ell = k)$ and $\delta_{\ell,k} = 0$ $(\ell \neq k)$. The operator $\langle \cdot \mid \cdot \rangle$ is unchanged from [4], Chapter 2.

Remark 1. *The sequence $s_\ell(x)$ satisfying (2) is called the Sheffer sequence for $(q(\tau), p(\tau))$, or $s_\ell(x)$ is Sheffer for $(g(\tau), p(\tau))$, which is usually denoted as $s_\ell(x) \sim (q(\tau), p(\tau))$. Remain aware that $q(\tau)$ and $p(\tau)$ should be an invertible series and a delta series, respectively.*

There are two forms of Sheffer sequences worth noting:

(i) *If $s_\ell(x) \sim (1, p(\tau))$, the $s_\ell(x)$ is said to be the* associated *sequence for $p(\tau)$, or $s_\ell(x)$ is* associated *with $p(\tau)$;*

(ii) *If $s_\ell(x) \sim (q(\tau), \tau)$, the $s_\ell(x)$ is said to be the* Appell *sequence for $q(\tau)$, or $s_\ell(x)$ is* Appell *for $q(\tau)$ (see [4], p. 17; see also [5]).*

If $s_\ell(x)$ is Sheffer for $(q(\tau), p(\tau))$, the Sheffer sequence $s_\ell(x)$ is generated by depending solely on the series $q(\tau)$ and $p(\tau)$. To emphasize this dependence, in [5], the $s_\ell(x)$ was represented by ${}_{[q,p]}s_\ell(x)$.

Amid various Sheffer sequences' characterizations, the following generating function is recalled (consult, for instance, [4], p. 18): The sequence $s_\ell(x)$ is Sheffer for $(q(\tau), p(\tau))$ if and only if:

$$\frac{1}{q(\bar{p}(\tau))}\, e^{x \bar{p}(\tau)} = \sum_{k=0}^{\infty} \frac{s_k(x)}{k!}\, t^k \tag{3}$$

for every x in $\mathbb{C}$, where $\bar{p}(\tau) = p^{-1}(\tau)$ is the inverse of composition of $p(\tau)$.

The particular polynomials of two variables are significant in view of an application. In addition, these polynomials facilitate the derivation of numerous valuable identities and aid in the introduction of new families of particular polynomials; see, for instance, [6–9]. The Laguerre-Sheffer polynomials ${}_L s_\ell(x, y)$ are generated by the following function (consult [10]):

$$\frac{1}{q(p^{-1}(\tau))} \exp\left(y p^{-1}(\tau)\right) C_0\left(x p^{-1}(\tau)\right) = \sum_{n=0}^{\infty} {}_L s_\ell(x, y) \frac{\tau^\ell}{\ell!}, \tag{4}$$

for all x, y in $\mathbb{C}$, where $C_0(x\tau)$ denotes the 0th-order Bessel-Tricomi function, which possesses the subsequent operational law:

$$C_0(\xi x) := \sum_{k=0}^{\infty} \frac{(-1)^k\, (\xi\, x)^k}{(k!)^2} = \exp(-\xi \hat{D}_x^{-1})\{1\}, \tag{5}$$

where

$$\hat{D}_x^{-n}\{1\} := \frac{x^n}{n!} \quad (n \in \mathbb{Z}_{\geq 0}). \tag{6}$$

Generally,

$$\hat{D}_x^{-\xi}\{p(x)\} = \frac{1}{\Gamma(\xi)} \int_0^x (x - \eta)^{\xi - 1}\, p(\eta)\, d\eta, \tag{7}$$

where Γ is the well-known Gamma function (consult, for example, [11], Section 1.1), which is a left-sided Riemann-Liouville fractional integral of order $\xi \in \mathbb{C}$ $(\Re(\xi) > 0)$ (see, for example, [12], Chapter 2). For some recent applications for geometric analysis, one may consult, for example, [13,14].

As in Remark 1, the case $q(\tau) = 1$ and the case $p(\tau) = \tau$ of the Laguerre-Sheffer polynomials ${}_L s_\ell(x, y)$ in (4) are called, respectively, the Laguerre-associated Sheffer sequence and the Laguerre-Appell sequence, and denoted, respectively, by ${}_L\mathfrak{s}_\ell(\mathrm{x},\mathrm{y})$ and ${}_LA_\ell(x,y)$ (consult [15]).

Remark 2. *For $\kappa \in \mathbb{Z}_{>0}$, let $\mathbb{C}^{\kappa\times\kappa}$ indicate the set of all κ by κ matrices whose entries are in $\mathbb{C}$. Let $\sigma(B)$ be the set of all eigenvalues of $B \in \mathbb{C}^{\kappa\times\kappa}$, which is said to be the* spectrum *of B. For $B \in \mathbb{C}^{\kappa\times\kappa}$, let $\alpha(B) := \max\{\Re(w) \mid w \in \sigma(B)\}$ and $\beta(B) := \min\{\Re(w) \mid w \in \sigma(B)\}$. If $\beta(B) > 0$, that is, $\Re(w) > 0$ for all $w \in \sigma(B)$, the matrix B is referred to as* positive stable.

For $B \in \mathbb{C}^{\kappa\times\kappa}$, its 2-norm is denoted by:

$$\|B\| = \sup_{\rho\neq 0} \frac{\|B\rho\|_2}{\|\rho\|_2},$$

where for any vector $\rho \in \mathbb{C}^\kappa$, $\|\rho\|_2 = \left(\rho^H \rho\right)^{1/2}$ is the Euclidean norm of ρ. Here ρ^H indicates the Hermitian matrix of ρ.

If $p(w)$ and $q(w)$ are holomorphic functions of the variable $w \in \mathbb{C}$, which are defined in an open set Λ of the plane $\mathbb{C}$, and R is a matrix in $\mathbb{C}^{\kappa\times\kappa}$ such that $\sigma(R) \subset \Lambda$, then from the matrix functional calculus's characteristics ([16], p. 558), one finds that $f(R)\,g(R) = g(R)\,f(R)$. Therefore, if Q in $\mathbb{C}^{\kappa\times\kappa}$ is another matrix with $\sigma(Q) \subset \Lambda$, such that $RQ = QR$, then $f(R)g(Q) = g(Q)f(R)$ (consult, for instance, [17,18]).

As the reciprocal of the Gamma function indicated by $\Gamma^{-1}(w) = 1/\Gamma(w)$ is an entire function of the variable $w \in \mathbb{C}$, for any R in $\mathbb{C}^{\kappa\times\kappa}$, the functional calculus of Riesz-Dunford reveals that the image of $\Gamma^{-1}(w)$ acting on R, symbolized by $\Gamma^{-1}(R)$, is a well-defined matrix (consult [16], Chapter 7).

Recently, the matrix polynomials of Gould-Hopper (GHMaP) $\mathrm{g}_n^\ell(x, y; C, E)$ were introduced by virtue of the subsequent generating function (consult [19]):

$$\sum_{n=0}^{\infty} \mathrm{g}_n^\ell(x, \mathrm{y}; C, E)\frac{\tau^n}{n!} = \exp(x\tau\sqrt{2C})\exp(E\,y\tau^\ell). \tag{8}$$

Here C, E are matrices in $\mathbb{C}^{\kappa\times\kappa}$ $(\kappa \in \mathbb{Z}_{>0})$ such that C is positive stable and an $\ell \in \mathbb{Z}_{>0}$. Consider the principal branch of $w^{\frac{1}{2}} = \exp\left(\frac{1}{2}\log w\right)$ defined on the domain $\Lambda := \mathbb{C} \setminus (-\infty, 0]$. Then, as in Remark 2, $\sqrt{C}$ is well-defined if $\sigma(C) \subset \Lambda$.

The polynomials $\mathrm{g}_n^\ell(x, y; C, E)$ are specified to be the series

$$\mathrm{g}_n^\ell(x, y; C, E) = \sum_{k=0}^{[\frac{n}{\ell}]} \frac{n!\,(\sqrt{2C})^{n-\ell k}E^k}{(n-\ell k)!\,k!} x^{n-\ell k} y^k. \tag{9}$$

As a result of the idea of monomiality, the majority of the features of generalized and conventional polynomials have been demonstrated to be readily derivable within a framework of operations. The monomiality principle is underpinned by Steffensen's [20] introduction of the idea of poweroid. Following that, Dattoli [21] reconstructed and elaborated the idea of monomiality (consult, for instance, [22]).

As per the monomiality principle, there are two operators $\hat{\mathrm{M}}$ and $\hat{\mathrm{P}}$ that operate on a polynomial set $\{q_\ell(x)\}_{\ell\in\mathbb{Z}_{>0}}$, termed the multiplicative and derivative operators, respectively. Then the polynomial set $\{q_\ell(x)\}_{\ell\in\mathbb{Z}_{>0}}$ is said to be quasi-monomial if it satisfies:

$$\hat{\mathrm{M}}\{q_\ell(x)\} = q_{\ell+1}(x), \quad \hat{\mathrm{P}}\{q_\ell(x)\} = \ell\, q_{\ell-1}(x), \quad q_0(x) = 1. \tag{10}$$

One easily finds from (10) that

$$\hat{\mathrm{M}}\hat{\mathrm{P}}\{q_\ell(x)\} = \ell\, q_\ell(x), \tag{11}$$

and

$$\hat{\mathrm{P}}\hat{\mathrm{M}}\{q_\ell(x)\} = (\ell+1)\, q_\ell(x). \tag{12}$$

A Weyl group structure of the operators $\hat{M}$ and $\hat{P}$ is shown by the relation of commutation:

$$[\hat{P}, \hat{M}] := \hat{P}\hat{M} - \hat{M}\hat{P} = \hat{1}, \tag{13}$$

where $\hat{1}$ is the identity operator.

As a result of $\hat{M}^m$ acting on $q_0(x)$, we may deduce the $q_m(x)$:

$$q_m(x) = \hat{M}^m\{q_0(x)\}. \tag{14}$$

The matrix polynomials of Gould-Hopper $g_m^\ell(x, y; C, E)$ are quasi-monomial with regard to the subsequent derivative and multiplicative operators [23]:

$$\hat{P}_g = (\sqrt{2C})^{-1} D_x, \tag{15}$$

and

$$\hat{M}_g = x\sqrt{2C} + \ell E y(\sqrt{2C})^{-(\ell-1)} D_x^{\ell-1}, \tag{16}$$

respectively, where $D_x := \frac{\partial}{\partial x}$.

The generalization ${}_\alpha F_\beta$ $(\alpha,\ \beta \in \mathbb{Z}_{\geq 0})$ of the hypergeometric series is given by (consult, for instance, [11], Section 1.5):

$$\begin{aligned} {}_\alpha F_\beta\left[\begin{matrix}\mu_1, \ldots, \mu_\alpha; \\ \nu_1, \ldots, \nu_\beta;\end{matrix}\ w\right] &= \sum_{n=0}^{\infty} \frac{(\mu_1)_n \cdots (\mu_\alpha)_n}{(\nu_1)_n \cdots (\nu_\beta)_n} \frac{w^n}{n!} \\ &= {}_\alpha F_\beta(\mu_1, \ldots, \mu_\alpha; \nu_1, \ldots, \nu_\beta; w), \end{aligned} \tag{17}$$

where $(\xi)_\eta$ indicates the Pochhammer symbol (for $\xi,\ \eta \in \mathbb{C}$) defined by

$$(\xi)_\eta := \frac{\Gamma(\xi+\eta)}{\Gamma(\xi)} = \begin{cases} 1 & (\eta = 0;\ \xi \in \mathbb{C} \setminus \{0\}), \\ \xi(\xi+1)\cdots(\xi+n-1) & (\eta = n \in \mathbb{Z}_{>0};\ \xi \in \mathbb{C}). \end{cases} \tag{18}$$

Here it is assumed that $(0)_0 := 1$, an empty product as 1, and that the variable w, the parameters of numerators $\mu_1, \ldots, \mu_\alpha$, and the parameters of denominators $\nu_1, \ldots, \nu_\beta$ are supposed to get complex values, provided that

$$(\nu_j \in \mathbb{C} \setminus \mathbb{Z}_{\leq 0};\ j = 1, \ldots, \beta). \tag{19}$$

Recall the well-known generalized binomial theorem (consult, for example, [24], p. 34):

$$(1-z)^{-\alpha} = \sum_{k=0}^{\infty} \frac{(\alpha)_k\, z^k}{k!} \quad (\alpha \in \mathbb{C};\ |z| < 1). \tag{20}$$

Recall the familiar beta function (consult, for instance, [11], p. 8):

$$B(\xi, \eta) = \begin{cases} \displaystyle\int_0^1 u^{\xi-1}(1-u)^{\eta-1}\, du & (\min\{\Re(\xi), \Re(\eta)\} > 0) \\ \dfrac{\Gamma(\xi)\Gamma(\eta)}{\Gamma(\xi+\eta)} & (\xi, \eta \in \mathbb{C} \setminus \mathbb{Z}_{\leq 0}). \end{cases} \tag{21}$$

Here we introduce the Gould-Hopper-Laguerre-Sheffer matrix polynomials (GHLSMaP), which are denoted by ${}_{gL}s_n^\ell(x, y, z; C, E)$, by convoluting the Laguerre-Sheffer polynomials ${}_L s_n(x, y)$ with the Gould-Hopper matrix polynomials $g_n^\ell(x, y; C, E)$. The polynomials ${}_{gL}s_n^\ell(x, y, z; C, E)$ are generated as in the following definition.

Definition 1. *The Gould-Hopper-Laguerre-Sheffer matrix polynomials* ${}_{gL}s_n^\ell(x, y, z; C, E)$ *are generated by the following function:*

$$F(\boldsymbol{x},\boldsymbol{y},\boldsymbol{z};C,E)(\tau):=\frac{1}{q(p^{-1}(\tau))}\exp\left[\boldsymbol{x}\sqrt{2C}p^{-1}(\tau)+E\boldsymbol{y}\left(p^{-1}(\tau)\right)^{\ell}\right]C_0\left(\boldsymbol{z}p^{-1}(\tau)\right)$$
$$=\sum_{n=0}^{\infty}{}_{gL}s_n^{\ell}(\boldsymbol{x},\boldsymbol{y},\boldsymbol{z};C,E)\frac{\tau^n}{n!}. \tag{22}$$

Here and in the sequel, the functions p, q, C_0 *are as in* (4); *the matrices* C, E *are as in* (8), (9), *or* (16); *the variables* $\boldsymbol{x},\boldsymbol{y},\boldsymbol{z}\in\mathbb{C}$.

In addition, to emphasize the invertible series q *and the delta series* p, *whenever necessary, the following notation is used:*

$${}_{gL}s_n^{\ell}(\boldsymbol{x},\boldsymbol{y},\boldsymbol{z};C,E)={}_{[q,p]}{}_{gL}s_n^{\ell}(\boldsymbol{x},\boldsymbol{y},\boldsymbol{z};C,E). \tag{23}$$

Further,

$${}_{g}s_n^{\ell}(\boldsymbol{x},\boldsymbol{y};C,E):={}_{gL}s_n^{\ell}(\boldsymbol{x},\boldsymbol{y},0;C,E) \tag{24}$$

is called the Gould-Hopper-Sheffer matrix polynomials.

Remark 3. *First we show how to derive the generating function in* (22). *In* (4), *replacing* $\boldsymbol{y}$ *by the multiplicative operator* $\hat{M}_g$ *in* (16), *and* $\boldsymbol{x}$ *by* $\boldsymbol{z}$, *we obtain*

$$F(\tau):=\frac{1}{q(p^{-1}(\tau))}C_0\left(\boldsymbol{z}p^{-1}(\tau)\right)$$
$$\times\exp\left[\left(\boldsymbol{x}\sqrt{2C}\,p^{-1}(\tau)+\ell E\boldsymbol{y}(\sqrt{2C})^{-(\ell-1)}\,p^{-1}(\tau)\,D_{\boldsymbol{x}}^{\ell-1}\right)\{1\}\right]. \tag{25}$$

Recall the Crofton-type identity (see, for instance, [25], p. 12; see also [26]:

$$f\left(\boldsymbol{x}+\ell\lambda\frac{d^{\ell-1}}{d\boldsymbol{x}^{\ell-1}}\right)\{1\}=\exp\left(\lambda\frac{d^{\ell}}{d\boldsymbol{x}^{\ell}}\right)\{f(\boldsymbol{x})\}, \tag{26}$$

with f *usually being an analytic function. Setting* $\ell=1$ *gives:*

$$f\left(\boldsymbol{x}+\lambda\right)\{1\}=\exp\left(\lambda\frac{d}{d\boldsymbol{x}}\right)\{f(\boldsymbol{x})\}. \tag{27}$$

Using (25) *in* (26), *we get*

$$F(\tau)=\frac{1}{q(p^{-1}(\tau))}C_0(\boldsymbol{z}p^{-1}(\tau))\exp\left(E\boldsymbol{y}(\sqrt{2C})^{-\ell}D_{\boldsymbol{x}}^{\ell}\right)\left\{\exp\left(\boldsymbol{x}\sqrt{2C}p^{-1}(\tau)\right)\right\}. \tag{28}$$

By performing the operation in (28), *with the aid of* (32), *we can readily find that* $F(\tau)$ *is identical to the* $F(\boldsymbol{x},\boldsymbol{y},\boldsymbol{z};C,E)(\tau)$ *in* (22).

Second, as in (ii), Remark 1, setting $p(\tau)=p^{-1}(\tau)=\tau$ *in* (22), *we get the generating function for the Gould-Hopper-Laguerre-Appell matrix polynomials (GHLAMaP)* ${}_{gL}\mathcal{C}_n^{\ell}(\boldsymbol{x},\boldsymbol{y},\boldsymbol{z};C,E)$ *in [27].*

Using Euler's integral for the Gamma function Γ (consult, for instance, Section 1.1 in [11], p. 218 in [24]), we get

$$b^{-\nu}=\frac{1}{\Gamma(\nu)}\int_0^{\infty}u^{\nu-1}e^{-bu}\,du\quad(\min\{\Re(\nu),\,\Re(b)\}>0). \tag{29}$$

Dattoli et al. [28] used (29) to obtain the following operator:

$$\left(\alpha-\frac{\partial}{\partial x}\right)^{-\nu}f(x)=\frac{1}{\Gamma(\nu)}\int_0^{\infty}u^{\nu-1}e^{-\alpha u}e^{u\frac{\partial}{\partial x}}\{f(x)\}\,du$$
$$=\frac{1}{\Gamma(\nu)}\int_0^{\infty}u^{\nu-1}e^{-\alpha u}f(x+u)\,du, \tag{30}$$

for the second equality of which (27) is employed.

The following definition introduces the extended matrix polynomials of Gould-Hopper-Laguerre-Sheffer (EGHLSMaP), which are indicated by ${}_{gL}s^{\ell}_{n,\nu}(\boldsymbol{x}, \boldsymbol{y}, \boldsymbol{z}; C, E; \eta)$

Definition 2. *Let $\Re(\eta) > 0$ and $\Re(\nu) > 0$. Then the extended Gould-Hopper-Laguerre-Sheffer matrix polynomials ${}_{gL}s^{\ell}_{n,\nu}(\boldsymbol{x}, \boldsymbol{y}, \boldsymbol{z}; C, E; \eta)$ are defined by*

$$ {}_{gL}s^{\ell}_{n,\nu}(\boldsymbol{x}, \boldsymbol{y}, \boldsymbol{z}; C, E; \eta) := \left(\eta - \boldsymbol{y}E\left(\sqrt{2C}\right)^{-\ell}\frac{\partial^{\ell}}{\partial \boldsymbol{x}^{\ell}}\right)^{-\nu}\left\{{}_{L}s_n(\boldsymbol{z}, \boldsymbol{x}\sqrt{2C})\right\}. \tag{31} $$

In this article, we aim to introduce the Gould-Hopper-Laguerre-Sheffer matrix polynomials via the use of a generating function. For these newly presented matrix polynomials, we investigate quasi-monomial features and related operational principles. We also explore the extended form of these novel hybrid special matrix polynomials and their properties using an integral transform. Finally, we provide many instances to demonstrate how the results presented here may be used.

2. Gould-Hopper-Laguerre-Sheffer Matrix Polynomials

The following lemma provides an easily-derivable operational identity.

Lemma 1. *Let ξ and η be constants independent of x. Also let $\ell \in \mathbb{Z}_{\geq 0}$. Then:*

$$ \exp\left(\xi\,\frac{d^{\ell}}{d\boldsymbol{x}^{\ell}}\right)\{e^{\eta\,\boldsymbol{x}}\} = \exp\left(\eta\,\boldsymbol{x} + \xi\,\eta^{\ell}\right). \tag{32} $$

In particular,

$$ \exp\left(\xi\,\frac{d}{d\boldsymbol{x}}\right)\{e^{\eta\,\boldsymbol{x}}\} = \exp(\eta\,\boldsymbol{x} + \xi\,\eta). \tag{33} $$

Proof.

$$ \exp\left(\xi\,\frac{d^{\ell}}{d\boldsymbol{x}^{\ell}}\right)\{e^{\eta\,\boldsymbol{x}}\} = \sum_{k=0}^{\infty}\frac{\xi^k}{k!}\,\frac{d^{\ell k}}{d\boldsymbol{x}^{\ell k}}\,e^{\eta\,\boldsymbol{x}} = e^{\eta\,\boldsymbol{x}}\sum_{k=0}^{\infty}\frac{\left(\xi\,\eta^{\ell}\right)^k}{k!} = \exp\left(\eta\,\boldsymbol{x} + \xi\,\eta^{\ell}\right). $$

□

The following theorem shows that the Gould-Hopper-Laguerre-Sheffer matrix polynomials ${}_{gL}s^{\ell}_{n}(\boldsymbol{x}, \boldsymbol{y}, \boldsymbol{z}; C, E)$ may be obtained by performing a suitable differential operation on the Laguerre-Sheffer polynomials ${}_{L}s_n(x, y)$ in (4) with some suitable substitutions of $\boldsymbol{x}$ and $\boldsymbol{y}$.

Theorem 1. *The following identity holds true:*

$$ {}_{gL}s^{\ell}_{n}(\boldsymbol{x}, \boldsymbol{y}, \boldsymbol{z}; C, E) = \exp\left(\boldsymbol{y}E\left(\sqrt{2C}\right)^{-\ell}D^{\ell}_{\boldsymbol{x}}\right)\left\{{}_{L}s_n\left(\boldsymbol{z}, \boldsymbol{x}\sqrt{2C}\right)\right\}. \tag{34} $$

Proof. Replacing $\boldsymbol{x}$ and $\boldsymbol{y}$ by $\boldsymbol{z}$ and $\boldsymbol{x}\sqrt{2C}$, respectively, in (4), we get

$$ \frac{1}{q(p^{-1}(\tau))}\,C_0\left(\boldsymbol{z}p^{-1}(\tau)\right)\,\exp\left(\boldsymbol{x}\sqrt{2C}p^{-1}(\tau)\right) = \sum_{n=0}^{\infty}{}_{L}s_n(\boldsymbol{z}, \boldsymbol{x}\sqrt{2C})\frac{\tau^n}{n!}. \tag{35} $$

Performing the operation $\exp\left[yE\left(\sqrt{2C}\right)^{-\ell}D_x^{\ell}\right]$ on both sides of (35), we obtain

$$\begin{aligned}
&\sum_{n=0}^{\infty}\exp\left[yE\left(\sqrt{2C}\right)^{-\ell}D_x^{\ell}\right]\left\{{}_{L}s_n(z,x\sqrt{2C})\right\}\frac{\tau^n}{n!}\\
&=\frac{1}{q(p^{-1}(\tau))}C_0\left(zp^{-1}(\tau)\right)\exp\left[yE\left(\sqrt{2C}\right)^{-\ell}D_x^{\ell}\right]\left\{\exp\left(x\sqrt{2C}p^{-1}(\tau)\right)\right\}\\
&=\sum_{n=0}^{\infty}{}_{gL}s_n^{\ell}(x,y,z;C,E)\frac{\tau^n}{n!},
\end{aligned} \tag{36}$$

for the second equality of which (22) and (32) are used. Finally, matching the coefficients of τ^n on the first and last power series in (36) gives the identity (34). □

Theorem 2. *The Gould-Hopper-Laguerre-Sheffer matrix polynomials* ${}_{gL}s_n^{\ell}(x,y,z;C,E)$ *are operationally represented by the Gould-Hopper-Sheffer matrix polynomials* ${}_{g}s_n^{\ell}(x,y;C,E)$*:*

$${}_{gL}s_n^{\ell}(x,y,z;C,E)=\exp\left[-\hat{D}_z^{-1}\left(\sqrt{2C}\right)^{-1}D_x\right]\left\{{}_{g}s_n^{\ell}(x,y;C,E)\right\}. \tag{37}$$

Proof. From (22) and (24), we have

$$\begin{aligned}
&\frac{1}{q(p^{-1}(\tau))}\exp\left[x\sqrt{2C}p^{-1}(\tau)+Ey\left(p^{-1}(\tau)\right)^{\ell}\right]\\
&=\sum_{n=0}^{\infty}{}_{gL}s_n^{\ell}(x,y;C,E)\frac{\tau^n}{n!}.
\end{aligned} \tag{38}$$

Performing the following operation $\exp\left[-\hat{D}_z^{-1}\left(\sqrt{2C}\right)^{-1}D_x\right]$ on each side of (38), and using (5) and (33), in the same way as in the argument of Theorem 1, one may find the desired identity (37). □

The following theorem reveals the quasi-monomial principle of the matrix polynomials of Gould-Hopper-Laguerre-Sheffer ${}_{gL}s_n^{\ell}(x,y,z;C,E)$.

Theorem 3. *The matrix polynomials* ${}_{gL}s_n^{\ell}(x,y,z;C,E)$ *gratify the following quasi-monomiality, with respect to the operators of multiplication and differentiation:*

$$\begin{aligned}
\hat{M}_{gLs}=&\left(x\sqrt{2C}-\hat{D}_z^{-1}+\ell Ey(\sqrt{2C})^{-(\ell-1)}D_x^{\ell-1}-\frac{q'((\sqrt{2C})^{-1}D_x)}{q((\sqrt{2C})^{-1}D_x)}\right)\\
&\times\frac{1}{p'((\sqrt{2C})^{-1}D_x)}
\end{aligned} \tag{39}$$

and

$$\hat{P}_{qLs}=p\left(\left(\sqrt{2C}\right)^{-1}D_x\right), \tag{40}$$

respectively.

Proof. Performing derivatives on each side of the first and second members in (22) about x, k times, we derive

$$\left((\sqrt{2C})^{-1}D_x\right)^{k}\{F(x,y,z;C,E)(\tau)\}=\left(p^{-1}(\tau)\right)^{k}F(x,y,z;C,E)(\tau)\quad(k\in\mathbb{Z}_{\geq 0}). \tag{41}$$

In particular,

$$\Big((\sqrt{2C})^{-1}\,D_x\Big)\{F(x,y,z;C,E)(\tau)\} = p^{-1}(\tau)\,F(x,y,z;C,E)(\tau). \tag{42}$$

Applying (41) to the series in (1a), we find

$$\sum_{k=0}^{\infty}\frac{p_k}{k!}\Big((\sqrt{2C})^{-1}\,D_x\Big)^k\{F(x,y,z;C,E)(\tau)\} = \sum_{k=0}^{\infty}\frac{p_k}{k!}\Big(p^{-1}(\tau)\Big)^k\{F(x,y,z;C,E)(\tau)\},$$

which implies

$$\begin{aligned} p\Big((\sqrt{2C})^{-1}\,D_x\Big)\{F(x,y,z;C,E)(\tau)\} &= p\Big(p^{-1}(\tau)\Big)\{F(x,y,z;C,E)(\tau)\} \\ &= \tau\,F(x,y,z;C,E)(\tau). \end{aligned} \tag{43}$$

Then, utilizing the identity (43) in (22), we get

$$\begin{aligned} &\sum_{n=1}^{\infty} p\Big((\sqrt{2C})^{-1}\,D_x\Big)\,{}_{gL}s_n^{\ell}(x,y,z;C,E)\frac{\tau^n}{n!} \\ &\quad= \sum_{n=1}^{\infty} {}_{gL}s_{n-1}^{\ell}(x,y,z;C,E)\frac{\tau^n}{(n-1)!}. \end{aligned} \tag{44}$$

Now, identifying the coefficients of τ^n on each side of (44), in view of (10), may prove the derivative operator (40).

Next, in view of (5), we have

$$\frac{d}{d\tau}C_0\Big(zp^{-1}(\tau)\Big) = \frac{d}{d\tau}\exp(-p^{-1}(\tau)\hat{D}_z^{-1})\{1\} = -\Big(p^{-1}(\tau)\Big)'\,\hat{D}_z^{-1}\,C_0\Big(zp^{-1}(\tau)\Big). \tag{45}$$

Then, taking (45) into account, differentiating (22) about τ, we get

$$\begin{aligned} &\sum_{n=0}^{\infty} {}_{gL}s_{n+1}^{\ell}(x,y,z;C,E)\frac{\tau^n}{n!} \\ &= \frac{1}{p'(p^{-1}(\tau))}\left(x\sqrt{2C} + \ell E y(\sqrt{2C})^{-(\ell-1)}D_x^{\ell-1} - \hat{D}_z^{-1} - \frac{q'(p^{-1}(\tau))}{q(p^{-1}(t))}\right) \\ &\quad\times \sum_{n=0}^{\infty} {}_{gL}s_n^{\ell}(x,y,z;C,E)\frac{\tau^n}{n!}. \end{aligned} \tag{46}$$

Finally, applying (42) to (46), in view of (10), we can prove the multiplicative operator (39). □

Remark 4. *If $p(\tau)$ is a delta series, then $p'(\tau)$ is an invertible series. Therefore, the reciprocal $1/p'(p^{-1}(\tau))$ is well-defined in* (46). □

Combining the multiplicative operator in (39) and the derivative operator in (40), such as (11)–(14), we can provide several matrix differential equations for the matrix polynomials of Gould-Hopper-Laguerre-Sheffer ${}_{gL}s_n^{\ell}(x,y,z;C,E)$. One uses (11) to illustrate one of them in the next theorem, whose proof is simple and overlooked.

Theorem 4. *The following differential equation holds true:*

$$\begin{aligned} &\left\{\left(x\sqrt{2C} - \hat{D}_z^{-1} + \ell E y(\sqrt{2C})^{-(\ell-1)}D_x^{\ell-1} - \frac{q'((\sqrt{2C})^{-1}D_x)}{q((\sqrt{2C})^{-1}D_x)}\right)\right. \\ &\quad\left.\times\frac{p((\sqrt{2C})^{-1}D_x)}{p'((\sqrt{2C})^{-1}D_x)} - n\right\}{}_{gL}s_n^{\ell}(x,y,z;C,E) = 0. \end{aligned} \tag{47}$$

The polynomials ${}_{gL}s_n^{\ell}(x, y, z; C, E)$ may yield numerous particular matrix polynomials as special cases, some of which are offered in Table 1.

Table 1. Particular cases of the polynomials ${}_{gL}s_n^{\ell}(x, y, z; C, E)$.

S. No.	Values of the Indices and Variables	Relation between ${}_{gL}s_n^{\ell}(x, y, z; C, E)$ and Its Special Case	Name of the Special Matrix Polynomials	Generating Functions
I.	$\ell = 2$	${}_{gL}s_n^2(x, y, z; C, E) = {}_{HL}s_n(x, y, z; C, E)$	3-Variable Hermite-Laguerre-Sheffer matrix polynomials (3VHLSMaP)	$\frac{1}{q(p^{-1}(\tau))} \exp\left((xp^{-1}(\tau)\sqrt{2C} + Ey(p^{-1}(\tau))^2\right) \times C_0(zp^{-1}(\tau)) = \sum_{n=0}^{\infty} {}_{HL}s_n(x, y, z; C, E)\frac{\tau^n}{n!}$
II.	$z = 0$	${}_{gL}s_n^{\ell}(x, y, 0; C, E) = {}_{g}s_n^{\ell}(x, y; C, E)$	Gould-Hopper-Sheffer-matrix polynomials (GHSMaP)	$\frac{1}{q(p^{-1}(\tau))} \exp\left(xp^{-1}(\tau)\sqrt{2C} + Ey(p^{-1}(\tau))^{\ell}\right) = \sum_{n=0}^{\infty} {}_{g}s_n^{\ell}(x, y; C, E)\frac{\tau^n}{n!}$
III.	$\ell = r-1$, $z = 0$	${}_{gL}s_n^{r-1}(x, y, 0; C, E) = {}_{U}s_n^{r}(x, y; C, E)$	Generalized Chebyshev-Sheffer matrix polynomials (GCSMaP)	$\frac{1}{q(p^{-1}(\tau))} \exp\left(xp^{-1}(\tau)\sqrt{2C} + Ey(p^{-1}(\tau))^{r-1}\right) = \sum_{n=0}^{\infty} {}_{U}s_n^{r}(x, y; C, E)\frac{\tau^n}{n!}$
IV.	$\ell = 2$, $z = 0$	${}_{gL}s_n^2(x, y, 0; C, E) = {}_{H}s_n(x, y; C, E)$	Hermite Kampé de Fériet-Sheffer matrix polynomials (HKdFSMaP)	$\frac{1}{q(p^{-1}(\tau))} \exp\left(xp^{-1}(\tau)\sqrt{2C} + Ey(p^{-1}(\tau))^2\right) = \sum_{n=0}^{\infty} {}_{H}s_n(x, y; C, E)\frac{\tau^n}{n!}$
V.	$z = 0, x \to y$, $y \to D_x^{-1}$	${}_{L}s_n^{\ell}(y, D_x^{-1}, 0; C, E) = {}_{L}s_n^{\ell}(x, y; C, E)$	Generalized Laguerre-Sheffer matrix polynomials (GLSMaP)	$\frac{1}{q(p^{-1}(\tau))} C_0\left(-Ex(p^{-1}(\tau))^{\ell}\right) \times \exp\left(yp^{-1}(\tau)\sqrt{2C}\right) = \sum_{n=0}^{\infty} {}_{L}s_n^{\ell}(x, y; C, E)\frac{\tau^n}{n!}$
VI.	$x = -D_x^{-1}$, $z = 0$	${}_{gL}s_n^{\ell}(-D_x^{-1}, y; C, E) = {}_{[\ell]L}s_n^{\ell}(x, y; C, E)$	2-Variable generalized Laguerre type Sheffer matrix polynomials (2VgLtSMaP)	$\frac{1}{q(p^{-1}(\tau))} C_0\left(xp^{-1}(\tau)\sqrt{2C}\right) \exp\left(Ey(p^{-1}(\tau))^{\ell}\right) = \sum_{n=0}^{\infty} {}_{[\ell]L}s_n(x, y; C, E)\frac{\tau^n}{n!}$
VII.	$y = 0, z \to x$, $x \to y$	${}_{gL}s_n^{\ell}(y, 0, x; C, E) = {}_{L}s_n(x, y; C)$	Laguerre-Sheffer matrix polynomials (LSaMP)	$\frac{1}{q(p^{-1}(\tau))} C_0(xp^{-1}(\tau)) \exp\left(yp^{-1}(\tau)\sqrt{2C}\right) = \sum_{n=0}^{\infty} {}_{L}s_n(x, y; C)\frac{\tau^n}{n!}$

Remark 5. *For the particular matrix polynomials demonstrated in Table 1, we may offer some properties corresponding to those in Theorems 1–4.*

We may get a variety of outcomes that correspond to the above-presented results by varying the invertible series $q(\tau)$ and the delta series $p(\tau)$. As in Remark 1, the following corollaries give the corresponding results to those in Theorems 3 and 4 for the associated and Appell polynomials.

Associated Polynomials

Corollary 1. *The associated polynomials ${}_{[1,p]}{}_{gL}s_n^{\ell}(x, y, z; C, E)$ satisfy the following quasi-monomiality with regard to the operators of multiplication and differentiation:*

$$ {}_{[1,p]}\hat{M}_{gLs} = \left(x\sqrt{2C} - \hat{D}_z^{-1} + \ell Ey(\sqrt{2C})^{-(\ell-1)}D_x^{\ell-1}\right)\frac{1}{p'((\sqrt{2C})^{-1}D_x)} \tag{48} $$

and

$$_{[1,p]}\hat{P}_{\mathbf{g}LS} = p\left(\left(\sqrt{2C}\right)^{-1} D_{\boldsymbol{x}}\right), \tag{49}$$

respectively.

Corollary 2. *The associated polynomials* $_{[1,p]}\mathbf{g}_{L}s_n^{\ell}(\boldsymbol{x}, \boldsymbol{y}, \boldsymbol{z}; C, E)$ *satisfy the following differential equation:*

$$\begin{aligned}&\left\{\left(\boldsymbol{x}\sqrt{2C} - \hat{D}_{\boldsymbol{z}}^{-1} + \ell E\boldsymbol{y}(\sqrt{2C})^{-(\ell-1)} D_{\boldsymbol{x}}^{\ell-1}\right)\right.\\ &\left.\quad\times \frac{p((\sqrt{2C})^{-1}D_{\boldsymbol{x}})}{p'((\sqrt{2C})^{-1}D_{\boldsymbol{x}})} - n\right\} {}_{[1,p]}\mathbf{g}_{L}s_n^{\ell}(\boldsymbol{x}, \boldsymbol{y}, \boldsymbol{z}; C, E) = 0.\end{aligned} \tag{50}$$

Appell Polynomials

Corollary 3. *The Appell polynomials* $_{[q(\tau),\tau]}\mathbf{g}_{L}s_n^{\ell}(\boldsymbol{x}, \boldsymbol{y}, \boldsymbol{z}; C, E)$ *gratify the following quasi- monomiality with respect to the operators of multiplication and differentiation:*

$$_{[q(\tau),\tau]}\hat{M}_{\mathbf{g}LS} = \left(\boldsymbol{x}\sqrt{2C} - \hat{D}_{\boldsymbol{z}}^{-1} + \ell E\boldsymbol{y}(\sqrt{2C})^{-(\ell-1)} D_{\boldsymbol{x}}^{\ell-1} - \frac{q'((\sqrt{2C})^{-1}D_{\boldsymbol{x}})}{q((\sqrt{2C})^{-1}D_{\boldsymbol{x}})}\right) \tag{51}$$

and

$$_{[q(\tau),\tau]}\hat{P}_{\mathbf{g}LS} = \left(\sqrt{2C}\right)^{-1} D_{\boldsymbol{x}}, \tag{52}$$

respectively.

Corollary 4. *The Appell polynomials* $_{[q(\tau),\tau]}\mathbf{g}_{L}s_n^{\ell}(\boldsymbol{x}, \boldsymbol{y}, \boldsymbol{z}; C, E)$ *gratify the following differential equation:*

$$\begin{aligned}&\left\{\left(\boldsymbol{x}\sqrt{2C} - \hat{D}_{\boldsymbol{z}}^{-1} + \ell E\boldsymbol{y}(\sqrt{2C})^{-(\ell-1)} D_{\boldsymbol{x}}^{\ell-1} - \frac{q'((\sqrt{2C})^{-1}D_{\boldsymbol{x}})}{q((\sqrt{2C})^{-1}D_{\boldsymbol{x}})}\right)\right.\\ &\left.\quad\times \sqrt{2C})^{-1}D_{\boldsymbol{x}} - n\right\} {}_{[q(\tau),\tau]}\mathbf{g}_{L}s_n^{\ell}(\boldsymbol{x}, \boldsymbol{y}, \boldsymbol{z}; C, E) = 0.\end{aligned} \tag{53}$$

3. Extended Gould-Hopper-Laguerre-Sheffer Matrix Polynomials

Fractional calculus is a well-established theory that is extensively employed in a broad variety of fields of science, engineering, and mathematics today. The use of integral transforms and operational procedures to new families of special polynomials is a reasonably effective technique (consult, for instance, [28]).

This section provides some properties for the extended Gould-Hopper-Laguerre-Sheffer matrix polynomials in (31).

Theorem 5. *Let* $\Re(\eta) > 0$ *and* $\Re(\nu) > 0$. *Then the following integral representation for the extended Gould-Hopper-Laguerre-Sheffer matrix polynomials* $_{\mathbf{g}L}s_{n,\nu}^{\ell}(\boldsymbol{x}, \boldsymbol{y}, \boldsymbol{z}; C, E; \eta)$ *holds true:*

$$\begin{aligned}&{}_{\mathbf{g}L}s_{n,\nu}^{\ell}(\boldsymbol{x}, \boldsymbol{y}, \boldsymbol{z}; C, E; \eta)\\ &\quad= \frac{1}{\Gamma(\nu)} \int_0^{\infty} e^{-\eta t} t^{\nu-1} {}_{\mathbf{g}L}s_n^{\ell}(\boldsymbol{x}, \boldsymbol{y}t, \boldsymbol{z}; C, E)\, dt.\end{aligned} \tag{54}$$

Proof. Let $\mathcal{L}$ be the left-sided member of (54). Using (29) and (31), we have

$$\mathcal{L} = \frac{1}{\Gamma(\nu)} \int_0^\infty e^{-\eta t} t^{\nu-1} \exp\left(yEt \left(\sqrt{2C}\right)^{-\ell} \frac{\partial^\ell}{\partial x^\ell} \right) \left\{ {}_L s_n(z, x\sqrt{2C}) \right\} dt$$
$$= \frac{1}{\Gamma(\nu)} \int_0^\infty e^{-\eta t} t^{\nu-1} \, {}_{gL}s_n^\ell(x, yt, z; C, E) \, dt, \tag{55}$$

the second equality of which follows from (34). □

The following theorem gives the generating function of the EGHLSMaP.

Theorem 6. *The following function generates the extended Gould-Hopper-Laguerre-Sheffer matrix polynomials* ${}_{gL}s_{n,\nu}^\ell(x, y, z; C, E; \eta)$:

$$\frac{\exp(x\sqrt{2C}p^{-1}(u)) C_0(zp^{-1}(u))}{q(p^{-1}(u))\{\eta - Ey(p^{-1}(u))^\ell\}^\nu} = \sum_{n=0}^\infty {}_{gL}s_{n,\nu}^\ell(x, y, z; C, E; \eta) \frac{u^n}{n!}. \tag{56}$$

Additionally, the following differential-recursive relation holds true:

$$\frac{\partial}{\partial \eta} \, {}_{gL}s_{n,\nu}^\ell(x, y, z; C, E; \eta) = -\nu \, {}_{gL}s_{n,\nu+1}^\ell(x, y, z; C, E; \eta). \tag{57}$$

Proof. Multiplying each member of (54) by $\frac{u^n}{n!}$ and adding over n, one derives

$$\sum_{n=0}^\infty {}_{gL}s_{n,\nu}^\ell(x, y, z; C, E; \eta) \frac{u^n}{n!}$$
$$= \sum_{n=0}^\infty \frac{1}{\Gamma(\nu)} \int_0^\infty e^{-\eta t} t^{\nu-1} {}_{gL}s_n^\ell(x, yt, z; C, E) \frac{u^n}{n!} \, dt. \tag{58}$$

Using (22) in the integrand of the right-sided member of (58) gives

$$\sum_{n=0}^\infty {}_{gL}s_{n,\nu}^\ell(x, y, z; C, E; \eta) \frac{u^n}{n!}$$
$$= \frac{C_0(z(p^{-1}(u))^\ell) \exp(x\sqrt{2C}p^{-1}(u))}{q(p^{-1}(u))\Gamma(\nu)} \int_0^\infty e^{-\{\eta - Ey(f^{-1}(u))^\ell\}t} t^{\nu-1} \, dt,$$

the right member of which, upon using (29), leads to the left-sided member of (56).

Differentiating each member of (56) about η, one may get (57). □

The following theorem reveals that the EGHLSMaP ${}_{gL}s_{n,\nu}^\ell(x, y, z; C, E; \eta)$ is an extension of the GHLSMaP ${}_{gL}s_n^\ell(x, y, z; C, E)$.

Theorem 7. *The following identities hold true:*

$$\frac{\exp(x\sqrt{2C}p^{-1}(u)) C_0(zp^{-1}(u))}{q(p^{-1}(u))} {}_1F_1\left(\nu; 1; Ey(p^{-1}(u))^\ell\right)$$
$$= \sum_{n=0}^\infty {}_{gL}s_{n,\nu}^\ell(x, \hat{D}_y^{-1}, z; C, E; 1)\{1\} \frac{u^n}{n!}; \tag{59}$$

$${}_{gL}s_n^\ell(x, y, z; C, E) = {}_{gL}s_{n,1}^\ell(x, \hat{D}_y^{-1}, z; C, E; 1)\{1\}. \tag{60}$$

Proof. Taking $\eta = 1$ and $y = \hat{D}_y^{-1}$ in (56), we get

$$G(\nu; t) := \frac{\exp(x\sqrt{2C}p^{-1}(u)) C_0(zp^{-1}(u))}{q(p^{-1}(u))} \left(1 - E\hat{D}_y^{-1}(p^{-1}(u))^\ell\right)^{-\nu} \{1\}. \tag{61}$$

Using (20), we obtain

$$\begin{aligned}\left(1 - E\hat{D}_y^{-1}(p^{-1}(u))^{\ell}\right)^{-\nu}\{1\} &= \sum_{n=0}^{\infty} \frac{(\nu)_n}{n!} E^n \left(p^{-1}(u)\right)^{\ell n} \hat{D}_y^{-n}\{1\} \\ &= \sum_{n=0}^{\infty} \frac{(\nu)_n\, E^n y^n (p^{-1}(u))^{\ell n}}{(1)_n\, n!} \\ &= {}_1F_1\left(\nu\,;\, 1;\, Ey(p^{-1}(u))^{\ell}\right),\end{aligned} \tag{62}$$

for the second and third equalities of which (6) and (17) are employed, respectively.

Now, setting the last expression of (62) in (61), in view of (56), we obtain (59).

Noting

$$ {}_1F_1\left(1\,;\, 1;\, Ey(p^{-1}(u))^{\ell}\right) = \exp\left(Ey\left(p^{-1}(u)\right)^{\ell}\right),$$

we find that the resulting $G(t;1)$ is the generating function of the Gould-Hopper-Laguerre-Sheffer matrix polynomials ${}_{gL}s_n^{\ell}(x, y, z; C, E)$ in (22). We therefore have

$$\sum_{n=0}^{\infty} {}_{gL}s_{n,1}^{\ell}(x, \hat{D}_y^{-1}, z; C, E; 1)\{1\} \frac{u^n}{n!} = \sum_{n=0}^{\infty} {}_{gL}s_n^{\ell}(x, y, z; C, E) \frac{u^n}{n!},$$

which, upon equating the coefficients of u^n, yields (60).

The identity (60) may be obtained as follows: Combining (31) and (34) gives

$$\begin{aligned}{}_{gL}s_n^{\ell}(x, y, z; C, E) =& \left(1 - \hat{D}_y^{-1} E\left(\sqrt{2C}\right)^{-\ell} D_x^{\ell}\right) \exp\left(yE\left(\sqrt{2C}\right)^{-\ell} D_x^{\ell}\right) \\ &\times \left\{{}_{gL}s_{n,1}^{\ell}(x, \hat{D}_y^{-1}, z; C, E; 1)\right\}.\end{aligned}$$

As in (62), we find

$$\exp\left(yE\left(\sqrt{2C}\right)^{-\ell} D_x^{\ell}\right) = \left(1 - \hat{D}_y^{-1} E\left(\sqrt{2C}\right)^{-\ell} D_x^{\ell}\right)^{-1}\{1\}.$$

□

Remark 6. *As in (ii), Remark 1, the Laguerre-Sheffer polynomials ${}_Ls_n(x, y)$ reduce to the Laguerre-Appell polynomials ${}_L\mathcal{A}_n(x, y)$ (see [15]). Additionally, taking $p^{-1}(u) = u$ in the generating Equation (56), we can get the generalized Gould-Hopper-Laguerre-Appell matrix polynomials ${}_{gL}\mathcal{A}_{n,\nu}^{\ell}(x, y, z; C, E; \eta)$ (see [27]).*

The following theorem reveals the quasi-monomial principle of the extended Gould-Hopper-Laguerre-Sheffer matrix polynomials ${}_{gL}s_{n,\nu}^{\ell}(x, y, z; C, E; \eta)$.

Theorem 8. *The matrix polynomials ${}_{gL}s_{n,\nu}^{\ell}(x, y, z; C, E; \eta)$ satisfy the following quasi-monomiality with regard to the operators of multiplication and differentiation:*

$$\begin{aligned}\hat{M}_{{}_{gL}s_\nu} =& \left(x\sqrt{2C} - \hat{D}_z^{-1} - \ell E y(\sqrt{2C})^{-(\ell-1)} D_\eta D_x^{\ell-1} - \frac{q'((\sqrt{2C})^{-1}D_x)}{q((\sqrt{2C})^{-1}D_x)}\right) \\ &\times \frac{1}{p'((\sqrt{2C})^{-1}D_x)}\end{aligned} \tag{63}$$

and

$$\hat{P}_{{}_{gL}s_\nu} = p\left(\left(\sqrt{2C}\right)^{-1} D_x\right), \tag{64}$$

respectively. Here $D_\eta := \frac{\partial}{\partial \eta}$.

Proof. From Theorem 3, we have

$$\left(x\sqrt{2C} - \hat{D}_z^{-1} + \ell E y (\sqrt{2C})^{-(\ell-1)} D_x^{\ell-1} - \frac{q'((\sqrt{2C})^{-1}D_x)}{q((\sqrt{2C})^{-1}D_x)} \right) \times \frac{1}{p'((\sqrt{2C})^{-1}D_x)} {}_{gL}s_n^{\ell}(x,y,z;C,E) = {}_{gL}s_{n+1}^{\ell}(x,y,z;C,E), \tag{65}$$

and

$$p\left(\left(\sqrt{2C}\right)^{-1} D_x \right) {}_{gL}s_n^{\ell}(x,y,z;C,E) = n\, {}_{gL}s_{n-1}^{\ell}(x,y,z;C,E). \tag{66}$$

Replacing y by yt in each member of (66), multiplying both members of the resultant identity by $\frac{1}{\Gamma(\nu)} e^{-\eta t} t^{\nu-1}$, and integrating each member of the last resultant identity with respect to t from 0 to ∞, with the aid of (54), one obtains

$$p\left(\left(\sqrt{2C}\right)^{-1} D_x \right) \left\{ {}_{gL}s_{n,\nu}^{\ell}(x,y,z;C,E;\eta) \right\} = n\, {}_{gL}s_{n-1,\nu}^{\ell}(x,y,z;C,E;\eta),$$

which proves (64).

Furthermore, replacing y by yt in both sides of (65), multiplying both members of the resultant identity by $\frac{1}{\Gamma(\nu)} e^{-\eta t} t^{\nu-1}$, and integrating both sides of the last resulting identity with respect to t from 0 to ∞, with the help of (54) and (57), one can derive

$$\hat{M}_{gLs_\nu} \left\{ {}_{gL}s_{n,\nu}^{\ell}(x,y,z;C,E;\eta) \right\} = {}_{gL}s_{n+1,\nu}^{\ell}(x,y,z;C,E;\eta).$$

This proves (63). □

As in Theorem 4, using the results in Theorem 8, a differential equation for the extended Gould-Hopper-Laguerre-Sheffer matrix polynomials ${}_{gL}s_{n,\nu}^{\ell}(x,y,z;C,E;\eta)$ can be given in Theorem 9.

Theorem 9. *The following differential equation holds true:*

$$\left\{ \left(x\sqrt{2C} - \hat{D}_z^{-1} - \ell E y (\sqrt{2C})^{-(\ell-1)} D_z D_x^{\ell-1} - \frac{q'((\sqrt{2C})^{-1}D_x)}{q((\sqrt{2C})^{-1}D_x)} \right) \times \frac{p((\sqrt{2C})^{-1}D_x)}{p'((\sqrt{2C})^{-1}D_x)} - n \right\} {}_{gL}s_{n,\nu}^{\ell}(x,y,z;C,E;\eta) = 0. \tag{67}$$

As in Table 1, Table 2 includes certain particular cases of the extended Gould-Hopper-Laguerre-Sheffer matrix polynomials ${}_{gL}s_{n,\nu}^{\ell}(x,y,z;C,E;\eta)$, among numerous ones.

Table 2. Special cases of the EGHLSMaP ${}_{gL}s_{n,\nu}^{\ell}(x,y,z;C,E;\eta)$.

S. No.	Values of the Indices and Variables	Name of the Hybrid Special Polynomials	Generating Function
I.	$\ell = 2$	3-Variable extended Hermite-Laguerre-Sheffer matrix polynomials (3VEHLSMaP)	$\frac{\exp\left((xp^{-1}(u)\sqrt{2C}\right)C_0\left(zp^{-1}(u)\right)}{q(p^{-1}(u))\left(\eta - Ey(p^{-1}(u))^2\right)^{\nu}} = \sum_{n=0}^{\infty} {}_{HL}s_{n,\nu}(x,y,z;C,E,\eta)\frac{\tau^n}{n!}$
II.	$z = 0$	Extended Gould-Hopper-Sheffer-matrix polynomials (EGHSMaP)	$\frac{\exp\left(xp^{-1}(u)\sqrt{2C}\right)}{q(p^{-1}(u))\left(\eta - Ey(p^{-1}(u))^{\ell}\right)^{\nu}} = \sum_{n=0}^{\infty} {}_{g}s_{n,\nu}^{\ell}(x,y;C,E,\eta)\frac{\tau^n}{n!}$

Table 2. *Cont.*

S. No.	Values of the Indices and Variables	Name of the Hybrid Special Polynomials	Generating Function
III.	$\ell = r-1$, $\boldsymbol{z} = 0$	Extended generalized Chebyshev-Sheffer matrix polynomials (EGCSMaP)	$\frac{\exp((\boldsymbol{x}p^{-1}(u)\sqrt{2C})}{q(p^{-1}(u))\left(\eta - E\boldsymbol{y}(p^{-1}(u))^{r-1}\right)^{\nu}} = \sum_{n=0}^{\infty} {}_{U}s^{r}_{n,\nu}(\boldsymbol{x}, \boldsymbol{y}; C, E, \eta)\frac{\tau^n}{n!}$
IV.	$\ell = 2$, $\boldsymbol{z} = 0$	Extended Hermite Kampé de Fériet-Sheffer matrix polynomials (EHKdFSMaP)	$\frac{\exp\left(\boldsymbol{x}p^{-1}(u)\sqrt{2C}\right)}{q(p^{-1}(u))\left(\eta - E\boldsymbol{y}(p^{-1}(u))^{2}\right)^{\nu}} = \sum_{n=0}^{\infty} {}_{H}s_{n,\nu}(\boldsymbol{x}, \boldsymbol{y}; C, E, \eta)\frac{\tau^n}{n!}$
V.	$z = 0, \boldsymbol{x} \to \boldsymbol{y}$, $\boldsymbol{y} \to D_{\boldsymbol{x}}^{-1}$	Extended generalized Laguerre-Sheffer matrix polynomials (EGLSMaP)	$\frac{C_0\left(-E\boldsymbol{x}(p^{-1}(u))^{\ell}\right)}{q(p^{-1}(u))\left(\eta - \boldsymbol{y}\sqrt{2C}p^{-1}(u)\right)^{\nu}} = \sum_{n=0}^{\infty} {}_{L}s^{\ell}_{n,\nu}(\boldsymbol{x}, \boldsymbol{y}; C, E, \eta)\frac{\tau^n}{n!}$
VI.	$\boldsymbol{x} = -D_{\boldsymbol{x}}^{-1}$, $\boldsymbol{z} = 0$	2-Variable extended generalized Laguerre type Sheffer matrix polynomials (2VEgLtSMaP)	$\frac{C_0\left(\boldsymbol{x}p^{-1}(u)\sqrt{2C}\right)}{q(p^{-1}(u))\left(\eta - E\boldsymbol{y}(p^{-1}(u))^{\ell}\right)^{\nu}} = \sum_{n=0}^{\infty} {}_{[\ell]L}s_{n,\nu}(\boldsymbol{x}, \boldsymbol{y}; C, E, \eta)\frac{\tau^n}{n!}$
VII.	$\boldsymbol{y} = 0, \boldsymbol{z} \to \boldsymbol{x}$, $\boldsymbol{x} \to \boldsymbol{y}$	Extended Laguerre-Sheffer matrix polynomials (ELSaMP)	$\frac{C_0\left(\boldsymbol{x}p^{-1}(u)\right)}{q(p^{-1}(u))\left(\eta - \boldsymbol{y}\sqrt{2C}p^{-1}(u)\right)^{\nu}} = \sum_{n=0}^{\infty} {}_{L}s_{n,\nu}(\boldsymbol{x}, \boldsymbol{y}; C, \eta)\frac{\tau^n}{n!}$

Remark 7. *As in (i), Remark 1, if* $q(\tau) = 1$, *the Laguerre-Sheffer polynomials* ${}_{L}s_n(x, y)$ *reduce to the Laguerre-associated Sheffer polynomials* ${}_{[1,p]L}s_n(\boldsymbol{x}, \boldsymbol{y})$. *The extended Gould-Hopper-Laguerre-Sheffer matrix polynomials* ${}_{gL}s^{\ell}_{n,\nu}(\boldsymbol{x}, \boldsymbol{y}, \boldsymbol{z}; C, E; \eta)$ *reduce to the extended Gould-Hopper-Laguerre-associated Sheffer matrix polynomials (EGHLASMaP)* ${}_{[1,p]}{}_{gL}s^{\ell}_{n,\nu}(\boldsymbol{x}, \boldsymbol{y}, \boldsymbol{z}; C, E; \eta)$. *The following corollary contains the results for EGHLASMaP corresponding to those in Theorems 5–9.* □

Corollary 5. (i) *Let* $\Re(\eta) > 0$ *and* $\Re(\nu) > 0$.

$$ {}_{[1,p]}{}_{gL}s^{\ell}_{n,\nu}(\boldsymbol{x}, \boldsymbol{y}, \boldsymbol{z}; C, E; \eta) = \frac{1}{\Gamma(\nu)} \int_0^{\infty} e^{-\eta u} u^{\nu-1} \, {}_{[1,p]}{}_{gL}s^{\ell}_{n}(\boldsymbol{x}, \boldsymbol{y}u, \boldsymbol{z}; C, E)\, du. \tag{68} $$

(ii) *The polynomials* ${}_{[1,p]}{}_{gL}s^{\ell}_{n,\nu}(\boldsymbol{x}, \boldsymbol{y}, \boldsymbol{z}; C, E; \eta)$ *are generated by means of the following function:*

$$ \frac{\exp(\boldsymbol{x}\sqrt{2C}p^{-1}(u))C_0(\boldsymbol{z}p^{-1}(u))}{p^{-1}(u)\left\{\eta - E\boldsymbol{y}(p^{-1}(u))^{\ell}\right\}^{\nu}} = \sum_{n=0}^{\infty} {}_{[1,p]}{}_{gL}s^{\ell}_{n,\nu}(\boldsymbol{x}, \boldsymbol{y}, \boldsymbol{z}; C, E; \eta)\frac{u^n}{n!}. \tag{69} $$

Additionally, the following differential-recursive relation holds true:

$$ \frac{\partial}{\partial \eta} {}_{[1,p]}{}_{gL}s^{\ell}_{n,\nu}(\boldsymbol{x}, \boldsymbol{y}, \boldsymbol{z}; C, E; \eta) = -\nu \, {}_{[1,p]}{}_{gL}s^{\ell}_{n,\nu+1}(\boldsymbol{x}, \boldsymbol{y}, \boldsymbol{z}; C, E; \eta). \tag{70} $$

(iii) *The matrix polynomials* ${}_{[1,p]}{}_{gL}s^{\ell}_{n,\nu}(\boldsymbol{x}, \boldsymbol{y}, \boldsymbol{z}; C, E; \eta)$ *gratify quasi-monomiality with regard to the following operators of multiplication and differentiation:*

$$ {}_{[1,p]}\hat{M}_{{}_{gL}s_{\nu}} = \left(\boldsymbol{x}\sqrt{2C} - \hat{D}_{\boldsymbol{z}}^{-1} - \ell E\boldsymbol{y}(\sqrt{2C})^{-(\ell-1)} D_{\eta} D_{\boldsymbol{x}}^{\ell-1}\right) \frac{1}{p'((\sqrt{2C})^{-1}D_{\boldsymbol{x}})} \tag{71} $$

and

$$ {}_{[1,p]}\hat{P}_{{}_{gL}s_{\nu}} = p\left(\left(\sqrt{2C}\right)^{-1} D_{\boldsymbol{x}}\right), \tag{72} $$

respectively.

(iv) *The following differential equation holds true:*

$$\left\{\left(x\sqrt{2C} - \hat{D}_z^{-1} - \ell Ey(\sqrt{2C})^{-(\ell-1)} D_z D_x^{\ell-1}\right) \times \frac{p((\sqrt{2C})^{-1}D_x)}{p'((\sqrt{2C})^{-1}D_x)} - n\right\}_{[1,p]} {}_{gL}s_{n,\nu}^{\ell}(x,y,z;C,E;\eta) = 0. \tag{73}$$

4. Remarks and Further Particular Cases

The ${}_1F_1$ in (59), which is called the confluent hypergeometric function or Kummer's function, is an important and useful particular case of ${}_\alpha F_\beta$ in (17). It also has various other notations (consult, for instance, [11], p. 70). For properties and identities of ${}_1F_1$, one may consult the monograph [29]. In this regard, in view of (59), one may offer a variety of identities for the ${}_{gL}s_{n,\nu}^{\ell}(x, \hat{D}_y^{-1}, z; C, E; 1)\{1\}$. In order to give a demonstration, the ${}_1F_1$ in (59) has the following integral representation (consult, for instance, [11], p. 70, Equation (46)):

$$\begin{aligned} &{}_1F_1\left(\nu\,;\,1;\,Ey(p^{-1}(u))^{\ell}\right) \\ &= \frac{1}{\Gamma(\nu)\Gamma(1-\nu)} \int_0^1 \eta^{\nu-1}(1-\eta)^{-\nu} \exp\left(Ey\left(p^{-1}(u)\right)^{\ell}\eta\right) d\eta \quad (0 < \Re(\nu) < 1). \end{aligned} \tag{74}$$

Further, using (35) and (59), with the aid of (21) and (74), one may readily get the following identity:

$$\begin{aligned} &{}_{[q(u),u]}{}_{gL}s_{n,\nu}^{\ell}(x, \hat{D}_y^{-1}, z; C, E; 1)\{1\} \\ &= \sum_{k=0}^{[\frac{n}{\ell}]} \frac{n!\,(\nu)_k}{(k!)^2\,(n-\ell k)!}\,(Ey)^k\; {}_{[q(u),u]}{}_{L}s_{n-\ell k}(z, x\sqrt{2C}). \end{aligned} \tag{75}$$

The hybrid matrix polynomials introduced in Sections 2 and 3, besides the demonstrated particular cases, may produce numerous other particular cases as well as corresponding properties. In this section, we combine the findings from Sections 2 and 3 with several well-known (or classical) polynomials to derive some related identities.

(a) The Hermite polynomials $H_n(x)$, which are generated by the following function (consult, for example, [30]):

$$\exp(2x\tau - \tau^2) = \sum_{n=0}^{\infty} H_n(x) \frac{\tau^n}{n!} \tag{76}$$

belongs to the Sheffer family by choosing

$$q(\tau) = e^{\tau^2/4}, \quad p(\tau) = \frac{\tau}{2}, \quad \text{and} \quad p^{-1}(\tau) = 2\tau \tag{77}$$

in (3).

For these choices of $q(\tau)$ and $p(\tau)$ in (22) and (56), the GHLSMaP ${}_{gL}s_n^{\ell}(x,y,z;C,E)$ and the EGHLSMaP ${}_{gL}s_{n,\nu}^{\ell}(x,y,z;C,E;\eta)$ are called (denoted) as the matrix polynomials of Gould-Hopper-Laguerre-Hermite (GHLHMaP) ${}_{gL}H_n^{\ell}(x,y,z;C,E)$ and the extended matrix polynomials of Gould-Hopper-Laguerre-Hermite (EGHLHMaP) ${}_{gL}H_{n,\nu}^{\ell}(x,y,z;C,E;\eta)$, respectively.

Some identities corresponding to those in Sections 2 and 3 are recorded in Tables 3 and 4.

Table 3. Results for the GHLHMaP ${}_{gL}H_n^{\ell}(x, y, z; C, E)$.

Results	Expressions
Generating function:	$\exp\left(2x\tau\sqrt{2C} + Ey(2\tau)^{\ell} - \tau^2\right)C_0(2z\tau) = \sum_{n=0}^{\infty} {}_{gL}H_n^{\ell}(x, y, z; C, E)\frac{\tau^n}{n!}$.
Multiplicative and derivative operators:	$\hat{M}_{gLH} = \left(x\sqrt{2C} - \hat{D}_z^{-1} + \frac{\ell E y}{(\sqrt{2C})^{(\ell-1)}}\frac{\partial^{\ell-1}}{\partial x^{\ell-1}} - \frac{\left(\sqrt{2C}\right)^{-1}D_x}{2}\right)2,$ $\hat{P}_{gLH} = \frac{\left(\sqrt{2C}\right)^{-1}D_x}{2}$.
Differential equation:	$\left(\left(x\sqrt{2C} - \hat{D}_z^{-1} + \frac{\ell E y}{(\sqrt{2C})^{(\ell-1)}}\frac{\partial^{\ell-1}}{\partial x^{\ell-1}} - \frac{\left(\sqrt{2C}\right)^{-1}D_x}{2}\right)\left(\sqrt{2C}\right)^{-1}D_x - n\right)$ $\times {}_{gL}H_n^{\ell}(x, y, z; C, E) = 0.$

Table 4. Results for the EGHLHMaP ${}_{gL}H_{n,\nu}^{\ell}(x, y, z; C, E; \alpha)$.

Results	Expressions
Generating function:	$\frac{\exp(2x\tau\sqrt{2C})C_0(2z\tau)}{e^{\tau^2}(\alpha - Ey(2\tau)^{\ell})^{\nu}} = \sum_{n=0}^{\infty} {}_{gL}H_{n,\nu}^{\ell}(x, y, z; C, E; \alpha)\frac{\tau^n}{n!}$.
Multiplicative and derivative operators:	$\hat{M}_{gLH_\nu} = \left(x\sqrt{2C} - \hat{D}_z^{-1} - \frac{\ell E y}{(\sqrt{2C})^{(\ell-1)}}\frac{\partial^{\ell}}{\partial\alpha\partial x^{\ell-1}} - \frac{\left(\sqrt{2C}\right)^{-1}D_x}{2}\right)2,$ $\hat{P}_{gLH_\nu} = \frac{\left(\sqrt{2C}\right)^{-1}D_x}{2}$.
Differential equation:	$\left(\left(x\sqrt{2C} - \hat{D}_z^{-1} + \frac{\ell E y}{(\sqrt{2C})^{(\ell-1)}}\frac{\partial^{\ell}}{\partial\alpha\partial x^{\ell-1}} - \frac{\left(\sqrt{2C}\right)^{-1}D_x}{2}\right)\left(\sqrt{2C}\right)^{-1}D_x - n\right)$ $\times {}_{gL}H_{n,\nu}^{\ell}(x, y, z; C, E; \alpha) = 0.$

(b) The truncated exponential polynomials $e_n(x)$, which are generated by the following function (consult, for example, [31], p. 596, Equation (4); see also [32]):

$$\frac{e^{x\tau}}{1-\tau} = \sum_{n=0}^{\infty} e_n(x)\frac{\tau^n}{n!} \tag{78}$$

belong to the Sheffer family by choosing $q(\tau) = \frac{1}{1-\tau}$ and $p(\tau) = \tau$. As in (a), the GHLSMaP ${}_{gL}s_n^{\ell}(x, y, z; C, E)$ and EGHLSMaP ${}_{gL}s_{n,\nu}^{\ell}(x, y, z; C, E; \eta)$ are called (denoted) as the Gould-Hopper-Laguerre-truncated exponential matrix polynomials (GHLTEMaP) ${}_{gL}e_n^{\ell}(x, y, z; C, E)$ and extended Gould-Hopper-Laguerre-truncated exponential matrix polynomials (EGHLTEMaP) ${}_{gL}e_{n,\nu}^{\ell}(x, y, z; C, E; \eta)$, respectively. As in (a), their properties are recorded in Tables 5 and 6.

Table 5. Results for the GHLTEMaP ${}_{gL}e_n^{\ell}(x, y, z; C, E)$.

Results	Expressions
Generating function:	$\frac{1}{1-t}\exp\left(xt\sqrt{2C} + Eyt^{\ell}\right)C_0(zt) = \sum_{n=0}^{\infty} {}_{gL}e_n^{\ell}(x, y, z; C, E)\frac{t^n}{n!}$.
Multiplicative and derivative operators:	$\hat{M}_{gLe} = x\sqrt{2C} - \hat{D}_z^{-1} + \frac{\ell E y}{(\sqrt{2C})^{(\ell-1)}}\frac{\partial^{\ell-1}}{\partial x^{\ell-1}} - \frac{1}{1-(\sqrt{2C})^{-1}D_x},$ $\hat{P}_{gLe} = \left(\sqrt{2C}\right)^{-1}D_x$.
Differential equation:	$\left(\left(x\sqrt{2C} - \hat{D}_z^{-1} + \frac{\ell E y}{(\sqrt{2C})^{(\ell-1)}}\frac{\partial^{\ell-1}}{\partial x^{\ell-1}} - \frac{1}{1-(\sqrt{2C})^{-1}D_x}\right)(\sqrt{2C})^{-1}D_x - n\right)$ $\times {}_{gL}e_n^{\ell}(x, y, z; C, E) = 0.$

Table 6. Results for the EGHLTEMaP ${}_{\mathtt{g}L}e^{\ell}_{n,\nu}(x, y, z; C, E; \eta)$.

Results	Expressions
Generating function:	$\frac{1}{1-u}\frac{\exp(xu\sqrt{2C})C_0(zu)}{(\alpha-Eyu^{\ell})^{\nu}} = \sum_{n=0}^{\infty} {}_{\mathtt{g}L}e^{\ell}_{n,\nu}(x, y, z; C, E; \alpha)\frac{u^n}{n!}$.
Multiplicative and derivative operators:	$\hat{M}_{\mathtt{g}Le_\nu} = x\sqrt{2C} - \hat{D}_z^{-1} - \frac{\ell E y}{(\sqrt{2C})^{(\ell-1)}}\frac{\partial^{\ell}}{\partial\alpha\partial x^{\ell-1}} - \frac{1}{1-(\sqrt{2C})^{-1}D_x}$, $\hat{P}_{\mathtt{g}Le_\nu} = \left(\sqrt{2C}\right)^{-1} D_x$.
Differential equation:	$\left(\left(x\sqrt{2C} - \hat{D}_z^{-1} - \frac{\ell E y}{(\sqrt{2C})^{(\ell-1)}}\frac{\partial^{\ell}}{\partial\alpha\partial x^{\ell-1}} - \frac{1}{1-(\sqrt{2C})^{-1}D_x}\right)(\sqrt{2C})^{-1}D_x - n\right)$ $\times {}_{\mathtt{g}L}e^{\ell}_{n,\nu}(x, y, z; C, E; \alpha) = 0$.

(c) The Mittag-Leffler polynomials $M_n(x)$, which are the member of associated Sheffer family and defined as follows (see [4]):

$$\left(\frac{1+\tau}{1-\tau}\right)^{x} = \sum_{n=0}^{\infty} M_n(x)\frac{\tau^n}{n!} \tag{79}$$

by choosing $q(\tau) = 1$ and $p(\tau) = \frac{e^{\tau}-1}{e^{\tau}+1}$. As in (a), the GHLASMaP ${}_{\mathtt{g}L}\mathfrak{s}^{\ell}_{n}(x, y, z; C, E)$ and the EGHLASMaP ${}_{\mathtt{g}L}\mathfrak{s}^{\ell}_{n,\nu}(x, y, z; C, E; \eta)$ are called (denoted) as the Gould-Hopper-Laguerre-Mittag-Leffler matrix polynomials (GHLMLMaP) ${}_{\mathtt{g}L}M^{\ell}_{n}(x, y, z; C, E)$ and the extended Gould-Hopper-Laguerre-Mittag-Leffler matrix polynomials (EGHLMLMaP) ${}_{\mathtt{g}L}M^{\ell}_{n,\nu}(x, y, z; C, E; \eta)$, respectively. As in (a) or (b), their properties are recorded in Tables 7 and 8.

Table 7. Results for the GHLMLMaP ${}_{\mathtt{g}L}M^{\ell}_{n}(x, y, z; C, E)$.

Results	Expressions
Generating function:	$\exp\left(x\ln\left(\frac{1+\tau}{1-\tau}\right)\sqrt{2C} + Ey\ln\left(\frac{1+\tau}{1-\tau}\right)^{\ell}\right)C_0\left(z\ln\left(\frac{1+\tau}{1-\tau}\right)\right)$ $= \sum_{n=0}^{\infty} {}_{\mathtt{g}L}M^{\ell}_{n}(x, y, z; C, E)\frac{\tau^n}{n!}$.
Multiplicative and derivative operators:	$\hat{M}_{\mathtt{g}LM} = \left(x\sqrt{2C} - \hat{D}_z^{-1} + \frac{\ell E y}{(\sqrt{2C})^{(\ell-1)}}\frac{\partial^{\ell-1}}{\partial x^{\ell-1}}\right)\frac{\left(e^{(\sqrt{2C})^{-1}D_x}+1\right)^2}{2\,e^{(\sqrt{2C})^{-1}D_x}}$, $\hat{P}_{\mathtt{g}LM} = \frac{e^{(\sqrt{2C})^{-1}D_x}-1}{e^{(\sqrt{2C})^{-1}D_x}+1}$.
Differential equation:	$\left(\left(x\sqrt{2C} - \hat{D}_z^{-1} + \frac{\ell E y}{(\sqrt{2C})^{(\ell-1)}}\frac{\partial^{\ell-1}}{\partial x^{\ell-1}}\right)\frac{e^{2(\sqrt{2C})^{-1}D_x}-1}{2\,e^{(\sqrt{2C})^{-1}D_x}} - n\right)$ $\times {}_{\mathtt{g}L}M^{\ell}_{n}(x, y, z; C, E) = 0$.

Table 8. Results for the EGHLMLMaP ${}_{\mathtt{g}L}M^{\ell}_{n,\nu}(x, y, z; C, E; \eta)$.

Results	Expressions
Generating function:	$\frac{\exp(x\ln(\frac{1+\tau}{1-\tau})\sqrt{2C})C_0(z\ln(\frac{1+\tau}{1-\tau}))}{\left(\alpha-Ey\ln(\frac{1+t}{1-t})^{\ell}\right)^{\nu}} = \sum_{n=0}^{\infty} {}_{\mathtt{g}L}M^{\ell}_{n,\nu}(x, y, z; C, E; \eta)\frac{\tau^n}{n!}$.
Multiplicative and derivative operators:	$\hat{M}_{\mathtt{g}LM_\nu} = \left(x\sqrt{2C} - \hat{D}_z^{-1} - \frac{\ell E y}{(\sqrt{2C})^{(\ell-1)}}\frac{\partial^{\ell}}{\partial\alpha\partial x^{\ell-1}}\right)\frac{\left(e^{(\sqrt{2C})^{-1}D_x}+1\right)^2}{2\,e^{(\sqrt{2C})^{-1}D_x}}$, $\hat{P}_{\mathtt{g}LM_\nu} = \frac{e^{(\sqrt{2C})^{-1}D_x}-1}{e^{(\sqrt{2C})^{-1}D_x}+1}$.
Differential equation:	$\left(\left(x\sqrt{2C} - \hat{D}_z^{-1} - \frac{\ell E y}{(\sqrt{2C})^{(\ell-1)}}\frac{\partial^{\ell}}{\partial\alpha\partial x^{\ell-1}}\right)\frac{e^{2(\sqrt{2C})^{-1}D_x}-1}{2\,e^{(\sqrt{2C})^{-1}D_x}} - n\right)$ $\times {}_{\mathtt{g}L}M^{\ell}_{n,\nu}(x, y, z; C, E; \eta) = 0$.

Numerous necessary and sufficient properties for Sheffer sequences, accordingly, associated sequences and Appell sequences have been developed (see [4], pp. 17–28). In addition to the identities in Corollaries 3 and 4, here, we record several identities for the

Appell polynomials ${}_{[q(\tau),\tau]}gLs_n^{\ell}(x, y, z; C, E)$ in the following corollary, without their proofs (see [4], pp. 26–28).

Corollary 6. *The following identities hold true:*

(a)

$$ {}_{[q(\tau),\tau]}gLs_n^{\ell}(x, y, z; C, E) = q\left(\left(\sqrt{2C}\right)^{-1} D_x\right)^{-1}\{x^n\}. \tag{80}$$

(b)

$$\begin{aligned} &{}_{[q(\tau),\tau]}gLs_n^{\ell}(x_1 + x_2, y, z; C, E) \\ &= \sum_{k=0}^{n} \binom{n}{k} {}_{[q(\tau),\tau]}gLs_{n-k}^{\ell}(x_1, y, z; C, E) \left(\sqrt{2C}\, x_2\right)^k. \end{aligned} \tag{81}$$

(c) (Conjugate representation)

$$\begin{aligned} &{}_{[q(\tau),\tau]}gLs_n^{\ell}(x, y, z; C, E) \\ &= \sum_{k=0}^{n} \binom{n}{k} \left[q\left(\left(\sqrt{2C}\right)^{-1} D_x\right)^{-1}\left\{x^{n-k}\right\}\right] x^k. \end{aligned} \tag{82}$$

5. Conclusions and Posing a Problem

The authors introduced a new class of polynomials, the Gould-Hopper-Laguerre-Sheffer matrix polynomials, using operational approaches. This new family's generating function and operational representations were then constructed. They are also understood in terms of quasi-monomiality. The authors also extended Gould-Hopper-Laguerre-Sheffer matrix polynomials and explored their characteristics using the integral transform. There were other instances for individual members of the aforementioned matrix polynomial family.

It should be highlighted that the polynomials presented and studied in this article are regarded to be novel, primarily because they cannot be obtained by modifying previously published findings and identities, as far as we have researched. Also, the new polynomials and their identities are potentially useful, particularly in light of the tables' demonstrations of some of their special instances.

Posing a problem: Provide some new instances (which are nonexistent from the literature) for those novel polynomials, such as Gould-Hopper matrix polynomials and Gould-Hopper-Laguerre-Sheffer matrix polynomials.

Author Contributions: Writing—original draft, T.N. and J.C.; writing—review and editing, T.N. and J.C. All authors have read and agreed to the published version of the manuscript.

Funding: The second-named author was supported by the Basic Science Research Program through the National Research Foundation of Korea (NRF) funded by the Ministry of Education (NRF-2020R111A1A01052440).

Institutional Review Board Statement: Not applicable.

Informed Consent Statement: Not applicable.

Acknowledgments: The authors are quite appreciative of the anonymous referees' helpful and supportive remarks, which helped to enhance this article.

Conflicts of Interest: The authors have no conflict of interest.

References

1. Abdalla, M.; Akel, M.; Choi, J. Certain matrix Riemann-Liouville fractional integrals associated with functions involving generalized Bessel matrix polynomials. *Symmetry* **2021**, *13*, 622. [CrossRef]
2. Khammash, G.S.; Agarwal, P.; Choi, J. Extended k-Gamma and k-Beta functions of matrix arguments. *Mathematics* **2020**, *8*, 1715. [CrossRef]
3. Sheffer, I.M. Some properties of polynomial sets of type zero. *Duke Math. J.* **1939**, *5*, 590–622. [CrossRef]

4. Roman, S. *The Umbral Calculus*; Academic Press: New York, NY, USA; Dover Publications, Inc.: Mineola, NY, USA, 2005.
5. Khan, N.U.; Aman, M.; Usman, T.; Choi, J. Legendre-Gould Hopper-based Sheffer polynomials and operational methods. *Symmetry* **2020**, *12*, 2051. [CrossRef]
6. Khan, S.; Nahid, T.; Riyasat, M. Partial derivative formulas and identities involving 2-variable Simsek polynomials. *Bol. Soc. Mat. Mex.* **2020**, *26*, 1–13. [CrossRef]
7. Khan, S.; Nahid, T.; Riyasat, M. Properties and graphical representations of the 2-variable form of the Simsek polynomials. *Vietnam J. Math.* **2022**, *50*, 95–109. [CrossRef]
8. Nahid, T.; Ryoo, C.S. 2-variable Fubini-degenerate Apostol-type polynomials. *Asian-Eur. J. Math.* **2021**, 2250092. [CrossRef]
9. Wani, S.A.; Khan, S.; Nahid, T. Gould-Hopper based Frobenius-Genocchi polynomials and their generalized form. *Afr. Mat.* **2020**, *31*, 1397–1408. [CrossRef]
10. Khan, S.; Raza, N. Monomiality principle, operational methods and family of Laguerre-Sheffer polynomials. *J. Math. Anal. Appl.* **2012**, *387*, 90–102. [CrossRef]
11. Srivastava, H.M.; Choi, J. *Zeta and q-Zeta Functions and Associated Series and Integrals*; Elsevier: Amsterdam, The Netherland; London, UK; New York, NY, USA, 2012.
12. Kilbas, A.A.; Srivastava, H.M.; Trujillo, J.J. *Theory and Applications of Fractional Differential Equations*; North-Holland Mathematical Studies; Elsevier: Amsterdam, The Netherlands, 2006; Volume 204.
13. Bohner, M.; Tunç, O.; Tunç, C. Qualitative analysis of Caputo fractional integro-differential equations with constant delays. *Comp. Appl. Math.* **2021**, *40*, 214. [CrossRef]
14. Motamedi, M.; Sohail, A. Geometric analysis of the properties of germanene using Lifson-Wershel potential function. *Int. J. Geom. Methods Mod. Phys.* **2020**, *17*, 2050031. [CrossRef]
15. Khan, S.; Al-Saad, M.W.; Khan, R. Laguerre-based Appell polynomials: Properties and applications. *Math. Comput. Model.* **2010**, *52*, 247–259. [CrossRef]
16. Dunford, N.; Schwartz, J.; *Linear Operators Part I*; Interscience: New York, NY, USA, 1963.
17. Jódar, L.; Cortés, J.C. Some properties of Gamma and Beta matrix functions. *Appl. Math. Lett.* **1998**, *11*, 89–93. [CrossRef]
18. Jódar, L.; Cortés, J.C. On the hypergeometric matrix function. *J. Comp. Appl. Math.* **1998**, *99*, 205–217. [CrossRef]
19. Çekim, B.; Aktaş, R. Multivariable matrix generalization of Gould-Hopper polynomials. *Miskolc Math. Notes* **2015**, *16*, 79–89. [CrossRef]
20. Steffensen, J.F. The poweroid, an extension of the mathematical notion of power. *Acta. Math.* **1941**, *73*, 333–366. [CrossRef]
21. Dattoli, G. Hermite-Bessel and Laguerre-Bessel functions: A by-product of the monomiality principle. *Adv. Spec. Funct. Appl.* **2000**, *1*, 147–164.
22. Dattoli, G. Generalized polynomials, operational identities and their applications. *J. Comput. Appl. Math.* **2000**, *118*, 111–123. [CrossRef]
23. Qamar, R.; Nahid, T.; Riyasat, M.; Kumar, N.; Khan, A. Gould-Hopper matrix-Bessel and Gould-Hopper matrix-Tricomi functions and related integral representations. *AIMS Math.* **2020**, *5*, 4613–4623. [CrossRef]
24. Srivastava, H.M.; Manocha, H.L. *A Treatise on Generating Functions*; Halsted Press: Chichester, NH, USA; John Wiley and Sons: Hoboken, NJ, USA, 1984.
25. Dattoli, G.; Ottaviani, P.L.; Torre, A.; Vázquez, L. Evolution operator equations: Integration with algebraic and finite-difference methods. Applications to physical problems in classical and quantum mechanics and quantum field theory. *Riv. Nuovo Cim.* **1997**, *20*, 3. [CrossRef]
26. Ricci, P.E.; Tavkhelidze, I. An introduction to operational techniques and special functions. *J. Math. Sci.* **2009**, *157*, 161–188. [CrossRef]
27. Nahid, T.; Khan, S. Construction of some hybrid relatives of Laguuerre-Appell polynomials associated with Gould-Hopper matrix polynomials. *J. Anal.* **2021**, *29*, 927–946. [CrossRef]
28. Dattoli, G.; Ricci, P.E.; Cesarano, C.; Vázquez, L. Special polynomials and fractional calculus. *Math. Comput. Model.* **2003**, *37*, 729–733. [CrossRef]
29. Slater, L.J. *Confluent Hypergeometric Functions*; Cambridge University Press: London, UK; New York, NY, USA, 1960.
30. Rainville, E.D. *Special Functions*; Macmillan Company: New York, NY, USA; Chelsea Publishing Company: Bronx, NY, USA, 1971.
31. Dattoli, G.; Cesarano, C.; Sacchetti, D. A note on truncated polynomials, *Appl. Math. Comput.* **2003**, *134*, 595–605. [CrossRef]
32. Nahid, T.; Alam, P.; Choi, J. Truncated-exponential-based Appell-type Changhee polynomials. *Symmetry* **2020**, *12*, 1588. [CrossRef]

fractal and fractional

MDPI

Article

Some New Harmonically Convex Function Type Generalized Fractional Integral Inequalities

Rana Safdar Ali [1], Aiman Mukheimer [2], Thabet Abdeljawad [2,3,4,*], Shahid Mubeen [5], Sabila Ali [1], Gauhar Rahman [6,*] and Kottakkaran Sooppy Nisar [7]

1 Department of Mathematics, The University of Lahore, Sargodha Campus, Sargodha 40100, Pakistan; rsafdar0@gmail.com (R.S.A.); sabila21imran@gmail.com (S.A.)
2 Department of Mathematics and General Sciences, Prince Sultan University, Riyadh 11586, Saudi Arabia; mukheimer@psu.edu.sa
3 Department of Medical Research, China Medical University, Taichung 40402, Taiwan
4 Department of Computer Sciences and Information Engineering, Asia University, Taichung 41354, Taiwan
5 Department of Mathematics, University of Sargodha, Sargodha 40100, Pakistan; smjhanda@gmail.com
6 Department of Mathematics, Hazara University, Mansehra 21120, Pakistan
7 Department of Mathematics, College of Arts and Sciences, Prince Sattam bin Abdulaziz University, Wadi Aldawaser 11991, Saudi Arabia; n.sooppy@psau.edu.sa
* Correspondence: tabdeljawad@psu.edu.sa (T.A.); gauhar55uom@gmail.com or drgauhar.rahman@hu.edu.pk (G.R.)

Abstract: In this article, we established a new version of generalized fractional Hadamard and Fejér–Hadamard type integral inequalities. A fractional integral operator (FIO) with a non-singular function (multi-index Bessel function) as its kernel and monotone increasing functions is utilized to obtain the new version of such fractional inequalities. Our derived results are a generalized form of several proven inequalities already existing in the literature. The proven inequalities are useful for studying the stability and control of corresponding fractional dynamic equations.

Keywords: bessel function; harmonically convex function; non-singular function involving kernel fractional operator; harmonically convex function; Hadamard inequality; Fejér–Hadamard inequality

MSC: 2010: 33C10; 11K70; 33B20; 52A41; 05B20; 26D07

Citation: Ali, R.S.; Mukheimer, A.; Abdeljawad, T.; Mubeen, S.; Ali, S.; Rahman, G.; Nisar, K.S. Some New Harmonically Convex Function Type Generalized Fractional Integral Inequalities. *Fractal Fract.* **2021**, *5*, 54. https://doi.org/10.3390/fractalfract5020054

Academic Editors: Palle Jorgensen and Themistocles M. Rassias

Received: 23 April 2021
Accepted: 2 June 2021
Published: 7 June 2021

1. Introduction

In the present era, fractional integral operators involving inequalities are widely derived by [1–4]. These fractional integral operators of any arbitrary real or complex order involve a different type of kernel. The field of fractional calculus has gained considerable importance among mathematicians and scientists due to its wide applications in sciences, engineering, and many other fields [5–9]. Hadamard and Fejér–Hadamard type inequalities have been discussed for many functions using different fractional operators with different kernels. Abbas and Farid [10] proposed the Hadamard and Fejér–Hadamard type integral inequalities for harmonically convex functions using the two-sided generalized fractional integral operator. Farid et al. [11,12] discussed these results in generalized form with an extended generalized Mittag–Leffler function. Hadamard and Fejér–Hadamard type inequalities are widely studied by the researchers [12–19]. The objective of this paper is to derive Hadamard, Fejér–Hadamard, and some other related type inequalities for the harmonically convex function via a generalized fractional operator with a nonsingular function as its kernel, which involves a multi-index Bessel function. For a recent related weighted fractional generalized approach, we refer to [20].

Hermite–Hadamard inequality and Fejér–Hadamard inequality are given by

Theorem 1 ([21–23]). *The inequality derived on the interval $I = [u, v] \subseteq \mathbb{R}$ called Hermite Hadamard inequality is given by*

$$\rho\left[\frac{u+v}{2}\right] \leq \frac{1}{v-u}\int_u^v \rho(t)dt \leq \frac{\rho(u)+\rho(v)}{2}, \tag{1}$$

where $u, v \in I$, with $u \neq v$ and $\rho : I \rightarrow R$ is a convex function.

Theorem 2 ([21,24,25]). *The Fejér–Hadamard inequality is defined for a convex function $\rho : I \rightarrow \mathbb{R}$ and for a function $\mu : I \rightarrow \mathbb{R}$, which is non-negative, integrable, and symmetric about $\frac{u+v}{2}$, defined by*

$$\rho\left[\frac{u+v}{2}\right]\int_u^v \mu(t)dt \leq \int_u^v \rho(t)\mu(t)dt \leq \left[\frac{\rho(u)+\rho(v)}{2}\right]\int_u^v \mu(t)dt, \tag{2}$$

where $u, v \in I$, with $u \neq v$.

Definition 1 ([21,26]). *A function $\rho : [u, v] \rightarrow \mathbb{R}$ is said to be convex if*

$$\rho\Big[tx + (1-t)y\Big] \leq t\rho(x) + (1-t)\rho(y) \tag{3}$$

holds for all $x, y \in [u, v]$ and $t \in [0, 1]$.

Definition 2 ([21,22]). *Let I be an interval of nonzero real numbers. Then a function $\rho : I \rightarrow \mathbb{R}$ is said to be harmonically convex if*

$$\rho\left[\frac{uv}{tu + (1-t)v}\right] \leq t\rho(v) + (1-t)\rho(u) \tag{4}$$

holds for all $u, v \in I$ and $t \in [0, 1]$.

Definition 3 ([21,27]). *A function $\rho : [u, v] \rightarrow \mathbb{R}$ where $I \subset \mathbb{R}$ contains nonzero real numbers is said to be harmonically symmetric about $\frac{u+v}{2uv}$ if*

$$\rho\left[\frac{1}{t}\right] = \rho\left[\frac{1}{\frac{1}{u}+\frac{1}{v}-t}\right]. \tag{5}$$

$t \in [u, v]$

Definition 4 ([28,29]). *The Pochammer's symbol is defined for $s \in N$ as*

$$(\mu)_s = \begin{cases} 1, & \text{for } s = 0, \mu \neq 0, \\ \mu(\mu+1)\cdots(\mu+s-1), & \text{for } s \geq 1, \end{cases} \tag{6}$$

where $\mu \in \mathbb{C}$.

Definition 5 ([30]). *The generalized multi-index Bessel function defined by Choi et al. as follows;*

$$J_{(\tau_j)_{m,\sigma}}^{(\gamma_j)_{m,\lambda}}(t) = \sum_{s=0}^{\infty} \frac{(\lambda)_{\sigma s}}{(s!)\prod_{j=1}^{m}\Gamma(\gamma_j s + \tau_j + 1)}(-t)^s, \tag{7}$$

where $\gamma_j, \tau_j, \lambda \in \mathbb{C}$,$j = 1, 2, 3\cdots m$, $\Re(\lambda) > 0$, $\Re(\tau_j) > -1$, $\sum_{j=1}^{m}\Re(\gamma_j) > \max(0 : \Re(\sigma) - 1)$, $\sigma > 0$.

We define the following generalized fractional integral with a nonsingular function (generalized multi-index Bessel function) as a kernel.

Definition 6. *The generalized fractional integral operators (left and right-sided) containing the multi-index Bessel function in its kernel are, respectively, defined by*

$$\left(\mathscr{T}^{(\gamma_j,\tau_j)_m}_{\lambda,\sigma,\varsigma;u^+}\rho\right)(z) = \int_u^z (z-t)^{\tau_j} J^{(\gamma_j)_m,\lambda}_{(\tau_j)_m,\sigma}(\varsigma(z-t)^{\gamma_j})\rho(t)dt \tag{8}$$

and

$$\left(\mathscr{T}^{(\gamma_j,\tau_j)_m}_{\lambda,\sigma,\varsigma;v^-}\rho\right)(z) = \int_z^v (t-z)^{\tau_j} J^{(\gamma_j)_m,\lambda}_{(\tau_j)_m,\sigma}(\varsigma(t-z)^{\gamma_j})\rho(t)dt, \tag{9}$$

where $\gamma_j, \tau_j, \lambda, \varsigma \in \mathbb{C}, j = 1,2,3\cdots m, \Re(\lambda) > 0, \Re(\tau_j) > -1, \sum_{j=1}^{m} \Re(\gamma_j) > max(0 : \Re(\sigma) - 1),$ $\sigma > 0$ *and* $\rho \in L[u,v]$, $t \in [u,v]$.

Remark 1. *1. If we put* $\varsigma = 0$, $m = 1$ *and replace* τ_j *by* $\tau_j - 1$, *it reduces to left and right-sided Riemann–Liouville fractional integral operator.*

2. Main Results

In this section, we present Hadamard, and Fejér–Hadamard type inequalities for harmonically convex functions by employing the new generalized fractional integral operators with a multi-index Bessel function as its kernel. We also establish a new version of inequalities by expressing the generalized fractional integral operator as the sum of two fractional integrals.

Theorem 3. *Let* $\theta, \psi : [a,b] \to \mathbb{R}$, $(0 < a < b,\ range(\psi) \subset [a,b])$ *be functions such that* $\theta \in L_1[a,b]$ *is a positive and harmonically convex function and* ψ *is differentiable and strictly increasing on [a,b], then for the integral operators defined in Definition 6, we have*

$$\begin{aligned}
\theta\Big(\frac{2\psi(a)\psi(b)}{\psi(a)+\psi(b)}\Big)&\Big(\mathscr{T}^{(\gamma_j,\tau_j)_m}_{\lambda,\sigma,\zeta;(\frac{1}{\psi(a)})^-}1\Big)\Big(\frac{1}{\psi(b)}\Big)\\
&\le \frac{1}{2}\Big[\{\mathscr{T}^{(\gamma_j,\tau_j)_m}_{\lambda,\sigma,\zeta;(\frac{1}{\psi(a)})^-}\theta\circ\mu\}\Big(\frac{1}{\psi(b)}\Big) + \{\mathscr{T}^{(\gamma_j,\tau_j)_m}_{\lambda,\sigma,\zeta;(\frac{1}{\psi(b)})^+}\theta\circ\mu\}\Big(\frac{1}{\psi(a)}\Big)\Big]\\
&\le \Big[\frac{\theta(\psi(a))+\theta(\psi(b))}{2}\Big]\Big(\mathscr{T}^{(\gamma_j,\tau_j)_m}_{\lambda,\sigma,\zeta;(\frac{1}{\psi(a)})^-}1\Big)\Big(\frac{1}{\psi(b)}\Big),
\end{aligned} \tag{10}$$

where $\mu(x) = \frac{1}{x}$ *for all* $x \in [\frac{1}{b}, \frac{1}{a}]$.

Proof. If θ is harmonically convex on $[a,b]$, for every $x,y \in [a,b]$, the following inequality holds

$$\theta\Big(\frac{2\psi(x)\psi(y)}{\psi(x)+\psi(y)}\Big) \le \frac{\theta(\psi(x))+\theta(\psi(y))}{2}. \tag{11}$$

Now, taking $\psi(x) = \dfrac{\psi(a)\psi(b)}{t\psi(b)+(1-t)\psi(a)}$ and $\psi(y) = \dfrac{\psi(a)\psi(b)}{t\psi(a)+(1-t)\psi(b)}$ in Equation (11), we have

$$2\theta\Big(\frac{2\psi(a)\psi(b)}{\psi(a)+\psi(b)}\Big) \le \theta\Big(\frac{\psi(a)\psi(b)}{t\psi(b)+(1-t)\psi(a)}\Big) + \theta\Big(\frac{\psi(a)\psi(b)}{t\psi(a)+(1-t)\psi(b)}\Big). \tag{12}$$

By multiplying by $(1-t)^{\tau_j}\mathrm{J}_{(\tau_j)_{m,\sigma}}^{(\gamma_j)_{m,\lambda}}(\zeta(1-t)^{\gamma_j})$ and then integrating over $[0,1]$, we get

$$2\theta\Big(\frac{2\psi(a)\psi(b)}{\psi(a)+\psi(b)}\Big)\int_0^1(1-t)^{\tau_j}\mathrm{J}_{(\tau_j)_{m,\sigma}}^{(\gamma_j)_{m,\lambda}}(\zeta(1-t)^{\gamma_j})dt$$
$$\le\int_0^1(1-t)^{\tau_j}\mathrm{J}_{(\tau_j)_{m,\sigma}}^{(\gamma_j)_{m,\lambda}}(\zeta(1-t)^{\gamma_j})\Big\{\theta\Big(\frac{\psi(a)\psi(b)}{t\psi(b)+(1-t)\psi(a)}\Big)+\theta\Big(\frac{\psi(a)\psi(b)}{t\psi(a)+(1-t)\psi(b)}\Big)\Big\}dt$$

$$2\theta\Big(\frac{2\psi(a)\psi(b)}{\psi(a)+\psi(b)}\Big)\Big[\frac{1}{\frac{1}{\psi(a)}-\frac{1}{\psi(b)}}\Big]\mathscr{T}^{(\gamma_j,\tau_j)_m}_{\lambda,\sigma,\zeta;(\frac{1}{\psi(a)})^-}1\Big(\frac{1}{\psi(b)}\Big)$$
$$\le\sum_{s=0}^{\infty}\frac{((\lambda)_{s\sigma}(-\zeta)^s}{(s!)\prod_{j=1}^m\Gamma(\gamma_j s+\tau_j+1)}\Big[\int_0^1(1-t)^{\tau_j+\gamma_j s}\theta\Big(\frac{\psi(a)\psi(b)}{t\psi(b)+(1-t)\psi(a)}\Big)dt \tag{13}$$
$$+\int_0^1(1-t)^{\tau_j+(\gamma_j s)}\theta\Big(\frac{\psi(a)\psi(b)}{t\psi(a)+(1-t)\psi(b)}\Big)dt\Big].$$

Solving the integrals involved in right side of inequality (13) by making substitution $\frac{1}{u}=\frac{\psi(a)\psi(b)}{t\psi(b)+(1-t)\psi(a)}$ in first integral and $\frac{1}{\psi(x)}=\frac{\psi(a)\psi(b)}{t\psi(a)+(1-t)\psi(b)}$ in the second integral, we have

$$\theta\Big(\frac{2\psi(a)\psi(b)}{\psi(a)+\psi(b)}\Big)\mathscr{T}^{(\gamma_j,\tau_j)_m}_{\lambda,\sigma,\zeta;(\frac{1}{\psi(a)})^-}1\Big(\frac{1}{\psi(b)}\Big)$$
$$\le\frac{1}{2}\Big[\mathscr{T}^{(\gamma_j,\tau_j)_m}_{\lambda,\sigma,\zeta;(\frac{1}{\psi(a)})^-}\theta\circ\mu\Big(\frac{1}{(\psi(b))}\Big)+\mathscr{T}^{(\gamma_j,\tau_j)_m}_{\lambda,\sigma,\zeta;(\frac{1}{\psi(b)})^+}\theta\circ\mu\Big(\frac{1}{(\psi(a))}\Big)\Big]. \tag{14}$$

To obtain the second part of the inequality, the harmonic convexity of θ, we have the following relation

$$\theta\Big(\frac{\psi(a)\psi(b)}{t\psi(b)+(1-t)\psi(a)}\Big)+\theta\Big(\frac{\psi(a)\psi(b)}{t\psi(a)+(1-t)\psi(b)}\Big)\le\theta(\psi(a))+\theta(\psi(b)). \tag{15}$$

Multiplying by $(1-t)^{\tau_j}\mathrm{J}_{(\tau_j)_{m,\sigma}}^{(\gamma_j)_{m,\lambda}}(\zeta(1-t)^{\gamma_j})$ and integrating over $[0,1]$ in Equation (15), we have

$$\int_0^1(1-t)^{\tau_j}\mathrm{J}_{(\tau_j)_{m,\sigma}}^{(\gamma_j)_{m,\lambda}}(\zeta(1-t)^{\gamma_j})\theta\Big(\frac{\psi(a)\psi(b)}{t\psi(b)+(1-t)\psi(a)}\Big)dt$$
$$+\int_0^1(1-t)^{\tau_j}\mathrm{J}_{(\tau_j)_{m,\sigma}}^{(\gamma_j)_{m,\lambda}}(\zeta(1-t)^{\gamma_j})\theta\Big(\frac{\psi(a)\psi(b)}{t\psi(a)+(1-t)\psi(b)}\Big)dt$$
$$\le\Big[\theta(\psi(a))+\theta(\psi(b))\Big]\mathscr{T}^{(\gamma_j,\tau_j)_m}_{\lambda,\sigma,\zeta;(\frac{1}{\psi(b)})^+}1\Big(\frac{1}{\psi(a)}\Big). \tag{16}$$

Solving the integrals involved in the left side of inequality (16) by making substitution $\frac{1}{u}=\frac{\psi(a)\psi(b)}{t\psi(b)+(1-t)\psi(a)}$ in first integral and $\frac{1}{v}=\frac{\psi(a)\psi(b)}{t\psi(a)+(1-t)\psi(b)}$ in the second integral, we obtain

$$\frac{1}{2}\Big[\mathscr{T}^{(\gamma_j,\tau_j)_m}_{\lambda,\sigma,\zeta;(\frac{1}{\psi(a)})^-}\theta\circ\mu\Big(\frac{1}{\psi(b)}\Big)+\mathscr{T}^{(\gamma_j,\tau_j)_m}_{\lambda,\sigma,\zeta;(\frac{1}{\psi(b)})^+}\theta\circ\mu\Big(\frac{1}{\psi(a)}\Big)\Big]$$
$$\le\Big[\frac{\theta(\psi(a))+\theta(\psi(b))}{2}\Big]\mathscr{T}^{(\gamma_j,\tau_j)_m}_{\lambda,\sigma,\zeta;(\frac{1}{\psi(b)})^+}1\Big(\frac{1}{(\psi(a))}\Big). \tag{17}$$

Combining (14) and (17), we get the desired result. □

Corollary 1. *If $\psi(x) = \frac{1}{x}$ in Theorem 3 then the following inequality holds*

$$\theta\left[\frac{2}{a+b}\right]\left(\mathscr{T}^{(\gamma_j,\tau_j)_m}_{\lambda,\sigma,\zeta;(\frac{1}{a})-}1\right)\left(\frac{1}{b}\right)$$
$$\leq \frac{1}{2}\left[\mathscr{T}^{(\gamma_j,\tau_j)_m}_{\lambda,\sigma,\zeta;(\frac{1}{(a)})-}\theta\circ\mu\left(\frac{1}{b}\right) + \mathscr{T}^{(\gamma_j,\tau_j)_m}_{\lambda,\sigma,\zeta;(\frac{1}{(b)})+}\theta\circ\mu\left(\frac{1}{a}\right)\right]$$
$$\leq \left[\frac{\theta\left(\frac{1}{a}\right)+\theta\left(\frac{1}{b}\right)}{2}\right]\left(\mathscr{T}^{(\gamma_j,\tau_j)_m}_{\lambda,\sigma,\zeta;(\frac{1}{b})+}1\right)\left(\frac{1}{(a)}\right). \tag{18}$$

Now, we derive the following Lemma before giving the next result.

Lemma 1. *Let $\theta,\psi : [a,b] \to \mathbb{R}, 0 < a < b, range(\psi) \subset [a,b]$ be functions such that θ is positive, $\theta \in L_1[a,b]$, and ψ is differentiable and strictly increasing. If θ is a harmonically convex function on [a,b] and satisfies $\theta\left(\frac{1}{\psi(x)}\right) = \theta\left(\frac{1}{\frac{1}{\psi(a)}+\frac{1}{\psi(b)}-\psi(x)}\right)$, we have*

$$\mathscr{T}^{(\gamma_j,\tau_j)_m}_{\lambda,\sigma,\zeta;(\frac{1}{\psi(b)})+}\theta\circ\mu\left(\frac{1}{\psi(a)}\right) = \frac{1}{2}\left[\mathscr{T}^{(\gamma_j,\tau_j)_m}_{\lambda,\sigma,\zeta;(\frac{1}{\psi(a)})-}\theta\circ\mu\left(\frac{1}{\psi(b)}\right) + \mathscr{T}^{(\gamma_j,\tau_j)_m}_{\lambda,\sigma,\zeta;(\frac{1}{\psi(b)})+}\theta\circ\mu\left(\frac{1}{\psi(a)}\right)\right]$$
$$= \mathscr{T}^{(\gamma_j,\tau_j)_m}_{\lambda,\sigma,\zeta;(\frac{1}{\psi(a)})-}\theta\circ\mu\left(\frac{1}{\psi(b)}\right), \tag{19}$$

where $\mu(x) = \frac{1}{x}$, $\forall\, x \in [\frac{1}{b}, \frac{1}{a}]$.

Proof. Consider

$$\mathscr{T}^{(\gamma_j,\tau_j)_m}_{\lambda,\sigma,\zeta;(\frac{1}{\psi(b)})+}\theta\circ\mu\left(\frac{1}{\psi(a)}\right)$$
$$= \int_{\frac{1}{\psi(b)}}^{\frac{1}{\psi(a)}}\left[\frac{\frac{1}{\psi(a)}-u}{\frac{1}{\psi(a)}-\frac{1}{\psi(b)}}\right]^{\tau_j} \mathrm{J}^{(\gamma_j)_m,\lambda}_{(\tau_j)_m,\sigma}\left(\zeta\left[\frac{\frac{1}{\psi(a)}-u}{\frac{1}{\psi(a)}-\frac{1}{\psi(b)}}\right]^{\gamma_j}\right)(\theta\circ\mu)u du. \tag{20}$$

Putting $u = \frac{1}{\psi(a)} + \frac{1}{\psi(b)} - \psi(x)$ and using $\theta\left(\frac{1}{\psi(x)}\right) = \theta\left(\frac{1}{\frac{1}{\psi(a)}+\frac{1}{\psi(b)}-\psi(x)}\right)$ in Equation (20), we have

$$\mathscr{T}^{(\gamma_j,\tau_j)_m}_{\lambda,\sigma,\zeta;(\frac{1}{\psi(b)})+}\theta\circ\mu\left(\frac{1}{\psi(a)}\right) = \mathscr{T}^{(\gamma_j,\tau_j)_m}_{\lambda,\sigma,\zeta;(\frac{1}{\psi(a)})-}\theta\circ\mu\left(\frac{1}{\psi(b)}\right). \tag{21}$$

By the addition of $\mathscr{T}^{(\gamma_j,\tau_j)_m}_{\lambda,\sigma,\zeta;(\frac{1}{\psi(b)})+}\theta\circ\mu\left(\frac{1}{\psi(a)}\right)$ in Equation (21) on both sides, we have the required result. □

Theorem 4. *Let $\theta,\psi,\eta : [a,b] \to \mathbb{R}$, $\left(0 < a < b, range(\psi), range(\eta) \subset [a,b]\right)$, be functions such that $\theta \in L_1[a,b]$ is a positive function, ψ is a differentiable and strictly increasing function and η is nonnegative and integrable and satisfies $\eta\left(\frac{1}{\psi(x)}\right) = \eta\left(\frac{1}{\frac{1}{\psi(a)}+\frac{1}{\psi(b)}-\psi(x)}\right)$, then the following inequality holds*

$$\theta\left(\frac{2\psi(a)\psi(b)}{\psi(a)+\psi(b)}\right)\left[\mathscr{T}^{(\gamma_j,\tau_j)_m}_{\lambda,\sigma,\zeta;(\frac{1}{\psi(b)})+}\eta\circ\mu\left(\frac{1}{(\psi(a))}\right) + \mathscr{T}^{(\gamma_j,\tau_j)_m}_{\lambda,\sigma,\zeta;(\frac{1}{\psi(a)})-}\eta\circ\mu\left(\frac{1}{\psi(b)}\right)\right]$$
$$\leq \left[\mathscr{T}^{(\gamma_j,\tau_j)_m}_{\lambda,\sigma,\zeta;(\frac{1}{\psi(a)})-}\theta\eta\circ\mu\left(\frac{1}{(\psi(b))}\right) + \mathscr{T}^{(\gamma_j,\tau_j)_m}_{\lambda,\sigma,\zeta;(\frac{1}{\psi(b)})+}\theta\eta\circ\mu\left(\frac{1}{(\psi(a))}\right)\right] \tag{22}$$
$$\leq \left(\frac{\theta(\psi(a))+\theta(\psi(b))}{2}\right)\left[\mathscr{T}^{(\gamma_j,\tau_j)_m}_{\lambda,\sigma,\zeta;(\frac{1}{\psi(a)})-}\eta\circ\mu\left(\frac{1}{\psi(b)}\right) + \mathscr{T}^{(\gamma_j,\tau_j)_m}_{\lambda,\sigma,\zeta;(\frac{1}{\psi(b)})+}\eta\circ\mu\left(\frac{1}{\psi(a)}\right)\right],$$

where $\mu(x) = \frac{1}{x}, \forall\, x \in [\frac{1}{b}, \frac{1}{a}], \theta\,\eta \circ \mu = (\theta \circ \mu)(\eta \circ \mu)$.

Proof. By using the harmonic convexity of θ, we have

$$2\theta\Big(\frac{2\psi(a)\psi(b)}{\psi(a)+\psi(b)}\Big) \leq \theta\Big(\frac{\psi(a)\psi(b)}{t\psi(b)+(1-t)\psi(a)}\Big) + \theta\Big(\frac{\psi(a)\psi(b)}{t\psi(a)+(1-t)\psi(b)}\Big). \tag{23}$$

By multiplying by$(1-t)^{\tau_j}\mathrm{J}_{(\tau_j)_{m,\sigma}}^{(\gamma_j)_{m,\lambda}}(\zeta(1-t)^{\gamma_j})\eta\Big(\frac{\psi(a)\psi(b)}{t\psi(b)+(1-t)\psi(a)}\Big)$ in Equation (23) and then integrating over the closed interval $[0,1]$, we have

$$\begin{aligned}
&2\theta\Big(\frac{2\psi(a)\psi(b)}{\psi(a)+\psi(b)}\Big)\int_0^1 (1-t)^{\tau_j}\mathrm{J}_{(\tau_j)_{m,\sigma}}^{(\gamma_j)_{m,\lambda}}(\zeta(1-t)^{\gamma_j})\eta\Big(\frac{\psi(a)\psi(b)}{t\psi(b)+(1-t)\psi(a)}\Big)dt \\
&\leq \int_0^1 (1-t)^{\tau_j}\mathrm{J}_{(\tau_j)_{m,\sigma}}^{(\gamma_j)_{m,\lambda}}(\zeta(1-t)^{\gamma_j})\eta\Big(\frac{\psi(a)\psi(b)}{t\psi(b)+(1-t)\psi(a)}\Big) \\
&\times \Big[\theta\Big(\frac{\psi(a)\psi(b)}{t\psi(b)+(1-t)\psi(a)}\Big) + \theta\Big(\frac{\psi(a)\psi(b)}{t\psi(a)+(1-t)\psi(b)}\Big)\Big]dt
\end{aligned} \tag{24}$$

$$\begin{aligned}
&2\theta\Big(\frac{2\psi(a)\psi(b)}{\psi(a)+\psi(b)}\Big)\sum_{s=0}^{\infty}\frac{(\lambda)_{\sigma s}(-\zeta)^s}{(s!)\prod_{j=1}^{m}\Gamma(\gamma_j s+\tau_j+1)}\int_0^1 (1-t)^{\tau_j+\gamma_j s}\eta(\frac{\psi(a)\psi(b)}{t\psi(b)+(1-t)\psi(a)})dt \\
&\leq \sum_{s=0}^{\infty}\frac{((\lambda)_{\sigma s}(-\zeta)^s}{(s!)\prod_{j=1}^{m}\Gamma(\gamma_j s+\tau_j+1)}\Big[\int_0^1 (1-t)^{\tau_j+\gamma_j s}\eta(\frac{\psi(a)\psi(b)}{t\psi(b)+(1-t)\psi(a)}) \\
&\times \theta(\frac{\psi(a)\psi(b)}{t\psi(b)+(1-t)\psi(a)})dt + \int_0^1 (1-t)^{\tau_j+(\gamma_j s)}\eta(\frac{\psi(a)\psi(b)}{t\psi(b)+(1-t)\psi(a)}) \\
&\times \theta(\frac{\psi(a)\psi(b)}{t\psi(a)+(1-t)\psi(b)})dt\Big].
\end{aligned} \tag{25}$$

By making a substitution of $\frac{1}{u} = \frac{\psi(a)\psi(b)}{t\psi(b)+(1-t)\psi(a)}$ in the first integral and $\frac{1}{\psi(x)} = \frac{\psi(a)\psi(b)}{t\psi(a)+(1-t)\psi(b)}$ in second integrals occurring at right side and $\frac{1}{u} = \frac{\psi(a)\psi(b)}{t\psi(b)+(1-t)\psi(a)}$ in the integral occurring at left side of inequality (25) and using $\eta\Big(\frac{1}{\psi(x)}\Big) = \eta\Big(\frac{1}{\frac{1}{\psi(a)}+\frac{1}{\psi(b)}-\psi(x)}\Big)$ we have

$$\begin{aligned}
&\theta\Big(\frac{2\psi(a)\psi(b)}{\psi(a)+\psi(b)}\Big)\Big[\mathscr{T}_{\lambda,\sigma,\zeta;(\frac{1}{\psi(b)})+}^{(\gamma_j,\tau_j)_m}\eta\circ\mu\Big(\frac{1}{\psi(a)}\Big) + \mathscr{T}_{\lambda,\sigma,\zeta;(\frac{1}{\psi(a)})-}^{(\gamma_j,\tau_j)_m}\eta\circ\mu\Big(\frac{1}{\psi(b)}\Big)\Big] \\
&\leq \Big[\mathscr{T}_{\lambda,\sigma,\zeta;(\frac{1}{\psi(a)})-}^{(\gamma_j,\tau_j)_m}\theta\eta\circ\mu\Big(\frac{1}{\psi(b)}\Big) + \mathscr{T}_{\lambda,\sigma,\zeta;(\frac{1}{\psi(b)})+}^{(\gamma_j,\tau_j)_m}\theta\eta\circ\mu\Big(\frac{1}{\psi(a)}\Big)\Big].
\end{aligned} \tag{26}$$

Now, we take

$$\theta\Big(\frac{\psi(a)\psi(b)}{t\psi(b)+(1-t)\psi(a)}\Big) + \theta\Big(\frac{\psi(a)\psi(b)}{t\psi(a)+(1-t)\psi(b)}\Big) \leq \theta(\psi(a)) + \theta(\psi(b)). \tag{27}$$

By multiplying $(1-t)^{\tau_j}\mathrm{J}^{(\gamma_j)_{m,\lambda}}_{(\tau_j)_{m,\sigma}}(\zeta(1-t)^{\gamma_j})\eta(\frac{\psi(a)\psi(b)}{t\psi(b)+(1-t)\psi(a)})$ in Equation (27) and then integrating over $[0,1]$, we get

$$\begin{aligned}&\int_0^1 (1-t)^{\tau_j}\mathrm{J}^{(\gamma_j)_{m,\lambda}}_{(\tau_j)_{m,\sigma}}(\zeta(1-t)^{\gamma_j})\eta\Big(\frac{\psi(a)\psi(b)}{t\psi(b)+(1-t)\psi(a)}\Big)\theta\Big(\frac{\psi(a)\psi(b)}{t\psi(b+(1-t)\psi(a))}\Big)dt\\&+\int_0^1 (1-t)^{\tau_j}\mathrm{J}^{(\gamma_j)_{m,\lambda}}_{(\tau_j)_{m,\sigma}}(\zeta(1-t)^{\gamma_j})\eta\Big(\frac{\psi(a)\psi(b)}{t\psi(b)+(1-t)\psi(a)}\Big)\theta\Big(\frac{\psi(a)\psi(b)}{t\psi(a)+(1-t)\psi(b)}\Big)dt\\&\leq(\theta(\psi(a))+\theta(\psi(b)))\int_0^1 (1-t)^{\tau_j}\mathrm{J}^{(\gamma_j)_{m,\lambda}}_{(\tau_j)_{m,\sigma}}(\zeta(1-t)^{\gamma_j})\eta\Big(\frac{\psi(a)\psi(b)}{t\psi(b)+(1-t)\psi(a)}\Big)dt.\end{aligned} \tag{28}$$

Solving the integrals involved in left side of inequality (28) by making substitution $\frac{1}{u}=\frac{\psi(a)\psi(b)}{t\psi(b)+(1-t)\psi(a)}$ in the first integral and $\frac{1}{\psi(x)}=\frac{\psi(a)\psi(b)}{t\psi(a)+(1-t)\psi(b)}$ in the second integral and $\frac{1}{u}=\frac{\psi(a)\psi(b)}{t\psi(b)+(1-t)\psi(a)}$ in the integral on the right side of the inequality and using $\eta\Big(\frac{1}{\psi(x)}\Big)=\eta\Big(\frac{1}{\frac{1}{\psi(a)}+\frac{1}{\psi(b)}-\psi(x)}\Big)$, we have

$$\begin{aligned}&\Big[\mathscr{T}^{(\gamma_j,\tau_j)_m}_{\lambda,\sigma,\zeta;(\frac{1}{\psi(a)})^-}\theta\eta\circ\mu\Big(\frac{1}{\psi(b)}\Big)+\mathscr{T}^{(\gamma_j,\tau_j)_m}_{\lambda,\sigma,\zeta;(\frac{1}{\psi(b)})^+}\theta\eta\circ\mu\Big(\frac{1}{\psi(a)}\Big)\Big]\\&\leq\Big(\frac{\theta(\psi(a))+\theta(\psi(b))}{2}\Big)\mathscr{T}^{(\gamma_j,\tau_j)_m}_{\lambda,\sigma,\zeta;(\frac{1}{\psi(b)})^+}\eta\circ\mu\Big(\frac{1}{\psi(a)}\Big)+\mathscr{T}^{(\gamma_j,\tau_j)_m}_{\lambda,\sigma,\zeta;(\frac{1}{\psi(a)})^-}\eta\circ\mu\Big(\frac{1}{\psi(b)}\Big).\end{aligned} \tag{29}$$

Combining (26) and (29), we have the required result. □

Theorem 5. *Let $\theta,\psi:[a,b]\to\mathbb{R},(0<a<b, range(\psi)\subset[a,b])$ be functions, such that $\theta\in L_1[a,b]$ is a positive and harmonically convex function and ψ is differentiable and strictly increasing, then the following inequality holds for the operators defined in Definition 6*

$$\begin{aligned}&\theta(\frac{2\psi(a)\psi(b)}{\psi(a)+\psi(b)})(\mathscr{T}^{(\gamma_j,\tau_j)_m}_{\lambda,\sigma,\zeta;(\frac{\psi(a)+\psi(b)}{2\psi(a)\psi(b)})^-}1)(\frac{1}{\psi(b)})\\&\leq\frac{1}{2}[(\mathscr{T}^{(\gamma_j,\tau_j)_m}_{\lambda,\sigma,\zeta;(\frac{\psi(a)+\psi(b)}{2\psi(a)\psi(b)})^-}\theta\circ\mu)(\frac{1}{\psi(b)})+(\mathscr{T}^{(\gamma_j,\tau_j)_m}_{\lambda,\sigma,\zeta;(\frac{\psi(a)+\psi(b)}{2\psi(a)\psi(b)})^+}\theta\circ\mu)(\frac{1}{\psi(a)})]\\&\leq[\frac{\theta(\psi(a))+\theta(\psi(b))}{2}](\mathscr{T}^{(\gamma_j,\tau_j)_m}_{\lambda,\sigma,\zeta;(\frac{\psi(a)+\psi(b)}{2\psi(a)\psi(b)})^+}1)(\frac{1}{\psi(a)}),\end{aligned} \tag{30}$$

where $\mu(x)=\frac{1}{x}\,\forall\, x\in[\frac{1}{b},\frac{1}{a}]$

Proof. We have

$$2\theta(\frac{2\psi(a)\psi(b)}{\psi(a)+\psi(b)})\leq\theta(\frac{\psi(a)\psi(b)}{t\psi(b)+(1-t)\psi(a)})+\theta(\frac{\psi(a)\psi(b)}{t\psi(a)+(1-t)\psi(b)}). \tag{31}$$

By multiplying $(1-t)^{\tau_j}\mathrm{J}^{(\gamma_j)_{m,\lambda}}_{(\tau_j)_{m,\sigma}}(\zeta(1-t)^{\gamma_j})$ on both sides and then integrating over $[\frac{1}{2},1]$, we get

$$\begin{aligned}
&2\theta(\frac{2\psi(a)\psi(b)}{\psi(a)+\psi(b)})\int_{\frac{1}{2}}^{1}(1-t)^{\tau_j}\mathrm{J}^{(\gamma_j)_{m,\lambda}}_{(\tau_j)_{m,\sigma}}(\zeta(1-t)^{\gamma_j})dt\\
&\leq\int_{\frac{1}{2}}^{1}(1-t)^{\tau_j}\mathrm{J}^{(\gamma_j)_{m,\lambda}}_{(\tau_j)_{m,\sigma}}(\zeta(1-t)^{\gamma_j})\Big\{\theta\Big(\frac{\psi(a)\psi(b)}{t\psi(b)+(1-t)\psi(a)}\Big)+\theta\Big(\frac{\psi(a)\psi(b)}{t\psi(a)+(1-t)\psi(b)}\Big)\Big\}dt\\
&2\theta(\frac{2\psi(a)\psi(b)}{\psi(a)+\psi(b)}\sum_{s=0}^{\infty}\frac{(\lambda)_{\sigma s}(-\zeta)^s}{(s!)\prod_{j=1}^{m}\Gamma(\gamma_j s+\tau_j+1)}\int_{\frac{1}{2}}^{1}(1-t)^{\tau_j+\gamma_j s}dt\\
&\leq\sum_{s=0}^{\infty}\frac{(\lambda)_{\sigma s}(-\zeta)^s}{(s!)\prod_{j=1}^{m}\Gamma(\gamma_j s+\tau_j+1)}\Big[\int_{\frac{1}{2}}^{1}(1-t)^{\tau_j+\gamma_j s}\theta\Big(\frac{\psi(a)\psi(b)}{t\psi(b)+(1-t)\psi(a)}\Big)dt\\
&+\int_{\frac{1}{2}}^{1}(1-t)^{\tau_j+(\gamma_j s)}\theta\Big(\frac{\psi(a)\psi(b)}{t\psi(a)+(1-t)\psi(b)}\Big)dt\Big].
\end{aligned}\tag{32}$$

Solving the integrals involved in the right side of inequality (32) by making a substitution of $\frac{1}{u}=\frac{\psi(a)\psi(b)}{t\psi(b)+(1-t)\psi(a)}$ in the first integral and $\frac{1}{v}=\frac{\psi(a)\psi(b)}{t\psi(a)+(1-t)\psi(b)}$ in the second integral as well as in the integral occurring at the left side of inequality (32), we have

$$\begin{aligned}
&\theta\Big(\frac{2\psi(a)\psi(b)}{\psi(a)+\psi(b)}\Big)\mathscr{T}^{(\gamma_j,\tau_j)_m}_{\lambda,\sigma,\zeta;(\frac{\psi(a)+\psi(b)}{2\psi(a)\psi(b)})-}1\Big(\frac{1}{\psi(b)}\Big)\\
&\leq\frac{1}{2}\Big[\mathscr{T}^{(\gamma_j,\tau_j)_m}_{\lambda,\sigma,\zeta;(\frac{\psi(a)+\psi(b)}{2\psi(a)\psi(b)})-}\theta\circ\mu\Big(\frac{1}{\psi(b)}\Big)+\mathscr{T}^{(\gamma_j,\tau_j)_m}_{\lambda,\sigma,\zeta;(\frac{\psi(a)+\psi(b)}{2\psi(a)\psi(b)})+}\theta\circ\mu\Big(\frac{1}{(\psi(a))}\Big)\Big].
\end{aligned}\tag{33}$$

To obtain the second part of inequality, the harmonic convexity of θ gives the following relation

$$\theta\Big(\frac{\psi(a)\psi(b)}{t\psi(b)+(1-t)\psi(a)}\Big)+\theta\Big(\frac{\psi(a)\psi(b)}{t\psi(a)+(1-t)\psi(b)}\Big)\leq\theta(\psi(a))+\theta(\psi(b)).\tag{34}$$

Multiplying by $(1-t)^{\tau_j}\mathrm{J}^{(\gamma_j)_{m,\lambda}}_{(\tau_j)_{m,\sigma}}(\zeta(1-t)^{\gamma_j})$ and integrating over $[\frac{1}{2},1]$, we get

$$\begin{aligned}
&\int_{\frac{1}{2}}^{1}(1-t)^{\tau_j}\mathrm{J}^{(\gamma_j)_{m,\lambda}}_{(\tau_j)_{m,\sigma}}(\zeta(1-t)^{\gamma_j})\theta\Big(\frac{\psi(a)\psi(b)}{t\psi(b)+(1-t)\psi(a)}\Big)dt\\
&+\int_{\frac{1}{2}}^{1}(1-t)^{\tau_j}\mathrm{J}^{(\gamma_j)_{m,\lambda}}_{(\tau_j)_{m,\sigma}}(\zeta(1-t)^{\gamma_j})\theta\Big(\frac{\psi(a)\psi(b)}{t\psi(a)+(1-t)\psi(b)}\Big)dt\\
&\leq(\theta(\psi(a))+\theta(\psi(b)))\mathscr{T}^{(\gamma_j,\tau_j)_m}_{\lambda,\sigma,\zeta;(\frac{\psi(a)+\psi(b)}{2\psi(a)\psi(b)})+}1\Big(\frac{1}{\psi(a)}\Big).
\end{aligned}\tag{35}$$

Simplify the integrals involved in the left side of inequality (35) by making a substitution of $\frac{1}{u}=\frac{\psi(a)\psi(b)}{t\psi(b)+(1-t)\psi(a)}$ in the first integral and $\frac{1}{v}=\frac{\psi(a)\psi(b)}{t\psi(a)+(1-t)\psi(b)}$ in the second integral, we have

$$\begin{aligned}
&\frac{1}{2}\Big[\mathscr{T}^{(\gamma_j,\tau_j)_m}_{\lambda,\sigma,\zeta;(\frac{\psi(a)+\psi(b)}{2\psi(a)\psi(b)})-}\theta\circ\mu\Big(\frac{1}{\psi(b)}\Big)+\mathscr{T}^{(\gamma_j,\tau_j)_m}_{\lambda,\sigma,\zeta;(\frac{\psi(a)+\psi(b)}{2\psi(a)\psi(b)})+}\theta\circ\mu\Big(\frac{1}{\psi(a)}\Big)\Big]\\
&\leq\Big[\frac{\theta(\psi(a))+\theta(\psi(b))}{2}\Big]\mathscr{T}^{\gamma_j,\tau_j)_m}_{\lambda,\sigma,\zeta;(\frac{\psi(a)+\psi(b)}{2\psi(a)\psi(b)})+}1\Big(\frac{1}{\psi(a)}\Big).
\end{aligned}\tag{36}$$

Combining (33) and (36), we have the result. □

Remark 2. *1. If $\psi(x) = x$, $m = 1$, $\varsigma = 0$ and τ_j is replaced by $\tau_j - 1$, it reduces to the result produced by Mehmet et al. [31]*

$$\theta\Big(\frac{2ab}{a+b}\Big) \leq \frac{\Gamma(\tau_j+1)}{2^{1-\tau_j}}\Big(\frac{ab}{b-a}\Big)^{\tau_j}\Big(I^{\tau_j}_{\frac{a+b}{2ab}-}\theta\circ\mu\big(\frac{1}{b}\big) + I^{\tau_j}_{\frac{a+b}{2ab}+}\theta\circ\mu\big(\frac{1}{a}\big)\Big) \leq \frac{\theta(a)+\theta(b)}{2}$$

where $\mu(x) = \frac{1}{x}\, \forall\, x \in [\frac{1}{b}, \frac{1}{a}]$

Lemma 2. *Let $\theta, \psi : [a,b] \to \mathbb{R}$, $(0 < a < b, range(\psi) \subset [a,b])$ be functions such that $\theta > 0$, $\theta \in L_1[a,b]$, and ψ is differentiable and strictly increasing. If θ is a harmonically convex function on [a,b] and satisfies $\theta\Big(\frac{1}{\psi(x)}\Big) = \theta\Big(\frac{1}{\frac{1}{\psi(a)}+\frac{1}{\psi(b)}-\psi(x)}\Big)$, we have*

$$\begin{aligned}
&\mathscr{T}^{(\gamma_j,\tau_j)_m}_{\lambda,\sigma,\zeta;(\frac{\psi(a)+\psi(b)}{2\psi(a)\psi(b)})+}\theta\circ\mu\Big(\frac{1}{\psi(a)}\Big)\\
&\quad= \frac{1}{2}\Big[\mathscr{T}^{(\gamma_j,\tau_j)_m}_{\lambda,\sigma,\zeta;(\frac{\psi(a)+\psi(b)}{2\psi(a)\psi(b)})-}\theta\circ\mu\Big(\frac{1}{\psi(b)}\Big) + \mathscr{T}^{(\gamma_j,\tau_j)_m}_{\lambda,\sigma,\zeta;(\frac{\psi(a)+\psi(b)}{2\psi(a)\psi(b)})+}\theta\circ\mu\Big(\frac{1}{\psi(a)}\Big)\Big]\\
&\quad= \mathscr{T}^{(\gamma_j,\tau_j)_m}_{\lambda,\sigma,\zeta;(\frac{\psi(a)+\psi(b)}{2\psi\psi(a)\psi(b)})-}\theta\circ\mu\Big(\frac{1}{\psi(b)}\Big)
\end{aligned} \tag{37}$$

where $\mu(x) = \frac{1}{x}$, $\forall\, x \in [\frac{1}{b}, \frac{1}{a}]$.

Proof. Consider

$$\begin{aligned}
&\mathscr{T}^{(\gamma_j,\tau_j)_m}_{\lambda,\sigma,\zeta;(\frac{\psi(a)+\psi(b)}{2\psi(a)\psi(b)})+}\theta\circ\mu\Big(\frac{1}{\psi(a)}\Big)\\
&\quad= \int_{\frac{\psi(a)+\psi(b)}{2\psi(a)\psi(b)}}^{\frac{1}{\psi(a)}}\Big[\frac{\frac{1}{\psi(a)}-u}{\frac{1}{\psi(a)}-\frac{1}{\psi(b)}}\Big]^{\tau_j} \mathrm{J}^{(\gamma_j)_{m,\lambda}}_{(\tau_j)_{m,\sigma}}\Big(\zeta\Big[\frac{\frac{1}{\psi(a)}-u}{\frac{1}{\psi(a)}-\frac{1}{\psi(b)}}\Big]^{\gamma_j}\Big)(\theta\circ\mu)u\,du.
\end{aligned} \tag{38}$$

Substituting $u = \frac{1}{\psi(a)} + \frac{1}{\psi(b)} - \psi(x)$ and using $\theta\Big(\frac{1}{\psi(x)}\Big) = \theta\Big(\frac{1}{\frac{1}{\psi(a)}+\frac{1}{\psi(b)}-\psi(x)}\Big)$ in Equation (38), we have

$$\mathscr{T}^{(\gamma_j,\tau_j)_m}_{\lambda,\sigma,\zeta;(\frac{\psi(a)+\psi(b)}{2\psi(a)\psi(b)})+}\theta\circ\mu\Big(\frac{1}{\psi(a)}\Big) = \mathscr{T}^{(\gamma_j,\tau_j)_m}_{\lambda,\sigma,\zeta;(\frac{\psi(a)+\psi(b)}{2\psi(a)\psi(b)})-}\theta\circ\mu\Big(\frac{1}{\psi(b)}\Big). \tag{39}$$

By the addition of $\mathscr{T}^{(\gamma_j,\tau_j)_m}_{\lambda,\sigma,\zeta;(\frac{\psi(a)+\psi(b)}{2\psi(a)\psi(b)})+}\theta\circ\mu\Big(\frac{1}{\psi(a)}\Big)$ in Equation (39) on both sides, we have the required result. □

Theorem 6. *Let $\theta, \psi, \eta : [a,b] \to \mathbb{R}$, $\Big(0 < a < b, range(\psi), range(\eta) \subset [a,b]\Big)$ be functions such that $\theta \in L_1[a,b]$ is a positive function, ψ is a differentiable, strictly increasing function and η is nonnegative and integrable and satisfies $\eta\Big(\frac{1}{\psi(x)}\Big) = \eta\Big(\frac{1}{\frac{1}{\psi(a)}+\frac{1}{\psi(b)}-\psi(x)}\Big)$, then the following inequality holds for the operators defined in Definition 6.*

$$\begin{aligned}
&\theta\Big(\frac{2\psi(a)\psi(b)}{\psi(a)+\psi(b)}\Big)\Big[\mathscr{T}^{(\gamma_j,\tau_j)_m}_{\lambda,\sigma,\zeta;(\frac{\psi(a)+\psi(b)}{2\psi(a)\psi(b)})+}\eta\circ\mu\Big(\frac{1}{\psi(a)}\Big) + \mathscr{T}^{(\gamma_j,\tau_j)_m}_{\lambda,\sigma,\zeta;(\frac{\psi(a)+\psi(b)}{2\psi(a)\psi(b)})-}\eta\circ\mu\Big(\frac{1}{\psi(b)}\Big)\Big]\\
&\leq \Big[\mathscr{T}^{(\gamma_j,\tau_j)_m}_{\lambda,\sigma,\zeta;(\frac{\psi(a)+\psi(b)}{2\psi(a)\psi(b)})-}\theta\eta\circ\mu\Big(\frac{1}{\psi(b)}\Big) + \mathscr{T}^{(\gamma_j,\tau_j)_m}_{\lambda,\sigma,\zeta;(\frac{\psi(a)+\psi(b)}{2\psi(a)\psi(b)})+}\theta\eta\circ\mu\Big(\frac{1}{\psi(a)}\Big)\Big]\\
&\leq \Big(\frac{\theta(\psi(a))+\theta(\psi(b))}{2}\Big)\Big[\mathscr{T}^{(\gamma_j,\tau_j)_m}_{\lambda,\sigma,\zeta;(\frac{\psi(a)+\psi(b)}{2\psi(a)\psi(b)})-}\eta\circ\mu\Big(\frac{1}{\psi(b)}\Big) + \mathscr{T}^{(\gamma_j,\tau_j)_m}_{\lambda,\sigma,\zeta;(\frac{\psi(a)+\psi(b)}{2\psi(a)\psi(b)})+}\eta\circ\mu\Big(\frac{1}{\psi(a)}\Big)\Big],
\end{aligned} \tag{40}$$

where $\mu(x) = \frac{1}{x}, \forall\, x \in [\frac{1}{b}, \frac{1}{a}], \theta\eta\circ\mu = (\theta\circ\mu)(\eta\circ\mu)$.

Proof. By the harmonic convexity of θ, we have

$$2\theta\Big(\frac{2\psi(a)\psi(b)}{\psi(a)+\psi(b)}\Big) \leq \theta\Big(\frac{\psi(a)\psi(b)}{t\psi(b)+(1-t)\psi(a)}\Big)+\theta\Big(\frac{\psi(a)\psi(b)}{t\psi(a)+(1-t)\psi(b)}\Big). \tag{41}$$

By multiplying $(1-t)^{\tau_j}\mathrm{J}^{(\gamma_j)_{m,\lambda}}_{(\tau_j)_{m,\sigma}}(\zeta(1-t)^{\gamma_j})\eta\Big(\frac{\psi(a)\psi(b)}{t\psi(b)+(1-t)\psi(a)}\Big)$ in the Equation (41) and then integrating over the closed interval $[\frac{1}{2},1]$, we have

$$\begin{aligned}
&2\theta\Big(\frac{2\psi(a)\psi(b)}{\psi(a)+\psi(b)}\Big)\int_{\frac{1}{2}}^{1}(1-t)^{\tau_j}\mathrm{J}^{(\gamma_j)_{m,\lambda}}_{(\tau_j)_{m,\sigma}}(\zeta(1-t)^{\gamma_j})\eta\Big(\frac{\psi(a)\psi(b)}{t\psi(b)+(1-t)\psi(a)}\Big)dt\\
&\leq \int_{\frac{1}{2}}^{1}(1-t)^{\tau_j}\mathrm{J}^{(\gamma_j)_{m,\lambda}}_{(\tau_j)_{m,\sigma}}(\zeta(1-t)^{\gamma_j})\eta\Big(\frac{\psi(a)\psi(b)}{t\psi(b)+(1-t)\psi(a)}\Big)\\
&\times\Big[\theta\Big(\frac{\psi(a)\psi(b)}{t\psi(b)+(1-t)\psi(a)}\Big)+\theta\Big(\frac{\psi(a)\psi(b)}{t\psi(a)+(1-t)\psi(b)}\Big)\Big]dt\\
&2\theta\Big(\frac{2\psi(a)\psi(b)}{\psi(a)+\psi(b)}\Big)\sum_{s=0}^{\infty}\frac{(\lambda)_{\sigma s}(-\zeta)^s}{(s!)\prod_{j=1}^{m}\Gamma(\gamma_j s+\tau_j+1)}\int_{\frac{1}{2}}^{1}(1-t)^{\tau_j+\gamma_j s}\eta(\frac{\psi(a)\psi(b)}{t\psi(b)+(1-t)\psi(a)})dt\\
&\leq\sum_{s=0}^{\infty}\frac{((\lambda)_{\sigma s}(-\zeta)^s}{(s!)\prod_{j=1}^{m}\Gamma(\gamma_j s+\tau_j+1)}\Big[\int_{\frac{1}{2}}^{1}(1-t)^{\tau_j+\gamma_j s}\eta(\frac{\psi(a)\psi(b)}{t\psi(b)+(1-t)\psi(a)})\\
&\times\theta(\frac{\psi(a)\psi(b)}{t\psi(b)+(1-t)\psi(a)})dt+\int_{\frac{1}{2}}^{1}(1-t)^{\tau_j+(\gamma_j s)}\eta(\frac{\psi(a)\psi(b)}{t\psi(b)+(1-t)\psi(a)})\\
&\times\theta(\frac{\psi(a)\psi(b)}{t\psi(a)+(1-t)\psi(b)})dt\Big].
\end{aligned}\tag{42}$$

By substituting $\frac{1}{u}=\frac{\psi(a)\psi(b)}{t\psi(b)+(1-t)\psi(a)}$ in the first integral and $\frac{1}{\psi(x)}=\frac{\psi(a)\psi(b)}{t\psi(a)+(1-t)\psi(b)}$ in the second integrals occurring at the right side and $\frac{1}{u}=\frac{\psi(a)\psi(b)}{t\psi(b)+(1-t)\psi(a)}$ in the integral occurring at left side of inequality (42), we have

$$\begin{aligned}
&\theta\Big(\frac{2\psi(a)\psi(b)}{\psi(a)+\psi(b)}\Big)\mathscr{T}^{(\gamma_j,\tau_j)_m}_{\lambda,\sigma,\zeta;(\frac{\psi(a)+\psi(b)}{2\psi(a)\psi(b)})^+}\eta\circ\mu\Big(\frac{1}{(\psi(a))}\Big)+\mathscr{T}^{(\gamma_j,\tau_j)_m}_{\lambda,\sigma,\zeta;(\frac{\psi(a)+\psi(b)}{2\psi(a)\psi(b)})^-}\eta\circ\mu\Big(\frac{1}{\psi(b)}\Big)\\
&\leq\Big[\mathscr{T}^{(\gamma_j,\tau_j)_m}_{\lambda,\sigma,\zeta;(\frac{\psi(a)+\psi(b)}{2\psi(a)\psi(b)})^-}\theta\eta\circ\mu\Big(\frac{1}{(\psi(b))}\Big)+\mathscr{T}^{(\gamma_j,\tau_j)_m}_{\lambda,\sigma,\zeta;(\frac{\psi(a)+\psi(b)}{2\psi(a)\psi(b)})^+}\theta\eta\circ\mu\Big(\frac{1}{(\psi(a))}\Big)\Big].
\end{aligned}\tag{43}$$

Now, we take

$$\theta\Big(\frac{\psi(a)\psi(b)}{t\psi(b)+(1-t)\psi(a)}\Big)+\theta\Big(\frac{\psi(a)\psi(b)}{t\psi(a)+(1-t)\psi(b)}\Big)\leq\theta(\psi(a))+\theta(\psi(b)). \tag{44}$$

By multiplying $(1-t)^{\tau_j}\mathrm{J}^{(\gamma_j)_{m,\lambda}}_{(\tau_j)_{m,\sigma}}(\zeta(1-t)^{\gamma_j})\eta(\frac{\psi(a)\psi(b)}{t\psi(b)+(1-t)\psi(a)})$ in Equation (44) and then integrating over $[\frac{1}{2},1]$, we get

$$\begin{aligned}
&\int_{\frac{1}{2}}^{1}(1-t)^{\tau_j}\mathrm{J}^{(\gamma_j)_{m,\lambda}}_{(\tau_j)_{m,\sigma}}(\zeta(1-t)^{\gamma_j})\eta\Big(\frac{\psi(a)\psi(b)}{t\psi(b)+(1-t)\psi(a)}\Big)\theta\Big(\frac{\psi(a)\psi(b)}{t\psi(b+(1-t)\psi(a))}\Big)dt\\
&+\int_{\frac{1}{2}}^{1}(1-t)^{\tau_j}\mathrm{J}^{(\gamma_j)_{m,\lambda}}_{(\tau_j)_{m,\sigma}}(\zeta(1-t)^{\gamma_j})\eta\Big(\frac{\psi(a)\psi(b)}{t\psi(b)+(1-t)\psi(a)}\Big)\theta\Big(\frac{\psi(a)\psi(b)}{t\psi(a)+(1-t)\psi(b)}\Big)dt\\
&\leq(\theta(\psi(a))+\theta(\psi(b)))\int_{\frac{1}{2}}^{1}(1-t)^{\tau_j}\mathrm{J}^{(\gamma_j)_{m,\lambda}}_{(\tau_j)_{m,\sigma}}(\zeta(1-t)^{\gamma_j})\eta\Big(\frac{\psi(a)\psi(b)}{t\psi(b)+(1-t)\psi(a)}\Big)dt.
\end{aligned}\tag{45}$$

Solving the integrals involved in left side of inequality (45) by making substitution $\frac{1}{u} = \frac{\psi(a)\psi(b)}{t\psi(b)+(1-t)\psi(a)}$ in the first integral and $\frac{1}{\psi(x)} = \frac{\psi(a)\psi(b)}{t\psi(a)+(1-t)\psi(b)}$ in the second integral and $\frac{1}{u} = \frac{\psi(a)\psi(b)}{t\psi(b)+(1-t)\psi(a)}$ in the integral on the right side of the inequality and using the above lemma and the condition $\frac{1}{u} = \frac{\psi(a)\psi(b)}{t\psi(b)+(1-t)\psi(a)}$, we have

$$\begin{aligned}
&\left[\mathscr{T}^{(\gamma_j,\tau_j)_m}_{\lambda,\sigma,\zeta;(\frac{\psi(a)+\psi(b)}{2\psi(a)\psi(b)})-}\theta\eta\circ\mu\Big(\frac{1}{(\psi(b))}\Big)+\mathscr{T}^{(\gamma_j,\tau_j)_m}_{\lambda,\sigma,\zeta;(\frac{\psi(a)+\psi(b)}{2\psi(a)\psi(b)})+}\theta\eta\circ\mu\Big(\frac{1}{(\psi(a))}\Big)\right]\\
&\qquad\le\Big(\frac{\theta(\psi(a))+\theta(\psi(b))}{2}\Big)\mathscr{T}^{(\gamma_j,\tau_j)_m}_{\lambda,\sigma,\zeta;(\frac{\psi(a)+\psi(b)}{2\psi(a)\psi(b)})+}\eta\circ\mu\Big(\frac{1}{(\psi(a))}\Big)\\
&\qquad+\mathscr{T}^{(\gamma_j,\tau_j)_m}_{\lambda,\sigma,\zeta;(\frac{\psi(a)+\psi(b)}{2\psi(a)\psi(b)})-}\eta\circ\mu\Big(\frac{1}{(\psi(b))}\Big).
\end{aligned}\tag{46}$$

Combining (43) and (46), we have the required result. □

3. Conclusion Remarks

In this article, we established Hadamard and Fejér–Hadamard type inequalities via a new generation of the generalized fractional integral operators (8) and (9) with a non-singular function (multi-index Bessel function) as its kernel for harmonically convex functions. It is concluded that many classical inequalities cited in the literature can be easily derived by employing certain conditions on generalized fractional integral operators (8) and (9). We believe that our formulated inequalities will be useful to investigate the stability of certain fractional controlled systems.

Author Contributions: Conceptualization, R.S.A., S.M. and S.A.; methodology, G.R. and K.S.N.; software, G.R. and K.S.N.; validation, A.M., T.A. and G.R.; formal analysis, K.S.N.; investigation, S.M. and K.S.N.; resources, S.M., G.R. and K.S.N.; data curation, S.M., and T.A.; writing—original draft preparation, R.S.A., S.M. and S.A.; writing—review and editing, T.A., G.R. and K.S.N.; visualization, A.M., K.S.N. and T.A.; supervision, T.A., G.R. and K.S.N.; project administration, A.M. and T.A.; funding acquisition, A.M. and T.A. All authors have read and agreed to the published version of the manuscript.

Funding: The authors Aiman Mukheimer and Thabet Abdeljawad would like to thank Prince Sultan University for funding this work through research group Nonlinear Analysis Methods in Applied Mathematics (NAMAM) group number RG-DES-2017-01-17.

Institutional Review Board Statement: Not applicable.

Informed Consent Statement: Not applicable.

Data Availability Statement: Not applicable.

Conflicts of Interest: The authors declare no conflict of interest.

References

1. Mehmood, S.; Farid, G.; Khan, K.A.; Yussouf, M. New Hadamard and Fejér–Hadamard fractional inequalities for exponentially m-convex function. *Eng. Appl. Sci. Lett.* **2020**, *3*, 45–55. [CrossRef]
2. Mumcu, I.; Set, E.; Akdemir, A.O. Hermite–Hadamard type inequalities for harmonically convex functions via Katugampola fractional integrals. *Miskolc Math. Notes* **2019**, *20*, 409–424. [CrossRef]
3. Rao, Y.; Yussouf, M.; Farid, G.; Pecaric, J.;Tlili, I. Further generalizations of Hadamard and Fejér–Hadamard fractional inequalities and error estimates. *Adv. Differ. Equ.* **2020**, *2020*, 1–14. [CrossRef]
4. Rashid, S.; Safdar, F.; Akdemir, A.O.; Noor, M.A.; Noor, K.I. Some new fractional integral inequalities for exponentially m-convex functions via extended generalized Mittag-Leffler function. *J. Inequalities Appl.* **2019**, *2019*, 1–17. [CrossRef]
5. Salim, T.O.; Faraj, A.W. A generalization of Mittag-Leffler function and integral operator associated with fractional calculus. *J. Fract. Calc. Appl.* **2012**, *3*, 1–13.
6. Sarikaya, M.Z.; Alp, N. On Hermite–Hadamard-Fejér type integral inequalities for generalized convex functions via local fractional integrals. *Open J. Math. Sci.* **2019**, *3*, 273–284. [CrossRef]

7. Sarikaya, M.Z.; Yildirim, H. On Hermite–Hadamard type inequalities for Riemann-Liouville fractional integrals. *Miskolc Math. Notes* **2016**, *17*, 1049–1059. [CrossRef]
8. Waheed, A.; Farid, G.; Rehman, A.U.; Ayub, W. k-Fractional integral inequalities for harmonically convex functions via Caputo k-fractional derivatives. *Bull. Math. Anal. Appl.* **2018**, *10*, 55–67.
9. Yaldiz, H.; Akdemir, A.O. Katugampola fractional integrals within the class of convex functions. *Turk. J. Sci.* **2018**, *3*, 40–50.
10. Abbas, G.; Farid, G. Hadamard and Fejér–Hadamard type inequalities for harmonically convex functions via generalized fractional integrals. *J. Anal.* **2017**, *25*, 107–119. [CrossRef]
11. Farid, G.; Rehman, A.U.; Mehmood, S. Hadamard and Fejér–Hadamard type integral inequalities for harmonically convex functions via an extended generalized Mittag-Leffler function. *J. Math. Comput. Sci.* **2018**, *8*, 630–643.
12. Farid, G.; Mishra, V.N.; Mehmood, S. Hadamard And Fejér–Hadamard Type Inequalities for Convex and Relative Convex Functions via an Extended Generalized Mittag-Leffler Function. *Int. J. Anal. Appl.* **2019**, *17*, 892–903.
13. Set, E.; Mumcu, I. Hermite–Hadamard-Fejér type inequalities for conformable fractional integrals. *Miskolc Math. Notes* **2019**, *20*, 475–488. [CrossRef]
14. Chen, H.; Katugampola, U.N. Hermite–Hadamard and Hermite–Hadamard-Fejér type inequalities for generalized fractional integrals. *J. Math. Anal. Appl.* **2017**, *446*, 1274–1291. [CrossRef]
15. Dahmani, Z. On Minkowski and Hermite–Hadamard integral inequalities via fractional integration. *Ann. Funct. Anal.* **2010**, *1*, 51–58. [CrossRef]
16. Ekinci, A.; Ozdemir, M.E. Some new integral inequalities via Riemann-Liouville integral operators. *Appl. Comput. Math.* **2019**, *18*, 288–295.
17. Farid, G.; Rehman, A.U.; Zahra, M. On Hadamard inequalities for relative convex functions via fractional integrals. *Nonlinear Anal. Forum* **2016**, *21*, 77–86.
18. Kang, S.M.; Abbas, G.; Farid, G.; Nazeer, W. A generalized Fejér–Hadamard inequality for harmonically convex functions via generalized fractional integral operator and related results. *Mathematics* **2018**, *6*, 122. [CrossRef]
19. Khan, H.; Tunc, C.; Baleanu, D.; Khan, A.; Alkhazzan, A. Inequalities for n-class of functions using the Saigo fractional integral operator. *Rev. Real Acad. Cienc. Exactas. Fis. Nat. Ser. Mat.* **2019**, *113*, 2407–2420. [CrossRef]
20. Mohammed, P.O.; Abdeljawad, T.; Kashuri, A. Fractional Hermite–Hadamard–Fejér Inequalities for a Convex Function with Respect to an Increasing Function Involving a Positive Weighted Symmetric Function. *Symmetry* **2020**, *12*, 1503. [CrossRef]
21. Qiang, X.; Farid, G.; Yussouf, M.; Khan, K.A.; Rahman, A.U. New generalized fractional versions of Hadamard and Fejér inequalities for harmonically convex functions. *J. Inequalities Appl.* **2020**, *2020*, 1–13. [CrossRef]
22. Iscan, I.; Wu, S. Hermite–Hadamard type inequalities for harmonically convex functions via fractional integrals. *Appl. Math. Comput.* **2014**, *238*, 237–244.
23. Ion, D.A. Some estimates on the Hermite–Hadamard inequality through quasi-convex functions. *Ann. Univ. Craiova Math. Comput. Sci. Ser.* **2007**, *34*, 82–87.
24. Fejér, L. Uber die fourierreihen, II. *Math. Naturwiss. Anz Ungar. Akad. Wiss* **1906**, *24*, 369–390.
25. Tseng, K.L.; Hwang, S.R.; Dragomir, S.S. Fejér-type inequalities (I). *J. Inequalities Appl.* **2010**, *2010*, 531976. [CrossRef]
26. Toader, G.H. Some generalizations of the convexity. In Proceedings of the Colloquium on Approximation and Optimization, Cluj-Napoca, Romania, 25–27 October 1984; pp. 329–338.
27. Kunt, M.; Iscan, I.; Gozutok, U. On new inequalities of Hermite–Hadamard-Fejér type for harmonically convex functions via fractional integrals. *SpringerPlus* **2016**, *5*, 635. [CrossRef]
28. Ali, R.S.; Mubeen, S.; Nayab, I.; Araci, S.; Rahman, G.; Nisar, K.S. Some Fractional Operators with the Generalized Bessel-Maitland Function. *Discret. Dyn. Nat. Soc.* **2020**, *2020*. [CrossRef]
29. Mubeen, S.; Ali, R.S.; Nayab, I.; Rahman, G.; Abdeljawad, T.; Nisar, K.S. Integral transforms of an extended generalized multi-index Bessel function. *AIMS Math.* **2020**, *5*, 7531–7547. [CrossRef]
30. Suthar, D.L.; Purohit, S.D.; Parmar, R.K. Generalized fractional calculus of the multiindex Bessel function. *Math. Nat. Sci.* **2017**, *1*, 26–32. [CrossRef]
31. Iscan, I.; Kunt, M.; Yazici, N. Hermite–Hadamard-Fejér type inequalities for harmonically convex functions via fractional integrals. *New Trends Math. Sci.* **2016**, *4*, 239. [CrossRef]

MDPI
St. Alban-Anlage 66
4052 Basel
Switzerland
Tel. +41 61 683 77 34
Fax +41 61 302 89 18
www.mdpi.com

Fractal and Fractional Editorial Office
E-mail: fractalfract@mdpi.com
www.mdpi.com/journal/fractalfract

www.ingramcontent.com/pod-product-compliance
Lightning Source LLC
LaVergne TN
LVHW070238100526
838202LV00015B/2146
* 9 7 8 3 0 3 6 5 4 9 0 5 7 *